Applied Probability and Statistics

BAILEY · The Elements of Stochastic ...
to the Natural Sciences

BAILEY · Mathematics, Statistics and Systems for Health

BARTHOLOMEW · Stochastic Models for Social Processes, *Second Edition*

BECK and ARNOLD · Parameter Estimation in Engineering and Science

BENNETT and FRANKLIN · Statistical Analysis in Chemistry and the Chemical Industry

BHAT · Elements of Applied Stochastic Processes

BLOOMFIELD · Fourier Analysis of Time Series: An Introduction

BOX · R. A. Fisher, The Life of a Scientist

BOX and DRAPER · Evolutionary Operation: A Statistical Method for Process Improvement

BOX, HUNTER, and HUNTER · Statistics for Experimenters: An Introduction to Design, Data Analysis, and Model Building

BROWN and HOLLANDER · Statistics: A Biomedical Introduction

BROWNLEE · Statistical Theory and Methodology in Science and Engineering, *Second Edition*

BURY · Statistical Models in Applied Science

CHAMBERS · Computational Methods for Data Analysis

CHATTERJEE and PRICE · Regression Analysis by Example

CHERNOFF and MOSES · Elementary Decision Theory

CHOW · Analysis and Control of Dynamic Economic Systems

CLELLAND, deCANI, BROWN, BURSK, and MURRAY · Basic Statistics with Business Applications, *Second Edition*

COCHRAN · Sampling Techniques, *Third Edition*

COCHRAN and COX · Experimental Designs, *Second Edition*

COX · Planning of Experiments

COX and MILLER · The Theory of Stochastic Processes, *Second Edition*

DANIEL · Application of Statistics to Industrial Experimentation

DANIEL · Biostatistics: A Foundation for Analysis in the Health Sciences

DANIEL and WOOD · Fitting Equations to Data

DAVID · Order Statistics

DEMING · Sample Design in Business Research

DODGE and ROMIG · Sampling Inspection Tables, *Second Edition*

DRAPER and SMITH · Applied Regression Analysis

DUNN · Basic Statistics: A Primer for the Biomedical Sciences, *Second Edition*

DUNN and CLARK · Applied Statistics: Analysis of Variance and Regression

ELANDT-JOHNSON · Probability Models and Statistical Methods in Genetics

FLEISS · Statistical Methods for Rates and Proportions

GALAMBOS · The Asymptotic Theory of Extreme Order Statistics

GIBBONS, OLKIN, and SOBEL · Selecting and Ordering Populations: A New Statistical Methodology

GNANADESIKAN · Methods for Statistical Data Analysis of Multivariate Observations

GOLDBERGER · Econometric Theory

GOLDSTEIN and DILLON · Discrete Discriminant Analysis

continued on back

The Theory of Statistical Inference

The Theory of Statistical Inference

SHELEMYAHU ZACKS

Professor of Mathematics and Statistics
Case Western Reserve University
Cleveland, Ohio

JOHN WILEY & SONS, INC.

New York · London · Sydney · Toronto

Library of Congress Catalog Card Number: 77-132227

ISBN 0-471-98103-6

Printed in the United States of America

10 9 8 7 6 5 4 3

To my wife Hanna

״אשת חיל מי ימצא ...״ (משלי, לא:10).

Preface

The theory of statistical inference has been developed during the last three decades to such an extent that even a great number of volumes will not suffice to cover every theorem and result of some importance. The library shelves are stacked every year with journals in which hundreds of papers present studies in the general field of statistical inference. Concurrently, many textbooks are published every year on the methods of statistical inference and on the basic theory. Generally these textbooks are written for the population of undergraduate and beginning graduate students. As a consequence of this general trend, there are only a few advanced books on the theory of statistical inference. In this book I attempt to provide the advanced graduate student and the researcher in mathematical statistics with some of the material that is in the literature but to a large extent has not yet been compiled in a book. To attain this objective I have chosen to write this book on an advanced level. The reader therefore has to be proficient in the theory of probability and the theory of statistical distribution functions and have some background in advanced stochastic processes for Chapters 9 and 10. In addition, general knowledge of advanced calculus, real variables, and functional analysis is helpful.

The book has ten chapters. The first chapter gives a general view of the material in the book. The following nine chapters deal with the topics under consideration. These chapters are to a large degree self-contained. One can read each chapter almost independently of the others. Chapter 2, however, should be considered as a basic one, and is desirable to read the material in this chapter first. Each chapter contains a number of problems for solution. The problems are neither uniformly difficult nor easy, and not all of them are of high statistical interest. All the problems, however, reflect the material in the various sections. For the convenience of the reader, a list of the references cited is provided at the end of each chapter.

I began writing this book during the spring semester of 1967. At that time I was working in the Department of Statistics at Kansas State University. I would like to acknowledge Professor H. C. Fryer, the chairman of that department, for providing me with excellent working conditions and some secretarial help. My sincere gratitude is extended also to Professors J. R. Blum and L. H. Koopmans who, as chairman of the Mathematics and

Statistics Department at the University of New Mexico, provided me with the necessary help and support.

I am also grateful to Professor Geoffry S. Watson who introduced my book to John Wiley and Sons for publication; to Professor Debabrata Basu for some interesting discussions of various topics of statistical inference; to Miss Beatrice Shube, editor in the Wiley-Interscience Division; and to Mrs. Arlene Conkle for an excellent typing of the manuscript.

SHELEMYAHU ZACKS

Cleveland, Ohio
November, 1970

Contents

Commonly Used Abbreviations

a.s.	:	almost surely
i.i.d.	:	independent identically distributed
r.v.	:	random variable
d.f.	:	distribution function
s.t.	:	such that
m.g.f.	:	moment generating function
U.M.V.U.	:	uniformly minimum variance unbiased estimator
L.M.V.U.	:	locally minimum variance unbiased
L.U.E.	:	linear unbiased estimator
B.L.U.E.	:	best linear unbiased estimator
L.S.E.	:	least squares estimator
M.L.E.	:	maximum likelihood estimator
M.S.E.	:	mean square error
B.A.N.	:	best asymptotically normal
C.A.N.	:	consistent asymptotically normal
U.C.A.N.	:	uniformly consistent asymptotically normal
A.O.V.	:	analysis of variance
U.M.P.	:	uniformly most powerful (test)
U.M.P.U.	:	uniformly most powerful unbiased (test)
U.M.A.	:	uniformly most accurate (confidence intervals)
U.M.A.U.	:	uniformly most accurate unbiased (confidence intervals)
S.P.R.T.	:	sequential probability ratio test
M.L.R.	:	monotone likelihood ratio
I.F.R.A.	:	increasing failure rate average
I.F.R.	:	increasing failure rate
D.F.R.	:	decreasing failure rate
D.F.R.A.	:	decreasing failure rate average
L.C.	:	log-convex

The Theory of Statistical Inference

CHAPTER 1

Synopsis

1.1. GENERAL INTRODUCTION

This book represents the author's attempt to summarize in an integrated form the major results in certain fields of the theory of statistical inference. There are virtually thousands of papers in the statistical literature on various topics of statistical inference, many of which have not yet been accounted for in textbooks. Results of many of these interesting papers are reported and presented here. We do not however, report all the studies which have been published throughout the years on the topics which we discuss. Only a relatively small number of papers are cited and studied in each chapter. The choice of which papers to discuss and what results should be presented is in itself a difficult decision. The author has probably committed the sin of overlooking some important results, or not reporting them. In the present book we concentrate on nine basic branches of the theory of statistical inference. These include the following: sufficient statistics; unbiased estimation; the efficiency of estimators under quadratic loss; maximum likelihood estimation; Bayes and minimax estimation; equivariant estimation; admissibility of estimators; testing statistical hypotheses; confidence and tolerance estimation. A chapter is devoted to each one of these nine subjects. In order to introduce the reader to the subject matter of each area, we provide introductory sections which summarize the material discussed in the corresponding chapters. These sections are designed to present a general view of what can be found in this book.

Important subject areas of statistical inference which are not discussed in this book are the following: nonparametric procedures of testing and estimation (only a few sections in the book deal with distribution free methods), and in particular we do not discuss the important area of robust procedures. The subject of sampling finite populations is discussed only briefly. There is no account of many of the recent important contributions to the theory of simultaneous testing of hypotheses and simultaneous confidence intervals estimation. (An introduction to this subject is found in the textbook by Miller [1].) Special multivariate techniques are not discussed, and in particular we do not treat the subjects of optimal design of experiments, optimal statistical control, or adaptive procedures, although in several chapters we do mention certain results from these areas. A reasonably complete account of all these areas would span several volumes. The subjects we treat in this book are, on the other hand, fundamental to all areas of statistical inference. Eight out of nine chapters of subject matter deal with problems of estimation, but only one chapter is devoted to testing statistical hypotheses. This fact should not be considered a reflection of the author's value judgment concerning the relative importance of the various areas of statistical inference. It merely reflects the fact that an excellent book is available on testing statistical hypotheses (Lehmann [2]), and we therefore present in one chapter only certain subjects that are either not discussed or only briefly mentioned in Lehmann's book. On the other hand, there is no book on the theory of statistical estimation that discusses the subjects mentioned above on the theoretical level of this book.

The optimality and admissibility of estimators and of test procedures are defined and discussed relative to a loss function, which expresses the "regret" for erroneous decisions in a quantitative manner. The types of loss functions used are the ones on which results are available in the literature. We do not discuss the question of the proper choice of a loss function. The same remark applies to the types of prior distributions applied in the various examples in this book. The choice of a prior distribution is not guided by any normative principle. We proceed now to a synopsis of the various chapters.

1.2. SUFFICIENT STATISTICS

The concept of sufficient statistics was introduced by Fisher [2] in 1922 and has been ever since a subject of many investigations. Chapter 2 is devoted to this important and basic notion of the theory of statistical inference. We start, in Section 2.1, with several examples of statistical models and their corresponding sufficient statistics. As originally indicated by Fisher, a statistic (a function of the observed sample values) is sufficient for the objectives of statistical inference if it contains, in a certain sense, all the "information" on the parent distribution (the distribution function according to which the sample values have been generated). In what sense do we use the word "information" here? We assume in the statistical model that the observed random variables have a certain joint distribution function which belongs to a specified family $\mathcal{F}$ of distribution functions. The actual (true) parent distribution is, however, unknown. Suppose that $T(X)$ is a statistic (a properly measurable function of the observed random variables X), such that the conditional expectation of *any* other statistic $Z(X)$, given $T(X)$, is independent of $\mathcal{F}$. $T(X)$ is called then a sufficient statistic for $\mathcal{F}$. If the sample value of $T(X)$ is known, the values of any other statistic $Z(X)$ do not add any further relevant statistical information on $\mathcal{F}$. Furthermore, we can say that a knowledge of the sample value of a sufficient statistic $T(X)$ provides the statistician with all the empirical information required for regenerating an equivalent sample. This can be done by the common methods of Monte Carlo simulation. Indeed, since the conditional joint distribution of the observed random variables $X_1, \ldots, X_n$, given $T(X)$, is independent of $\mathcal{F}$, we can generate for each value of $T(X)$ a sample $Y_1, \ldots, Y_n$ whose conditional distribution, given $T(X)$, is like that of $X_1, \ldots, X_n$. Thus one method of verifying whether $T(X)$ is a sufficient statistic is to determine the conditional distribution of $X_1, \ldots, X_n$, given $T(X)$. This method could very often be difficult and laborious.

Fisher [2] and later Neyman [1] provided a simple criterion with which we can generally determine whether a family $\mathcal{F}$ of distribution functions admits a nontrivial sufficient statistic, and what is the form of this statistic. This criterion is provided by the celebrated Neyman-Fisher factorization theorem. A rigorous measure-theoretic proof of the Neyman-Fisher factorization theorem was given only in 1949 by Halmos and Savage [1], and in some generalization by Bahadur [2]. We provide in Section 2.2 some preliminary measure-theoretic framework for the proof of the factorization theorem, that is presented in Section 2.3. Conditions are given for the existence of sufficient statistics. The usefulness of the theorem is illustrated with examples.

Section 2.4 is devoted to the subject of minimal sufficient statistics. The notion is explained first by a simple example, Example 2.6, in which we show the relevance of the correspondence between random variables, or statistics, and the sample space partitioning. The idea of minimal sufficient statistics is then explained in terms of the associated contours in the sample space, which contain the contours of all other sufficient statistics. A minimal sufficient statistic partitions the sample space to the smallest number of contours. The theory of Lehmann and Scheffé [1], published in 1950, is then discussed. This theory establishes a method of constructing the contours of minimal sufficient statistics (if exist). It also provides criteria for deciding whether a nontrivial sufficient statistic exists. [A trivial sufficient statistic is either the whole sample $(X_1, \ldots, X_n)$ or the corresponding order statistic $(X_{(1)}, \ldots, X_{(n)})$, $X_{(1)} \leqslant \cdots \leqslant X_{(n)}$.] As we prove in Theorems 2.4.1 and 2.4.4, the Lehmann-Scheffé method of constructing minimal sufficient statistics is valid if the (dominated) family of probability measures is countable; or if the corresponding family of density functions is separable, in an $\mathfrak{L}_1$-metric, $\delta(f, g) = \int |f(x) - g(x)| \, \mu(dx)$. These conditions are quite mild. [Whenever we have a parametric family of distribution functions for random variables (vectors) with a parameter space, which is an open set in a Euclidean k-space absolutely continuous with respect to the linear Lebesgue measure, the separability condition holds.] Bahadur [2] extended the above results and proved that whenever the family $\mathcal{F}$ of probability measures is dominated the log-likelihood function is a minimal sufficient statistic [see (2.4.26)]. This result has many interesting applications. In particular see Godambe [1] for a formulation of this result in the theory of sampling finite populations.

Minimal sufficient statistics are closely related to the exponential family of distribution functions. This relationship is studied in Section 2.5 where we present a theory developed by Dynkin [1] and published in 1951. This theory prevails only in cases of families of probability measures which are regular in Dynkin's special sense. In Dynkin's theory the log-likelihood function plays an important role. Each probability measure P in a regular family $\mathcal{F}$ is associated with a log-likelihood function $g(x, P)$, where the domain of the variable x is an interval Δ on the real line. We consider the linear space $\mathcal{L}(\mathcal{G}, \Delta)$ generated by the class $\mathcal{G} = \{g(x, P); P \in \mathcal{F}\}$. The dimension of this linear space gives the dimension of the minimal sufficient statistic. Whenever $\mathcal{L}(\mathcal{G}, \Delta)$ is an infinite dimensional space, the only sufficient statistic is a trivial one. On the other hand, if the dimension of $\mathcal{L}(\mathcal{G}, \Delta)$ is $r = k + 1$ and the sample size n is greater than k, the family of distribution functions is a k-parameter exponential (the Koopmans-Darmois) type. Several examples exhibit the main result. An interesting result of this theory is that whenever a family $\mathcal{F}$ of probability measures consists of mixtures of probability measures the only sufficient statistic is a trivial one (see Example 2.11). The implication of this is that sufficient statistics are very sensitive to the assumptions on which the statistical models are based. For example, if $\mathcal{F}$ consists of all probability measures corresponding to the normal $\mathcal{N}(\theta, 1)$ distributions, $-\infty < \theta < \infty$, a minimal sufficient statistic is $T_n = \sum_{i=1}^{n} X_i$. On the other hand, if $\mathcal{F}$ consists of probability measures with corresponding distribution functions $F_{\alpha,\theta}(x) = \alpha\Phi(x - \theta) + (1 - \alpha)\,\Phi(x)$, where α, $0 < \alpha < 1$, is known or unknown and $\Phi(x)$ is the $\mathcal{N}(0, 1)$ distribution function, then the only sufficient statistic is a trivial one. This shows that even if α is very close to 1, but still smaller than 1, we cannot reduce the data without loss of information. From a pragmatic point of view, however, if α is very close to 1 so that probability measure of Borel sets B contributed by $\alpha\,\Phi(x)$ is negligible, we could approximate the mixtures $F_{\alpha,\theta}(x)$ by $F(x - \theta)$ and use the sufficient statistic T_n. Another implication from the main result of Section 2.5 is that in many statistical models that are widely used in applications (e.g., those with the Wiebull distributions, various types of the Pearson system, Laplace distributions with unknown location parameters) the only sufficient statistics are the trivial ones. This is a source of many complications in estimation and in testing of hypotheses, which will be further discussed later on.

Section 2.6 deals with completeness and sufficiency. The properties of completeness and sufficiency are of fundamental importance in the theory of uniformly minimum risk unbiased estimation, which is discussed in Chapter 3. Several theorems are proven in this section concerning the conditions under which families of the induced distribution functions of sufficient statistics are complete. Several examples are provided.

In Section 2.7 we present the theory of sufficiency and invariance. The results of this theory play a basic role in the theory of minimax estimation (Chapter 6), equivariant estimation (Chapter 7), invariant testing of hypotheses (Chapter 9), and so forth. The main problem studied in this section is the following: Under what conditions does the maximal invariant reduction of a sufficient statistic, with respect to a group of transformations $\mathcal{G}$, yield an equivalent statistic to the one obtained by a sufficiency reduction of a maximal invariant statistic? The conditions are stated in a theorem of Stein (Theorem 2.7.1), proven recently by Hall, Wijsman, and Ghosh [1]. The theory of sufficiency and invariance can be used in many cases to prove that certain statistics are independent. Among the problems of Section 2.11 are several that can be easily solved by using the results of this section.

Section 2.8 concerns the property of transitivity of a sequence of sufficient statistics. The results of this section have important applications in the theory of sequential estimation and sequential testing of hypotheses.

Section 2.9 is devoted to the notion of sufficient experiments, introduced by Blackwell [1] in 1951. This notion has various applications in the theory of optimum design of experiments. Such applications have been shown by DeGroot [3].

Section 2.10 discusses the role of sufficient statistics in Bayesian analysis. The results of this section are well known and appear in various books and papers. In particular we mention the book of Raiffa and Schlaifer [1] and the interesting application of the notion of Bayes sufficiency in the theory of sampling finite populations by Godambe [3].

1.3. UNBIASED ESTIMATION

Chapter 3 is devoted to the problem of unbiased estimation of a functional $g: \mathcal{F} \to E^{(k)}$, where $E^{(k)}$, $k \geqslant 1$, is the Euclidean k-space. In other words, a function g, whose domain is a specified family $\mathcal{F}$ of probability measures, is considered. The family $\mathcal{F}$ could be a parametric or a nonparametric one, and the function g ascribes each P of $\mathcal{F}$ a real or a vector valued parameter $g(P)$ (e.g., the first k moments of P, if exist). The objective is to estimate the point $g(P)$ on the basis of the observed random variables $X_1, X_2, \ldots$. The class of all estimators $\hat{g}(X)$, which are statistics on $\mathfrak{X} = \mathfrak{X}_1 \times \mathfrak{X}_2 \times \cdots$, mapping $\mathfrak{X}$ to the range of $g(P)$, say Θ, and satisfying

$$E_P\{\hat{g}(X)\} = g(P), \qquad \text{all } P \in \mathcal{F},$$

is called the *class of unbiased estimators of g, relative to* $\mathcal{F}$.

Lehmann [2], p. 11, defines an unbiased point estimator in a more general manner. He considers a loss function $L(g(P), \hat{g}(X))$ and says that $\hat{g}$ is an unbiased estimator if,

$$E_P\{L(g(P), \hat{g}(X))\} \leqslant E_P\{L(g(P), \tilde{g}(X))\}$$

for all $P \in \mathcal{F}$, where $\tilde{g}(X)$ is any other estimator of $g(P)$. It is easy to verify that the two definitions coincide when the loss function $L(g, \hat{g})$ is quadratic, that is, in the real case $L(g, \hat{g}) = \lambda(g)(\hat{g} - g)^2$, $0 < \lambda(g) < \infty$. In this book we adopted a more common approach, in which an unbiased estimator is any estimator whose expectation, under P, is equal to $g(P)$ *for all* $P \in \mathcal{F}$. This definition is independent of any particular choice of a loss function $L(g, \hat{g})$. Moreover, if we adopt Lehmann's more general definition we may find, in most cases of interest, that an unbiased estimator does not exist. In this book we call an estimator $\hat{g}$ that minimizes the risk function $E_P\{L(g(P), \hat{g}(X))\}$, at all $P \in \mathcal{F}$, a *uniformly minimum risk estimator.*

As will be shown in examples (see in particular Chapter 7) uniformly minimum risk estimators (if exist) are not necessarily unbiased in the classical sense. An estimator $\hat{g}(X)$ is called a *uniformly minimum risk unbiased* if it minimizes the risk $E_P\{L(g(P), \hat{g}(X))\}$ uniformly in $P \in \mathcal{F}$, with respect to the class of *unbiased* estimators only. As mentioned above, uniformly minimum risk unbiased estimators do not coincide, necessarily, with uniformly minimum risk estimators.

In Section 3.1 we formulate the general theory of uniformly minimum risk unbiased estimators for quadratic or convex loss functions when the family $\mathcal{F}$ of probability measures is a parametric one, that is, when $\mathcal{F} = \{P_\theta; \theta \in \Theta\}$ where Θ is some open interval (rectangle) in a k-dimensional Euclidean space $E^{(k)}$. We prove the celebrated Blackwell-Rao-Lehmann-Scheffé theorem in a general framework for quadratic loss functions of the form

$$L(g, \hat{g}) = (\hat{g} - g)'\mathbf{A}(\hat{g} - g),$$

where $\mathbf{A}$ is a $k \times k$ positive definite matrix. The main result is then generalized to cases with any loss function $L(g, \hat{g})$, which is convex in $\hat{g}$ for each g. As shown in this section, the existence of a uniformly minimum risk unbiased estimator depends to a large extent on whether the family $\mathcal{F}$ admits a complete sufficient statistic.

Section 3.2 presents several examples of the method of deriving uniformly minimum risk unbiased estimators in various models. In these examples we restrict attention to real parameter cases with squared-error loss functions, that is, $L(g, \hat{g}) = (\hat{g} - g)^2$. Since $\hat{g}$ is unbiased, the risk function $E_P\{(\hat{g}(X) - g(P))^2\}$ is the variance of $\hat{g}(X)$. Thus the best unbiased estimators for this loss function are those that are uniformly minimum variance unbiased (U.M.V.U.).

In Section 3.3 we discuss locally minimum variance unbiased (L.M.V.U.) estimators for cases in which U.M.V.U. estimators do not exist. An unbiased estimator $\hat{g}(X, \theta_0)$ is a minimum variance unbiased estimator at $\theta = \theta_0$ if $\mathrm{Var}_{\theta_0}\{\hat{g}(X; \theta_0)\} \leqslant \mathrm{Var}_{\theta_0}\{\tilde{g}(X)\}$, where $\tilde{g}(X)$ is any unbiased estimator of $g(\theta)$. We restrict attention to parametric cases. A more general definition is given for cases of vector valued parameters $g(\theta)$. We start with a theorem that establishes a necessary and sufficient condition for an unbiased estimator $\hat{g}$ to be L.M.V.U. at $\theta = \theta_0$. Many examples are available in the literature of statistical models for which no U.M.V.U. estimators exist but one can construct L.M.V.U. estimators that attain a zero variance at one or more values of θ. We show such an example (Example 3.9) from Sethuraman [1], in which an L.M.V.U. estimator attains a zero variance at an infinite sequence of θ values. The question raised is whether we can provide a general method of constructing L.M.V.U. estimators that attain a zero variance at one or more points of θ. We present such a method, apparently attributable to Takeuchi and published by Morimoto and Sibuya [1]. This method is limited to models of absolutely continuous distribution functions whose support is an open interval $(\theta, b(\theta))$, where $b(\theta)$ is a differentiable function. By the method presented we can construct the required estimator in a recursive manner on intervals of Θ. Example 3.8 illustrates this method. In Section 3.3 we also provide a theorem of Stein [3] that establishes necessary and sufficient conditions, in terms of the likelihood ratio function, for an estimator $\hat{g}$ to be L.M.V.U. at $\theta = \theta_0$. The theorem is restricted to dominated families of probability measures, with (generalized) density functions satisfying certain regularity conditions in terms of some linear operators on an L^2-space. In these cases we show that Stein's theorem is a generalization of the previous theorem of this section. Whenever the theorem is applicable it also yields the variance of the L.M.V.U. estimator, in terms of the linear operators used.

Kitagawa [1] provided a general theory of linear translatable operators on function spaces that can yield in certain statistical models U.M.V.U. estimators. Washio, Morimoto, and Ikeda [1] applied the theory of Kitagawa to the case of 1-parameter exponential families of distribution functions. We know that in this case a U.M.V.U. estimator exists, since a minimal sufficient statistic is complete. The main theorem (Theorem 3.4.2) from

Kitagawa provides a general method of using linear translatable operators to derive a U.M.V.U. estimator in cases of exponential families. Stein's theorem, which we discussed previously, is a special case of Kitagawa's theorem in the 1-parameter exponential case. Two examples illustrate the method on a family of normal distributions with a known variance.

In Sections 3.5 and 3.6 we provide methods of deriving U.M.V.U. estimators (if exist) of location and of scale parameters. These methods were developed in several different studies and unified in a general framework by Tate [1]. Section 3.5 is partitioned into two subsections. The first one deals with location parameters of the truncation type. In this case the density functions are generally of the form $f(X; \theta) = k(\theta) f(X) I_{(\theta,a)}(X)$ or $f(X; \theta) = k(\theta) f(X) I_{(b,\theta)}(X)$, where $I_A(X)$ is the indicator function of the set A. In other words, the location parameter is a point θ (unknown) which is at the left or right limit of the support interval of $f(X; \theta)$. The constants a and b are known and can be ∞ or $-\infty$. In the second part of Section 3.5 we discuss a method of deriving the U.M.V.U. estimator when the location parameter is of the translation type, that is, $f(X; \theta) = f(X - \theta)$. In this case we can use, under mild regularity conditions, the theory of bilateral Laplace transforms to derive U.M.V.U. estimators. In Section 3.6 we discuss the derivation of U.M.V.U. estimators when the density functions depend on a scale parameter, that is, $f(X; \sigma) = (1/\sigma) f(X/\sigma)$, $0 < \sigma < \infty$. It is shown that if the complete sufficient statistic is a homogeneous function of $(X_1, \ldots, X_n)$, of rank α, $\alpha \neq 0$, then under conditions of existence a U.M.V.U. estimator of a function $g(\theta)$ can be obtained by using certain Mellin transforms (see Theorem 3.6.1). The use of the Laplace and the Mellin transforms enables us, in certain cases, to overcome some difficulties which we may face in using the basic method prescribed by the Blackwell-Rao-Lehmann-Scheffé theorem. This is especially important when we try to estimate some nontrivial functions of θ. The use of tables of the Laplace and Mellin transforms may give us the required estimator right away, while the Blackwell-Rao theorem requires that we first find an unbiased estimator of $g(\theta)$, say $\tilde{g}(X)$, and then determine $\hat{g}(t) = E\{\tilde{g}(X) \mid T(X) = t\}$, where $T(X)$ is a complete sufficient statistic. This method could be quite laborious. In Example 3.13 we illustrate the use of the transform theory on a problem of estimating certain functions $g(\alpha, \theta)$, where $X_1, \ldots, X_n$ have a common 2-parameter (location α and scale θ) negative exponential distribution.

Section 3.7 is devoted to distribution free models. Such models are prevalent in various areas of application such as sampling from finite population or nonparametric procedures, in which the researcher does not wish to commit the model to too many assumptions. We present in this section the basic theory of u-statistics as developed by Halmos [1] in 1946. We prove that the order statistic $(X_{(1)}, \ldots, X_{(n)})$, where $X_{(1)} \leqslant \cdots \leqslant X_{(n)}$, is a sufficient statistic for the family of all joint distributions of independent and identically distributed (i.i.d.) random variables. This result, combined with the result of Section 2.6 concerning the completeness of the order statistic, implies that the u-statistics are U.M.V.U. estimators of functions $g(P)$ on $\mathfrak{T}$.

Linear models occupy a very significant place in the theory and applications of statistical methodology. There are hundreds of papers that have been published in the area of linear statistical inference. Many of these papers have been accounted for in several books on this subject (see Rao [4], Graybill [2], Anderson [1], etc.). In Section 3.8 we provide a short development of the main results in the area of linear unbiased estimation. The section is partitioned into several subsections. A general introduction is followed by a concise treatment of linear models of a full rank. This is followed by a discussion of the conditions of (unbiased) estimability of linear functions. This subsection leads us into a subsection which discusses generalized least squares estimators. We employ here the generalized inverse operator to obtain the module of all consistent solutions of singular normal equations. An application of the theory of generalized least squares estimators to the field of fractional weighing designs is the subject of the fourth subsection. In the fifth subsection we treat linear models of incomplete rank with nuisance parameters. These models are of wide use in fractional replications of factorial experiments. Unbiased generalized least squares estimators of a subgroup of preassigned parameters are obtained only under randomized designs. The reader can find more details in the studies of Ehrenfeld and Zacks [1], Zacks [1; 11], and Lentner [1]. The final part of Section 3.8 studies the conditions under which best linear unbiased estimators (B.L.U.E.) coincide with least squares estimators (L.S.E.), when the covariance (error) matrix is nonsingular but is not $\sigma^2 I$ and the rank of the design matrix could be incomplete. We start with a simple case discussed by Magness and McGuire [1], in which the role of the eigenvectors of the covariance (error) matrix is nicely displayed. We then prove a theorem of Zyskind [1] that generalizes the results of Magness and McGuire.

Section 3.9 deals with the subject of best unbiased linear combinations of order statistics. We show how the theory of least squares can be applied to obtain B.L.U.E.'s based on order statistics. Several important applications of this theory and special cases are given in Sarhan and Greenberg [1].

1.4. THE EFFICIENCY OF ESTIMATORS UNDER QUADRATIC LOSS

In the fourth chapter we introduce a measure of efficiency of estimators under quadratic loss, based on the celebrated Cramér-Rao lower bound for the variance of unbiased estimators. To explain it, suppose that θ is a real parameter. Let $\hat{\theta}$ be an estimator of θ, not necessarily unbiased. Then if the quadratic loss function is $L(\theta, \hat{\theta}) = \lambda(\theta)(\hat{\theta} - \theta)^2$, $0 < \lambda(\theta) < \infty$, the risk function associated is

$$R(\theta, \hat{\theta}) = \lambda(\theta)E_\theta\{(\hat{\theta} - \theta)^2\} = \lambda(\theta)[\text{Var}_\theta \{\hat{\theta}\} + B^2(\theta)],$$

where $B(\theta) = E_\theta\{\hat{\theta}\} - \theta$ is the bias of $\hat{\theta}$ at θ. In order to measure the efficiency of the estimator $\hat{\theta}$ it is necessary to derive a lower bound, as a function of θ, for the risk functions $R(\theta, \tilde{\theta})$ of all point estimators $\tilde{\theta}$. We can then express the efficiency of the estimator $\hat{\theta}$ as a function of θ, which is the ratio of the lower bound of all possible values of $R(\theta, \tilde{\theta})$ at θ to the actual risk $R(\theta, \hat{\theta})$ attained by $\hat{\theta}$. That is, if $\rho(\theta)$ designates this lower bound then

$$\text{rel eff}\,(\hat{\theta}; \theta) = \frac{\rho(\theta)}{R(\theta, \hat{\theta})}, \quad \theta \in \Theta.$$

This measure of efficiency is in a sense absolute in contrast to relative measures of efficiency, in which we compare the risk function of one estimator with that of another. In this chapter we study the various forms the lower bound $\rho(\theta)$ may attain when θ is a point in a finite-dimensional Euclidean space. If $\hat{\theta}$ is a biased estimator of θ, then $E\{\hat{\theta}\} = \theta + B(\theta)$. That is, $\hat{\theta}$ is an unbiased estimator of $g(\theta) = \theta + B(\theta)$. Thus we focus attention on the lower bounds of unbiased estimators of functions $g(\theta)$ of θ. From these lower bounds we can easily derive the lower bound $\rho(\theta)$ for the quadratic risk functions.

In Section 4.1 we discuss the Cramér-Rao lower bound for variances of unbiased estimators of a function $g(\theta)$ in the case of a real parameter θ. This lower bound, in a slightly different form, was established in two separate papers by C. R. Rao [1] and H. Cramér [1] in 1945 and 1946, respectively. The derivation of the Cramér-Rao lower bounds requires that certain regularity conditions be imposed on the family of distribution functions, on the function $g(\theta)$, and on its unbiased estimator $\hat{g}(X)$. We list these regularity conditions and show their relevancy for the various derivations. An important concept is also defined in this section, namely the *Fisher information* function.

Let $I_n(\theta)$ designate the Fisher information function associated with an observed sample value of $\mathbf{X}_n = (X_1, \ldots, X_n)$. This information function is defined as

$$I_n(\theta) = E_\theta\left\{\left[\frac{\partial}{\partial\theta}\sum_{i=1}^{n} \log f(X_i; \theta)\right]^2\right\},$$

where $f(X; \theta)$ is the common density function of the observed random variables. The celebrated Cramér-Rao inequality for the variance of any unbiased estimator $\hat{g}(\mathbf{X}_n)$ of $g(\theta)$ is

$$\text{Var}_\theta \{\hat{g}(\mathbf{X}_n)\} \geqslant \frac{(g'(\theta))^2}{I_n(\theta)}.$$

If we relate the variance of each unbiased estimator of $g(\theta)$ to the Cramér-Rao lower bound we obtain the commonly used efficiency function

$$E_g(\theta; \hat{g}) = \frac{(g'(\theta))^2}{I_n(\theta)\operatorname{Var}_\theta\{\hat{g}(X_n)\}}.$$

From the Cramér-Rao inequality we easily obtain, for any estimator of θ satisfying certain regularity conditions, that its mean square error $E_\theta\{(\hat{\theta}(X_n) - \theta)^2\}$ is bounded below by the right-hand side of the inequality

$$E_\theta\{[\hat{\theta}(X_n) - \theta]^2\} \geqslant B_n^2(\theta) + \frac{(1 + B_n'(\theta))^2}{I_n(\theta)},$$

where $B_n(\theta) = E_\theta\{\hat{\theta}(X_n)\} - \theta$ is the bias of $\hat{\theta}(X_n)$. This special form of the Cramér-Rao inequality is applied in Chapter 8 to determine the admissibility of certain estimators (see Section 1.8 for more explanations). Thus the right-hand side of this special form of the Cramér-Rao inequality is a lower bound for the mean square error of all estimators of θ having the same bias function, $B_n(\theta)$. However, if we compare the efficiency of biased and unbiased estimators, or of estimators having a different bias function, then the right-hand side of the Cramér-Rao inequality cannot provide a common comparison level. For this purpose, and since $B_n(\theta) \to 0$ as $n \to \infty$ whenever $\lim_{n\to\infty} E_\theta[\hat{\theta}(X_n) - \theta]^2 = 0$ [a property called squared error consistency of $\hat{\theta}(X_n)$], a widely used measure of relative efficiency, especially for large sample situations, is

$$\text{rel eff}(\theta; \hat{\theta}) = \frac{1}{I_n(\theta)E_\theta\{[\hat{\theta}(X_n) - \theta]^2\}}.$$

In Section 4.2 we derive a system of more stringent lower bounds for the variance of unbiased estimators. These lower bounds were introduced by Bhattacharyya [1] in 1946. As illustrated in Example 4.1, there are many cases of U.M.V.U. estimators whose variances do not attain the Cramér-Rao lower bound except asymptotically (as $n \to \infty$). We prove in Section 4.1 that a necessary and sufficient condition for a variance of an estimator to attain the Cramér-Rao lower bound is that the estimator have a (sampling) distribution of the exponential type. If this is not the case, then, as proven in a theorem of Fend [1], if the U.M.V.U. estimator is a polynomial of degree k, and if its variance has not attained the Bhattacharyya lower bound of order j, $1 \leqslant j \leqslant k - 1$, it attains the Bhattacharyya lower bound of order k. There are, however, U.M.V.U. estimators with variance functions which do not attain any of the Bhattacharyya lower bounds.

Section 4.3 is devoted to multiparameter extensions of the Cramér-Rao inequality. When $\hat{g}(X_n)$ is a vector valued unbiased estimator of $g(\theta) = (g,(\theta), \ldots, g_k(\theta))$, we consider the whole covariance matrix of $\hat{g}(X_n)$, say

$\Sigma(\hat{g}(X_n))$. An unbiased estimator $\hat{g}_1$ having a covariance matrix Σ_1 is said to be efficient at least as $\hat{g}_2$ with covariance Σ_2 if $\Sigma_2 - \Sigma_1$ is a non-negative definite matrix. This is denoted by $\Sigma_2 \geqslant \Sigma_1$. The question is whether one can generalize the Cramér-Rao inequality to matrix inequalities as above. It is shown that if $D(\theta) = \| [\partial g_i(\theta)/\partial \theta_j];\ i, j = 1, \ldots, k \|$ is a matrix of partial derivatives of $g(\theta)$ and $I(\theta)$ is the Fisher information matrix, that is,

$$I(\theta) = \left\| E\left\{\frac{\partial}{\partial \theta_i} \sum_{r=1}^{n} \log f(X_r; \theta) \cdot \frac{\partial}{\partial \theta_j} \sum_{r=1}^{n} \log f(X_r; \theta)\right\};\ i, j = 1, \ldots, k \right\|,$$

the covariance matrix $\Sigma(\hat{g}(X_n))$ of any unbiased estimator of $g(\theta)$ satisfies the inequality

$$\Sigma(\hat{g}(X_n)) \geqslant D(\theta) I^{-1}(\theta)\, D'(\theta).$$

We provide also, as special cases, inequalities on the determinants of the covariance matrices, which are called the *generalized variances.*

Section 4.4 is devoted to the sequential analog of the Cramér-Rao inequality. The main result of this section was established by Wolfowitz [1]. In this section we prove the celebrated Wald theorem concerning the expected value of the sum of a random number N of i.i.d. random variables. The version of the Wald theorem proven in this section is the common one, assuming that $E\,|X| < \infty$ and $EN < \infty$. Some generalizations of this theorem are discussed in Section 9.7.4.

In Section 4.5 we discuss the asymptotic efficiency, in large sample cases, of sequences of consistent estimators (i.e., $\hat{\theta}(X_n) \xrightarrow{P} \theta$). The notion of asymptotic efficiency is defined in the sense of Bahadur [4]. The starting point is the probability that a sequence of estimators will attain estimate values in a close neighborhood of the true value of the parameter θ. (Instead of a parameter θ we can consider a function $g(P)$, $P \in \mathfrak{T}$, in nonparametric models.) By definition, all consistent estimators $\hat{\theta}(X_n)$ are associated with intervals $(\hat{\theta}(X_n) - \delta, \hat{\theta}(X_n) + \delta)$ whose probabilities of covering the true value of θ converge to 1 as $n \to \infty$, for arbitrarily small values of δ; that is,

$$\lim_{n \to \infty} P_\theta[|\theta - \hat{\theta}(X_n)| < \delta] = 1, \quad \text{all } \delta > 0.$$

However, some consistent estimators have coverage probabilities which converge to 1 at a faster rate than other consistent estimators. Moreover, the convergence of the coverage probabilities may not be uniform in θ. Chernoff [1] established that certain efficient estimators (see Section 1.5 for more details) have coverage probabilities that converge to 1 at an exponential rate. This rate depends on the amount of information available in the sample for discriminating between different parameter values. The information function used here is the *Kullback-Leibler* information function $I(\theta, \varphi)$ (see:

Kullback [1]). Bahadur [4] extended the results of Chernoff and developed a general theory of asymptotic efficiency of consistent estimators whose asymptotic distribution is normal. We prove in Section 4.5 the main results of Bahadur. It is interesting to notice that Bahadur's criterion of asymptotic efficiency is reduced in regular cases, as shown in Theorem 4.5.2, to an asymptotic generalization of the Cramér-Rao inequality. In Section 4.5 we also discuss the phenomenon of super-efficiency.

1.5. MAXIMUM LIKELIHOOD ESTIMATORS

The method of maximum likelihood estimation has been in common use for many years. It was promoted by R. A. Fisher [3] in his classical 1925 paper. This method serves also as a basis for large sample procedures of testing hypotheses and of confidence interval estimation. In contrast to statements in many textbooks, maximum likelihood estimation is not a universally good procedure and should not be used in a dogmatic fashion. As we show in examples, there are various cases in which maximum likelihood estimators are inefficient compared with unbiased or other types of estimator. On the other hand, if the family of distributions under consideration satisfies certain regularity conditions, maximum likelihood estimators are best asymptotically normal (B.A.N.) estimators, in a sense that we explain later in this section. The advantage is that in many cases it is very simple to derive the maximum likelihood estimators. Moreover, due to an invariance property established in Section 5.1, if $\hat{\theta}$ is an M.L.E. of θ then $g(\hat{\theta})$ is an M.L.E. of $g(\theta)$, for any function $g(\theta)$ of θ (not necessarily one-to-one). This invariance property is not shared by unbiased estimators or by other types of estimators that are discussed in later chapters of the book.

For a specified statistical model, given the sample values of $X_1, \ldots, X_n$, that is, $(x_1, \ldots, x_n)$, the likelihood function is a function on the parameter space Θ or on the family $\mathcal{F}$ of probability measures which, for a given value of θ (respectively P), is proportional to the (probability) density function of $(X_1, \ldots, X_n)$ at $(x_1, \ldots, x_n)$. The factor of proportionality may depend on $(x_1, \ldots, x_n)$. Thus if $f(x_1, \ldots, x_n; \theta)$ is the joint (probability) density function of $(X_1, \ldots, X_n)$ under θ, the likelihood function of θ, given $(x_1, \ldots, x_n) = \mathbf{x}_n$ is

$$L(\theta; \mathbf{x}_n) = k(\mathbf{x}_n) f(x_1, \ldots, x_n; \theta), \qquad \theta \in \Theta.$$

As mentioned in Section 2.4, the transformation $\mathbf{X}_n \to L(\theta; \mathbf{X}_n)$ is a minimal sufficient statistic. An M.L.E. of θ is defined as a point $\hat{\theta}_n$ in Θ, if such a point exists, for which $L(\theta; \mathbf{X}_n)$ attains its supremum. Roughly speaking, one could say that the probability of observing a value in a close neighborhood of the sample value is maximized when the state of nature θ is an M.L.E.

If the family $\mathcal{F}$ admits a sufficient statistic, then obviously an M.L.E. is a function of the minimal sufficient statistic. An M.L.E. is thus a necessary statistic. We provide in Section 5.1 examples that show that an M.L.E. is not necessarily a sufficient statistic. An M.L.E. is always a minimal sufficient statistic if the family $\mathcal{F}$ of probability measures is a k-parameter exponential type. This is also the case in certain families of translation parameter distributions of the truncation type. This fact led to the common mistake, stated in several books, that a maximum likelihood estimator is sufficient.

In Section 5.2 we mention several computation routines. There are cases, even if $\mathcal{F}$ is of the exponential type, in which the determination of an M.L.E. is not simple. For example, consider the family of location parameter Cauchy distributions, or the family of scale parameter extreme value distributions. The numerical procedures, in cases where the likelihood function is twice differentiable in the neighborhood of the maximum, are based on the Newton-Raphson iterative procedure. Various modifications are mentioned that are designed to stabilize the process of successive approximations. However, we should remember that these methods are not universal. Each case requires good judgment and insight for choosing the appropriate method.

Section 5.3 is devoted to the consistency of maximum likelihood estimators. As shown by examples, not all M.L.E.'s are consistent, in the sense that $\hat{\theta}(X_n) \to \theta$ in probability or almost surely as the sample size n grows. We establish the conditions for consistency of M.L.E.'s. These conditions are stated in terms of the information functions for discriminating between distributions. These are the functions defined in (5.3.1), which are extensions of the Kullback-Leibler information function for discriminating between two distributions. We discuss also an extension of Hannan [1] to discrete cases.

In Section 5.4 we discuss the asymptotic efficiency of M.L.E.'s. This discussion is in terms of the theoretical framework developed in Section 4.5. The main result is that, under the required regularity conditions and if the M.L.E. is consistent, it attains the Bahadur efficiency; that is, its asymptotic effective variance is equal to the Cramér-Rao lower bound. We also mention Rao's definition [3] of second-order efficiency of estimators, with an indication that in many cases M.L.E. estimators are second-order efficient estimators. This concept serves to discriminate between M.L.E.'s and other types of estimators that are asymptotically efficient, but whose rate of convergence to the asymptotic variance is too slow.

In Section 5.5 we discuss the properties of best asymptotically normal (B.A.N.) estimators. Best asymptotically normal estimators are those consistent estimators whose asymptotic distribution is normal, with the inverse of the Fisher information matrix as a covariance matrix. We prove first a theorem of Walker [1] which establishes conditions under which asymptotically normal estimators have variances which are greater or equal to the inverse of the Fisher information function. We then indicate which one of these conditions does not hold in Example 4.6 of a super-efficient estimator. This theorem is then generalized to a k-parameter case. The third theorem gives the condition under which the M.L.E. is a B.A.N. estimator. Finally we provide a method of converting a consistent estimator (having a certain rate of convergence in probability) to a B.A.N. estimator. This method is applied in an example for obtaining a B.A.N. estimator of the location parameter of a Cauchy distribution.

In Section 5.6 we study the asymptotic properties of the M.L.E. from a decision-theoretic point of view. We consider loss functions that are bounded and that can be expanded in the neighborhood of the true parameter point into a quadratic loss, plus a negligible term. Then we ask under what conditions the M.L.E. is asymptotically optimal in the sense that the risk function for the M.L.E. estimator converges, as the sample size n grows to infinity, to the asymptotic variance of the M.L.E. An example of Commins [1] is given to show that the asymptotic risk function, using the M.L.E., could diverge to infinity, whereas the asymptotic variance is finite. A condition is given under which the asymptotic risk coincides with the asymptotic variance of an M.L.E.

The last section is devoted to a few remarks concerning maximum likelihood estimation on the basis of groups and censored data. Several references to studies in this area are mentioned.

1.6. BAYES AND MINIMAX ESTIMATORS

It is rarely the case that an experiment is planned without any prior information on the statistical properties of the variables to be observed. The classical methods of unbiases and of maximum likelihood estimation do not provide the means to incorporate such prior knowledge. In the sixth chapter we present the theory of Bayesian estimation, which gives us the general method of incorporating prior information with the values observed in the experiment. The basic difference between the philosophy of the Bayesian and non-Bayesian is that the Bayesian considers the parameter of the distribution as a random variable, whereas the non-Bayesian regards it as a fixed point. This difference is often a source of endless controversy about the validity of the Bayesian methodology. We do not treat the problem of Bayesian estimation from a normative point of view, but concentrate only on the study of the various properties of the theory. In the course of this study we often ask questions which dogmatic adherents to the Bayesian approach consider irrelevant.

The most important element in the Bayesian approach is the specification of a distribution function on the parameter space, which is called a *prior distribution*. In many actual applications we find that the Bayesian method is used with prior distributions that have been selected by pragmatic criteria rather than by their intrinsic relationship to the prior information available. Nevertheless the resulting Bayes estimators are often "good" estimators and compare well with other types of estimators. The Bayesian methodology is generally a very effective one. This is illustrated in many examples as well as discussed in later chapters.

In Section 6.1 we present the general structure of Bayes estimators. In addition to the prior distribution that should be specified, the Bayes estimator for a particular statistical model depends strongly on the loss function assumed. We adopt the general definition that a Bayes estimator (not necessarily unique) is a statistic (if exists) that minimizes the posterior risk function, given the observed sample values. Accordingly, after a sample has been observed, the prior distribution of the parameters is converted into a posterior distribution. A choice of an estimator is then associated with the expected loss under the posterior distribution. We also show, in Section 6.1, that whenever the loss function is bounded, a Bayes estimator also minimizes the expected loss under the prior distribution, namely, the prior risk. This is not necessarily the case when the loss function is unbounded. As shown by Girshick and Savage [1] (see Example 6.2), a certain prior distribution may have, with a squared-error loss, an infinite prior risk for all estimators, whereas its posterior risk may have, for some estimator, a finite risk.

In Section 6.2 we study the properties of Bayes estimators under quadratic and convex loss functions (unbounded). A simple integrability condition assures that the prior risk will be minimized, with a finite risk, by the same estimator minimizing the posterior risk. In addition to a theorem of Girshick and Savage [1] we provide in this section a theorem of DeGroot and Rao [1], which characterizes Bayes estimators for convex loss functions symmetric about the origin.

In Section 6.3 we discuss the notion of *generalized* Bayes estimators. Generalized Bayes estimators are defined as regular (pointwise) limits of sequences of proper Bayes estimators, which can be formally expressed as Bayes estimators against some (generally improper) prior distributions. These estimators are also called improper- or pseudo-Bayes. In Section 6.3 we present the theory, developed by Sacks [1], concerning generalized Bayes estimators of the parameter of a 1-parameter exponential family. Conditions are given for the existence of a sigma finite measure on the measurable space $(\Theta, \mathcal{T})$, with respect to which a limit of proper Bayes estimators can be expressed as a formal Bayes estimator.

Section 6.4 is devoted to the asymptotic behavior of Bayes estimators.

The main problem of this section is irrelevant from the purely Bayesian point of view. However, it has an important bearing in a more general theoretical framework. If a non-Bayesian applies the Bayesian procedure with some prior distribution that he has chosen by some pragmatic criteria, it is relevant for him to ask whether the Bayes estimator is a consistent estimator; that is, under what conditions Bayes estimators converge strongly or weakly to the "true" parameter point. A strong convergence of a sequence of Bayes estimators is illustrated in Example 6.6. In this example we derive a sequence of Bayes estimators for the probability density function of a Poisson distribution. For squared-error loss and prior negative exponential distribution the Bayes estimator of the Poisson density $p(i; \lambda)$, $i = 0, 1, \ldots,$ is the density function of a negative binomial distribution. It is easy to show that these negative binomial densities converge a.s. to the Poisson densities. The main results of Freedman [1] concerning the consistency of Bayes estimators of the probability density functions of discrete distributions are presented. References are given to further studies in this area. We also provide the heuristic arguments of Lindley [1], who has shown that under the regularity conditions generally imposed on an M.L.E. it could be asymptotically (as the sample size increases) equivalent to a Bayes estimator for a squared-error loss.

In Section 6.5 we discuss minimax estimators. The standard theorems are proven. These theorems were established by Hodges and Lehmann [1], Blyth [1], and Girshick and Savage [1] in 1950–1951. These theorems are widely used in the statistical application of the minimax theory. We do not emphasize the game-theoretic point of view. Various examples illustrate the possible applications.

As explained in Section 6.5, the minimax approach in statistics is warranted when there is only a very vague prior knowledge of the parameters and we are willing to adopt the (very pessimistic) objective of optimizing against the least favorable prior distribution of the parameters. In many instances there exists a substantial amount of prior information that is, however, incomplete in the Bayesian sense. The statistician is not ready to specify a prior distribution but is willing to incorporate more prior information than the minimax criterion allows. Several procedures for such an incorporation of incomplete prior information are discussed in Section 6.6. We start with an example from linear models with nuisance parameters. (This example is a highly simplified version of the randomized fractional replication model studied by Ehrenfeld and Zacks [1].) In this example we show how an additional prior information, even if given in a very crude form, can lead to an improvement upon the unrestricted minimax estimation. This example is followed by a discussion of minimax estimation which is restricted to a certain subclass of prior distributions. We present the main results of Blum and Rosenblatt [3] and prove a convergence theorem. The criterion adopted by Blum and Rosenblatt [3] is the one suggested by Hodges and Lehmann [3] for restricted minimax estimation. Hodges and Lehmann argue that if the statistician cannot prescribe a prior distribution he may often be ready to say that, with a probability (subjective) p, the prior distribution is $\xi_0(\theta)$, whereas with probability $(1 - p)$ it is known only that the prior distribution $\xi(\theta)$ is restricted to a specific family $\mathcal{H}$. Thus the criterion set by Hodges and Lehmann for these cases of restricted minimax estimation is to choose an estimator $\hat{\theta}$ which minimizes

$$p \int R(\theta, d)\xi_0(\mathrm{d}\theta) + (1 - p) \sup_{\xi \in \mathcal{H}} \int R(\theta, d)\xi(\mathrm{d}\theta).$$

In Section 6.7 we present a theory developed by Kudō [2] on partial prior information and parametric sufficiency. In certain estimation problems we do not have to specify the prior distribution of the basic parameters but possibly only the prior probabilities of certain events. The question raised by Kudō [2] is how to characterize the structures of models for which only a partial prior information is available. The partial prior information is specified in terms of the prior probabilities of some partition $\{\Theta_1, \ldots, \Theta_k\}$ of Θ. Let $\pi_1, \ldots, \pi_k$ be the prior probabilities of these mutually exclusive and exhaustive sets. Let d be an estimator of θ and $R(\theta, d)$ the associated risk function of d. Kudō adopted the principle of minimizing the *mean-max* risk of d, which is

$$\sum_{j=1}^{k} \pi_j \sup_{\theta \in \Theta_j} R(\theta, d).$$

If each coset of the partition contains only a single point of Θ, then a minimal mean-max estimator reduces to a corresponding Bayes estimator. If the partition is trivial, that is, $\{\phi, \Theta\}$, then the minimal mean-max estimator coincides with the (unrestricted) minimax estimator. Kudō defines the notion of parametric sufficiency of a partial prior information. This notion could be very well illustrated by considering a problem of testing hypotheses, as a special case of an estimation problem, in the following sense. Suppose that we consider the parametric function

$$g(\theta) = \begin{cases} 1, & \text{if} \quad \theta \in \Theta_1, \\ 0, & \text{if} \quad \theta \in \Theta_0, \end{cases}$$

where $\{\Theta_0, \Theta_1\}$ is a partition of Θ. Then obviously for a Bayes estimator of $g(\theta)$ with a zero-one loss it is sufficient to specify the prior probabilities of Θ_0 and Θ_1. The actual prior distribution on Θ is irrelevant. There are many prior distributions which yield the same pair of prior probabilities. The partial prior information given by the prior probabilities of Θ_0 and Θ_1

is a parametric sufficient information, in the sense of Kudō, for this particular estimation problem.

Section 6.8 is devoted to a concise review of some basic studies of sequential minimax estimation. In particular we discuss some general results of Stein and Wald [1], Wolfowitz [3], Blyth [1], and Kiefer [1]. We again set the theoretical framework for sequential estimation and define the appropriate risk functions and the minimax criteria. The characteristic result which we discuss is that, quite generally, the minimax sequential estimation procedure is a fixed sample size procedure. This is in contrast to the general characteristic of Bayes sequential procedures.

In Section 6.9 we discuss empirical Bayes estimation procedures, which have been introduced by Robbins [2]. These procedures are appropriate when we have an estimation problem that repeats at an (infinite) sequence of epochs. The empirical Bayes procedures are adaptive procedures in their nature. We assume that at each epoch the observation is determined from a distribution $F(x; \theta)$, with a parameter value θ which is a sample realization of an unobservable random variable having an unknown prior distribution. The empirical Bayes procedure does not require specification of the prior distribution, only the family of prior distributions to which the actual one belongs. Then by a properly chosen sequence of adaptive estimators we can approach asymptotically the Bayes risk, which corresponds to the "true" prior distributions. Robbins [2] has shown how these sequences of empirical Bayes estimators can be constructed and what conditions are sufficient to guarantee the required convergence.

Finally Section 6.10 is devoted to a recent contribution of Bickel and Yahav [1; 2] to the theory of asymptotic pointwise optimal Bayes sequential procedures. The problem discussed is how to determine stopping rules for sequential estimation procedures that yield posterior risk functions which converge a.s. (in large sample situations) to the Bayes risk functions. General stopping rules are specified in terms of the expected incremental risk reduction due to continuation of sampling versus the expected incremental sampling cost. It is proven that the two general types of stopping rules specified are asymptotically pointwise optimal in the Bayesian sense.

1.7. EQUIVARIANT ESTIMATION

In Section 2.7 we discussed the fundamental problem of invariance theory, namely, the problem of sufficiently invariant statistics, with respect to a group of transformations $\mathcal{G}$. In Chapter 7 we study the properties of equivariant estimation procedures, which are based on sufficiently invariant statistics. The important feature of equivariant estimators is that they have risk functions corresponding to properly defined invariant loss functions, which assume constant values on the orbits of the induced group $\bar{\mathcal{G}}$ operating on the parameter space. This property has some important implications on the theory of minimum risk equivariant estimation and on minimax theory. In Section 7.1 we start with a discussion of the general structure of equivariant estimators for specified groups of transformations and invariant loss functions. We also discuss the role of invariant randomized estimators. Section 7.2 is devoted to equivariant estimators of a location parameter for families of distribution functions depending on a location parameter of the translation type. The celebrated Pitman estimator is derived. This estimator yields the minimum risk equivariant estimator of a location parameter with respect to the group of real translations when the loss function is a squared-error one. We provide also a theorem of Girshick and Savage [1] in which we prove that the Pitman estimator is not only a minimum risk equivariant for a squared-error loss but also a minimax estimator (in the class of all estimators).

In Section 7.3 we present the group-theoretic structural model that is relevant to our discussion. The theory of the structural model has been developed by Fraser in a series of papers. We refer in particular to Fraser's recent book [9]. The main framework of the structural model, for general locally compact topological groups of transformations, is presented through a special case of a multivariate normal distribution with zero mean and a general covariance matrix. In this special case we consider the group of nonsingular linear transformations. We illustrate the important notions of invariant differentials along orbits of $\mathcal{G}$ in the space of the statistic and the corresponding left and right invariant differentials along the orbits of $\bar{\mathcal{G}}$ in the parameter space. These invariant differential elements are special cases of the more general invariant Haar measures, which are also discussed. We present in this section a concise introduction to the general theory of the Haar integral on locally compact groups. The reader is referred to the book of Nochbin [1] on this subject.

In Section 7.4 we discuss the theory of fiducial minimum risk equivariant estimators. The notion of fiducial estimation was introduced to the theory of statistics by Fisher [4], who, however, failed to connect this notion with the group structure. Fisher's definition of fiducial distributions and of fiducial estimators is appropriate for the special cases which he treated. However, in later attempts to treat various other cases by the method proposed by Fisher many controversial results were obtained. As shown by Fraser [6], Hora and Buehler [1; 2], Barnard [1], and others, if the notion of fiducial distributions is properly embedded in the group invariant structural model, meaningful results are obtainable. We further show that, restricted to group structural models, fiducial estimators coincide with minimum risk equivariant estimators for invariant loss functions and can be *formally* expressed as improper Bayes estimators against prior invariant measures. The Pitman estimators for the location and for scale parameters are special cases.

In Section 7.5 we treat the *modified* minimax principle and a generalized version of the Hunt-Stein theorem [1]. The modified minimax principle was introduced by Wesler [1]. According to this principle we should determine an estimator which minimizes the maximal risk value on each orbit of $\bar{\mathcal{G}}$ at Θ. Equivariant estimators with invariant loss functions assume, as proven in Section 7.1, equal risk values on each of the parameter orbits of $\bar{\mathcal{G}}$. Thus equivariant estimators can be compared by comparing one value on each of their parameter orbits. In particular, if a uniformly minimum risk equivariant estimator is available, it is obviously a modified minimax estimator among the equivariant estimators. The question is Under what conditions does a uniformly minimum risk equivariant estimator coincide with a modified minimax estimator in the unrestricted class? An answer to this question is provided by the generalized Hunt-Stein theorem, which is proven in this section.

Section 7.6 deals with the problem of sequential equivariant estimators. We present the results of Kiefer [2]. It is shown that if, in addition to the assumptions of the generalized Hunt-Stein theorem, three additional assumptions are imposed, the optimal sequential equivariant procedures are fixed sample size ones.

In Section 7.7 we discuss the problem of equivariant estimators when nuisance parameters are available. In these cases uniformly minimum risk equivariant estimators do not generally exist. Bayes equivariant estimators can play an important role in these cases. Bayes equivariant estimators are equivariant estimators which minimize the Bayes risk against some prior distribution over the subspace of the nuisance parameters. Bayes equivariant estimators are generally not Bayes estimators. They coincide, however, with certain minimal risk fiducial estimators. The general theory and examples of statistical interest are discussed in this last section of the chapter.

1.8. ADMISSIBILITY OF ESTIMATORS

The notion of the admissibility of decision functions in the context of game theory was introduced by von Neumann and Morgenstern [1]. Wald [6] discussed this notion in terms of statistical decision theory. Many papers have been written in the last two decades on this subject, which occupies a central part in the theory of statistical inference. In this chapter we present some of the major results in this area.

The notion of admissibility, which is defined formally in Section 8.1, imposes a partial ordering on the class of all estimators (decision functions) relative to the particular risk function under consideration. An estimator,

d_1, is said to be strictly dominated by another estimator, d_2, if the risk function of d_2 never exceeds that of d_1 and assigns smaller risk values to d_2 at some points of θ. Any estimator strictly dominated by another estimator is labeled "inadmissible." Clearly, relative to a specified risk function, we can restrict attention only to admissible estimators. Procedures of estimation labeled inadmissible can be discarded. The problem is that generally the class of all estimators, for a given estimation problem, is so rich that it is difficult to characterize *all* the admissible estimators. It is sometimes very difficult even to determine whether a given estimator is admissible or not. We try to determine a subclass of estimators such that every estimator outside this class is strictly dominated by an estimator within the subclass. If such a subclass exists, it is called "complete." All the admissible estimators belong to the smallest complete class (if it is not empty). Generally the determination of a complete class does not characterize just the admissible estimators, since there are often inadmissible estimators in complete classes.

In Section 8.1 we present the basic theory of admissibility. We start with several well-known theorems (Blyth [1]), which are illustrated with examples. These theorems are followed by a theorem of Stein [4], which provides sufficient conditions for admissibility of an estimator with a general loss function. The notion of strict admissibility is also discussed. We prove (Stein [4]) that if an estimator d is admissible and the class $\mathfrak{D}$ of estimators is weakly compact in a Wald's sense, then d is strictly admissible. This result is followed by a necessary and sufficient condition of Stein [4] for admissibility. Two theorems on complete classes are cited.

In Section 8.2 we discuss the admissibility of estimators when the loss function is quadratic. We start with a theorem of Hodges and Lehmann [2], which gives a sufficient condition for the admissibility of regular estimators whose mean square error attain the Cramér-Rao lower bound. As shown in Example 8.4, it is generally difficult to verify that the condition of this theorem is satisfied. When an estimator has a 1-parameter exponential distribution and the natural parameter space for the family of densities is the whole real line, we have admissibility. This is proved in a theorem by Girshick and Savage [1]. Their result has been extended by Karlin [4], who has also provided conditions for almost admissibility. An inadmissible estimator is called almost admissible if the set of parameter points on which it is dominated has a Lebesgue measure zero. We then prove a theorem of Karlin [4] which provides a sufficient condition of almost admissibility of an estimator having an exponential type of distribution and a natural parameter space that is *not* the whole real line.

Section 8.3 is devoted to the results of Katz [1] on admissibility and minimaxity in truncated cases under quadratic loss. These results generalize some results of Karlin, mentioned above. In this section we also restrict attention to the 1-parameter exponential family. A general form of an admissible estimator is given. A generalization to general 1-parameter families of density functions is given in Theorem 8.3.2.

In Section 8.4 we discuss the admissibility of the Pitman translation equivariant estimator of a single location parameter for squared-error loss functions. The main result proven in this section is attributable to Stein [7]. An example from Perng [1] is provided to show that generally we cannot relax the conditions of Stein's theorem. These conditions can be relaxed only in certain special cases. An analog of the Pitman estimator, for a bilinear convex loss function, is derived. We present a theorem of Fox and Rubin [1] which states that this estimator is admissible.

Section 8.5 is devoted to two examples from Stein [5; 8]. The first example proves the inadmissibility of the sample mean as an estimator of the mean vector μ of an $\mathcal{N}(\mu, I)$ distribution, when the dimension of μ is $p \geqslant 3$. The second example shows that the best equivariant estimator of the variance of a normal distribution (unknown mean) is inadmissible. These examples raise the question, Under what conditions are the best equivariant estimators admissible or inadmissible? A partial answer to this question is given in Section 8.6, which is devoted to a general theory of admissibility of an equivariant estimator of a location parameter. The results of Farrel [1] and Cohen [1] are discussed. Farrel's results deal with a single location parameter, whereas those of Cohen deal with a vector valued location parameter.

In Section 8.7 we discuss admissibility in the sequential case. The framework is according to Brown [2]. Section 8.8 treats a very interesting problem posed by Linnik and solved by Kagan [1]. The problem asks, With respect to which subclass of estimator, and under what conditions, is the sample mean $\bar{X}_n$ an admissible estimator? As shown by Kagan, Linnik, and Rao [1], the sample mean $\bar{X}_n$ is an admissible estimator of a single location parameter for a squared-error loss if and only if the family of distributions is normal. However, if we restrict attention only to *linear* estimators then the sample mean is an admissible estimator of a single location parameter for any distribution function depending on a single location parameter. Under which family of distributions is the sample mean an admissible estimator if the class contains all polynomial unbiased estimators of degree $k, k \geqslant 1$? An answer to this question is given in Theorem 8.8.3.

Section 8.9 is devoted to the subject of admissibility of estimation procedures when sampling is from finite populations. We outline the general unified theory of sampling finite populations, according to Godambe [5], and discuss the main results of Godambe [4], Joshi [1; 4], and Godambe and Joshi [1].

1.9. TESTING STATISTICAL HYPOTHESES

The subject of testing statistical hypotheses is not represented in this book in accordance with its importance or the number of papers available in the literature on testing hypotheses. Whereas seven chapters are devoted to topics of point estimation, only one chapter presents material on testing hypotheses. The reason for this is twofold. First, the basic theory of testing statistical theory is provided in Lehmann's [2], with which most of the readers are familiar. There is no need to rewrite that material. Ferguson [1] also gives several chapters on testing statistical hypotheses, in which we find more material, especially on sequential procedures for testing hypotheses. The second reason is that some of the results discussed in the preceding seven chapters can be applied easily to problems of testing hypotheses by selecting an appropriate loss function. This point has been discussed in Section 1.1. Accordingly, we devote the chapter on testing hypotheses to the theory of Bayes and minimax procedures. We start in Section 9.1 with a general formulation of Bayes and minimax tests of multiple hypotheses in the fixed sample case. We also prove two general minimax theorems of Wald [6]. In addition we illustrate in examples that the test procedures are monotone procedures, in the sense that the sample space of the sufficient statistic can be partitioned into intervals such that the same hypothesis is accepted whenever the sufficient statistic realizes a value in a certain interval. The general theory of monotone tests, which has been developed by Rubin and Karlin [1] and Karlin [1; 2; 3], is discussed in Section 9.2. There we provide the conditions under which a certain subclass of monotone procedures constitutes a minimal complete class.

Section 9.3 deals with empirical Bayes procedures of testing hypotheses. The subject of empirical Bayes procedures for estimation was discussed in Chapter 6. In Section 9.3 we discuss the results of Robbins [4] and Samuel [1] on empirical Bayes procedures for testing multiple composite hypotheses. In particular we provide a theorem of Robbins, which yields a method of estimating the unknown prior distribution of the parameter under consideration.

In Section 9.4 we present the basic theory of Bayes sequential test procedures. We establish the general functional equation which the minimal Bayes expected risk function satisfies. We then give the conditions for approximating this function by a sequence of minimal Bayes expected risk functions of truncated sequential procedures. The minimal risk functions and the Bayes stopping rules for the truncated problems can be determined, at least numerically, by the dynamic programming method. The determination of the Bayes sequential stopping rules for a specified prior distribution is generally very difficult.

Section 9.5 is devoted to the celebrated Wald sequential probability ratio tests (S.P.R.T.) of two simple hypotheses. We discuss in this section the general characteristics of this very widely used procedure. In Section 9.6 we prove that the Wald S.P.R.T. minimizes the expected sample size, with respect to all procedures for which the error probabilities (of Type I and of Type II) do not exceed those of the S.P.R.T.

As shown in Section 4, the determination of a Bayes sequential stopping rule for a specified prior distribution is generally a very difficult and tedious process (if at all manageable). For this reason various attempts at determination have been made and methods of approximation are offered in the literature. One of these approximating methods is based on the novel approach of first embedding the random sequence of the observed random variables (or their sufficient statistics) in a continuous stochastic process, and then approximating the optimal stopping rule of the original problem by one appropriate for the stochastic process. In certain cases it is somewhat easier to determine the characteristics of continuous stopping time analog than the characteristics of the stopping variables of the sequential procedures. Such cases are brought forth in Section 9.7. We start with a description of one of the earliest studies on continuous time analogs for sequential procedures, the study by Dvoretzky, Kiefer, and Wolfowitz [1]. We follow this with a presentation of the results of Epstein and Sobel [1] on sequential life tests. A third study presented is that of DeGroot [2]. In this study a minimax sequential symmetric S.P.R.T. for deciding about the sign of the mean of a normal distribution is developed. Here too the minimax sequential stopping time is constructed for a Wiener process, which is analogous to the original process; one hopes that it will approximate closely the corresponding rule for the observed sequence (if one exists). In the fourth part of Section 9.7 we discuss some recent results on the extension of the celebrated Wald theorem to Wiener processes.

Section 9.8 presents further ideas and techniques of solving the Bayes sequential testing problem for multiple hypotheses. This technique is based on results of Whittle [1]. A solution of the optimal Bayes stopping rule problem is given for the case of two simple hypotheses and symmetric loss structures (the same constant loss due to the two possible types of error). In developing the solution we show the need for the adoption of the "principle of merging risks." The formulation of the solution follows that of Shiryaev [1].

A more general approach, which reduces the problem of determining a Bayes stopping rule to a generalized Stephan free boundary problem, is discussed in Section 9.9. We start with a result of Skorokhod [1] concerning the embedding of certain supermartingale sequences in Wiener processes. Then we show that for a Wiener process the minimal Bayes expected risk function should satisfy, within the continuation region, a parabolic partial differential equation which, after some transformations, is reduced to the famous heat equation. On the boundary of the continuation region and outside, the minimal Bayes expected risk coincides with the Bayes (posterior) risk function arising from erroneous decisions (since sampling terminates there). Furthermore, using a "smooth-pasting" condition we show that along the boundary of the continuation region an additional condition should hold; certain partial derivatives should match. Thus the problem of determining the optimal boundaries of the continuation region of a Bayes sequential procedure is reduced to a free boundary problem of parabolic partial differential equations. These results are due to Chernoff [1] and Mikhalevich [1]. Several other papers are cited in the paper of Grigelionis and Shiryaev [1].

The last section of the chapter is devoted to large sample approximations to the Bayes sequential stopping rule. We discuss in particular the results of Kiefer and Sacks [1], which generalize earlier results of Chernoff [3], A. E. Albert [1], and Bessler [1]. We present in some detail the elegant characterization of the asymptotic shapes of the Bayes continuation region for families of exponential type distributions, developed by Shwarz [1].

1.10. CONFIDENCE AND TOLERANCE INTERVALS

A confidence interval for a certain parameter of a distribution function is, roughly speaking, an interval in the parameter space determined by the sufficient statistic of the observed sample and having the property that the probability that it contains the "true" parameter point is at least a preassigned value, say γ; γ is called the confidence level. (We use the term interval in a general sense here.) Tolerance intervals are intervals in the sample space of the observed random variables, which are determined by the sufficient statistics and have the property that, with a confidence level γ, the probability measure of the interval according to the "true" distribution is at least β. We also discuss β-expectation tolerance intervals which are defined later. Procedures of confidence interval and of tolerance interval estimation are used to a large extent. In this chapter we present some of the theory in this field, which has been developed in a series of papers in the last thirty years. A short account of the material available in the various sections of Chapter 10 follows.

In Section 10.1 we present the theory of uniformly most accurate confidence intervals for the 1-parameter case. We start with the well-known basic theory, which relates most powerful test functions to most accurate confidence intervals. Uniformly most accurate confidence intervals for discrete cases of M.L.R. families are discussed and illustrated.

Section 10.2 is devoted to confidence intervals in the presence of nuisance parameters. We discuss and study uniformly most accurate unbiased confidence intervals. An example is provided from the field of reliability estimation.

In Section 10.3 we discuss the expected length of confidence intervals (defined in the case of one-sided confidence intervals as the expected value of the confidence limit). We start with some results of Pratt [2]. Pratt has shown that the criterion of minimizing the expected length of confidence intervals is related to the criterion of minimizing the probability of covering false values. We show that in cases of two-sided uniformly most accurate unbiased (U.M.A.U.) confidence intervals the expected interval length is also minimal. This conclusion does not hold in cases of one-sided uniformly most accurate confidence intervals. This phenomenon is illustrated by examples of Pratt [2] and Madansky [1]. The cause for this shortcoming is analyzed, and criteria for minimizing the expected excess and the expected shortage of confidence limits, suggested by Harter [1], are presented and analyzed.

Section 10.4 presents the basic theory of tolerance intervals. There we discuss two kinds of tolerance intervals, (β, γ) tolerance intervals and β-expectation tolerance intervals. The first kind of tolerance intervals are those for which, with confidence probability γ, the probability measure of the interval is at least β. The other kind of tolerance intervals are those for which the expected probability measure of the interval is at least β. Examples are offered of the intervals obtained for the same statistical model according to these two criteria. We also define the notion of uniformly most accurate (β, γ) tolerance intervals and provide a method of constructing uniformly most accurate one-sided tolerance intervals in cases of M.L.R. families of distribution functions. In addition, the discrete case according to Zacks [10] is treated. Goodman and Madansky [1] discussed the notion of *most stable* tolerance intervals. Fraser and Guttman [1] and Guttman [1] discussed uniformly most powerful (invariant) β-expectation tolerance intervals. These notions are defined and discussed in Section 10.4.

Section 10.5 deals with distribution free confidence and tolerance intervals. We concentrate mainly on families of absolutely continuous distributions on the real line. An important subfamily of distributions for which interesting results are available is the subfamily of increasing failure rate (I.F.R.) distributions. In this area, results of Hanson and Koopmans [1] and of Barlow and Proschan [1] are developed.

Section 10.6 is devoted to Bayes and fiducial confidence and tolerance limits. In our formulation we follow Lindley [2] and Aitchison [1]. In contrast to the non-Bayesian confidence and tolerance limits, in the Bayesian framework we consider posterior probability of intervals in the parameter space, given the observed values of the sufficient statistic. Thus the probability confidence statements in the Bayesian framework are entirely different in meaning and in definition from those in the non-Bayesian framework. We illustrate these differences with several examples. We show also that for certain prior distributions we obtain Bayes confidence and tolerance intervals, which are analogous to the non-Bayesian confidence and tolerance intervals. Fiducial confidence and tolerance limits are discussed too. As in Chapter 7, we restrict the definition of the fiducial distribution to cases in which the structural group model applies.

The last four sections of the chapter are devoted to fixed-width confidence intervals. In Section 10.7 we present three cases. The first case is the well-known case of estimating the mean of a normal distribution when the variance is unknown. We prove the well-known property that no fixed-width confidence interval is available when the sample has a fixed size. (This was first proven, apparently, by Dantzig [1].) Then we present the Stein two-stage sampling procedure, under which the fixed-width interval centered at the sample mean is a confidence interval with the specified confidence level. The second case treated is that of fixed-width confidence intervals for the vth moment of an I.F.R. distribution. The third case treats an m-dependent Gaussian sequence with an unknown m. It is shown that there exists no stopping rule (for any sequential procedure) which is finite with probability one and which yields a fixed-width confidence interval for the mean. In Section 10.8 we discuss necessary and sufficient conditions for the existence of single-stage, multi-stage, or sequential sampling procedures which yield fixed-width confidence intervals for a certain real parameter. In Sections 10.9 and 10.10 we present some of the major results concerning asymptotically efficient fixed-width confidence intervals. Section 10.9 is devoted entirely to the normal case, whereas Section 10.10 deals with the general case.

CHAPTER 2

Sufficient Statistics

2.1. INTRODUCTION AND EXAMPLES

One of the most important objectives in the primary stage of statistical analysis is to process the observed data and transform it to a form most suitable for decision making. This primary data processing generally reduces the size of the original sets of sample values to a relatively small number of statistics. It is desired, however, that no information relevant to the decision process will be lost in this primary data reduction. As shown later in the chapter, not all families of distribution functions allow such a reduction without losing information. On the other hand, there are families of distribution functions for which all the set of sample values can be reduced to a real valued statistic. The theory of sufficient statistics provides us with the necessary criteria for the characteristization of families of distribution functions and the corresponding transformations, which yield sufficiently informative nontrivial statistics.

A rigorous definition and characterization of the notion of *sufficient statistics* is introduced in Section 2.3, following a brief review of the background in probability theory. In this section we discuss the ideas of sufficiency only in a heuristic manner. We then exhibit the main principle by a series of examples involving various families of distribution functions.

Let $\mathbf{X}$ be an observed set of random variables representing the sample data. The statistical model specifies the sample space of $\mathbf{X}$ and the family of distribution functions $\mathfrak{F}$ to which the distribution of $\mathbf{X}$ belongs. The crux of the problem is that the analyst does not know the particular distribution from which the observations are generated. He has to infer from the observed values certain properties of the parent distribution. The question is whether there exists a statistic $T(\mathbf{X})$ which is simpler than $\mathbf{X}$ and yet sufficiently informative. The main formulation of the concept "sufficient statistic" is based on the assertion that if the conditional distribution of $\mathbf{X}$, given T, is independent of the actual (nonconditional) distribution of $\mathbf{X}$, then T is sufficiently informative for $\mathfrak{F}$. Indeed, if the conditional distribution law of $\mathbf{X}$ given T can be specified independently of $\mathcal{F}$, and the value of T is given, we can generate a sample $\mathbf{X}'$ which has the same distribution as that of $\mathbf{X}$ by using a table of random numbers. Thus if a statistic T has the property mentioned above, the sample values do not provide information on $\mathcal{F}$ more than T does; T is then called a sufficient statistic. We now present several examples.

Example 2.1. Let $X_1, \ldots, X_n$ be i.i.d. random variables having a Poisson distribution with parameter λ. Let $T_n = \sum_{i=1}^n X_i$. As is well known, the distribution of T_n is a Poisson with mean $n\lambda$. We verify now that T_n is a sufficient statistic in the above sense. For this purpose it is sufficient to prove that the conditional distribution of $X_1, \ldots, X_{n-1}$, given T_n, is independent of λ. Indeed the joint density of $X_1, \ldots, X_{n-1}, T_n$ is

$$(2.1.1) \qquad f^\lambda(x_1, \ldots, x_{n-1}, t) = e^{-\lambda n}\lambda^t\left[\prod_{i=1}^{n-1} x_i!\left(t - \sum_{i=1}^{n-1} x_i\right)!\right]^{-1},$$

for all $x_i = 0, 1, \ldots$ $(i = 1, 2, \ldots, n-1)$, $\sum_{i=1}^{n-1} x_i \leqslant t$, and $t = 0, 1, \ldots$. Furthermore the density of T_n is

$$(2.1.2) \qquad f^\lambda(t) = \frac{e^{-\lambda n}(\lambda n)^t}{t!}, \quad t = 0, 1, \ldots.$$

Hence the conditional density of $X_1, \ldots, X_{n-1}$, given T_n, is

(2.1.3)

$$g(x_1, \ldots, x_{n-1} \mid t) = \frac{f^\lambda(x_1, \ldots, x_{n-1}, t)}{f^\lambda(t)}$$

$$= \frac{t!}{\prod_{i=1}^{n-1} x_i!\,(t - \sum_{i=1}^{n-1} x_i)!}\frac{1}{n^t}, \quad \text{for} \begin{cases} x_i = 0, 1, \ldots, \\ \sum_{i=1}^{n-1} x_i \leqslant t. \end{cases}$$

The density function, (2.1.3), is a multinomial density, independent of λ. Therefore the conditional distribution of any statistic $\phi(X_1, \ldots, X_{n-1})$, given T_n, is independent of λ. ■

Example 2.2. Let $X_1, \ldots, X_n$ be i.i.d. random variables having an $\mathcal{N}(\mu, 1)$ distribution, $-\infty < \mu < \infty$. The joint distribution of the vector $\mathbf{X} = (X_1, \ldots, X_n)'$ is the multivariate $\mathcal{N}(\mu\mathbf{1}_n, I_n)$. We show now that the sample mean $\bar{x} = (1/n)\mathbf{1}_n'\mathbf{X}$ is a sufficient statistic.

Make the transformation $\mathbf{Y} = \mathbf{HX}$, where $\mathbf{H}$ is an orthogonal matrix whose last row vector is $(1/\sqrt{n})\mathbf{1}_n'$ and the other $n - 1$ row vectors are arbitrary. The transformation $\mathbf{H}\colon \mathbf{X} \to \mathbf{Y}$ is known in a particular form as the Helmert transformation (see Kendall and Buckland [1], p. 126). The distribution of $\mathbf{Y}$ is $\mathcal{N}(\mu\mathbf{H}\mathbf{1}_n, I_n)$ and, as is easily verified, $\mathbf{H}\mathbf{1}_n = (\mathbf{0}', \sqrt{n})'$.

Thus $Y_1, \ldots, Y_{n-1}$ are i.i.d., having an $\mathcal{N}(0, 1)$ distribution independent of Y_n, which has a $\mathcal{N}(\sqrt{n}\,\mu, 1)$ distribution. Hence the conditional distribution of $Y_1, \ldots, Y_{n-1}$, given $Y_n = \sqrt{n}\,\bar{X}$, is $\mathcal{N}(\mathbf{0}, I_{n-1})$. We therefore conclude that the conditional distribution of any measurable function of $\mathbf{X}$ given $\bar{X}$ is independent of μ, and the statistic is sufficient for the family of normal distributions $\mathcal{N}(\mu, 1)$; $-\infty < \mu < \infty$. ■

Example 2.3. Let $X_1, \ldots, X_n$ be i.i.d. random variables having a 2-parameter negative exponential density

$$f(x; \alpha, \beta) = \begin{cases} 0 & \text{if } X < \alpha, \\ \dfrac{1}{\beta} \exp\left\{-\dfrac{x - \alpha}{\beta}\right\} & \text{if } x \geqslant \alpha, \end{cases} \tag{2.1.4}$$

where $-\infty < \alpha < \infty$, $0 < \beta < \infty$; α is called the location parameter and β, the scale parameter. This statistical model is prevalent in systems reliability theory, where X denotes the life length of a system. The location parameter (here $0 \leqslant \alpha < \infty$) is the minimal life of a system. Let $X_{(1)} \leqslant \cdots \leqslant X_{(n)}$ denote the order statistic of the given sample. The order statistic corresponds in a life testing of a given system to the time points of failure of independent and identical systems. Let $T^*_{n-1} = \sum_{i=2}^{n} X_{(i)}$. We show in this example that $(X_{(1)}, T^*_{n-1})$ is a pair of sufficient statistics. To verify this assertion, consider the transformations

$$\begin{aligned} Y_1 &= nX_{(1)}, \\ Y_2 &= (n-1)[X_{(2)} - X_{(1)}], \\ &\ \vdots \\ Y_n &= X_{(n)} - X_{(n-1)}. \end{aligned} \tag{2.1.5}$$

The joint density of $Y_1, \ldots, Y_n$ is

$$\begin{aligned} f(y_1, \ldots, y_{n-1}) \\ = \beta^{-n} \exp\left(-\frac{1}{\beta}\sum_{i=1}^{n} y_i + \frac{n\alpha}{\beta}\right), \qquad n\alpha \leqslant y_1, 0 \leqslant y_2, \ldots, y_n \leqslant \infty. \end{aligned} \tag{2.1.6}$$

We make a further transformation:

$$\begin{aligned} U_1 &= Y_2, \\ U_2 &= Y_2 + Y_3, \\ &\ \vdots \\ U_{n-2} &= Y_2 + Y_3 + \cdots + Y_{n-1}, \\ T_{n-1} &= Y_2 + Y_3 + \cdots + Y_{n-1} + Y_n. \end{aligned} \tag{2.1.7}$$

Deriving the joint density of $Y_1, U_1, \ldots, U_{n-2}, T_{n-1}$ from (1.1.6) and integrating out $U_1, \ldots, U_{n-2}$ we find that the marginal joint density of (Y_1, T_{n-1}) is

(2.1.8)
$$g(y, t) = \beta^{-1} \exp\left\{-\frac{y - n\alpha}{\beta}\right\} \times \frac{t^{n-2}}{(n-2)!\,\beta^{n-1}} \exp\left\{-\frac{t}{\beta}\right\}, \qquad n\alpha \leqslant y \leqslant \infty, 0 \leqslant t \leqslant \infty.$$

We conclude that Y_1 and T_{n-1} are independent and that $Y_1 + n\alpha$ has a negative exponential distribution (location at zero) with expectation β, and T_{n-1} has a gamma distribution, $\mathcal{G}(1/\beta, n-1)$.† Moreover, from the joint density of $Y_1, U_1, \ldots, U_{n-2}, T_{n-1}$ and (2.1.8) we find that the conditional joint density of $u_1, u_2, \ldots, u_{n-2}$, given (Y_1, T_{n-1}), is

(2.1.9)
$$h(u_1, \ldots, u_{n-2} \mid y, t) = \frac{(n-2)!}{t^{n-2}}, \qquad 0 \leqslant u_1 \leqslant u_2 \leqslant \cdots \leqslant u_{n-2} \leqslant t.$$

Thus (Y_1, T_{n-1}) is a pair of sufficient statistics and, equivalently, since $\sum_{i=1}^{n} X_{(i)} = \sum_{i=1}^{n} Y_i$, the pair $(X_{(1)}, T_{n-1}^*)$ is a sufficient statistic.

It is important to notice in the above examples that if the values of the sufficient statistics are known, one can generate, with the aid of tables of random numbers, a sample having a distribution P equal to the distribution of the original sample. Indeed, since the conditional distribution of any other statistic, given the value of the sufficient statistic, is independent of the class P of possible probability measures (and is thus known), the value of the observed random variable can be generated by the common Monte Carlo procedure (see Shreider [1]) from this conditional distribution. To demonstrate this, consider again the case of Example 2.3. We are given the values of $X_{(1)}$, say x, and T_{n-1}^*, say t^*. By the simple transformation $Y_1 = nX_{(1)}$ and $T_{n-1} = X_{(1)} + T_{n-1}^* - Y_1 = T_{n-1}^* - (n-1)X_{(1)}$ we find $y = nx$ and $t = t^* - (n-1)x$. Given the pair (y, t) we generate $u_1, \ldots, u_{n-2}$ according to (2.1.9). The density (2.1.9) is that of $(n-2)$ randomly distributed points on the interval $[0, t]$. Accordingly, we first generate $n-2$ values of i.i.d. random variables having a uniform distribution over the interval $[0, t]$. Given these randomly generated values we obtain $u_1, \ldots, u_{n-2}$ by ordering the sample values. The values $X_{(1)} \leqslant \cdots \leqslant X_{(n)}$ are determined by transformations inverse to (2.1.7) and (2.1.5). Finally, since $\{X_{(1)}, \ldots, X_{(n)}\}$ is a sufficient statistic and $f(X_1, \ldots, X_n \mid X_{(1)}, \ldots, X_{(n)}) = 1/n!$, we choose at random, with probability $1/n!$, a permutation of $(X_{(1)}, \ldots, X_{(n)})$. This

† We denote by $\mathcal{G}(\lambda, p)$ the gamma distribution with a scale parameter λ, $0 < \lambda < \infty$, and expectation p/λ.

permutation has the same joint distribution as that of the original sample. From the viewpoint of probability theory, and thus for all purposes of statistical inference, the original sample and the one generated by the process described here are equivalent. ■

2.2. STATISTICS, SUBFIELDS AND CONDITIONAL EXPECTATION

In this section we review the measure-theoretic notions required for further treatment of the subject. The reader who has insufficient preparation for studying the material presented here is advised to consider the books of either Loève [2], or Halmos [1], or Royden [1]. A complementary treatment of the subject is provided by examples that are given at the end of the chapter among the problems for solution.

Given a basic probability space $(\mathcal{X}, \mathcal{B}, P)$, $P \in \mathcal{P}$, associated with the observed random variable X, we consider a mapping or transformation $Y: \mathcal{X} \to \mathcal{Y}$, where $\mathcal{Y}$ is contained in a Euclidean space. Let $\mathcal{B}_Y$ be the smallest Borel sigma-field on $\mathcal{Y}$.

For every Borel set $B \in \mathcal{B}$ we denote by $Y(B)$ the set of all $\mathcal{Y}$ points that are obtained by mapping the points of B via Y, that is,

$$(2.2.1) \qquad Y(B) = \{y; y = Y(x), x \in B\}.$$

Let C be a Borel set in $\mathcal{B}_Y$. The set of all points in $\mathcal{X}$ whose Y images are in C is the Y-inverse image of C, $Y^{-1}(C)$, that is,

$$(2.2.2) \qquad Y^{-1}(C) = \{x; Y(x) \in C\}.$$

A mapping $Y: (\mathcal{X}, \mathcal{B}) \to (\mathcal{Y}, \mathcal{B}_Y)$ is called $\mathcal{B}$-measurable if $Y^{-1}(C) \in \mathcal{B}$ for all $C \in \mathcal{B}_Y$. A $\mathcal{B}$ measurable transformation $Y: (\mathcal{X}, \mathcal{B}) \to (\mathcal{Y}, \mathcal{B}_Y)$ is called a *statistic* if it does not depend on the class of probability measures $\mathcal{P}$; for example, let X be a real-valued random variable, $\mathcal{X}$, the entire real line, $\mathcal{B}$, the smallest Borel sigma-field on the real line. Let $\mathcal{P}$ be the class of all normal distributions $\{\mathcal{N}(\mu, \sigma^2); -\infty < \mu < \infty, 0 < \sigma^2 < x\}$. Then $Y = X^2$ is a statistic, whereas $Z = X^2/\sigma^2$, $0 < \sigma^2 < \infty$, is a $\mathcal{B}$-measurable transformation but not a statistic.

Let Y be any $\mathcal{B}$-measurable transformation of $(\mathcal{X}, \mathcal{B}) \to (\mathcal{Y}, \mathcal{B}_Y)$. Let $Y^{-1}(\mathcal{B}_Y)$ denote the collection of all inverse images of Borel sets $C \in \mathcal{B}_Y$. Since every $Y^{-1}(C) \in \mathcal{B}$, $Y^{-1}(\mathcal{B}_Y) \subset \mathcal{B}$. In other words, the collection of sets of x values, $Y^{-1}(\mathcal{B}_Y)$, contains only Borel sets of $\mathcal{B}$. It may, however, contain a smaller number of Borel sets than $\mathcal{B}$ does. A simple example to illustrate this point is the following: as before, let $\mathcal{X}$ be the entire real line and $\mathcal{B}$, the Borel sigma-field over $\mathcal{X}$. Every open interval in $\mathcal{X}$ belongs to $\mathcal{B}$ (together with many more sets). Consider the transformation $Y = X^2$.

Thus $\mathcal{Y}$ is the non-negative part of the real line; $\mathcal{B}_Y$ is the Borel sigma-field over $[0, \infty)$. Let $C = (\frac{1}{4}, 1)$; $C \in \mathcal{B}_Y$. The inverse image of C is $Y^{-1}(C) = (-1, -\frac{1}{2}) \cup (\frac{1}{2}, 1)$. Obviously $(-1, -\frac{1}{2}) \cup (\frac{1}{2}, 1)$ belongs to $\mathcal{B}$. However, $\mathcal{B}$ contains also the individual open intervals $(-1, -\frac{1}{2})$ and $(\frac{1}{2}, 1)$, whereas $Y^{-1}(\mathcal{B}_Y)$ does not. The reason is that $Y = X^2$ is *not* a 1:1 transformation. We can show that whenever $Y: (\mathcal{X}, \mathcal{B}) \to (\mathcal{Y}, \mathcal{B}_Y)$ is 1:1 then $Y^{-1}(\mathcal{B}_Y) = \mathcal{B}$. We can easily show that $Y^{-1}(\mathcal{B}_Y)$ for any measurable transformation Y is itself a sigma-field. We therefore call it a *sigma-subfield* of $\mathcal{B}$.

A sigma-finite measure μ on $(\mathcal{X}, \mathcal{B})$ is a non-negative set function:

1. $\mu(\phi) = 0$, where ϕ denotes the null set.
2. If $B_1, B_2, \ldots \in \mathcal{B}$, $B_i \cap B_j = \phi$ all $i \neq j$, then $\mu(\bigcup_{i=1}^{\infty} B_i) = \sum_{i=1}^{\infty} \mu(B_i)$.
3. There exists a partition† $\{B_1, \ldots, B_n, \ldots\}$ of $\mathcal{X}$ such that $\mu(B_i) < \infty$ for all B_i $(i = 1, 2, \ldots)$ in this partition.

The following is a simple example of a sigma-finite measure. Let $(\mathcal{X}, \mathcal{B})$ be the entire real line with the Borel sigma-field on it. For every Borel set $B \in \mathcal{B}$, let $\mu(B) =$ number of integers in B. It is easy to verify that μ satisfies all three specified properties; μ is not a finite measure.

Let μ and ν be two sigma-finite measures on the same measure space $(\mathcal{X}, \mathcal{B})$; ν is said to be *absolutely continuous* with respect to μ if $\mu(B) = 0$ implies $\nu(B) = 0$. We designate this relationship to $\nu \ll \mu$. If μ and ν are two sigma-finite measures such that $\nu \ll \mu$ and $\mu \ll \nu$, they are called *equivalent*. The Radon-Nikodym theorem (see [21]) states: *Let ν and μ be sigma-finite measures on $(\mathcal{X}, \mathcal{B})$. Then $\nu \ll \mu$ if and only if there exists a $\mathcal{B}$-measurable non-negative function f on $\mathcal{X}$ (we denote $f \in \mathcal{B}$) so that*

$$\nu(B) = \int_B f(x)\mu(\mathrm{d}x), \qquad \text{all } B \in \mathcal{B}. \tag{2.2.3}$$

The function $f(x)$ is called the *Radon-Nikodym derivative* of ν with respect to μ, that is,

$$f(x) = \frac{\nu(\mathrm{d}x)}{\mu(\mathrm{d}x)} \quad \text{a.s. } [\mu]. \tag{2.2.4}$$

The function $f(x)$ in (2.2.4) is determined up to a μ-null set; we denote this by a.s. $[\mu]$. There might be different versions of f, which vary only on a set of μ-measure zero; for example, let $\mathcal{X}$ be the entire real line, $\mathcal{B}$, the Borel sigma-field on $\mathcal{X}$. Let μ be the sigma-finite measure on $(\mathcal{X}, \mathcal{B})$ which assign to every Borel set $B \in \mathcal{B}$ the number of non-negative integers lying in B. Let ν on $(\mathcal{X}, \mathcal{B})$ be the Poisson probability measure. If $B \in \mathcal{B}$ is such that

† A collection of subsets of $\mathcal{X}$ is called a partition of $\mathcal{X}$ if all the subsets are mutually exclusive and their union is $\mathcal{X}$.

$\mu(B) = 0$ then $\nu(B) = 0$. Hence $\nu \ll \mu$ and we have

$$\nu(B) = \int_B e^{-\lambda} \frac{\lambda^x}{x!} \mu(\mathrm{d}x), \quad \text{every } B \in \mathcal{B}. \tag{2.2.5}$$

$0 < \lambda < \infty$. The Radon Nikodym derivative $d\nu/d\mu$ is usually taken, as in (2.2.5), to be

$$f(x) = \begin{cases} e^{-\lambda} \dfrac{\lambda^x}{x!}, & x = 0, 1, \ldots, \\ 0, & \text{otherwise.} \end{cases} \tag{2.2.6}$$

But on the Borel sets of μ-measure zero we can consider for f any arbitrary finite version; for example, the polygonal function

$$f^*(x) = \begin{cases} f(x), & \text{if } x \text{ is an integer,} \\ w(x)f([x]) + (1 - w(x))f([x] + 1), & \text{otherwise,} \end{cases} \tag{2.2.7}$$

where $[x]$ denotes the largest integer not exceeding x and $w(x) = [x] + 1 - x$ is an equivalent version of the Radon-Nikodym derivative, that is, $f^*(x) = f(x)$ a.s. $[\mu]$.

We now discuss the transformation induced on the class of probability measures $\mathcal{P}$ by a statistic Y. As previously denoted, the basic probability space is $(\mathcal{X}, \mathcal{B}, P)$, where P belongs to a specified class $\mathcal{P}$. Let Y be a $\mathcal{B}$-measurable statistic $Y: (\mathcal{X}, \mathcal{B}) \to (\mathcal{Y}, \mathcal{B}_Y)$. The class $\mathcal{P}$ is transformed to a class $\mathcal{P}^Y$ of induced probability measures P^Y on $(\mathcal{Y}, \mathcal{B}_Y)$ defined by

$$P^Y(G) = P[Y^{-1}(G)], \quad \text{all } G \in \mathcal{B}_Y. \tag{2.2.8}$$

Define the sigma-finite measure on $\mathcal{B}_Y$,

$$Q^Y[G] = P[B_0 \cap Y^{-1}(G)], \quad B_0 \in \mathcal{B}, G \in \mathcal{B}_Y, \quad P \in \mathcal{P}.$$

Obviously $Q^Y \ll P^Y$, hence by the Radon-Nikodym theorem there exists a non-negative function $g_P(y; B_0)$ on $\mathcal{Y}$ such that

$$\begin{aligned} Q^Y[G] &= P[B_0 \cap Y^{-1}(G)] \\ &= \int_G g_P(y; B_0) P^Y(\mathrm{d}y), \quad \text{every } G \in \mathcal{B}_Y. \end{aligned} \tag{2.2.9}$$

The function $g_P(y; B_0)$ is called the *conditional probability* of B_0, given $\{Y = y\}$, and is determined uniquely up to a P^Y-null set. We also denote $g_P(y; B_0) \equiv P[B_0 \mid Y = y]$. The conditional probability function $g_P(y; B_0)$ may or may not depend on the original probability measure P. It depends on the particular properties of the statistic Y. We generalize (2.2.9) by defining the conditional expectation function.

Let f be a non-negative function on $\mathfrak{X}$, which is $\mathscr{B}$-measurable and P-integrable. Define the sigma-finite measure on $(\mathfrak{X}, \mathscr{B})$

$$\nu_P(B) = \int_B f(x)P(\mathrm{d}x), \quad B \in \mathscr{B}. \tag{2.2.10}$$

In the general theory we allow $\nu_P(\mathfrak{X}) = \infty$, provided ν_P is a sigma-finite measure. It is implied from (2.2.10) and from the definition of the Lebesgue-Stieltjes integral that $\nu_P \ll P$. Furthermore let Y be a $\mathscr{B}$-measurable statistic, $Y: (\mathfrak{X}, \mathscr{B}) \to (\mathfrak{Y}, \mathscr{B}_Y)$, and let P^Y be the induced probability measure on $(\mathfrak{Y}, \mathscr{B}_Y)$. Let $\nu_P{}^Y$ be the induced measure on $(\mathfrak{Y}, \mathscr{B}_Y)$ defined by $\nu_P{}^Y(G) = \nu_P(Y^{-1}(G))$ for all $G \in \mathscr{B}_Y$. We immediately obtain $\nu_P{}^Y \ll P^Y$, since $\nu_P \ll P$. Hence by the Radon-Nikodym theorem there exists a non-negative function g_P on $\mathfrak{Y}$, $g_P \in \mathscr{B}_Y$, such that

$$\nu_P{}^Y(G) = \int_G g_P(y)P^Y(\mathrm{d}y). \tag{2.2.11}$$

On the other hand, since $\nu_P{}^Y(G) = \nu_P(Y^{-1}(G))$ we obtain from (2.2.10) and (2.2.11) that

$$\int_G g_P(y)P^Y(\mathrm{d}y) = \int_{Y^{-1}(G)} f(x)P(\mathrm{d}x), \quad G \in \mathscr{B}_Y. \tag{2.2.12}$$

The function $g_P(y)$ is called the *conditional expectation* of $f(X)$, given $\{Y = y\}$, and is designated also by $E_P\{f(X) \mid Y = y\}$. Since Y is defined on $\mathfrak{X}$, g_P on $\mathfrak{Y}$ is the composite function $g_P Y$ on $\mathfrak{X}$. Thus we consider the conditional expectation $E_P\{f(X) \mid Y = y\}$ a function on $\mathfrak{X}$, which assumes constant values on the cosets $A(y) = \{x;\ Y(x) = y\}$. The conditional expectation function $E_P\{f(X) \mid Y = y\}$ is measurable with respect to the sigma-subfield $Y^{-1}(\mathscr{B}_Y)$ and is determined uniquely up to sets of P-measure zero.

When the function f on $\mathfrak{X}$ may assume negative values, we extend the definition of the conditional expectation by making the decomposition $f(x) = f^+(x) - f^-(x)$, where $f^+(x) = \max\{0, f(x)\}$ and $f^-(x) = \max\{0, -f(x)\}$. We then define

$$E_P\{f(X) \mid Y = y\} = E_P\{f^+(X) \mid Y = y\} - E_P\{f^-(X) \mid Y = y\}, \tag{2.2.13}$$

provided at least one of the terms on the right-hand side is finite.

The conditional probability of a Borel set $B \in \mathscr{B}$, given $\{Y = y\}$, is obtained from (2.2.13) by the function

$$f(X) = I_B(X) = \begin{cases} 1, & \text{if } X \in B, \\ 0, & \text{otherwise.} \end{cases}$$

Example 2.4. Let $X_1, \ldots, X_n$ be random variables having a symmetric joint distribution, in the sense that if $\mathcal{B}$ is the Borel sigma-field over the Euclidean n-space, and $B \in \mathcal{B}$, then

$$P[(X_1, \ldots, X_n) \in B] = P[(X_{i_1}, \ldots, X_{i_n}) \in B]$$

for every permutation $(i_1, \ldots, i_n)$ of $(1, 2, \ldots, n)$. Let $f(X_1, \ldots, X_n)$ be any $\mathcal{B}$-measurable and P-integrable function. Let $Y = (X_{(1)}, \ldots, X_{(n)})$, where $X_{(1)} \leqslant \cdots \leqslant X_{(n)}$, be the order statistic. We show that

$$E\{f(X_1, \ldots, X_n) \mid Y\} = \frac{1}{n!} \sum_{(i_1, \ldots, i_n)} f(X_{i_1}, \ldots, X_{i_n}),$$

where the sum ranges over all permutations.

The transformation Y transforms the sample space $\mathcal{X} \subset E^{(n)}$ to a corresponding set $\mathcal{Y}$ of ordered vectors in $E^{(n)}$. Let $\mathcal{B}_Y$ be the smallest sigma-field over $\mathcal{Y}$. Every Borel set B in $Y^{-1}(\mathcal{B}_Y)$ is invariant to permutation of the coordinates of its points; that is, if $(x_1, \ldots, x_n) \in B$ then $(x_{i_1}, \ldots, x_{i_n}) \in B$ for all permutations $(i_1, \ldots, i_n)$ of $(1, 2, \ldots, n)$. Let C be any Borel set in $\mathcal{B}_Y$. Then

$$\int_{Y^{-1}(G)} f(x_1, \ldots, x_n) P(\mathrm{d}x_1, \ldots, \mathrm{d}x_n) = \int_{Y^{-1}(G)} f(x_{i_1}, \ldots, x_{i_n}) P(\mathrm{d}x_1, \ldots, \mathrm{d}x_n),$$

for all permutations $(i_1, \ldots, i_n)$ of $(1, 2, \ldots, n)$, since by the symmetry of P, $P(\mathrm{d}x_1, \ldots, \mathrm{d}x_n) = P(\mathrm{d}x_{i_1}, \ldots, \mathrm{d}x_{i_n})$ for all permutations, and $Y^{-1}(G) \in Y^{-1}(\mathcal{B}_Y)$. Therefore

$$\int_{Y^{-1}(C)} f(x_1, \ldots, x_n) P(\mathrm{d}x_1, \ldots, \mathrm{d}x_n) = \int_{Y^{-1}(C)} \left[\frac{1}{n!} \sum f(x_{i_1}, \ldots, x_{i_n})\right] P(\mathrm{d}x_1, \ldots, \mathrm{d}x_n),$$

where the sum in the right-hand side ranges over all permutations. The function $1/n! \sum f(x_{i_1}, \ldots, x_{i_n})$ is symmetric, and can be written as $g(y_1, \ldots, y_n)$, where $y_i = x_{(i)}$ $(i = 1, \ldots, n)$ is the ith largest coordinate. Thus, since

$$\int_{Y^{-1}(C)} \left\{\frac{1}{n!} \sum f(x_{i_1}, \ldots, x_{i_n})\right\} P(\mathrm{d}x_1, \ldots, \mathrm{d}x_n) = \int_G g(y_1, \ldots, y_n) P^Y(\mathrm{d}y_1, \ldots, \mathrm{d}y_n)$$

for every Borel set C in $\mathcal{B}_Y$, then according to (2.12),

$$E\{f(X_1, \ldots, X_n) \mid Y\} = \frac{1}{n!} \sum f(X_{i_1}, \ldots, X_{i_n}).$$

It is important to notice that this conditional expectation does not depend on P, as long as P is a symmetric probability measure with respect to permutations. In the more general case a similar result can be obtained if P is absolutely continuous with respect to a sigma-finite measure μ on $(\mathcal{X}, \mathcal{B})$, where μ is symmetric. In this case let $h(x_1, \ldots, x_n) = P(dx_1, \ldots, dx_n)/\mu(dx_1, \ldots, dx_n)$ a.s. $[P]$ be the Radon-Nikodym derivative; then we have for every $\mathcal{B}$-measurable and P-integrable function on $\mathcal{X}$ that

$$E\{f(X_1, \ldots, X_n) \mid Y\} = \frac{\sum f(X_{i_1}, \ldots, X_{i_n}) h(X_{i_1}, \ldots, X_{i_n})}{\sum h(X_{i_1}, \ldots, X_{i_n})} \text{ a.s. } [P].$$

2.3. DEFINITION OF SUFFICIENT STATISTICS AND THE NEYMAN-FISHER FACTORIZATION THEOREM

A statistic $T: (\mathcal{X}, \mathcal{B}) \rightarrow (\mathcal{T}, \mathcal{F})$ is called a *sufficient statistic* for $\mathcal{P}$ if, for every two Borel sets $B \in \mathcal{B}$, $F \in \mathcal{F}$,

$$P[B \cap T^{-1}(F)] = \int_F P[B \mid T = t] \, P^T(dt) \tag{2.3.1}$$

where $P[B \mid T = t]$ is independent of the class of probability measures $\mathcal{P}$.

As proven in Example 2.4, if the family $\mathcal{P}$ consists of symmetric probability measures then the order statistic is a sufficient statistic for $\mathcal{P}$, since the conditional expectation of every measurable and integrable function, given the order statistic, is independent of $\mathcal{P}$. Accordingly, whenever a sample consists of n observations on independent random variables (not necessarily of the same distribution) the order statistic is the least reduction of the original data which does not lose information on $\mathcal{P}$. The examples of sufficient statistics given in Section 2.1 were following in essence the above definition of sufficiency. We can, however, apply a more convenient criterion for determining whether a statistic is sufficient. This is the famous Neyman-Fisher factorization criterion (see Neyman [1]; Fisher [2]). To demonstrate this factorization criterion, we consider again the case of i.i.d. random variables $X_1, \ldots, X_n$ having a normal distribution of the family $\mathcal{P} = \{\mathcal{N}(\theta, 1); -\infty < \theta < \infty\}$. As shown in Example 2.1, the sample mean $\bar{X}_n$ is a sufficient statistic. Now the joint density of $X_1, \ldots, X_n$ is

(2.3.2)

$$\begin{aligned} f(x_1, \ldots, x_n; \theta) &= (2\pi)^{-n/2} \exp\left\{-\frac{1}{2}\sum_{i=1}^{n}(x_i - \theta)^2\right\} \\ &= (2\pi)^{-n/2} \exp\left\{-\frac{1}{2}\sum_{i=1}^{n}(x_i - \bar{X}_n)^2\right\} \cdot \exp\left\{-\frac{n}{2}(\bar{X}_n - \theta)^2\right\}. \end{aligned}$$

Thus the joint density is factored into a product of two non-negative measurable functions, one of which does not depend on the unknown parameter θ and the other of which depends on $x_1, \ldots, x_n$ only through the sufficient statistic. This is the essence of the factorization criterion. In this section we prove it rigorously, following Halmos and Savage [1]. We start with several lemmas.

Lemma 2.3.1. *Let* $T\colon (\mathfrak{X}, \mathfrak{B}, P) \to (\mathfrak{T}, \mathfrak{F}, P^T)$ *be* $\mathfrak{B}$*-measurable and let* f *be a* $\mathfrak{B}$*-measurable function on* $\mathfrak{X}$*. A necessary and sufficient condition for the existence of an* $\mathfrak{F}$*-measurable function* g *on* $\mathfrak{T}$ *such that* $f(x) = g(T(x))$ *is that* $f \in T^{-1}(\mathfrak{F})$.

Proof. Since T is $\mathfrak{B}$-measurable, $T^{-1}(F) \in \mathfrak{B}$ for all $F \in \mathfrak{F}$. Hence $T^{-1}(F) \in T^{-1}(\mathfrak{F})$ for all $F \in \mathfrak{F}$. Let g be an $\mathfrak{F}$-measurable transformation, $g\colon (\mathfrak{T}, \mathfrak{F}, P^T) \to (\mathfrak{Z}, \mathcal{A}, P^{gT})$, and let $f(X) = gT(X)$. Let $B = \{x; gT(x) \in A\}$, $A \in \mathcal{A}$. Then $A = gT(B)$ and $T(B) = g^{-1}(A) \in \mathfrak{F}$. It follows that $B = T^{-1}g^{-1}(A) \in T^{-1}(\mathfrak{F})$. This proves the necessity. To prove the sufficiency of the condition, let t_0 be an arbitrary point in $\mathfrak{T}$ and define $B_0 = T^{-1}(t_0)$. Let $x_0 \in B_0$, $A_0 = \{x; f(x) = f(x_0)\}$. Since $f \in T^{-1}(\mathfrak{F})$, then $F = T(A_0) \in \mathfrak{F}$. Furthermore $x_0 \in B_0$ and $T(B_0) = t_0$ imply that $t_0 \in F$. Therefore B_0 is included in $T^{-1}(F) = A_0$. Thus f assumes a constant value on B_0, and consequently the relationship $g(t_0) = f(x_0)$ defines a function g on $\mathfrak{T}$. The uniqueness of g follows from the uniqueness of the transformation $T\colon \mathfrak{X} \to \mathfrak{T}$. Finally, if C is any set in the Borel field on the range of g, $g^{-1}(C) \in \mathfrak{F}$. Hence g is $\mathfrak{F}$-measurable.

Lemma 2.3.2. *Let* T *be a* $\mathfrak{B}$*-measurable transformation,* $T\colon (\mathfrak{X}, \mathfrak{B}, P) \to (\mathfrak{T}, \mathfrak{F}, P^T)$*. If* g *is (real valued) measurable* $\mathfrak{F}$ *and integrable* P^T *then, for every* $F \in \mathfrak{F}$,

(2.3.3) $$\int_F g(y) P^T(\mathrm{d}y) = \int_{T^{-1}(F)} gT(x)\, P(\mathrm{d}x).$$

Proof. According to the previous lemma $gT \in T^{-1}(\mathfrak{F})$. Moreover, since g is P^T-integrable, we obtain that the usual steps of constructing the integral in the left-hand side of (2.3.3) lead to the integral on the right-hand side of (2.3.3).

Lemma 2.3.3. *Let* T *be a* $\mathfrak{B}$*-measurable transformation;* $T\colon (\mathfrak{X}, \mathfrak{B}, P) \to (\mathcal{T}, \mathfrak{F}, P^T)$*. Let* $g \geqslant 0$ *be* $\mathfrak{F}$*-measurable and* P^T*-integrable. Define a sigma-finite measure* ν_P *on* $(\mathfrak{X}, \mathfrak{B})$ *by* $\mathrm{d}\nu_P = gT\,\mathrm{d}P$*; then* $d\nu_P{}^T = g\,\mathrm{d}P^T$.

Proof. For every $F \in \mathfrak{F}$,

(2.3.4) $$\nu_P{}^T(F) = \nu_P(T^{-1}(F)) = \int_{T^{-1}(F)} gT(x)\, P(\mathrm{d}x).$$

By Lemma 2.3.2.,

$$\int_{T^{-1}(F)} gT(x)\, P(\mathrm{d}x) = \int_F g(t)\, P^T(\mathrm{d}t). \tag{2.3.5}$$

Finally, this lemma is implied from (2.3.4) and (2.3.5).

Lemma 2.3.4. *Let T be a $\mathcal{B}$-measurable transformation, $T: (\mathcal{X}, \mathcal{B}, P) \to (\mathcal{T}, \mathcal{F}, P^T)$. Let f, gT and g be defined on $\mathcal{X}$ and $\mathcal{T}$ and measurable $\mathcal{B}$, $T^{-1}(\mathcal{F})$, and $\mathcal{F}$, respectively. Then*

$$E_P\{f(X)\, gT(X) \mid T(X) = t\} = g(t)\, E_P\{f(X) \mid T(X) = t\} \text{ a.s. } [P]. \tag{2.3.6}$$

The proof of this simple lemma is left to the reader.

Definition. (i) *Let $\mathcal{P}$ denote a family (class) of sigma-finite measures and μ a sigma-finite measure, all defined on the same space* $(\mathcal{X}, \mathcal{B})$. *We say that $\mathcal{P}$ is dominated by μ, $\mathcal{P} \ll \mu$, if $P \ll \mu$ for all $P \in \mathcal{P}$.* (ii) *Let $\mathcal{P}$ and $\mathcal{M}$ be two families of sigma-finite measures on* $(\mathcal{X}, \mathcal{B})$. *We say that $\mathcal{P}$ is dominated by $\mathcal{M}$, $\mathcal{P} \ll \mathcal{M}$, if $\mu(B) = 0$ for all $\mu \in \mathcal{M}$ implies that $P(B) = 0$ for all $P \in \mathcal{P}$.* (iii) *$\mathcal{P}$ is said to be* equivalent *to $\mathcal{M}$, $\mathcal{P} \equiv \mathcal{M}$, if $\mathcal{P} \ll \mathcal{M}$ and $\mathcal{M} \ll \mathcal{P}$.*

Any countable family of sigma-finite measures $\mathcal{P} = \{(P_1, P_2, \ldots\}$ is equivalent to a sigma-finite measure

$$\lambda(B) = \sum_{n=1}^{\infty} 2^{-n} P_n(B), \quad B \in \mathcal{B}. \tag{2.3.7}$$

Indeed, since for every $B \in \mathcal{B}$, $P_n(B) \geqslant 0$ for all $n = 1, 2, \ldots$, $\lambda(B) = 0$ if and only if $P_n(B) = 0$ for all $n = 1, 2, \ldots$. Thus $\lambda \equiv \mathcal{P}$. It should be remarked that, without loss of generality, the equivalent measure λ can be taken as a finite measure, since if $\lambda(B)$ is not finite consider

$$\tilde{\lambda}(B) = \frac{\lambda(A_i \cap B)}{\lambda(A_i)}, \quad B \in \mathcal{B},$$

where $\{A_1, A_2, \ldots\}$ is a partition of $\mathcal{X}$ such that $\lambda(A_i) > 0$, $\lambda(A_i) < \infty$, for all $i = 1, 2, \ldots$. The equivalent measure λ is said to be *dense* in $\mathcal{P}$. We now give a more general definition.

Let $\mathcal{P}$ be a family of probability measures on a measure space $(\mathcal{X}, \mathcal{B})$. A subfamily $\mathcal{P}^*$ of $\mathcal{P}$, consisting of a countable number of probability measures, is called *dense* in $\mathcal{P}$ if and only if $\mathcal{P}^*$ is equivalent to $\mathcal{P}$ (see Sverdrup [25] and Bahadur [1; 2]).

For the above case of a countable number of measures the following mixture can be generally used as a dense measure:

$$\lambda(B) = \sum_{n=1}^{\infty} w_n P_n(B), \qquad B \in \mathcal{B}, \quad w_n \geqslant 0, \quad \sum_{n=1}^{\infty} w_n = 1. \tag{2.3.8}$$

For the sake of further discussion, assume that the family $\mathcal{F}$ is dominated by the sigma-finite measure λ. Let $f_P(x) = P(dx)/\lambda(dx)$ a.s. be the (generalized) *density function* of P (the Radon-Nikodym derivative of P with respect to λ). The *support* of P is the set

$$K_P = \{x; f_P(x) > 0\}. \tag{2.3.9}$$

A *kernel* K is a subset of $\mathfrak{X}$ such that, for some P in $\mathcal{F}$, $K \subset K_P$ and $P[K] > 0$. A *chain* is a union of disjoint kernels. Since $\mathcal{F} \ll \lambda$, the λ-measure of every kernel K is positive. We now prove that every dominated family of probability measures has a dense subfamily.

Lemma 2.3.5. (*Halmos and Savage* [13].) *If a family of probability measures $\mathcal{F}$ on $(\mathfrak{X}, \mathfrak{B})$ is dominated by a sigma-finite measure λ, then there exists a countable subfamily equivalent to P.*

Proof. Let $C_1, C_2, \ldots$ be a sequence of chains such that $\lim_{n\to\infty} \lambda(C_n) = \sup \lambda(C)$. Let $C^* = \bigcup_{n=1}^{\infty} C_n$. Then $\lambda(C^*) = \sup \lambda(C)$; C^* is a chain and is thus a countable union of disjoint kernels $K_1, K_2, \ldots$. Let P_n be a probability measure in $\mathcal{F}$ such that $P_n(K_n) > 0$, $n = 1, 2, \ldots$. We show that the countable subfamily $\mathcal{F}^* = \{P_1, P_2, \ldots\}$ is dense in P. Since $\mathcal{F}^* \subset \mathcal{F}$, the relationship $\mathcal{F}^* \ll \mathcal{F}$ obviously holds. It remains to establish the relationship $\mathcal{F} \ll \mathcal{F}^*$. In order to do this we have to prove that if $B \in \mathfrak{B}$ is such that $P_n(B) = 0$ for *all* $P_n \in \mathcal{F}^*$, then $P(B) = 0$ for all $P \in \mathcal{F}$. We prove first that, for every $P \in \mathcal{F}$, $P[B - C^*] = 0$. By negation, if $P[B - C^*] > 0$ for some P of $\mathcal{F}$ then, if K_P designates the corresponding support of P, $P[(B - C^*) \cap \bar{K}_P] = 0$, where $\bar{K}_P$ is the complement of K_P in $\mathfrak{X}$. Thus $P[(B - C^*) \cap K_P] > 0$. Therefore $(B - C^*) \cap K_P$ is a kernel of P and $\lambda((B - C^*) \cap K_P) > 0$. Furthermore $(B - C^*) \cap K_P$ is disjoint with C^*. Therefore $(B - C^*) \cap K_P \cup C^*$ is a chain with $\lambda[(B - C^*) \cap K_P \cup C^*] = \lambda[(B - C^*) \cap K_P] + \lambda(C^*) > \lambda(C^*)$. This contradicts the assumption that C^* is a chain of a maximal λ-measure. We conclude that $P[B - C^*] = 0$ for all $P \in \mathcal{F}$.

We have to show now that $P[B \cap C^*] = 0$ for all $P \in \mathcal{F}$. Since $P_n(B) = 0$ for every $P_n \in \mathcal{F}^*$, we have

$$P_n(B \cap K_n) = \int_{B \cap K_n} f_n(x)\lambda(dx) = 0, \quad \text{for every } P_n \in \mathcal{F}^*. \tag{2.3.10}$$

Since $B \cap K_n$ is a subset of the support of P_n, $f_n(x) > 0$ for each $x \in B \cap K_n$. Hence (2.3.10) implies that $\lambda(B \cap K_n) = 0$ for all $n = 1, 2, \ldots$ and $\lambda(B \cap C^*) = \sum_{n=1}^{\infty} \lambda(B \cap K_n) = 0$. Finally, since $\mathcal{F} \ll \lambda$, $P[B \cap C^*] = 0$ for all $P \in \mathcal{F}$. This completes the proof.

Theorem 2.3.1. (Halmos and Savage [13].) *Let $(\mathfrak{X}, \mathfrak{B}, P)$ be a probability space; P is a probability distribution belonging to a family $\mathcal{F}$, which is dominated* *by a sigma-finite measure μ. Let T be a statistic transforming $(\mathcal{X}, \mathcal{B}, \mathcal{P})$ to $(\mathcal{Y}, \mathcal{F}, P^T)$. A necessary and sufficient condition for the sufficiency of T with respect to $\mathcal{P}$ is the existence of a probability measure λ dense in $\mathcal{P}$ such that $P(\mathrm{d}x)/\lambda(\mathrm{d}x) \in T^{-1}(\mathcal{F})$ for all $P \in \mathcal{P}$.*

Remark. The condition of the theorem means that the generalized density of each P, with respect to λ, is a function of $T(X)$. That is, there exists a function g_P, which may depend on P, such that $P(\mathrm{d}x)/\lambda(\mathrm{d}x) = g_P T(x)$ a.s. $[P]$.

Proof. (i) Suppose that T is a sufficient statistic for the family $\mathcal{P}$. Thus, according to the definition of the sufficiency of a statistic,

$$P[B \cap T^{-1}(F)] = \int_F P[B \mid T = y] P^T(\mathrm{d}y),$$

for every $B \in \mathcal{B}$ and every $F \in \mathcal{F}$, where the conditional probability $P[B \mid T = y]$ does not depend on the probability measures in P.

According to the previous lemma, since $\mathcal{P} \ll \mu$ there exists a countable subfamily $\{P_1, P_2, \ldots\} \subset \mathcal{P}$ which is dense in $\mathcal{P}$. Define the probability measure

$$\lambda(B) = \sum_{n=1}^{\infty} 2^{-n} P_n(B), \quad B \in \mathcal{B}. \tag{2.3.11}$$

The quantity λ is equivalent to $\{P_1, P_2, \ldots\}$ and hence dense in $\mathcal{P}$. Thus for every $B \in \mathcal{B}$, $F \in \mathcal{F}$,

$$\begin{aligned} \lambda(B \cap T^{-1}(F)) &= \sum_{n=1}^{\infty} 2^{-n} \int_F P[B \mid T = y] P_n{}^T(\mathrm{d}y) \\ &= \int_F P[B \mid T = y] \lambda^T(\mathrm{d}y) \\ &= \int_{T^{-1}(F)} P[B \mid T(x)] \lambda(\mathrm{d}x), \end{aligned} \tag{2.3.12}$$

where

$$\lambda^T(F) = \lambda(T^{-1}(F)) = \sum_{n=1}^{\infty} 2^{-n} P_n(T^{-1}(F)) = \sum_{n=1}^{\infty} 2^{-n} P_n{}^T(F).$$

Thus according to (2.3.12), $P[B \mid T(X)]$ is also the conditional λ-probability of B, given T.

Let $f_P(x) = P(\mathrm{d}x)/\lambda(\mathrm{d}x)$ a.s. be the Radon-Nikodym derivative of P, with respect to λ. This is the generalized density of P with respect to λ. We show now that, for every $P \in \mathcal{P}$, $f_P(x) = g_P T(x)$.

For each $B \in \mathcal{B}$ we have

$$P[B] = \int P[B \mid T(x)]P(\mathrm{d}x) \tag{2.3.13}$$
$$= \int P[B \mid T(x)]f_P(x)\lambda(\mathrm{d}x)$$
$$= \int P[B \mid T = y]E_\lambda\{f_P(X) \mid T = y\}\lambda^T(\mathrm{d}y),$$

where the function $E_\lambda\{f_P(X) \mid T = y\} \in \mathcal{F}$. Thus

$$P[B] = \int E_\lambda\{I_B(X)f_P(X) \mid T = y\}\lambda^T(\mathrm{d}y) \tag{2.3.14}$$
$$= \int_B E_\lambda\{f_P(X) \mid T(x)\}\lambda(\mathrm{d}x),$$

where $E_\lambda\{f_P(X) \mid T(X)\} \in T^{-1}(\mathcal{F})$. Thus by Lemma 2.3.1 there exists a function g_P on y such that $E_\lambda\{f_P(X) \mid T(X)\} = g_P T(X)$ a.s. It follows from (2.3.14) that, for every $B \in \mathcal{B}$,

$$\int_B f_P(x)\lambda(\mathrm{d}x) = \int_B g_P T(x)\lambda(\mathrm{d}x). \tag{2.3.15}$$

Hence $f_P(x) = g_P T(x)$ a.s. Thus $f_P(x) \in T^{-1}(\mathcal{F})$.

(ii) We now establish the sufficiency of the condition. Assume that $\mathrm{d}P/\mathrm{d}\lambda \in T^{-1}(\mathcal{F})$. By Lemma 2.3.1 there exists a function g_P on $\mathcal{Y}$ such that $f_P(X) = g_P T(X)$ a.s., where $f_P = \mathrm{d}P/\mathrm{d}\lambda$. Thus

$$\begin{aligned} \mathrm{d}P &= g_P T\, d\lambda, \\ \mathrm{d}P^T &= g_P\, \mathrm{d}\lambda^T. \end{aligned} \tag{2.3.16}$$

Let B be any Borel set in $\mathcal{B}$. Define $\mathrm{d}\nu_P = I_B\, \mathrm{d}P$. From (2.3.16) we obtain

$$\nu_P{}^T(\mathrm{d}y) = E_\lambda\{I_B(X)g_P T(X) \mid T = y\}\lambda^T(\mathrm{d}y) \text{ a.s.} \tag{2.3.17}$$

Furthermore, since $g_P T \in T^{-1}(\mathcal{F})$,

$$E_\lambda\{I_B(X)g_P T(X) \mid T = y\} = g_P(y)E_\lambda\{I_B(X) \mid T = y\} \quad \text{a.s} \tag{2.3.18}$$

We thus arrive at

$$\nu_P{}^T(\mathrm{d}y) = g_P(y)E_\lambda\{I_B(X) \mid T = y\}\lambda^T(\mathrm{d}y) \tag{2.3.19}$$
$$= P_\lambda\{B \mid T = y\}g_P(y)\lambda^T(\mathrm{d}y) \quad \text{a.s.,}$$

where $P_\lambda\{B \mid T = y\}$ is independent of $\mathcal{P}$. Finally, since for every $B \in \mathcal{B}$ and $F \in \mathcal{F}$,

$$P[B \cap T^{-1}(F)] = \int_{T^{-1}(F)} I_B(x) P(\mathrm{d}x) = \int_{T^{-1}(F)} \nu_P(\mathrm{d}x) = \int_F P_\lambda\{B \mid T = y\} P^T(\mathrm{d}y), \tag{2.3.20}$$

where, as in (2.3.19), $P_\lambda\{B \mid T = y\}$ is independent of $\mathcal{P}$. Hence T is a sufficient statistic for $\mathcal{P}$. (Q.E.D.)

The following is a simple example for the Theorem 2.3.1. Let $X_1, \ldots, X_n$ be i.i.d. random variables having an $\mathcal{N}(\theta, 1)$ distribution. The joint distribution function is absolutely continuous with respect to the Lebesgue linear measure $\mu(\mathrm{d}X_1, \ldots, \mathrm{d}X_n) = \mathrm{d}X_1 \cdots \mathrm{d}X_n$. A probability measure of $\mathcal{P}$, dense in $\mathcal{P}$, is obtained by a mixture of any countable subfamily of $\mathcal{P}$. In particular, we can take

$$\lambda(\mathrm{d}x_1, \ldots, \mathrm{d}x_n) = (2\pi)^{-n/2} \exp\left\{-\frac{1}{2}\sum_{i=1}^{n} x_i^2\right\} \mathrm{d}x_1 \cdots \mathrm{d}x_n.$$

Thus for every $P \in \mathcal{P}$,

$$\frac{P(\mathrm{d}X_1, \ldots, \mathrm{d}X_n; \theta)}{\lambda(\mathrm{d}X_1, \ldots, \mathrm{d}X_n)} = \exp\left\{n\theta\left(\bar{X}_n - \frac{\theta}{2}\right)\right\}, \qquad -\infty < \theta < \infty.$$

The sufficient statistic for $\mathcal{P}$ is the sample mean $\bar{X}_n$. Thus we see that the generalized density of every $P \in \mathcal{P}$, with respect to λ, is a function of the sufficient statistic $\bar{X}_n$.

In the following theorem, called the *Neyman-Fisher* factorization theorem, we show that if T is a sufficient statistic, the density function of every probability measure P of $\mathcal{P}$ can be factored into two functions. One function in this product is independent of P, and the function which depends on P is a function of $T(X)$. Although the factorization property was suggested by both Neyman and Fisher [1], the theorem has been proved in the required generality only in 1949 by Halmos and Savage [1], as a corollary from the previous theorem. We provide here Halmos and Savage's proof and discuss the theorem later with examples. A certain deficiency is pointed out and a more general statement of the theorem, from Bahadur [1], is given without a proof.

Theorem 2.3.2. (The Factorization Theorem.) *Let $\mathcal{P}$ be a family of probability measures on $(\mathcal{X}, \mathcal{B})$ dominated by μ, that is, $\mathcal{P} \ll \mu$; and let $f_P = \mathrm{d}P/\mathrm{d}\mu$ a.s.*

$[\mu]$ for $P \in \mathcal{P}$. A statistic T transforming $(\mathcal{X}, \mathcal{B}, P)$ into $(\mathcal{Y}, \mathcal{F}, P^T)$ is sufficient for $\mathcal{P}$ if and only if

$$f_P(x) = g_P T(x) \cdot h(x) \text{ a.s. } [\mu] \tag{2.3.21}$$

where

(i) $g_P T \geqslant 0$, $h \geqslant 0$;

(ii) $h \in \mathcal{B}$ *and* $g_P T \in T^{-1}(\mathcal{F})$;

(iii) h *and* $g_P T$ *are* μ*-integrable;*

(iv) $h(X) = 0$ *only on a* $\mathcal{P}$*-null set.*

Proof. *Necessity.* If T is a sufficient statistic for $\mathcal{P}$ there exists, according to Theorem 2.3.1., a probability measure λ dense in $\mathcal{P}$ such that, for every $P \in \mathcal{P}$, $\mathrm{d}P/\mathrm{d}\lambda \in T^{-1}(\mathcal{F})$. Hence $g_P T \in T^{-1}(\mathcal{F})$. Furthermore let f_P denote the generalized density of $P \in \mathcal{P}$, that is, $f_P = \mathrm{d}P/\mathrm{d}\mu$. Then since

$$f_P = \frac{\mathrm{d}P}{\mathrm{d}\lambda} \cdot \frac{\mathrm{d}\lambda}{\mathrm{d}\mu} \text{ a.s. } [P], \tag{2.3.22}$$

we have

$$f_P(x) = g_P T(x) \cdot h(x) \text{ a.s. } [\mu], \tag{2.3.23}$$

where $h(X) = \lambda(\mathrm{d}X)/\mu(\mathrm{d}X) \geqslant 0$. We show now that $h(X)$ vanishes on sets of x values for which $P = 0$ for all $P \in \mathcal{P}$ (i.e., $\mathcal{P}$-null sets of $\mathcal{X}$). Let $N \in \mathcal{B}$ be such that $P(N) = 0$ for all $P \in \mathcal{P}$. Since λ is dense in $\mathcal{P}$, $\lambda(N) = 0$. Thus

$$\lambda(N) = \int_N h(x)\mu(\mathrm{d}x) = 0. \tag{2.3.24}$$

If $\mu(N) = 0$, we define $h(X) = 0$ on N. If, on the other hand, $\mu(N) > 0$, (2.3.24) implies that $h(X) = 0$ on N.

Sufficiency. Suppose that (2.3.21) holds. Define a sigma-finite measure by $\mathrm{d}\lambda = h\,\mathrm{d}\mu$. Since $h(X) = 0$ on the set $N = \{x; f_P(x) = 0\}$ for all $P \in \mathcal{P}$, $\lambda \ll \mu$. On the other hand, if $B \in \mathcal{B}$ is such that $\lambda(B) = 0$, then

$$\begin{aligned} P[B] &= \int_B g_P T(x) h(x) \mu(\mathrm{d}x) \\ &= \int_B g_P T(x) \lambda(\mathrm{d}x) = 0, \quad \text{all } P \in \mathcal{P}. \end{aligned}$$

It follows that $\mathcal{P} \ll \lambda$. Therefore $\mathcal{P} \equiv \lambda$. Finally, by Lemma 2.3.1,

$$g_P T(x) = \frac{P(\mathrm{d}x)}{\lambda(\mathrm{d}x)} \in T^{-1}(\mathcal{F}). \tag{2.3.25}$$

Hence, according to Theorem 2.3.1, T is sufficient for $\mathcal{P}$. (Q.E.D.)

In Example 2.2 we are given a class $\mathcal{P}$ of n-variate normal distributions $\mathcal{P} = \{\mathcal{N}(\theta 1_n, I_n); -\infty < \theta < \infty\}$ over the Euclidean n-space $\mathcal{B}$, the common Borel sigma-field on $\mathcal{X}$. In this case, the class $\mathcal{P}$ is dominated by any one of its members. For a sigma-finite measure μ dominating $\mathcal{P}$, we can take, for example, the probability measure $\mathcal{N}(\mathbf{0}_n, I_n)$, that is,

$$\mu(\mathrm{d}x_1, \ldots, \mathrm{d}x_n) = (2\pi)^{-n/2} \exp\left\{-\frac{1}{2}\sum_{i=1}^{n} x_i^2\right\} \mathrm{d}x_1 \cdots \mathrm{d}x_n.$$

Consider now the factorization (2.3.2), for which we write

$$\begin{aligned} T(X_1, \ldots, X_n) &= \bar{X}_n, \\ h(X_1, \ldots, X_n) &\equiv 1, \\ g_\theta(T(X_1, \ldots, X_n)) &= \exp\left\{n\theta\left(\bar{X}_n - \frac{\theta}{2}\right)\right\}. \end{aligned} \tag{2.3.26}$$

The functions h and $g_0 T$ are non-negative and $\mathcal{B}$-measurable. Moreover, if $\mathcal{F}$ denotes the Borel sigma-field on the real line, $T^{-1}(\mathcal{F})$ designates the smallest sigma-field generated by the sets $\{(X_1, \ldots, X_n); \bar{X}_n \leqslant a\}$, where $-\infty < a < \infty$. It follows immediately that $g_0 T \in T^{-1}(\mathcal{F})$ for all $-\infty < \theta < \infty$. Finally both $h(X_1, \ldots, X_n)$ and $g_\theta T(X_1, \ldots, X_n)$ are μ-integrable, since $h(X_1, \ldots, X_n) \equiv 1$ and $\int g_\theta(T(X))\mu(\mathrm{d}X) = 1$ for all $(X_1, \ldots, X_n) \in \mathcal{X}$. Thus the factorization equation, (2.3.2), and Theorem 2.3.2. imply that $T(X_1, \ldots, X_n) = \bar{X}_n$ is a sufficient statistic for $\mathcal{P}$.

The proof of the sufficiency of a statistic, according to the factorization given by Theorem 2.3.2, is not always simple or convenient. A difficulty in using Theorem 2.3.2. arises, for example, in the following case.

Example 2.5. Let $X_1, \ldots, X_n$ be i.i.d. random variables having a uniform distribution on $[0, \theta]$, $0 < \theta < \infty$. We denote a random variable having such a distribution law by $R(0, \theta)$. Again let $\mathcal{X}$ denote the Euclidean n-space, $\mathcal{B}$ the Borel sigma-field on $\mathcal{X}$. The class $\mathcal{P} = \{\mathcal{R}^{(n)}(0, \theta); 0 < \theta < \infty\}$ of the joint distributions of the i.i.d. $X_1, \ldots, X_n$ is dominated by the Lebesgue linear measure $\mu(\mathrm{d}X_1, \ldots, \mathrm{d}X_n) = \mathrm{d}X_1 \cdots \mathrm{d}X_n$. Let $X_{(1)} = \min\{X_1, \ldots, X_n\}$ and $X_{(n)} = \max\{X_1, \ldots, X_n\}$. The joint density of $X_1, \ldots, X_n$, with respect to μ, is

$$f(X_1, \ldots, X_n; \theta) = \begin{cases} \theta^{-n}, & \text{if } 0 \leqslant X_{(1)} \leqslant X_{(n)} \leqslant \theta, \\ 0, & \text{otherwise.} \end{cases} \tag{2.3.27}$$

Define,

$$h(X_1, \ldots, X_n) = \begin{cases} 1, & \text{if } X_{(1)} \geqslant 0, \\ 0, & \text{otherwise,} \end{cases} \tag{2.3.28}$$

and

$$g_\theta(t) = \begin{cases} \theta^{-n}, & t \leqslant \theta, \\ 0, & \text{otherwise.} \end{cases} \tag{2.3.29}$$

Then the joint density of $X_1, \ldots, X_n$ is

$$f(X_1, \ldots, X_n; \theta) = h(X_1, \ldots, X_n) g_\theta(X_{(n)}). \tag{2.3.30}$$

There is a tendency to infer from this factorization that $X_{(n)}$ is a sufficient statistic. Although $X_{(n)}$ is indeed a sufficient statistic, the function $h(X_1, \ldots, X_n)$ does not satisfy (iii) of Theorem 2.3.2, since

$$\int h(X_1, \ldots, X_n)\mu(\mathrm{d}X_1, \ldots, \mathrm{d}X_n) = \infty, \tag{2.3.31}$$

It is easy to verify that $T(X_1, \ldots, X_n) = \max\{X_1, \ldots, X_n\}$ is a sufficient statistic for $\mathfrak{F}$. Indeed, if we make the transformation

$$Y_i = X_{(i)}, \qquad i = 1, 2, \ldots, n,$$

where $X_{(1)} \leqslant X_{(2)} \leqslant \cdots \leqslant X_{(n)}$ are the ordered statistics of the sample, we obtain that the joint density of $(Y_1, \ldots, Y_n)$ is

$$f(y_1, \ldots, y_n; \theta) = \frac{n!}{\theta^n}, \qquad 0 \leqslant y_1 \leqslant \cdots y_n \leqslant \theta. \tag{2.3.32}$$

Furthermore the density function of $T(X_1, \ldots, X_n) = Y_n$ is

$$h(t; \theta) = \frac{n}{\theta^n} t^{n-1}, \qquad 0 \leqslant t \leqslant \theta. \tag{2.3.33}$$

Finally, the conditional density of $Y_1, \ldots, Y_{n-1}$, given $\{T = t\}$, is

(2.3.34)

$$f(y_1, \ldots, y_{n-1} \mid t) = \frac{(n-1)!}{t^{n-1}}, \qquad 0 \leqslant y_1 \leqslant \cdots \leqslant y_{n-1} \leqslant t, 0 < t < \infty.$$

This density is independent of θ, and thus T is a sufficient statistic. It would be helpful if we could still use the factorization theorem, Theorem 2.3.2, for the purpose of verifying that $T(X_1, \ldots, X_n) = \max\{X_1, \ldots, X_n\}$ is a sufficient statistic. This can be attained by modifying the joint density under consideration. The class $\mathfrak{F} = \{\mathcal{R}^{(n)}(0, \theta); 0 < \theta < \infty\}$ is dominated by a measure λ dense in $\mathfrak{F}$, whose density is the mixture

$$\lambda(\mathrm{d}X_1, \ldots, \mathrm{d}X_n) = \sum_{j=1}^{\infty} 2^{-j} g_j(X_{(n)})\, \mathrm{d}X_1, \ldots, \mathrm{d}X_n. \tag{2.3.35}$$

Let $\phi(X_1, \ldots, X_n; \theta)$ designate the generalized density of $X_1, \ldots, X_n$ with respect to λ. We have

$$\phi(X_1, \ldots, X_n; \theta) = h(X_1, \ldots, X_n) \cdot g_\theta^*(X_{(n)}), \tag{2.3.36}$$

where

$$g_\theta^*(X_{(n)}) = \frac{g_\theta(X_{(n)})}{\sum_{j=1}^{\infty} 2^{-j} g_j(X_{(n)})}. \tag{2.3.37}$$

We now show that the factorization in (2.3.36) satisfies all the four conditions of Theorem 2.3.2. Conditions (i), (ii), and (iv) are verified immediately. The integrability condition, (iii), is also satisfied with respect to λ. Indeed,

$$\int h(X_1, \ldots, X_n)\lambda(\mathrm{d}X_1, \ldots, \mathrm{d}X_n) = \sum_{j=1}^{\infty} 2^{-j} \int f(X_1, \ldots, X_n; j)\, \mathrm{d}X_1 \cdots \mathrm{d}X_n = \sum_{j=1}^{\infty} 2^{-j} = 1. \tag{2.3.28}$$

Furthermore

$$\int g_\theta^*(X_{(n)})\lambda(\mathrm{d}X_1, \ldots, \mathrm{d}X_n) \leqslant 1, \tag{2.3.39}$$

Bahadur [1] has modified the above factorization theorem so that a factorization as simple as (2.3.30) is sufficient. Bahadur's version of the factorization theorem follows.

Theorem 2.3.3. (Bahadur.) *Let* $(\mathcal{X}, \mathcal{B}, P)$ *be a probability space. Let* T *be a statistic,* $T\colon (\mathcal{X}, \mathcal{B}, P) \to (\mathcal{Y}, \mathcal{F}, P^T)$. *Let* $\mathcal{P}$ *be a dominated family of probability measures,* $\mathcal{P} \ll \mu$. *A necessary and sufficient condition for* T *to be a sufficient statistic for* $\mathcal{P}$ *is the existence of non-negative functions* h *on* $\mathcal{X}$ *and* g_P *on* $\mathcal{Y}$ *such that*

(i) h *is* $\mathcal{B}$*-measurable;*
(ii) $g_P T$ *is* $T^{-1}(\mathcal{F})$*-measurable for each* $P \in \mathcal{P}$;
(iii) *for each* $P \in \mathcal{P}$,

$$P(\mathrm{d}x) = h(x) g_P T(x) \mu(\mathrm{d}x) \text{ a.s. } [\mathcal{P}].$$

We do not reproduce the proof of this version of the factorization theorem, since it is similar in many respects to the proof of Theorem 2.3.2. The reader is referred to Bahadur's paper [1].

2.4. THE EXISTENCE AND CONSTRUCTION OF MINIMAL SUFFICIENT STATISTICS

We start the discussion on minimal sufficient statistics with an example through which a few of the ideas will be conveyed.

Example 2.6. Consider the case of n independent binomial experiments. Let X_i $(i = 1, \ldots, n)$ be i.i.d. random variables representing the results of the n experiments; $X_i = 1$ if the i-th experiment is successful and $X_i = 0$ otherwise. The sample space $\mathfrak{X}$ is a discrete one and contains the 2^n points $(X_1, X_2, \ldots, X_n)$, $X_i = 0, 1$, all $i = 1, \ldots, n$. The joint density of $(X_1, \ldots, X_n)$ is

(2.4.1)

$$f(X_1, \ldots, X_n; \theta) = \begin{cases} \theta^{t_n}(1-\theta)^{n-t_n}, & X_i = 0, 1,\ t_n = \sum_{i=1}^{n} X_i,\ 0 < \theta < 1, \\ 0, & \text{otherwise.} \end{cases}$$

According to the factorization theorem, the sample point $S_1 = (X_1, \ldots, X_n)$ is a sufficient statistic for $\mathcal{F} = \{P_\theta; 0 < \theta < 1\}$, where P_θ denotes the probability measure associated with (2.4.1). Moreover, according to the same theorem, the statistics $S_2 = (X_1 + X_2, X_3, \ldots, X_n)$, $S_3 = (X_1 + X_2 + X_3, X_4, \ldots, X_n), \ldots, S_{n-1} = (X_1 + \cdots + X_{n-1}, X_n)$, and $S_n = \sum_{i=1}^n X_i$ are all sufficient for $\mathcal{F}$. However, it is intuitively clear that if S_n is a sufficient statistic for $\mathcal{F}$ there is no sense in giving any of the other sufficient statistics mentioned above, since a sample equivalent to the original may be regenerated from the knowledge of S_n only. The statistic S_n is called minimal sufficient. It is a function of all the other sufficient statistics mentioned above. The term minimal sufficient can be explained in terms of the partitions of the sample space $\mathfrak{X}$ associated with the above statistics. The statistic S_1 partitions $\mathfrak{X}$ into 2^n disjoint an exhaustive sets, each of which contains one of the original points $(x_1, \ldots, x_n)$. The statistic S_2 partitions $\mathfrak{X}$ into $3 \times 2^{n-2}$ sets; 2^{n-2} sets contain one point of the form $(0, 0, x_3, \ldots, x_n)$, 2^{n-2} sets contain two points of the form $\{(1, 0, x_3, \ldots, x_n), (0, 1, x_3, \ldots, x_n)\}$, and 2^{n-2} sets contain one point of the form $(1, 1, x_3, \ldots, x_n)$. We remark that S_2 is a function of S_1, say, $T(S_1)$ which adds the first two components of S_1 and leaves the other components unchanged. We further notice that each set of the partition induced by S_1 is included in one of the sets of the partition induced by S_2 but not vice versa. In a similar manner we notice that $S_3 = T(S_2)$ partitions $\mathfrak{X}$ into $4 \times 2^{n-3}$ sets: 2^{n-3} sets of the form $\{(0, 0, 0, x_4, \ldots, x_n)\}$ and 2^{n-3} sets of the form $\{(1, 0, 0, x_4, \ldots, x_n), (0, 1, 0, x_4, \ldots, x_n), (0, 0, 1, x_4, \ldots, x_n)\}$, each having 3 points, and so on. The partitions associated with S_i and S_j where $i < j$ have $(i + 1)2^{n-i}$ and $(j + 1)2^{n-j}$ sets. For every $1 \leqslant i < j \leqslant n$ and every $n \geqslant 2$ we have $(j + 1)2^{n-j} \leqslant (i + 1)2^{n-i}$. Thus each of the sufficient statistics S_i $(i = 1, \ldots, n)$ is associated with a partition ζ_i, and if $i < j$ then the number of sets in ζ_i is not smaller than that of ζ_j. Each of the sets in a given partition ζ is called a *contour* of the statistic S. The number of different contours is determined by the number of different points in the range of S. All the contours in a given partition ζ are disjoint.

If $S^* = T(S)$ is a function of S then each contour of the partition ζ associated with S is included in a contour of the partition ζ^* of S^*. In the present example the contours of the partition ζ_n associated with the sufficient statistic S_n include the contours of all the finer partitions ζ_j $(j = 1, \ldots, n = 1)$ associated with S_j $(j = 1, \ldots, n - 1)$. Indeed, there are $(n + 1)$ contours in ζ_n; each of these contours can be characterized as $K_{n,\nu} = \{(X_1, \ldots, X_n); X_i = 0, 1, \sum_{i=1}^{n} X_i = \nu\}, \nu = 0, \ldots, n$. There are $2n$ contours in the partition ζ_{n-1}; these contours can be characterized as

$$K_{n-1,\nu}(x_n) = \left\{(x_1, \ldots, x_n); X_i = 0, 1, \sum_{i=1}^{n-1} x_i = \nu\right\}, \qquad \nu = 0, \ldots, n - 1,$$

and we have the inclusion relationship; for every $\nu = 0, 1, \ldots, n - 1$,

$$K_{n-1,\nu}(x_n) \subset \begin{cases} K_{n,\nu}, & \text{if } x_n = 0, \\ K_{n,\nu+1}, & \text{if } x_n = 1. \end{cases}$$

Similarly we can show that each of the contours of ζ_{n-2} is included in one of the contours of ζ_{n-1}, etc. The statistic S_n is considered a minimal sufficient statistic, in the sense that the number of contours it forms is minimal. ■

Let $(\mathfrak{X}, \mathfrak{B}, P)$ be a probability space, and $\mathfrak{F}$ a family of probability measures. Let μ be a sigma-finite measure on $(\mathfrak{X}, \mathfrak{B})$ such that $\mathfrak{F} \ll \mu$. A sufficient statistic T for $\mathfrak{F}$ is called minimal sufficient if, given any other sufficient statistic S for $\mathfrak{F}$ and a contour $K_S(s) = \{x: T(x) = t_s\}$, there exists a μ-null set N and a contour $K_T(t_s) = \{x; T(x) = t_s\}$ such that on $\mathfrak{X} - N$, $K_S(s) \subset K_T(t_s)$.

We now present a method of constructing minimal sufficient statistics that utilizes this notion of contour inclusion. This method was first suggested by Lehmann and Scheffé [1].

As before, we assume that the class $\mathfrak{F}$ of probability measures on $(\mathfrak{X}, \mathfrak{B})$ is dominated by a sigma-finite measure μ, that is, $\mathfrak{F} \ll \mu$. Let $\Psi = \{g_P; P \in \mathfrak{F}\}$ be the corresponding class of density functions, that is, $g_P = dP/d\mu$. Now we introduce a partition of $\mathfrak{X}$ in terms of the density functions in Ψ. A partition of $\mathfrak{X}$ can be described in terms of the cosets of $\mathfrak{X}$. We define a *coset* of a point x_0 (of $\mathfrak{X}$) to be the set of all points x in $\mathfrak{X}$ having a certain relationship $R(x, x_0)$ with x_0. For example, suppose $\mathfrak{X}$ is the plane, that is, $\mathfrak{X} = \{(x_1, x_2); -\infty \leqslant x_i \leqslant \infty, i = 1, 2\}$. Let $x^0 = (x^0{}_1, x_2{}^0)$. The collection of all points x of $\mathfrak{X}$ lying on the straight line passing through x_0 and through the origin $(0, 0)$ is a coset of x_0. Another example of a coset of x_0 is the set of all points on the circle whose radius is $\rho = |x_0|$.

We assume that the relationship $R(x, x_0)$, which determines the coset of x_0, is transitive; that is, if $R(x, x_0)$ and $R(x_0, x_1)$ then $R(x, x_1)$. This means that if a point x lies in a certain coset of x_0 and a point y lies in the same coset of x_0, then x lies in the coset of y, given by the relationship $R(x, y)$.

Let $C(x_0)$ designate the coset of x_0, which is determined by a relationship $R(x, x_0)$. If $C(x_0)$ does not exhaust $\mathfrak{X}$, let x_1 be a point of $\mathfrak{X}$ not in $C(x_0)$. Construct the coset $C(x_1)$. Obviously $C(x_0)$ and $C(x_1)$ are disjoint. If $C(x_0) \cup C(x_1) \neq \mathfrak{X}$, construct a coset $C(x_2)$, where $x_2 \in \mathfrak{X} - (C(x_0) \cup C(x_1))$. In this manner we proceed until we obtain a partition of $\mathfrak{X}$, being a collection of disjoint cosets exhausting $\mathfrak{X}$.

Consider now the following partitioning operation. Let x_0 be a point of $\mathfrak{X}$ and let the coset $C(x_0)$ be the set

$$C(x_0) = \left\{x;\ x \in \mathfrak{X},\ \frac{g_P(x)}{g_P(x_0)} = k(x, x_0) \qquad \text{for all } P \in \mathcal{F}\right\}, \tag{2.4.2}$$

where the function $k(x, x_0)$ is independent of $\mathcal{F}$. If this partitioning operation is applied to a family of density functions Ψ we denote by $\zeta(\Psi)$ the resulting partition of $\mathfrak{X}$.

We illustrate the partitioning according to (2.4.2) by referring back to Example 2.6. The family of density functions in Example 2.6 is given by (2.4.1), where $0 < \theta < 1$. Take any point x_0 in $\mathfrak{X}$, which is an n-tuple $(x_1^0, \ldots, x_n^0)$. The ratio of densities in (2.4.2) is then, for all $x \in \mathfrak{X}$,

$$\frac{f(x_1, \ldots, x_n; \theta)}{f(x_1^0, \ldots, x_n^0; \theta)} = \left(\frac{\theta}{1-\theta}\right)^{\sum_{i=1}^n (x_i - x_i^0)} \qquad 0 < \theta < 1. \tag{2.4.3}$$

The ratio (2.4.3) is independent of θ if and only if $\sum_{i=1}^n x_i = \sum_{i=1}^n x_i^0$. Thus the coset of x^0, $C(x^0)$, consists of all the n-tuples in $\mathfrak{X}$ whose components' sum is equal to $S_n(\mathbf{x}_0) = \sum_{i=1}^n x_i^0$. The partition of $\mathfrak{X}$ according to the cosets defined by (2.4.2) is equivalent to that determined by the minimal sufficient statistic S_n of Example 2.6.

Example 2.7. Let $X_1, \ldots, X_n$ be i.i.d. random variables having a normal distribution law $\mathcal{N}(\theta, \sigma^2)$, $-\infty < \theta < \infty$, $0 < \sigma^2 < \infty$. Let $\mathfrak{X}$ be the n-dimensional Euclidean space $X = (X_1, \ldots, X_n)'$ and $\mathcal{B}$ the usual Borel sigma-field on $\mathfrak{X}$. Let $\mathcal{F}$ be the class of all n-variate normal distributions $\mathcal{N}(\theta 1_n, \sigma^2 I_n)$. The ratio of density points in the determination of cosets according to (2.4.2) is

$$\frac{f(X_1, \ldots, X_n; \theta, \sigma^2)}{f(X_1^0, \ldots, X_n^0; \theta, \sigma^2)} \tag{2.4.4}$$

$$= \exp\left\{-\frac{1}{2\sigma^2}\left[\sum_{i=1}^n X_i^2 - \sum_{i=1}^n (X_i^0)^2\right] + \frac{\theta}{\sigma^2}\left[\sum_{i=1}^n X_i - \sum_{i=1}^n X_i^0\right]\right\}.$$

This ratio is independent of (θ, σ^2) if and only if $\sum_{i=1}^n X_i^2 = \sum_{i=1}^n (X_i^0)^2$ *and* $\sum_{i=1}^n X_i = \sum_{i=1}^n X_i^0$. The partition of $\mathfrak{X}$ obtained by determining the cosets according to (2.4.4) is equivalent to the partition of $\mathfrak{X}$ induced by the sufficient statistic $T_n = (\sum X_i, \sum X_i^2)$. As will be proven later, T_n is a minimal sufficient statistic for the family of normal distributions, when both the expectation θ and the variance σ^2 are unknown. ■

The measure-theoretic difficulty in proving that the partition, induced by the construction of cosets according to (2.4.2), corresponds to that of a minimal sufficient statistic is due to the fact that when the family $\mathfrak{F}$ is not denumerable the partition of $\mathfrak{X}$ might be unmeasurable. This undesirable property results when for every $P \in \mathfrak{F}$, the density function $g_P(x)$ is only essentially unique. On a set N_P of P measure zero there may be various different determinations of $g_P(x)$. The problem is that when $\mathfrak{F}$ is nondenumerable the union $\bigcup_{P\in\mathfrak{F}} N_P$ is not necessarily measurable, and if measurable it might have a positive measure. Following Lehmann and Scheffé [1], we start proving the minimal sufficiency of the partition induced by the likelihood ratio operation by restricting to countable families $\mathfrak{F}$. Then by introducing the notion of separability, we show that if $\mathfrak{F}$ is contained in a Euclidean space, the likelihood ratio operation yields sufficient statistics even if $\mathfrak{F}$ is nondenumerable.

Theorem 2.4.1. *Let* $(\mathfrak{X}, \mathfrak{B}, P)$ *be a probability space,* $\mathfrak{F}$ *a countable family of probability measures. Furthermore, let* μ *be a sigma-finite measure such that* $\mathfrak{F} \ll \mu$ *and let* $\Psi = \{g_P; P \in \mathfrak{F}\}$ *be a family of particular determination of density functions of* $\mathfrak{F}$ *with respect to* μ. *If the partitioning operation* (2.4.2) *is applied to* Ψ, *the resulting partition of* $\mathfrak{X}$ *is* $\mathfrak{B}$*-measurable and is associated with a minimal sufficient statistic for* $\mathfrak{F}$. *The partitions of* $\mathfrak{X}$ *induced by applying the operation* (2.4.2) *on two different determinations of* Ψ *are equivalent in the strong sense.*†

Proof. Since $\mathfrak{F}$ is countable, let $\Psi = \{g_i(x): i = 1, 2, \ldots\}$ denote a particular determination of density functions. Let $\zeta(\Psi)$ denote the partition of $\mathfrak{X}$ induced by the operation (2.4.2) on Ψ. We first prove that $\zeta(\Psi)$ is $\mathfrak{B}$-measurable. Let $C_i^0 = \{x; x \in \mathfrak{X}, g_i(x) = 0\}$ $i = 1, 2, \ldots$. Since $g_i \in \mathfrak{B}$ so is C_i^0. Hence $C^0 = \bigcap_{i=1}^{\infty} C_i^0$ is $\mathfrak{B}$-measurable. Let x^0 be a point of $\mathfrak{X}$, $x^0 \notin C^0$. There exists a least integer $I = I(x^0)$ such that $g_I(x^0) > 0$. We define

$$C_i(x^0) = \left\{x; x \in \mathfrak{X}, \frac{g_i(x)}{g_I(x)} = \frac{g_i(x^0)}{g_I(x^0)}\right\}. \tag{2.4.5}$$

Since $g_i(x)$ and $g_I(x)$ are $\mathfrak{B}$-measurable, so is their quotient in $\mathfrak{X} - C_I^0$. Thus the sets $C_i(x^0)$, $i = 1, 2, \ldots$, are $\mathfrak{B}$-measurable. Therefore $C(x^0) = \bigcap_{i=1}^{\infty} C_i(x^0)$ is $\mathfrak{B}$-measurable. The set $C(x^0)$ is the coset of x^0 in $\mathfrak{X}$, as defined by the operation (2.4.2). This completes the proof of the measurability of $\zeta(\Psi)$.

If ϕ' is a family of density functions of $\mathfrak{F}$ determined differently from Ψ then, for each $i = 1, 2, \ldots, g_i(x) \neq g_i'(x)$ only on a null set N_i' of $\mathfrak{X}$,

† Two different partitions of $\mathfrak{X}$, say ζ and ζ', are called equivalent in the strong sense if there exists a null set N for $\mathfrak{F}$ such that ζ and ζ' coincide on $\mathfrak{X} - N$.

that is, $\mu(N_i') = 0$. Since a countable union of μ-null sets is μ-null, we obtain that Ψ coincides with ϕ' on $\mathfrak{X} - \bigcup_{i=1}^{\infty} N_1' \equiv \mathfrak{X} - N$. Hence $\zeta(\Psi)$ coincides with $\zeta(\phi)$ on $\mathfrak{X} - N$. This means that $\zeta(\psi)$ and $\zeta(\phi')$ are strongly equivalent. It remains to show that $\zeta(\psi)$ is strongly equivalent to a partition induced by a minimal sufficient statistic. From each coset C in $\zeta(\psi)$ choose a point x^0 by some choice function χ on ζ, that is,

$$x^0 = \chi(C), \qquad C \in \zeta(\psi), \qquad x^0 \in C. \tag{2.4.6}$$

Furthermore, define the statistic

$$T(x) = x^0 \quad \text{if} \quad x \in C \quad \text{and} \quad \chi(C) = x^0. \tag{2.4.7}$$

The partition of $\mathfrak{X}$ induced by T is strongly equivalent to $\zeta(\psi)$. Furthermore, for every $x \in C$,

$$\begin{aligned} g_i(x) &= k(x, x^0)g_i(x^0) \\ &= k(x, T(x))g_i(T(x)), \qquad i = 1, 2, \ldots; \end{aligned} \tag{2.4.8}$$

$k(x, T(x)) = h(x)$ is non-negative and positive on $\mathfrak{X} - C^0$. We define $h(x) = 0$ on C^0. Also $g_i(x^0) \geqslant 0$. One can also show that $h(x)$ and $g_i(T(x))$ are $\mathfrak{B}$-measurable ($i = 1, 2, \ldots$). (See details in [1].) Hence according to (2.4.8) and the factorization theorem, Theorem 2.3.3., $T(x)$ is a sufficient statistic. We show now that $T(x)$ is a minimal sufficient statistic for $\mathfrak{F}$. Let T' be a sufficient statistic for $\mathfrak{F}$. According to the factorization theorem, there exists a determination of density functions

$$g_i'(x) = h'(x)g_i'(T'(x)), \qquad i = 1, 2, \ldots, \tag{2.4.9}$$

where h' and $g'T'$ are non-negative and $\mathfrak{B}$-measurable. For each $i = 1, 2, \ldots$ the determination $g_1'(x)$ given by (2.4.9) differs from the determination $g_i(x)$ given by (2.4.8) at most on a μ-null set N_i'. Let $N' = \bigcup_{i=1}^{\infty} N'$; N' is μ-null. Let $N_0 = \{x; h'(x) = 0\}$. We have

$$P_i[N_0] = \int_{N_0} h'(x)g_i'(T'(x))\mu(dx) = 0, \quad \text{all } i = 1, 2, \ldots. \tag{2.4.10}$$

Hence N_0 is a μ-null set. Let $N'' = N_0 \cup N'$; N'' is a μ-null set. On $\mathfrak{X} - N''$, $g_i'(x) = g_i(x)$ for all $i = 1, 2, \ldots,$ *and* $h(x) > 0$. We show now that on $\mathfrak{X} - N''$ every contour of $T'(x)$ is included in a contour of $T(x)$. Let $x^0 \in \mathfrak{X} - N''$ and let

$$C'(x^0) = \{x; x \in \mathfrak{X} - N'' \quad \text{and} \quad T'(x) = T'(x^0)\} \tag{2.4.11}$$

That is, $C'(x^0)$ is a contour of $T'(x)$ through x^0 in $\mathfrak{X} - N''$. For all x in $C'(x^0)$,

$$\begin{aligned} g_i(x) = g_i'(x) &= h'(x)g_i'(T'(x)) \\ &= h'(x)g_i'(x^0) \\ &= k(x, x^0)g_i(x^0), \end{aligned} \tag{2.4.12}$$

where $k(x, x^0) = h'(x)/h'(x^0)$, $h'(x^0) > 0$. Thus according to (2.4.2), $C'(x^0) \subset C(x^0)$, which is a contour of $T(x)$ through x^0 on $\mathfrak{X} - N''$. This proves that $T(x)$ is a minimal sufficient statistic and completes the proof of the theorem. (Q.E.D.)

We extend the existence theorem of minimal sufficient statistics for countable families of probability measures $\mathcal{P}$ to general (not necessarily countable) families. For this purpose we introduce the notion of *separability* of $\mathcal{P}$. Let $(\mathfrak{X}, \mathcal{B}, \mu)$ be a given measure (probability) space. Consider the class $\mathcal{L}_1$ of μ-integrable functions on $\mathfrak{X}$; that is,

$$\mathcal{L}_1 = \left\{ f;\ \int |f(x)|\ \mu(\mathrm{d}x) < \infty \right\}. \tag{2.4.13}$$

Let f and g be two functions in $\mathcal{L}_1$. The distance between f and g is defined to be

$$\delta(f, g) = \int |f(x) - g(x)|\ \mu(\mathrm{d}x). \tag{2.4.14}$$

We say that a sequence of functions $\{f_n\}$ in $\mathcal{L}_1$ *converge in the mean* to g in $\mathcal{L}_1$ if

$$\lim_{n\to\infty} \delta(f_n, g) = 0. \tag{2.4.15}$$

Let $\psi = \{f_\theta;\ \theta \in \Theta\}$ be a family of $\mathcal{L}_1$ functions indexed by θ in Θ. Let $\Psi_0 = \{f_n;\ n = 1, 2, \ldots\}$ be a countable subfamily of Ψ. We say that Ψ_0 is *dense* in Ψ if, for every f_θ in Ψ, there exists a subsequence of functions in Ψ_0 such that

$$\lim_{j\to\infty} \delta(f_{n_j}, f_\theta) = 0. \tag{2.4.16}$$

A family Ψ of functions in $\mathcal{L}_1$ is said to be *separable* if it contains a countable subfamily Ψ_0 dense in Ψ. In our restriction to a sample space $\mathfrak{X}$, which is a Euclidean space, to $\mathcal{B}$ being a Borel sigma-field on $\mathfrak{X}$, and to μ being the Lebesgue measure on $(\mathfrak{X}, \mathcal{B})$, we guarantee that if $\mathcal{P}$ is a family of probability measures on $(\mathfrak{X}, \mathcal{B})$ and $\mathcal{P} \ll \mu$ and $\Psi = \{g_P(x);\ P \in \mathcal{P}\}$, then in any determination of density functions for $\mathcal{P}$, $g_P = \mathrm{d}P/\mathrm{d}\mu$, the family Ψ is *separable* (see Lehmann and Scheffé [17]). The following theorem extends Theorem 2.4.1 to cases where $\mathcal{P}$ is separable. We proceed with the following lemma.

Lemma 2.4.2. *If a statistic T on $(\mathfrak{X}, \mathcal{B}, \mathcal{P})$ is sufficient for the family $\mathcal{P}$; if it is a minimal sufficient statistic for a subfamily $\mathcal{P}_1 \subset \mathcal{P}$; and if every null set of $\mathcal{P}_1$ is a null set of $\mathcal{P}$, then T is a minimal sufficient statistic for $\mathcal{P}$.*

Proof. First we mention that if a statistic is sufficient for a family of probability measures $\mathcal{P}$ it is sufficient for any subfamily $\mathcal{P}_1 \subset \mathcal{P}$. Thus let T'

be any sufficient statistic for $\mathcal{P}$ on $(\mathcal{X}, \mathcal{B}, P)$. To show that T is a minimal sufficient statistic it is enough to prove the existence of a $\mathcal{P}$-null set N such that every contour of T' is included in a contour of T on $\mathcal{X} - N$; T' is sufficient for $\mathcal{P}_1$, $\mathcal{P}_1 \subset \mathcal{P}$. But T is minimal sufficient for $\mathcal{P}_1$. Therefore there exists a $\mathcal{P}_1$-null set N_1 such that every contour of T' is included in a contour of T on $\mathcal{X} - N_1$. Finally, since N_1 is also a $\mathcal{P}$-null set, we obtain the required result.

Theorem 2.4.2. *If the family of probability measures $\mathcal{P}$ on $(\mathcal{X}, \mathcal{B})$ yields density functions g_P, $P \in \mathcal{P}$, with respect to a sigma-finite measure μ; and if the family $\Psi = \{g_P; P \in \mathcal{P}\}$ is separable, there exists a minimal sufficient statistic T for $\mathcal{P}$ which may be constructed by applying the partition $\zeta(\Psi_0)$ [see (2.4.2)] to any countable subfamily Ψ_0 dense in Ψ.*

Proof. Let Ψ_1 be a particular determination of a countable subfamily of densities, $\Psi_1 \subset \Psi$, which is dense in Ψ. According to Theorem 2.4.1, the statistic T associated with the partition $\zeta(\Psi_1)$ is a minimal sufficient for $\mathcal{P}_1$, where $\mathcal{P}_1$ is the subfamily of probability measures associated with Ψ_1. If we prove that the statistic T is sufficient for $\mathcal{P}$ and that every $\mathcal{P}_1$-null set N is also a $\mathcal{P}$-null set, then the theorem is proven.

Let $P_0 \in \mathcal{P}$ (arbitrary) and $g_0 = dP_0/d\mu$. The separability of Ψ implies that there exists a sequence of densities $\{g_n; n = 1, 2, \ldots\}$ in Ψ_1 such that

$$\lim_{n\to\infty} \delta(g_n, g_0) = 0.$$

Let N_1 be a $\mathcal{P}_1$-null set. Hence for every g_n in the above sequence

$$\int_{N_1} g_n(x)\, \mu(dx) = 0, \qquad n = 1, 2, \ldots. \tag{2.4.17}$$

Hence from the separability of Ψ we obtain that, for any $\epsilon > 0$ and all $n \geqslant N(\epsilon, g_0)$, $\delta(g_n, g_0) < \epsilon$. Therefore

$$\begin{aligned}\int_{N_1} g_0(x)\, \mu(dx) &= \left| \int_{N} g_0(x)\, \mu(dx) - \int_{N_1} g_n(x)\, \mu(dx) \right| \\ &\leqslant \int_{N_1} |g_0(x) - g_n(x)|\, \mu(dx) \leqslant \delta(g_n, g_0) < \epsilon.\end{aligned} \tag{2.4.18}$$

Since g_0 is an arbitrary element of Ψ, N_1 is also $\mathcal{P}$-null. It remains to show that T is a sufficient statistic for $\mathcal{P}$; T is minimal sufficient for $\mathcal{P}_1$. Let $T: (\mathcal{X}, \mathcal{B}, P) \to (\mathcal{Y}, \mathcal{F}, P^T)$. Thus, for every $P_n \in \mathcal{P}_1$, $n = 1, 2, \ldots$, and every $B \in \mathcal{B}$, $F \in \mathcal{F}$,

$$P_n[B \cap T^{-1}(F)] = \int_F P[B \mid T = y] P_n{}^T(dy), \tag{2.4.19}$$

where the conditional probability function $P[B \mid T = y]$ is independent of $\mathcal{P}_1$ (since T is sufficient for $\mathcal{P}_1$). This conditional probability function $P[B \mid T(x)]$ as a function of $\mathcal{X}$ is $T^{-1}(\mathcal{F})$-measurable; we denote it by $\pi(x; B)$.

The integral in (2.4.19) can be written as:

$$P_n[B \cap T^{-1}(F)] = \int_{T^{-1}(F)} \pi(x; B)P_n(dx) = \int_{T^{-1}(F)} \pi(x; B)g_n(x)\,\mu(dx), \tag{2.4.20}$$

where $g_n = dP_n/d\mu$. Let P_0 be an arbitrary probability measure in $\mathfrak{T}$, having a density g_0 with respect to μ. Let $\{\tilde{g}_n; n = 1, 2, \ldots\}$ be a sequence of densities in Ψ_1 which converge in the mean to g_0. Thus for any given $\epsilon > 0$ and sufficiently large n, $n \geqslant N(\epsilon, g_0)$,

$$|P_n[B \cap T^{-1}(F)] - P_0[B \cap T^{-1}(F)]| \leqslant \int_{B\cap T^{-1}(F)} \left|g_n(x) - g_0(x)\right| \mu(dx) \leqslant \int \left|g_n(x) - g_0(x)\right| \mu(dx) < \epsilon. \tag{2.4.21}$$

Furthermore, denoting by $\pi_0(x; B)$ the conditional probability of B given $T(X)$, under $\mathfrak{T}_1$, we have

$$\begin{aligned}|P_0[B \cap T^{-1}(F)] - P_n[B \cap T^{-1}(F)]| &= \left|\int_{T^{-1}(F)} [\pi_0(x; B)g_0(x) - \pi(x; B)g_n(x)]\mu(dx)\right| \\ &\geqslant \Bigg|\left|\int_{T^{-1}(F)} \pi_0(x; B)(g_0(x) - g_n(x))\mu(dx)\right| \\ &\quad - \left|\int_{T^{-1}(F)} (\pi(x; B) - \pi_0(x; B))g_n(x)\mu(dx)\right|\Bigg|.\end{aligned}$$

But

$$\left|\int_{T^{-1}(F)} \pi_0(x; B)(g_0(x) - g_n(x))\mu(dx)\right| \leqslant \int_{T^{-1}(F)} \pi_0(x; B)\left|g_0(x) - g_n(x)\right| \mu(dx) < \epsilon.$$

Hence for every $n \geqslant N(\epsilon)$,

$$\left| \int_{T^{-1}(F)} (\pi_0(x; B) - \pi(x; B))g_n(x)\mu(\mathrm{d}x) \right| < 2\epsilon; \tag{2.4.22}$$

otherwise (2.4.21) is contradicted. Finally, since (2.4.22) holds for every $B \in \mathcal{B}$ and every $F \in \mathcal{F}$, $\pi_0(x; B) = \pi(x; B)$ a.s. Thus the conditional probability of B given $T(X)$ is independent of the particular distribution P_0 in $\mathcal{T}$, hence $T(X)$ is sufficient for $\mathcal{T}$. (Q.E.D.)

Example 2.8. Let $(\mathfrak{X}, \mathcal{B}, P)$ be a probability space, $\mathfrak{X}$ a Euclidean space, $\mathcal{B}$ the common Borel sigma-field on $\mathfrak{X}$, and $\mathcal{T}$ a family of k-parameter exponential type densities (the Darmois-Koopman family) given by

$$f(x; \theta_1, \ldots, \theta_k) = h(x) \exp\left\{ \sum_{i=1}^{k} \theta_i U_i(x) + V(\theta_1, \ldots, \theta_k) \right\}, \tag{2.4.23}$$

where $\theta_1, \ldots, \theta_k$ are real parameters. Let x_0 be any point of $\mathfrak{X}$ such that $f(x_0; \theta_1, \ldots, \theta_k) > 0$. Then the coset of x_0 according to the partitioning operation (2.4.2) is the set of all x in $\mathfrak{X}$ for which

$$\frac{f(x; \theta_1, \ldots, \theta_k)}{f(x_0; \theta_1, \ldots, \theta_k)} = \frac{h(x)}{h(x_0)} \exp\left\{ \sum_{i=1}^{k} \theta_i (U_i(x) - U_i(x_0)) \right\} = k(x, x_0)$$

independently of $(\theta_1, \ldots, \theta_k)$. Let Θ be the parameter space,† namely,

$$\Theta = \{(\theta_1, \ldots, \theta_k), \quad (\theta_1, \ldots, \theta_k) \text{ belongs to a certain specified set in the Euclidean } k\text{-space}\}. \tag{2.4.24}$$

If Θ contains k linearly independent vectors, the dimension of Θ is k. In this case (2.4.23) holds for all $(\theta_1, \ldots, \theta_k) \in \Theta$ if and only if $U_1(x) = U_i(x_0)$ for all $i = 1, \ldots, k$. Hence, since (2.4.23) is symmetric in $(U_1(x), \ldots, U_k(x))$, the ordered vector $(U_{(1)}(x), \ldots, U_{(k)}(x))$ is a minimal sufficient statistic. We remark in this connection that, in terms of the reduction of the sample data, if $X_1, X_2, \ldots, X_n$ are i.i.d. random variables and the likelihood ratio partition (2.4.2) yields a k-dimensional vector $(U_{(1)}(x), \ldots, U_{(k)}(x))$, $(k \geqslant n)$, as a minimal sufficient statistic, then the minimal sufficient statistic is equivalent to the trivial sufficient statistic $(X_{(1)}, X_{(2)}, \ldots, X_{(n)})$. The only significant reduction of data is attained when the sample size n is greater than k. This statement will be proven in the next section, where we establish that families of distribution functions having certain regularity conditions and admitting sufficient statistics are either of the k-parameter exponential type, or their minimal sufficient statistics are the trivial ones.

† Θ is called the natural parameter space. As proven by Lehmann [2, p. 51], Θ is convex.

We continue now with Example 2.8. Let $X_1, \ldots, X_n$ be a sample (i.i.d.) from a normal distribution $\mathcal{N}(\xi_1, \sigma_1^2)$ and $Y_1, \ldots, Y_n$ a sample (i.i.d.) from a normal distribution $\mathcal{N}(\xi_2, \sigma_2^2)$, $-\infty < \xi_1, \xi_2 < \infty$, $0 < \sigma_1^2, \sigma_2^2 < \infty$. The joint density of the two samples can be written as

$$(2.4.25)\quad f(x_1, \ldots, x_n, y_1, \ldots, y_n; \xi_1, \xi_2, \sigma_1, \sigma_2) = (2\pi)^{-n}\sigma_1^{-n}\sigma_2^{-n} \cdot \exp\left\{\frac{\xi_1}{\sigma_1^2}\sum_{i=1}^n x_i - \frac{1}{2\sigma_1^2}\sum_{i=1}^n x_i^2 + \frac{\xi_2}{\sigma_2^2}\sum_{i=1}^n y_i - \frac{1}{2\sigma_2^2}\sum_{i=1}^n y_i^2 - \frac{n}{2}\left(\frac{\xi_1^2}{\sigma_1^2} + \frac{\xi_2^2}{\sigma_2^2}\right)\right\}.$$

Redefine the parameters

$$\theta_1 = \frac{\xi_1}{\sigma_1^2};\quad \theta_2 = \frac{\xi_2}{\sigma_2^2};\quad \theta_3 = -\frac{1}{2\sigma_1^2};\quad \theta_4 = -\frac{1}{2\sigma_2^2}.$$

We can easily verify that the joint density, (2.4.25), is of the general 4-parameter exponential type, (2.4.22). In the general case, the rank of

$$\Theta = \{(\theta_1, \ldots, \theta_4); -\infty < \theta_1, \theta_2 < \infty; -\infty < \theta_3, \theta_4 < 0\}$$

is 4 and a minimal sufficient statistic is $T_4 = (\sum_{i=1}^n X_i, \sum_{i=1}^n Y_i, \sum_{i=1}^n X_i^2, \sum_{i=1}^n Y_i^2)$. Equivalently, the set of the two sample means and the two sample variances is a sufficient statistic. It should be remarked that even if $\xi_1 = \xi_2$ but $\sigma_1^2 \neq \sigma_2^2$, Θ is still 4-dimensional and a minimal sufficient statistic is T_4. On the other hand, if $\xi_1 \neq \xi_2$ but $\sigma_1^2 = \sigma_2^2$, the dimension of Θ is reduced to 3 and a minimal sufficient statistic is reduced to $T_3 = \{\sum_{i=1}^n X_i, \sum_{i=1}^n Y_i, \sum_{i=1}^n (X_i^2 + Y_i^2)\}$. ■

Bahadur treated the problem of minimal sufficient statistics in a more abstract measure-theoretic manner and did not have to assume that the class of densities corresponding to $\mathcal{F}$ is separable. The only requirement of Bahadur's theorems is that $\mathcal{F}$ is dominated (see [1]). In other words, Bahadur extended the result of Lehmann and Scheffé and proved that whenever the family $\mathcal{F}$ is dominated, there exists a minimal sufficient statistic. The minimal sigma-field over the class of all Borel sets of the form

$$(2.4.26)\quad B_P(r) = \{x; g_P(x) < r\},\quad 0 \leqslant r \leqslant \infty,\quad \text{for all } P \in \mathcal{F},$$

where $g_P = dP/d\mu$, is a sigma-subfield of $\mathcal{B}$ induced by a minimal sufficient statistic.

In Example 2.8 we have considered a family of density functions of the k-parameter exponential type. We have seen that if the dimension of the parameter space Θ is k, a minimal sufficient statistic is $(U_1(x), \ldots, U_k(x)) \equiv U(x)$. It is interesting to notice that in this case the dimensionality of $U(x)$ does not change if n i.i.d. observations are performed, and $n > k$. Indeed, if $X_1, X_2, \ldots, X_n$, $(n > k)$, are i.i.d. random variables having a k-parameter exponential density, the joint density of the sample is

$$(2.4.27)\quad f(x_1, \ldots, x_n; \theta_1, \ldots, \theta_k) = \prod_{i=1}^{n} h(x_i) \cdot \exp\left\{\sum_{i=1}^{k} \theta_i \left(\sum_{j=1}^{n} U_i(x_j)\right) + nV(\theta_1, \ldots, \theta_k)\right\},$$

and a minimal sufficient statistic is $(\sum_{j=1}^{n} U_1(x_j), \ldots, \sum_{j=1}^{n} U_k(x_j))$, again of dimensionality not greater than k. We see in the next section a strong connection between the dimensionality of a minimal sufficient statistic and the form of the corresponding family of density functions. The exponential family of densities will play an important role there. Here we present an example showing that a minimal sufficient statistic is not necessarily of the same dimensionality at different regions of the sample space. This example was provided by Barankin and Katz [1].

Example 2.9. Let X be a random variable having a density function

$$(2.4.28)\quad f(x; \sigma) = \begin{cases} (1 + \sigma\sqrt{(2\pi)})^{-1} \exp\left\{-\dfrac{1}{2\sigma^2} x^2\right\}, & \text{if } x < 0, \\ (1 + \sigma\sqrt{(2\pi)})^{-1}, & \text{if } 0 \leqslant x \leqslant 1, \\ (1 + \sigma\sqrt{(2\pi)})^{-1} \exp\left\{-\dfrac{1}{2\sigma^2} (x - 1)^2\right\}, & \text{if } 1 \leqslant x, \end{cases}$$

$0 < \sigma < \infty$. Let X_1 and X_2 be independent random variables having density functions $f(x; \sigma_1)$ and $f(x; \sigma_2)$, respectively, $0 < \sigma_1, \sigma_2 < \infty$. The joint density of X_1 and X_2 assumes nine different forms on the following sets:

$$\begin{aligned}
A_1 &= \{(x_1, x_2); x_1 < 0, \text{ and } x_2 < 0\} \\
A_2 &= \{(x_1, x_2); x_1 < 0, \text{ and } 0 \leqslant x_2 \leqslant 1\} \\
A_3 &= \{(x_1, x_2); x_1 < 0, \text{ and } 1 < x_2\} \\
A_4 &= \{(x_1, x_2); 0 \leqslant x_1 \leqslant 1, \text{ and } x_2 < 0\} \\
A_5 &= \{(x_1, x_2); 0 \leqslant x_1 \leqslant 1, \text{ and } 0 \leqslant x_2 \leqslant 1\} \\
A_6 &= \{(x_1, x_2); 0 \leqslant x_1 \leqslant 1, \text{ and } 1 < x_2\} \\
A_7 &= \{(x_1, x_2); 1 < x_1, \text{ and } x_2 < 0\} \\
A_8 &= \{(x_1, x_2); 1 < x_1, \text{ and } 0 \leqslant x_2 \leqslant 1\} \\
A_9 &= \{(x_1, x_2); 1 < x_1, \text{ and } 1 < x_2\}.
\end{aligned}$$

Using the operation (2.4.2) of cosets determination, it is easy to verify that a minimal sufficient statistic is

$$(2.4.29)\quad T(X_1, X_2) = \begin{cases} (X_1, X_2), & \text{if } (X_1, X_2) \in A_1 \cup A_3 \cup A_7 \cup A_9 \\ X_1^2, & \text{if } (X_1, X_2) \in \text{int } A_2 \cup \text{int } A_8 \\ X_2^2, & \text{if } (X_1, X_2) \in \text{int } A_4 \cup \text{int } A_6 \\ 0, & \text{if } (X_1, X_2) \in \text{int } A_5 \end{cases}$$

where int A denotes the interior of the set A. We see that the minimal sufficient statistic $T(X_1, X_2)$ is 2-dimensional on four subsets of $\mathfrak{X}$, 1-dimensional on another four subsets of $\mathfrak{X}$, and of zero dimensionality on the interior of the square $\{0 < x_1 < 1 \text{ and } 0 < x_2 < 1\}$. ■

Barankin and Katz [1] have provided some basic results concerning sufficient statistics having locally and globally minimal dimension. They also related the properties of sufficient statistics of minimal dimension to those of minimal statistics in the functional sense, which were studied in the previous section. Further results in this direction were given by Shimizu [1] and others.

2.5. MINIMAL SUFFICIENT STATISTICS AND THE EXPONENTIAL FAMILY

As mentioned earlier, there is an intrinsic relationship between families of distribution functions admitting nontrivial sufficient statistics and the exponential family of distributions. In this section we present Dynkin's theory, which was published in [1].

Let $(\mathfrak{X}, \mathfrak{B}, P)$ be a probability space and $\mathfrak{F}$ a family of probability measures admitting sufficient statistics. A statistic T on $\mathfrak{X}$ is called *necessary* if it is a function of every sufficient statistic S for $\mathfrak{F}$.

In other words, each contour of a sufficient statistic is included in a contour of a necessary statistic. From the functional definition of a minimal sufficient statistic it is immediately implied that a minimal sufficient statistic is both necessary and sufficient. There is no redundancy in the information given by a necessary statistic, but it may not be all the information available; for example, let $X_1, \ldots, X_n$ be i.i.d. random variables having a probability measure P belonging to a family $\mathfrak{F}$ admitting a sufficient statistic. The vector $(X_{(1)}, \ldots, X_{(n)}) \equiv \mathbf{X}$ of the order statistic is a trivial sufficient statistic. Suppose now that $S = (\sum_{i=1}^n X_i, \sum_{i=1}^n X_i^2)$ is a minimal sufficient statistic for $\mathfrak{F}$. Although the trivial sufficient statistic X provides all the information in the sample concerning $\mathfrak{F}$ it is, as previously discussed, redundant in the sense that the same information can be furnished by a 2-dimensional statistic S.

On the other hand, the necessary statistics $T_1 = \sum_{i=1}^n X_i$ and $T_2 = \sum_{i=1}^n X_i^2$ are individually insufficient for $\mathfrak{F}$. One must provide the information on $\mathfrak{F}$ by $S = (T_1, T_2)$.

A family of probability measures $\mathfrak{F}$ is called *regular in the sense of Dynkin* [1] if it satisfies the following regularity conditions:

(i) $\mathfrak{F} = \{P_\theta; \theta \in \Theta\}$ is a parametric family, where Θ is an open subset of a Euclidean k-space. $\mathfrak{F}$ is dominated by a sigma-finite measure μ.

(ii) If $f(x;\theta) = P_\theta(\mathrm{d}x)/\mu(\mathrm{d}x)$ then $f(x;\theta) > 0$ for all $x \in \mathfrak{X}$ and all $\theta \in \Theta$.

(iii) The class of density functions $\{f(x;\theta); \theta \in \Theta\}$ corresponding to $\mathcal{F}$ is such that, for each $\theta \in \Theta, f(x;\theta)$ is continuously differentiable with respect to x.

[Dynkin assumes only that $f(x;\theta)$ are piecewise smooth (see [1]). As shown by Brown in [1] we have to require the stronger assumption (iii)].

Choose θ_0 in Θ and define the function

$$g(x;\theta) = \log f(x;\theta) - \log f(x;\theta_0),\ x \in \mathfrak{X},\ \theta \in \Theta. \tag{2.5.1}$$

According to regularity assumption, (ii), $g(x;\theta)$ is well defined. For a given $X = x$ the function $g(x;\theta)$ ranging over its domain Θ is known as the *log-likelihood function* (see Fraser [1]). For a single observation $X = x$, the maximal reduction that can be made without losing sufficiency is provided by the transformation $x \to g(x;\theta)$. Dynkin calls the random function $g(X;\theta)$, which is defined on $\mathfrak{X} \times \Theta$, a statistic. The reader may not be used to such a general approach, but there are many advantages in adopting this outlook. Since we are interested in the properties of log-likelihood functions on Θ, we consider the variable, x in $g(x;\theta)$ as a parameter. Furthermore a function $g(x;\theta)$ will be called a sufficient statistic if it depends on x only through the sufficient statistic $T(X)$. For example, consider the k-parameter exponential family, with densities given by (2.4.23). The log-likelihood functions corresponding to these density functions are

$$\begin{aligned} g(X;\theta) = {} & (u_1(X), \ldots, u_k(X))'(\theta_1 - \theta_1^0, \ldots, \theta_k - \theta_k^0) \\ & + V(\theta) - V(\theta^0), \end{aligned} \tag{2.5.2}$$

where $\theta = (\theta_1, \ldots, \theta_k)'$. We see that $g(X;\theta)$ depends on X only through the minimal sufficient statistic $(u_1(X), \ldots, u_k(X))$.

Lemma 2.5.1. (Dynkin.) *A necessary and sufficient statistic for a regular family $\mathcal{F}$ is the function $g(X;\theta)$, relating to each $X = x$ the function $g(x;\theta)$ over Θ.*

Proof. To show that the statistic $g(X;\theta)$ is sufficient, write

$$f(x;\theta) = f(x;\theta_0)\exp\{g(x;\theta)\}, \qquad x \in \mathfrak{X}, \qquad \theta \in \Theta. \tag{2.5.3}$$

Both $f(x;\theta_0)$ and $\exp\{g(x;\theta)\}$ are $\mathcal{B}$-measurable positive functions. Hence by Bahadur's factorization theorem (Theorem 2.3.3) $g(X;\theta)$ is a sufficient statistic.

We prove now that $g(X;\theta)$ is a necessary statistic. Let $S(X)$ on $(\mathfrak{X}, \mathcal{B}, P)$ be any sufficient statistic for $\mathcal{F}$. By the factorization theorem, there exist

$\mathcal{B}$-measurable, non-negative functions $h(x)$ and $k(S(x); \theta)$ such that

$$f(x; \theta) = h(x)k(S(x); \theta) \quad \text{a.s. } [P_\theta], \quad \text{and all } \theta \in \Theta. \tag{2.5.4}$$

Hence

$$g(x; \theta) = \log K(S(x); \theta) - \log K(S(x); \theta_0) \qquad \text{a.s. } [\mathcal{F}], \tag{2.5.5}$$

since the set of x values on which $h(x) = 0$ has a P-measure zero for every $P \in \mathcal{F}$. Thus $g(x; \theta)$ is a function of every sufficient statistic for $\mathcal{F}$, hence a necessary statistic for $\mathcal{F}$. (Q.E.D.)

Corollary 2.5.1. *If $X_1, X_2, \ldots, X_n$ are i.i.d. random variables whose common density belongs to a regular family, then a necessary and sufficient statistic for the sample is*

$$g(X_1, \ldots, X_n; \theta) = \sum_{i=1}^{n} g(X_i; \theta). \tag{2.5.6}$$

Minimal sufficiency under sampling falls thus into the domain of transforms over linear function spaces, which transfer sets of n functions $g(X_1; \theta), \ldots, g(X_n; \theta)$ into the function $\sum_{i=1}^{n} g(X_i; \theta)$ over Θ. We restrict attention to the case of a 1-dimensional random variable X, that is, $\mathcal{X}$ will be the real line. Whenever n random variables $X_1, X_2, X_3, \ldots, X_n$ are considered it is assumed that they are independent and identically distributed.

Let Δ designate an open interval on the real line independent of Θ. For a given $\mathcal{F} = \{P_\theta; \theta \in \Theta\}$ over $(\mathcal{X}, \mathcal{B})$ we denote by $\mathcal{G}$ the class of log-likelihood functions when θ ranges over Θ; that is,

$$\mathcal{G} = \{g(x; \theta); \theta \in \Theta, x \in \mathcal{X}\}. \tag{2.5.7}$$

Let $\mathcal{L}(\mathcal{G}, \Delta)$ designate the minimal linear space of functions on Δ, including all the log-likelihood functions in $\mathcal{G}$ and the unit function $\phi_0(x) \equiv 1$. For example, consider the k-parameter exponential family with density functions as in (2.4.23). This is a regular family, and the log-likelihood functions are

$$g(x; \theta) = \sum_{i=1}^{k} c_i(\theta)U_i(x) + c_0(\theta), \theta \in \Theta, x \in \mathcal{X}. \tag{2.5.8}$$

Here $\Delta = \mathcal{X}$, and we can consider every function $g(x; \theta)$, for a fixed θ, as a linear combination of the $(k + 1)$ functions $\phi_0(x) \equiv 1$ and $u_1(x), u_2(x), \ldots, u_k(x)$. Thus the linear space $\mathcal{L}(\mathcal{G}, \Delta)$ is the space of functions spanned by $(1, u_1(x), \ldots, u_k(x))$. If a sample of n random variables $X_1, \ldots, X_n$ is under consideration, $\mathcal{L}(\mathcal{G}, \Delta)$ is a linear space of functions on $\Delta \subset E^{(n)}$, which includes all the log-likelihood functions $g(x_1, \ldots, x_n; \theta)$. The functions in $\mathcal{L}(\mathcal{G}, \Delta)$ also can be vector valued.

Theorem 2.5.2. (Dynkin.) *Let $X_1, \ldots, X_n$ be i.i.d. random variables having a distribution belonging to a family $\mathcal{F}$ which is regular on $(\mathcal{G}, \Delta)$. Let $\mathcal{L}(\mathcal{G}, \Delta)$ be the corresponding linear space spanned by $\mathcal{G}$, of dimensionality $k + 1$ (not* excluding $k = \infty$). *Then*

(i) *if $n \leqslant k$, any sufficient statistic for $\mathcal{F}$ is trivial;*
(ii) *if $n > k$ and the functions $(1, \phi_1(x), \ldots, \phi_k(x))$ constitute a base of $\mathcal{L}(\mathcal{G}, \Delta)$, then the system of functions*

$$S_i(X_1, \ldots, X_n) = \sum_{j=1}^{n} \phi_i(X_j), \qquad i = 1, \ldots, k, \tag{2.5.9}$$

is functionally independent and forms a necessary and sufficient statistic for $\mathcal{F}$.

Proof. (i) Suppose the sample size n is smaller than k. First we recall that the order statistic $(X_{(1)}, \ldots, X_{(n)})$, where $X_{(i)} \leqslant X_{(j)}$ for all $i < j$, is a trivial sufficient statistic. We show now that the order statistic is a necessary statistic. Since $\mathcal{L}(\mathcal{G}, \Delta)$ is of dimensionality $(k + 1)$ we have, for every $\theta \in \Theta$,

$$g(x_{(j)}; \theta) = c_0(\theta) + \sum_{i=1}^{k} c_i(\theta)\phi_i(x_{(j)}), \qquad j = 1, \ldots, n, \tag{2.5.10}$$

where $(1, \phi_1(x), \ldots, \phi_k(x))$ is a base of $\mathcal{L}(\mathcal{G}, \Delta)$. Furthermore, if $S(X_{(j)})$ is any function of $X_{(j)}$ in $\mathcal{L}(\mathcal{G}, \Delta)$, then for each θ in Θ

$$S(X_{(j)}) = b_0(\theta) + \sum_{i=1}^{k} b_i(\theta)\phi_i(X_{(j)}), \qquad j = 1, \ldots, n. \tag{2.5.11}$$

Without loss of generality, assume that $X_{(1)} < X_{(2)} < \cdots < X_{(n)}$. Since $n \leqslant k$, the n column vectors of the matrix

$$\mathbf{F}(\mathbf{X}) = \begin{bmatrix} 1 & 1 & \cdots & 1 \\ \phi_1(X_{(1)}) & \phi_1(X_{(2)}) & \cdots & \phi_1(X_{(n)}) \\ \cdot & \cdot & & \cdot \\ \cdot & \cdot & & \cdot \\ \cdot & \cdot & & \cdot \\ \phi_k(X_{(1)}) & \phi_k(X_{(2)}) & \cdots & \phi_k(X_{(n)}) \end{bmatrix}$$

are linearly independent. Let

$$g(X; \theta) = (g(X_{(1)}; \theta), \ldots, g(X_{(n)}; \theta))'$$

and

$$S(X) = (S(X_{(1)}), \ldots, S(X_{(n)}))'.$$

The matrix $\mathbf{F}(\mathbf{X})(\mathbf{F}(\mathbf{X}))'$ is non-singular, where $(\mathbf{F}(\mathbf{X}))'$ designates the transpose of $\mathbf{F}(\mathbf{X})$. Hence according to (2.5.10) and (2.5.11),

$$(S(X))' = (g(X; \theta))'(\mathbf{F}(\mathbf{X}))'\mathbf{B}(\boldsymbol{\theta}; \mathbf{X})\mathbf{F}(\mathbf{X}), \tag{2.5.12}$$

where $\mathbf{B}(\boldsymbol{\theta}; \mathbf{X})$ is a $(k + 1) \times (k + 1)$ matrix such that

$$(2.5.13) \quad (b_0(\theta), \ldots, b_k(\theta))' = (c_0(\theta), \ldots, c_k(\theta))'\mathbf{F}(\mathbf{X})(\mathbf{F}(\mathbf{X}))'\mathbf{B}(\boldsymbol{\theta}; \mathbf{X}).$$

Finally, according to Lemma 2.5.1, $g(X; \theta)$ is a necessary and sufficient statistic. But, according to (2.5.12), the trivial sufficient statistic $(X_{(1)}, \ldots, X_{(n)})$ is a function of $g(X; \theta)$ and hence is a necessary statistic. This completes the proof of (i).

(ii) Suppose that $k < n$ and let $\{1, \phi_1(X), \ldots, \phi_\theta(X)\}$ be a base of $\mathcal{L}(\mathcal{G}, \Delta)$. According to (2.5.6) and (2.5.10), the log-likelihood function of the joint density of $(X_1, \ldots, X_n)$ is, for each $\theta \in \Theta$,

$$(2.5.14) \qquad g(X_1, \ldots, X_n; \theta) = \sum_{i=1}^{k} c_i(\theta)\left(\sum_{j=1}^{n} \phi_i(X_j)\right) + nc_0(\theta).$$

According to Lemma 2.5.1, $g(X_1, \ldots, X_n)$ is a sufficient and necessary statistic. Hence

$$S(X_1, \ldots, X_n) = \left(\sum_{j=1}^{n} \phi_1(X_j), \sum_{j=1}^{n} \phi_2(X_j), \ldots, \sum_{j=1}^{n} \phi_k(X_j)\right)$$

is a sufficient statistic. The necessity of $S(X_1, \ldots, X_n)$ is implied from (i).

We show now that $S_i(X_1, \ldots, X_n) = \sum_{j=1}^{n} \phi_i(X_j)$, $i = 1, \ldots, k$, are functionally independent. First we notice that

$$(2.5.15) \qquad \frac{\partial}{\partial X_j} S_i(X_1, \ldots, X_n) = \frac{\mathrm{d}}{\mathrm{d}X} \phi_i(X), \qquad i = 1, \ldots, k.$$

According to the regularity conditions on $\mathcal{F}$, all the functions $\phi_i(X)$ are continuously differentiable on the closure of Δ. If $S_i(X_1, \ldots, X_n)$ $(i = 1, \ldots, k)$ are not linearly independent but there exists some linear relationship $F(S_1(X_1, \ldots, X_n), \ldots, S_k(X_1, \ldots, X_n)) \equiv 0$, then, for k different points $X_1, \ldots, X_k \in \mathfrak{X}$, we have

$$(2.5.16) \qquad \frac{\partial(S_1, \ldots, S_k)}{\partial(X_1, \ldots, X_k)} = \det\{\|\phi_i'(X_j)\colon i, j = 1, \ldots, k\|\} = 0,$$

where $\phi_i'(X)$ is the derivative of $\phi_i(X)$ and $\|\phi_i'(X_j); i, j = 1, \ldots, k\|$ is a $k \times k$ matrix of the derivatives of ϕ_i at the points X_j. The notation $\det\{A\}$ denotes the determinant of A. We show now that (2.5.16) implies the linear dependence of $\{1, \phi_1(X), \ldots, \theta_k(X)\}$. This is proven by induction on k. Suppose we have shown that (2.5.16) with $(k - 1)$ functions implies that $\{1, \phi_1(X), \ldots, \phi_{k-1}(X)\}$ are dependent. Substitute in (2.5.16) with k functions $x_i = x_i^0$ $(i = 1, \ldots, k - 1)$ and $x_k = y$. Expanding the determinant in (2.5.16) by the elements of the last row, we obtain

$$(2.5.17) \qquad A_1\phi_1'(y) + \cdots + A_k\phi_k'(y) = 0, \qquad y \in \Delta,$$

where $A_1, \ldots, A_k$ do not depend on y and $A_k \neq 0$. Equation 2.5.17 implies that

$$\phi(y) = A_1\phi_1(y) + \cdots + A_k\phi_k(y), \qquad y \in \Delta, \tag{2.5.18}$$

is a constant over Δ. (The functions in $\mathfrak{L}(\mathfrak{G}, \Delta)$ are continuously differentiable on Δ.) Therefore $\phi_k(y)$ depends on $(\phi_1(y), \ldots, \phi_{k-1}(y))$. (Q.E.D.)

We now introduce the concept of the *rank* of a family of probability measures. Let $\mathfrak{F}$ be a regular family of probability measures. Let $\mathfrak{L}(\mathfrak{G}, \Delta)$ be the corresponding linear space of functions, generated by the set of log-likelihood functions $\mathfrak{G}$. If the dimensionality of $\mathfrak{L}(\mathfrak{G}, \Delta)$ is $r + 1$ we say that the *rank* of $\mathfrak{F}$ is r.

A function $f(X)$ is called *piecewise smooth* on a set Δ if there exists a subset Δ^* of Δ such that $f(X)$ is continuously differentiable on Δ^* and Δ is the closure of Δ^*.

Theorem 2.5.3. (Dynkin.) *If the family $\mathfrak{F}$ of probability measures on $(\mathfrak{X}, \mathfrak{B})$ is regular and has a finite rank r on the interval Δ, then the density $f(x; \theta)$ of X can be written in the form*

$$f(x; \theta) = h(x) \exp\left\{\sum_{i=1}^{r} \phi_i(x)c_i(\theta) + c_0(\theta)\right\}, \qquad x \in \Delta, \tag{2.5.19}$$

$\theta \in \Theta$, *where the functions $\phi_1(x), \ldots, \phi_r(x)$ are piecewise smooth on Δ and the systems of functions $\{1, \phi_1(x), \ldots, \phi_r(x)\}$ and $\{1, c_1(\theta), \ldots, c_r(\theta)\}$ are linearly independent.*

Proof. Let $\{1, \phi_1(x), \ldots, \phi_r(x)\}$ be a base for $\mathfrak{L}(\mathfrak{G}, \Delta)$. Then for a given $X = x$ we can write the log-likelihood function in the form

$$g(x; \theta) = \sum_{i=1}^{r} c_i(\theta)\phi_i(x) + c_0(\theta), \quad \theta \in \Theta, \quad x \in \Delta. \tag{2.5.20}$$

Hence

$$f(x; \theta) = f(x; \theta_0) \exp\left\{\sum_{i=1}^{r} c_i(\theta)\phi_i(x) + c_0(\theta)\right\}. \tag{2.5.21}$$

Writing $h(x) = f(x; \theta_0)$, we obtain the exponential type density (2.5.19). It remains to prove that the system $\{1, c_1(\theta), \ldots, c_r(\theta)\}$ is linearly independent. Indeed, if the functions in this system are dependent we can write

$$c_r(\theta) = b_0 + b_1c_1(\theta) + \cdots + b_{r-1}c_{r-1}(\theta), \quad b_{r-1} \neq 0. \tag{2.5.22}$$

Substituting (2.5.22) in (2.5.20) we have

$$g(x; \theta) = \sum_{i=1}^{r-1} c_i(\theta)\phi_i^*(x) + c_0(\theta) + b_0\phi_r(x), \tag{2.5.23}$$

where

$$\phi_i^*(x) = \phi_i(x) + b_i\phi_r(x), \qquad i = 1, \ldots, r-1. \tag{2.5.24}$$

Finally, since $g(x; \theta_0) = 0$ we have

(2.5.25)

$$g(x; \theta) = g(x; \theta) - g(x; \theta_0) = \sum_{i=1}^{r} \phi_i^*(x)[c_i(\theta) - c_i(\theta_0)] + [c_0(\theta) - c_0(\theta_0)].$$

This implies that the dimension of $\mathcal{L}(\mathcal{G}, \Delta)$ is less than $r + 1$. But this contradicts the statement that $\{1, \phi_1(X), \ldots, \phi_r(X)\}$ forms a base for $\mathcal{L}(\mathcal{G}, \Delta)$.

(Q.E.D.)

We have proven in this section that whenever a family of distribution functions is regular and has a finite rank r, its minimal sufficient statistic is either the trivial one or an r-dimensional vector. Furthermore, whenever the rank r of a regular family of distributions $\mathcal{F}$ is finite, the density functions of $\mathcal{F}$ are of the r-parameter exponential type. In Example 2.9 we showed a case where the minimal sufficient statistic is a vector of varying dimensionality on different subsets of the sample space $\mathcal{X}$. But there the family $\mathcal{F}$ is not regular. Indeed, the density functions $f(X; \theta)$ are not differentiable at $X = 0$ and at $X = 1$ for all θ, $-\infty < \theta < \infty$. In the following example we present a regular family of densities but whose rank is either finite or infinite, depending whether the value of a certain parameter is known.

Example 2.10. Let $\mathcal{X} = \{x;\ 0 \leqslant x < \infty\}$, $\mathcal{B}$ be the Borel sigma-field on $\mathcal{X}$, and $\mathcal{F}$ be the class of probability measures with corresponding densities (of the Weibull distributions)

$$f(x; \beta, \lambda, \mu) = \begin{cases} 0, & \text{if } x \leqslant 0 \\ K x^{\beta} e^{-\lambda x^{\mu}}, & \text{if } x > 0, \end{cases} \tag{2.5.26}$$

$$0 < \beta < \infty; \qquad 0 < \lambda < \infty; \qquad \theta \leqslant \mu \leqslant \infty.$$

This family is regular over the interval $\Delta = (0, \infty)$.

Case 1. The unknown is μ, β and λ are known. The log-likelihood function is

$$g(x; \beta, \lambda, \mu) = -\lambda x^{\mu} + \log \frac{K}{K_0} + \lambda_0 x^{\mu_0}. \tag{2.5.27}$$

It is apparent that $\mathcal{L}(\mathcal{G}, \Delta)$ is a linear space of functions generated by all the functions X^{μ} $(0 \leqslant \mu \leqslant \infty)$, and therefore of infinite dimension $(r = \infty)$. Hence when μ is unknown there are no nontrivial sufficient statistics. For every sample of finite size n, we have to consider for a sufficient statistic the order statistic $(X_{(1)}, \ldots, X_{(n)})$.

Case 2. The known is μ, λ and β are unknown. In this case $Y = X^{\mu}$ has a gamma distribution law $\mathcal{G}(\lambda, 1 + \beta/\mu)$. The log-likelihood function is

$$(2.5.28)\quad g(X; \beta, \lambda, \mu) = \log \frac{K}{K_0} + (\beta - \beta_0) \log X - (\lambda - \lambda_0) X^{\mu}.$$

In this case the rank of the family $\mathcal{F}$ is $r = 2$, and the necessary and sufficient statistic, based on a sample of $X_1, \ldots, X_n$ i.i.d., is

$$S_n = \left(\sum_{i=1}^{n} \log X_i, \sum_{i=1}^{n} X_i^{\mu} \right)$$

If μ is known, β unknown, λ known, then the rank of the family is $r = 1$ and the necessary-sufficient statistic is $T_1 = \sum_{i=1}^{n} \log X_i$. When μ and β are known but λ unknown, the rank is $r = 1$ and the necessary sufficient is $T_2 = \sum_{i=1}^{n} X_i^{\mu}$. ■

Example 2.11. Let $\mathcal{F}$ be a family of mixtures of two normal distributions $\mathcal{N}(\xi_1, \sigma_1^2)$ and $\mathcal{N}(\xi_2, \sigma_2^2)$ with mixing probabilities θ and $(1 - \theta)$, respectively, where $-\infty < \xi_1, \xi_2 < \infty$; $0 < \sigma_1, \sigma_2 < \infty$; $0 < \theta < 1$. This family is regular over the whole real line $\Delta = (-\infty, \infty)$. The density functions for this family are

$$(2.5.29)\quad f(X; \xi_1, \xi_2, \sigma_1, \sigma_2, \theta) = \frac{\theta}{\sigma_1\sqrt{(2\pi)}} \exp\left\{-\frac{1}{2\sigma_1^2}(X - \xi_1)^2\right\}$$

$$+ (1 - \theta) \cdot \frac{1}{\sigma_2\sqrt{(2\pi)}} \exp\left\{-\frac{1}{2\sigma_2^2}(X - \xi_2)^2\right\}, \qquad -\infty < X < \infty.$$

When all the five parameters are known the rank of the family is $r = 0$, but if at least one of the five parameters is unknown the rank is $\gamma = \infty$, and the only sufficient statistics are the trivial ones. ■

For further investigation of sufficiency and the exponential family see Denny [1, 2, 3, 4, 5].

We conclude by remarking that the main result of this section is based on the assumption that the domain Δ over which the log-likelihood functions are defined is independent of the parameter space Θ. In certain cases in which the domain depends on Θ, for example, the rectangular distributions $\mathcal{R}(0, \theta)$; $0 < \theta < \infty$, there might be a nontrivial sufficient statistic while the family $\mathcal{F}$ is not of the exponential type. Such families of distribution functions will be discussed in Section 3.5. Statistical models which assume that the support of the distribution functions depend on the parameters, θ, are called *selection models*. Studies of sufficiency under selection model have been done by several people. In particular see Fraser [1], Brunk [1], and Morimoto and Sibuya [1].

2.6. SUFFICIENCY AND COMPLETENESS

The completeness of sufficient statistics in certain statistical models is an important property, playing an essential role in the theory of best unbiased estimation and of best unbiased test procedures. In this section we introduce the definitions of a complete family of distribution functions and the completeness of a statistic; we establish conditions under which the minimal sufficient statistics in the exponential family are complete; and we study the general relationship between completeness and minimality of sufficient statistics. We furthermore discuss the conditions under which pairs of complete statistics constitute complete statistics. Finally, we prove that the order statistic is a complete sufficient statistic in the nonparametric case, where the family of distributions consists of *all* absolutely continuous distributions (with respect to the Lebesgue measure).

A family of distribution functions $\{F_\theta; \theta \in \Theta\}$ on $(\mathcal{X}, \mathcal{B})$ is called *complete* if, for every $\mathcal{B}$-measurable statistic φ on $\mathcal{X}_0$

$$\int \varphi(x) F_\theta(\mathrm{d}x) = 0 \quad \text{for all} \quad \theta \in \Theta \tag{2.6.1}$$

implies that $\varphi(x) = 0$ a.s. The following are examples of several complete families of distribution functions.

Example 2.12. (i) We show that the family of binomial distributions with densities, $0 < \theta < 1$,

$$f(x; \theta) = \begin{cases} \binom{n}{x} \theta^x (1-\theta)^{n-x}, & x = 0, 1, \ldots, n, \\ 0, & \text{otherwise}, \end{cases} \tag{2.6.2}$$

is complete. If $\varphi(x)$ is a real function such that

$$\sum_{x=0}^{n} \binom{n}{x} \varphi(x) \theta^x (1-\theta)^{n-x} = 0, \tag{2.6.3}$$

for all θ in (0, 1), then $\varphi(x)$ is necessarily zero for all $x = 0, \ldots, n$; that is, $\varphi(x) = 0$ a.s. $[\mathcal{F}]$. Indeed, the function $\sum_{x=0}^{n} \binom{n}{x} \varphi(x)\theta^x(1-\theta)^{n-x}$ is a polynomial in θ, of degree n. Such a polynomial has at most n different roots. But if this polynomial assumes the value zero for all θ, its coefficients should be identically zero. Thus the family of binomial distributions is complete.

(ii) Consider the family of normal distributions $\mathcal{N}(\xi, \sigma^2)$, $-\infty < \xi < \infty$, $0 < \sigma < \infty$. Let $\varphi(x)$ be an integrable function satisfying

$$\frac{1}{\sigma\sqrt{(2\pi)}} \int_{-\infty}^{\infty} \varphi(x) \exp\left\{-\frac{1}{2\sigma^2}(x-\xi)^2\right\} \mathrm{d}x = 0 \tag{2.6.4}$$

for all (ξ, σ). This can be written in the form

$$\int_{-\infty}^{\infty} \exp\left\{\frac{\xi}{\sigma^2} x\right\} \cdot \varphi(x) \exp\left\{-\frac{1}{2\sigma^2} x^2\right\} dx = 0 \tag{2.6.5}$$

for all (ξ, σ). Equation 2.6.5 says that the bilateral Laplace transform of $\varphi(x) \exp\{-(1/2\sigma^2)x^2\}$ is zero for all values of $\tau = \xi/\sigma^2$, $-\infty < \tau < \infty$. This implies that

$$\varphi(x) \exp\left\{-\frac{1}{2\sigma^2} x^2\right\} = 0 \text{ a.s. } [\mathcal{T}].$$

In other words, $\varphi(x) = 0$ a.s. $[\mathcal{T}]$.

The following is an example of an incomplete family of distributions. Let $\{F_\theta; \theta \in \Theta\}$ be the family of joint distributions of two i.i.d. random variables X, Y. Let $\varphi(X)$ be any integrable function with respect to the common distribution of X and Y such that $0 < \text{Var}_\theta \{\varphi(X)\} < \infty$ all $\theta \in \Theta$. However

$$\int [\varphi(x) - \varphi(y)] F_\theta(dx) F_\theta(dy) = 0, \qquad \text{all } \theta \in \Theta. \tag{2.6.6}$$

But $P_\theta[\varphi(X) \neq \varphi(Y)] > 0$ for all $\theta \in \Theta$; otherwise X and Y are not independent. This means that the class $\{F_\theta; \theta \in \Theta\}$ is not complete. ■

A statistic S on $(\mathfrak{X}, \mathfrak{B}, F_\theta)$, $\theta \in \Theta$, to $(\zeta, \mathcal{F}, F_\theta^{S})$ is called *complete* if the induced family of distribution functions $\{F_\theta^S; \theta \in \Theta\}$ is complete.

We remark in this context that, as is often the case, a family of distribution functions $\{F_\theta; \theta \in \Theta\}$ for $(\mathfrak{X}, \mathfrak{B})$ is incomplete, whereas an induced family $\{F_\theta^S; \theta \in \Theta\}$ might be complete. For example, as mentioned before, if X_1 and X_2 are i.i.d. such as $\mathcal{N}(\xi, \sigma^2)$, $-\infty < \xi < \infty$, $0 < \sigma < \infty$, the joint distribution of (X_1, X_2), namely, $\mathcal{N}(\xi I_2, \sigma^2 I_2)$, is incomplete. The family of the induced distributions of the statistic $S = X_1 + X_2$, namely, $\{\mathcal{N}(2\xi, 2\sigma^2)$, $-\infty < \xi < \infty, 0 < \sigma < \infty\}$, is, however, a complete one. In the following lemma we provide the condition for a sufficient statistic in a k-parameter exponential family to be a complete statistic. This lemma has many applications in parameteric methods of statistical inference. In the proof of this lemma we follow Lehmann [2], p. 132.

Theorem 2.6.1. *Let $X_1, \ldots, X_n$ be i.i.d. random variables having a density function, with respect to a sigma-finite measure μ on the real line, of the k-parameter exponential type*

$$f(x; \theta) = h(x) \exp\left\{\sum_{i=1}^{k} \theta_i U_i(x) + V(\theta)\right\}, \quad x \in \Delta, \quad \theta \in \Theta.$$

Let $U = (\sum_{j=1}^{n} U_1(X_j), \ldots, \sum_{j=1}^{n} U_k(X_j))'$. Then U is a complete sufficient statistic provided Θ contains a k-dimensional rectangle.

Proof. The sufficiency and necessity of U was established in the previous sections. We have shown that if Θ contains k-linearly independent vectors then U is minimal sufficient. We prove now that the condition that Θ contains a k-dimensional rectangle implies that U is a complete statistic. Without loss of generality, assume that Θ contains the rectangle $I = \{\theta; -a \leqslant \theta_j \leqslant a,$ all $j = 1, \ldots, k\}$ for some $0 < a < \infty$. If this is not the case, a simple reparametrization reduces the parameter space Θ to Θ^* containing I. To prove the completeness of U we have to show that $\mathcal{F}^u$ is such that, for every measurable function f on $\mathcal{U}$ (the sample space of U) for which

$$\int f(u_1, \ldots, u_k) P_\theta(du_1, \ldots, du_k) = 0 \quad \text{for all} \quad \theta \in \Theta, \tag{2.6.7}$$

we have $f(u_1, \ldots, u_k) = 0$ a.s. $[\mathcal{F}^u]$. Let

$$f(u_1, \ldots, u_k) = f^+(u_1, \ldots, u_k) - f^-(u_1, \ldots, u_k).$$

From (2.6.7) we imply that

(2.6.8)

$$\int f^+(u_1, \ldots, u_k) P_\theta(du_1, \ldots, du_k) = \int f^-(u_1, \ldots, u_k) P_\theta(du_1, \ldots, du_k),$$

for all $\theta \in I$. Furthermore, writing

$$P_\theta(du_1, \ldots, du_k) = \exp\left\{\sum_{i=1}^{k} \theta_i u_i\right\} \nu_\theta(du_1, \ldots, du_k) \tag{2.6.9}$$

and defining the sigma finite-measures

$$\begin{aligned} \nu_\theta^+(du_1, \ldots, du_k) &= f^+(u_1, \ldots, u_k)\nu_\theta(du_1, \ldots, du_k), \\ \nu_\theta^-(du_1, \ldots, du_k) &= f^-(u_1, \ldots, u_k)\nu_\theta(du_1, \ldots, du_k), \end{aligned} \tag{2.6.10}$$

we obtain

$$\int \exp\left\{\sum_{i=1}^{k} \theta_i u_i\right\} \nu_\theta^+(du_1, \ldots, du_k) = \int \exp\left\{\sum_{i=1}^{k} \theta_i u_i\right\} \nu_\theta^-(du_1, \ldots, du_k), \tag{2.6.11}$$

for all $\theta \in I$. In particular for $\theta = 0$ we obtain

$$\int \nu_0^+(du_1, \ldots, du_k) = \int \nu_0^-(du_1, \ldots, du_k). \tag{2.6.12}$$

Without loss of generality we can assume that the common value of these two integrals is 1. (If these integrals equal to C, $0 < C < \infty$, we divide $f(u_1, \ldots, u_k)$ by the common value C.) Moreover, for every $\theta \in \Theta$,

$$\nu_\theta^\pm(du_1, \ldots, du_k) = K(\theta)\nu_0^\pm(du_1, \ldots, du_k). \tag{2.6.13}$$

Hence

$$\int \exp\left\{\sum_{i=1}^{k}\theta_i u_i\right\}\nu_0^+(\mathrm{d}u_1,\ldots,\mathrm{d}u_k) = \int \exp\left\{\sum_{i=1}^{k}\theta_i u_i\right\}\nu_0^-(\mathrm{d}u_1,\ldots,\mathrm{d}u_k), \tag{2.6.14}$$

for all $\theta \in I$. If $h(u_1, \ldots, u_k)$ is any bounded measurable function, the integral

$$\varphi(Z_1, \ldots, Z_k) = \int h(u_1, \ldots, u_k)\exp\left\{\sum_{i=1}^{k} Z_i u_i\right\}\mu(\mathrm{d}u_1, \ldots, \mathrm{d}u_k)$$

is an analytic function of the complex variables $Z_j = \theta_j + it_j$ $(j = 1, \ldots, k)$ over the strips $-a \leqslant \theta_j \leqslant a$, $-\infty \leqslant t_j \leqslant \infty$. In the above case $\nu_0^{\pm}$ $(\mathrm{d}u, \ldots, \mathrm{d}u_k)$ are probability measures. Taking $h(u_1, \ldots, u_k) \equiv 1$, we obtain that $\int \exp\{\sum_{i=1}^k Z_i u_i\}\nu_0^{\pm}$ $(\mathrm{d}u_1, \ldots, \mathrm{d}u_k)$ are analytic functions of $Z_1, \ldots, Z_k$ over the strip $-a \leqslant \theta_j \leqslant a$, $-\infty \leqslant t_j \leqslant \infty$. The analyticity of these integrals permits us to extend the above equality over the whole strip and in particular to imply that, for all real $(t_1, \ldots, t_k)$,

(2.6.15)

$$\int \exp\left\{i\sum_{j=1}^{k} t_j u_j\right\}\nu_0^+(\mathrm{d}u_1,\ldots,\mathrm{d}u_k) = \int \exp\left\{i\sum_{j=1}^{k} t_j u_j\right\}\nu_0^-(\mathrm{d}u_1,\ldots,\mathrm{d}u_k).$$

This means that the characteristic functions of ν_0^+ and of ν_0^- coincide. From the unicity of the correspondence between distribution functions and their characteristic functions (see Loève [2]) it follows that ν_0^+ $(\mathrm{d}u_1, \ldots, \mathrm{d}u_k) = \nu_0^-$ $(\mathrm{d}u_1, \ldots, \mathrm{d}u_k)$ a.s. $[\mathfrak{F}^u]$. Finally, the definition of $\nu_0^{\pm}(u_1, \ldots, u_k)$ and the last result imply that

$$f(u_1, \ldots, u_k) = f^+(u_1, \ldots, u_k) - f^-(u_1, \ldots, u_k) = 0 \quad \text{a.s. } [\mathfrak{F}^u].$$

(Q.E.D.)

The joint distribution functions of two independent statistics on the same probability space $(\mathfrak{X}, \mathfrak{B}, F_\theta)$, $\theta \in \Theta$, constitute a complete family, provided the marginal distributions are elements of complete families and at least one of the statistics is *strongly complete* in the following sense (see Fraser [5], p. 26).

A family of distribution functions $\{F_0; \theta \in \Theta\}$ is called *strongly complete* if there exists a measure $\mathcal{M}$ on $(\Theta, \mathcal{C})$, where $\mathcal{C}$ is the Borel sigma-field on Θ such that for every $\Theta^* \subset \Theta$ for which $\mathcal{M}(\Theta - \Theta^*) = 0$,

$$\int \varphi(x)\, F_\theta(\mathrm{d}x) = 0, \quad \text{all } \theta \text{ in } \Theta^*, \tag{2.6.16}$$

implies that $\varphi(X) = 0$ a.s. $[P_\theta]$ for every $\theta \in \Theta$. Obviously strong completeness implies the completeness of a family of distribution functions.

Lemma 2.6.2. (Fraser [5].) *Let $(\mathfrak{X}, \mathfrak{B}, P)$ be a probability space. Let T and T' be two independent statistics on $(\mathfrak{X}, \mathfrak{B})$. If $\{F_\theta^T; \theta \in \Theta\}$ is complete and*

$\{F_\theta^{T'}; \theta \in \Theta'\}$ is strongly complete, the family of joint distributions

$$\{F_{\theta,\tau}^{T,T'}; (\theta, \tau) \in \Theta \times \Theta'\}$$

is complete.

Proof. It is necessary to prove that,

$$\iint h(y, y')F_\theta^{T}(dy)F_\tau^{T'}(dy') = 0 \quad \text{for all} \quad (\theta, \tau) \tag{2.6.17}$$

implies that $h(y, y') = 0$ a.s. $[\mathfrak{F}^{T,T'}]$. According to Fubini's theorem we can write the double integral of (2.6.17) in the form

$$\int F_\theta^{T}(dy)\int h(y, y')F_\tau^{T'}(dy') = 0, \tag{2.6.18}$$

for all $(\theta, \tau) \in \Theta \times \Theta'$. Since T is a complete statistic, it implies that

$$g(y, \tau) = \int h(y, y')F_\tau^{T'}(dy') = 0 \quad \text{a.s. } [\mathfrak{F}^{T}] \tag{2.6.19}$$

and for almost every τ $[\mathcal{M}]$. Finally, since T' is strongly complete, (2.6.19) implies that $h(y, y') = 0$ a.s. $[\mathfrak{F}^{T,T'}]$. (Q.E.D.)

Example 2.13. Lemma 2.6.3 can be applied to prove the completeness of the minimal sufficient statistics $(\bar{X}, Q)$, $Q = \sum_{i=1}^{n} (X_i - \bar{X})^2$, in the case where $X_1, \ldots, X_n$ are i.i.d. like $\mathcal{N}(\xi, \sigma^2)$, $-\infty < \xi < \infty$, $0 < \sigma < \infty$. Both $\bar{X}$ and Q are strongly complete. We show this for Q; $Q \sim \sigma^2\chi^2[n-1]$. Thus the density function of Q is

(2.6.20)

$$f^Q(x; \sigma^2) = \left[(2\sigma^2)^{(n-1)/2}\Gamma\left(\frac{n-1}{2}\right)\right]^{-1} x^{(n-3)/2} \exp\left\{-\frac{x}{2\sigma^2}\right\}, \qquad 0 \leqslant x \leqslant \infty.$$

Let $\mathfrak{D}$ denote the parameter space of σ^2; that is, $\mathfrak{D} = \{\sigma^2; 0 < \sigma^2 < \infty\}$. Let Γ be the Borel sigma-field on $\mathfrak{D}$ and $g_\lambda(\sigma^2)$ a negative exponential density with expectation λ^{-1} on $\mathfrak{D}$. Suppose that with respect to the measure m on $(\mathfrak{D}, \Gamma)$, induced by $g_\lambda(\sigma^2)$, for every $\mathfrak{D}^* \subset \mathfrak{D}$ for which $m(\mathfrak{D} - \mathfrak{D}^*) = 0$,

$$\int_0^\infty \varphi(x)f^Q(x; \sigma^2)\, dx = 0 \quad \text{for all} \quad \sigma^2 \in \mathfrak{D}^*. \tag{2.6.21}$$

Integrating (2.6.21) with respect to $g_\lambda(\sigma^2)$ and applying Fubini's theorem, we obtain

$$\lambda\int_0^\infty \frac{\varphi(x)x^{(n-3)/2}}{\Gamma[(n-1)/2]}\left\{\int_0^\infty e^{-(\lambda+x)\theta}\theta^{(n-1)/2}\, d\theta\right\} dx = 0 \tag{2.6.22}$$

where $\theta = (2\sigma^2)^{-1}$. Since

$$\int_0^\infty e^{-(\lambda+x)\theta}\theta^{(n-1)/2}\,d\theta > 0 \quad \text{for all} \quad 0 \leqslant x < \infty,$$

we obtain from (2.6.22)

$$\varphi(x)\cdot x^{(n-3)/2} = 0 \quad \text{a.s. } [m], \tag{2.6.23}$$

Finally, since $g_\lambda(\sigma^2)$ assigns a positive density for each $0 < \sigma^2 < \infty$, (2.6.23) implies that $\varphi(x) = 0$ a.s. $[\mathcal{T}^Q]$. This completes the proof that Q is strongly complete. ■

Lemma 2.6.3. *Let S be a statistic on $(\mathfrak{X}, \mathcal{B}, F_\theta)$, $\theta \in \Theta$, to $(\zeta, \mathcal{F}, P_\theta^S)$. If $\{P_\theta^S; \theta \in \Theta\}$ is a complete family, and if S is sufficient for $\{F_\theta; \theta \in \Theta\}$, then S is necessary.*

Proof. Let T be any other sufficient statistic for $\{F_\theta;\ \theta \in \Theta\}$ such that $T(X) = h(S(x))$. Consider the function $\psi(S) = S - E\{S \mid T\}$; $\psi(S)$ is $\mathcal{B}_S$-measurable, but since T is a sufficient statistic $\psi(S)$ is independent of F_θ, $\theta \in \Theta$. Finally, since

$$\int \psi(s)P_\theta^S(ds) = 0 \quad \text{for all} \quad \theta \in \Theta, \tag{2.6.24}$$

we obtain from the completeness of S that $\psi(S) = 0$ a.s. $[F_\theta^S;\ \theta \in \Theta]$. This implies that S is $\mathcal{B}_T$-measurable. In other words, the contours of S are included in the contours of any other sufficient statistic for $\{F_\theta;\ \theta \in \Theta\}$. Thus S is necessary (minimal). (Q.E.D.)

We remark in this connection that although the completeness of a sufficient statistics implies its minimality, the converse is not true. *A minimal sufficient statistic is not necessarily a complete one.* Indeed, as we have demonstrated in Example 2.8, in the case of two samples from normal distributions with equal means, that is, $\mathcal{N}(\xi, \sigma_1^2)$ and $\mathcal{N}(\xi, \sigma_2^2)$, the minimal sufficient statistic is $(\bar{X}, \bar{Y}, Q_1, Q_2)$ where $\bar{X}$ and $\bar{Y}$ are the sample means and Q_1, Q_2 are the sample sums of squares of deviations. This minimal sufficient statistic is obviously incomplete.

We now consider the very extensive family of all absolutely continuous distributions and prove that the order statistic of the sample $T = (X_{(1)}, \ldots, X_{(n)})$ is a complete statistic for this general family. This result is very important for distribution free methods.

We remark first that the partition of the sample space induced by the order statistic $T(X_1, \ldots, X_n)$ is equivalent to the partition induced by the statistic

$$S(X_1, \ldots, X_n) = \left(\sum_{i=1}^n X_i, \sum_{i=1}^n X_i^2, \ldots, \sum_{i=1}^n X_i^n\right). \tag{2.6.25}$$

We can show that the system of equations

$$\sum_{i=1}^{n} X_i^{\alpha} = c_{\alpha}, \qquad (\alpha = 1, \ldots, n), \tag{2.6.26}$$

has at most $n!$ solutions, all of which belong to the same coset of an order statistic.

Let $S_\alpha = \sum_{i=1}^{m} X_i^{\alpha}$, $\alpha = 1, \ldots, n$. Let $\mathcal{B}_S$ be the sigma-subfield induced by $S = (S_1, \ldots, S_n)$, and let $h(S_1, \ldots, S_n)$ be a $\mathcal{B}_S$-measurable function. Let $\zeta^{(n)}$ denote the sample space of $(S_1, \ldots, S_n)$ and $F^{S_1,\ldots,S_n}(x_1, \ldots, x_n)$ the joint distribution function of $(S_1, \ldots, S_n)$. We denote by Ψ^S the class of all distributions $F^{S_1,\ldots,S_n}$, where $F \in \Psi$. Let $h(S_1, \ldots, S_n)$ be a $\mathcal{B}_S$-measurable function such that

$$\int h(x_1, \ldots, x_n) F^{S_1,\ldots,S_n}(\mathrm{d}x_1, \ldots, \mathrm{d}x_n) = 0 \tag{2.6.27}$$

for all $F^{S_1,\ldots,S_n} \in \Psi^S$. We wish to show that $h(S_1, \ldots, S_n) = 0$ a.s. $[\Psi^S]$. Consider a subfamily of absolutely continuous distributions $\tilde{\Psi}^S$ such that if $F^{S_1,\ldots,S_n} \in \tilde{\Psi}^S$ then

$$\begin{aligned} F^{S_1,\ldots,S_n}&(\mathrm{d}x_1, \ldots, \mathrm{d}x_n) \\ &= c(\theta_1, \ldots, \theta_n) g(x_1, \ldots, x_n) \exp\left\{-\sum_{j=1}^{n} \theta_j x_j\right\} \mathrm{d}x_1 \cdots \mathrm{d}x_n, \end{aligned} \tag{2.6.28}$$

Ψ^S is the parametric subfamily of Ψ, of the exponential type. Since $E\{h(S_1, \ldots, S_n)\} = 0$ for all $F^S \in \Psi^S$, we have in particular

$$E_\theta\{h(S_1, \ldots, S_n)\} = 0 \quad \text{for all} \quad \theta \in \Theta \tag{2.6.29}$$

where Θ is the range of parameters θ corresponding to the exponential distributions in $\tilde{\Psi}^S$. Finally,

$$\begin{aligned} &E_\theta\{h(S_1, \ldots, S_n\} \\ &= c(\theta_1, \ldots, \theta_n) \int h(x_1, \ldots, x_n) g(x_1, \ldots, x_n) \exp\left\{-\sum_{j=1}^{n} \theta_j x_j\right\} \mathrm{d}x_1 \cdots \mathrm{d}x_n \end{aligned} \tag{2.6.30}$$

is the Laplace transform of $h(s_1, \ldots, s_n) g(s_1, \ldots, s_n)$. By the unicity of the Laplace transform, (2.6.30) implies that $h(S_1, \ldots, S_n) g(S_1, \ldots, S_n) = 0$ a.s. $[\tilde{\Psi}^S]$, and since $g(S_1, \ldots, S_n) > 0$ a.s. $[\tilde{\Psi}^S]$ we conclude that $h(S_1, \ldots, S_n) = 0$ a.s. $[\tilde{\Psi}^S]$. Thus $(S_1, \ldots, S_n)$ is a complete sufficient statistic for the subfamily of absolutely continuous distributions of the exponential type. It remains to show that $(S_1, \ldots, S_n)$ is complete relative to the family of all absolutely continuous distributions. This is concluded from the following lemma. (See Fraser [5], p. 26.) We remind the reader that a set N is said to be null with respect to a family Ψ of distribution functions if

$$\int_N F(dx) = 0 \quad \text{for all} \quad F \in \Psi. \tag{2.6.31}$$

Lemma 2.6.4. *Let $\tilde{\Psi}$ and Ψ be families of distribution functions, $\tilde{\Psi} \subset \Psi$. The completeness of $\tilde{\Psi}$ implies the completeness of Ψ if, for every set, N, null with respect to $\tilde{\Psi}$, and every distribution function $G \in \Psi - \tilde{\Psi}$,*

$$\int_N G(dx) = 0. \tag{2.6.32}$$

Proof. We have to prove that if h is an unbiased estimator of zero, that is,

$$\int h(x)F(dx) = 0 \quad \text{for all} \quad F \in \Psi, \tag{2.6.33}$$

then $h(X) = 0$ a.s. $[\Psi]$. Since $\tilde{\Psi} \subset \Psi$ and $\tilde{\Psi}$ is complete, we have for every h satisfying (2.6.33) that $h(X) = 0$ a.s. $[\tilde{\Psi}]$. But condition (2.6.32) implies that $h(X) = 0$ a.s. $[\Psi]$. Hence Ψ is complete. (Q.E.D.)

Thus Lemma 2.6.4 and the completeness of the family of absolutely continuous distribution functions of the exponential type, combined with the property that the family of *all* absolutely continuous distributions is dominated by the subfamily of absolutely continuous distributions of the exponential type, imply the following theorem.

Theorem 2.6.5. *The order statistic $T(X_1, \ldots, X_n)$ is complete with respect to the family of all distribution functions over the Euclidean n-space, $E^{(n)}$, provided that each random variable X_i $(i = 1, \ldots, n)$ has the same absolutely continuous distribution function.*

The completeness of the order-statistic for the family of all discrete distribution functions was proven by Halmos (see Fraser [5], p. 29).

2.7. SUFFICIENCY AND INVARIANCE

In many statistical decision problems it is reasonable to confine attention to procedures that are invariant with respect to a certain group of transformations. For instance, if we are concerned about the variance of a distribution, the variance is a translation-invariant parameter and it is justified in many cases to demand that the decision procedure concerning the variance of the distribution be independent of the origin of measurements. The restriction to invariant statistics reduces the sample space of the original observations. In terms of sigma-fields, the original Borel sigma-field $\mathcal{B}$ on

$\mathcal{X}$ is reduced to a sigma-field $\mathcal{B}_I$ of invariant sets (the terminology will be rigorously defined later on). Further reduction of the statistics, and of the corresponding sigma-fields can be attempted by considering sufficient statistics over the set of invariant statistics. These statistics, called invariantly sufficient, provide all the information on the family of distributions given by invariant statistics. The following example will further clarify the issue.

Example 2.14. Let $(X_1, \ldots, X_n)$ be i.i.d. random variables having a common normal distribution $\mathcal{N}(\xi, \sigma^2)$, $-\infty < \xi < \infty$, $0 < \sigma < \infty$. We wish to decide about the value of the variance σ^2, and we confine attention to translation invariant procedures.

A translation of $\mathbf{X} = (X_1, \ldots, X_n)'$ is an addition of a constant c, $-\infty < c < \infty$, to each observation X_i $(i = 1, \ldots, n)$. A translation-invariant statistic is a $\mathcal{B}$-measurable function f on $\mathcal{X}$ such that

$$f(X_1, \ldots, X_n) = f(X_1 + c, \ldots, X_n + c) \quad \text{for all} \quad -\infty < c < \infty.$$

As will be shown later on, every invariant statistic is a function of a statistic, called maximal invariant. A maximal invariant statistic in the present case is $(X_1 - X_n, \ldots, X_{n-1} - X_n)$. Let $Y_i = X_i - X_n$ $(i = 1, \ldots, n-1)$ and $\mathbf{Y} = (Y_1, \ldots, Y_{n-1})'$. The transformation $\mathbf{X} \to \mathbf{Y}$ reduces the dimension of the statistic and restricts the treatment of the data to translation-invariant procedures. If n is larger than 2 we can reduce the data further by considering minimal sufficient statistics for the class $\mathcal{F}^{\mathbf{Y}}$ of the probability measures of $\mathbf{Y}$. As we shall exhibit, the minimal sufficient statistic for $\mathcal{F}^{\mathbf{Y}}$, which is also an invariant statistic, is 1-dimensional.

It is immediate to verify that

$$\mathbf{Y} \sim N(\mathbf{0}, \sigma^2(\mathbf{I}_{n-1} + \mathbf{J}_{n-1})),$$

where $\mathbf{J}_k = \mathbf{1}_k\mathbf{1}_k'$ is a $k \times k$ matrix with all elements equal to 1. The inverse of $\mathbf{I}_{n-1} + \mathbf{J}_{n-1}$ is

$$(\mathbf{I}_{n-1} + \mathbf{J}_{n-1})^{-1} = I_{n-1} - \frac{1}{n}\mathbf{J}_{n-1}.$$

Therefore the density function of $\mathbf{Y}$ is

$$f(y; \sigma^2) = (2\pi)^{-(n-1)/2}\sigma^{-(n-1)}[\det\{\mathbf{I}_{n-1} + \mathbf{J}_{n-1}\}]^{-1/2} \times \exp\left\{-\frac{1}{2\sigma^2}\mathbf{Y}'\left(\mathbf{I}_{n-1} - \frac{1}{n}\mathbf{J}_{n-1}\right)\mathbf{Y}\right\}.$$

Thus $Q = \mathbf{Y}'\left(\mathbf{I}_{n-1} - \frac{1}{n}\mathbf{J}_{n-1}\right)\mathbf{Y}$ is a minimal sufficient statistic for the class of probability measures having the above densities $f(y; \sigma^2)$, $0 < \sigma^2 < \infty$. Writing this invariantly sufficient statistic Q in terms of the original vector of observations $\mathbf{X}$ we have

$$Q = \mathbf{X}'\left(\mathbf{I}_n - \frac{1}{n}\mathbf{J}_n\right)\mathbf{X} = \sum_{i=1}^{n}(X_i - \bar{X})^2$$

where $\bar{X} = (1/n)1_n'X$ is the sample mean. We now show that the statistic Q can be obtained more easily by another route of operations. A minimal sufficient statistic for $\mathcal{F}^X$ is the pair $(\bar{X}, Q)$. If we apply a translation transformation $X_i \to X_i + c$, $-\infty < c < \infty$, the minimal sufficient statistic $(\bar{X}, Q)$ is transformed to $(\bar{X} + c, Q)$. In particular, making the transformation $X_i \to X_i - \bar{X}$ $(i = 1, \ldots, n)$ we obtain $(\bar{X}, Q) \to (0, Q)$. Hence, Q is a maximal translation invariant over the sample space of $(\bar{X}, Q)$. As was previously shown, Q is invariantly sufficient. The two routes of operation yield in this example the same statistic Q. ∎

The general problem of sufficiency and invariance is to establish conditions on the family of distribution functions so that the two routes of reduction will yield the same statistic. We present in the present section the theory developed by Hall, Wijsman, and Ghosh [11].

Let $(\mathcal{X}, \mathcal{B}, \mathcal{F})$ be a probability space; the sample space $\mathcal{X}$ is an open set in a finite-dimensional Euclidean space; $\mathcal{B}$ is the Borel sigma-field on $\mathcal{X}$; and $\mathcal{F}$ is a family of probability measures P on $(\mathcal{X}, \mathcal{B})$.

Let $\mathcal{G}$ be a group of 1:1 transformations of $\mathcal{X}$ onto $\mathcal{X}$; that is, $g\mathcal{X} = \mathcal{X}$ for all g in $\mathcal{G}$. The transformations on $\mathcal{F}$ defined by $\bar{g}P = P^{gX}$ induce a group of 1:1 transformations $\bar{\mathcal{G}}$. We assume that $\mathcal{F}$ is such that for every $P \in \mathcal{F}$ and $\bar{g} \in \bar{\mathcal{G}}$, $\bar{g}P \in \mathcal{F}$. This is denoted by $\bar{g}\mathcal{F} = \mathcal{F}$.

For example, suppose $\mathcal{X}$ is the real line and $\mathcal{F}$ the class of all location and scale parameter distributions $F[(x - \xi)/\sigma]$; $-\infty < \xi < \infty, 0 < \sigma < \infty$. $\mathcal{G}$ is the group of affine transformations

$$g_{\alpha,\beta}X = \alpha X + \beta, \quad 0 < \alpha < \infty, \quad -\infty < \beta < \infty.$$

The induced group $\bar{\mathcal{G}}$ is the group operating on the parameters so that

$$\bar{g}_{\alpha,\beta}(\xi, \sigma) = (\alpha\xi + \beta, \alpha\sigma)$$

Thus we have for every $-\infty < b < \infty$,

$$P^{(\xi,\sigma)}[X \in (-\infty, b)] = P^{(\alpha\xi+\beta,\alpha\sigma)}[X \in (-\infty, \alpha b + \beta)]. \tag{2.7.1}$$

More generally, for every Borel set $B \in \mathcal{B}$, if gB denotes set of image points of x in B, that is, $gB = \{y; y = gx \text{ and } x \in B\}$, we have

$$P[X \in B] = \bar{g}P[X \in gB]. \tag{2.7.2}$$

An *invariant statistic* on $(\mathcal{X}, \mathcal{B})$, f, is a $\mathcal{B}$-measurable function on $\mathcal{X}$ such that

$$f(x) = f(gx) \quad \text{for all } x \in \mathcal{X}, \quad \text{all } g \in \mathcal{G}. \tag{2.7.3}$$

An *almost invariant* statistic is a $\mathcal{B}$-measurable function f on $\mathcal{X}$ such that

$$f(x) = f(gx) \text{ a.s. } [P] \quad \text{for all } P \in \mathcal{F}, \quad \text{all } g \in \mathcal{G}. \tag{2.7.4}$$

The group of transformations $\mathcal{G}$ induces a partition of $\mathcal{X}$ to cosets or *orbits*, where an orbit of x_0, $x_0 \in \mathcal{X}$, relative to $\mathcal{G}$ is the set

$$G(x_0) = \{x; g(x_0) = x, \quad g \in \mathcal{G}\}. \tag{2.7.5}$$

From the definition of invariant statistic and of orbit we immediately conclude that an invariant statistic f assumes the same value for all x belonging to the same orbit. It may be the case, however, that f assumes the same constant value on several orbits. For example, $f =$ constant on $\mathcal{X}$ is an invariant function which assumes the same value on all orbits of $\mathcal{G}$. If an invariant statistic assumes a different value on each orbit it is called *maximal invariant.* In other words, a statistic $U(x)$ is maximal invariant if and only if $U(x_1) = U(x_2)$ implies that $x_2 = gx_1$ for some $g \in \mathcal{G}$. Maximal invariant statistics are not unique. Every invariant statistic is a function of a maximal invariant statistic.

Example 2.15 (Lehmann [2,] p. 21.) Let $\mathcal{X}$ be the n-dimensional Euclidean space $x = (x_1, \ldots, x_n)$. Let $\mathcal{G}_1$ be the group of translations; that is, $\mathcal{G}_1 = \{g_c; g_c x_i = x_i + c, i = 1, \ldots, n\}$. A maximal invariant of $\mathcal{X}$ with respect to $\mathcal{G}_1$ is the set

$$U_1 = \{(y_1, \ldots, y_{n-1}); y_i = x_i - x_n, i = 1, \ldots, n-1, x \in \mathcal{X}\}.$$

We designate the vector of difference $(x_1 - x_n, x_2 - x_n, \ldots, x_{n-1} - x_n)$ by $U(x)$. We now show that $U(x)$ is maximal invariant. Indeed, if $x^{(1)}$ and $x^{(2)}$ are two points of $\mathcal{X}$ and $U(x^{(1)}) = U(x^{(2)})$, it is readily obtained that $x^{(2)} = x^{(1)} + (x_n^{(2)} - x_n^{(1)})\mathbf{1}_n$. Hence $x^{(2)}$ and $x^{(1)}$ lie on the same orbit. In the same manner we can also show that $(x_1 - x_2, x_3 - x_2, \ldots, x_n - x_2)$ is a maximal invariant.

Let $\mathcal{G}_2$ be the group of transformations

$$\mathcal{G}_2 = \{g_\alpha; g_\alpha x = \alpha x, 0 < \alpha < \infty\}.$$

Then a maximal invariant with respect to $\mathcal{G}_2$, assuming that $x_n \neq 0$, is

$$U_2 = \left\{u(x_1, \ldots, x_n) = \left(\frac{x_1}{x_n}, \ldots, \frac{x_{n-1}}{x_n}\right), x \in \mathcal{X}\right\}.$$

Let $\mathcal{G} = \mathcal{G}_2 \cdot \mathcal{G}_1 = \{g = g_2 g_1; g_2 \in \mathcal{G}_2, g_1 \in \mathcal{G}_1\}$. We choose first g_1 in $\mathcal{G}_1$ and transform x to $g_1 x$; then we choose g_2 in $\mathcal{G}_2$ and transform $g_1 x$ to $g_2 g_1 x$.

Assuming that $x_n \neq x_{n-1}$, a maximal invariant of $\mathcal{X}$ is

$$\mathcal{U} = \left\{u(x_1, \ldots, x_n) = \left(\frac{x_1 - x_n}{x_{n-1} - x_n}, \frac{x_2 - x_n}{x_{n-1} - x_n}, \ldots, \frac{x_{n-2} - x_n}{x_{n-1} - x_n}\right)\right\}. \quad \blacksquare$$

Let U be a transformation of the sample space $\mathcal{X}$ to the space $\mathcal{U}$ determined by a maximal invariant statistic. Let $\mathcal{B}_I$ be the sigma-subfield induced by U. $\mathcal{B}_I \subset \mathcal{B}$. Every $\mathcal{G}$-invariant statistic f is measurable $\mathcal{B}_I$. The Borel sets B of $\mathcal{B}_I$ are invariant sets in the sense that $gB = B$ for all $B \in \mathcal{B}_I$ and all $g \in \mathcal{G}$. Let $\mathcal{F}^u$ be the induced family of probability measures. Thus we have

$$u: (\mathcal{X}, \mathcal{B}, \mathcal{F}) \to (\mathcal{U}, \mathcal{B}_I, \mathcal{F}^u). \tag{2.7.6}$$

Let S_I be a sufficient statistic for $\mathcal{F}^u$ on $\mathcal{U}$. S_I induces a sigma-subfield of $\mathcal{B}_I$, namely $\mathcal{B}_{SI}$. By the definition of a sufficient statistic, for every invariant statistic $f_I \in \mathcal{B}_I$ any particular determination of the conditional probability $E\{f_I \mid S_I\}$ is independent of $\mathcal{F}^I$ and is equivalent to a $\mathcal{B}_{SI}$ statistic, say f_{SI}. The statistic S_I is called *invariantly sufficient*.

Let S be a sufficient statistic for $\mathcal{F}$ and $\mathcal{B}_S$ the corresponding sigma-subfield, $\mathcal{B}_S \subset \mathcal{B}$. Let $\mathcal{G}_S$ be the corresponding group of transformations. In terms of Example 2.13, the group of transformations $\mathcal{G}_S$ is the set of transformations g_S: $(\bar{X}, Q) \to (\bar{X} + c, Q)$ whenever $x_i \to x_i + c$ for all $i = 1, \ldots, n$. Let $U_S(S)$ be a statistic reducing every S to a certain maximal invariant $U_S(\cdot)$ with respect to $\mathcal{G}_S$. Let $\mathcal{V}$ denote the sample space of $U_S(S)$. The problem raised in the present section is the following:

Under what conditions is the sigma-subfield induced by $U_S(\cdot)$ equivalent to the sigma-subfield $\mathcal{B}_{SI}$ of the invariantly sufficient statistic?

The following scheme represents the two possible routes of transformations leading to $(\mathcal{V}, \mathcal{B}_{SI}, \mathcal{F}^{SI})$:

$$\begin{array}{ccc} (\mathcal{X}, \mathcal{B}, \mathcal{F}) & \xrightarrow{S} & (\mathcal{S}, \mathcal{B}_S, \mathcal{F}^S) \\ \downarrow U & & \downarrow U \\ (\mathcal{U}, \mathcal{B}_I, \mathcal{F}^I) & \xrightarrow{S} & (\mathcal{V}, \mathcal{B}_{SI}, \mathcal{F}^{SI}) \end{array}$$

Theorem 2.7.1. (Stein.) *Either one of the three conditions A, B, or C implies that the statistic $V = U_S(S)$, reducing a sufficient statistic S for $\mathcal{F}$ to a maximal invariant with respect to $\mathcal{G}_S$, is invariantly sufficient.*

Condition A.

(i) $g\mathcal{B}_S = \mathcal{B}_S$ *for all* $g \in \mathcal{G}$.
(ii) *If* $f_S \in \mathcal{B}_S$ *and is almost invariant, there exists an invariant function* $f_{SI} \in \mathcal{B}_{SI}$ *equivalent to* f_S*; that is,* $f_{SI}(X) = f_S(X)$ a.s. $[\mathcal{F}]$.

Condition B. There exists an invariant conditional probability measure, $P[B \mid S] = P[gB \mid g_S S]$ *for all* $B \in \mathcal{B}$ *and* $g \in \mathcal{G}$.

Condition C. $\mathcal{X}$ *is an n-dimensional Euclidean space,* $\mathcal{B}$ *the Borel sigma-field on* $\mathcal{X}$, $\mathcal{P} = \{P_\theta; \theta \in \Theta\}$ *where* Θ *is an arbitrary index set, and with respect to a Lebesque measure* $\mu(dx_1, \ldots, dx_n)$, P_θ *has a density*

$$f(x; \theta) = h(x) g_\theta(S(x)), \quad x \in \mathcal{X}, \tag{2.7.7}$$

whose factors $h(x)$ *and* $g_\theta(S(x))$ *satisfy the conditions of the factorization theorem; Theorem 2.3.3. Furthermore, suppose that there is an open set* $B_{SI} \in \mathcal{B}_{SI}$ *of* $\mathcal{P}$*-measure 1 such that on* B_{SI}

(i) *each* $g \in \mathcal{G}$ *is continuously differentiable and the Jacobian depends on* $S(x)$;
(ii) *for each* $g \in \mathcal{G}$, $S(x) = S(x')$ *implies* $S(gx) = S(gx')$;
(iii) $S(x)$ *is continuously differentiable, and the matrix* $\|[\partial S_j(x)/\partial x_i]; j = 1, \ldots, k, i = 1, \ldots, n\|$ *is of rank k* (*S is k-dimensional*);
(iv) *for each* $g \in \mathcal{G}$, $h(g(x))/h(x)$ *depends only on* $S(x)$.

The proof of Stein's theorem will proceed through a sequence of lemmas, among which we provide several examples. In order to prove Stein's theorem we have to show that given a maximal invariant statistic on the space $\mathcal{S}$ of the sufficient statistic, namely, $U_S(S)$, every determination of the conditional expectation of an invariant function f_I is equivalent to some function of $\mathcal{B}_{SI}$ and is thus independent of $\mathcal{P}$. We first prove that Condition A implies the conclusions of Stein's theorem.

We introduce the notation $E\{f \mid \mathcal{B}_S\}$ for the conditional expectation of f given any statistic which is $\mathcal{B}_S$ measurable. This conditional expectation operator is a slight generalization of the operator $E\{f \mid S\}$ treated so far. The fact is that, given any two statistics S_1 and S_2 which are $\mathcal{B}_S$-measurable, $E\{f \mid S_1\} = E\{f \mid S_2\}$ a.s. $[P]$ for every $P \in \mathcal{P}$. It is therefore immaterial which particular statistic in $\mathcal{B}_S$ we choose for conditioning the expectation operator. We assume that all functions under consideration are P-integrable for all $P \in \mathcal{P}$.

Lemma 2.7.1. *If under Condition A(i)* f_I *is invariant, then any version of* $E\{f_I \mid \mathcal{B}_S\}$ *is almost invariant.*

Proof. The transformations g in $\mathcal{G}$ produce an isomorphism between $(\mathcal{X}, \mathcal{B}, \mathcal{P})$ and $(g\mathcal{X}, g\mathcal{B}, \bar{g}\mathcal{P})$. If f is any $\mathcal{B}$-measurable and P-integrable function, let $\tilde{g}$ be the transformation of f induced by g; that is, $\tilde{g}f(x) = f(gx)$ for all $x \in \mathcal{X}$; $\tilde{g}f$ is a $\bar{g}P$-integrable function. Consider an invariant function $f_I(x)$ and the sigma-subfield $\mathcal{B}_S$. Then since $\tilde{g}f_I(x) = f_I(gx) = f_I(x)$ for all

x and all $g \in \mathcal{G}$,

(2.7.8) $$E\{\tilde{g}f_I(X) \mid \mathcal{B}_S\} = E\{f_I(X) \mid \mathcal{B}_S\} = f_{SI}(X) \text{ a.s. } [\bar{g}\mathcal{F}],$$

where $f_{SI}(X)$ is some $\mathcal{B}_S$-measurable function. Since $f_{SI}(X)$ is also invariant, we obtain

(2.7.9) $$\begin{aligned} \tilde{g}f_{SI}(X) = \tilde{g}E\{f_I(X) \mid \mathcal{B}_S\} &= f_{SI}(X) \\ &= E\{f_I(X) \mid \mathcal{B}_S\} \end{aligned}$$

a.s. $[\bar{g}\mathcal{F}]$. Hence $E\{f_I(X) \mid \mathcal{B}_S\}$ is almost invariant. (Q.E.D.)

Condition A(ii) and Theorem 2.7.1 imply that $f_{SI}(X)$ is equivalent to a $\mathcal{B}_{SI}$-measurable function. We introduce now the notion of sufficient subfield, which is a straightforward generalization of the notion of sufficient statistic.

A sigma-subfield $\mathcal{B}_S$ is called sufficient for $\mathcal{F}$ if, for any $\mathcal{B}$-measurable and P-integrable function $f\,(P \in \mathcal{F})$, any version of the conditional expectation $E\{f \mid \mathcal{B}_S\}$ is independent of $\mathcal{F}$.

Lemma 2.7.2. *Under Condition A $\mathcal{B}_{SI}$ is sufficient for $\mathcal{F}^I$.*

Proof. We must prove that if $f_I \in \mathcal{B}_I$, there exists a $\mathcal{B}_{SI}$-measurable function f_{SI} such that

(2.7.10) $$E_P\{f_I(X) \mid \mathcal{B}_{SI}\} = f_{SI}(X) \quad \text{a.s. } [P], \quad P \in \mathcal{F}^I.$$

Let $F_S \in \mathcal{B}_S$ be any version of $E\{f_I \mid \mathcal{B}_S\}$, $\mathcal{B}_{SI} \subset \mathcal{B}_S$. Hence by the property of the iterated conditional expectations,

(2.7.11) $$E_P\{f_I(X) \mid \mathcal{B}_{SI}\} = E_P\{F_S(X) \mid \mathcal{B}_{SI}\} \quad \text{a.s. } [P], \quad P \in \mathcal{F}^I.$$

By Condition A(ii) and Lemma 2.7.1 there exists a $\mathcal{B}_{SI}$-measurable function f_{SI} equivalent to F_S. Substituting this function (2.7.11), we obtain

(2.7.12) $$\begin{aligned} E_P\{f_I(X) \mid \mathcal{B}_{SI}\} &= E_P\{f_{SI}(X) \mid \mathcal{B}_{SI}\} \\ &= f_{SI}(X) \quad \text{a.s. } [P] \quad \text{all } P \in \mathcal{F}_I. \end{aligned}$$

Lemma 2.7.2 proves that Condition A implies the conclusion of Stein's theorem. Hall, Wijsman, and Ghosh [1] prove that $\mathcal{B}_S$ and $\mathcal{B}_I$ are *conditionally independent*, given $\mathcal{B}_{SI}$. That is, for any $f_I \in \mathcal{B}_I$ and $f_S \in \mathcal{B}_S$,

(2.7.13) $$E\{f_I(X)f_S(X) \mid \mathcal{B}_{SI}\} = E\{f_I(X) \mid \mathcal{B}_{SI}\}E\{f_S(X) \mid \mathcal{B}_{SI}\} \quad \text{a.s.}$$

(Q.E.D.)

The following example is provided in [11].

Example 2.16. Let X_1, X_2 be independent random variables identically distributed like $\mathcal{N}(\theta, 1)$, $-\infty < \theta < \infty$. A sufficient and necessary statistic for $\mathcal{F}$ is $S = X_1 + X_2$.

Consider the group of sign transformations $\mathcal{G} = \{g^+, g^-\}$, where $g^+(X) = X$ and $g^-(X) = -X$. The induced group of transformations on $\mathcal{T}$ is $\bar{\mathcal{G}} = \{\bar{g}^+, \bar{g}^-\}$; $\bar{g}^{\pm}(\theta) = \pm\theta$. Furthermore $\bar{\mathcal{G}} = \mathcal{G}_S$, where $\mathcal{G}_S$ is the induced group of transformations on the sample space $\mathcal{S}$ of the sufficient statistic. A maximal invariant of (X_1, X_2) with respect to $\mathcal{G}$ is $U(X_1, X_2) = (|X_1|, |X_2|, \text{sgn}\,(X_1), \text{sgn}\,(X_2))$ where $\text{sgn}\,(X) = 1$ if $X \geqslant 0$ and $= -1$ if $X < 0$. Indeed, if (X_1, X_2) and (X_1', X_2') are such that $U(X_1, X_2) = U(X_1', X_2')$, then $(X_1', X_2') = g^+(X_1, X_2)$ or $(X_1', X_2') = g^-(X_1, X_2)$, which means that (X_1', X_2') and (X_1, X_2) lie on the same orbit.

A maximal invariant of θ is $U(\theta) = |\theta|$. A maximal invariant of S is $U_S(S) = |X_1 + X_2|$. Stein's theorem assures that the conditional distribution of $U(X_1, X_2)$, given $U_S(S) = |X_1 + X_2|$, does not depend on θ (is distribution free). ■

We now consider Condition B of Stein's theorem. We first formulate this condition in more rigorous terms.

Condition B. There exists a $\mathcal{B}_{SI}$-measurable set B_{SI} of $\mathcal{T}$-measure 1 and a real valued function Q on $\mathcal{B} \times \mathcal{X}$, $Q(B, x)$, which is zero for all $x \notin B_{SI}$ such that

(*i*) for every $x \in B_{SI}$, $Q(B, x)$ is a probability measure on $\mathcal{B}$;
(*ii*) for every $B \in \mathcal{B}$, $Q(B, x)$ is a version of $P[B \mid \mathcal{B}_S]$;
(*iii*) for every $x \in \mathcal{X}$, $B \in \mathcal{B}$ and $g \in \mathcal{G}$, $Q(gB; gx) = Q(B, x)$.

Lemma 2.7.3. *If Condition B holds, then the conclusion of Lemma* 2.7.2 *holds and $\mathcal{B}_I$ is conditionally independent of $\mathcal{B}_S$ given $\mathcal{B}_{SI}$.*

Proof. We remark first that Condition A(*i*) is implied by Condition B. Indeed, $Q(B, x)$, $B \in \mathcal{B}$, $x \in B_{SI}$, generates a sufficient subfield of $\mathcal{B}_S$, differing from $\mathcal{B}_S$ only in null sets and satisfying Condition A(ii) (see Hall, Wijsman, and Ghosh [1], p. 605).

A possible version of $E\{f \mid \mathcal{B}_S\}$ is

$$E\{f \mid \mathcal{B}_S\} = \int f(y) Q(dy, x), \qquad x \in B_{SI}. \tag{2.7.14}$$

Hence if $f(X)$ is invariant and $Q(B, X)$ satisfies Condition B, $E\{f(X) \mid \mathcal{B}_S\}$ is invariant. Therefore, since $E\{f \mid \mathcal{B}_S\}$ is $\mathcal{B}_S$-measurable, it is a $\mathcal{B}_{SI}$-measurable function. This is the conclusion of Lemma 2.7.2. (Q.E.D.)

The example given by Hall, Wijsman and Ghosh [1] is as follows.

Example 2.17. Let $X_1, \ldots, X_n$ be i.i.d. random variable with an unknown distribution. Let $\mathcal{G}$ be the group of transformations $g(X_1, \ldots, X_n) = (h(X_1), \ldots, h(X_n))$ where $h(x)$ is strictly monotone, continuous, and maps the real line onto itself. A sufficient statistic is the order statistic $(X_{(1)}, \ldots, X_{(n)})$ (the trivial statistic). Let $\pi(X)$ denote a permutation of the coordinates of $x = (x_1, \ldots, x_n)$.

$\mathcal{B}_S$ is the sigma-subfield generated by sets $B \in \mathcal{B}$ and having the property that if $X \in B$ then $\pi(X) \in B$.

Given x and a Borel set $B \in \mathcal{B}$, let $m_B(x)$ be the number of distinct points $\pi(x)$ that are in B. Define $Q(B, x) = m_B(x)/m(x)$, $Q(B, x)$ satisfies Condition B, with $B_{SI} = \mathcal{X}$. ■

To show that Condition C implies the conclusion of Stein's theorem, the following lemma is proven by Hall, Wijsman and Ghosh [1].

Lemma 2.7.4. *Condition C implies Condition B and therefore the conclusions of Stein's theorem.*

We do not provide here the proof of this lemma, which is quite involved. We present only an example, taken also from Hall, Wijsman and Ghosh [1].

Example 2.18. Let $X_1, \ldots, X_n$ $(n \geqslant 3)$ be i.i.d. random variables having a normal distribution $\mathcal{N}(\xi, \sigma^2)$; both ξ and σ are unknown. Let $\theta = (\xi, \sigma)$, $S = (T_1, T_2)$, $T_1 = \sum_{i=1}^n X_i$, $T_2 = \sum_{i=1}^n X_i^2$. According to (2.7.6) we write $h(x) = 1$ and

$$g_\theta(S(X)) = (2\pi)^{-n/2}\sigma^{-n} \exp\left\{-\frac{1}{2\sigma^2} T_2 + \frac{\xi}{\sigma^2} T_1 - \frac{n\xi^2}{2\sigma^2}\right\}. \tag{2.7.15}$$

Let $\mathcal{G}$ be the group of transformations $X_i \to cX_i$ $(i = 1, \ldots, n)$, $0 < c < \infty$. The matrix $\mathbf{D}(\mathbf{X}) = \|\partial T_i(X)/\partial X_j;\ i = 1, 2;\ j = 1, \ldots, n\|$ is of rank 2, unless all the X's are equal (null-event). All the assumptions specified in Condition C hold, and a maximal invariant on S can be taken as the Student t-statistic; namely; $U_S(T_1, T_2) = T_1[T_2 - (T_1^2/n)]^{-1/2}$. ■

2.8. SUFFICIENCY AND TRANSITIVITY FOR SEQUENTIAL MODELS

In a typical sequential model an infinite sequence of possible experiments $(\mathcal{E}_1, \mathcal{E}_2, \ldots)$ is specified. A sequence $(X_1, X_2, \ldots)$ of random variables is associated with the sequence of possible experiments. We denote by $\mathcal{X}$ the sample space corresponding to this sequence of random variables and by $\mathcal{X}^{(n)}, n = 1, 2, \ldots$, the sample space of the first n random variables $(X_1, \ldots, X_n)$. It is assumed here that $\mathcal{X}^{(n)}$ $(n = 1, 2, \ldots)$ are Euclidean spaces. Let $\mathcal{B}^{(n)}$ designate the Borel sigma-field on $\mathcal{X}^{(n)}$ and $\mathcal{F}^{(n)}$ denote the family of probability measures on $(\mathcal{X}^{(n)}, \mathcal{B}^{(n)})$. We have the inclusion relationship $\mathcal{B}^{(1)} \subset \mathcal{B}^{(2)} \subset \cdots$. A sequential procedure provides a set of stopping rules $\{R_n(X_1, \ldots, X_n);\ n = 1, 2, \ldots\}$ which are $\mathcal{B}^{(n)}$-measurable statistics, assigning to $(X_1, \ldots, X_n)$ an integer value so that if $R_n(X_1, \ldots, X_n) = n$, we terminate sampling after the n-th observation. Otherwise, X_{n+1} is observed.

Assume that for each $n \geqslant 1$ the families $\mathcal{P}^{(n)}$ of probability measures on $(\mathcal{X}^{(n)}, \mathcal{B}^{(n)})$ admit sufficient statistics [it is sufficient to assume that for each n, $\mathcal{P}^{(n)}$ is dominated by a sigma-finite measure $\mu^{(n)}$]. Let $S_n(X_1, \ldots, X_n)$ be a (minimal) sufficient statistic for $\mathcal{P}^{(n)}$. The sequence $(S_1, S_2, \ldots)$ is called a *sufficient sequence* for the sequential model.

We now define the notion of transitivity of a sequence of random variables associated with a sequential model. This notion was introduced by Bahadur [1]. A sequence of random variables $\{T_n(X_1, \ldots, X_n); n = 1, 2, \ldots\}$ such that $T_n \in \mathcal{B}^{(n)}$ is called a *transitive* sequence for the sequential model if for any $\mathcal{B}_0^{(n)}$-measurable and $\mathcal{P}^{(n)}$-integrable function f on $\mathcal{X}^{(n)}$,

(2.8.1)
$$E_{P^{(n)}}\{f(X_1, \ldots, X_n) \mid \mathcal{B}^{(n-1)}\} = E_{P^{(n)}}\{f(X_1, \ldots, X_n) \mid \mathcal{B}_0^{(n-1)}\} \quad \text{a.s. } [P^{(n)}],$$

all $P^{(n)} \in \mathcal{P}^{(n)}$, $n = 1, 2, \ldots$, where $\mathcal{B}_0^{(n)}$ is the sigma subfield generated by T_n. In particular, if $\{T_n; n = 1, 2, \ldots\}$ is a transitive sequence for a sequential model $\{(\mathcal{X}^{(n)}, \mathcal{B}^{(n)}, \mathcal{P}^{(n)}); n = 1, 2, \ldots\}$, any version of the conditional distribution of T_n given $(X_1, X_2, \ldots, X_{n-1})$ depends only on $T_{n-1}(X_1, \ldots, X_{n-1})$ and is thus equivalent to the conditional distribution of T_n given T_{n-1} (for all $n = 1, 2, \ldots$).

Bahadur [1] has shown that in sequential decision problems attention can be confined to sequential decision rules (including stopping rules and action rules) which depend at each n only on S_n, provided the sequence $(S_1, S_2, \ldots)$ is sufficient and transitive for the sequential model. The question arising is; *Under what condition is a sufficient sequence also a transitive one?* The following theorem of Wijsman (see [1]) gives the condition in terms of the sigma-subfields $\mathcal{B}_S^{(n)}$ induced by S_n at each n. The sequence $\{\mathcal{B}_S^{(n)}; n = 1, 2, \ldots\}$ is called transitive if $\{S_n; n = 1, 2, \ldots\}$ is a transitive sequence.

Theorem 2.8.1. (Wijsman.) *The sequence* $\{\mathcal{B}_S^{(n)}\ n = 1, 2, \ldots\}$, *where* $\mathcal{B}_S^{(n)} \subset \mathcal{B}^{(n)}$ *for each* $n \geqslant 1$, *is transitive for* $\{\mathcal{B}^{(n)}; n = 1, 2, \ldots\}$ *if and only if for each* n, $n \geqslant 1$, $\mathcal{B}^{(n)}$ *and* $\mathcal{B}_S^{(n+1)}$ *are conditionally independent given* $\mathcal{B}_S^{(n)}$.

Proof. We denote by $\mathcal{B}_1 \vee \mathcal{B}_3$ the smallest sigma-field containing both $\mathcal{B}_1$ and $\mathcal{B}_3$. It is proven in Loève [1] that $\mathcal{B}_1$ and $\mathcal{B}_2$ are conditionally independent, given $\mathcal{B}_3$, if and only if for every $f_2 \in \mathcal{B}_2$,

$$E\{f_2(X) \mid \mathcal{B}_1 \vee \mathcal{B}_3\} = E\{f_2(X) \mid \mathcal{B}_3\} \text{ a.s.} \tag{2.8.2}$$

Let f be $\mathcal{B}_S^{(n+1)}$ measurable. Thus according to (2.8.2) we have for all $P \in \mathcal{P}$,

(2.8.3)
$$E_P\{f(X_1, \ldots, X_{n+1}) \mid \mathcal{B}^{(n)} \vee \mathcal{B}_S^{(n)}\} = E_P\{f(X_1, \ldots, X_{n+1}) \mid \mathcal{B}_S^{(n)}\} \quad \text{a.s. } [P],$$

if and only if $\mathcal{B}^{(n)}$ and $\mathcal{B}_S^{(n+1)}$ are conditionally independent, given $\mathcal{B}_S^{(n)}$.

But since $\mathcal{B}_S^{(n)} \subset \mathcal{B}^{(n)}$ we have $\mathcal{B}^{(n)} \vee \mathcal{B}_S^{(n)} = \mathcal{B}^{(n)}$. Substituting $\mathcal{B}^{(n)}$ for $\mathcal{B}^{(n)} \vee \mathcal{B}_S^{(n)}$ in (2.8.3) and comparing to (2.8.1) we obtain the proof of the theorem. (Q.E.D.)

Example 2.19. Let $X = (X_1, X_2, \ldots)$ where $X_1, X_2, \ldots$ are i.i.d. random variables, each of which is distributed over the real line and has a distribution function with a k-parameter exponential density. As was previously shown, the minimal sufficient statistic for $\mathcal{F}^{(n)}$ is $S_n = (\sum_{j=1}^n U_1(X_j), \ldots, \sum_{j=1}^n U_k(X_j))$. Furthermore, for every $n \geqslant 1$ we have $S_{n+1} = S_n + (U_1(X_{n+1}), \ldots, U_k(X_{n+1}))$. Since the dependence of S_{n+1} on $(X_1, \ldots, X_n)$ is only through S_n, and since X_{n+1} is independent of $\mathcal{B}^{(n)}$ and in particular of $\mathcal{B}_S^{(n)}$, it follows that $\mathcal{B}^{(n)}$ and $\mathcal{B}_S^{(n+1)}$ are conditionally independent, given $\mathcal{B}_S^{(n)}$. Hence for the k- (finite, but arbitrary) parameter exponential family, the sufficient sequence $(S_1, S_2, \ldots)$ is transitive. ■

To conclude this section we denote a few words to the problem of invariantly sufficient statistics for sequential models. Let $\mathcal{G}$ be a group of transformations on the sequential model $(\mathcal{X}, \mathcal{B}, \mathcal{F})$, which is 1:1 onto $g\mathcal{X} = \mathcal{X}$, $g\mathcal{B} = \mathcal{B}$, and $\tilde{g}\mathcal{F} = \mathcal{F}$. We further assume that each $g \in \mathcal{G}$ induces a transformation $g^{(n)}$ on $\mathcal{X}^{(n)}$. If g transforms $x = (x_1, x_2, \ldots)$ componentwise, then $g^{(n)}(X_1, \ldots, X_n) = (g_1x_1, g_2x_2, \ldots, g_nx_n)$. Let $\mathcal{G}^{(n)}$ denote the induced group of transformations on $\mathcal{X}^{(n)}$. Furthermore, let $U^{(n)}$ denote a maximal invariant on $\mathcal{X}^{(n)}$ with respect to $\mathcal{G}^{(n)}$.

The *principle of invariance for sequential cases* stipulates that the statistics at the n-th stage be based on $U^{(n)}$. A sufficiency reduction of each $U^{(n)}$, $n \geqslant 1$, results in a sequence $(V_1, V_2, \ldots)$ of elements in which, for each fixed n, V_n is invariantly sufficient. Such a sequence is called an *invariantly sufficient sequence* for the sequential model. The restriction to sequential rules based on an invariantly sufficient sequence is justified as long as the sequence is transitive. As proven by Wijsman [1], it is sufficient to verify the transitivity of the sufficient sequence $(S_1, S_2, \ldots)$. Then, for each $n \geqslant 1$, if $\mathcal{F}^{(n)}$ and S_n satisfy the conditions of Stein's theorem, we make a maximal invariant reduction of S_n to V_n. The resulting sequence $(V_1, V_2, \ldots)$ is invariantly sufficient and transitive.

2.9. THE COMPARISON OF EXPERIMENTS AND SUFFICIENT STATISTICS

The notion of *sufficient experiments* was introduced by Blackwell [1] and studied further by LeCam [3], DeGroot [3], and others. DeGroot [3] utilized the notion of sufficient experiments for attaining optimal design of experiments. We define the relation of sufficiency between two experiments in the following manner.

Let $\mathcal{E}_1$ be an experiment resulting in a random variable X_1, and let $\mathcal{E}_2$ be an experiment resulting in a random variable X_2. If from an observation on X_1 and a use of a table of random numbers one can generate a random variable distributed like X_2, then the experiment $\mathcal{E}_1$ is said to be sufficient for $\mathcal{E}_2$. Accordingly, if an experiment $\mathcal{E}_1$ is sufficient for an experiment $\mathcal{E}_2$, there is no justification in performing $\mathcal{E}_2$.

Example 2.20. An experimental design problem in which the notion of sufficient experiments can be utilized for obtaining an optimal design is the allocation problem for a random effect analysis of variance (Model II; see Scheffé [1], p. 221). In such a design problem, n experiments are performed for the purpose of estimating a certain effect. The experiments can be chosen from a certain set Ω. Let $\mathcal{E}$ be an element of Ω. If the experiment $\mathcal{E}$ is performed, a random variable X is observed. The distribution law of X, given a parameter b associated with $\mathcal{E}$, is $\mathcal{N}(b, \sigma^2)$. The parameter σ^2 is known, but b unknown. In the random effect model b is assumed to be a random variable having a normal distribution $\mathcal{N}(\xi, \tau^2)$, τ^2 known. The objective is to estimate ξ. Consider two different designs.

Design I. An experiment $\mathcal{E}$ is chosen from Ω and repeated n times independently.

Design II. n different experiments $\mathcal{E}_1, \mathcal{E}_2, \ldots, \mathcal{E}_n$ are chosen from Ω independently, and each experiment is repeated only once.

Let $X_1, \ldots, X_n$ be the random variables observed in Design I and $Y_1, \ldots, Y_n$ the random variables observed in Design II. The sufficient statistics for these designs are the respective sample means $\bar{X}$ and $\bar{Y}$. If, as assumed, the value of τ^2 is known, then Design II is sufficient for Design I. In other words, Design II contains all the information which Design I may contain. Moreover, if the value of $\bar{Y}$ is given, with the aid of random numbers we can generate an observation having the same distribution as $\bar{X}$. Indeed, an observation on a random variable Z having a distribution $\mathcal{N}(0, \tau^2(1 - 1/n))$ can be generated from a table of random numbers, independently of $\bar{Y}$. Define then $\bar{X}^* = \bar{Y} + Z$. It is immediately verified that $\bar{X}^* \sim \mathcal{N}(\xi, \tau^2 + \sigma^2/n)$ and hence $\bar{X}^* \sim \bar{X}$. On the other hand, if the experiments are performed according to Design I and the value of $\bar{X}$ is observed, we cannot generate, without performing experiments according to Design II, a random variable that is distributed like $\bar{Y}$. The reason is that the variance of $\bar{X}$ is uniformly greater than that of $\bar{Y}$ for all (σ^2, τ^2). Thus Design II yields sufficient experiments for those variables of Design I, but not vice versa. Design II is therefore better than Design I. We now present Blackwell's definition of sufficient experiments (see [4]). We first define the notion of stochastic transformation from $(\mathcal{X}, \mathcal{B}, P)$ to $(\mathcal{Y}, \mathcal{F}, Q)$.

A *stochastic transformation* from $(\mathfrak{X}, \mathfrak{B}, P)$ to $(\mathfrak{Y}, \mathfrak{F}, Q)$ is a non-negative function $\pi(C \mid x)$ on $\mathfrak{F} \times \mathfrak{X}$ such that

(*i*) for each $x \in \mathfrak{X}$, $\pi(C \mid x)$ is a probability measure on $\mathfrak{F}$;
(*ii*) for each $C \in \mathfrak{F}$, $\pi(C \mid x)$ is a $\mathfrak{B}$-measurable function on $\mathfrak{X}$.

Definition. *Let* $(\mathfrak{X}, \mathfrak{B}, P_\theta)$, $\theta \in \Theta$, *be a probability space associated with an experiment* $\mathcal{E}_1$. *Let* $(\mathfrak{Y}, \mathfrak{F}, Q_\theta)$, $\theta \in \Theta$, *be a probability space associated with an experiment* $\mathcal{E}_2$. *Experiment* $\mathcal{E}_1$ *is said to be* sufficient *for experiment* $\mathcal{E}_2$ *if there exists a stochastic transformation* $\pi(\cdot \mid \cdot)$ *from* $(\mathfrak{X}, \mathfrak{B}, P_\theta)$ *to* $(\mathfrak{Y}, \mathfrak{F}, Q_\theta)$ *such that, for every* $\theta \in \Theta$ *and all* $C \in \mathfrak{F}$,

$$Q_\theta(C) = \int \pi(C \mid x) P_\theta(\mathrm{d}x). \tag{2.9.1}$$

In Example 2.20 we presented each of the experiments (Design I or Design II) by the sample means $\bar{X}$ and $\bar{Y}$. In this case both $\mathfrak{X}$ and $\mathfrak{Y}$ are the real line, $\mathfrak{B} = \mathfrak{F}$ is the Borel sigma-field on the real line, $\Theta = \{(\xi, d^2); -\infty < \xi < \infty; 0 < d < \infty\}$, where $d^2 = \tau^2 + \sigma^2/n$ for experiments of Design I, and $d^2 = (1/n)(\tau^2 + \sigma^2)$ for experiments of Design II. Since τ^2 is known, the stochastic transformation $\pi(\cdot \mid \cdot)$ on $\mathfrak{B} \times \mathfrak{Y}$ is

$$\pi(B \mid y) = \left(2\pi\left(1 - \frac{1}{n}\right)\tau^2\right)^{-1/2} \int_B \exp\left\{-\frac{1}{2\tau^2(1-(1/n))}(x-y)^2\right\} \mathrm{d}x. \tag{2.9.2}$$

We now present the definition of *sufficient combined experiments.* Let $\mathcal{E}_1$ denote an experiment with an associated probability space $(\mathfrak{X}_1, \mathfrak{B}_1, P_1)$ and $\mathcal{E}_2$ an experiment with a probability space $(\mathfrak{X}_2, \mathfrak{B}_2, P_2)$. In a combined experiment $\mathcal{E}_1$ and $\mathcal{E}_2$ are performed *independently* and the pair of random variables (X_1, X_2) is observed. The probability space associated with the combined experiment is the product space $(\mathfrak{X}_1 \times \mathfrak{X}_2, \mathfrak{B}_1 \times \mathfrak{B}_2, P_1 \times P_2)$ where $\mathfrak{X}_1 \times \mathfrak{X}_2$ is the sample space of all pairs (x_1, x_2) in which $x_1 \in \mathfrak{X}_1$ and $x_2 \in \mathfrak{X}_2$. $\mathfrak{B}_1 \times \mathfrak{B}_2$ is the smallest sigma-field over the collection of all cross products $B_1 \times B_2 = \{(x_1, x_2); x_1 \in B_1, x_2 \in B_2\}$ and $B_1 \in \mathfrak{B}_1$, $B_2 \in \mathfrak{B}_2$. The product measure on $\mathfrak{B}_1 \times \mathfrak{B}_2$ is

$$P(B_1, B_2) = P(B_1)Q(B_2), \quad B_1 \in \mathfrak{B}_1, \quad B_2 \in \mathfrak{B}_2. \tag{2.9.3}$$

We prove the following property of the combination of sufficient experiment.

Lemma 2.9.1. *Let* $\mathcal{E}_1, \mathcal{E}_2, \mathcal{E}_3, \mathcal{E}_4$ *be experiments with associated probability spaces* $(\mathfrak{X}_1, \mathfrak{B}_1, P_1)$, $(\mathfrak{X}_2, \mathfrak{B}_2, P_2)$, $(\mathfrak{X}_3, \mathfrak{B}_3, P_3)$, *and* $(\mathfrak{X}_4, \mathfrak{B}_4, P_4)$, *respectively. Suppose that* $\mathcal{E}_1$ *is sufficient for* $\mathcal{E}_2$ *and* $\mathcal{E}_3$ *is sufficient for* $\mathcal{E}_4$. Then the combined experiment $\mathcal{E}_1 \times \mathcal{E}_3$ *is sufficient for the combined experiment* $\mathcal{E}_2 \times \mathcal{E}_4$.

Proof. According to the definition of sufficient experiments, there exists a stochastic transformation $\pi_1(\cdot \mid \cdot)$ on $\mathcal{B}_2 \times \mathcal{X}_1$, transforming $(\mathcal{X}_1, \mathcal{B}_1, P_1) \rightarrow (\mathcal{X}_2, \mathcal{B}_2, P_2)$ and such that

$$P_2[C] = \int \pi_1(C \mid x)P_1(\mathrm{d}x), \quad \text{all} \quad C \in \mathcal{B}_2. \tag{2.9.4}$$

Similarly there exists a stochastic transformation $\pi_2(\cdot \mid \cdot)$ on $\mathcal{B}_4 \times \mathcal{X}_3$ such that

$$P_4[D] = \int \pi_2(D \mid y)P_3(\mathrm{d}y), \quad \text{all} \quad D \in \mathcal{B}_4. \tag{2.9.5}$$

Let C be a Borel set in $\mathcal{B}_2$ and D a Borel set in $\mathcal{B}_4$. Let $Q(C, D)$ be the probability measure on $\mathcal{B}_2 \times \mathcal{B}_4$. We have

$$\begin{aligned} Q(C, D) &= \int I_{C\times D}(x_2, x_4)P_2(\mathrm{d}x_2)P_4(\mathrm{d}x_4) \\ &= \int Q[C \times D \mid x_1, x_3]P_1(\mathrm{d}x_1)P_3(\mathrm{d}x_3), \end{aligned} \tag{2.9.6}$$

where $Q(C \times D \mid x_1, x_3)$ is, for a fixed $C \times D$, a $\mathcal{B}_1 \times \mathcal{B}_3$-measurable function, and for every $(x_1, x_3) \in \mathcal{X}_1 \times \mathcal{X}_3$ it is a version of the conditional probability of $C \times D$ given $\mathcal{B}_1 \times \mathcal{B}_3$. On the other hand, according to the definition of the product measure $Q(C, D)$ we have

$$Q(C, D) = P_2(C)P_4(D) = \int \pi_1(C \mid x)\pi_2(D \mid y)P_1(\mathrm{d}x)P_3(\mathrm{d}y). \tag{2.9.7}$$

The comparison of (2.9.6) and (2.9.7) yields

$$Q(C \times D \mid x_1, x_3) = \pi_1(C \mid x_1)\pi_2(D \mid x_3) \text{ a.s.} \tag{2.9.8}$$

We thus define a stochastic transformation $\pi(\cdot \mid \cdot)$ on $(\mathcal{B}_2 \times \mathcal{B}_4) \times (\mathcal{X}_1 \times \mathcal{X}_3)$ by

$$\pi(C \times D \mid x_1, x_3) = \pi_1(C \mid x_1)\pi_2(D \mid x_3) \text{ a.s.,} \tag{2.9.9}$$

all $C \in \mathcal{B}_2$, $D \in B_4$, $x_1 \in \mathcal{X}_1$, and $x_3 \in \mathcal{X}_3$. It is easy to verify that (2.9.9) is indeed a stochastic transformation. Finally, according to (2.9.7) $\mathcal{E}_1 \times \mathcal{E}_3$ is a sufficient combined experiment for $\mathcal{E}_2 \times \mathcal{E}_4$. (Q.E.D.)

Corollary 2.9.1. *If* $\mathcal{E}_i^{(1)}$ $(i = 1, \ldots, r)$ *are sufficient experiments for* $\mathcal{E}_i^{(2)}$ $(i = 1, \ldots, r)$, *then* $\mathcal{E}_1^{(1)} \times \cdots \times \mathcal{E}_r^{(1)}$ *is a sufficient combined experiment for* $\mathcal{E}_1^{(2)} \times \cdots \times \mathcal{E}_r^{(2)}$.

DeGroot [3] utilizes these results on sufficient and combined sufficient experiments to derive very interesting theorems concerning optimal allocation of experiments for statistical models similar to the random effects model presented in Example 2.20. Another interesting example elaborated by DeGroot treats the problem of optimal allocation of binomial experiments. Suppose there is available a population (or a lot) of coins. Let Z be the probability that a coin falls "heads." We assume that Z may differ from one coin to another. Z is treated as an unobserved random variable. The observed random variables are the results of tossing coins chosen from that population. Suppose that the objective of the experiment is to estimate the proportion of coins in that population having a bias in favor of "heads," that is, to estimate $P[Z > \frac{1}{2}]$. The theory derived by DeGroot provides results necessary for the optimal design of sampling of coins from that population, on the basis of assumptions concerning the chance behavior of the unobserved random variable Z.

2.10. SUFFICIENT STATISTICS AND POSTERIOR DISTRIBUTIONS—THE BAYESIAN DEFINITION OF SUFFICIENCY

Let $(\mathcal{X}, \mathcal{B}, P)$ be a probability space corresponding to a fixed sample size statistical model. That is, $\mathcal{X}$ is a finite-dimensional Euclidean space, $\mathcal{B}$ the Borel sigma-field on $\mathcal{X}$, $P \in \mathcal{P}$, where P is a family of probability measures on $(\mathcal{X}, \mathcal{B})$. We shall further assume that the probability measures P of $\mathcal{P}$ are indexed by a parameter θ (real or vector valued). We denote by Θ the parameter space and assume that Θ is an open set of a k-dimensional Euclidean space ($k \geqslant 1$). The Bayesian approach is to consider the parameter Θ of P as a random variable. We therefore construct a probability space for θ, namely $(\theta, \mathcal{C}, H)$ where $\mathcal{C}$ is a Borel sigma-field over Θ and H a probability measure over $(\theta, \mathcal{C})$. H is called a the *prior* probability measure of θ. The prior measure H belongs to a certain specified class of prior measures $\mathcal{H}$. Further assumption which we have to impose on our statistical model is that both $\mathcal{P} = \{P_\theta;\ \theta \in \mathcal{X}\}$ and $\mathcal{H}$ are dominated by some sigma-finite measures μ and ζ, respectively. Thus for every given θ, we denote by $f(x; \theta)$ a version of the density of P with respect to μ on $(\mathcal{X}, \mathcal{B})$, and by $h(\theta)$ a version of the prior density of H with respect to ζ. A version of the joint density of (X, θ) over $(\mathcal{X} \times \Theta, \mathcal{B} \times \mathcal{C})$ is

$$g(x, \theta) = f(x; \theta)h(\theta), \quad x \in \mathcal{X}, \quad \theta \in \Theta. \tag{2.10.1}$$

According to Bayes' theorem, a version of the conditional density of θ given $\{X = x\}$, called the *posterior* density of θ, is

$$h(\theta \mid X = x) = \frac{f(x; \theta)h(\theta)}{\displaystyle\int_\Theta f(x; \tau)H(\mathrm{d}\tau)}, \qquad \theta \in \Theta, \tag{2.10.2}$$

for every $x \in \mathcal{X}$ for which

$$f(x) = \int_{\Theta} f(x; \tau)h(\tau)\zeta(\mathrm{d}\tau) > 0.$$

The set of all x values N_ζ for which $f(x) = 0$ has a probability 0 with respect to almost all the probability measures P_θ, $\theta \in \Theta$. Indeed,

$$P_\zeta[N_\zeta] = \int_{N_\zeta} f_\zeta(x)\mu(\mathrm{d}x) = \int_{\Theta}\left(\int_{N_\zeta} f(x; \theta)\mu(\mathrm{d}x)\right) h(\theta)\zeta(\mathrm{d}\theta) = 0.$$

Hence

$$P_\theta[N_\zeta] = \int_{N_\zeta} f(x; \theta)\mu(\mathrm{d}x) = 0 \quad \text{a.s. } [H].$$

Therefore on the set N_ζ we can define $h(\theta \mid x) \equiv 0$.

If Y is a statistic on $(\mathcal{X}, \mathcal{B}, P_\theta)$ to $(\mathcal{Y}, \mathcal{F}, P_\theta^Y)$, the probability measures P_θ, $\theta \in \Theta$ are transformed to P_θ^Y, and we denote by $f^Y(y; \theta)$ a version of the density of P_θ^Y. Thus given the value of $Y(X)$, the posterior density of θ given $\{Y(X) = y\}$ becomes

$$h^Y(\theta \mid Y = y) = \frac{f^Y(y; \theta)h(\theta)}{\int_{\Theta} f^Y(y; \theta)h(\theta)\zeta(\mathrm{d}\theta)}, \qquad \theta \in \Theta. \tag{2.10.3}$$

The following is a *Bayesian* definition of a sufficient statistic (see Raiffa and Schlaifer [1], p. 32).

Definition. *Given the product probability space* $(\mathcal{X} \times \Theta, \mathcal{B} \times \mathcal{C}, \mathcal{P} \times \mathcal{H})$, *a statistic* $S(X)$ *on* $(\mathcal{X}, \mathcal{B}, \mathcal{P})$ *to* $(\mathcal{S}, \mathcal{B}_S, \mathcal{P}^S)$ *is called* Bayesian sufficient *for* $\mathcal{H}$ *if*

$$h^S(\theta \mid S(X)) = h(\theta \mid X) \text{ a.s. } [\mu], \quad \text{for all} \quad H \in \mathcal{H}. \tag{2.10.4}$$

From the Bayesian point of view, a sufficient statistic yields a posterior density of θ equivalent to the posterior density of θ, given the original observation (sample).

Example 2.21. Let $X_1, X_2, \ldots, X_n$ be i.i.d. random variables having a common normal distribution $\mathcal{N}(\theta_1, \theta_2^2)$, both θ_1 and θ_2 unknown. Here $\Theta = \{(\theta_1, \theta_2);\ -\infty < \theta_1 < \infty;\ 0 < \theta_2 < \infty\}$. That is, Θ is the "upper" half of the Euclidean plane. The density function of $X_1, \ldots, X_n$ given $\theta = (\theta_1, \theta_2)$ is

$$f(X_1, \ldots, X_n; \theta_1, \theta_2) = (2\pi)^{-n/2}\theta_2^{-n} \exp\left\{-\frac{1}{2\theta_2^2}\sum_{i=1}^{n}(X_i - \bar{X})^2 - \frac{n}{2\theta_2^2}(\bar{X} - \theta_1)^2\right\}, \tag{2.10.5}$$

where

$$\bar{X} = \frac{1}{n}\sum_{i=1}^{n} X_i.$$

Assuming that θ_1 and θ_2 have a prior density $h(\theta_1, \theta_2)$, we obtain that the posterior density of (θ_1, θ_2) given the original sample $(X_1, \ldots, X_n)$ is

$$(2.10.6)\quad h(\theta_1, \theta_2 \mid X_1, \ldots, X_n) = \frac{\theta_2^{-n} h(\theta_1, \theta_2) \exp\{-(n/2\theta_2^2)(\bar{X} - \theta_1)^2 - (1/2\theta_2^2) \sum_{i=1}^n (X_i - \bar{X})^2\}}{\displaystyle\int_{-\infty}^{\infty} d\theta_1 \int_0^{\infty} d\theta_2 \cdot \theta_2^{-n} h(\theta_1, \theta_2) \times \exp\{-(n/2\theta_2^2)(\bar{X} - \theta_1)^2 - (1/2\theta_2^2) \sum_{i=1}^n (X_i - \bar{X})^2\}}$$

We see in (2.10.6) that the posterior density of (θ_1, θ_2) depends on the original sample $(X_1, \ldots, X_n)$ only through the statistics $(\bar{X}, \sum_{i=1}^n (X_i - \bar{X})^2)$, which are the minimal sufficient statistics in the non-Bayesian sense. But $\bar{X}$ and $Q = \sum_{i=1}^n (X_i - \bar{X})^2$ are independent statistics having the density functions

$$(2.10.7)\quad f^{\bar{X}}(x; \theta_1, \theta_2) = \frac{n^{1/2}}{\sqrt{2\pi}} \theta_2^{-1} \exp\left\{-\frac{n}{2\theta_2^2}(x - \theta_1)^2\right\}, \qquad -\infty < x < \infty$$

and

$$(2.10.8)\quad f^Q(q; \theta_2) = \left[\Gamma\left(\frac{n-1}{2}\right)(2\theta_2^2)^{(n-1)/2}\right]^{-1} q^{(n-3)/2} \exp\left\{-\frac{1}{2\theta_2^2} q\right\}, \qquad 0 \leqslant q \leqslant \infty.$$

From (2.10.7) and (2.10.8) we obtain the posterior density of (θ_1, θ_2) given $(\bar{X}, Q)$, namely,

$$(2.10.9)\quad h(\theta_1, \theta_2 \mid \bar{X}, Q) = \frac{\theta_2^{-n} h(\theta_1, \theta_2) \exp\{-(n/2\theta_2^2)(\bar{X} - \theta_1)^2 - (1/2\theta_2^2)Q\}}{\displaystyle\int_{-\infty}^{\infty} d\theta_1 \int_0^{\infty} d\theta_2 \cdot \theta_2^{-n} h(\theta_1, \theta_2) \exp\{(n/2\theta_2^2)(\bar{X} - \theta_1)^2 - (1/2\theta_2^2)Q\}}.$$

Comparing (2.10.6) with (2.10.9) we obtain

$$(2.10.10)\qquad h(\theta_1, \theta_2 \mid \bar{X}, Q) = h(\theta_1, \theta_2 \mid X_1, \ldots, X_n) \text{ a.s.}$$

Thus $(\bar{X}, Q)$ are also Bayesian-sufficient statistics. This is not a coincidence. ■

We prove now that the Bayesian and non-Bayesian definitions of sufficiency are equivalent.

Theorem 2.10.1. *Given any Bayesian model with a product probability space* $(\mathcal{X} \times \Theta, \mathcal{B} \times \mathcal{C}, \mathcal{P} \times \mathcal{H})$, *a statistic* $S\colon (\mathcal{X}, \mathcal{B}, \mathcal{P}) \to (\mathcal{S}, \mathcal{B}_S, \mathcal{P}^S)$ *is Bayesian sufficient for* $\mathcal{H}$ *if and only if it is sufficient for* $\mathcal{P}$.

Proof. (i) We prove first that if S is a (non-Bayesian) sufficient statistic for $\mathcal{F}$, it is a Bayesian sufficient statistic. Indeed, if S is a (non-Bayesian) sufficient statistic for $\mathcal{F}$, then by the factorization theorem (Theorem 2.3.3) there exists a non-negative $\mathcal{B}$-measurable function $k(x)$ and a non-negative function $g(S(x); \theta)$ which is, for a fixed $\theta \in \Theta$, $\mathcal{B}_S$-measurable such that

$$f(x; \theta) = k(x)g(S(x); \theta) \quad \text{a.s.} \quad [\mathcal{F}]; \qquad (2.10.11)$$
$$k(x) = 0 \text{ only on a } \mathcal{F}\text{-null set.}$$

Hence the posterior density of θ, given X, is

$$h(\theta \mid X) = \frac{h(\theta)g(S(X); \theta)}{\int_\Theta h(\theta)g(S(X); \theta)\zeta(d\theta)}, \qquad \theta \in \Theta. \qquad (2.10.12)$$

The induced density of $S(X)$ can be written, according to (2.10.11), in the form

$$f^S(x; \theta) = J(x)g(x; \theta), \quad x \in \mathcal{S}, \quad \theta \in \Theta, \qquad (2.10.13)$$

where $J(x)$ is a non-negative constant on the cosets $\{S(x) = s\}$. Therefore the posterior density of θ, given $S(X) = s$, is

$$h(\theta \mid S(x) = s) = \frac{h(\theta)g(s; \theta)}{\int_\Theta h(\theta)g(s; \theta)\zeta(d\theta)}, \qquad \theta \in \Theta. \qquad (2.10.14)$$

It follows that $h(\theta \mid X) = h(\theta \mid S(X))$ a.s. Hence S is a Bayesian sufficient statistic.

(*ii*) We prove now that if $S: (\mathcal{X}, \mathcal{B}, \mathcal{F}) \rightarrow (\mathcal{S}, \mathcal{B}_S, \mathcal{F}^S)$ is a Bayesian sufficient statistic, then it is non-Bayesian sufficient for $\mathcal{F}$. According to the definition of Bayesian sufficiency, the posterior density of θ given $\{X = x\}$, namely, $h(\theta \mid x)$, is, for a fixed θ, a $\mathcal{B}_S$-measurable function. Let θ be such that $h(\theta) > 0$ and let θ_0 be fixed with $h(\theta_0) > 0$. From the definition, (2.10.2), of the posterior density $h(\theta \mid X)$ we have

$$\log \frac{h(\theta \mid X)}{h(\theta_0 \mid X)} - \log \frac{f(X; \theta)}{f(X; \theta_0)} + \log \frac{h(\theta)}{h(\theta_0)}, \qquad (2.10.15)$$

where $g(X, \theta) = \log f(X, \theta)/f(X, \theta_0)$ is the log-likelihood function and for any fixed θ is $\mathcal{B}_S$-measurable. Hence we obtain from (2.10.15) that, for every θ and θ_0 such that $h(\theta) > 0$, $h(\theta_0) > 0$,

$$\log f(X, \theta) = \log f(X, \theta_0) + g(S(X), \theta), \qquad (2.10.16)$$

where

$$g(S(X), \theta) = \log \frac{h(\theta \mid X)}{h(\theta_0 \mid X)} - \log \frac{h(\theta)}{h(\theta_0)}$$

is a $\mathcal{B}_S$-measurable function. Finally, letting $k(X) = f(X, \theta_0)$, which is non-negative and $\mathcal{B}$-measurable, we obtain

$$f(X, \theta) = k(X) \exp\{g(S(X), \theta)\} \text{ a.s.} \tag{2.10.17}$$

Thus by Theorem 2.3.3 $S(X)$ is a (non-Bayesian) sufficient statistic for $\mathcal{F}$.

(Q.E.D)

PROBLEMS

Section 2.1

The solution of the following problems is expected to follow the method exhibited in the example of Section 2.1.

1. Let $X_1, \ldots, X_n$ be i.i.d. random variables having a common rectangular distribution $\mathcal{R}(\theta_1, \theta_2)$; $-\infty < \theta_1 < \theta_2 < \infty$; that is, the density function of X_1 is

$$f(x; \theta_1, \theta_2) = \begin{cases} (\theta_2 - \theta_1)^{-1}, & \theta_1 \leqslant X \leqslant \theta_2 \\ 0, & \text{otherwise.} \end{cases}$$

Find a nontrivial sufficient statistic.

2. Suppose that $X_1, \ldots, X_n$ are i.i.d. random variables with a common normal distribution $\mathcal{N}(\mu, \sigma^2)$; $-\infty < \mu < \infty, 0 < \sigma < \infty$. Show that $(\sum_{i=1}^n X_i, \sum_{i=1}^n X_i^2)$ is a sufficient statistic.

3. Let $X_1, \ldots, X_n$ be i.i.d. random variables with a common log-normal distribution; that is, $\ln X_1 \sim N(\mu, \sigma^2)$. What is a nontrivial sufficient statistic?

4. Suppose that $X_1, \ldots, X_n$ are i.i.d. random variables with a common distribution F belonging to the class $\mathcal{F}$ of *all* absolutely continuous distributions. Show that the order statistic $X_{(1)} \leqslant \cdots \leqslant X_{(n)}$ is a sufficient statistic. (As will be shown later on, the order statistic cannot be further reduced to attain another sufficient statistic for $\mathcal{F}$.)

5. Let $X_1, \ldots, X_n$ be i.i.d. random variables having a common $\mathcal{G}(\lambda, p)$ distribution. Show that $T_n = \sum_{i=1}^n X_i$ is a sufficient statistic.

6. The random variables $X_1, \ldots, X_n$ are i.i.d. with a common Laplace distribution with density

$$f(x; \sigma) = \frac{1}{2\sigma} \exp\left\{\frac{-|x|}{\sigma}\right\}, \quad -\infty < x < \infty, \quad 0 < \sigma < \infty.$$

Given the value of $T_n = \sum_{i=1}^n |X_i|$, how will you proceed to generate a random sample equivalent to $X_1, \ldots, X_n$?

7. Prove that if $X_1, \ldots, X_n$ are i.i.d. like a binomial random variable $B(k; \theta)$, $0 < \theta < 1$, and k known, then $T_n = \sum_{i=1}^n X_i$ is a sufficient statistic.

8. Let $X_1, \ldots, X_n$ be i.i.d. like a discrete random variable with a geometric density

$$f(x; \theta) = \theta(1 - \theta)^x, \quad x = 0, 1, \ldots; \quad 0 < \theta < 1.$$

Let $T_n = \sum_{i=1}^n X_i$. Find the conditional joint density of $(X_1, \ldots, X_{n-1})$ given T_n. Is T_n a sufficient statistic? How would you generate a sample of n i.i.d. geometric random variables by simulation methods if the value of T_n is given?

Section 2.2

9. Let $(\Omega, \mathcal{F}, P)$ be a probability space, and let $X: \Omega \to \mathcal{R}$ be a random variable.

(i) What is the sigma-subfield $\mathcal{B}$ induced by X?
(ii) What is the probability measure induced by X?
(iii) Suppose that X has an $\mathcal{N}(0, 1)$ distribution function. What is the corresponding probability measure P on $(\Omega, \mathcal{F})$?

10. Let X be a random variable with a corresponding probability space $(\mathcal{X}, \mathcal{B}, P)$, where $\mathcal{X}$ is the real line, $\mathcal{B}$ the Borel sigma-field on $\mathcal{X}$. Suppose that $\mathcal{T}$ is the class of probability measures corresponding to the family of gamma distributions $\mathcal{G}(1/\theta, \nu)$, $0 < \theta < \infty$, $0 < \nu < \infty$. Let Q be the probability measure on $(\mathcal{X}, \mathcal{B})$ induced by the $\mathcal{N}(0, 1)$ distribution.

(i) Is $P \ll Q$ for each $P \in \mathcal{T}$?
(ii) Write the Radon-Nikodym derivative $P(\mathrm{d}x)/Q(\mathrm{d}x)$ for $P \in \mathcal{T}$.
(iii) Is $Q \ll P$ for any $P \in \mathcal{T}$?

11. Let $B(n; \theta)$ designate a binomial r.v. corresponding to n Bernoulli trials, with a probability of success θ, $0 < \theta < 1$. Write the density of $B(n; \theta)$ in terms of the density of a Poisson r.v. $P(\lambda)$, $0 < \lambda < \infty$; that is, if $B(\cdot \mid n, \theta)$ designates the binomial probability measure and $P(\cdot \mid \lambda)$ the Poisson probability measure, what is $B(\mathrm{d}x \mid n, \theta)/P(\mathrm{d}x \mid \lambda)$? Are the families $\{B(\cdot \mid n, \theta);\ 0 < \theta < 1\}$ and $\{P(\cdot \mid \lambda); 0 < \lambda < \infty\}$ equivalent?

12. Consider Example 2.3 of section 2.1. Let $\mathcal{B}^{(n)}$ be the Borel sigma-field generated by $(X_1, \ldots, X_n)$.

(i) What is the sigma-subfield generated by the sufficient statistic $(X_{(1)}, T_{n-1}^*)$?
(ii) Let $f(X_1, \ldots, X_n)$ be a $\mathcal{B}^{(n)}$ measurable function such that $E\{|f(\mathbf{X})|\} < \infty$. What is $E\{f(\mathbf{X}) \mid (X_{(1)}, T_{n-1}^*)\}$?

13. Let X_1, X_2 be i.i.d. random variables having a common Poisson distribution $P(\lambda)$. Let $T = X_1 + X_2$. What is $P[X_1 \leqslant x \mid T = t]$?

Section 2.3

14. Let $\mathcal{T}$ be the family of probability measures on $(\mathcal{X}, \mathcal{B})$ corresponding to the family of $\mathcal{N}(\theta, \sigma^2)$, $-\infty < \theta < \infty$, $0 < \sigma < \infty$, distributions. Give several examples of subfamilies, $\mathcal{T}^* \subset \mathcal{T}$, which are dense in $\mathcal{T}$.

15. Let N.B. (ψ, ν) designate a negative binomial distribution with parameters (ψ, ν); $0 < \nu < \infty, 0 < \psi < 1$.† Consider the family $\mathcal{F}_\nu = \{\text{N.B. } (\psi, \nu); 0 < \psi < 1, 0 < \nu < \infty\}$.

† The corresponding density function is

$$f(x; \psi, \nu) = \frac{\Gamma(\nu + x)}{\Gamma(x + 1)\Gamma(\nu)} (1 - \psi)^\nu \psi^x, \quad x = 0, 1, \ldots.$$

with a known value ν.

(i) Show a subfamily dense in $\mathcal{F}_v$.
(ii) Apply Theorem 2.3.1 to show that if $X_1, \ldots, X_n$ are i.i.d. with a common distribution in $\mathcal{F}_v$, then $T_n = \sum_{i=1}^n X_i$ is a sufficient statistic.

16. Let $Y_i \sim N(\alpha + \beta X_i, \sigma^2)$, $i = 1, \ldots, n$, be independent, where $-\infty < \alpha, \beta < \infty$, $0 < \sigma < \infty$ are unknown parameters and $(X_1, \ldots, X_n)$ are known constants. Apply the Neyman-Fisher factorization theorem to prove that $(\sum_{i=1}^n Y_i, \sum_{i=1}^n X_i Y_i, \sum_{i=1}^n Y_i^2)$ is a sufficient statistic.

17. Let $\{(X_i, Y_i), i = 1, \ldots, n\}$ be n independent random vectors having a common bivariate distribution $\mathcal{N}\left(\begin{pmatrix}\theta_1\\ \theta_2\end{pmatrix}, \begin{pmatrix}\sigma_1^2 & \rho\sigma_1\sigma_2\\ \cdot & \sigma_2^2\end{pmatrix}\right)$, $-\infty < \theta_1, \theta_2 < \infty$; $0 < \sigma_1, \sigma_2 < \infty$; $-1 \leqslant \rho \leqslant 1$. Apply the Neyman-Fisher factorization theorem to prove that

$$(\sum X_i, \sum X_i^2, \sum X_i Y_i, \sum Y_i, \sum Y_i^2)$$

is a sufficient statistic.

18. Let $\{X_{ij}; i = 1, \ldots, I, j = 1, \ldots, J\}$ be independent random variables and

$$X_{ij} = \mu + a_i + e_{ij}; \qquad i = 1, \ldots, I, \ \ j = 1, \ldots, J,$$

where $a_1, \ldots, a_I$ are i.i.d. distributed like $N(0, \tau^2)$ and e_{ij} are i.i.d. (independent of $\{a_1, \ldots, a_I\}$) with $e_{ij} \sim N(0, \sigma^2)$. The parameters μ, τ, σ are unknown: $-\infty < \mu < \infty$, $0 < \tau, \sigma < \infty$. Show that $(X.., S_e, S_a)$ is a sufficient statistic where

$$S_e = \sum_{i=1}^{I}\sum_{j=1}^{J} (X_{ij} - \bar{X}_i)^2,\ \bar{X}_i = \frac{1}{J}\sum_{j=1}^{J} X_{ij},\ S_a = J\sum_{i=1}^{I} (\bar{X}_i - X..)^2,\ X.. = \frac{1}{IJ}\sum_{i=1}^{I}\sum_{j=1}^{J} X_{ij}$$

(The statistical model under consideration is known as the Model II of Analysis of Variance.)

Section 2.4

19. Let $X_1, \ldots, X_n$ be i.i.d. like a binomial random variable $B(m; \theta)$. Consider the family of distributions $\{B(m; \theta); m = 0, 1, \ldots, 0 < \theta < 1\}$. Both parameters m and θ are unknown. Show that the probability measure $\lambda(dx) = \sum_{0=m}^{\infty} 2^{-m} B(dx \mid m, \frac{1}{2})$ dominates $\mathcal{F}$ but a minimal sufficient statistic is the trivial one; $(X_{(1)}, \ldots, X_{(n)})$, $X_{(1)} \leqslant \cdots \leqslant X_{(n)}$.

20. Show that the sufficient statistics discussed in Problems 15–18 are minimal.

21. Let $X_1, \ldots, X_n$ be i.i.d. random variables having an absolutely continuous distribution with a density function

$$f(x; \theta) = \begin{cases} \dfrac{2}{\theta}(\theta - x), & 0 \leqslant x \leqslant \theta, \quad 0 < \theta < \infty. \\ 0, & \text{otherwise} \end{cases}$$

What is a minimal sufficient statistic?

Section 2.5

22. Let $X_1, \ldots, X_n$ be independent random variables having normal distributions; that is,

$$X_i \sim N(\mu_i(J), 1), \qquad i = 1, \ldots, n,$$

where

$$\mu_i = \begin{cases} 0, & \text{if } i \leqslant J, \\ \mu, & \text{if } i > J \end{cases} \qquad -\infty < \mu < \infty,$$

where J is a discrete random variable with a (probability) density function

$$P[J = j] = \begin{cases} \dfrac{1}{n}, & \text{if } j = 1, \ldots, n, \\ 0, & \text{otherwise.} \end{cases}$$

(i) Is the family of distributions under consideration regular in the sense of Dynkin?
(ii) What is the rank of $\mathfrak{F}$?
(iii) What is a minimal sufficient statistic?

(Distributions of this kind are found when one observes a normal process which is subjected to a shift in the mean at an unknown time point.)

23. Let $X_1, \ldots, X_n$ be i.i.d. having a common distribution with a density function

$$f(x; \alpha) = \exp\{-(x - \alpha) - e^{-(x-\alpha)}\}, \qquad -\infty < x < \infty.$$

The parameter α is real; $-\infty < \alpha < \infty$. [The distribution with density $f(x; 0)$ is known as the "first asymptotic distribution of extreme statistics."]

(i) Is the family $\mathfrak{F}$ regular in the sense of Dynkin?
(ii) Find the rank of $\mathfrak{F}$ and a (nontrivial) minimal sufficient statistic.

24. Consider the family of Cauchy distributions with densities

$$f(x; \alpha, \sigma) = \frac{1}{\pi}\left[1 + \left(\frac{x - \alpha}{\sigma}\right)^2\right]^{-1}, \qquad -\infty < x < \infty;$$

$-\infty < \alpha < \infty, 0 < \sigma < \infty$. Show that the only sufficient statistic for this family, in a sample of size n, $n \geqslant 1$, is the trivial one.

Section 2.6

25.(i) Let $X_1, \ldots, X_n$ be i.i.d. having a common rectangular distribution $\mathcal{R}(0, \theta)$, $0 < \theta < \infty$. Let $X_{(1)} \leqslant \cdots \leqslant X_{(n)}$. Prove that $X_{(n)}$ is a minimal sufficient statistic by showing that it is a complete sufficient statistic.
(ii) If the common distribution is $\mathcal{R}(\theta_1, \theta_2)$, $-\infty < \theta_1 < \theta_2 < \infty$, then $(X_{(1)}, X_{(n)})$ is a complete sufficient statistic and therefore minimal.
(iii) If $X_1 \sim \mathcal{R}(\theta, 3\theta)$, $-\infty < \theta < \infty$, then $(X_{(1)}, X_{(n)})$ is a minimal sufficient statistic, but it is incomplete.

26. Consider the family of distribution functions specified in Problem 18. Is the minimal sufficient statistic $(X_{..}, S_e, S_a)$ a complete one?

Section 2.7

27. Consider Example 2.3 (Section 2.1). Let $S_X = \sum_{i=2}^{m} (X_{(i)} - X_{(1)})$. Let $\mathcal{G}$ be the group of real translation.

(i) Show that $T_X = (X_{(1)}, S_X)$ is a minimal sufficient statistic.
(ii) What is the maximal invariant $U(T_X)$ on the sample space of T_X?
(iii) Applying (2.7.13) prove that $X_{(1)}$ and $U(X) = (X_{(2)} - X_{(1)}, \ldots, X_{(n)} - X_{(1)})$ are conditionally independent given S_X.
(iv) Does the conditional expectation $E\{U(X) \mid S_X\}$ depend on (α, σ)?
(v) Prove that $X_{(1)}$ and S_X are independent.

28. Let $X_1, \ldots, X_n$ be i.i.d. random variables distributed like $N(\mu, \sigma^2)$, $-\infty < \mu < \infty$, $0 < \sigma < \infty$. Using the group $\mathcal{G}$ of affine transformations in conjunction with (2.7.13), prove that the sample mean is independent of the sample variance.

29. Let $X_1, \ldots, X_n$ and $Y_1, \ldots, Y_n$ be independent random variables; $X_i \sim N(\mu, \sigma_1^2)$ and $Y_i \sim N(\mu, \sigma_2^2)$, $i = 1, \ldots, n$; $-\infty < \mu < \infty; 0 < \sigma_1, \sigma_2 < \infty$. Let $\bar{X}, \bar{Y}, S_X{}^2, S_X{}^2$ be the sample means and sample variances, respectively.

(i) Apply the Stein theorem to prove that $(S_X{}^2/(\bar{X} - \bar{Y})^2, S_X{}^2/(\bar{X} - \bar{Y})^2)$ is invariantly sufficient.
(ii) Show that the distribution of this invariantly sufficient statistic depends only on $\rho = \sigma_1{}^2/\sigma_2{}^2$.

Section 2.8

30. As a continuation of Problem 29, consider a sequential model based on $\{(\mathcal{X}^{(n)} \times \mathcal{Y}^{(n)}, \mathcal{F}_n, \mathcal{P}_n); n \geqslant 2\}$ where $\mathcal{X}^{(n)}$ is the sample space of the first n observations on X; $\mathcal{Y}^{(n)}$ that of the first n observations on Y; $\mathcal{F}_n$ the smallest Borel sigma-field generated by $(X_1, \ldots, X_n; Y_1, \ldots, Y_n)$; and $\mathcal{P}_n$ the corresponding family of probability measures induced by the joint distribution of $(X_1, \ldots, X_n; Y_1, \ldots, Y_n)$. Let $\{(\bar{X}_n, \bar{Y}_n, S^2_{X,n}, S^2_{Y,n}); n \geqslant 2\}$ be the sequence of sufficient statistics. Let

$$\hat{\mu}(\bar{X}_n, \bar{Y}_n, S^2_{X,n}, S^2_{Y,n}) = \frac{\bar{X}_n S^2_{Y,n} + \bar{Y} S^2_{X,n}}{s^2_{X,n} + s^2_{Y,n}}.$$

Determine

$$E\{\hat{\mu}(\bar{X}_n, \bar{Y}_n, S^2_{X,n}, S)^2_{Y,n} \mid \mathcal{F}_{n-1}\}, \qquad n \geqslant 3.$$

31. Let $X_1, \ldots, X_n$ be a sequence of i.i.d. random variables having a common log-normal distribution; that is, $\log X_1 \sim N(\mu, \sigma^2)$; $-\infty < \mu < \infty$, $0 < \sigma < \infty$. Let $\mathcal{B}_n$ be the Borel sigma-field generated by $(X_1, \ldots, X_n)$. Let $Y_i = \log X_i$, $i = 1, \ldots, n$; $\bar{Y}_n = \sum_i Y_i/n$, $Q_n = \sum_{i=1}^{n} (Y_i - \bar{Y}_n)^2$. Determine

$$E\left\{\exp\left\{\bar{Y}_{n+1} + \frac{1}{2(n+1)} Q_{n+1}\right\} \middle| \mathcal{B}_n\right\}.$$

Section 2.9

32. Let $Y(x) \sim N(\alpha + \beta x, \sigma^2)$, $\beta \neq 0$, $-\infty < \alpha < \infty$, $0 < \sigma < \infty$; $2n$ independent experiments can be performed. In each experiment a point x is chosen in the interval $[0, 1]$ and the corresponding value of $Y(x)$ is observed. Show that the combined experiment in which n observations are performed at $x = 0$ and n observations at $x = 1$ is a sufficient experiment.

33. Let o_1, o_2 be two objects whose weights are ω_1 and ω_2, respectively. A balance weighing apparatus of the chemical type is available. The observations on this apparatus are normally distributed with mean ξ and a known variance σ^2, where ξ is the sum of weights of objects put on the left pan. In order to estimate the weights ω_1 and ω_2 two combined experiments are considered, $\mathcal{E}_1 \times \mathcal{E}_2$ and $\mathcal{E}_3 \times \mathcal{E}_4$; $\mathcal{E}_1$ is the experiment in which o_1 is weighed individually (say on the right pan); $\mathcal{E}_2$ is the experiment in which o_2 is weighed individually. In o_3 both objects are put on one pan. In $\mathcal{E}_4$ one object is put on the right pan and one on the left pan. Prove that $\mathcal{E}_3 \times \mathcal{E}_4$ is a sufficient experiment for $\mathcal{E}_1 \times \mathcal{E}_2$.

Section 2.10

34. Let $X_1, \ldots, X_n$ be i.i.d. random variables having a common $\mathcal{N}(\mu, \sigma^2)$ distribution; $-\infty < \mu < \infty$, $0 < \sigma < \infty$. Let $\theta_1 = \mu$, $\theta_2 = 1/2\sigma^2$. Assume the prior distribution $H(\theta_1, \theta_2)$ such that the conditional distribution of θ_1 given θ_2 is $\mathcal{N}(0, 1/2\theta_2)$; and θ_2 has the gamma distribution $\mathcal{G}(\tau, \nu)$. Find the posterior distribution of (μ, σ^2) given $(X_1, \ldots, X_n)$.

35. (i) Let X be a discrete random variable having a hypergeometric distribution, with a (probability) density function

$$f(x \mid N, M, n) = \begin{cases} \binom{M}{x}\binom{N-M}{n-x} \Big/ \binom{N}{n}, & x = 0, 1, \ldots, n, \\ 0, & \text{otherwise,} \end{cases}$$

where N and n are specified positive integers, $1 \leq n \leq N$, and $M = 0, 1, \ldots, N$.

Assume that the prior distribution of M is the discrete uniform, that is,

$$h(M) = \begin{cases} (N+1)^{-1}, & M = 0, \ldots, N, \\ 0, & \text{otherwise.} \end{cases}$$

Find the posterior density of M, given X.

(ii) What is the posterior density of M if its prior distribution is the binomial $\mathcal{B}(N; \theta)$, $0 < \theta < 1$.

(iii) If $X_1, \ldots, X_k$ are i.i.d. random variables having a common hypergeometric distribution, with parameters (N, M, n), and M has a prior density as in (i), what is the posterior density of M, given $(X_1, \ldots, X_k)$?

(iv) As in (iii), what is the posterior density of M if its prior density is specified in (ii)?

36. Let $X_1, X_2, \ldots$ be a sequence of independent random variables having an $\mathcal{N}(\mu_n, 1)$ distribution; $\mu_1 = \cdots = \mu_J = 0$ and $\mu_{J+1} = \mu_{J+2} = \cdots = \mu$. Both J and μ are unknown and $-\infty < \mu < \infty$; $J = 0, 1, \ldots$. Both μ and J are given a prior distribution of two independent random variables. The prior distribution of μ is the normal, $\mathcal{N}(0, \tau^2)$, and that of J is the geometric, N.B. $(\psi, 1)$, $0 < \psi < 1$.

(i) Determine the posterior density of J, given $(X_1, \ldots, X_n)$.
(ii) Determine the posterior density of μ, given $(X_1, \ldots, X_n)$.

REFERENCES

Bahadur [1], [2]; Barankin and Katz [1]; Blackwell [1]; Brown [1]; Dynkin [1]; DeGroot [3]; Fisher [2]; Fraser [5], [7]; Hall [1]; Halmos [2]; Halmos and Savage [1]; Kendall and Buckland [1]; LeCam [3]; Lehmann [2]; Lehmann and Scheffé [1]; Loève [2]; Morimoto and Sibuya [1]; Neymann [1]; Raiffa and Schlaifer [1]; Royden [1]; Scheffé[1]; Shimizu [1]; Shreider [1]; Sverdrup [1].

CHAPTER 3

Unbiased Estimation

3.1. THE PARAMETRIC CASE—GENERAL THEORY

In the parametric case the statistical model consists of a triplet $(\mathcal{X}, \mathcal{B}, F_\theta)$, where $\mathcal{X}$ is the sample space of the observed random variable (it could be a vector of observations); $\mathcal{B}$ is the Borel sigma-field on $\mathcal{X}$: and $\{F_\theta; \theta \in \Theta\}$ is a *parametric* family of distribution functions. We are interested in the unbiased estimation of the function $g(\theta)$ mapping Θ (a subset of a k-dimensional Euclidean space) into Ω (a subset of an r-dimensional Euclidean space). *An unbiased estimator of $g(\theta)$ is a statistic φ mapping $(\mathcal{X}, \mathcal{B}, F_\theta)$ into $(\Omega, \mathcal{F}, H_\theta)$ and satisfying*

(3.1.1) $$\int_{\mathcal{X}} \varphi(x)\, dF_\theta(x) = g(\theta), \quad \text{for all} \quad \theta \in \Theta.$$

A function $g(\theta)$ admitting an unbiased estimator is called *estimable*. This term is especially used in the theory of linear models. In many cases, as shown in the following example, the family of distribution functions $\{F_\theta; \theta \in \Theta\}$ of the observed random variable is not complete, and there are many possible unbiased estimators of the same function $g(\theta)$. In such cases we look for an optimal unbiased estimator, in a sense that is explained later.

Example 3.1. Let $X = (X_1, \ldots, X_n)'$ where X_i $(i = 1, \ldots, n; n \geqslant 2)$ are i.i.d. random variables having a common normal distribution law $\mathcal{N}(\xi, \sigma^2)$. Here Θ is the upper half-plane, $\{(\xi, \sigma^2); -\infty < \xi < \infty, 0 < \sigma^2 < \infty\}$.

Let $g(\xi, \sigma^2) = (\xi, \sigma^2)$; g is vector valued, and maps Θ onto itself. It is simple to prove that the estimator

(3.1.2) $$\varphi(X_1, \ldots X_n) = (X_1, \tfrac{1}{2}(X_1 - X_2)^2)$$

is unbiased. Indeed, $EX_1 = \xi$ and $E\{\frac{1}{2}(X_1 - X_2)^2\} = \sigma^2$ for all values of (ξ, σ^2). The estimator (3.1.2) is, however, very inefficient since it disregards the information on ξ given by $X_2, \ldots, X_n$ and the information on σ^2 given by $X_3, \ldots, X_n$. As we show later on, if the family of distribution functions $\{F_\theta; \theta \in \Theta\}$ admits a sufficient statistic S, and if S is also necessary, the most efficient unbiased estimator of $g(\theta)$ should be $\mathcal{B}_S$-measurable. In other words, the estimator should be a function of all the information on $\{F_\theta; \theta \in \Theta\}$ provided by the sample. From this point of view, it will be proved that the unbiased estimator

$$(3.1.3) \qquad \varphi^*(X_1, \ldots, X_n) = \left(\frac{1}{n}\sum_{i=1}^{n} X_i, \frac{1}{n-1}\sum_{i=1}^{n}(X_i - \bar{X})^2\right),$$

where $\bar{X} = (1/n)\sum_{i=1}^{n} X_i$ is the "best" unbiased estimator of (ξ, σ^2).

Consider now a function $g(\xi, \sigma^2)$ which maps Θ into $[0, 1]$, namely, $g(\xi, \sigma^2) = \Phi(-\xi/\sigma)$ where

$$\Phi(u) = \frac{1}{\sqrt{2\pi}}\int_{-\infty}^{u} \exp\{-\tfrac{1}{2}x^2\}\, dx, \qquad -\infty < u < \infty,$$

is the standard normal integral (the distribution function of $N(0, 1)$). A trivial and inefficient unbiased estimator of $g(\xi, \sigma^2)$ is

$$(3.1.4) \qquad \hat{p}(X_1, \ldots, X_n) = \begin{cases} 1, & \text{if } X_1 \leqslant 0, \\ 0, & \text{otherwise.} \end{cases}$$

Indeed,

$$(3.1.5) \qquad E_{(\xi,\sigma^2)}\{\hat{p}(X_1, \ldots, X_n)\} = P_{(\xi,\sigma^2)}[X_1 \leqslant 0] = \Phi\left(-\frac{\xi}{\sigma}\right)$$

for all (ξ, σ^2). This is an inefficient estimator since it disregards the information on $\Phi(-\xi/\sigma)$ given by $X_2, \ldots, X_n$. As proved by Lieberman and Resnikoff [1], the "best" unbiased estimator of $\Phi(-\xi/\sigma)$ is

$$(3.1.6) \quad p^*(X_1, \ldots, X_n) = \begin{cases} I_{z(\bar{X},Q)}\left(\frac{n}{2} - 1, \frac{n}{2} - 1\right), & 0 \leqslant z(\bar{X}, Q) \leqslant 1, \\ 1, & z(\bar{X}, Q) > 1, \\ 0, & z(\bar{X}, Q) < 0, \end{cases}$$

where

(i)

$$Q = \sum_{i=1}^{n}(X_i - \bar{X})^2,$$

(ii) $I_z(p, q)$ is the incomplete beta function ratio, for positive finite arguments (p, q), being

$$(3.1.7) \qquad I_z(p, q) = \frac{1}{B(p, q)}\int_0^z u^{p-1}(1 - u)^{q-1}\, du, \qquad 0 \leqslant z \leqslant 1$$

and

(iii)

$$z(\bar{X}, Q) = \frac{1}{2}\left[1 - \frac{\bar{X}}{Q^{1/2}}\left(\frac{n}{n-1}\right)^{1/2}\right].$$

As in the previous case the best unbiased estimator of $\Phi(-\xi/\sigma)$ is a function of the minimal sufficient statistic $(\bar{X}, Q)$. ■

We now consider the risk function of an unbiased estimator associated with a general quadratic loss function. We present the theory in a general context to suit cases of vector valued functions $g(\theta)$. Cases of real valued $g(\theta)$ are obtained as simple corollaries.

Let $g(\theta)$ be an r-dimensional vector mapping Θ into Ω. Let $\mathbf{A}$ be an $r \times r$ *positive definite* matrix and φ a statistic on $(\mathfrak{X}, \mathfrak{B})$ into $(\Omega, \mathfrak{F})$. *A quadratic loss function* for estimation is

$$L(\varphi(X), g(\theta)) = (\varphi(X) - g(\theta))'\mathbf{A}(\varphi(X) - g(\theta)), \tag{3.1.8}$$

defined on $\Omega \times \Omega$.

Lemma 3.1.1. *Let $(\mathfrak{X}, \mathfrak{B}, F_\theta)$, $\theta \in \Theta$, be a probability space for a parametric model. Let $\varphi(X)$ be an estimator of $g(\theta)$. Let S be any statistic mapping $(\mathfrak{X}, \mathfrak{B}, F_\theta) \to (\mathcal{S}, \mathfrak{B}_S, F_\theta^S)$ and $L(\varphi(X), g(\theta))$ a quadratic loss function. Then*

$$E_\theta\{L(\varphi(X), g(\theta))\} = E_\theta\{L(\hat{g}_\theta(S), g(\theta))\} + E_\theta\{E_\theta\{L(\varphi(X), \hat{g}_\theta(S)) \mid \mathfrak{B}_S\}\}, \qquad \theta \in \Theta, \tag{3.1.9}$$

where

$$\hat{g}_\theta(S) = E_\theta\{\varphi(X) \mid \mathfrak{B}_S\}.$$

Proof. According to (3.1.8),

$$\begin{aligned} L(\varphi(X), g(\theta)) = {} & [\varphi(X) - \hat{g}_\theta(S)]'\mathbf{A}[\varphi(X) - \hat{g}_\theta(S)] \\ & + 2[\varphi(X) - \hat{g}_\theta(S)]'\mathbf{A}[\hat{g}_\theta(S) - g(\theta)] \\ & + [\hat{g}_\theta(S) - g(\theta)]'\mathbf{A}[\hat{g}_\theta(S) - g(\theta)]. \end{aligned} \tag{3.1.10}$$

Since $\hat{g}_\theta(S)$ is $\mathfrak{B}_S$-measurable and $E_\theta\{\varphi(X) - \hat{g}_\theta(S) \mid \mathfrak{B}_S\} = 0$ a.s. for all $\theta \in \Theta$, we obtain from the iterated expectation law

$$\begin{aligned} E_\theta\{L(\varphi(X), g(\theta))\} &= E_\theta\{E_\theta\{L(\varphi(X), g(\theta)) \mid \mathfrak{B}_S\}\} \\ &= E_\theta\{E_\theta\{L(\varphi(X), \hat{g}_\theta(S)) \mid \mathfrak{B}_S\}\} \\ &\quad + E_\theta\{L(\hat{g}_\theta(S), g(\theta))\}. \end{aligned} \tag{3.1.11}$$

(Q.E.D.)

From this lemma we immediately obtain the celebrated Blackwell-Lehmann-Scheffé theorem, which is the most basic theorem in the theory of unbiased estimation.

Theorem 3.1.1. (Blackwell-Lehmann-Scheffé.) *If* $\mathcal{B}_S$ *is a sufficient subfield for* $\{F_\theta; \theta \in \Theta\}$ *and* $\varphi(X)$ *is an unbiased estimator of* $g(\theta)$, *then* $\hat{g}(S) = E\{\varphi(X) \mid \mathcal{B}_S\}$ *is an unbiased estimator of* $g(\theta)$; *and if* $L(\varphi(X), g(\theta))$ *is a quadratic loss function* (3.1.8) *then*

$$E_\theta\{L(\hat{g}(S), g(\theta))\} \leqslant E_\theta\{L(\varphi(X), g(\theta))\}, \tag{3.1.12}$$

for all $\theta \in \Theta$. *Equality holds if and only if* $\varphi(X) \in \mathcal{B}_S$.

Proof. Since $\mathcal{B}_S$ is a sufficient subfield, $\hat{g}(S)$ is independent of θ and is thus an estimator of $g(\theta)$. The unbiasedness of $\hat{g}(S)$ is implied from the unbiasedness of $\varphi(X)$ and the law of the iterated expectation. Furthermore, according to Lemma 3.1.1 and since $L(\varphi(X), \hat{g}(S))$ is non-negative, $E\{L(\varphi(X), \hat{g}(S)) \mid \mathcal{B}_S\} \geqslant 0$. This and (3.1.9) imply (3.1.12). Equality in (3.1.12) holds if and only if $\varphi(X)$ is $\mathcal{B}_S$-measurable, in which case $\varphi(X) = \hat{g}(S)$ a.s. and $L(\varphi(X), \hat{g}(S)) > 0$ only on a null set. (Q.E.D.)

The univariate version of Theorem 3.1.1 is known as the Blackwell-Rao theorem (see Fraser [5], p. 57). From Theorem 3.1.1 we learn about the strong relationship between minimum risk estimators (for quadratic loss) and sufficient statistics. It is clear that if a family of distribution functions admits a necessary and sufficient statistic which is not the trivial sufficient statistic, then the minimum risk unbiased estimator should be a function of the minimal sufficient statistics. This is implied from the following lemma.

Lemma 3.1.2. *If* $\mathcal{B}_S$ *is sufficient but not necessary for* $\mathfrak{F}$, *and if there exists a necessary and sufficient subfield* $\mathcal{B}_{S^*}$, *then for every unbiased estimator* $\varphi(X)$ *of* $g(\theta)$ *and a quadratic loss function*

$$E_\theta\{L(E(\varphi(X) \mid \mathcal{B}_{S^*}), g(\theta))\} < E_\theta\{L(E(\varphi(X) \mid \mathcal{B}_S), g(\theta))\} \tag{3.1.13}$$

for all $\theta \in \Theta$.

The existence of a minimal sufficient statistic is a necessary but not sufficient condition for the existence of a uniformly minimum risk unbiased estimator. As an example of such a deficiency, consider the case of two random samples $\{x_1, \ldots, x_n\}$ and $\{y_1, \ldots, y_n\}$ from normal distributions having a common mean ξ but unequal variances. As shown in Example 2.8, a minimal sufficient statistic in this case is $(\bar{X}_n, Q_n(X), \bar{Y}_n, Q_n(Y))$ where $\bar{X}_n$, $\bar{Y}_n$ are the corresponding sample means and $Q_n(X)$ and $Q_n(Y)$ are the respective sample sums of squares of deviations about the means. If the variance ratio $\rho = \sigma_Y^2/\sigma_X^2$ is unknown, there exists no unbiased estimator which has a minimum risk uniformly in $(\xi, \sigma_X^2, \sigma_Y^2)$. The existence of locally optimal unbiased estimators of the common mean ξ will be shown in Section 3.3. A study of this estimation problem for small samples, using additional criterions of optimality, was carried on by Zacks

[4] and will be reported in another chapter. The source of the problem in this case is that the minimal sufficient statistic is not a complete one. Indeed, $E\{\bar{X}_n - \bar{Y}_n\} = 0$ for all $(\xi, \sigma_x, \sigma_y^2)$ but $\bar{X}_n \neq \bar{Y}_n$ a.s. The role of a complete sufficient statistic is given in the following theorem.

Theorem 3.1.2. (Lehmann-Scheffé.) *Let* $(\mathfrak{X}, \mathfrak{B}, F_\theta)$, $\theta \in \Theta$, *be a probability space. Suppose that* S: $(\mathfrak{X}, \mathfrak{B}, F_\theta) \to (\mathfrak{S}, \mathfrak{F}, F_\theta^S)$ *is a complete sufficient statistic for* $\{F_\theta;\ \theta \in \Theta\}$. *Then if a function* $g:\Theta \to \Omega$ *admits an unbiased estimator it has a uniformly minimum risk* (*quadratic loss*) *unbiased estimator, which is* $\mathfrak{B}_S$*-measurable and essentially unique.*

Proof. Since $g(\theta)$ is estimable there exists an unbiased estimator $\varphi(X)$ of $g(\theta)$. Consider the unbiased estimator $\hat{g}(S) = E\{\varphi(X) \mid \mathfrak{B}_S\}$. According to Theorem 3.3.1, the risk of $\hat{g}(S)$ for any $\theta \in \Theta$ is not larger than the risk of $\varphi(X)$; and if $\varphi(X) \notin \mathfrak{B}_S$ then the risk of $\hat{g}(S)$ is strictly smaller than that of $\varphi(X)$. Since $\varphi(X)$ is arbitrary, $\hat{g}(S)$ is a minimum risk estimator. Furthermore this result does not depend on θ; hence $\hat{g}(S)$ is a uniformly minimum risk unbiased estimator. Finally, if $U(S)$ is another uniformly minimum risk unbiased estimator, then

$$\int_S (\hat{g}(s) - U(s))\, dF_\theta^{(S)}(s) = 0, \quad \text{for all} \quad \theta \in \Theta. \tag{3.1.14}$$

Therefore by the completeness of $\{F_\theta^{(S)};\ \theta \in \Theta\}$, $\hat{g}(S) = U(S)$ a.s. $[F_\theta^{(S)};\ \theta \in \Theta]$. This establishes the essential uniqueness of $\hat{g}(S)$. (Q.E.D.)

We generalize the results of this section concerning the minimum risk unbiased estimators for cases where the loss function is convex in $(X_1, \ldots, X_n)$ for each $\theta \in \Theta$. A function $f(X_1, \ldots, X_n)$ on a Euclidean n-space $E^{(n)}$ is called *convex* if, for any two points $\mathbf{X}_n = (X_1, \ldots, X_n)$ and $\mathbf{X}'_n = (X'_1, \ldots, X'_n)$ and any $0 \leqslant \alpha \leqslant 1$, we have

$$f(\alpha \mathbf{X}_n + (1 - \alpha)\mathbf{X}'_n) \leqslant \alpha f(\mathbf{X}_n) + (1 - \alpha) f(\mathbf{X}'_n). \tag{3.1.15}$$

A well-known theorem on convex functions (see Fraser [5], p. 54) states that through any point $\mathbf{X}_0 \in E^{(n)}$ passes a hyperplane

$$l(\mathbf{X}; \mathbf{X}_0) = f(\mathbf{X}_0) + (\nabla f(\mathbf{X}_0))'(\mathbf{X} - \mathbf{X}_0) \tag{3.1.16}$$

which lies entirely below the convex function $f(\mathbf{X})$ and coincides with $f(\mathbf{X})$ at $\mathbf{X}_0$. The quantity $l(\mathbf{X}; \mathbf{X}_0)$ is called a supporting hyperplane. The gradient of $f(\mathbf{X})$ at $\mathbf{X}_0$ is designated by $\nabla f(\mathbf{X}_0)$, when the partial derivatives $(\partial/\partial X_j) f(X_1, \ldots, X_n)$ $(j = 1, \ldots, n)$ exist in some neighborhood of $\mathbf{X}_0$. If some of the partial derivatives do not exist the corresponding component of $\nabla f(\mathbf{X}_0)$ is taken to be some value between the left and the right partial derivatives of $f(\mathbf{X})$ at $\mathbf{X}_0$.

Lemma 3.1.3. *Let $X_1, \ldots, X_n$ be random variables having a joint distribution function F. Let $f(\mathbf{X}_n)$ be a convex function on $E^{(n)}$, integrable with respect to F, and suppose that $E\{\mathbf{X}_n\}$ exists* (finite). *Then*

$$f(E\{\mathbf{X}_n\}) \leqslant E\{f(\mathbf{X}_n)\}, \tag{3.1.17}$$

where

$$\begin{aligned} E\{\mathbf{X}_n\} &= \left(\int X_1 \, \mathrm{d}F(X_n), \ldots, \int X_n \, \mathrm{d}F(X_n)\right)', \\ E\{f(\mathbf{X}_n)\} &= \int f(\mathbf{X}_n) \, \mathrm{d}F(\mathbf{X}_n). \end{aligned} \tag{3.1.18}$$

Proof. Let $l_0(\mathbf{X}_n)$ be a supporting hyperplane for $f(\mathbf{X}_n)$ through $E\{\mathbf{X}_n\}$. Then for every $\mathbf{X}_n \in E^{(n)}$,

$$f(\mathbf{X}_n) \geqslant l(\mathbf{X}_n) = f(E\{\mathbf{X}_n\}) + (\nabla f(E\{\mathbf{X}_n\}))'(\mathbf{X}_n - E\{\mathbf{X}_n\}). \tag{3.1.19}$$

The components of $\nabla f(E\{\mathbf{X}_n\})$ are finite and $E\{\mathbf{X}_n - E\{\mathbf{X}_n\}\} = 0$. Hence since $E\{f(\mathbf{X}_n)\} \geqslant E\{l(\mathbf{X}_n)\} = f(E\{\mathbf{X}_n\})$, (3.1.17) is proven. (Q.E.D.)

We now generalize the Blackwell-Rao-Lehmann-Scheffé theorem for cases with convex loss functions. A loss function $L(\varphi(\mathbf{X}_n), g(\theta))$ is called convex if it is a convex function of $\mathbf{X}_n$ for each $\theta \in \Theta$.

Theorem 3.1.3. *Let $X_1, \ldots, X_n$ be i.i.d. random variables on $(\mathfrak{X}, \mathfrak{B})$ having a nondegenerate common distribution F_θ, $\theta \in \Theta$. Let $g(\theta)$ be an estimable parametric function on Θ, and let $U(\mathbf{X}_n)$ be an unbiased estimator of $g(\theta)$. Let $S(\mathbf{X}_n)$ be a complete sufficient statistic for $\mathfrak{F} = [F_\theta; \theta \in \Theta]$, inducing a sigma-subfield $\mathfrak{B}_S$. Then $\hat{g}(S(\mathbf{X}_n)) = E\{U(\mathbf{X}_n) \mid S(\mathbf{X}_n)\}$ is an unbiased estimator of $g(\theta)$. Furthermore if $L(\varphi(\mathbf{X}_n), g(\theta))$ is a convex loss function then*

$$E_\theta\{L(\hat{g}(S(\mathbf{X}_n)), g(\theta))\} \leqslant E_\theta\{L(U(\mathbf{X}_n), g(\theta))\} \tag{3.1.20}$$

for all $\theta \in \Theta$. Assuming that $L(\cdot, g(\theta))$ is a nonconstant function for each $\theta \in \Theta$, equality in (3.1.20) *holds if and only if $U(\mathbf{X}_n)$ is $\mathfrak{B}_S$-measurable. Moreover $\hat{g}(S(\mathbf{X}_n))$ is essentially unique.*

Proof. The unbiasedness of $\hat{g}(S(\mathbf{X}_n))$ and its essential uniqueness are implied by the hypothesis that $S(\mathbf{X}_n)$ is a complete sufficient statistic. To prove the inequality (3.1.20) we remark first that $\hat{g}(S(\mathbf{X}_n)) \in \mathfrak{B}_S$. By the law of the iterated expectation we write

$$E_\theta\{L(\varphi(\mathbf{X}_n), g(\theta))\} = E_\theta\{E\{L(\varphi(\mathbf{X}_n), g(\theta)) \mid S(\mathbf{X}_n)\}\}. \tag{3.1.21}$$

Furthermore, according to the convexity of the loss function and Lemma 3.1.3,

$$L(\hat{g}(S(\mathbf{X}_n)), g(\theta)) \leqslant E\{L(\varphi(\mathbf{X}_n), g(\theta)) \mid S(\mathbf{X}_n)\} \text{ a.s. } [\mathfrak{F}^S], \tag{3.1.22}$$

for each $\theta \in \Theta$. Inequality 3.1.22 implies (3.1.20). Finally, if $U(\mathbf{X}_n) \in \mathfrak{B}_S$ then $U(\mathbf{X}_n) = \hat{g}(S(\mathbf{X}_n))$ a.s. $[\mathfrak{F}^S]$, and hence equality holds in (3.1.20). On the other hand, if equality holds in (3.1.20) then

$$(3.1.23)\quad L(\hat{g}(S(\mathbf{X}_n)), g(\theta)) = E\{L(U(\mathbf{X}_n), g(\theta)) \mid S(\mathbf{X}_n)\} \text{ a.s. } [\mathcal{F}^S].$$

Since $L(U(\mathbf{X}_n), g(\theta))$ is a convex function of $\mathbf{X}_n$ for each $\theta \in \Theta$ and $\hat{g}(S(\mathbf{X}_n)) = E\{U(\mathbf{X}_n) \mid S(\mathbf{X}_n)\}$, (3.1.23) can hold if all the distributions in $\mathcal{F}$ are degenerate, which is an excluded case, or if $L(U(\mathbf{X}_n), g(\theta))$ is a constant independent of $\mathbf{X}_n$, which is an excluded case, or when $U(\mathbf{X}_n) \in \mathcal{B}_S$. (Q.E.D.)

In the following section we present several examples in which the uniformly minimum risk unbiased estimator is derived.

3.2. SEVERAL EXAMPLES OF MINIMUM VARIANCE UNBIASED ESTIMATORS IN THE PARAMETRIC CASE

In this section we present a few cases where uniformly minimum risk unbiased estimators exist. We confine our attention to real parameters and a squared-error loss function, and thus the risk associated with an unbiased estimator is its variance. Hence we derive uniformly minimum variance unbiased (U.M.V.U.) estimators.

Example 3.2. We return to the unbiased estimation of $\Phi(-\xi/\sigma)$ in the normal $\mathcal{N}(\xi, \sigma^2)$ case, which was considered in Example 3.1. As shown in Chapter 2, the minimal sufficient statistic $(\bar{X}, Q)$ is complete. Hence if we prove that the estimator (3.1.6) is unbiased we can conclude, from the Lehmann-Scheffé theorem, that it is the essentially unique U.M.V.U. estimator. To show the unbiasedness of (3.1.6) we first use a result of Ellison [1]. Let $\beta((n-2)/2, (n-2)/2)$ denote a random variable having a beta distribution; that is,

$$(3.2.1)\quad P\left[\beta\left(\frac{n}{2}-1, \frac{n}{2}-1\right) \leqslant x\right] = \begin{cases} 0, & x < 0, \\ I_x\left(\frac{n}{2}-1, \frac{n}{2}-1\right), & 0 \leqslant x \leqslant 1, \\ 1, & x > 1. \end{cases}$$

Ellison [1] proved that if $\beta((n-2)/2, (n-2)/2)$ is independent of $\chi^2[n-1]$ then the distribution law of $(2\beta((n-2)/2, (n-2)/2) - 1)(\chi^2[n-1])^{1/2}$ is like that of a standard normal random variable.

Since $Q^{1/2} \sim \sigma(\chi^2[n-1])^{1/2}$ we have

$$(3.2.2)\quad p^*(X_1, \ldots, X_n) = \begin{cases} 0, & z(\bar{X}, Q) \leqslant 0 \\ I_{Z(\bar{X},Q)}\left(\frac{n}{2}-1, \frac{n}{2}-1\right), & 0 \leqslant z(\bar{X}, Q) \leqslant 1 \\ 1, & z(\bar{X}, Q) \geqslant 1 \end{cases}$$

$$= P\left\{\beta\left(\frac{n}{2}-1, \frac{n}{2}-1\right) \leqslant z(\bar{X}, Q) \mid (\bar{X}, Q)\right\},$$

where $z(\bar{X}, Q) = \frac{1}{2}[1 - (\bar{X}/Q^{1/2})(n/(n-1))^{1/2}]$, $\beta((n-2)/2, (n-2)/2)$ and $(\bar{X}, Q)$ are independent. Thus from Ellison's result ((3.2.2)) and the definition of $z(\bar{X}, Q)$, we obtain

$$(3.2.3)\quad E\{p^*(X_1, \ldots, X_n)\}$$
$$= E\left\{P\left\{\beta\left(\frac{n}{2}-1, \frac{n}{2}-1\right) \leqslant \frac{1}{2}\left[1 - \frac{\bar{X}}{Q^{1/2}}\left(\frac{n}{n-1}\right)^{1/2}\right] \,\Big|\, (\bar{X}, Q)\right\}\right\}$$
$$= P\left\{\sigma\left(2 \cdot \beta\left(\frac{n}{2}-1, \frac{n}{2}-1\right) - 1\right)(\chi^2[n-1])^{1/2} \leqslant -\bar{X}\left(\frac{n}{n-1}\right)^{1/2}\right\}$$
$$= P\left\{N_1(0, \sigma^2) + N_2\left(\xi\left(\frac{n}{n-1}\right)^{1/2}, \sigma^2\frac{1}{n-1}\right) \leqslant 0\right\},$$

where $N_1(0, \sigma^2)$ and $N_2(\xi[n/(n-1)]^{1/2}, \sigma^2/(n-1))$ are independent normally distributed random variables. Finally, since

$$(3.2.4)\quad N_1(0, \sigma^2) + N_2\left(\xi\left(\frac{n}{n-1}\right)^{1/2}, \frac{\sigma^2}{n-1}\right) \sim N\left(\xi\left(\frac{n}{n-1}\right)^{1/2}, \sigma^2\frac{n}{n-1}\right),$$

we deduce from (3.2.3) that

$$(3.2.5)\quad E\{p^*(X_1, \ldots, X_n)\} = P\left\{N\left(\xi\left(\frac{n}{n-1}\right)^{1/2}, \sigma^2\frac{n}{n-1}\right) \leqslant 0\right\}$$
$$= \Phi\left(-\frac{\xi}{\sigma}\right).$$

This proves the unbiasedness of (3.1.6). ■

Example 3.3. Let $X_1, \ldots, X_n$ be i.i.d. random variables having a common rectangular distribution on (θ_1, θ_2), that is, $\mathcal{R}(\theta_1, \theta_2)$. We derive the essentially unique U.M.V.U. estimators of θ_1 and θ_2.

Case 1. When θ_1 is known, without loss of generality we assume $\theta_1 = 0$. If $\theta_1 \neq 0$ we let $Y_i = X_i - \theta_1$ $(i = 1, \ldots, n)$. Then $Y_i \sim \mathcal{R}(0, \theta)$, where $\theta = \theta_2 - \theta_1$. As established in Chapter 2, Example 2.5, $\max(X_1, \ldots, X_n)$ is a sufficient statistic for $\mathcal{F} = \{\mathcal{R}(0, \theta); 0 < \theta < \infty\}$. Furthermore, as is easy to check, $T_n = \max(X_1, \ldots, X_n)$ is a minimal (necessary) sufficient. The density of T_n, given θ, is given in (2.3.33). Let $F_\theta(t)$ denote the distribution function of T_n and suppose that $\psi(T)$ is any statistic, measurable $\mathcal{B}_T$, such that

$$(3.2.6)\quad \int_0^\theta \psi(t)\, \mathrm{d}F_\theta(t) = \frac{n}{\theta^n}\int_0^\theta \psi(t)t^{n-1}\, \mathrm{d}t = 0$$

for all $0 < \theta < \infty$. It follows that $\psi(t) = 0$ for almost all $0 < t < \infty$. Hence T_n is a complete sufficient statistic. To derive the essentially unique

U.M.V.U. estimator of θ we employ the following device. It is first noticed that $E_\theta\{2X_1\} = \theta$ for all $0 < \theta < \infty$. Hence $\varphi(X_1, \ldots, X_n) = 2X_1$ is an unbiased estimator of θ. Thus according to the Lehmann-Scheffé theorem,

$$\hat{g}(T_n) = E\{2X_1 \mid T_n\} \tag{3.2.7}$$

is an essentially unique U.M.V.U. estimator of θ. The conditional density of X_1 given T_n is, as given by Patil and Wani [1],

$$f(x \mid t) = \begin{cases} \left(1 - \dfrac{1}{n}\right)\dfrac{1}{t}, & 0 < x < t, \\ \dfrac{1}{n}, & x = t. \end{cases} \tag{3.2.8}$$

Indeed, since $X_1, \ldots, X_n$ are i.i.d., the symmetry of the model implies that $P[X_1 = T_n \mid T_n] = 1/n$ and $P[X_1 < T_n \mid T_n] = (1 - (1/n))$. Furthermore, given $\{T_n = t, X_1 < T_n\}$, the uniform distribution of X_1 implies that the conditional marginal distribution of X_1 is uniform on $(0, t)$. This leads to (3.2.8). One can establish (3.2.8) also by formal rigorous computation. The conditional expectation of X_1 given T_n is

$$E\{X_1 \mid T_n = t\} = \left(1 - \frac{1}{n}\right) \cdot \frac{t}{2} + \frac{t}{n}. \tag{3.2.9}$$

Hence $\hat{g}(T_n)$ is

$$\hat{g}(T_n) = \left(1 - \frac{1}{n}\right) T_n + \frac{2}{n} T_n = \left(1 + \frac{1}{n}\right) T_n. \tag{3.2.10}$$

It is instructive to compare the variance of the U.M.V.U. estimator $\hat{g}(T_n)$ to that of an unbiased estimator based on the sample mean, say $U(X_1, \ldots, X_n) = 2\bar{X}_n$. The variance of $2\bar{X}_n$ is

$$\text{Var}_\theta \{2\bar{X}_n\} = \frac{\theta^2}{3n}, \qquad 0 < \theta < \infty. \tag{3.2.11}$$

The variance of $\hat{g}(T_n)$ is, on the other hand,

$$\text{Var}_\theta \{\hat{g}(T_n)\} = \frac{\theta^2}{n(n+2)}, \qquad 0 < \theta < \infty. \tag{3.2.12}$$

The two estimates $\hat{g}(T_n)$ and $2\bar{X}_n$ coincide when $n = 1$. In this case (3.2.11) and (3.2.12) are equal. For every $n > 1$ the variance of $\hat{g}(T_n)$ is smaller than that of $2\bar{X}_n$, and the difference becomes excessive for large values of n.

Case 2. Both θ_1 and θ_2 are unknown. In this case the complete sufficient statistic is $(X_{(1)}, X_{(n)})$ where $X_{(1)} = \min_{1 \leq i \leq n} \{X_i\}$ and $X_{(n)} = \max_{1 \leq i \leq n} \{X_i\}$. The marginal densities of $X_{(1)}$ and of $X_{(n)}$ are

$$(3.2.13)\qquad f_{\theta_1,\theta_2}^{X(1)}(x) = \frac{n}{(\theta_2-\theta_1)^n}(\theta_2 - x)^{n-1}, \qquad \theta_1 \leqslant x \leqslant \theta_2,$$
$$= 0, \qquad \text{otherwise,}$$

and

$$(3.2.14)\qquad f_{\theta_1,\theta_2}^{X(n)}(y) = \frac{n}{(\theta_2-\theta_1)^n}(y - \theta_1)^{n-1}, \qquad \theta_1 \leqslant y \leqslant \theta_2,$$
$$= 0, \qquad \text{otherwise.}$$

Simple integration yields the following:

$$(3.2.15)\qquad E_{\theta_1,\theta_2}\{X_{(1)}\} = \frac{n}{n+1}\theta_1 + \frac{1}{n+1}\theta_2,$$

and

$$(3.2.16)\qquad E_{\theta_1,\theta_2}\{X_{(n)}\} = \frac{1}{n+1}\theta_1 + \frac{n}{n+1}\theta_2.$$

From (3.2.15) and (3.2.16) we obtain the essentially unique U.M.V.U. estimators of θ_1 and θ_2 which are

$$(3.2.17)\qquad \hat{\theta}_1(X_1, \ldots, X_n) = \frac{1}{n-1}[nX_{(1)} - X_{(n)}]$$

and

$$(3.2.18)\qquad \hat{\theta}_2(X_1, \ldots, X_n) = \frac{1}{n-1}[nX_{(n)} - X_{(1)}]. \qquad \blacksquare$$

Example 3.4. *Estimation of Poisson probabilities* (Glasser [1]). Let $X_1, \ldots, X_n$ be i.i.d. random variables having a common Poisson distribution $P(\lambda)$, $0 < \lambda < \infty$. The quantity λ is unknown, and we wish to estimate unbiasedly the density function

$$(3.2.19)\qquad P(k; \lambda) = e^{-\lambda}\frac{\lambda^k}{k!}, \qquad k = 0, 1, \ldots$$

A complete sufficient statistic for $\mathfrak{F} = \{P(\lambda); 0 < \lambda < \infty\}$ is the sample total $T_n = \sum_{i=1}^n X_i$. Thus according to the Lehmann-Scheffé theorem the conditional probability of $\{X_1 = k\}$, given T_n, is the essentially unique U.M.V.U. estimator of $P(k; \lambda)$. As we derived in Chapter 2, Example 2.1, the conditional joint distribution of $(X_1, \ldots, X_{n-1})$, given $\{T_n = t\}$, is the multinomial distribution with density as in (2.1.3). Hence the conditional probability of $\{X_1 = k\}$ given T_n is

$$(3.2.20)\qquad \hat{P}(k; T_n) = \binom{T_n}{k}\left(\frac{1}{n}\right)^k\left(1 - \frac{1}{n}\right)^{T_n-k}, \qquad k = 0, \ldots, T_n$$
$$= 0, \qquad \text{otherwise.}$$

It is simple to verify that $E_\lambda\{\hat{P}(k; T_n)\} = P(k; \lambda)$ for all $0 < \lambda < \infty$.

Glasser [1] proves that $\hat{P}(k; T_n)$ is the essentially unique U.M.V.U. estimator of $P(k; \lambda)$ in another manner. Since Glasser's approach has some instructive importance and adds more to the example we present it too. We start with the remark that a function $g(\lambda)$ is estimable if and only if it can be expressed as a power function in integral non-negative powers of λ. Indeed, let $T_n \sim P(n\lambda)$ and let $U(T_n)$ be an unbiased estimator of $g(\lambda)$ where $g(\lambda) = \sum_{j=0}^{\infty} a_j \lambda^j$. Thus

$$E_\lambda\{U(T_n)\} = e^{-\lambda n} \sum_{t=0}^{\infty} \frac{n^t \lambda^t}{t!} U(t) = \sum_{t=0}^{\infty} a_t \lambda^t. \tag{3.2.21}$$

Equation 3.2.21 shows that only power series in λ are estimable. In particular, the U.M.V.U. estimator of λ^i is

$$U_i(T_n) = \begin{cases} 0, & \text{if } T_n < i, \\ \dfrac{T_n!}{(T_n - i)!\, n^i}, & \text{if } T_n \geqslant i. \end{cases} \tag{3.2.22}$$

Thus expanding $P(k; \lambda)$ in powers of λ we have

$$P(k; \lambda) = \frac{1}{k!} \sum_{j=0}^{\infty} (-1)^j \frac{\lambda^{j+k}}{j!}, \qquad k = 0, 1, \ldots. \tag{3.2.23}$$

Denoting by $U_{j+k}(T_n)$ the U.M.V.U. estimator of λ^{j+k} we obtain, by substituting in (3.2.23),

$$\hat{P}(k; T_n) = \frac{1}{k!} \sum_{j=0}^{T_n - k} (-1)^j \frac{T_n!}{(T_n - j - k)!\, j!\, n^{j+k}} = \begin{cases} \dbinom{T_n}{k} \left(\dfrac{1}{n}\right)^k \left(1 - \dfrac{1}{n}\right)^{T_n - k}, & 0 \leqslant k \leqslant T_n, \\ 0, & \text{otherwise}. \end{cases} \tag{3.2.24}$$

The function $P(0; \lambda) = e^{-\lambda}$ has an important role in reliability theory. It yields the probability of no failure in a Poisson process during a specified time period. The properties of its U.M.V.U. estimator $\hat{P}(0; T_n)$ were studied by Zacks and Even [1]. We give an account of the results of this paper later on, when we compare the efficiency of various methods of estimation. ■

Example 3.5. *Unbiased estimation of reliability* (Basu [1]). Let $X_1, \ldots, X_n$ be i.i.d. random variables representing the life length of a system and having a common distribution function $F_\theta(x)$; $\theta \in \Theta$. The reliability function for the given system is defined in different manners, depending on the objective of the life-testing experiment. One of the common definitions is

$$R_\theta(\tau) = P_\theta[X \geqslant \tau], \qquad \theta \in \Theta. \tag{3.2.25}$$

Let $I_t(X)$ denote the indicator function of the interval $[t, \infty)$. An unbiased estimator of $R_\theta(\tau)$ is $I_\tau(X)$. This is a trivial estimator but, according to the Lehmann-Scheffé theorem, if $S_n(X_1, \ldots, X_n)$ denotes a complete sufficient statistic for $\{F_\theta;\ \theta \in \Theta\}$, then

$$\hat{R}(\tau; S_n) = E\{I_\tau(X_1) \mid S_n(X_1, \ldots, X_n)\} \tag{3.2.26}$$

is the essentially unique U.M.V.U. estimator of $R_\theta(\tau)$.

Case 1. *The gamma distribution.* $X_1, \ldots, X_n$ are i.i.d. having a gamma distribution $\mathcal{G}(1/\theta, p)$, p known, $0 < \theta < \infty$. A complete sufficient statistic is $S_n = \sum_{i=1}^n X_i$. We now derive the conditional density of X_1 given S_n. Let $S_{n-1}^* = \sum_{i=2}^n X_i$; X_1 is independent of S_{n-1}^*; hence the joint density of X_1 and S_{n-1}^* is

$$f_\theta^{X_1, S^*_{n-1}}(x, s) = \frac{x^{p-1}e^{-x/\theta}}{\theta^p \Gamma(p)} \cdot \frac{s^{(n-1)p-1}e^{-s/\theta}}{\theta^{(n-1)p}\Gamma((n-1)p)}, \tag{3.2.27}$$

for $0 \leqslant x \leqslant \infty$, $0 \leqslant s \leqslant \infty$. Letting $S_n = X_1 + S_{n-1}^*$ we have from (3.2.27) that

$$f_\theta^{X_1, S_n}(x, t) = \frac{1}{\theta^{np}\Gamma(p)\Gamma((n-1)p)}\, x^{p-1}(t-x)^{(n-1)p-1}e^{-t/\theta}, \tag{3.2.28}$$

$0 \leqslant x \leqslant t \leqslant \infty$. Therefore the conditional density of X_1, given S_n, is

$$f^{X_1|S_n}(x \mid s) = \frac{(n-1)p}{\Gamma(p)\cdot s} \cdot \left(\frac{x}{s}\right)^{p-1}\left(1 - \frac{x}{s}\right)^{(n-1)p-1}, \tag{3.2.29}$$

$0 \leqslant x \leqslant s$. Finally, the U.M.V.U. estimator of $R_\theta(\tau)$ is

$$\hat{R}(\tau; S_n) = \begin{cases} 0, & \text{if } S_n \leqslant \tau, \\ \displaystyle\int_\tau^{S_n} f^{X1|S_n}(x \mid S_n)\, dx, & \text{if } S_n > \tau. \end{cases} \tag{3.2.30}$$

Or

$$\hat{R}(\tau; S_n) = \begin{cases} 0, & \text{if } S_n \leqslant \tau, \\ 1 - I_{\tau/S_n}(p, (n-1)p), & \text{if } S_n > \tau. \end{cases} \tag{3.2.31}$$

In the special case of negative exponential distribution ($p = 1$) we obtain the formula

$$\hat{R}(\tau; S_n) = \begin{cases} 0, & \text{if } S_n \leqslant \tau, \\ \left(1 - \dfrac{\tau}{S_n}\right)^{n-1}, & \text{if } S_n > \tau. \end{cases} \tag{3.2.32}$$

Case 2. The truncated 2-parameter exponential distribution. This case is prevalent in life testing when the failures follow a Poisson process (see Epstein and Sobel [1]). We conduct the experiment on n identical systems. A system which fails is not replaced. Immediately after the r-th failure $(1 \leqslant r \leqslant n)$ the experiment is terminated. Let $X_{(1)} \leqslant X_{(2)} \leqslant \cdots \leqslant X_{(r)}$ denote the time points of failure. We can show that this order statistic is equivalent to the first r order statistics in a sample of n i.i.d. random variables having a 2-parameter exponential density

$$f(x; \alpha, \beta) = \begin{cases} 0, & \text{if } x < \alpha, \\ \dfrac{1}{\beta} \exp\left\{-\dfrac{x-\alpha}{\beta}\right\}, & \text{if } x \geqslant \alpha, \end{cases} \tag{3.2.33}$$

where $0 < \beta < \infty$ and $-\infty < \alpha < \infty$. As a generalization of the result in Example 2.3, we can show that a complete sufficient statistic in the present case is $S_{n,r} = (X_{(1)}, \sum_{i=2}^{r} X_{(i)} + (n-r)X_{(r)})$. The essentially unique U.M.V.U. estimator of $R_\theta(\tau)$ is in this case

(3.2.34)
$$\hat{R}(\tau; S_{n,r}) = \frac{n-1}{n}\left[\left(1 - \frac{\tau - X_{(1)}}{\sum_{i=2}^{r} X_{(i)} + (n-r)X_{(r)} - (n-1)X_{(1)}}\right)^{+}\right]^{r-2},$$

where $a^+ = \max(a, 0)$. ■

Example 3.6. (Olkin and Pratt [1].) Let $(X_1, Y_1), \ldots, (X_n, Y_n)$ be i.i.d. random vectors having a joint bivariate normal distribution with an expectation vector (ξ, η) and a covariance matrix:

$$\Sigma = \begin{pmatrix} \sigma_1^2 & \rho\sigma_1\sigma_2 \\ \rho\sigma_1\sigma_2 & \sigma_2^2 \end{pmatrix}, \tag{3.2.35}$$

$0 < \sigma_1^2, \sigma_2^2 < \infty$, $-1 \leqslant \rho \leqslant 1$, where ρ is the coefficient of correlation. We consider the problem of the unbiased estimation of ρ when all the five parameters are unknown. The complete sufficient statistic is the statistic $T_n = (\sum_{i=1}^n X_i, \sum_{i=1}^n X_i^2, \sum_{i=1}^n Y_i, \sum_{i=1}^n Y_i^2, \sum_{i=1}^n X_i Y_i)$. Letting $\nu = n - 1$, the density function of the sample correlation coefficient

$$r = \frac{\sum_{i=1}^n (X_i - \bar{X})(Y_i - \bar{Y})}{\{\sum_{i=1}^n (X_i - \bar{X})^2 \cdot \sum_{i=1}^n (Y_i - \bar{Y})^2\}^{1/2}} \tag{3.2.36}$$

is (see Anderson [1], p. 69)

$$f(r \mid \rho) = \frac{2^{\nu-2}}{\pi\Gamma(\nu-1)}(1-\rho^2)^{\nu/2}(1-r^2)^{(\nu-3)/2}\sum_{k=0}^{\infty}\Gamma^2\left(\frac{\nu+k}{2}\right)\frac{(2\rho)^k}{k!}. \tag{3.2.37}$$

It is well known that the sample correlation coefficient r is a *biased* estimator of ρ for every $0 < \nu < \infty$. We seek an unbiased estimator of ρ which is a function of r, say $G(r)$. For $G(r)$ to be unbiased the following equation should be satisfied for all $-1 < \rho < 1$:

(3.2.38)

$$\frac{2^{\nu-2}}{\pi\Gamma(\nu-1)} \cdot \sum_{k=0}^{\infty} \Gamma^2\left(\frac{\nu+k}{2}\right) \frac{(2\rho)^k}{k!} \int_{-1}^{1} G(r)(1-r^2)^{(\nu-3)/2}\, dr = (1-\rho^2)^{-\nu/2}\rho.$$

Furthermore, expanding $(1-\rho^2)^{-\nu/2}$ as a power series and comparing the coefficients of equal powers of ρ in the right- and left-hand sides of (3.2.38) we obtain, for every $j = 0, 1, \ldots,$

(3.2.39)

$$\int_0^1 G(r)(1-r^2)^{(\nu-3)/2} r^{2j+1}\, dr = \frac{\pi\Gamma(\nu-1)\Gamma(2j+2)}{2^{\nu+2j-1}\Gamma^2((\nu+2j+1)/2)} \cdot \frac{\Gamma((\nu+2j)/2)}{\Gamma(\nu/2)\Gamma(j+1)}.$$

Making the transformation $r = \exp\{-\frac{1}{2}y\}$, the left-hand side of (3.2.39) is transformed into

$$\frac{1}{2}\int_0^{\infty} G(e^{-\frac{1}{2}y})(1-e^{-y})^{(\nu-3)/2} e^{-jy} e^{-y}\, dy.$$

The Laplace transform (if exists) of an integrable function $f(y)$, to be designated by $\mathfrak{L}\{f(y); s\}$, is the function

(3.2.40)
$$f^*(s) = \int_0^{\infty} e^{-sy} f(y)\, dy, \qquad \mathfrak{Re}\{s\} > s_0,$$

defined on the complex plane (see Widder [1]). Thus from (3.2.39) we obtain that, for every $j = 0, 1, \ldots,$

(3.2.41)

$$\mathfrak{L}\{G(e^{-\frac{1}{2}y})(1-e^{-y})^{(\nu-3)/2}e^{-y}; j\} = \frac{\pi\Gamma(\nu-1)\Gamma(2j+2)\Gamma\left(\frac{\nu}{2}+j\right)}{2^{\nu-1+2j}\Gamma^2\left(\frac{\nu}{2}+j+\frac{1}{2}\right)\Gamma\left(\frac{\nu}{2}\right)\Gamma(j+1)}.$$

The theory of the Laplace transform establishes the unique correspondence between integrable functions on $(0, \infty)$ and their Laplace transforms. Denote the inverse of the Laplace transform $f^*(s)$ by $f(X) = \mathfrak{L}^{-1}(f^*(s); X)$. We determine the inverse function $\mathfrak{L}^{-1}(f^*(s); X)$ by aid of tables of integral transforms [1]. The right-hand side of (3.2.41) can be further simplified by the Legendre identity

(3.2.42)
$$\sqrt{\pi}\,\Gamma(2p) = 2^{2p-1}\Gamma(p)\Gamma(p+\tfrac{1}{2}), \qquad 0 < p < \infty$$

(see Olkin [1]). We can then write the right-hand side of (3.2.41) in the form

$$\Gamma\left(\frac{\nu-1}{2}\right)\frac{\Gamma(\frac{3}{2}+j)\Gamma\left(\frac{\nu}{2}+j\right)}{\Gamma^2\left(\frac{\nu+1}{2}+j\right)},$$

which is, according to the tables (see Bateman Manuscript Project [1] p. 262(7)), the Laplace transform of

$$(3.2.43)\quad \psi(y) = e^{-3/2y}(1-e^{-y})^{(\nu-1)/2-1}F\left(\frac{1}{2};\frac{1}{2};\frac{\nu-1}{2};1-e^{-y}\right),$$

where $F(.\,;.\,;.\,;.)$ is the hypergeometric function

$$(3.2.44)\qquad F(\alpha;\beta;\gamma;x) = \sum_{k=0}^{\infty}\frac{\Gamma(\alpha+k)\Gamma(\beta+k)\Gamma(\gamma)}{\Gamma(\alpha)\Gamma(\beta)\Gamma(\gamma+k)}\cdot\frac{x^k}{k!}$$

Thus from (3.2.41) and (3.2.43) we obtain the unbiased estimator of ρ, which is

$$(3.2.45)\qquad G(r) = rF\left(\frac{1}{2};\frac{1}{2};\frac{\nu-1}{2};1-r^2\right).$$

Olkin and Pratt [1] give a table of the values of $F(\frac{1}{2};\frac{1}{2};(\nu-1)/2;1-r^2)$ for different combinations of ν and r. Finally, since $(\bar{X},\bar{Y},\sum X_i^2,\sum Y_i^2,\sum X_iY_i)$ is a complete sufficient statistic, $G(r)$ given by (3.2.45) is the minimum variance unbiased estimator of ρ. ■

3.3. LOCALLY MINIMUM VARIANCE UNBIASED ESTIMATORS

In the previous section we have presented examples of U.M.V.U. estimators. The essential element for the existence of a U.M.V.U. estimator is the completeness of the family of distributions in the given statistical model. In the absence of completeness there will not exist a U.M.V.U. estimator. Locally minimum variance unbiased (L.M.V.U.) estimators may, however, exist, as will be shown in the material that follows.

If $g(\theta)$ is a real valued parameter and $\varphi(X)$ is an unbiased estimator of $g(\theta)$, then $\varphi(X)$ is L.M.V.U. at θ_0 if, given any other unbiased estimator of $g(\theta)$, say $\varphi^*(X)$,

$$(3.3.1)\qquad \text{Var}_{\theta_0}\{\varphi(X)\} \leqslant \text{Var}_{\theta_0}\{\varphi^*(X)\}.$$

The definition of an L.M.V.U. estimator can be formulated in a more general fashion for vector valued unbiased estimators in the following manner.

Let $(\mathcal{X}, \mathcal{B}, F_\theta)$, $\theta \in \Theta$, be a given probability space and $\mathbf{g}(\theta)$ a vector valued estimable function mapping Θ into an open set Ω of an r-dimensional Euclidean space. Denote by $\mathbf{\Sigma}_\theta(\boldsymbol{\varphi})$ the variance-covariance matrix of an estimator $\boldsymbol{\varphi}(X)$ of $\mathbf{g}(\theta)$ under θ. An estimator $\boldsymbol{\varphi}^0(X)$ is called locally minimum variance unbiased (L.M.V.U.) at θ_0 if, given any unbiased estimator $\boldsymbol{\varphi}(X)$ of $\mathbf{g}(\theta)$, $\mathbf{\Sigma}_{\theta_0}(\boldsymbol{\varphi}) - \mathbf{\Sigma}_{\theta_0}(\boldsymbol{\varphi}^0)$ is a non-negative definite matrix.

In a similar manner we can define the notion of a locally minimum risk unbiased estimator for any appropriate risk function. In this section we confine attention to the case of estimable real valued functions $g(\theta)$.

Theorem 3.3.1. *Let $(\mathcal{X}, \mathcal{B}, F_\theta)$, $\theta \in \Theta$, be a given probability space. Let g be an estimable function mapping Θ into an open interval of the real line. An unbiased estimator φ_0 of $g(\theta)$ is L.M.V.U. at $\theta_0 \in \Theta$ if and only if $\text{cov}_{\theta_0}(\varphi_0(X), f(X)) = 0$ where $f(X)$ is any unbiased estimator of* 0. $\text{Var}_{\theta_0}\{f\} < \infty$.

Proof. Without loss of generality we assume that X is a minimal sufficient statistic (if exists). If not, then the Blackwell-Rao theorem implies that φ_0 cannot be L.M.V.U. at $\theta = \theta_0$. Furthermore, we assume that the family of distributions of X, $\{F_\theta(x);\ \theta \in \Theta\}$, is incomplete. Otherwise by the Blackwell-Rao-Lehmann-Scheffé theorem, $\varphi_0(X)$ is a U.M.V.U. estimator $f(X) = 0$ a.s., and the theorem is trivially true.

Thus let $f(X)$ be any nontrivial unbiased estimator of 0. We prove that $\text{cov}_{\theta_0}(\varphi_0(X), f(X)) = 0$ is a necessary and sufficient condition for $\varphi_0(X)$ to be an L.M.V.U. estimator at $\theta = \theta_0$.

(i) *Necessity*. The proof is by negation. Assume that $\varphi_0(X)$ is L.M.V.U. at $\theta = \theta_0$ and that $\text{cov}_{\theta_0}(\varphi_0(X), f(X)) > 0$. Define the unbiased estimator $\varphi_1(X) = \varphi_0(X) + \lambda f(X)$ where $-2[\text{cov}_{\theta_0}(\varphi_0(X), f(X))/\text{Var}_{\theta_0}\{f(X)\}] < \lambda < 0$. The variance of $\varphi_1(X)$ at θ_0 is

(3.3.2)

$$\text{Var}_{\theta_0}\{\varphi_1(X)\} = \text{Var}_{\theta_0}\{\varphi_0(X)\} + 2\lambda\,\text{cov}_{\theta_0}(\varphi_0(X), f(X)) + \lambda^2\,\text{Var}_{\theta_0}\{f(X)\}.$$

According to the above restriction on λ,

$$(3.3.3)\quad 2\lambda\,\text{cov}_{\theta_0}(\varphi_0(X), f(X)) + \lambda^2\,\text{Var}_{\theta_0}\{f(X)\} = 2\lambda\,\text{cov}_{\theta_0}(\varphi_0(X), f(X))\left\{1 + \lambda\frac{\text{Var}_{\theta_0}\{f(X)\}}{2\,\text{cov}_{\theta_0}(\varphi_0(X), f(X))}\right\} < 0.$$

Hence $\text{Var}_{\theta_0}\{\varphi_1(X)\} < \text{Var}_{\theta_0}\{\varphi_0(X)\}$, which contradicts the assumption. In a similar way we show that if $\text{cov}_{\theta_0}(\varphi_0(X), f(X)) < 0$ then $\varphi_0(X)$ is not an L.M.V.U. estimator at $\theta = \theta_0$.

(ii) *Sufficiency*. Assume that $\text{cov}_{\theta_0}(\varphi_0(X), f(X)) = 0$ for every unbiased estimator of $0, f(X)$. Let $\varphi_1(X)$ be any unbiased estimator of $g(\theta)$ and define

$U(X) = \varphi_0(X) - \varphi_1(X)$. The estimator $U(X)$ is an unbiased estimator of 0. Therefore

$$\text{(3.3.4)} \quad \operatorname{cov}_{\theta_0}(\varphi_0(X), U(X)) = \operatorname{cov}_{\theta_0}(\varphi_0(X), \varphi_0(X) - \varphi_1(X)) = \operatorname{Var}_{\theta_0}\{\varphi_0(X)\} - \operatorname{cov}_{\theta_0}(\varphi_0(X), \varphi_1(X)) = 0.$$

By the Schwarz inequality,

$$\text{(3.3.5)} \quad \operatorname{cov}_{\theta_0}(\varphi_0(X), \varphi_1(X)) \leqslant [\operatorname{Var}_{\theta_0}\{\varphi_0(X)\} \cdot \operatorname{Var}_{\theta_0}\{\varphi_1(X)\}]^{1/2}.$$

Equations 3.3.4 and 3.3.5 imply that

$$\text{(3.3.6)} \quad \operatorname{Var}_{\theta_0}\{\varphi_0(X)\} \leqslant \operatorname{Var}_{\theta_0}\{\varphi_1(X)\}.$$

Hence $\varphi_0(X)$ is an L.M.V.U. estimator at $\theta = \theta_0$. (Q.E.D.)

In the following example we apply the result of the present theorem to the case of estimating a common mean of two normal distributions.

Example 3.7. Let $X_1, \ldots, X_n$ be i.i.d. random variables having an $\mathcal{N}(\mu, \sigma_1^2)$ distribution law, $-\infty < \mu < \infty$ and $0 < \sigma_1^2 < \infty$. Let $Y_1, \ldots, Y_n$ be i.i.d. random variables having an $\mathcal{N}(\mu, \sigma_2^2)$ distribution law, $0 < \sigma_2^2 < \infty$. The variables $Y_1, \ldots, Y_n$ are mutually independent of $X_1, \ldots, X_n$. The two normal distributions have the same expectation μ. Neither parameter is known nor is the variance ratio $\rho = \sigma_2^2/\sigma_1^2$.

A minimal sufficient statistic is, as shown in Example 2.8, $T_n = (\bar{X}_n, Q_n(X), \bar{Y}_n, Q_n(Y))$, where $Q_n(X)$ and $Q_n(Y)$ are the sample sum of squares of deviations for X and Y, respectively.

We now show that $\hat{\mu}(\rho_0) = (\rho_0\bar{X} + \bar{Y})/(1 + \rho_0)$ is L.M.V.U. at all (σ_1^2, σ_2^2) such that $\sigma_2^2/\sigma_1^2 = \rho_0$. This is verified by Theorem 3.3.1. Let $f \equiv f(T_n)$ be any statistic which is an unbiased estimator of zero. Non-trivial unbiased estimators of zero exist, since the family of distributions of T_n is incomplete (see Example 2.8). Thus

$$\operatorname{cov}\left(\frac{\rho_0\bar{X} + \bar{Y}}{1 + \rho_0}, f(T_n)\right) = \frac{\rho_0}{1 + \rho_0}\operatorname{cov}(\bar{X}, f) + \frac{1}{1 + \rho_0}\operatorname{cov}(\bar{Y}, f)$$

Furthermore, since $\bar{Y} \sim N(\mu, (\sigma^2\rho/n))$ and $\bar{X} \sim N(\mu, (\sigma^2/n))$, $\operatorname{cov}(\bar{Y}, f) = \operatorname{cov}(\bar{X}, f)\rho^{1/2}$; hence

$$\text{(3.3.7)} \quad \operatorname{cov}\left(\frac{\rho_0\bar{X} + \bar{Y}}{1 + \rho_0}, f\right) = \frac{\sigma\rho_0^{1/2}(1 + \rho_0)^{1/2}}{\sqrt{n}(1 + \rho_0)}\operatorname{cov}(U, f)$$

where $U \sim N(0, 1)$. Since $f(T_n)$ is an unbiased estimator of zero it should be translation invariant. That is,

$$f(\bar{X}_n, \bar{Y}_n, Q_n \mid A, Q_n(Y) = f^*(\bar{X}_n - \bar{Y}_n; Q_n(X), Q_n(Y)).$$

Moreover, since $Q_n(X)$ and $Q_n(Y)$ are independent of $\bar{X}_n - \bar{Y}_n$ and are strongly complete,

$$E_\theta\{f^*(\bar{X}_n - \bar{Y}_n; Q_n(X), Q_n(Y)) \mid Q_n(X), Q_n(Y)\} = 0 \quad \text{a.s.}$$

for all

$$\theta = (\sigma_1^2, \sigma_2^2).$$

This in turn implies that

$$f^*(\bar{X}_n - \bar{Y}_n; \phi_n(X), \phi_n(Y)) = -f^*(-(\bar{X}_n - \bar{Y}_n); \phi_n(X), \phi_n(Y)) \quad \text{a.s.}$$

Finally, since $\bar{X}_n - \bar{Y}_n$ has a symmetric distribution

$$\begin{aligned} \text{(3.3.8)} \quad \operatorname{cov}(u, f(T_n)) &= E\{uf^*(\bar{X}_n - \bar{Y}_n; Q_n(X), Q_n(Y)\} \\ &= E\{\tfrac{1}{2}f^*(|\bar{X}_n - \bar{Y}_n|; Q_n(X), Q_n(Y))E\{u \mid |\bar{X}_n - \bar{Y}_n|\} \\ &\quad + \tfrac{1}{2}f^*(-|\bar{X}_n - \bar{Y}_n|; Q_n(X), Q_n(Y)) \\ &\quad \times E\{u \mid |\bar{X}_n - \bar{Y}_n|\}\} = 0. \end{aligned}$$

This proves that $\hat{\mu}(\rho_0)$ is a locally minimum variance unbiased estimator. Moreover, since this optimal property holds for all ρ_0, $\hat{\mu}(\rho)$ is a U.M.V.U. estimator, whenever ρ is known. ■

In certain cases one can construct L.M.V.U. estimators having a zero variance at a certain parameter point θ_0. The construction of such estimators together with some very interesting examples is discussed in an article by Morimoto and Sibuya [1]. (They credit Professor K. Takeuchi of Tokyo University with the main idea.) To present such a case, consider a family of densities whose supporting intervals depend on a location (selection) parameter θ and are of the form $(\theta, b(\theta))$, where $-\infty < \theta < b(\theta) < \infty$ and $b(\theta)$ is a nondecreasing function of θ, *everywhere differentiable*. Such a family of densities is specified by

$$\text{(3.3.9)} \quad f(x; \theta) = \begin{cases} \dfrac{f(x)}{F(\theta)}, & \theta \leqslant x \leqslant b(\theta), \qquad -\infty < \theta < b(\theta) < \infty, \\ 0, & \text{otherwise,} \end{cases}$$

in which $F(\theta) = \int_\theta^{b(\theta)} f(x)\,dx$. The objective is to construct for such a family of densities an unbiased estimator of a function $g(\theta)$ such that its variance is zero for a given point θ_0. We start the construction of such an estimator on the basis of one observation X (a sample of size $n = 1$). The function $g(\theta)$ is assumed to be absolutely continuous. Since θ_0 is specified, we first define

$$\text{(3.3.10)} \qquad \varphi_0(X) = g(\theta_0) \quad \text{for all} \quad \theta_0 \leqslant X \leqslant b(\theta_0).$$

This is consistent with the requirement of unbiasedness, since

$$E_{\theta_0}\{\varphi_0(X)\} = g(\theta_0).$$

Moreover

$$\text{Var}_{\theta_0}\{\varphi_0(X)\} = 0. \tag{3.3.11}$$

To assure that $\varphi_0(X)$ is unbiased we have to satisfy the condition

$$\int_{\theta}^{b(\theta)} \varphi_0(x) f(x)\, dx = g(\theta) \cdot F(\theta) \quad \text{for all} \quad -\infty < \theta < \infty. \tag{3.3.12}$$

Differentiating (3.3.12) with respect to θ, we obtain for almost all x

$$b'(x)\varphi_0(b(x))f(b(x)) - \varphi_0(x)f(x) = [g(x)F(x)]'. \tag{3.3.13}$$

Since $(g(x)F(x))'$ is determined almost everywhere, $b'(x) \geqslant 0$ exists for all x and the function $\varphi_0(x)$ is determined recursively, starting from the interval $(\theta_0, b(\theta_0))$ on which $\varphi_0(x) = g(\theta_0)$. For any point $\theta_0 < x < b(\theta_0)$, $b(x) \geqslant b(\theta_0)$, and if $b'(x) > 0$ we determine $\varphi_0(b(x))$ uniquely from (3.3.13). Similarly we determine the values of $\varphi_0(x)$ for values of $x < \theta_0$, starting from a value of x such that $\theta_0 \leqslant b(x) \leqslant b(\theta_0)$. In this manner the unbiased estimator $\varphi_0(X)$ is determined recursively for all x.

If a sample of n i.i.d. random variables $X_1, \ldots, X_n$ is given with a common density of the family (3.3.9), we determine the L.M.V.U. estimator by determining the conditional expectation of $\varphi_0(X_1)$, given the minimal sufficient statistic $(X_{(1)}, X_{(n)})$ where $X_{(1)}$ and $X_{(n)}$ denote, as previously, the sample minimum and the sample maximum. This conditional expectation is given by

(3.3.14)

$$\begin{aligned} \hat{\varphi}(X_{(1)}, X_{(n)}) &= E\{\varphi_0(X_1) \mid X_{(1)}, X_{(n)}\} \\ &= \frac{1}{n}\left[\varphi_0(X_{(n)}) + \varphi_0(X_{(1)})\right] + \frac{n-2}{n} \frac{\int_{X_{(1)}}^{X_{(n)}} \varphi_0(y) f(y)\, dy}{\int_{X_{(1)}}^{X_{(n)}} f(y)\, dy} \end{aligned}$$

Example 3.8. Let $X_1, X_2, \ldots, X_n$ be i.i.d. random variables having a location (selection) parameter density of the form

$$f(x; \theta) = \begin{cases} \dfrac{1}{\theta(\theta - 1)}, & \theta \leqslant x \leqslant \theta^2, \\ 0, & \text{otherwise,} \end{cases} \tag{3.3.15}$$

where $1 < \theta < \infty$. We wish to estimate $g(\theta) = e^{-\theta}$. According to the previous notation, $f(x) \equiv 1$, $b(x) = x^2$, and

$$F(\theta) = \theta(\theta - 1), \quad 1 < \theta < \infty. \tag{3.3.16}$$

Suppose that we have to construct an L.M.V.U. estimator with zero variance at $\theta_0 = 2$. We thus set

$$\varphi_0(X) \equiv e^{-2}, \quad 2 \leqslant X \leqslant 4.$$

According to (3.3 13) we have

$$(3.3.17) \qquad 2X\varphi_0(X^2) - \varphi_0(X) = -(X^2 - 3X + 1)e^{-X}.$$

Thus for all values of $4 \leqslant X \leqslant 16$ we obtain

$$(3.3.18) \quad \varphi_0(X) = \tfrac{1}{2}[e^{-2} - (X - 3X^{1/2} + 1)e^{-X^{1/2}}]X^{-1/2}, \qquad 4 < X \leqslant 16.$$

The function $\varphi_0(X)$ is discontinuous at $X = 4$ since $\varphi_0(4) = e^{-2}$ and $\lim_{X \to 4+} \varphi_0(X) = \frac{1}{2}e^{-2}$. From (3.3.18) we can obtain the values of $\varphi_0(X)$ for $16 < X \leqslant 256$, etc. For values of $1 \leqslant X < 2$ we have

$$(3.3.19) \quad \varphi_0(X) = 2Xe^{-2} + (X^2 - 3X + 1)e^{-X}, \qquad \sqrt{2} \leqslant X < 2.$$

The function $\varphi_0(X)$ is discontinuous at $X = 2$. Indeed, $\varphi_0(2) = e^{-2}$ but $\lim_{X \to 2-} \varphi_0(X) = 3e^{-2}$. In a similar manner we obtain from (3.3.13) and (3.3.19) the values of $\varphi_0(X)$ for $2^{1/4} \leqslant X < 2^{1/2}$, etc. ■

The estimator $\varphi_0(X)$ defined above in a piecewise manner is the L.M.V.U. estimator for a sample of size $n = 1$. If $n > 1$ the L.M.V.U. estimator is obtained from $\varphi_0(X)$ and (3.3.14). Morimoto and Sibuya extend the above result (see [1], Theorem 5) and exhibit a method of constructing L.M.V.U. estimators with zero variances over an infinite sequence of θ values. They provide and also modify examples of Sethuraman [1] and Takeuchi [1]. The modified example of Sethuraman, for the case of two observations, is the following.

Example 3.9. Let X_1, X_2 be i.i.d. random variables having a rectangular distribution over $(\theta, 2\theta)$, $0 < \theta < \infty$. We provide an unbiased estimator of θ which has zero variance for *all* $\theta_k = \theta_0 2^k$, $k = 0, \pm 1, \pm 2, \ldots$, where θ_0 is a finite prescribed constant.

Let $X_{(1)} \leqslant X_{(2)}$ be the order statistic. The L.M.V.U. estimator of θ having zero variance at all points of the sequence $\{\theta_0 2^k, k = 0, \pm 1, \pm 2, \ldots\}$ is

(3.3.20)

$$\varphi_0(X_{(1)}, X_{(2)}) = \begin{cases} \theta_0 2^k, & \text{if } \theta_0 2^k \leqslant X_{(1)} \leqslant X_{(2)} \leqslant \theta_0 2^{k+1}, \\ \frac{3}{2}X_{(1)} - \frac{1}{2}X_{(2)} + 3 \cdot 2^{k-1}\theta_0, & \text{if } \theta_0 2^k < X_{(1)} \leqslant 2^{k+1}\theta_0 \leqslant X_{(2)}. \end{cases}$$

Obviously, if $\theta = \theta_0 2^k$ $(k = 0, \pm 1, \pm 2, \ldots)$, $\theta_0 2^k \leqslant X_{(1)} \leqslant X_{(2)} \leqslant \theta_0 2^{k+1}$ and $\varphi_0(X_{(1)}, X_{(2)}) = \theta_0 2^k$, which has a zero variance. If $\theta_0 2^k < \theta < \theta_0 2^{k+1}$ then

$$\begin{aligned}(3.3.21) \quad E_\theta\{\varphi_0(X_{(1)}, X_{(2)})\} &= \theta_0 2^k P_\theta\{\theta \leqslant X_{(1)} \leqslant X_{(2)} \leqslant \theta_0 2^{k+1}\} \\ &+ \theta_0 2^{k+1} P_\theta\{\theta_0 2^{k+1} \leqslant X_{(1)} \leqslant X_{(2)} \leqslant 2\theta\} \\ &+ \int_{(\theta \leqslant x \leqslant 2^{k+1}\theta_0 \leqslant y \leqslant 2\theta)} [\tfrac{1}{2}x - \tfrac{1}{2}y + 3 \cdot 2^{k-1}\theta_0]\, dP_\theta[X_{(1)} \leqslant x, X_{(2)} \leqslant y].\end{aligned}$$

The joint density of $(X_{(1)}, X_{(2)})$ is

$$f_{X_{(1)},X_{(2)}}(x, y) = \begin{cases} \dfrac{2}{\theta^2}, & \theta \leqslant x \leqslant y \leqslant 2\theta, \\ 0, & \text{otherwise.} \end{cases} \tag{3.3.22}$$

Thus

$$\begin{aligned} P_\theta\{\theta \leqslant X_{(1)} \leqslant X_{(2)} \leqslant \theta_0 2^{k+1}\} \\ = \frac{2}{\theta^2} \int_\theta^{\theta_0 2^{k+1}} \mathrm{d}x \int_x^{\theta_0 2^{k+1}} \mathrm{d}y = 1 - \frac{\theta_0}{\theta} 2^{k+2} + \frac{\theta_0^{\,2}}{\theta^2} 2^{2k+2}. \end{aligned} \tag{3.3.23}$$

Similarly,

$$P_\theta\{\theta_0 2^{k+1} \leqslant X_{(1)} \leqslant X_{(2)} \leqslant 2\theta\} = 4 - \frac{\theta_0}{\theta} 2^{k+3} + \frac{\theta_0^{\,2}}{\theta} 2^{2k+2}. \tag{3.3.24}$$

Furthermore,

$$\begin{aligned} \int_{\{\theta \leqslant x \leqslant \theta_0 2^{k+1} \leqslant y \leqslant 2\theta\}} (\alpha x + \beta y + \gamma \cdot 2^{k-1}\theta_0) \, \mathrm{d}P[X_{(1)} \leqslant x, X_{(2)} \leqslant y] \\ = \alpha\left(\frac{\theta_0^{\,2}}{\theta} 2^{2k+3} - \frac{\theta_0^{\,3}}{\theta^2} 2^{3k+3} - 2\theta + \theta_0 2^{k+1}\right) \\ + \beta\left(\frac{\theta_0^{\,2}}{\theta} 2^{2k+2} - \frac{\theta_0^{\,3}}{\theta^2} 2^{3k+3} - 4\theta + \theta_0 2^{k+3}\right) \\ + \gamma\left(3 \frac{\theta_0^{\,2}}{\theta} \theta^{2k+2} - \frac{\theta_0^{\,3}}{\theta^2} 2^{3k+2} - \theta_0 2^{k+1}\right). \end{aligned} \tag{3.3.25}$$

Substituting (3.3.23), (3.3.24), and (3.3.25),

(3.3.26)

$$\begin{aligned} \theta_0 2^k P_\theta\{\theta \leqslant X_{(1)} \leqslant X_{(2)} \leqslant \theta_0 2^{k+1}\} + \theta_0 2^{k+1} P_\theta\{\theta_0 2^{k+1} \leqslant X_{(1)} \leqslant X_{(2)} \leqslant 2\theta\} \\ + \int_{\{\theta \leqslant x \leqslant 2^{k+1}\theta_0 \leqslant y \leqslant 2\theta\}} (\alpha x + \beta y + \gamma \cdot 2^{k-1}\theta_0) \, \mathrm{d}P_\theta[X_{(1)} \leqslant x, X_{(2)} \leqslant y] = \theta, \end{aligned}$$

we conclude that α, β, and γ should satisfy the equations

$$\begin{cases} \alpha + 2\beta = -\tfrac{1}{2}, \\ 2\alpha + 2\beta + \gamma = 3. \end{cases} \tag{3.3.27}$$

There are infinitely many possible solutions. If we set $\gamma = 3$ then we have the particular solution $\alpha = \frac{3}{2}$, $\beta = -\frac{1}{2}$, $\gamma = 3$. These coefficients appear in (3.3.20). Equation 3.3.27 show that there are infinitely many different unbiased estimators of θ having zero variances at $\theta = \theta_0 2^k$, $k = 0, \pm 1, \pm 2, \ldots$. ∎

The method we have presented for constructing an L.M.V.U. estimator having a zero variance at a certain point of the parameter space is restricted to selection models with densities of the type (3.3.9). If we do not restrict attention of this family of truncation parameter distributions we may face a case where no L.M.V.U. estimator with zero variance exists. The following example was given by Stein [3].

Example 3.10. Consider a family of density $\{f(x;\theta), \theta = 0, 1\}$ where

$$f(x;0) = \begin{cases} 1, & 0 < x < 1, \\ 0, & \text{otherwise,} \end{cases}$$

and

$$f(x;1) = \begin{cases} \tfrac{1}{2}x^{-1/2}, & 0 < x < 1, \\ 0, & \text{otherwise.} \end{cases}$$

We wish to estimate $g(\theta) = \theta$. The estimator, based on one observation,

$$\varphi(X) = \begin{cases} h, & \text{if } \tfrac{1}{2} - \delta \leqslant X \leqslant \tfrac{1}{2} \\ -h, & \text{if } \tfrac{1}{2} < X \leqslant \tfrac{1}{2} + \delta \end{cases}, \qquad 0 < \delta < \tfrac{1}{2},$$

is unbiased for a proper choice of h. Under $\theta = 0$, $E_0\{\varphi(X)\} = \delta h - \delta h = 0$ for all $0 < h < \infty$. We have to choose h so that under $\theta = 1$, $E_1\{\varphi(X)\} = 1$. We have

$$\text{(3.3.28)} \qquad \begin{aligned} E_1\{\varphi(X)\} &= \frac{h}{2}\int_{1/2-\delta}^{1/2} x^{-1/2}\,dx - \frac{h}{2}\int_{1/2}^{1/2+\delta} x^{-1/2}\,dx \\ &= -h[(\tfrac{1}{2} - \delta)^{1/2} + (\tfrac{1}{2} + \delta)^{1/2}] + h\sqrt{2}. \end{aligned}$$

Hence we have to choose

$$\text{(3.3.29)} \qquad h = [\sqrt{2} - [(\tfrac{1}{2} - \delta)^{1/2} + (\tfrac{1}{2} + \delta)^{1/2}]]^{-1}.$$

The variance of $\varphi(X)$ at $\theta = 0$ is

$$\text{(3.3.30)} \qquad \begin{aligned} \text{Var}_0\{\varphi(X)\} &= 2\delta h^2 \\ &= \frac{2\delta}{[\sqrt{2} - ((\tfrac{1}{2} - \delta)^{1/2} + (\tfrac{1}{2} + \delta)^{1/2})]^2}; \end{aligned}$$

$\text{Var}_0\{\varphi(X)\}$ can be made arbitrarily small, but it does not attain the value 0 and there is no L.M.V.U. estimator at $\theta = 0$. The only estimator with a zero variance at $\theta = 0$ is $g(X) \equiv 0$. But this estimator is obviously not unbiased. Stein [3] analyzed this situation and gave sufficient conditions for general families of densities to have an essentially unique L.M.V.U. estimator of a parametric function $g(\theta)$ at $\theta = \theta_0$. We reproduce Stein's theorem. ■

Given a probability space $(\mathfrak{X}, \mathcal{B}, F_\theta)$, $\theta \in \Theta$, we assume that $\{F_\theta\}$ is dominated by a sigma-finite measure μ on $(\mathfrak{X}, \mathcal{B})$. Let $f(x \mid \theta)$ designate the generalized density of F_θ with respect to μ. Define the probability measure

$$\nu(B) = \int_B f(x \mid \theta_0)\, d\mu(x), \qquad B \in \mathcal{B}. \tag{3.3.31}$$

The quantity $\nu(B)$ is the probability of $B \in \mathcal{B}$ under $\theta = \theta_0$, where θ_0 is a point of Θ at which we wish to minimize the variance of an unbiased estimator. Introduce the likelihood ratio function

$$R(x \mid \theta) = \frac{f(x \mid \theta)}{f(x \mid \theta_0)}, \qquad x \in \mathfrak{X}, \qquad \theta \in \Theta, \tag{3.3.32}$$

and furthermore define on $\Theta \times \Theta$

$$\begin{aligned} A(\theta_1, \theta_2) &= \int R(x \mid \theta_1) R(x \mid \theta_2)\, d\nu(x) \\ &= E_{\theta_2}\{R(X \mid \theta_1)\}. \end{aligned} \tag{3.3.33}$$

We assume that

$$A(\theta, \theta) = E_\theta\{R(X \mid \theta)\} < \infty, \quad \text{for all } \theta \in \Theta. \tag{3.3.34}$$

This implies (by using the Schwarz inequality) that $A(\theta_1, \theta_2) < \infty$ for all $\theta_1, \theta_2 \in \Theta$. If the condition that $A(\theta, \theta) < \infty$ for all $\theta \in \Theta$ is violated there may not exist an L.M.V.U. estimator of θ. This is the case in Example 3.10 where for $\theta_0 = 0$ we have $R(X \mid 0) \equiv 1$ and $R(X \mid 1) = \frac{1}{2}X^{-1/2}$. Hence $A(0, 0) = 1$ but $A(1, 1) = E_1\{\frac{1}{2}X^{-1/2}\} = \infty$. Indeed, there is no L.M.V.U. estimator at $\theta = \theta_0$.

The requirement that $A(\theta, \theta) < \infty$ for all $\theta \in \Theta$ is sufficient but not necessary. This could be demonstrated in the case of Example 3.7, where an essentially unique L.M.V.U. estimator of the common mean of two normal distributions exists but the condition $A(\theta, \theta) < \infty$ is not satisfied for all $\theta \in \Theta$.

Let L^2 denote the class of all measurable functions $\phi(x)$ on $\mathfrak{X}$ such that

$$\int \phi^2(x)\, d\nu(x) < \infty.$$

Furthermore, let $\mathcal{G}$ denote the class of all functions on Θ, expressible in the form

$$\psi(\theta) = \int \phi(x) R(x \mid \theta)\, d\nu(x), \qquad \theta \in \Theta.$$

According to (3.3.34), $A(\theta, \theta) \in \mathcal{G}$. A functional T on $\mathcal{G}$ is a function assigning to each function $\psi(\theta)$ in $\mathcal{G}$ a real value. As, for example, if (Θ, Γ) is a measure space and $\lambda(\theta)$ a probability measure on (Θ, Γ), then the expectation of $\psi(\theta)$ with respect to $\lambda(\theta)$ is a functional on $\mathcal{G}$.

Theorem 3.3.2. (Stein [3].) *If $R(x \mid \theta)$ is finite for all $\theta \in \Theta$ and almost all $x \in \mathfrak{X}$; if $A(\theta, \theta) < \infty$ for all $\theta \in \Theta$; and if there exists an unbiased estimator of $g(\theta)$, then there exists an L.M.V.U. estimator $\varphi_0(X)$ of $g(\theta)$ at $\theta = \theta_0$. Moreover, if* $\text{Var}_{\theta_0}\{\varphi_0(X)\} < \infty$ *then φ_0 is an essentially unique L.M.V.U. estimator at θ_0. Finally, a statistic φ on $(\mathfrak{X}, \mathfrak{B})$ is an L.M.V.U. estimator of $g(\theta)$ at $\theta = \theta_0$ if and only if there exists a real valued functional T on $\mathcal{G}$ for which*

(i) *$TA(\theta, \theta) = g(\theta_1)$ for each $\theta_1 \in \Theta$;*
(ii) *$T \int \phi(x)R(x \mid \theta)\, d\nu(x) = \int \phi(x)\varphi(x)\, d\nu(x)$ for all $\phi \in L^2$.*

The minimum variance of φ at $\theta = \theta_0$ is

$$\text{Var}_{\theta_0}\{\varphi(X)\} = Tg(\theta) - (g(\theta_0))^2. \tag{3.3.35}$$

The proof of Theorem 3.3.2 is omitted and the reader is referred to Stein [3] for a complete proof. We would rather discuss the theorem and its implications.

We first remark that (ii) is an extension of the necessary and sufficient condition of Theorem 3.3.1, which states that an unbiased estimator $\varphi(X)$ is L.M.V.U. at $\theta = \theta_0$ if and only if $\text{cov}_{\theta_0}\{\varphi(X), \phi(X)\} = 0$ for any unbiased estimator of zero, $\phi(X)$. Condition (ii) implies that if $E_0\{\varphi(X)\} = \psi(\theta)$ and $\varphi(X)$ is L M.V.U. at $\theta = \theta_0$ then

$$\text{cov}_{\theta_0}\{\phi(X), \varphi(X)\} = T\psi(\theta) - \psi(\theta_0)g(\theta_0).$$

Letting $\phi(X) = \varphi(X)$ a.s. we obtain $\text{Var}_{\theta_0}\{\varphi(X)\}$ as a special case of $\text{cov}_{\theta_0}\{\phi(X), \varphi(X)\}$.

Possible functionals T on the class $\mathcal{G}$ of $\psi(\theta)$ functions are, for example, the following:

1. The integral of $\psi(\theta)$ with respect to an additive set function, defined on the smallest Borel field on Θ, and such that for almost all x values

$$\int R(x \mid \theta)\, |d\lambda(\theta)| < \sum c_k R(x \mid \theta_k) < \infty,$$

for some finite collection of θ_k values in Θ and satisfying

$$\int A(\theta, \theta_1)\, d\lambda(\theta) = g(\theta_1) \quad \text{for all} \quad \theta_1 \in \Theta.$$

Such a set function could be, for example, a (prior) distribution of θ (considered as random variable).

$$T\psi(\theta) = \sum_{n=0}^{m} \left[\frac{\partial^n}{\partial \theta^n}\psi(\theta)\right]_{\theta=\theta_0}, \tag{3.3.36}$$

which is applicable if Θ is a set of real values and $\psi(\theta)$ is differentiable m times. The use of these operators is justified under conditions which can be derived as corollaries from Theorem 3.3.2. For the differential operator $T\psi(\theta)$ we have the following theorem.

Theorem 3.3.3 (Stein [3].) *Suppose that $R(x \mid \theta)$ is finite for all θ, almost all x, and $A(\theta, \theta) < \infty$ for all θ. Let Θ be a set of real parameters. Furthermore,*

(i) *for some m (either a positive integer or ∞), $R(x \mid \theta)$ is, for almost all x, differentiable m times with respect to θ at θ_0;*

(ii) *for each $n < m$ there exists a finite collection of $\theta_{n,k}$ values and positive constants $\tau_{n,k}$ so that*

$$\left|\frac{R^{(n)}(x \mid \theta_0 + \delta) - R^{(n)}(x \mid \theta_0)}{\delta}\right| \leqslant \sum_k c_{n,k} R(x \mid \theta_{n,k})$$

for δ sufficiently close to zero, and almost all x;

(iii) *there exist constants a_n such that*

$$g(\theta_1) = \sum_{n=0}^{m} a_n \left[\frac{\partial^n}{\partial \theta^n} A(\theta, \theta_1)\right]_{\theta=\theta_0} \quad \textit{for all} \quad \theta_1 \in \Theta;$$

(iv) *there exists a finite collection θ_k and positive constants τ_k such that, for almost all x,*

$$\sum_{n=0}^{m} \left| a_n \left[\frac{\delta^n}{\partial \theta^n} R(x \mid \theta)\right]_{\theta=\theta_0} \right| \leqslant \sum c_k R(x \mid \theta_k).$$

Then the L.M.V.U. estimator of θ at θ_0 is

$$\varphi^*(X) = \sum_{n=0}^{m} a_n \left[\frac{\partial^n}{\partial \theta^n} R(X \mid \theta)\right]_{\theta=\theta_0}, \tag{3.3.37}$$

with a minimum variance of

$$\operatorname{Var}_{\theta_0}\{\varphi^*(X)\} = \sum_{n=0}^{m} a_n \left[\frac{\partial^n}{\partial \theta^n} g(\theta)\right]_{\theta=\theta_0} - g^2(\theta_0). \tag{3.3.38}$$

This theorem is proven on the basis of the previous theorem. We must only show that the functional T defined by

$$T \int \phi(x) R(x \mid \theta)\, d\nu(x) = \sum_{n=0}^{m} a_n \left[\frac{\partial^n}{\partial \theta^n} \int \phi(x) R(x \mid \theta)\, d\nu(x)\right]_{\theta=\theta_0} \tag{3.3.39}$$

satisfies (i) and (ii) of Theorem 3.3.2. Condition (iv) permits differentiation on the right-hand side of (3.3.39) under the integral sign. With (iii) we imply that the L.M.V.U. estimator at $\theta = \theta_0$ is given by (3.3.37).

In the next section we present a theory in which translatable linear operators (functionals on $\mathcal{G}$) are applied to derive minimum variance unbiased estimators in the 1-parameter exponential case. The differential operator defined by (3.3.39) is used in Example 3.7 of Section 3.4 for estimating a polynomial $g(\theta) = b_0 + b_1\theta + \cdots + b_r\theta^r$ in the $\mathcal{N}(\theta, 1)$ case. The reader will find a very interesting application of the functional defined by (3.3.36) in Stein's paper [3]. We refer also to Barankin [1] for further study of L.M.V.U. estimators. In particular, Barankin attained interesting lower bounds on the possible variances of L.M.V.U. estimators. His results will be discussed later.

3.4. MINIMUM VARIANCE UNBIASED ESTIMATION IN THE CASE OF A 1-PARAMETER EXPONENTIAL FAMILY OF DENSITIES

In this section we present the theory of unbiased estimation for the exponential family, which has been developed by Kitagawa [1] and applied by Washio, Morimoto, and Ikeda [1]. Kitagawa's theory utilizes the special general form of the exponential type of densities in the context of the theory of bilateral Laplace transforms and the theory of bounded linear translatable operators to derive minimum variance unbiased estimators.

Let $X_1, X_2, \ldots, X_n$ be i.i.d. random variables having a 1-parameter exponential distribution with a density function of the general form

$$f(x; \theta) = h(\theta) \exp\{-\theta U(x) + W(x)\} \tag{3.4.1}$$

with respect to a sigma-finite measure μ on the real line. The parameter θ belongs to an open interval Θ on the real line. We confine attention in this exposition to cases of distribution functions which are absolutely continuous with respect to the Lebesgue linear measure μ on the real line. A treatment of cases with discrete random variables is given in Washio, Morimoto, and Ikeda's paper [1].

A basic assumption of the theory discussed in their paper is that the set of x values over which $\exp\{W(x)\} > 0$ is *independent* of the parameter θ. This restriction will not allow us to apply the theory to cases where the parameter is a location parameter of a truncation type, like the one in (2.1.4) for negative exponential density (substitute there $\beta = 1$ and $\alpha = \tau$). A theory treating these cases was given by Tate [1] and is presented in Section 3.5.

To sketch the first part of the main theorem, let U be a complete sufficient statistic for a 1-parameter exponential family having a density of the form (3.4.1). Let $\nu(u)$ be a sigma-finite measure defined by $d\nu(u) = \exp\{\psi(u)\}\, du$ where $\psi(u)$ is properly obtained from $W(x)$ by making the transformation $U(x) = u$. Let $F^u(u; \theta)$ designate the distribution function of the complete sufficient statistic U. It is easy to check that

$$\mathrm{d}F^u(u;\theta) = h(\theta)e^{-\theta u}\,\mathrm{d}\nu(u), \qquad \theta \in \Theta. \tag{3.4.2}$$

In other words, $h(\theta)e^{-\theta u}$ is the generalized density of U with respect to $\nu(u)$. Hence

$$h(\theta) = \left(\int_{-\infty}^{\infty} e^{-\theta u}\,\mathrm{d}\nu(u)\right)^{-1}, \qquad \theta \in \Theta. \tag{3.4.3}$$

Thus $h^{-1}(\theta)$ is the bilateral Laplace-Stieltjes transform of $\nu(u)$.

Let $g(\theta)$ be a real valued function mapping Θ into an open interval Ω. A statistic $\varphi(u)$ is an unbiased estimator of $g(\theta)$ if and only if

$$\int_{-\infty}^{\infty} \varphi(u)e^{-\theta u}\,\mathrm{d}\nu(u) = \frac{g(\theta)}{h(\theta)} \quad \text{for all} \quad \theta \in \Theta. \tag{3.4.4}$$

Let $\omega(u)\,\mathrm{d}u = \varphi(u)\,\mathrm{d}\nu(u)$. Under certain conditions, which are specified in the following theorem, $\omega(u)$ can be determined as the inverse bilateral Laplace-Stieltjes transform of $g(\theta)/h(\theta)$; $\varphi(U)$ is then determined by a simple conversion of the inverse transform. Such an approach was already exhibited in Example 3.6.

Theorem 3.4.1. *Let U be a complete sufficient statistic for a 1-parameter exponential family. Assume*

(i) $\int_{-\infty}^{\infty} ue^{-\theta u+\psi(u)}\,\mathrm{d}u$ *converges for all $\theta \in \Theta'$, where $\Theta \subset \Theta'$;*
(ii) $g(z)$, $z = \theta + it$, *is an analytic function in a strip σ whose real part is Θ;*
(iii) *the characteristic function of u, that is, $\phi_\theta(t) = \int_{-\infty}^{\infty} e^{itu}h(\theta)e^{-\theta u}\,\mathrm{d}\nu(u)$, satisfies*

$$\int_{-\infty}^{\infty} |\phi_\theta(-t)g(\theta + it)|\,\mathrm{d}t < \infty, \qquad \theta \in \Theta, \tag{3.4.5}$$

and

$$\lim_{|t|\to\infty} \phi_\theta(-t)g(\theta + it) = 0 \tag{3.4.6}$$

uniformly in every subinterval of Θ.

Then the essentially unique minimum risk (convex loss) unbiased estimator of $g(\theta)$ is

$$\varphi(U) = e^{-\psi(U)}\omega(U), \tag{3.4.7}$$

where

$$\omega(u) = \frac{1}{2\pi i}\int_{s-i\infty}^{s+i\infty} e^{zu}\frac{g(z)}{h(z)}\,\mathrm{d}z, \qquad s \in \Theta, \tag{3.4.8}$$

and u belongs to the set $\mathfrak{D}$ on which $\exp\{\psi(u)\} > 0$.

Proof. It is sufficient to establish the unbiasedness of $\varphi(U)$ given by (3.4.7). The rest follows from the Blackwell-Rao-Lehmann-Scheffé theorem. Condition (i) implies that $h^{-1}(\theta + it)$ is analytic over the strip σ. Thus we can write

$$h^{-1}(\theta + it) = \int_{\mathfrak{D}} \exp\{-(\theta + it)u + \psi(u)\}\, du = \frac{\phi_\theta(-t)}{h(\theta)}. \tag{3.4.9}$$

Hence

$$\frac{g(\theta + it)}{h(\theta + it)} = \frac{\phi_\theta(-t)g(\theta + it)}{h(\theta)}. \tag{3.4.10}$$

Conditions (ii) and (iii) imply, according to Widder [1], p. 265, that

$$g(\theta) = h(\theta)\int_{\mathfrak{D}} e^{-\theta u}\omega(u)\, du, \qquad \theta \in \Theta, \tag{3.4.11}$$

where $\omega(u)$ is the inverse Laplace transform of $g(z)/h(z)$, given by (3.4.8). Finally, from (3.4.7) and (3.4.11) we obtain

$$g(\theta) = h(\theta)\int_{\mathfrak{D}} \varphi(u) \exp\{-\theta u + \psi(u)\}\, du \tag{3.4.12}$$

for all $\theta \in \Theta$. This proves that $\varphi(U)$ is an unbiased estimator of $g(\theta)$. (Q.E.D.)

The next theorem (Kitagawa's) gives conditions for the use of bounded linear translatable operators in the derivation of U.M.V.U. estimators. This theorem yields Stein's theorems (Theorems 3.3.2 and 3.3.3) as particular cases. We introduce first the required notions from functional analysis.

Let L^p $(1 \leqslant p < \infty)$ denote the space of all functions $f(x)$ on the real line, satisfying

$$\int_{-\infty}^{\infty} |f(x)|^p\, dx < \infty. \tag{3.4.13}$$

Consider the norm $\|f\|_p$ on L^p as

$$\|f\|_p = \left(\int_{-\infty}^{\infty} |f(x)|^p\, dx\right)^{1/p}. \tag{3.4.14}$$

Let $\mathfrak{B}$ denote the Borel sigma-field on the real line and $\mathfrak{B}^*$ a system of *bounded* measurable sets of $\mathfrak{B}$ having a finite Lebesgue measure. A mapping σ of $\mathfrak{B}^*$ is such that if $M_1, M_2 \in \mathfrak{B}^*$ and $M_1 \subset M_2$, then $\sigma M_1 \subset \sigma M_2$, and vice versa. Furthermore, for any $M \in \mathfrak{B}^*$, there exists an $M_1 \in \mathfrak{B}^*$ such that $\sigma M \subset M_1$.

A *bounded linear translatable operator* Λ_u transforming $L^p \to L^p$ satisfies the following conditions:

1. It is an additive operator; that is, for any complex values τ_1 and τ_2.

$$\Lambda_u\{\tau_1 f + \tau_2 f\} = c_1\Lambda_u\{f\} + \tau_2\Lambda_u\{f\} \tag{3.4.15}$$

for almost all u, $-\infty < u < \infty$.

2. Denoting by T^α the translation $T^\alpha\{f(u)\} = f(u + \alpha)$, the operators T^α and Λ_u are *commutative*; that is,

$$T^\alpha\Lambda_u\{f\} = \Lambda_u\{T^\alpha f\} \quad \text{a.e.} \tag{3.4.16}$$

3. The operator Λ_u is *bounded* in the following sense: There exists a mapping σ^Λ of $\mathfrak{B}^*$, associated with Λ. To each $M \in \mathfrak{B}^*$ there is a positive (real) constant $\tau(M, \Lambda)$ such that, for all $f \in L^p$,

$$\|\Lambda_u\{f\}\|_{p,M} \leqslant \tau(M, \Lambda)\, \|f\|_{p,\sigma^\Lambda M}, \qquad M \in \mathfrak{B}^*, \tag{3.4.17}$$

where

$$\|f\|_{p,M} = \left(\int_M |f(x)|^p \, dx\right)^{1/p}.$$

Theorem 3.4.2. (Kitagawa [1].) *If there exists a bounded linear translatable operator* Λ_u *satisfying*

$$\Lambda_u\{e^{\theta u}\} = g(\theta)e^{\theta u}, \qquad \theta \in \Theta, \tag{3.4.18}$$

the existence of both members and the equality

$$\int_{-\infty}^{\infty} \Lambda_u\left\{\frac{e^{(\theta+it)u}}{h(\theta + it)}\right\} dt = \Lambda_u\left\{\int_{-\infty}^{\infty} \frac{e^{(\theta+it)u}}{h(\theta + it)}\, dt\right\} \tag{3.4.19}$$

are affirmed; then the uniformly minimum risk (*convex loss*) *unbiased estimator of* $g(\theta)$ *is*

$$\varphi(U) = e^{-\psi(U)}\Lambda_u\{e^{\psi(U)}\}. \tag{3.4.20}$$

Proof. According to (3.4.8) and (3.4.18),

$$\omega(u) = \frac{1}{2\pi i}\int_{s-i\infty}^{s+i\infty} \frac{\Lambda_u\{e^{zu}\}}{h(z)}\, dz, \qquad u \in \mathfrak{D}, \qquad s \in \Theta. \tag{3.4.21}$$

Furthermore, (3.4.19) and (3.4.21) imply that

$$\omega(u) = \Lambda_u\left\{\frac{1}{2\pi i}\int_{s-i\infty}^{s+i\infty} \frac{e^{zu}}{h(z)}\, dz\right\}, \qquad u \in \mathfrak{D}, \qquad s \in \Theta. \tag{3.4.22}$$

Finally, since $1/h(z)$ is the Laplace transform of $e^{\psi(u)}$, (3.4.22) implies that $\omega(u) = \Lambda_u\{e^{\psi(u)}\}$. Substituting this result in (3.4.7) we arrive at (3.4.20).

(Q.E.D.)

The following corollaries are implied from Theorem 3.4.1 and 3.4.2.

Corollary 3.4.1. *Let $\varphi_k(U)$, $k = 1, 2, \ldots$, be the essentially unique minimum risk estimator of $g_k(\theta)$. If the conditions of Theorems 3.3.1 and 3.3.2 are satisfied by all $g_k(\theta)$, and if*

(4.3.23) $$g(\theta + it) = \sum_{k=1}^{\infty} a_k g_k(\theta + it), \qquad \theta \in \Theta,$$

converges uniformly in t for every fixed $\theta \in \Theta$, where $\{a_k\}$ is a sequence of constants, then

(3.4.24) $$\varphi(U) = \sum_{k=1}^{\infty} a_k \varphi_k(U)$$

is the essentially unique minimum risk unbiased estimator of $g(\theta)$.

This result was used implicitly in Example 3.6.

Corollary 3.4.2. *Let U be a complete sufficient statistic having a density function $f(u; \theta)$ given by (3.4.12). For a given non-negative integer k assume that $f(u; \theta)$ possesses the first $(k + 1)$ derivatives with respect to u, all belonging to $L^1(-\infty, \infty)$. Then, if (iii) of Theorem 3.4.1 holds, the essentially unique minimum risk unbiased estimator of $g(\theta) = \theta^k$ is*

(3.4.25) $$\varphi(U) = \exp\{\psi(U)\}\left[\frac{\partial^k}{\partial u^k} e^{\psi(u)}\right]_{u=U}$$

The existence of the first $(k + 1)$ derivatives of $f(u; \theta)$ guarantees that (i) and (ii) of Theorem 3.4.1 are satisfied. Together with (iii) of that theorem we have the implication of the theorem. Furthermore, let $\Lambda_u\{f(u)\} = (\partial^k/\partial u^k)\{f(u)\}$ where Λ_u is a bounded translatable linear operator and (3.4.18) is obviously satisfied. Moreover, the requirements that all the first $(k + 1)$ derivatives of $f(u; \theta)$ will belong to $L^1(-\infty, \infty)$ implies that (3.4.19) is satisfied. This is sufficient for (3.4.25).

We now provide an example in which Corollary 3.4.2 can be applied. In this example we treat the problem of estimating ξ^k in the $N(\xi, 1)$ case, which has been discussed previously.

Example 3.7. Let $X_1, X_2, \ldots, X_n$ be i.i.d. random variables having an $\mathcal{N}(\xi, 1)$ distribution. The complete sufficient statistic is the sample mean $\bar{X}_n$. We seek a statistic $\varphi_k(\bar{X}_n)$ which is a U.M.V.U. estimator of $g(\xi) = \xi^k$,

$k \geqslant 1$. Now $\bar{X}_n \sim N(\xi, n^{-1})$. Thus the density function of $\bar{X}_n$ is

$$(3.4.26) \qquad f^{\bar{X}_n}(x; \xi) = \frac{\sqrt{n}}{\sqrt{(2\pi)}} \exp\left\{-\frac{n}{2}(x-\xi)^2\right\}$$
$$= h(\xi)e^{-\xi U + \psi(U)},$$

where $h(\xi) = (\sqrt{n}/\sqrt{(2\pi)}) \exp\{-(n/2)\xi^2\}$, $U = -n\bar{X}_n$, $\psi(U) = -(1/2n)U^2$. Thus according to Corollary 3.4.2, the U.M.V.U. estimator of ξ^k is

$$(3.4.27) \qquad \varphi(\bar{X}_n) = \exp\{-\psi(-n\bar{X}_n)\}\left[\frac{\partial^k}{\partial u^k} \exp\{\psi(u)\}\right]_{u=-n\bar{X}_n}$$
$$= \exp\left\{\frac{n}{2}\bar{X}_n^2\right\}\left[\frac{\partial^k}{\partial u^k} \exp\left\{-\frac{1}{2n}u^2\right\}\right]_{u=-n\bar{X}_n}.$$

It is easy to check that (3.4.27) yields $\varphi_1(\bar{X}_n) = \bar{X}_n$ for $k = 1$ and $\varphi_2(\bar{X}_n) = \bar{X}_n - 1/n$ for $k = 2$. Indeed $E\{\bar{X}_n\} = \xi$ and $E\{\bar{X}_n^2 - 1/n\} = \xi^2$. Generally, the k-th order derivative of $e^{-(1/2n)u^2}$ involves the Hermite polynomials, defined by

$$(3.4.28) \qquad H_k(x) = (-1)^k e^{\frac{1}{2}x^2} \frac{d^k}{dx^k}\{e^{-\frac{1}{2}x^2}\}.$$

The Hermite polynomials $H_k(x)$ are tabulated (see Cramér [2], p. 133). In terms of the Hermite polynomials we have

$$(3.4.29) \qquad \varphi_k(\bar{X}_n) = \left(-\frac{1}{\sqrt{n}}\right)^k H_k(-\sqrt{(n)}\,\bar{X}_n).$$

The expression of the U.M.V.U. estimator of ξ^k as a function of the Hermite polynomial $H_k(-\sqrt{(n)}\,\bar{X}_n)$ simplifies also the derivation of the variance of these estimators. Indeed, for any $k \geqslant l$,

(3.4.30)

$$\operatorname{cov}(\varphi_k(\bar{X}_n), \varphi_l(\bar{X}_n)) = \left(-\frac{1}{\sqrt{n}}\right)^{k+l} \operatorname{cov}(H_k(-\sqrt{(n)}\,\bar{X}_n), H_l(-\sqrt{(n)}\,\bar{X}_n)).$$

Denote again $U = -n\bar{X}_n$. The distribution of U is like that of $\mathcal{N}(-n\xi, n)$ and we have

$$(3.4.31) \qquad E_\xi\left\{H_k\left(\frac{U}{\sqrt{n}}\right)H_l\left(\frac{U}{\sqrt{n}}\right)\right\}$$
$$= \frac{1}{\sqrt{(2\pi n)}} \int_{-\infty}^{\infty} H_k\left(\frac{u}{\sqrt{n}}\right)H_l\left(\frac{u}{\sqrt{n}}\right) \exp\left\{-\frac{u^2}{2n} - \xi u - \frac{n}{2}\xi^2\right\} du.$$

Making the transformation $Z = (u/\sqrt{n}) + \sqrt{(n)}\,\xi$, we obtain

(3.4.32)

$$E_\xi\left\{H_k\left(\frac{U}{\sqrt{n}}\right)H_l\left(\frac{U}{\sqrt{n}}\right)\right\} = \frac{1}{\sqrt{(2\pi)}}\int_{-\infty}^{\infty} H_k(z - \sqrt{(n)}\,\xi)H_l(z - \sqrt{(n)}\,\xi)e^{-\frac{1}{2}z^2}\,dz.$$

Let $H_k^{(j)}(u) = (\partial^j/\partial u^j)H_k(u)$. Then it is simple to verify that

$$H_k^{(j)}(u) = k(k-1)\cdots(k-j+1)H_{k-j}(u) \tag{3.4.33}$$

and

$$\begin{aligned} H_k(z - \lambda) &= \sum_{j=0}^{k}\frac{1}{j!}(-\lambda)^j H_k^{(j)}(z) \\ &= \sum_{j=0}^{k}(-\lambda)^j\binom{k}{j}H_{k-j}(z). \end{aligned} \tag{3.4.34}$$

Furthermore, letting $\lambda = \sqrt{(n)}\,\xi$, we obtain

$$\begin{aligned} &E_\lambda\left\{H_k\left(\frac{U}{\sqrt{n}}\right)H_l\left(\frac{U}{\sqrt{n}}\right)\right\} \\ &= \sum_{j=0}^{k}\sum_{i=0}^{l}(-\lambda)^{j+i}\binom{k}{j}\binom{l}{i}\cdot\frac{1}{\sqrt{(2\pi)}}\int_{-\infty}^{\infty}H_{k-j}(u)H_{l-i}(u)e^{-\frac{1}{2}u^2}\,du. \end{aligned} \tag{3.4.35}$$

The Hermite polynomials have the following important orthogonality property:

$$\frac{1}{\sqrt{(2\pi)}}\int_{-\infty}^{\infty}H_{k-j}(u)H_{l-i}(u)e^{-\frac{1}{2}u^2}\,du = \begin{cases} 0, & k-j \neq l-i, \\ (k-j)!, & k-j = l-i. \end{cases} \tag{3.4.36}$$

Substituting (3.4.35) in (3.4.36),

$$E_\lambda\left\{H_k\left(\frac{U}{\sqrt{n}}\right)H_l\left(\frac{U}{\sqrt{n}}\right)\right\} = \sum_{i=0}^{l}(-\lambda)^{k-l+2i}\binom{k}{k-l+i}\binom{l}{i}(l-i)!. \tag{3.4.37}$$

Finally, for every $k \geqslant l$,

(3.4.38)

$$\begin{aligned} \operatorname{cov}(\varphi_k(\bar{X}_n), \varphi_l(\bar{X}_n)) &= \left(-\frac{1}{\sqrt{n}}\right)^{k+l}E_\xi\left\{H_k\left(\frac{U}{\sqrt{n}}\right)H_l\left(\frac{U}{\sqrt{n}}\right)\right\} - \xi^{k+l} \\ &= \frac{\xi^{k-l}}{n^l}\sum_{i=0}^{l}n^i\xi^{2i}\binom{k}{k-l+i}\binom{l}{i}(l-i)! - \xi^{k+l}. \end{aligned}$$

In particular, for $k = l$ we obtain the variance of $\varphi_l(\bar{X}_n)$, namely,

$$\operatorname{Var}\{\varphi_l(\bar{X}_n)\} = \frac{1}{n^l}\sum_{i=0}^{l}n^i\xi^{2i}\binom{l}{i}^2(l-i)! - \xi^{2l}. \tag{3.4.39}$$

The following is another example of interest for the normal case, in which we apply Theorem 3.4.1.

Example 3.8. As in Example 3.7, let $X_1, \ldots, X_n$ be i.i.d. like $N(\xi, 1)$. We wish to derive the U.M.V.U. estimator of the density function; that is,

$$(3.4.40) \quad g(x; \xi) = \frac{1}{\sqrt{(2\pi)}} \exp\{-\tfrac{1}{2}(x - \xi)^2\}, \qquad -\infty < x < \infty,$$

and from that estimator, say $\varphi(x; \bar{X}_n)$, to obtain the U.M.V.U. estimator of

$$(3.4.41) \quad P_k(\tau; \xi) = P_\xi[\max(X_1, \ldots, X_k) \leqslant \tau], \qquad -\infty < \tau < \infty.$$

We start with the estimation of $g(x; \xi)$. Applying Theorem 3.4.1, we show that the U.M.V.U. estimator of $g(x; \xi)$ is, whenever the sample size is $n \geqslant 2$,

(3.4.42)

$$\varphi(x; \bar{X}_n) = \left[2\pi\left(1 - \frac{1}{n}\right)\right]^{-\frac{1}{2}} \exp\left\{-\frac{1}{2}\frac{(x - \bar{X}_n)^2}{1 - 1/n}\right\}, \qquad -\infty < x < \infty.$$

As in Example 3.7, let $U = -n\bar{X}_n$; $U \sim N(-n\xi, n)$ and its density is given by $f^U(u, \xi) = h(\xi)\exp\{-\xi u + \psi(u)\}$, where $\psi(u) = -(u^2/2n)$ and $h(\xi) = (2\pi n)^{-\frac{1}{2}}\exp\{-(n/2)\xi^2\}$. Furthermore, the characteristic function of U is

$$(3.4.43) \qquad \phi_\xi^U(t) = \exp\{-in\xi t - \tfrac{1}{2}nt^2\}.$$

Therefore for a fixed $-\infty < x < \infty$,

(3.4.44)

$$\phi_\xi^U(-t)g(x; \xi + it) = g(x; \xi)\cdot\exp\{-\tfrac{1}{2}(n-1)t^2 + it[(n-1)\xi + x]\}.$$

From this we obtain that, for every $n \geqslant 2$ and a fixed $-\infty < x < \infty$,

$$(3.4.45) \quad \begin{aligned} &\lim_{|t|\to\infty} \phi_\xi^u(-t)g(x; \xi + it) = g(x; \xi); \\ &\lim_{|t|\to\infty} \exp\{-\tfrac{1}{2}(n-1)t^2 + it[(n-1)\xi + x]\} = 0. \end{aligned}$$

Moreover,

(3.4.46)

$$\begin{aligned} \int_{-\infty}^{\infty} |\phi_\xi^U(-t)g(x; \xi + it)|\,dt &= g(x; \xi)\int_{-\infty}^{\infty} \exp\{-\tfrac{1}{2}(n-1)t^2\}\,dt \\ &= g(x; \xi)\left(\frac{2\pi}{n-1}\right)^{\frac{1}{2}} < \infty. \end{aligned}$$

Both (3.4.45) and (3.4.46) do not exist when $n = 1$, and therefore the requirement that $n \geqslant 2$ is essential to satisfy (iii) of Theorem 3.4.1. The function $h(z) = \exp\{-(n/2)z^2\}(2\pi n)^{-\frac{1}{2}}$ is an entire function and obviously analytic.

Similarly, the function $g(x; z)$ is analytic for each fixed $-\infty < x < \infty$, which satisfies (ii) of the theorem. Finally, (i) is satisfied since the distribution of U possesses all moments. Thus Theorem 3.4.1 is applicable, and we have

$$\begin{aligned}
\text{(3.4.47)} \quad \omega(u) &= \frac{1}{2\pi i}\int_{-i\infty}^{i\infty} e^{zu} \exp\{-\tfrac{1}{2}(x - z)^2\}\sqrt{(2\pi n)} \exp\left\{\frac{n}{2} z^2\right\} dz \\
&= \left(\frac{n}{2\pi}\right)^{1/2} \int_{-\infty}^{\infty} \exp\left\{-\tfrac{1}{2}(x - it)^2 - \frac{n}{2} t^2 + itu\right\} dt \\
&= \frac{1}{\sqrt{(2\pi)}}\left(\frac{n}{n-1}\right)^{1/2} e^{-1/2 x} \cdot \exp\left\{-\frac{(x + u)^2}{2(n - 1)}\right\}.
\end{aligned}$$

Finally, substituting $-n\bar{X}_n$ for U and multiplying by $\exp\{u^2/2n\}$ we obtain (3.4.42).

From (3.4.42) we immediately obtain that

$$\text{(3.4.48)} \qquad \hat{P}_1(\tau; \bar{X}_n) = \int_{-\infty}^{\tau} \varphi(x; \bar{X}_n)\, dx, \qquad -\infty < \tau < \infty.$$

Indeed, using Fubini's theorem we have, for each $-\infty < \tau < \infty$,

$$\begin{aligned}
\text{(3.4.49)} \qquad E\{\hat{P}_1(\tau; \bar{X}_n)\} &= \int_{-\infty}^{\infty} dP[\bar{X}_n \leqslant t] \int_{-\infty}^{\tau} \varphi(x; t)\, dx \\
&= \int_{-\infty}^{\tau} dx \int_{-\infty}^{\infty} \varphi(x; t)\, dP[\bar{X}_n \leqslant t] \\
&= \int_{-\infty}^{\tau} g(x; \xi)\, dx = P[N(\xi, 1) \leqslant \tau].
\end{aligned}$$

Finally, substituting (3.4.42) and (3.4.47), we obtain the well-known result

(3.4.50)

$$\begin{aligned}
\hat{P}_1(\tau; \bar{X}_n) &= \frac{1}{\sqrt{(2\pi)}\,(1 - 1/n)^{1/2}} \int_{-\infty}^{\tau} \exp\left\{-\frac{1}{2(1 - 1/n)}(x - \bar{X}_n)\right\} dx \\
&= \Phi\left(\frac{\tau - \bar{X}_n}{(1 - 1/n)^{1/2}}\right).
\end{aligned}$$

The U.M.V.U. estimator $\hat{P}_1(\tau; \bar{X}_n)$ of $\Phi(\tau - \xi)$ can be also derived by the following consideration. Ellison [8] has shown that if $X \sim N(\xi, \sigma^2)$ and $\Phi(u)$ is the standard normal integral then

$$\text{(3.4.51)} \qquad E\{\Phi(X)\} = \Phi\left(\frac{\xi}{\sqrt{(1 + \sigma^2)}}\right).$$

Writing

$$W = \frac{\tau - \bar{X}_n}{(1 - 1/n)^{1/2}} \sim N\left(\frac{\tau - \xi}{(1 - 1/n)^{1/2}}, \frac{1}{n - 1}\right),$$

we have

$$(3.4.52)\quad E\left\{\Phi\left(\frac{\tau - \bar{X}_n}{(1 - 1/n)^{1/2}}\right)\right\} = \Phi\left(\frac{(\tau - \xi)/(1 - 1/n)^{1/2}}{(1 + 1/(n + 1))^{1/2}}\right) = \Phi(\tau - \xi).$$

We conclude the present example with the derivation of the variances of $\varphi(x; \bar{X}_n)$ and $\hat{P}_1(\tau; \bar{X}_n)$. Formally, for every fixed $-\infty < x < \infty$,

$$(3.4.53)\quad \text{Var}\,\{\varphi(x; \bar{X}_n)\}$$
$$= E\{\varphi^2(x; \bar{X}_n)\} - g^2(x; \xi)$$
$$= E\left\{\left[2\pi\left(1 - \frac{1}{n}\right)\right]^{-1} \exp\left[-\frac{1}{1 - 1/n}(x - \bar{X}_n)^2\right]\right\} - g^2(x; \xi).$$

To find this expression we notice that

$$(3.4.54)\quad E\left\{\left[2\pi\left(1 - \frac{1}{n}\right)\right]^{-1} \exp\left\{-\frac{1}{1 - 1/n}(x - \bar{X}_n)^2\right\}\right\}$$
$$= \left[2\left(\pi\left(1 - \frac{1}{n}\right)\right)^{1/2}\right]^{-1} E\left\{\frac{\sqrt{2}}{\sqrt{(2\pi(1 - 1/n))}} \exp\left\{-\frac{2}{2(1 - 1/n)}(x - \bar{X}_n)^2\right\}\right\}.$$

Furthermore, consider the following result from the theory of normal distributions; namely, if $Y \mid W \sim N(W, \sigma^2)$ and $W \sim N(\xi, \tau^2)$, then the marginal distribution of Y is like that of $N(\xi, \sigma^2 + \tau^2)$. The expression under the expectation sign in (3.4.54) is the conditional density function of a normal random variable $N(\bar{X}_n, (n - 1)/2n)$, given $\bar{X}_n$. Moreover, since $\bar{X}_n \sim N(\xi, n^{-1})$ and the expectation of a conditional density is a marginal density, we obtain

$$(3.4.55)\quad E\left\{\frac{\sqrt{2}}{\sqrt{(2\pi(1 - 1/n))}} \exp\left\{-\frac{1}{1 - 1/n}(x - \bar{X}_n)^2\right\}\right\}$$
$$= \frac{(n^2 - 1)^{1/2}}{(4\pi n)^{1/2}} \exp\left\{-\frac{n}{n^2 - 1}(x - \xi)^2\right\}.$$

Substituting (3.4.55) in (3.4.54) and (3.4.53) we obtain, for each x, $-\infty < x < \infty$,

$$(3.4.56)\quad \text{Var}\,\{\varphi(x; \bar{X}_n)\}$$
$$= \frac{1}{2\pi} \exp\{-(x - \xi)^2\}\left[\tfrac{1}{2}(n + 1)^{1/2} \exp\left\{\left(1 - \frac{n}{n^2 - 1}\right)(x - \xi)^2\right\} - 1\right].$$

The variance of $\hat{P}_1(\tau; \bar{X}_n)$ is found in the following manner:

$$(3.4.57)\quad \text{Var}\,\{\hat{P}_1(\tau; \bar{X}_n)\} = E\{\hat{P}_1{}^2(\tau; \bar{X}_n)\} - \Phi^2(\tau - \xi).$$

Extending Ellison's result (see Zacks and Even [8]) we have

(3.4.58)
$$E\{\hat{P}_1^2(\tau; \bar{X}_n)\} = E\left\{\Phi^2\left(\frac{\tau - \bar{X}_n}{(1 - 1/n)^{1/2}}\right)\right\}$$
$$= E\left\{P\left[U_1 \leqslant \frac{\tau - \bar{X}_n}{(1 - 1/n)^{1/2}}, U_2 \leqslant \frac{\tau - \bar{X}_n}{(1 - 1/n)^{1/2}} \,\Big|\, \bar{X}_n\right]\right\},$$

where U_1 and U_2 are i.i.d. standard normal variables, independent of $\bar{X}_n$. As previously, let $W = (\tau - \bar{X}_n)(1 - 1/n)^{-1/2}$. Since

$$W \sim N((\tau - \xi)(1 - 1/n)^{-1/2}, (n - 1)^{-1}),$$

the expectation in (3.4.58) is the joint probability

(3.4.59)
$$E_{\bar{X}_n}\{\hat{P}_1^2(\tau; \bar{X}_n)\} = P(Z_1 \leqslant 0, Z_2 \leqslant 0),$$

where

(3.4.60)
$$\begin{bmatrix} Z_1 \\ Z_2 \end{bmatrix} \sim N\left[\begin{pmatrix} (\xi - \tau)\left(1 - \dfrac{1}{n}\right)^{-1/2} \\ (\xi - \tau)\left(1 - \dfrac{1}{n}\right)^{-1/2} \end{pmatrix}, \begin{pmatrix} \dfrac{n}{n-1} & \dfrac{1}{n-1} \\ \dfrac{1}{n-1} & \dfrac{n}{n-1} \end{pmatrix}\right].$$

Hence

(3.4.61)
$$\operatorname{Var}\{\hat{P}_1(\tau; \bar{X}_n)\} = \Phi\left(\tau - \xi, \tau - \xi; \frac{1}{n}\right) - \Phi^2(\tau - \xi),$$

where $\Phi(a, b; \rho)$ is the bivariate standard normal integral at (a, b), with a correlation coefficient ρ. The values of $\Phi(a, b; \rho)$ can be read in the tables of the bivariate normal distribution function. ■

Washio, Morimoto, and Ikeda extend Theorems 4.3.1 and 4.3.2 and the derived corollaries to the case of 2-parameter exponential families. The extension is quite straightforward and will not be reproduced here. We proceed now to consider the use of integral transforms in cases of location and scale parameters, where the supporting set of the distribution may depend on the location parameter.

3.5. UNBIASED ESTIMATION OF LOCATION PARAMETERS

Statistical models with density functions depending on location parameters are classified into two classes: a class of *truncation* models, termed also *selection* models, and a class of *translation* models. In the present section we study the use of integral transforms and related techniques for the construction of unbiased estimators.

3.5.1. LOCATION PARAMETERS OF THE TRUNCATION (SELECTION) TYPE

Following Tate [1], we call a density function a *Type I truncation parameter density* if it is of the form

$$f(x;\theta) = k_1(\theta)h_1(x), \qquad a < \theta \leqslant x < b, \tag{3.5.1}$$

for some $-\infty \leqslant a < b \leqslant \infty$. A truncation parameter density is said to be of *Type II* if

$$f(x;\theta) = k_2(\theta)h_2(x), \qquad a < x \leqslant \theta \leqslant b, \tag{3.5.2}$$

where $-\infty \leqslant a < b \leqslant \infty$.

In connection with these types of density functions we imply from the assumption that (3.5.1) and (3.5.2) are density functions the following:

1. Both $h_1(x)$ and $h_2(x)$, being non-negative, are absolutely continuous over (θ, b) and (a, θ), respectively.
2.

$$\left.\begin{aligned} k_1(\theta) &= \left\{\int_\theta^b h_1(x)\,\mathrm{d}x\right\}^{-1} \\ k_2(\theta) &= \left\{\int_a^\theta h_2(x)\,\mathrm{d}x\right\}^{-1} \end{aligned}\right\}, \quad a < \theta < b. \tag{3.5.3}$$

3. Both $k_1(\theta)$ and $k_2(\theta)$ are everywhere differentiable.

Let $X_1, \ldots, X_n$ be i.i.d. random variables. If their common density function has a truncation parameter of Type I, then $X_{(1)} = \min\limits_{1\leqslant i\leqslant n} \{X_i\}$ is a complete sufficient statistic. If the common density is of Type II then the complete sufficient statistic is $X_{(n)} = \max\limits_{1\leqslant i\leqslant n} \{X_i\}$. The density function of $X_{(1)}$ in the Type I case is

$$f_\theta^{X_{(1)}}(x) = nk_1^{\,n}(\theta)h_1(x)\left\{\int_x^b h_1(t)\,\mathrm{d}t\right\}^{n-1}, \qquad \theta \leqslant x \leqslant b. \tag{3.5.4}$$

Suppose we wish to estimate the function $g(\theta)$ in the Type I case. We assume that $g(\theta)$ is absolutely continuous and estimable. We search first for an unbiased estimator of $g(\theta)$ based on one observation, say $\varphi(X_1)$, having finite variances for all $a < \theta < b$. Then by applying the Blackwell-Rao-Lehmann-Scheffé theorem, we obtain the essentially unique U.M.V.U.

estimator of $g(\theta)$ as

$$\hat{g}(X_{(1)}) = E\{\varphi(X_1) \mid X_{(1)}\} \tag{3.5.5}$$

$$= \frac{1}{n}\varphi(X_{(1)}) + \left(1 - \frac{1}{n}\right)\frac{\int_{X_{(1)}}^{b} \varphi(y)h_1(y)\,dy}{\int_{X_{(1)}}^{b} h_1(y)\,dy}.$$

To derive the unbiased estimator $\varphi(X_1)$ we have to find a statistic φ satisfying

$$k_1(\theta)\int_{\theta}^{b} \varphi(x)h_1(x)\,dx = g(\theta) \quad \text{for all} \quad a < \theta < b. \tag{3.5.6}$$

Differentiating (3.5.6) with respect to θ we obtain

$$k_1'(\theta)\int_{\theta}^{b} \varphi(x)h_1(x)\,dx - k_1(\theta)\varphi(\theta)h_1(\theta) = g'(\theta). \tag{3.5.7}$$

Moreover, from (3.5.6)

$$\int_{\theta}^{b} \varphi(x)h_1(x)\,dx = \frac{g(\theta)}{k_1(\theta)} \quad \text{for all} \quad a < \theta < b. \tag{3.5.8}$$

Hence substituting in (3.5.7) we obtain

$$\varphi(x) = \frac{1}{k_1^2(x)h_1(x)}[k_1'(x)g(x) - k_1(x)g'(x)] \tag{3.5.9}$$

for all $a < x < b$. Equation 3.5.9 is the formula of an unbiased estimator of $g(\theta)$ based on one observation.

Furthermore, from (3.5.3) and (3.5.6) we have

$$g(X_{(1)}) = \frac{\int_{X_{(1)}}^{b} \varphi(y)h_1(y)\,dy}{\int_{X_{(1)}}^{b} h_1(y)\,dy}. \tag{3.5.10}$$

Substituting (3.5.9) and (3.5.10) in (3.5.5) we obtain Theorem 3.5.1.

Theorem 3.5.1. (Tate.) *Let $X_1, \ldots, X_n$ be i.i.d. random variables with a truncation density of Type I over some finite interval (a, b). Let $g(\theta)$ be an absolutely continuous function over (a, b). Then the essentially unique minimum risk (convex loss) unbiased estimator of $g(\theta)$ is*

$$\hat{g}(X_{(1)}) = g(X_{(1)}) - \frac{g'(X_{(1)})}{nk_1(X_{(1)})h_1(X_{(1)})}. \tag{3.5.11}$$

The following is a simple example in which Theorem 3.5.1 is applied.

Example 3.9. (i) Let $X_1, \ldots, X_n$ be i.i.d. random variables having a common rectangular distribution $\mathcal{R}(-\theta, 0)$ where $0 < \theta < a < \infty$. In this case

$h_1(x) \equiv 1$ and $k_1(\theta) = \theta^{-1}$. If we wish to estimate θ, we have $g(\theta) = \theta$ and according to (3.5.11), the U.M.V.U. estimator of θ is

$$\hat{\theta}(X_{(1)}) = X_{(1)}\left(1 - \frac{1}{n}\right). \tag{3.5.12}$$

This result is analogous to Case 1 of Example 3.3. Formula 3.5.12 is easy to obtain by straightforward derivation from the distribution of $X_{(1)}$ in the $\mathcal{R}(-\theta, 0)$ case. However, if $g(\theta)$ is a more complicated function a straightforward approach will not be easy, whereas (3.5.11) gives the result immediately. For example, in case $g(\theta) = e^{-\theta}\theta^{1/2}$ we obtain from (3.5.11) that the U.M.V.U. estimator of $g(\theta)$ is

$$\hat{g}(X_{(1)}) = e^{X_{(1)}}|X_{(1)}|^{1/2}\left\{1 + \frac{|X_{(1)}|}{n}\left(1 - \frac{1}{2\,|X_{(1)}|}\right)\right\}. \tag{3.5.13}$$

(ii) Let $X_1, \ldots, X_n$ be i.i.d. random variables having a truncated Pareto distribution with a density function

$$f_\theta(x) = \frac{(1/\theta)(\theta/x)^2}{1 - (\theta/b)}, \qquad 0 < \theta < x < b. \tag{3.5.14}$$

Here

$$k_1(\theta) = \frac{\theta}{1 - (\theta/b)}, \tag{3.5.15}$$
$$h_1(x) = x^{-2}.$$

Suppose we wish to estimate $g(\theta) = \theta$. Theorem 3.5.1 yields that the U.M.V.U estimator of θ is

$$\hat{\theta}(X_{(1)}) = X_{(1)} - \frac{1}{n}X_{(1)}\left(1 - \frac{X_{(1)}}{b}\right). \tag{3.5.16}$$ ■

In many statistical models of truncation parameters, the interval of possible θ values is unbounded. Theorem 3.5.1 holds for the case of $b = \infty$ provided $(k'(x)g(x) - k_1(x)g'(x))/h_1(x)$ is integrable over any finite subinterval of (θ, ∞). This condition permits the derivation of (3.5.9) as outlined before. Furthermore, if $\hat{g}_b(X_{(1)})$ designates the unbiased estimator for the case of a finite interval (a, b), and if $\hat{g}_b(X_{(1)})$ is bounded by an integrable function $G(X_{(1)})$ having a second moment, then the Lebesgue dominated convergence theorem implies that

$$\hat{g}_\infty(X_{(1)}) = \lim_{b\to\infty} \hat{g}_b(X_{(1)}). \tag{3.5.17}$$

is an unbiased estimator of $g(\theta)$. Indeed,

$$E\{\hat{g}_b(X_{(1)})\} = n\int_\theta^b \hat{g}_b(y)h_1(y)k_{1,b}^n(\theta)\left\{\int_y^b h_1(x)\,dx\right\}^{n-1} dy. \tag{3.5.18}$$

Since $\hat{g}_b(X_{(1)})$ is an unbiased estimator for the case of $a < \theta < b$ we have

$$(3.5.19)\qquad n\int_\theta^b \hat{g}_b(y)h_1(y)\left\{\int_y^b h_1(x)\,dx\right\}^{n-1} = k_{1,b}^{-n}(\theta)\,g(\theta)$$

where $k_{1,b}^{-n}(\theta) = (\int_\theta^b h_1(x)\,dx)^n$. Letting $b \to \infty$, we have

$$(3.5.20)\qquad \lim_{b\to\infty} k_{1,b}^{-n}(\theta)g(\theta) = \frac{g(\theta)}{k_{1,\infty}^n(\theta)}.$$

Moreover, $(\int_x^\infty h_1(y)\,dy)^n \leqslant (\int_\theta^\infty h_1(y)\,dy)^n = k_{1,\infty}^{-n}(\theta)$ for all $\theta \leqslant x < \infty$. Hence

$$(3.5.21)\qquad k_{1,\infty}^n(\theta)\left\{\int_x^\infty h_1(y)\,dy\right\}^n \leqslant 1, \qquad \theta \leqslant x < \infty.$$

Applying the Lebesgue dominated convergence theorem,

$$(3.5.22)\qquad g(\theta) = \lim_{b\to\infty}\left[nk_{1,b}^n(\theta)\int_\theta^b \hat{g}_b(y)h_1(y)\left\{\int_y^b h_1(x)\,dx\right\}^{n-1} dy\right]$$
$$= \int_\theta^\infty \left(\lim_{b\to\infty}\hat{g}_b(y)\right)nk_{1,\infty}^n(\theta)h_1(y)\left(\int_y^\infty h_1(x)\,dx\right)^{n-1} dy.$$

This proves the unbiasedness of $\hat{g}_\infty(X_{(1)})$. Using this result we can verify, for example, that $X_{(1)}(1 - 1/n)$ is the U.M.V.U. estimator of the truncation parameter θ in the case $f_\theta(x) = \theta/x^2$ when $0 < \theta \leqslant x < \infty$.

A theorem similar to 3.5.1 can be formulated for the case of a truncation parameter density function of Type II.

Theorem 3.5.2. *If $X_1, \ldots, X_n$ are i.i.d. random variables having a density function of Type II over a finite interval (a, b), and if $g(\theta)$ is absolutely continuous over (a, b), then the essentially unique minimum risk (convex loss) unbiased estimator of $\xi(\theta)$ is*

$$(3.5.23)\qquad \hat{g}(X_{(n)}) = g(X_{(n)}) + \frac{g'(X_{(n)})}{nk_2(X_{(n)})h_2(X_{(n)})}.$$

Theorem 3.5.2 can be extended to the case of $a = -\infty$ in a fashion similar to that above. We conclude the discussion of unbiased estimation of truncation parameters with a statement concerning the estimation of the distribution function $g(z; \theta) = P[X \leqslant z \mid \theta]$. According to Theorems 3.5.1 and 3.5.2, let $X_1, \ldots, X_n$ be i.i.d. random variables having a density function of Type I or Type II. Since $g(z; \theta)$ is absolutely continuous over (a, b), and since an unbiased estimator of $g(z; \theta)$ based on one observation is the indicator function $I_{(a,z)}(X)$, we conclude that the essentially unique uniformly minimum risk unbiased estimator of $g(z;\theta)$ is

$$(3.5.24)\quad \hat{g}(z;X_{(1)}) = \begin{cases} 0, & a < z \leqslant X_{(1)} < b, \\ \dfrac{1}{n} + \left(1 - \dfrac{1}{n}\right)\dfrac{\int_{X_{(1)}}^{z} h_1(y)\,dy}{\int_{X_{(1)}}^{b} h_1(y)\,dy}, & a < X_{(1)} < z \leqslant b, \end{cases}$$

for densities of Type I and

$$(3.5.25)\quad \hat{g}(z;X_{(n)}) = \begin{cases} \left(1 - \dfrac{1}{n}\right)\dfrac{\int_{a}^{z} h_2(y)\,dy}{\int_{a}^{X_{(n)}} h_2(y)\,dy}, & a < z \leqslant X_{(n)} < b, \\ 1, & a < X_{(n)} < z < b. \end{cases}$$

for densities of Type II. Thus in the case where $X_1, \ldots, X_n$ are i.i.d. like $\mathcal{R}(0,\theta)$, $0 < \theta < b$, we have

$$(3.5.26)\qquad g(z;\theta) = P[X \leqslant z \mid \theta] = \begin{cases} 0, & z \leqslant 0, \\ \dfrac{z}{\theta}, & 0 < z \leqslant \theta, \\ 1, & z > 1. \end{cases}$$

According to (3.2.25) the U.M.V.U. estimator of $g(z;\theta)$ is

$$(3.5.27)\qquad \hat{g}(z;X_{(n)}) = \begin{cases} 0, & z \leqslant 0, \\ \left(1 - \dfrac{1}{n}\right)\dfrac{z}{X_{(n)}}, & 0 < z \leqslant X_{(n)}, \\ 1, & X_{(n)} < z. \end{cases}$$

3.5.2. LOCATION PARAMETER OF THE TRANSLATION TYPE

It is reasonable to consider in the case of location parameters of the translation type unbiased estimators having the *translation* property. A statistic $U(X_1, \ldots, X_n)$ is said to have the translation property if, for all $-\infty < \tau < \infty$,

$$(3.5.28)\qquad U(X_1 + c, \ldots, X_n + c) = c + U(X_1, \ldots, X_n).$$

We restrict our attention in this section to such unbiased estimators.

If $X_1, \ldots, X_n$ are i.i.d. having a density function $f_\theta(x) = f(x - \theta)$, and if $U(X_1, \ldots, X_n)$ is a statistic having the translation property (3.5.28), then the density of U is of the type $g_\theta{}^U(u) = g(u - \theta)$. Thus for an arbitrary function $\xi(\theta)$ we consider the equation

$$(3.5.29)\qquad \int_{-\infty}^{\infty} \psi(u) g(u - \theta)\,du = \xi(\theta).$$

The integral transform of $\psi(u)$ in (3.5.29) is of the convolution type, and if bilateral Laplace transforms of $\psi(u)$ and $g(z)$ exist, we have the well-known result that the bilateral Laplace transform of a convolution $\psi * g$ is the product of the transforms; that is,

(3.5.30)

$$\int_{-\infty}^{\infty} e^{-s\theta}\left\{\int_{-\infty}^{\infty} \psi(-u)g(-u+\theta)\,du\right\} d\theta = \int_{-\infty}^{\infty} e^{-su}\psi(-u)\,du \cdot \int_{-\infty}^{\infty} e^{-sz}g(z)\,dz.$$

Thus, replacing u by $-u$ and θ by $-\theta$ in equation (3.5.29) and using relationship (3.5.30) we arrive at

$$\mathcal{B}\{\psi(-u); s\}\mathcal{B}\{g(u); s\} = \mathcal{B}\{\xi(-u); s\}, \tag{3.5.31}$$

where $\mathcal{B}\{f(x); s\}$ designates the bilateral Laplace transform of $f(x)$ at s; that is,

$$\mathcal{B}\{f(x); s\} = \int_{-\infty}^{\infty} e^{-sx}f(x)\,dx. \tag{3.5.32}$$

Let $\mathcal{B}^{-1}\{\psi(s); x\}$ designate the inverse of $\mathcal{B}\{f(x); s\} = \psi(s)$, which is determined uniquely almost everywhere (up to a set of a Lebesgue measure zero). We have proven the following theorem.

Theorem 3.5.3. (Tate.) *Let $X_1, X_2, \ldots, X_n$ be i.i.d. random variables having a translation parameter density $f(x - \theta)$, $\theta \in \Theta$. Let $U(X_1, \ldots, X_n)$ be a statistic having the translation property and $\xi(\theta)$ a function having a bilateral Laplace transform. If there exists an unbiased estimator $\varphi(U)$ of $\xi(\theta)$, having a bilateral Laplace transform, it is determined uniquely a.e. by*

$$\varphi(U) = \mathcal{B}^{-1}\left\{\frac{\mathcal{B}\{\xi(-x); s\}}{\mathcal{B}\{g(x); s\}}; -U\right\}, \tag{3.5.33}$$

where $g(u - \theta)$ is the density function of U.

In the following examples we have been able to use The Bateman Manuscript Project [1], tables of integral transforms to determine the inverse transforms of certain functions. This is the case in many problems. We very rarely have to find the inverse integral transform analytically. We also remark in this connection that the unbiased estimators constructed according to (3.5.33) are not necessarily U.M.V.U. ones.

Suppose that we wish to estimate the distribution function of X, that is, $F(x; \theta) = P[X \leqslant x \mid \theta]$. We show that there exists an unbiased estimator of $F(x; \theta)$ based on a statistic U having the translation property. We notice that for every $-\infty < x < \infty$,

$$\begin{aligned}(3.5.34)\qquad \mathcal{B}\{F(x;-y);s\} &= \int_{-\infty}^{\infty} e^{-sy}F(x;-y)\,dy \\ &= \int_{-\infty}^{\infty} e^{-sy}\left\{\int_{-\infty}^{x} f(y+z)\,dz\right\}dy \\ &= \int_{-\infty}^{\infty} f(\omega)\left\{\int_{\omega-x}^{\infty} e^{-sy}\,dy\right\}d\omega \\ &= \frac{e^{sx}}{s}\,\mathcal{B}\{f(y);s\},\end{aligned}$$

in which interchange of integrals is affirmed by Fubini's theorem. Substituting (3.5.34) in (3.5.33) we obtain that an unbiased estimator of $F(x;\theta)$ is given by

$$(3.5.35)\qquad \hat{F}(x;U) = \mathcal{B}^{-1}\left\{\frac{e^{sx}\mathcal{B}\{f(y);s\}}{s\mathcal{B}\{g(y);s\}};-U\right\}.$$

Finally, noticing that for each $-\infty < x < \infty$

$$(3.5.36)\qquad \frac{e^{sx}}{s} = \mathcal{B}\{I_{(-x,\infty)}(y);s\},$$

where $I_A(y)$ is the indicator function of the set A; and that the bilateral Laplace transform of the convolution $f*g$ is the product of the respective transforms, we obtain

$$\begin{aligned}(3.5.37)\qquad \hat{F}(x;U) &= \left[\mathcal{B}^{-1}\left\{\frac{\mathcal{B}\{f(y);s\}}{\mathcal{B}\{g(y);s\}};U\right\} * I_{(-x,\infty)}(U)\right]_{u=-U} \\ &= \int_{-\infty}^{\infty} \mathcal{B}^{-1}\left\{\frac{\mathcal{B}\{f(y);s\}}{\mathcal{B}\{g(y);s\}};u\right\} I_{(-\infty,x-U)}(u)\,du \\ &= \int_{-\infty}^{x-U} \mathcal{B}^{-1}\left\{\frac{\mathcal{B}\{f(y);s\}}{\mathcal{B}\{g(y);s\}};u\right\}du.\end{aligned}$$

Example 3.10. In the present example we derive by means of (3.5.37) the U.M.V.U. estimator of the distribution function in the normal case $N(\xi, 1)$, $-\infty < \xi < \infty$. Let $\varphi(x) = (2\pi)^{-1/2}e^{-1/2x^2}$, $-\infty < x < \infty$. The density function of $X \sim N(\xi, 1)$ is $\varphi(x-\xi)$. We take for U the complete sufficient statistic, that is, $U = \bar{X}_n$. The density function of U is then $g(u-\xi)$, where $g(u) = n^{1/2}(2\pi)^{-1/2}e^{-nu^2/2}$. The bilateral Laplace transform of the function $n(x;\sigma) = (2\pi\sigma^2)^{-1/2}\exp\{-(1/2\sigma^2)x^2\}$ is

$$(3.5.38)\qquad \mathcal{B}\{n(x;\sigma);s\} = \exp\left\{\frac{s^2}{2}\cdot\sigma^2\right\},\qquad -\infty < s < \infty.$$

Thus

$$\frac{\mathcal{B}\{\varphi(x); s\}}{\mathcal{B}\{g(x); s\}} = \exp\left\{\frac{s^2}{2}\left(1 - \frac{1}{n}\right)\right\}, \qquad -\infty < s < \infty. \tag{3.5.39}$$

Since the bilateral Laplace transform has a unique inverse almost everywhere we obtain from (3.5.38) that

$$\begin{aligned}\mathcal{B}^{-1}\left\{\frac{\mathcal{B}\{\varphi(x); s\}}{\mathcal{B}\{g(x); s\}}; u\right\} &= \mathcal{B}^{-1}\left\{\exp\left\{\frac{s^2}{2}\left(1 - \frac{1}{n}\right)\right\}; u\right\} \\ &= \left(\left[2\pi\left(1 - \frac{1}{n}\right)\right]^{1/2}\right)^{-1} \exp\left\{-\left[2\left(1 - \frac{1}{n}\right)\right]^{-1} u^2\right\}, \qquad -\infty < u < \infty.\end{aligned} \tag{3.5.40}$$

Finally, we obtain from (3.5.37) that the U.M.V.U. estimator of $F(x; \xi) = \Phi(x - \xi)$ is $\Phi((x - \bar{X}_n)/((1 - 1/n)^{1/2}))$. This result was previously derived in Example 3.8. ∎

We now consider cases in which the translation parameter can be estimated unbiasedly by a function of the minimal value in the sample $X_{(1)}$. We first remark that if $U(X_1, \ldots, X_n)$ is a statistic having the translation property (3.5.28) with a density function $g(u - \theta)$, $-\infty < \theta < \infty$, and if $\eta_0 = E_0\{U(X_1, \ldots, X_n)\}$ designates the expected value of $U(X_1, \ldots, X_n)$ with respect to the density $g(u)$ at $\theta = 0$, then

$$\phi(U) = U - \eta_0 \tag{3.5.41}$$

is an unbiased estimator of θ, provided η_0 exists. Indeed, the expectation of $U(X_1, \ldots, X_n)$ under any $-\infty < \theta < \infty$ is

$$\begin{aligned}E_\theta\{U(X_1, \ldots, X_n)\} &= \int_{-\infty}^{\infty} u g(u - \theta)\, du \\ &= \theta + \int_{-\infty}^{\infty} z g(z)\, dz = \theta + \eta_0.\end{aligned} \tag{3.5.42}$$

Thus the existence of η_0 implies the existence of $E_\theta\{U\}$ for all $-\infty < \theta < \infty$. Moreover, since U has the translation property, $U(X_1, \ldots, X_n) - \eta_0 = U(Y_1, \ldots, Y_n)$, where $Y_i = X_i - \eta_0$ $(i = 1, \ldots, n)$. The density function of Y_i is $f(y + \eta_0 - \theta)$, and that of $U^* = U(Y_1, \ldots, Y_n)$ is $g(\mu + \eta_0 - \theta)$, $-\infty < \theta < \infty$. Finally,

$$\begin{aligned}E_\theta\{\phi(U)\} &= \int_{-\infty}^{\infty} u g(u + \eta_0 - \theta)\, du \\ &= \theta - \eta_0 + \int_{-\infty}^{\infty} u g(u)\, du \\ &= \theta, \quad \text{for all} \quad -\infty < \theta < \infty.\end{aligned} \tag{3.5.43}$$

The minimal value in a sample $X_{(1)}$ is a statistic having the translation property (3.5.28). If $f(x - \theta)$ is the density of X then the density of $X_{(1)}$ in a sample of n is

$$g_{\theta}^{X(1)}(u; \theta, n) = nf(x - \theta)[1 - F(x - \theta)]^{n-1}, \tag{3.5.44}$$

which can be written as $g(u - \theta)$ where $F(x)$ is the distribution function of X under $\theta = 0$. The expected value of $X_{(1)}$ under $\theta = 0$ is

$$E_0\{X_{(1)}\} = n\int_{-\infty}^{\infty} xf(x)[1 - F(x)]^{n-1}\, dx. \tag{3.5.45}$$

Generally, the crux of the problem is to prove that $E_0\{X_{(1)}\}$ exists and find its value. In certain particular cases, and especially if the density function $f(x)$ is positive only on a set bounded below, the task may be easier. For example, in the negative exponential case, that is,

$$f(x - \theta) = \begin{cases} e^{-(x-\theta)}, & x \geqslant \theta, \\ 0, & \text{otherwise}, \end{cases}$$

$-\infty < \theta < \infty$, we have seen that in a sample of n i.i.d. random variables, $X_{(1)} \sim \theta + G(n, 1)$.† Hence $E\{X_{(1)}\} = \theta + 1/n$. A general theorem given by Tate [39] concerning the existence of $E\{X_{(1)}^k\}$, $k = 1, 2, \ldots$, in a sample of size n, when the density function $f(x)$ is positive only for non-negative values of x is the following:

If $f(x) = 0$ for $x < 0$, and

$$f(x) = 0\left(\frac{1}{x^{\alpha+1+k/n}}\right) \tag{3.5.46}$$

as $x \to \infty$ for some $\alpha > 0$ (k, n are positive integers) *then $E\{X_{(1)}^k\}$ exists where $X_{(1)}$ is the minimal value in a sample of n observations.*

Example 3.11. (i) Let $X_1, \ldots, X_n$ be i.i.d. having a density function

$$f(x) = \exp\{-(e^x - (1 + x))\}, \qquad 0 < x < \infty. \tag{3.5.47}$$

The corresponding distribution function is

$$F(x) = \begin{cases} 1 - \exp\{1 - e^x\}, & 0 < x < \infty, \\ 0, & x \leqslant 0. \end{cases}$$

This distribution function is known as the first asymptotic distribution of the smallest value (see Sarhan and Greenberg [1], p. 65). In this case we obtain

† The quantity $G(\lambda, p)$ designates an r.v. having a $\mathcal{G}(\lambda, p)$ distribution law.

from (3.5.45) by integration in parts

$$E\{X_{(1)}\} = \int_0^\infty [1 - F(x)]^n \, dx = \int_0^\infty \exp\{n - ne^x\} \, dx. \tag{3.5.48}$$

Letting $y = e^x - 1$ we obtain

$$E\{X_{(1)}\} = \int_0^\infty (1 + y)^{-1} e^{-ny} \, dy. \tag{3.5.49}$$

The exponential integral $Ei(-\alpha)$ is defined as

$$Ei(-\alpha) = -\int_\alpha^\infty \frac{1}{z} e^{-z} \, dz. \tag{3.5.50}$$

Tables of the exponential integral are given by Jahnke and Emde [1]. Making a change of variables $z = 1 + y$, we obtain that for every $n \geqslant 1$ $E\{X_{(1)}\} = -e^n Ei(-n)$. Finally, from (3.5.41) and (3.5.50) we imply that an unbiased estimator of θ, when $X_1, \ldots, X_n$ are i.i.d. having a density function $f(x - \theta)$, $-\infty < \theta < \infty$, where $f(x)$ is (3.5.47), is

$$\hat{\theta}_n(X_{(1)}) = X_{(1)} + nEi(-n). \tag{3.5.51}$$

The joint density function of $X_1, \ldots, X_n$ is

$$f(x_1, \ldots, x_n; \theta) = \exp\left\{-\frac{1}{e^\theta}\sum_{i=1}^n e^{x_i} - n(\theta - 1) + \sum_{i=1}^n x_i\right\}. \tag{3.5.52}$$

Hence by the Neyman-Fisher factorization theorem, $S_n = \sum_{i=1}^n e^{X_i}$ is a sufficient statistic. Moreover, as shown in Section 3.2, $S_n = \sum_{i=1}^n e^{X_i}$ is a minimal sufficient statistic. Thus a U.M.V.U. estimator should be a function of S_n. The quantity $\hat{\theta}_n(X_{(1)})$ is not a function of S_n, and hence is not a U.M.V.U. estimator. The problem is that $S_n = \sum_{i=1}^n e^{X_i}$ does not have the translation property (3.5.28) and therefore the U.M.V.U. estimator of θ does not have the translation property. We do not expect $\hat{\theta}_n(X_{(1)})$ to be a minimum variance unbiased estimator even among the estimators based on statistics having the translation property, since the sample mean $\bar{X}_n$ has the translation property and is a more informative estimator concerning θ than $X_{(1)}$. (Compare $\bar{X}_n$ and $X_{(1)}$ to the minimal sufficient statistic S_n.) ∎

3.6. UNBIASED ESTIMATION OF SCALE PARAMETERS

A density function $f(x; \rho)$ of a random variable X is called a *scale-parameter density* if

$$f(x; \rho) = \rho f(\rho x), \quad \text{for all } -\infty < x < \infty, \tag{3.6.1}$$

where $f(x)$ is a density function and $0 < \rho < \infty$. For example, the density function of a gamma random variable $G(\lambda, p)$ depends on a scale parameter $0 < \lambda < \infty$; that of a normal random variable $N(\mu, \sigma^2)$ depends on a scale parameter $0 < \sigma < \infty$. Following Tate [1], we consider in this section unbiased estimators of functions of scale parameters, which depend on the sample values $\{X_1, \ldots, X_n\}$ through a homogeneous statistic $Y = H_\alpha(X_1, \ldots, X_n)$ of a known degree $\alpha \neq 0$. An example of such a statistic is $H_\alpha(X_1, \ldots, X_n) = \sum_{i=1}^n X_i^\alpha$. We now show that the density function of such a homogeneous statistic Y is of the general form

$$g(y; \rho) = \rho^\alpha g(\rho^\alpha y), \qquad -\infty < y < \infty. \tag{3.6.2}$$

Indeed, according to (3.6.1) the density function of ρX is $f(x)$. Hence if $Y(\rho) = H_\alpha(\rho X_1, \ldots, \rho X_n)$ then the density function of $Y(\rho)$ is $g(y)$. Furthermore, since $H_\alpha(X_1, \ldots, X_n)$ is a homogeneous statistic of degree α, $Y = \rho^{-\alpha} Y(\rho)$. Hence the density function of Y under ρ, $0 < \rho < \infty$, is (3.6.2).

Let $\xi(\rho)$ be a given function of ρ and $\varphi(Y)$ an unbiased estimator of $\xi(\rho)$. The following equation must be satisfied:

$$\int_{-\infty}^{\infty} \varphi(y)\rho^\alpha g(\rho^\alpha y)\, dy = \xi(\rho), \quad \text{for all} \quad 0 < \rho < \infty. \tag{3.6.3}$$

Lemma 3.6.1. *If $X_1, \ldots, X_n$ are i.i.d. random variables, having a scale parameter density $\rho f(\rho x)$, and if $Y = H_\alpha(X_1, \ldots, X_n)$ is a homogeneous statistic of rank $\alpha \neq 0$, then an unbiased estimator of $\xi(\rho) = \rho^r$ is*

$$\varphi(Y) = \frac{Y^{-r/\alpha}}{E_1\{Y^{-r/\alpha}\}}, \tag{3.6.4}$$

for all values of r and $\alpha \neq 0$ for which $E_1\{Y^{-r/\alpha}\}$ exists.

Proof. We have to show that

$$E_\rho\{\varphi(Y)\} = \frac{E_\rho\{Y^{-r/\alpha}\}}{E_1\{Y^{-r/\alpha}\}} = \rho^r, \quad \text{for all} \quad 0 < \rho < \infty. \tag{3.6.5}$$

Assuming that $E_1\{Y^{-r/\alpha}\}$ exists, we can immediately imply that $E_\rho\{Y^{-r/\alpha}\}$ exists for every $0 < \rho < \infty$. Indeed, change the variable y in the integral expression for $E_1\{Y^{-r/\alpha}\}$ to $z = \rho^{-\alpha} y$. Finally,

$$\frac{E_\rho\{Y^{-r/\alpha}\}}{E_1\{Y^{-r/\alpha}\}} = \frac{\int_{-\infty}^{\infty} y^{-r/\alpha}\rho^\alpha g(\rho^\alpha y)\, dy}{\int_{-\infty}^{\infty} y^{-r/2} g(y)\, dy} = \rho^r. \tag{3.6.6}$$

(Q.E.D.)

We remark that $\varphi(Y)$ given by (3.6.4) is not necessarily a uniformly minimum risk estimator, unless $Y = H_\alpha(X_1, \ldots, X_n)$ is a complete sufficient statistic.

Example 3.12. The statistic $Y = \sum_{i=1}^{n} X_i^{\alpha}$, $\alpha \geqslant 1$, is a complete sufficient statistic for the case where $X_1, \ldots, X_n$ are i.i.d. having a Weibull density

$$f_\rho^X(x) = \alpha \rho^\alpha x^{\alpha-1} \exp\{-\rho^\alpha x^\alpha\}, \qquad 0 < x < \infty, \tag{3.6.7}$$

$0 < \rho < \infty$; α is known. Letting $\theta = \rho^\alpha$ and $Z = X^\alpha$ it is easy to check that the density of Z is that of $G(\theta, 1)$. Hence $Y = \sum_{i=1}^{n} X_i^\alpha \sim G(\theta, n)$, and whenever $n > r/\alpha$, $\varphi(Y) = Y^{-r/\alpha}/\Gamma(n - r/\alpha)$ is the U.M.V.U. estimator of ρ^r. ∎

The Mellin transform of a function $f(x)$, if exists, is defined as

$$\mathcal{M}[f(x); s] = \int_0^\infty x^{s-1} f(x)\, dx, \qquad s_0 < \mathcal{Re}\{s\} < s_1. \tag{3.6.8}$$

The inverse transform will be designated by $\mathcal{M}^{-1}[\varphi(s); x]$.

Theorem 3.6.1. (Tate.) *Let $X_1, \ldots, X_n$ be i.i.d. random variables having a scale-parameter density $\rho f(\rho x)$, $0 < \rho < \infty$. Let $Y = H_\alpha(X_1, \ldots, X_n)$ be a non-negative, homogeneous statistic with density $\rho^\alpha g(\rho^\alpha x)$, $\alpha \neq 0$. Assume that both $g(x)$ and $\xi(\rho)$ have Mellin transforms. If there exists an unbiased estimator $\phi(Y)$ of $\xi(\rho)$, for which the Mellin transform exists, then $\phi(Y)$ is determined uniquely a.e. by*

$$\phi(Y) = \frac{1}{Y} \mathcal{M}^{-1}\left\{\frac{\mathcal{M}[(1/x)\xi(x^{1/\alpha}); s]}{\mathcal{M}[g(x); s]}; \frac{1}{Y}\right\}. \tag{3.6.9}$$

Proof. The starting point is (3.6.3). Changing ρ into $\rho^{1/\alpha}$ and making the transformation $z = 1/y$, we obtain

$$\int_0^\infty \frac{1}{z} \varphi\left(\frac{1}{z}\right) g\left(\frac{\rho}{z}\right) \cdot \frac{dz}{z} = \frac{1}{\rho} \xi(\rho^{1/\alpha}). \tag{3.6.10}$$

It is easy to verify that the Mellin transform of the left-hand side of (3.6.10) is the product $\mathcal{M}\{(1/x)\varphi(1/x); s\}\mathcal{M}\{g(x); s\}$. Thus the hypotheses of the theorem imply that

$$\mathcal{M}\left\{\frac{1}{x}\varphi\left(\frac{1}{x}\right); s\right\} = \frac{\mathcal{M}\{(1/x)\xi(x^{1/\alpha}); s\}}{\mathcal{M}\{g(x); s\}}, \qquad s_0 < \mathcal{Re}\{s\} < s. \tag{3.6.11}$$

Inverting (3.6.11) and substituting $x = 1/Y$ we obtain the unbiased estimator $\phi(Y)$. (Q.E.D.)

Suppose X is a *non-negative* r.v. having a scale-parameter density $\rho f(\rho x)$, $0 \leqslant x \leqslant \infty$, $0 < \rho < \infty$. We consider the problem of unbiased estimation of the density function; that is, we set $\xi(\rho; z) = \rho f(\rho z)$ and assume that

$\rho f(\rho z)$ possesses a Mellin transform. We then have

$$\text{(3.6.12)} \qquad \mathcal{M}\left\{\frac{1}{\rho}\,\xi(\rho^{1/\alpha}; z); s\right\} = \frac{\alpha}{z^{1+\alpha(s-1)}}\,\mathcal{M}\{f(x); \alpha(s-1)+1\}.$$

Substituting (3.6.12) in (3.6.9) we obtain that an unbiased estimator of $\rho f(\rho z)$, $-\infty < z < \infty$, based on a non-negative homogeneous statistic of degree $\alpha \neq 0$, is,

$$\text{(3.6.13)} \qquad \hat{f}(z; Y) = \frac{\alpha}{Y}\,\mathcal{M}^{-1}\left\{\frac{\mathcal{M}\{f(x); \alpha(s-1)+1\}}{z^{\alpha(s-1)+1}\mathcal{M}\{g(x); s\}}; \frac{1}{Y}\right\}.$$

Using this result we can easily obtain an unbiased estimator of the distribution function of X; namely, $F(x; \rho) = P_\rho[X \leqslant x]$. This is obtained by integrating $\hat{f}(z; Y)$ from $-\infty$ to x. In the following example we show the application of the theory presented here for estimating functions of the parameters of scale and location in the case of 2-parameter negative exponential densities. Some of the results were derived by Epstein and Sobel [1].

Example 3.13. Let $X_1, \ldots, X_n$ be i.i.d. random variables having a 2-parameter negative exponential density $f_{\theta,\rho}(x) = \rho \exp\{-\rho(x-\theta)\}$, $\theta \leqslant x \leqslant \infty$. Let $X_{(1)} \leqslant X_{(2)} \leqslant \cdots \leqslant X_{(n)}$ be the order statistic. As shown in Chapter 2, Example 2.3, the pair $(X_{(1)}, T^*_{n-1})$, where $T^*_{n-1} = \sum_{i=2}^n X_{(i)}$, constitutes a minimal sufficient statistic. We can, equivalently, consider functions of $(X_{(1)}, \bar{X}_n - X_{(1)})$. As shown in Example 2.3, $X_{(1)}$ and $Y = \bar{X}_n - X_{(1)}$ are independent; $X_{(1)} \sim \theta + G(n\rho, 1)$ and $Y = \bar{X}_n - X_{(1)} \sim G(n\rho, n-1)$. Thus the joint density function of $X_{(1)}$ and Y is

$$\text{(3.6.14)} \qquad f_{\theta,\rho}(x, y) = \frac{(n\rho)^n}{(n-2)!}\,y^{n-2}\exp\{-n\rho(x+y-\theta)\},$$
$$\theta \leqslant x < \infty, \qquad 0 \leqslant y \leqslant \infty.$$

Let $\xi(\theta, \rho)$ be a given estimable function. If $\phi(X_{(1)}, Y)$ is an unbiased estimator of $\xi(\theta, \rho)$ it should satisfy the equation

(3.6.15)
$$\int_\theta^\infty \int_0^\infty \phi(x, y)y^{n-2}\exp\{-n\rho(x+y)\}\,dx\,dy = \frac{(n-2)!\,\xi(\theta, \rho)}{(n\rho)^n}\,e^{-n\rho\theta},$$

for all $-\infty < \theta < \infty$, and all $0 < \rho < \infty$. We assume the following:

(i) $\xi(\theta, \rho)$ has a partial derivative with respect to θ;
(ii) the unbiased estimator $\phi(X_{(1)}, Y)$ is continuous in $X_{(1)}$ almost everywhere and has a Laplace transform with respect to Y.

Then writing

$$(3.6.16)\quad \int_{\theta}^{\infty}\int_{0}^{\infty} \phi(x, y)y^{n-2} \exp\{-n\rho(x + y)\}\, dx\, dy = \int_{\theta}^{\infty} e^{-n\rho x}\, dx \mathfrak{L}\{\phi(x, y)y^{n-2}; n\rho\}$$

and differentiating (3.6.15) partially with respect to θ, we obtain the equation

$$(3.6.17)\quad \mathfrak{L}\{\phi(\theta, y)y^{n-2}; n\rho\} = \frac{(n-2)!}{(n\rho)^{n-1}}\left(\xi(\theta, \rho) - \frac{1}{n\rho}\cdot\frac{\partial}{\partial\theta}\xi(\theta, \rho)\right).$$

Hence the essentially unique, uniformly minimum risk (convex loss) unbiased estimator of $\xi(\theta, \rho)$ is given by

$$(3.6.18)\quad \phi(X_{(1)}, Y) = \frac{(n-2)!}{Y^{n-2}}\mathfrak{L}^{-1}\left\{\frac{\xi(\theta, s/n)}{s^{n-1}} - \frac{(\partial/\partial\theta)\xi(\theta, s/n)}{s^n}; Y\right\}_{\theta = X_{(1)}}.$$

Formula (3.6.18) yields the following unbiased estimators:

$\xi(\theta, \rho)$	$\phi(X_{(1)}, Y)$
$\theta^r,\ r \geqslant 0$	$X_{(1)}^r - \frac{r}{n-1} YX_{(1)}^{r-1}$
$\rho^r,\ r < n-1$	$\frac{(nY)^r\Gamma(n-1)}{\Gamma(n-r-1)}$
$\theta + \frac{1}{\rho}\ln\left[\frac{1}{1-p}\right]$ (p-th fractile, $0 < p < 1$)	$X_{(1)} - \frac{Y}{n-1}[1 + n\ln(1-p)]$
$P_{\theta,\rho}[X \leqslant Z] = (1 - e^{-\rho(Z-\theta)})$,	$\begin{cases} 0, & Z < X_{(1)} \\ 1 - \left(1 - \frac{1}{n}\right)\left(1 - \frac{Z - X_{(1)}}{nY}\right)^{n-2}, & \frac{Z - X_{(1)}}{n} < Y \\ 1, & Y < \frac{Z - X_{(1)}}{n} \end{cases}$

■

3.7. DISTRIBUTION FREE UNBIASED ESTIMATION

In this section we consider the problem of a distribution free unbiased estimation. In all the examples of the previous sections we assumed that the sample $X_1, \ldots, X_n$ of n i.i.d. random variables is generated according to a specified distribution law, such as normal, gamma, binomial, Poisson, etc. In other words, the statistical model specified the functional form of the distribution function of the observed random variable up to the value of its parameters. In this section we consider the problem of unbiased estimation when the functional form of the distribution function is not specified. We assume, as before, that $X_1, \ldots, X_n$ are i.i.d. real valued random variables and that the class of distribution functions on $(\mathfrak{X}, \mathcal{B})$, say $\mathcal{F}$, contains all the absolutely continuous distribution functions, or all the discrete distribution functions, but we do not specify their functional form. We wish to estimate unbiasedly a certain function of the distribution function F. For example, if we assume that $E\,|X| < \infty$, we may wish to estimate on the basis of the observed values of $X_1, \ldots, X_n$ the expectation of X, say $\xi = E\{X\}$. It is well known that the sample mean $\bar{X}_n = (1/n) \sum_{i=1}^{n} X_i$ is an unbiased estimator of ξ. Is $\bar{X}_n$ the essentially unique U.M.V.U. estimator of ξ? Suppose that $\text{Var}\,\{X\} \equiv \sigma^2 < \infty$. Is the sample variance

$$s^2 = (n-1)^{-1} \sum_{i=1}^{n} (X_i - \bar{X}_n)^2$$

an essentially unique U.M.V.U. estimator of σ^2? These are examples of distribution free estimation problems. A basic theory of distribution free unbiased estimation was given by Halmos [1] in 1946. In this section we follow the exposition of Fraser [5], pp. 135–147. The reader is advised to see also Lehmann [1].

There is an eminent symmetry in the statistical model, since the observations on $X_1, \ldots, X_n$ are independent and *identically* distributed. There is no significance to the order of the observation within the sample. We thus wish to estimate a function $g(F)$ in such a manner that the value of the estimate $\varphi(X_1, \ldots, X_n)$ will be invariant under permutations of $(X_1, \ldots, X_n)$. As we show later on, unbiased estimators which are invariant under permutations of $(X_1, \ldots, X_n)$ are U.M.V.U.

Given a sample $x_1, \ldots, x_n$ of i.i.d. random variables having a common distribution function F of a class Ψ, let $\mathfrak{X}^{(n)}$ denote the sample space of $(X_1, \ldots, X_n)$ and $\mathcal{B}^{(n)}$ the Borel sigma-field on $\mathfrak{X}^{(n)}$. We introduce the *order statistic* on $\mathfrak{X}^{(n)}$,

$$t(x_1, \ldots, x_n) = (x_{(1)}, \ldots, x_{(n)}), \tag{3.7.1}$$

where $x_{(1)} \leqslant x_{(2)} \leqslant \cdots \leqslant x_{(n)}$. The order statistic $t(\mathbf{X}_n)$ partitions $\mathfrak{X}^{(n)}$ into cosets of $\mathbf{X}_n$ values such that if $\mathbf{X}_n^0 = (X_1^0, \ldots, X_n^0)$ is a point of $\mathfrak{X}^{(n)}$, then the coset of $\mathbf{X}_n^0$ is the set of all permutations of the components of $\mathbf{X}_n^0$. Every coset in $\mathfrak{X}^{(n)}$ contains *at most* $n!$ points. The largest number of points in a coset is obtained whenever all the components of its elements are different.

A function $f(X_1, \ldots, X_n)$ is called *symmetric* if it is invariant under permutations of $(X_1, \ldots, X_n)$. A symmetric function may assume different values on different cosets of the order statistics but does not change its value on each of these cosets. We now show that the order statistic is a sufficient statistic for the family of all distribution functions of n i.i.d. random variables.

Let A be any Borel set in $\mathcal{B}^{(n)}$, and let B be any symmetric set; that is,

$$B = \{\mathbf{X}_n;\ \pi(\mathbf{X}_n) \in B \quad \text{for all permutations } \pi\} \tag{3.7.2}$$

where $\pi(\mathbf{X}_n)$ denotes a permutation of $\mathbf{X}_n$. Consider the probability of $A \cap B$ under $F \in \Psi$. We have, for all $F \in \Psi$,

$$P_F\{A \cap B\} = \int_B I_A(x_1, \ldots, x_n) \prod_{i=1}^{n} \mathrm{d}F(x_i), \tag{3.7.3}$$

where $I_A(x_1, \ldots, x_n)$ is the indicator function of A. Since $\prod_{i=1}^n \mathrm{d}F(x_i) = \prod_{j=1}^n \mathrm{d}F(x_{i_j})$ for every permutation $(x_{i_1}, \ldots, x_{i_n})$ of $(x_1, \ldots, x_n)$, and since B is a symmetric set, we obtain

$$P_F(A \cap B) = \int_B I_A(x_{i_1}, \ldots, x_{i_n}) \prod_{j=1}^{n} \mathrm{d}F(x_{i_j}), \tag{3.7.4}$$

for all permutations $(x_{i_1}, \ldots, x_{i_n})$ of X_n, and all $F \in \Psi$. Therefore

$$P_F(A \cap B) = \int_B \left[\sum_{\{(i_1, \ldots, i_n);\, 1 \leqslant i_1 < \cdots < i_n \leqslant n\}} \frac{I_A(x_{i_1}, \ldots, x_{i_n})}{n!} \right] \prod_{j=1}^{n} \mathrm{d}F(x_j) \tag{3.7.5}$$

for all $F \in \Psi$. Since $I_A(x_{i_1}, \ldots, x_{i_n})$ is the indicator function of $(x_{i_1}, \ldots, x_{i_n})$, the function

$$m(A; \{x_1, \ldots, x_n\}) = \sum_{\{(i_1, \ldots, i_n)\}} I_A(x_{i_1}, \ldots, x_{i_n}) \tag{3.7.6}$$

yields the number of different permutations of $(x_1, \ldots, x_n)$ which belong to A. Finally, from (3.7.5) $m(A; \{x_1, \ldots, x_n\})/n!$ is the conditional probability of the Borel set $A \in \mathcal{B}^{(n)}$, given $t(x_1, \ldots, x_n)$. This conditional probability is independent of F in Ψ, and hence *the order statistic $t(X_1, \ldots, X_n)$ is a sufficient statistic for the class Ψ of all distribution functions of n i.i.d. random variables.*

In Section 3.2 we have proven the completeness of the order statistic. We now utilize the completeness and sufficiency of the order statistic for the derivation of essentially unique U.M.V.U. estimators in the distribution free case.

Let $g(F)$ be a function on Ψ to an open subset Ω of a Euclidean k-space. That is, $g(F)$ is either real or vector valued. As before, $g(F)$ is called *estimable* if there exists a statistic φ on $(\mathfrak{X}^{(n)}, \mathcal{B}^{(n)}, P)$ such that

$$\int_{\mathfrak{X}^{(n)}} \varphi(x_1, \ldots, x_n)\, \mathrm{d}F(x_1, \ldots, x_n) = g(F) \quad \text{for all} \quad F \in \Psi. \tag{3.7.7}$$

The *degree* $m, m \geqslant 1$, of an estimable function $g(F)$ is the least sample size for which there exists an unbiased estimator of $g(F)$. For example, suppose that Ψ is the family of all distribution functions having a second moment, and suppose that

$$\int_{\mathfrak{X}} x \, dF_0(x) \neq 0 \quad \text{for some } F_0 \text{ in } \Psi.$$

Let $g_1(F) = E_F\{X\}$ and $g_2(F) = \sigma^2(F)$ be the expectation and the variance of X. The function $g_1(F)$ is estimable of degree $m = 1$; $g_2(F)$ is estimable of degree $m = 2$. One cannot estimate $\sigma^2(F)$ unbiasedly by one observation only. At least two i.i.d. observations are required. [For estimating $\sigma^2(F)$ the i.i.d. requirement can be replaced by a weaker one, such as, identically distributed and uncorrelated.]

A *kernel* for an estimable function of degree m is an unbiased estimator, which is a function of the least possible sample size. Returning to the previous example, consider the function $g_2(F) \equiv \sigma^2(F)$ and let $X_1, \ldots, X_n$, $n \geqslant m$, be a sample of i.i.d. random variables. The degree of $g_2(F)$ is $m = 2$. Kernels for $g_2(F)$ are, for example,

$$f(X_i, X_j) = X_i^2 - X_i X_j, \qquad i = 1, \ldots, n; i \neq i. \tag{3.7.8}$$

These kernels are not symmetric since $f(X_i, X_j) \neq f(X_j, X_i)$. However, there always exist symmetric kernels. In the present example, a symmetric kernel will be

$$\begin{aligned} f^*(X_i, X_j) &= \tfrac{1}{2}[f(X_i, X_j) + f(X_j, X_i)] \\ &= \tfrac{1}{2}(X_i - X_j)^2; \quad i = 1, \ldots, n; j \neq i. \end{aligned} \tag{3.7.9}$$

More generally, let $g(F)$ be an estimable function of degree m; let $X_1, \ldots, X_n$ be a sample of $n \geqslant m$ i.i.d. random variables; and let $f(X_{i_1}, \ldots, X_{i_n})$ be a kernel for $g(F)$. Then

$$f_s(X_{i_1}, \ldots, X_{i_m}) = \frac{1}{m!} \sum_{\text{per.}} f(X_{i_1}, \ldots, X_{i_m}) \tag{3.7.10}$$

is a *symmetric kernel*, where the summation is over all the $m!$ permutations of the components of $(X_{i_1}, \ldots, X_{i_m})$.

Based on the sample of n observations $(n \geqslant m)$, the symmetric kernels $f_s(X_{i_1}, \ldots, X_{i_m})$ may assume at most $\binom{n}{m}$ different values, which depend on the particular choice of m variates $(X_{i_1}, \ldots, X_{i_m})$ out of $(X_1, \ldots, X_n)$. A *U-statistic* corresponding to a given symmetric kernel, for an estimable function of degree m is the average of all the possible values of $f_s(X_{i_1}, \ldots, X_{i_m})$, namely,

$$U(X_1, \ldots, X_n) = \frac{1}{\binom{n}{m}} \sum_C f_s(X_{i_1}, \ldots, X_{i_m}), \tag{3.7.11}$$

where,

$$C = \{(i_1, \ldots, i_m);\ 1 \leqslant i_1 < \cdots < i_m \leqslant n\} \tag{3.7.12}$$

is the set of all $\binom{n}{m}$ different m-tuples of indices chosen from $\{1, 2, \ldots, n\}$.

Returning to the example of estimating the variance $\sigma^2(F)$ given a sample of n i.i.d. random variables $X_1, \ldots, X_n$, we have seen that a symmetric kernel for $\sigma^2(F)$ is $f_s(X_{i_1}, X_{i_2}) = \frac{1}{2}(X_{i_1} - X_{i_2})^2$, where $i_1 = 1, 2, \ldots, n$; $i_2 \neq i_1$. The corresponding U-statistic is

$$\begin{aligned} U(X_1, \ldots, X_n) &= \frac{1}{\binom{n}{2}} \sum_{i_1 < i_2} \tfrac{1}{2}(X_{i_1} - X_{i_2})^2 \\ &= \frac{1}{n-1} \sum_{i=1}^{n} (X_i - \bar{X})^2, \end{aligned} \tag{3.7.13}$$

where $\bar{X} = \sum_{i=1}^{n} X_i/n$. Thus the U-statistic for an unbiased estimation of $\sigma^2(F)$ is the common sample variance.

The U-statistic is a symmetric function depending on all the n observations. Since the U-statistic is a constant on each of the cosets of $\mathfrak{X}^{(n)}$ under the group of permutations, the conditional expectation of a U-statistic given the order statistic is equal a.s. to the U-statistic. Thus we obtain the following theorem from the Lehmann-Scheffé theorem.

Theorem 3.7.1. *Let Ψ be a family of all absolutely continuous or all discrete distributions. Let g be an estimable function on Ψ. Let φ be a statistic on $(\mathfrak{X}^{(n)}, \mathfrak{B}^{(n)})$ which is an unbiased estimator of g, and let U be the corresponding U-statistic. Then U is an unbiased estimator of g. Furthermore, for any convex loss function, the risk function of U is uniformly smaller or equal to that of φ; that is,*

$$R(U, g(F)) \leqslant R(\varphi, g(F)) \quad \text{for all } F \in \Psi, \tag{3.7.14}$$

Equality holds if and only if $\varphi(X) = U(X)$ a.s. Moreover, if U^* is any other unbiased estimator of $g(F)$ satisfying

$$R(U^*, g(F)) \leqslant R(U, g(F)) \quad \text{for all } F \in \Psi, \tag{3.7.15}$$

then $U^*(X) = U(X)$ a.s.

Theorem 3.7.1 states that a U-statistic as an unbiased estimator of a function $g(F)$, relative to the family of *all* absolutely continuous or discrete distributions, is an essentially unique uniformly minimum risk (convex loss) unbiased estimator. The following example (see Fraser [5], p. 164, and Lehmann [1]) is based on the theory developed in this section.

Example 3.14. Let Ψ be the class of all absolutely continuous distributions on the real line. A distance function $\Delta(F, G)$ is defined on $\Psi \times \Psi$ to be

$$\Delta(F, G) = \int_{-\infty}^{\infty} [F(x) - G(x)]^2 \, \mathrm{d}\, \frac{F(x) + G(x)}{2}, \tag{3.7.16}$$

where F and G belong to Ψ. We can show that $\Delta(F, G) = 0$ if and only if $F = G$ (see Fraser [5], p. 164).

Given a sample of n_1 i.i.d. random variables $X_1, \ldots, X_{n_1}$ having a common distribution F and an independent sample of n_2 i.i.d. random variables $Y_1, \ldots, Y_{n_2}$ having a common distribution G, we wish to estimate the distance $\Delta(F, G)$.

We prove first that

$$\begin{aligned} g(F, G) \equiv P[\{\max(X_1, X_2) < \min(Y_1, Y_2)\} \cup \{\max(Y_1, Y_2) \\ < \min(X_1, X_2)\}] = \tfrac{1}{3} + 2\Delta(F, G). \end{aligned} \tag{3.7.17}$$

These events, $\{\max(X_1, X_2) < \min(Y_1, Y_2)\}$ and $\{\max(Y_1, Y_2) < \min(X_1, X_2)\}$, are disjoint. Hence

$$\begin{aligned} g(F, G) = P[\max(X_1, X_2) < \min(Y_1, Y_2)] \\ + P[\max(Y_1, Y_2) < \min(X_1, X_2)] \end{aligned} \tag{3.7.18}$$

Moreover,

$$P[\max(X_1, X_2) \leqslant z] = F^2(z), \tag{3.7.19}$$

and

$$P[\min(Y_1, Y_2) \geqslant z] = (1 - G(z))^2. \tag{3.7.20}$$

Therefore

$$\begin{aligned} g(F, G) &= \int_{-\infty}^{\infty} (1 - G(z))^2 \, \mathrm{d}F^2(z) + \int_{-\infty}^{\infty} (1 - F(z))^2 \, \mathrm{d}G^2(z) \\ &= 2 + \int_{-\infty}^{\infty} \mathrm{d}F^2(z)G^2(z) - 4\int_{-\infty}^{\infty} F(z)G(z) \, \mathrm{d}(F(z) + G(z)) \\ &= 3 - 2\int_{-\infty}^{\infty} [(F(z) + G(z))^2 - (F(z) - G(z))^2] \, \mathrm{d}\, \frac{F(z) + (G)}{2} \\ &= 3 - 8\int_{-\infty}^{\infty} \left(\frac{F(z) + G(z)}{2}\right)^2 \mathrm{d}\, \frac{F(z) + G(z)}{2} \\ &\quad + 2\int_{-\infty}^{\infty} (F(z) + G(z))^2 \, \mathrm{d}\, \frac{F(z) + G(z)}{2}. \end{aligned} \tag{3.7.21}$$

Finally, since the distribution of $(F(z) + G(z))/2$ is uniform on $(0, 1)$,

$$\int_{-\infty}^{\infty} \left(\frac{F(z) + G(z)}{2}\right)^2 d\,\frac{F(z) + G(z)}{2} = \int_0^1 u^2\, du = \tfrac{1}{3}. \tag{3.7.22}$$

Substituting in (3.7.21) we obtain (3.7.17). We proceed now with the derivation of the U.M.V.U. estimator of $\Delta(F, G)$. Define the indicator function

$$\phi(X_1, X_2, Y_1, Y_2) = \begin{cases} 1, & \text{if } \max(X_1, X_2) < \min(Y_1, Y_2) \\ & \quad \text{or } \max(Y_1, Y_2) < \min(X_1, X_2) \\ 0, & \text{otherwise.} \end{cases} \tag{3.7.23}$$

The function $\phi(X_1, X_2, Y_1, Y_2)$ is an unbiased estimator of $g(F, G)$; (3.7.23) is a kernel for $g(F, G)$. According to Theorem 3.7.1 the U-statistic

$$\frac{1}{\binom{n_1}{2}} \cdot \frac{1}{\binom{n_2}{2}} \sum_{i_1 < i_2} \sum_{j_1 < j_2} \phi(X_{i_1}, X_{i_2}, Y_{j_1}, Y_{j_2}) \tag{3.7.24}$$

is the essentially unique U.M.V.U. estimator of $\frac{1}{3} + 2\Delta(F, G)$. Thus the essentially unique U.M.V.U. estimator of the distance function $\Delta(F, G)$ is

$$\Delta(\mathbf{X}_{n_1}, \mathbf{Y}_{n_2}) = \frac{1}{2\binom{n_1}{2}\binom{n_2}{2}} \sum_{i_1 < i_2} \sum_{j_1 < j_2} \phi(X_{i_1}, X_{i_2}, Y_{j_1}, Y_{j_2}) - \tfrac{1}{3}. \tag{3.7.25}$$

3.8. LINEAR UNBIASED ESTIMATION

The class of linear unbiased estimators (L.U.E.) of a certain function $g(P)$ is the family of all unbiased estimators of $g(P)$, which are linear functions of the observed random variables. The problem of linear unbiased estimation is twofold. First, Is the class of linear unbiased estimators nonempty? Second, What is the optimal linear unbiased estimator? The practice is to call an estimator "*best linear unbiased estimator*" (B.L.U.E.) if the variance of that estimator is minimal with respect to all L.U.E.'s uniformly for all P in a given family $\mathcal{T}$ of probability measures. The following is an example of a B.L.U.E.

Example 3.15. Let $\mathbf{X}$ be an n-dimensional random vector such that $E\{\mathbf{X}\} = \theta\mathbf{1}$, where θ is an unknown real scalar, $-\infty < \theta < \infty$; $\mathbf{1}' = (1, 1, \ldots, 1)$; and the covariance matrix of $\mathbf{X}$, $\mathbf{\Sigma}$, is assumed to be known and nonsingular. Let f be the family of all n-variate distribution functions such that $E\{\mathbf{X}\} = \theta\mathbf{1}$ and $\mathbf{\Sigma} = E\{(\mathbf{X} - \theta\mathbf{1})(\mathbf{X} - \theta\mathbf{1})'\}$. What is the B.L.U.E. of θ? Let $\boldsymbol{\lambda}$ be an

n-dimensional vector (known). A linear estimator of θ is any linear function $\boldsymbol{\lambda}'\mathbf{X}$; $\boldsymbol{\lambda}$ provides an L.U.E. if and only if $\boldsymbol{\lambda}'\mathbf{1} = 1$. Indeed,

$$E\{\boldsymbol{\lambda}'\mathbf{X}\} = \theta\boldsymbol{\lambda}'\mathbf{1}. \tag{3.8.1}$$

Hence from the definition of an unbiased estimator, the class of all L.U.E.'s of θ is represented by all the vectors $\boldsymbol{\lambda}$ lying on the hyperplane $\boldsymbol{\lambda}'\mathbf{1} = 1$. The variance of a general linear estimator $\boldsymbol{\lambda}'\mathbf{X}$ is

$$\text{Var}\,\{\boldsymbol{\lambda}'\mathbf{X}\} = \boldsymbol{\lambda}'\boldsymbol{\Sigma}\boldsymbol{\lambda}. \tag{3.8.2}$$

Thus the problem of finding the B.L.U.E. is equivalent to the problem of finding a vector $\boldsymbol{\lambda}^0$ in the hyperplane $\boldsymbol{\lambda}'\mathbf{1} = 1$, which minimizes $\boldsymbol{\lambda}'\boldsymbol{\Sigma}\boldsymbol{\lambda}$. Employing the method of the Lagrange multipliers, we easily find that the B.L.U.E. of θ is

$$\hat{\theta} = \frac{\mathbf{1}'\boldsymbol{\Sigma}^{-1}\mathbf{X}}{\mathbf{1}'\boldsymbol{\Sigma}^{-1}\mathbf{1}}. \tag{3.8.3}$$

The following special cases are well known:

(i) If $\boldsymbol{\Sigma} = \sigma^2\mathbf{I}$, where $\mathbf{I}$ is the $n \times n$ identity matrix, then $\hat{\theta} = \bar{X} = \sum_{i=1}^n X_i/n$, namely the sample mean.

(ii) If $\boldsymbol{\Sigma} = \text{diag}\,(\sigma_1^2, \ldots, \sigma_n^2)$, where $\text{diag}\,(a_1, \ldots, a_n)$ designates a diagonal matrix with the diagonal elements $\{a_1, \ldots, a_n\}$, then $\boldsymbol{\Sigma}^{-1} = \text{diag}\,\{\sigma_1^{-2}, \ldots, \sigma_n^{-2}\}$ and

$$\hat{\theta} = \frac{\sum_{i=1}^n X_i/\sigma_i^2}{\sum_{i=1}^n 1/\sigma_i^2}. \tag{3.8.4}$$

■

This example shows that B.L.U. estimation may yield in parametric cases very inefficient estimators. Consider, for example, the parametric case, in which $X_1, \ldots, X_n$ are i.i.d. having a rectangular distribution over $[0, 2\theta]$, $0 < \theta < \infty$. According to (i) above, the B.L.U.E. of θ is the sample mean $\bar{X}_n$. But we have shown in Example 3.3 that the sample mean is an inferior estimator of θ. On the other hand, there are parametric cases where the B.L.U.E. coincides with the M.V.U.E. Such a case occurs when $\mathbf{X} \sim N(\theta\mathbf{1}, \boldsymbol{\Sigma})$.

Best linear unbiased estimators are in wide usage, especially in the analysis of linear statistical models. We give in this section the general theory of best linear unbiased estimation in the linear statistical models. The literature on this subject is too vast for a complete account of all the available results. We present a short summary of the main results. The interested reader is referred for complementary details, examples, and results to the texts of Scheffé [1], Ch. 1, 2; Graybill [2], Ch. 5, 6; and Rao [7], Ch. 4.

3.8.1. THE GENERAL LINEAR STATISTICAL MODEL OF A FULL RANK

The general linear statistical model of a full rank represents an n-dimensional (observed) random vector Y in the form

$$Y = \mathbf{X}\boldsymbol{\beta} + \boldsymbol{\epsilon}, \tag{3.8.5}$$

where $\mathbf{X}$ is an $n \times p$ matrix of known coefficients, called *the design matrix;* $\boldsymbol{\beta}$ is a $p \times 1$ vector of unknown parameters; and $\boldsymbol{\epsilon}$ is an $n \times 1$ random vector satisfying

$$E\{\boldsymbol{\epsilon}\} = \mathbf{0}, \qquad E\{\boldsymbol{\epsilon}\boldsymbol{\epsilon}'\} = \sigma^2\mathbf{I}, \tag{3.8.6}$$

$0 < \sigma^2 < \infty$, where σ^2 is generally unknown. This linear model is said to be of a *full rank* if rank $(\mathbf{X}) = p$. We are concerned with the estimation of the vector of unknown parameters $\boldsymbol{\beta}$, or some linear function of it, say $\boldsymbol{\lambda}'\boldsymbol{\beta}$. We should notice that the linear model (3.8.5)–(3.8.6) represents a wider class of linear models in which $E\{\boldsymbol{\epsilon}\} = \mathbf{0}$ but $E\{\boldsymbol{\epsilon}\boldsymbol{\epsilon}'\} = \sigma^2\mathbf{V}$, with $\mathbf{V}$ being a *known* positive definite symmetric matrix; σ^2 unknown. But this general model can be reduced to (3.8.5)–(3.8.6). Indeed, since $\mathbf{V}$ is positive definite there exists a nonsingular $n \times n$ matrix $\mathbf{D}$ such that $\mathbf{V} = \mathbf{D}\mathbf{D}'$. Make the transformation $\mathbf{Y}^* = \mathbf{D}^{-1}\mathbf{Y}$. Then we have

$$\mathbf{Y}^* = \mathbf{X}^*\boldsymbol{\beta} + \boldsymbol{\epsilon}^*,$$

where $\mathbf{X}^* = \mathbf{D}^{-1}\mathbf{X}$; $E\{\boldsymbol{\epsilon}^*\} = \mathbf{0}$; and $E\{\boldsymbol{\epsilon}^*\boldsymbol{\epsilon}^{*\prime}\} = \sigma^2\mathbf{D}^{-1}\mathbf{V}(\mathbf{D}^{-1})' = \sigma^2\mathbf{I}$. Replacing $\mathbf{Y}$ by $\mathbf{Y}^*$ and $\mathbf{X}$ of (3.8.5) by $\mathbf{X}^*$, the general model is reduced to (3.8.5)–(3.8.6). If the general model is of full rank, so is the reduced model. Thus there is no loss in generality in assuming that $E\{\boldsymbol{\epsilon}\boldsymbol{\epsilon}') = \sigma^2\mathbf{I}$. If $\mathbf{V}$ is singular we cannot apply the present theory and should consider a more general theory. (See Zyskind [1].)

The p column vectors of $\mathbf{X}$ are n-dimensional and lie in a p-dimensional linear manifold (subspace). Thus for every $\boldsymbol{\beta}$, the $n \times 1$ vector $\boldsymbol{\eta} = \mathbf{X}\boldsymbol{\beta}$ lies in a p-dimensional linear manifold generated by the column vectors of $\mathbf{X}$. Let Ξ_p designate this manifold and let $\{\boldsymbol{\xi}^{(1)}, \ldots, \boldsymbol{\xi}^{(p)}\}$ be an orthonormal basis of Ξ_p. Then every vector $\boldsymbol{\eta} = \mathbf{X}\boldsymbol{\beta}$ can be presented as a linear combination $\boldsymbol{\eta} = \sum_{i=1}^{p} c_i\boldsymbol{\xi}^{(i)}$. A vector $\hat{\boldsymbol{\eta}} \in \Xi_p$ is of minimal distance to the given vector $\mathbf{Y}$ if it is the orthogonal projection of $\mathbf{Y}$ on Ξ_p.

Let $\hat{\boldsymbol{\beta}}$ be such that $\hat{\boldsymbol{\eta}} = \mathbf{X}\hat{\boldsymbol{\beta}}$. This vector $\hat{\boldsymbol{\beta}}$ is called the *least-squares* estimator of $\boldsymbol{\beta}$. The distance between $\mathbf{Y}$ and $\mathbf{X}\hat{\boldsymbol{\beta}}$ is minimal. To find $\hat{\boldsymbol{\beta}}$ we notice that since $\mathbf{X}\hat{\boldsymbol{\beta}}$ is the orthogonal projection of $\mathbf{Y}$ on Ξ_p,

$$\boldsymbol{\xi}^{(i)\prime}(\mathbf{Y} - \mathbf{X}\hat{\boldsymbol{\beta}}) = 0 \quad \text{for all } i = 1, \ldots, p. \tag{3.8.7}$$

Hence since every column vector of $\mathbf{X}$ is a linear combination of the orthogonal vector $\boldsymbol{\xi}^{(i)}$, we imply that $\hat{\boldsymbol{\beta}}$ satisfies the equation

$$\mathbf{X}'(\mathbf{Y} - \mathbf{X}\hat{\boldsymbol{\beta}}) = 0. \tag{3.8.8}$$

Or, in the *normal equations* form,

$$\mathbf{X}'\mathbf{X}\hat{\boldsymbol{\beta}} = \mathbf{X}'\mathbf{Y}. \tag{3.8.9}$$

Finally, since $\mathbf{X}'\mathbf{X}$ is nonsingular, the least-squares estimator (L.S.E.) of $\boldsymbol{\beta}$ is

$$\hat{\boldsymbol{\beta}} = (\mathbf{X}'\mathbf{X})^{-1}\mathbf{X}'\mathbf{Y}. \tag{3.8.10}$$

The L.S.E. (3.8.10) is obviously unique. We prove now that for any linear combination $\boldsymbol{\lambda}'\boldsymbol{\beta}$, $\lambda'\hat{\boldsymbol{\beta}}$ is the B.L.U.E., provided the model is of a full rank.

Theorem 3.8.1. (Gauss-Markov.) *Let* $\mathbf{Y} = X\boldsymbol{\beta} + \boldsymbol{\epsilon}$ *be a linear model of a full rank, like* (3.8.5)–(3.8.6). *Let* $\boldsymbol{\lambda}'\boldsymbol{\beta}$ *be an arbitrary linear function of* $\boldsymbol{\beta}$. *Then the L.S.E.* $\boldsymbol{\lambda}'\hat{\boldsymbol{\beta}}$ *is* B.L.U.E. *of* $\boldsymbol{\lambda}'\boldsymbol{\beta}$ *where* $\hat{\boldsymbol{\beta}}$ *is given by* (3.8.10).

Proof. To show that $\boldsymbol{\lambda}'\hat{\boldsymbol{\beta}}$ is an unbiased estimator of $\boldsymbol{\lambda}'\boldsymbol{\beta}$ we notice that (3.8.6) implies

$$\begin{aligned} E\{\hat{\boldsymbol{\beta}}\} &= (\mathbf{X}'\mathbf{X})^{-1}\mathbf{X}'E\{\mathbf{X}\boldsymbol{\beta} + \boldsymbol{\epsilon}\} \\ &= \boldsymbol{\beta}, \quad \text{for all } \boldsymbol{\beta}. \end{aligned} \tag{3.8.11}$$

Furthermore, if $\mathbf{AY}$ is any unbiased estimator of $\boldsymbol{\beta}$ then $\mathbf{AX} = \mathbf{I}$. We show now that, for every vector $\boldsymbol{\lambda}$, Var $\{\boldsymbol{\lambda}'\boldsymbol{\beta}\} \leqslant$ Var $\{\boldsymbol{\lambda}'\mathbf{AY}\}$. For this it is sufficient to show that $\boldsymbol{\Sigma}(\mathbf{AY}) - \boldsymbol{\Sigma}(\hat{\boldsymbol{\beta}})$ is non-negative definite where $\boldsymbol{\Sigma}(\mathbf{Z})$ designates the covariance matrix of $\mathbf{Z}$. Let $\mathbf{Q} = \mathbf{A} - \mathbf{S}^{-1}\mathbf{X}'$ where $\mathbf{S} = \mathbf{X}'\mathbf{X}$. Thus

$$\boldsymbol{\Sigma}(\mathbf{AY}) = \boldsymbol{\Sigma}(\mathbf{QY}) + \boldsymbol{\Sigma}(\hat{\boldsymbol{\beta}}) + 2\sigma^2\mathbf{QXS}^{-1}. \tag{3.8.12}$$

But since $\mathbf{AY}$ is unbiased, that is, $\mathbf{AX} = \mathbf{I}$, we have

$$\begin{aligned} \mathbf{QXS}^{-1} &= (\mathbf{A} - \mathbf{S}^{-1}\mathbf{X}')\mathbf{XS}^{-1} \\ &= \mathbf{AXS}^{-1} - \mathbf{S}^{-1} = 0. \end{aligned} \tag{3.8.13}$$

Finally, since $\boldsymbol{\Sigma}(\mathbf{QY}) = \sigma^2\mathbf{QQ}'$ is non-negative definite, the theorem is proven. (Q.E.D.)

It is simple to verify that the variance of the B.L.U.E. $\boldsymbol{\lambda}'\hat{\boldsymbol{\beta}}$ is

$$\text{Var}\,\{\boldsymbol{\lambda}'\hat{\boldsymbol{\beta}}\} = \sigma^2\boldsymbol{\lambda}'S^{-1}\boldsymbol{\lambda}. \tag{3.8.14}$$

We conclude this section with the presentation of an unbiased estimator of the unknown variance parameter σ^2.

Lemma 3.8.1. *An unbiased estimator of* σ^2 *for a linear statistical model of a* *full rank is*

$$\hat{\sigma}^2 = \frac{1}{n-p} \mathbf{Y}'(\mathbf{I} - \mathbf{X}(\mathbf{X}'\mathbf{X})^{-1}\mathbf{X}')\mathbf{Y}. \tag{3.8.15}$$

Proof. The matrix $\mathbf{I} - \mathbf{X}\mathbf{S}^{-1}\mathbf{X}'$ is idempotent; that is, $(\mathbf{I} - \mathbf{X}\mathbf{S}^{-1}\mathbf{X}')^2 = (\mathbf{I} - \mathbf{X}\mathbf{S}^{-1}\mathbf{X}')$. Furthermore $\mathbf{X}'(\mathbf{I} - \mathbf{X}(\mathbf{X}'\mathbf{X})^{-1}\mathbf{X}')\mathbf{X} = 0$. Hence

$$\begin{aligned} E\{\mathbf{Y}'(\mathbf{I} - \mathbf{X}\mathbf{S}^{-1}\mathbf{X}')\mathbf{Y}\} &= \text{tr}\,\{\mathbf{\Sigma}((\mathbf{I} - \mathbf{X}\mathbf{S}^{-1}\mathbf{X}')\mathbf{Y})\} \\ &= \sigma^2\, \text{tr}\,\{(\mathbf{I} - \mathbf{X}\mathbf{S}^{-1}\mathbf{X}')\} \\ &= (n-p)\sigma^2. \end{aligned} \tag{3.8.16}$$

Hence $E\{\hat{\sigma}^2\} = \sigma^2$. (Q.E.D.)

3.8.2. CONDITIONS FOR THE ESTIMABILITY OF LINEAR FUNCTIONS

We are concerned with the unbiased estimation of a linear function $\boldsymbol{\lambda}'\boldsymbol{\beta}$, where the linear model (3.8.5)–(3.8.6) is not necessarily of a full rank. Thus we assume that rank $(\mathbf{X}) = r \leqslant p$. The case of $r = p$ was treated previously. As will be shown, not every linear function $\boldsymbol{\lambda}'\boldsymbol{\beta}$ has an unbiased estimator if rank $(\mathbf{X}) = r < p$. As is proven in the following theorem, a linear function $\boldsymbol{\lambda}'\boldsymbol{\beta}$ has an unbiased estimator if and only if $\boldsymbol{\lambda}$ belongs to the linear manifold spanned by the rows of $\mathbf{X}$. We denote this linear manifold, whose dimension is r, by $\mathcal{V}_r$. We also mention that a linear function $\boldsymbol{\lambda}'\boldsymbol{\beta}$ is called *estimable* if there exists an unbiased estimator $\mathbf{L}'\mathbf{Y}$ of $\boldsymbol{\lambda}'\boldsymbol{\beta}$.

Theorem 3.8.2. *Consider the linear model* (3.8.5)–(3.8.6), *where* rank $(\mathbf{X}) = r \leqslant p$. *A linear function* $\boldsymbol{\lambda}'\boldsymbol{\beta}$ *is estimable if and only if*

$$\text{rank}\,(\mathbf{X}') = \text{rank}\,(\mathbf{X}' \,\vdots\, \boldsymbol{\lambda}). \tag{3.8.17}$$

Proof. (i) Suppose that $\boldsymbol{\lambda}'\boldsymbol{\beta}$ is estimable. Then there exists a vector $\mathbf{L} = (l_1, \ldots, l_n)'$ such that

$$\mathbf{X}'\mathbf{L} = \boldsymbol{\lambda}.$$

Hence $\boldsymbol{\lambda}$ is a linear combination of the row vectors of $\mathbf{X}$ and belongs therefore to $\mathcal{V}_r$. In other words, (3.8.17) is satisfied.

(ii) Suppose that (3.8.17) holds. Then there exists a vector $\mathbf{L} = (l_1, \ldots, l_n)'$ such that $\boldsymbol{\lambda}' = \mathbf{L}'\mathbf{X}$. $\mathbf{L}'\mathbf{Y}$ is then an unbiased estimator of $\boldsymbol{\lambda}'\boldsymbol{\beta}$. Indeed,

$$E\{\mathbf{L}'\mathbf{Y}\} = \mathbf{L}'\mathbf{X}\boldsymbol{\beta} = \boldsymbol{\lambda}'\boldsymbol{\beta}.$$

(Q.E.D.)

Lemma 3.8.2. *Consider the linear model* (3.8.5)–(3.8.6) *with rank $r \leqslant p$. Let $\boldsymbol{\lambda}'\boldsymbol{\beta}$ be estimable. Let Ξ_r be the linear manifold spanned by the column vectors of* $\mathbf{X}$. *Then there exists a unique vector $\mathbf{L}^* \in \Xi_r$ such that $\mathbf{L}^{*\prime}\mathbf{Y}$ is an unbiased* *estimator of* $\boldsymbol{\lambda}'\boldsymbol{\beta}$. *Furthermore, if* $\mathbf{L}'\mathbf{Y}$ *is any unbiased estimator of* $\boldsymbol{\lambda}'\boldsymbol{\beta}$ *then* $\mathbf{L}^*$ *is the projection of* $\mathbf{L}$ *on* Ξ_r*; that is,*

$$\mathbf{L}^* = \text{Proj}_{\Xi_r}\{\mathbf{L}\}. \tag{3.8.18}$$

Proof. Since $\boldsymbol{\lambda}'\boldsymbol{\beta}$ is estimable, there exists a vector $\mathbf{L} = (l_1, \ldots, l_n)'$ such that $E\{\mathbf{L}'\mathbf{Y}\} = \boldsymbol{\lambda}'\boldsymbol{\beta}$. Let $\mathbf{L} = \mathbf{L}^* + \mathbf{K}$ where $\mathbf{L}^* = \text{Proj}_{\Xi_r}\{\mathbf{L}\}$ and $\mathbf{K}$ is orthogonal to $\mathbf{L}^*$. Hence $\mathbf{K}'\mathbf{X} = \mathbf{0}$, and we have

$$\begin{aligned} \boldsymbol{\lambda}'\boldsymbol{\beta} &= E\{(\mathbf{L}^* + \mathbf{K})'\mathbf{Y}\} \\ &= (\mathbf{L}^* + \mathbf{K})'\mathbf{X}\boldsymbol{\beta} \\ &= \mathbf{L}^{*\prime}\mathbf{X}\boldsymbol{\beta} = E\{\mathbf{L}^{*\prime}\mathbf{Y}\}. \end{aligned} \tag{3.8.19}$$

The quantity $\mathbf{L}^*\mathbf{Y}$ is therefore an unbiased estimator of $\boldsymbol{\lambda}'\boldsymbol{\beta}$. We show now that $\mathbf{L}^*$ is unique in Ξ_r. If $\mathbf{A} \in \Xi_r$ and $E\{\mathbf{A}'\mathbf{Y}\} = \boldsymbol{\lambda}'\boldsymbol{\beta}$ then

$$(\mathbf{A} - \mathbf{L}^*)'\mathbf{X}\boldsymbol{\beta} = 0. \tag{3.8.20}$$

$\mathbf{X}\boldsymbol{\beta} \in \Xi_r$ and is arbitrary. Hence $\mathbf{A} = \mathbf{L}^*$. (Q.E.D.)

Theorem 3.8.3. (Gauss-Markov.) *Consider the linear model* (3.8.5)–(3.8.6) *with rank* $r \leqslant p$. *Every estimable linear function* $\boldsymbol{\lambda}'\boldsymbol{\beta}$ *has a unique B.L.U.E., given by* $\boldsymbol{\lambda}'\hat{\boldsymbol{\beta}}$, *where* $\hat{\boldsymbol{\beta}}$ *is the L.S.E. of* $\boldsymbol{\beta}$, *being any consistent solution of the normal equations*

$$\mathbf{X}'\mathbf{X}\hat{\boldsymbol{\beta}} = \mathbf{X}'\mathbf{Y}.$$

Proof. $\boldsymbol{\lambda}'\boldsymbol{\beta}$ is estimable. Let $\mathbf{L}'\mathbf{Y}$ be an unbiased estimator of $\boldsymbol{\lambda}'\boldsymbol{\beta}$. We have shown in Lemma 3.8.2 that the unique unbiased estimator $\mathbf{L}^{*\prime}\mathbf{Y}$ of $\boldsymbol{\lambda}'\boldsymbol{\beta}$ with $\mathbf{L}^* \in \Xi_r$ is the one where $\mathbf{L}^* = \text{Proj}_{\Xi_r}\{\mathbf{L}\}$. The variance of $\mathbf{L}'\mathbf{Y}$ is

$$\text{Var}\{\mathbf{L}'\mathbf{Y}\} = \sigma^2\mathbf{L}'\mathbf{L}.$$

Since $\mathbf{L}^*$ is orthogonal to $(\mathbf{L} - \mathbf{L}^*)$, we have

$$\mathbf{L}'\mathbf{L} = \mathbf{L}^{*\prime}\mathbf{L}^* + (\mathbf{L} - \mathbf{L}^*)'(\mathbf{L} - \mathbf{L}^*).$$

Hence

$$\text{Var}\{\mathbf{L}^{*\prime}\mathbf{Y}\} = \sigma^2\mathbf{L}^{*\prime}\mathbf{L} \leqslant \text{Var}\{\mathbf{L}'\mathbf{Y}\}.$$

Thus $\mathbf{L}^{*\prime}\mathbf{Y}$ is the unique B.L.U.E. of $\boldsymbol{\lambda}'\boldsymbol{\beta}$. It remains to show that $\mathbf{L}^{*\prime}\mathbf{Y} = \boldsymbol{\lambda}'\hat{\boldsymbol{\beta}}$. The quantity Ξ_r is the linear manifold spanned by the column vectors of $\mathbf{X}$. Let $\mathbf{X}\hat{\boldsymbol{\beta}}$ be the orthogonal projection of $\mathbf{Y}$ on Ξ_r. As previously shown, $\hat{\boldsymbol{\beta}}$ satisfies the normal equations; that is,

$$\mathbf{X}'\mathbf{X}\boldsymbol{\beta} = X'\mathbf{Y}.$$

Since $\mathbf{L}^* \in \Xi_r$, $\mathbf{L}^{*\prime}(\mathbf{Y} - \mathbf{X}\hat{\boldsymbol{\beta}}) = 0$. Hence $\mathbf{L}^{*\prime}\mathbf{Y} = \mathbf{L}^{*\prime}\mathbf{X}\hat{\boldsymbol{\beta}}$. Moreover, since $\mathbf{L}^{*\prime}\mathbf{Y}$ is unbiased, $\mathbf{L}^{*\prime}\mathbf{X} = \boldsymbol{\lambda}'$. Hence $\mathbf{L}^{*\prime}\mathbf{Y} = \boldsymbol{\lambda}'\hat{\boldsymbol{\beta}}$. (Q.E.D.)

3.8.3. GENERALIZED LEAST-SQUARES ESTIMATORS

According to Theorem 3.8.2, if $\boldsymbol{\lambda}'\boldsymbol{\beta}$ is estimable, then any consistent solution of the normal equations $\mathbf{X}'\mathbf{X}\boldsymbol{\beta} = \mathbf{X}'\mathbf{Y}$, say $\hat{\boldsymbol{\beta}}$, yields a B.L.U.E., $\boldsymbol{\lambda}'\hat{\boldsymbol{\beta}}$ of $\boldsymbol{\lambda}'\boldsymbol{\beta}$. We now outline a method of arriving at such an estimator. When the linear model is of a full rank, p, there is a unique solution $\hat{\boldsymbol{\beta}}$ given by the L.S.E. (3.8.10). When the rank of $\mathbf{X}'\mathbf{X}$ is $r < p$ the least-squares solution is not unique. There is a $(p - r)$-dimensional linear manifold of solutions to the normal equations. Each one of these solutions is called a *generalized least-squares estimator*. The literature on generalized least-squares estimators is very rich. We shall give here a very brief account of the method and an example pertaining to the design and analysis of factorial experiment. The reader is referred to Rao [7], Ch. 1, 4, and Rao [4], for further details.

A matrix $\mathbf{S}^-$ of order $m \times n$ is called a *weakly generalized inverse*, to be designated shortly as a *g-inverse*, of an $n \times m$ matrix $\mathbf{S}$ if, for any given $n \times 1$ vector $\mathbf{Y}$, $\mathbf{S}^-\mathbf{Y}$ is a consistent solution of the equation $\mathbf{S}\mathbf{X} = \mathbf{Y}$. A g-inverse $\mathbf{S}^-$ of $\mathbf{S}$, if exists, satisfies

$$\mathbf{S}\mathbf{S}^-\mathbf{S} = \mathbf{S}. \tag{3.8.21}$$

Furthermore, if $\mathbf{S}^-$ is a g-inverse of $\mathbf{S}$ and $\mathbf{H} = \mathbf{S}^-\mathbf{S}$ then

(i) $\mathbf{H}$ is an idempotent matrix, that is, $\mathbf{H}^2 = \mathbf{H}$;
(ii) $\mathbf{S}\mathbf{H} = \mathbf{S}$ and rank $(\mathbf{S})$ = rank $(\mathbf{H})$ = trace $(\mathbf{H})$;
(iii) a general solution of the consistent equation $\mathbf{S}\mathbf{X} = \mathbf{Y}$ is

$$\hat{\mathbf{X}} = \mathbf{S}^-\mathbf{Y} + (\mathbf{H} - \mathbf{I})\mathbf{Z} \tag{3.8.22}$$

where $\mathbf{Z}$ is arbitrary;
(iv) $\boldsymbol{\lambda}'\mathbf{X}$ is unique for all $\mathbf{X}$ satisfying $\mathbf{S}\mathbf{X} = \mathbf{Y}$ if and only if $\boldsymbol{\lambda}'\mathbf{H} = \boldsymbol{\lambda}'$;
(v) $\mathbf{S}^-$ exists, and rank $(\mathbf{S}^-) \geqslant$ rank $(\mathbf{S})$. Moreover, there exists $\mathbf{S}^-$ such that rank $(\mathbf{S}^-) = \min(m, n)$ irrespective of rank $(\mathbf{S})$. Such a g-inverse is of maximal rank. We construct it in the following manner.

Given $\mathbf{S}$ of order $m \times n$, $m < n$, we define the $n \times n$ matrix

$$\mathbf{S}^* = \begin{bmatrix} \mathbf{S} \\ \hline \mathbf{0} \end{bmatrix},$$

where $\mathbf{0}$ is a zero matrix of order $(n - m) \times n$. By elementary transformations on the rows of $\mathbf{S}^*$ we reduce it to its Hermite canonical form. In other words, there exists a nonsingular matrix $\mathbf{C}$ of order $n \times n$ such that

$$\mathbf{C}\mathbf{S}^* = \left[\begin{array}{c|c} \mathbf{I}_r & \mathbf{K} \\ \hline 0 & 0 \end{array}\right],$$

where r = rank $(\mathbf{S})$. Let $\mathbf{H} = \mathbf{C}\mathbf{S}^*$. It is easy to check that $\mathbf{H}$ is idempotent.

Hence $\mathbf{CS^*CS^*} = \mathbf{H}$ and, since $\mathbf{C}$ is nonsingular, $\mathbf{S^*CS^*} = \mathbf{C}^{-1}\mathbf{H}$. Moreover,

$$\begin{aligned} \mathbf{S^*H} &= \mathbf{S^*CS^*CS^*} \\ &= \mathbf{C}^{-1}\mathbf{CS^*} \\ &= \mathbf{S^*}. \end{aligned} \tag{3.8.23}$$

Let $\mathbf{S}^-$ be the $n \times m$ matrix comprising of the first m column vectors of $\mathbf{C}$; that is,

$$\mathbf{C} = (\mathbf{S}^- \mid \mathbf{D}). \tag{3.8.24}$$

Then

$$\begin{aligned} \mathbf{S^*CS^*} &= \begin{bmatrix} \mathbf{S} \\ \mathbf{0} \end{bmatrix} (\mathbf{S}^- \mid \mathbf{D}) \begin{bmatrix} \mathbf{S} \\ \mathbf{0} \end{bmatrix} \\ &= \begin{bmatrix} \mathbf{SS^-S} \\ \mathbf{0} \end{bmatrix}. \end{aligned} \tag{3.8.25}$$

The comparison of (3.8.23) and (3.8.25) implies that $\mathbf{SS^-S} = \mathbf{S}$; that is, $\mathbf{S}^-$ so defined is the g-inverse of $\mathbf{S}$. Finally, since $\mathbf{C}$ is nonsingular all the m column vectors of $\mathbf{S}^-$ are linearly independent. Hence rank $(\mathbf{S}^-) = m$ irrespective of rank $(\mathbf{S})$. In the following example we show how this theory of generalized least-squares estimators obtained by a g-inverse of $\mathbf{S} = \mathbf{X'X}$ is utilized to solve certain problems connected with the design of weighing experiments.

3.8.4. APPLICATION IN WEIGHING DESIGNS

In this subsection we discuss the papers of K. S. Banerjee [1] concerning the design of fractional weighing designs. This paper shows how the notion of a g-inverse can be exploited in the design and analysis of experiments.

A number of objects, p, are subject to a weighing experiment. Let $\beta_1, \ldots, \beta_p$ be their unknown weights. The weighing experiment is of the chemical type, in which a two-pan scale is available. For each weighing operation the experimenter specifies which of the p objects will be placed on the right-hand pan, which on the left-hand pan, and which will not be weighted. Such a specification is represented by a p-dimensional row vector $\mathbf{X}$ consisting of $+1$, 0, and -1 elements. The quantity $X_i = +1$ if the i-th object ($i = 1, \ldots, p$) is placed on the right pan; equals -1 if it is placed on the left pan; and equals 0 otherwise. A sequence of n weighing operations is thus presented by an $n \times p$ design matrix $\mathbf{X}$. Such a weighing design is called *singular* if rank of $\mathbf{X} < p$. Hotelling [1] showed that an optimal nonsingular weighing design of the chemical type for the estimation of $\boldsymbol{\beta}' = (\beta_1, \ldots, \beta_p)$ is given by a

Hadamard matrix, $\mathbf{C}$, of order $p \times p$, if it exists. A Hadamard matrix of order p is a matrix $\mathbf{C}$, all of whose elements are $+1$ or -1, and satisfies

$$\mathbf{C}'\mathbf{C} = \mathbf{C}\mathbf{C}' = p\mathbf{I}.$$

A Hadamard matrix $\mathbf{C}$ of order p does not exist for every p (see Banerjee [1]). If $p = 2^m$ $(m = 1, 2, \ldots)$ the Hadamard matrix $\mathbf{C}$ exists and represents a 2^m factorial design. A Hadamard matrix exists not only for $p = 2^m$ but for all $p \equiv 2$ (Mod. 4). In this example we assume that p is such that a Hadamard weighing design matrix $\mathbf{C}$ exists. A weighing design of n weighing operations, chosen from the rows of $\mathbf{C}$ where $1 \leqslant n < p$, is called a *fractional weighing design* (F.W.D.) of order n. The linear model associated with an F.W.D. of order n is the following one: Let $\mathbf{G}$ be an $n \times p$ matrix consisting of the n rows of the Hadamard matrix $\mathbf{C}$, corresponding to the weighing operations chosen. Let $\mathbf{Y}_n$ be the corresponding vector of observations. Then

$$\mathbf{Y}_n = \mathbf{G}\boldsymbol{\beta} + \boldsymbol{\epsilon}_n, \tag{3.8.26}$$

where $\boldsymbol{\epsilon}_n$ is a random vector satisfying $E\{\boldsymbol{\epsilon}_n\} = \mathbf{0}$ and $E\{\boldsymbol{\epsilon}_n\boldsymbol{\epsilon}_n'\} = \sigma^2\mathbf{I}_n$. The objective is to estimate unbiasedly a specified linear function $\boldsymbol{\lambda}'\boldsymbol{\beta}$. If the weighing design is of a full rank, that is, $n = p$, $\mathbf{G} = \mathbf{C}$, and the B.L.U.E. of $\boldsymbol{\lambda}'\boldsymbol{\beta}$ is the L.S.E.,

$$\boldsymbol{\lambda}'\hat{\boldsymbol{\beta}} = \frac{1}{p}\,\boldsymbol{\lambda}'\mathbf{C}'\mathbf{Y}. \tag{3.8.27}$$

The variance of this B.L.U.E. is

$$\text{Var}\,\{\boldsymbol{\lambda}'\hat{\boldsymbol{\beta}}\} = \frac{\sigma^2}{p}\,\boldsymbol{\lambda}'\boldsymbol{\lambda}. \tag{3.8.28}$$

We know that certain linear functions are estimable even if the design is not of a full rank. The questions raised are the following:

(i) What linear functions $\boldsymbol{\lambda}'\boldsymbol{\beta}$ are estimable by a fractional weighing design of order $1 \leqslant n < p$; and which fractional design should be chosen?

(ii) Is the variance of the L.S.E. of $\boldsymbol{\lambda}'\boldsymbol{\beta}$ estimable by a F.W.D. of order $n < p$ equal to or greater than that of an L.S.E. corresponding to a design of a full rank?

If the variance of the B.L.U.E. of a function $\boldsymbol{\lambda}'\boldsymbol{\beta}$ estimable by an F.W.D. of order $n < p$ is equal to that of a B.L.U.E. in a full rank design, we save $(p - n)$ weighing operations which are redundant in the given problem. Consider an F.W.D. of order $n < p$. Let $\mathbf{G}$ be its design matrix. Without loss of generality, assume that $\mathbf{G}$ consists of the first n row vectors of $\mathbf{C}$.

(A permutation of the row vectors of $\mathbf{C}$, which does not alter its essential properties, can always reduce a given model to the case under consideration.) Partition the matrix $\mathbf{C}$ of full rank in the following manner:

$$\mathbf{C}' = \left[\begin{array}{c}\mathbf{G}\\ \hline \mathbf{G}^*\end{array}\right] = \left[\begin{array}{c|c}\mathbf{C}_{11} & \mathbf{C}_{12}\\ \hline \mathbf{C}_{21} & \mathbf{C}_{22}\end{array}\right], \tag{3.8.29}$$

where $\mathbf{C}_{11}$ of rank r is nonsingular and $\mathbf{C}_{22}$ of rank $(n - r)$ is nonsingular. Furthermore, since $\mathbf{CC}' = p\mathbf{I}$, we have

$$\mathbf{C}_{11}\mathbf{C}_{12}' + \mathbf{C}_{12}\mathbf{C}_{22}' = 0. \tag{3.8.30}$$

The linear model for such an F.W.D. yields the normal equations

$$\begin{aligned}\left[\begin{array}{c|c}\mathbf{C}_{11}'\mathbf{C}_{11} & \mathbf{C}_{11}'\mathbf{C}_{12}\\ \hline \mathbf{C}_{12}'\mathbf{C}_{11} & \mathbf{C}_{12}'\mathbf{C}_{12}\end{array}\right]\boldsymbol{\beta} &= \left[\begin{array}{c|c}\mathbf{S}_{11} & \mathbf{S}_{12}\\ \hline \mathbf{S}_{21} & \mathbf{S}_{22}\end{array}\right]\boldsymbol{\beta}\\ &= \left[\begin{array}{c}\mathbf{C}_{11}'\\ \hline \mathbf{C}_{12}'\end{array}\right]\mathbf{Y}_n,\end{aligned} \tag{3.8.31}$$

where $\mathbf{Y}_n$ is the observed vector of n weighing operations. The general solution to this equation is

$$\hat{\boldsymbol{\beta}} = \mathbf{S}^-\left[\begin{array}{c}\mathbf{C}_{11}'\\ \hline \mathbf{C}_{12}'\end{array}\right]\mathbf{Y}_n + (\mathbf{I} - \mathbf{H})\mathbf{Z}, \tag{3.8.32}$$

where $\mathbf{Z}$ is an arbitrary (nonrandom) vector; $\mathbf{S}^-$ is a g-inverse of

$$\mathbf{S} = \left[\begin{array}{c|c}\mathbf{S}_{11} & \mathbf{S}_{12}\\ \hline \mathbf{S}_{21} & \mathbf{S}_{22}\end{array}\right];$$

and $\mathbf{H} = \mathbf{S}^-\mathbf{S}$. Due to the special properties of the $\mathbf{C}$ matrix and because of (3.8.30), Banerjee showed that the matrix $\mathbf{H}$ can be written as

$$\mathbf{H} = \left[\begin{array}{c|c}\mathbf{I}_r & \mathbf{H}_{12}\\ \hline \mathbf{0} & \mathbf{0}\end{array}\right], \tag{3.8.33}$$

where

$$\mathbf{H}_{12} = -\mathbf{C}_{21}'(\mathbf{C}_{22}')^{-1}. \tag{3.8.34}$$

Furthermore, the g-inverse $\mathbf{S}^-$ is given by

$$\mathbf{S}^- = \mathbf{P}'\,\boldsymbol{\Delta}_S^-\mathbf{P}, \tag{3.8.35}$$

where $\boldsymbol{\Delta}_S$ is a diagonal matrix obtained by the diagonalization $\boldsymbol{\Delta}_S = \mathbf{PSP}'$, where

$$\mathbf{P} = \left[\begin{array}{c|c}\mathbf{I}_r & 0\\ \hline -\mathbf{H}_{12}' & \mathbf{I}_{p-r}\end{array}\right]. \tag{3.8.36}$$

Furthermore, if $\boldsymbol{\Delta}_S = \text{diag}\,\{d_1, d_2, \ldots, d_r, 0, \ldots, 0\}$ then

$$\boldsymbol{\Delta}_S^- = \text{diag}, \{d_1^{-1} d_2^{-1}, \ldots, d_r^{-1}, 0, \ldots, 0\}.$$

A necessary and sufficient condition for the estimability of $\boldsymbol{\lambda}'\boldsymbol{\beta}$ under an F.W.D. of order n and rank $r \leqslant n$ is that $\boldsymbol{\lambda}'\mathbf{H} = \boldsymbol{\lambda}'$. This condition, combined with (3.8.33)–(3.8.34), implies that if $\boldsymbol{\lambda}' = (\boldsymbol{\lambda}_r', \boldsymbol{\lambda}^{*\prime}_{p-r})$ then $\boldsymbol{\lambda}_r$ could be arbitrary; but a necessary condition for the estimability of $\boldsymbol{\lambda}'\boldsymbol{\beta}$ is that

$$\boldsymbol{\lambda}^*_{p-r} = -\mathbf{C}_{22}^{-1}\mathbf{C}_{21}\boldsymbol{\lambda}r. \tag{3.8.37}$$

This necessary condition has the following implication on the choice of weighing operations included in the design. Suppose we are given a linear function $\boldsymbol{\lambda}'\boldsymbol{\beta}$, with a known functional $\boldsymbol{\lambda}$. We are asked whether there exists an F.W.D. under which $\boldsymbol{\lambda}'\boldsymbol{\beta}$ is estimable and what the order of this design is. We transform $\boldsymbol{\lambda}$ into

$$\boldsymbol{\zeta} = \frac{1}{p}\,\mathbf{C}\boldsymbol{\lambda}. \tag{3.8.38}$$

If certain components of $\boldsymbol{\zeta}$ are zero, the corresponding row vectors of $\mathbf{C}$ can be omitted from the design. To show this suppose, without loss of generality, that the first r components of $\boldsymbol{\zeta}$ are nonzero and the last $(p - r)$ components of $\boldsymbol{\zeta}$ are zero $(1 \leqslant r \leqslant p)$. Decomposing $\mathbf{C}$ as in (3.8.29), the hypothesis that the last $(p - r)$ components of $\boldsymbol{\zeta}$ are zero implies that

$$(\mathbf{C}_{21} \mid \mathbf{C}_{22})\begin{pmatrix} \boldsymbol{\lambda}_r \\ \boldsymbol{\lambda}^*_{p-r} \end{pmatrix} = \mathbf{0}. \tag{3.8.39}$$

This, however, implies (3.8.37). Thus if certain components of $\boldsymbol{\zeta}$ are zero, the omission of the corresponding row vectors of $\mathbf{C}$ yields an F.W.D. under which $\boldsymbol{\lambda}'\boldsymbol{\beta}$ is estimable. It should be remarked that an F.W.D. which omits a row vector of $\mathbf{C}$, whose inner product with $\boldsymbol{\lambda}$ is nonzero, destroys the estimability property of $\boldsymbol{\lambda}'\boldsymbol{\beta}$. A way to overcome this difficulty is to consider also randomized F.W.D. We shall not discuss these procedures here. The interested reader is referred to Zacks [2]. We conclude the present example by showing that if a function $\boldsymbol{\lambda}'\boldsymbol{\beta}$ is estimable under an F.W.D. of order r, $1 \leqslant r \leqslant p$, and if the corresponding linear model is of rank r, then the variance of its least-squares estimator $\boldsymbol{\lambda}'\hat{\boldsymbol{\beta}}$, where $\hat{\boldsymbol{\beta}}$ is given by (3.8.32), is equal to the variance of $\boldsymbol{\lambda}'\hat{\boldsymbol{\beta}}$ attainable in a weighing design of a full rank, p. Indeed, the variance of $\boldsymbol{\lambda}'\hat{\boldsymbol{\beta}}$, where $\hat{\boldsymbol{\beta}}$ is a generalized least-squares solution is, in the present case of an F.W.D.,

$$\text{Var}\,\{\boldsymbol{\lambda}'\hat{\boldsymbol{\beta}}\} = \sigma^2\boldsymbol{\lambda}'\mathbf{S}^-\boldsymbol{\lambda}, \tag{3.8.40}$$

where $\mathbf{S}^-$ is the particular g-inverse of $\mathbf{S}$. But, the condition that $\boldsymbol{\lambda}' = \boldsymbol{\lambda}'\mathbf{H}$, where $\mathbf{H} = \mathbf{S}^{-}\mathbf{S}$, implies

$$\text{(3.8.41)} \qquad \boldsymbol{\lambda}'\hat{\boldsymbol{\beta}} = \boldsymbol{\lambda}'\mathbf{S}^{-}\mathbf{G}'\mathbf{Y}_r = \boldsymbol{\lambda}_r'\mathbf{C}_{11}^{-1}\mathbf{Y}_r.$$

The simplification in (3.8.41) is a consequence of the following: $\mathbf{S} = \mathbf{G}'\mathbf{G}$; $\mathbf{H} = \mathbf{S}^{-}\mathbf{S} = \mathbf{S}^{-}\mathbf{G}'\mathbf{G} = \mathbf{S}^{-}\mathbf{G}'(\mathbf{C}_{11} \mid \mathbf{C}_{12})$. Hence

$$\boldsymbol{\lambda}' = \boldsymbol{\lambda}'\mathbf{H} = (\boldsymbol{\lambda}'\mathbf{S}^{-}\mathbf{G}'\mathbf{C}_{11} \mid \boldsymbol{\lambda}'\mathbf{S}^{-}\mathbf{G}'\mathbf{C}_{12}).$$

It follows that $\boldsymbol{\lambda}'\mathbf{S}^{-}\mathbf{G}'\mathbf{C}_{11} = \boldsymbol{\lambda}_r'$ and $\boldsymbol{\lambda}'\mathbf{S}^{-}\mathbf{G}' = \boldsymbol{\lambda}_r'\mathbf{C}_{11}^{-1}$. Thus from (3.8.41) we imply that

$$\text{(3.8.42)} \qquad \operatorname{Var}\{\boldsymbol{\lambda}'\hat{\boldsymbol{\beta}}\} = \sigma^2\boldsymbol{\lambda}_r'\mathbf{C}_{11}^{-1}\boldsymbol{\lambda}_r.$$

Using the fact that $\mathbf{C}^{-1} = (1/p)\mathbf{C}'$ and (3.8.30), it is easy to verify that

$$\text{(3.8.43)} \qquad \mathbf{S}_{11}^{-1} = \frac{1}{p}[\mathbf{I}_r + \mathbf{C}_{21}'(\mathbf{C}_{22}\mathbf{C}_{22}')^{-1}\mathbf{C}_{21}].$$

Substituting (3.8.43) in (3.8.42) we obtain that the variance of the B.L.U.E· of $\boldsymbol{\lambda}'\boldsymbol{\beta}$ under an F.W.D. of order r and rank r, $1 \leqslant r \leqslant p$, is

$$\text{(3.8.44)} \qquad \operatorname{Var}\{\boldsymbol{\lambda}'\hat{\boldsymbol{\beta}}\} = \frac{\sigma^2}{p}[\boldsymbol{\lambda}_r'\boldsymbol{\lambda}_r + \boldsymbol{\lambda}_r'\mathbf{C}_{21}'(\mathbf{C}_{22}\mathbf{C}_{22}')^{-1}\mathbf{C}_{21}\boldsymbol{\lambda}_r].$$

On the other hand, in a weighing design of a full rank

$$\text{(3.8.45)} \qquad \operatorname{Var}\{\boldsymbol{\lambda}'\hat{\boldsymbol{\beta}}\} = \frac{\sigma^2}{p}\boldsymbol{\lambda}'\boldsymbol{\lambda},$$

but since $\boldsymbol{\lambda}_{p-r}' = -\mathbf{C}_{22}^{-1}\mathbf{C}_{21}\boldsymbol{\lambda}_r$ we have

$$\text{(3.8.46)} \qquad \operatorname{Var}\{\boldsymbol{\lambda}'\hat{\boldsymbol{\beta}}\} = \frac{\sigma^2}{p}(\boldsymbol{\lambda}_r' \mid (-\mathbf{C}_{22}^{-1}\mathbf{C}_{21}\boldsymbol{\lambda}_r)')\begin{pmatrix}\boldsymbol{\lambda}_r \\ \hline -\mathbf{C}_{22}^{-1}\mathbf{C}_{21}\boldsymbol{\lambda}_r\end{pmatrix} = \frac{\sigma^2}{p}[\boldsymbol{\lambda}_r'\boldsymbol{\lambda}_r + \boldsymbol{\lambda}_r'\mathbf{C}_{21}'(\mathbf{C}_{22}\mathbf{C}_{22}')^{-1}\mathbf{C}_{21}\boldsymbol{\lambda}_r],$$

which is identical to (3.8.44). This proves the required result.

3.8.5. INCOMPLETE RANK WITH NUISANCE PARAMETERS

Consider a linear model of full rank p. Suppose that the vector of parameter $\boldsymbol{\beta}$ consists of two subvectors: $\boldsymbol{\alpha}_r$ of order $r \times 1$ and $\boldsymbol{\beta}_{p-r}^*$ of order $p - r$. We are interested in estimating the subvector $\boldsymbol{\alpha}_r$. Is there a fractional design consisting of r row vectors of the original $\mathbf{X}$ ($1 \leqslant r < p$) and having a linear unbiased estimator of α? The answer is obviously negative, for the following reason. Consider a linear model $\mathbf{Y} = \mathbf{X}\boldsymbol{\beta} + \boldsymbol{\epsilon}$, where $\mathbf{X}$ is of a full rank p, $E\{\boldsymbol{\epsilon}\} = \mathbf{0}$, $E\{\boldsymbol{\epsilon}\boldsymbol{\epsilon}'\} = \sigma^2\mathbf{I}$. Let $\mathbf{X}^*_{(r)}$ be a matrix consisting of r row vectors of $\mathbf{X}$, $1 \leqslant r \leqslant p$. The rank of $\mathbf{X}^*_{(r)}$ is r. Let $\boldsymbol{\beta}' = (\boldsymbol{\alpha}'_r, \boldsymbol{\beta}^{*\prime}_{p-r})$. The linear model associated with the fractional design $\mathbf{X}^*_{(r)}$ is

$$\mathbf{Y}^*_r = \mathbf{X}^*_{(r)}\begin{pmatrix}\boldsymbol{\alpha}_r \\ \hdashline \boldsymbol{\beta}^*_{p-r}\end{pmatrix} + \boldsymbol{\epsilon}^*_r, \tag{3.8.47}$$

where $E\{\boldsymbol{\epsilon}^*_r\} = \mathbf{0}$, $E\{\boldsymbol{\epsilon}^*_r\boldsymbol{\epsilon}^{*\prime}_r\} = \sigma^2\mathbf{I}_r$. Write $\mathbf{X}^*_{(r)} = (\mathbf{C} \mid \mathbf{D})$, where $\mathbf{C}$ is of order $r \times r$. Assume that rank $\{\mathbf{C}\} = r$. A linear estimator $\mathbf{A}\mathbf{Y}^*_r$ is an unbiased estimator of $\boldsymbol{\alpha}_r$ if and only if

$$(\mathbf{AC} - \mathbf{I}_r)\boldsymbol{\alpha}_r + \mathbf{AD}\boldsymbol{\beta}^*_{p-r} = \mathbf{0}, \quad \text{for all } (\boldsymbol{\alpha}_r, \boldsymbol{\beta}^*_{p-r}). \tag{3.8.48}$$

But there exists no matrix $\mathbf{A}$ independent of $(\boldsymbol{\alpha}_r, \boldsymbol{\beta}^*_{p-r})$ which satisfies (3.8.48).

We may be interested in the form of the generalized least-squares estimator of $\boldsymbol{\alpha}_r$. Let

$$\mathbf{S} = \left[\begin{array}{c:c}\mathbf{C}'\mathbf{C} & \mathbf{C}'\mathbf{D} \\ \hdashline \mathbf{D}'\mathbf{C} & \mathbf{D}'\mathbf{D}\end{array}\right]$$

and let $\mathbf{S}^-$ be a g-inverse of $\mathbf{S}$. The matrix $\mathbf{Q} = \mathbf{C}'\mathbf{C}$ is nonsingular. A nonsingular matrix $\mathbf{P}$ which reduces $\mathbf{S}$ to its Hermite canonical form $\mathbf{H}$ is

$$\mathbf{P} = \left[\begin{array}{c:c}\mathbf{Q}^{-1} & \mathbf{0} \\ \hdashline \mathbf{E} & \mathbf{I}_{p-r}\end{array}\right], \tag{3.8.49}$$

where $\mathbf{E} = -\mathbf{D}'(\mathbf{C}')^{-1}$. It is easy to verify that

$$\mathbf{H} = \mathbf{PS} = \left[\begin{array}{c:c}\mathbf{I}_r & \mathbf{C}^{-1}\mathbf{D} \\ \hdashline \mathbf{0} & \mathbf{0}\end{array}\right] \tag{3.8.50}$$

and that $\mathbf{SPS} = \mathbf{S}$. Thus $\mathbf{S}^- = \mathbf{P}$ is a g-inverse of $\mathbf{S}$ of maximal rank p. Finally, the normal equations associated with the linear model (3.8.47) are

$$\mathbf{S}\begin{pmatrix}\hat{\boldsymbol{\alpha}}_r \\ \hdashline \hat{\boldsymbol{\beta}}^*_{p-r}\end{pmatrix} = \begin{pmatrix}\mathbf{C}' \\ \hdashline \mathbf{D}'\end{pmatrix}\mathbf{Y}^*_r. \tag{3.8.51}$$

A generalized least-squares estimator of $(\boldsymbol{\alpha}'_r, \boldsymbol{\beta}^*_{p-r})$ is of the form

$$\begin{bmatrix}\hat{\boldsymbol{\alpha}}_r \\ \hat{\boldsymbol{\beta}}^*_{p-r}\end{bmatrix} = \mathbf{S}^-\begin{pmatrix}\mathbf{C}' \\ \hdashline \mathbf{D}'\end{pmatrix}\mathbf{Y}^*_r + (\mathbf{I}_p - \mathbf{H})\mathbf{Z}, \tag{3.8.52}$$

where $\mathbf{Z}$ is arbitrary. Simple manipulations yield the linear manifold of estimators

$$\begin{pmatrix}\hat{\boldsymbol{\alpha}}_r \\ \hat{\boldsymbol{\beta}}^*_{p-r}\end{pmatrix} = \begin{pmatrix}\mathbf{C}^{-1}(\mathbf{Y}^*_r - \mathbf{D}\mathbf{Z}^*_{p-r}) \\ \mathbf{Z}^*_{p-r}\end{pmatrix}, \tag{3.8.53}$$

where $\mathbf{Z}^*_{p-r}$ is a subvector consisting of the last $p - r$ components of $\mathbf{Z}$. Equation 3.8.53 implies immediately that an unbiased estimator of $(\boldsymbol{\alpha}_r, \boldsymbol{\beta}^*_{p-r})$ exists if and only if $\boldsymbol{\beta}^*_{p-r}$ is known and is given by substituting $\mathbf{Z}^*_{p-r} = \boldsymbol{\beta}^*_{p-r}$ in (3.8.53). But in this case the whole linear model is reduced to one of a full rank r, and there is no problem in finding the B.L.U.E. of $\boldsymbol{\alpha}_r$. The expectation of the generalized L.S.E. $\hat{\boldsymbol{\alpha}}_r$ in (3.8.53) is

$$E\{\hat{\boldsymbol{\alpha}}_r\} = \boldsymbol{\alpha}_r + \mathbf{C}^{-1}\mathbf{D}(\boldsymbol{\beta}^*_{p-r} - \mathbf{Z}^*_{p-r}). \tag{3.8.54}$$

If there exists a p-dimensional vector $\boldsymbol{\xi}_p$, with non-negative coefficients, which is orthogonal to the $(p - r)$ column vectors of $\mathbf{X}$ corresponding to the components of $\boldsymbol{\beta}^*_{p-r}$, we can combine a *randomized* fractional design of rank r, whose r row vectors are chosen from $\mathbf{X}$ at random. Such a randomized fractional design may yield an unbiased estimator of $\boldsymbol{\alpha}_r$. (See Zacks [2].)

3.8.6. CONDITIONS UNDER WHICH B.L.U.E. AND SIMPLE L.S.E. COINCIDE

Consider the general linear model $\mathbf{Y} = \mathbf{X}\boldsymbol{\beta} + \boldsymbol{\epsilon}$ of a full rank, where the covariance matrix of $\boldsymbol{\epsilon}$ is $E\{\boldsymbol{\epsilon}\boldsymbol{\epsilon}'\} = \sigma^2\mathbf{V}$, $\mathbf{V}$ a known non-negative definite matrix. As mentioned in Section 3.8.1, if $\mathbf{V}$ is nonsingular and $\mathbf{V} = \mathbf{D}\mathbf{D}'$ we make the transformation $\mathbf{Y}^* = \mathbf{D}^{-1}\mathbf{Y}$ and the B.L.U.E. of $\boldsymbol{\beta}$ is the L.S.E. of the reduced model, namely,

$$\hat{\boldsymbol{\beta}} = (\mathbf{X}'\mathbf{V}^{-1}\mathbf{X})^{-1}\mathbf{X}'\mathbf{V}^{-1}\mathbf{Y}, \tag{3.8.55}$$

having a covariance matrix of

$$\boldsymbol{\Sigma}\{\hat{\boldsymbol{\beta}}\} = \sigma^2(\mathbf{X}'\mathbf{V}^{-1}\mathbf{X})^{-1}. \tag{3.8.56}$$

Even if $\mathbf{V}$ is nonsingular it might be a difficult task to invert $\mathbf{V}$. In any case, the least-squares estimator

$$\tilde{\boldsymbol{\beta}} = (\mathbf{X}'\mathbf{X})^{-1}\mathbf{X}'\mathbf{Y} \tag{3.8.57}$$

is simpler than the B.L.U.E. $\hat{\boldsymbol{\beta}}$. By definition of the B.L.U.E. $\hat{\boldsymbol{\beta}}$, $\boldsymbol{\Sigma}(\tilde{\beta}) - \boldsymbol{\Sigma}(\hat{\boldsymbol{\beta}}) \geqslant 0$ (non-negative definite). In certain cases, the two estimators coincide. We investigate in this subsection the conditions for the equivalence of the B.L.U.E. and the L.S.E.

We start with a relatively simple case discussed by Magness and McGuire [1]. Consider the linear regression

$$\mathbf{Y} = \beta\mathbf{X} + \boldsymbol{\epsilon}, \tag{3.8.58}$$

in which β is an unknown scalar; $\mathbf{Y}$ $\mathbf{X}$ and $\boldsymbol{\epsilon}$ are $p \times 1$ vectors; $\mathbf{X}'\mathbf{X} = 1$; and the random vector $\boldsymbol{\epsilon}$ has mean $\mathbf{0}$ and a nonsingular covariance matrix $\mathbf{R}$.

Furthermore, $E\epsilon_i^2 = 1$ for all $i = 1, \ldots, p$. Matrix $\mathbf{R}$ is therefore a correlation matrix. Let $\lambda_1, \ldots, \lambda_p$ be the eigenvalues of $\mathbf{R}$ (all of which are positive) and $\boldsymbol{\psi}_1, \ldots, \boldsymbol{\psi}_p$ the associated eigenvectors. We have $\mathbf{R}\boldsymbol{\psi}_i = \lambda_i\boldsymbol{\psi}_i$ $(i = 1, \ldots, p)$ and $\boldsymbol{\psi}_i'\boldsymbol{\psi}_j = 0$ if $i \neq j$. The set of orthonormal eigenvectors $\boldsymbol{\psi}_1, \ldots, \boldsymbol{\psi}_p$ constitutes a basis to the p-dimensional vector space, and we can write

$$\mathbf{X} = \sum_{i=1}^{p} c_i\boldsymbol{\psi}_i.$$

Since $\mathbf{X}'\mathbf{X} = 1$ we have $\sum_{i=1}^{p} c_i^2 = 1$. The B.L.U.E. of β is

$$\hat{\beta} = (\mathbf{X}'\mathbf{R}^{-1}\mathbf{X})^{-1}\mathbf{X}'\mathbf{R}^{-1}\mathbf{Y}. \tag{3.8.59}$$

The variance of $\hat{\beta}$ is

$$\begin{aligned} \text{Var}\,\{\hat{\beta}\} &= (\mathbf{X}'\mathbf{R}^{-1}\mathbf{X})^{-1} \\ &= \frac{1}{(\sum_{i=1}^{p} c_i\boldsymbol{\psi}_i)'\mathbf{R}^{-1}(\sum_{j=1}^{p} c_j\boldsymbol{\psi}_j)}. \end{aligned} \tag{3.8.60}$$

But since $\boldsymbol{\psi}_1, \ldots, \boldsymbol{\psi}_p$ are also the eigenvectors of $\mathbf{R}^{-1}$ with eigenvalues λ_i^{-1} $(i = 1, \ldots, p)$ we have

$$\sum_{i=1}^{p} c_i\mathbf{R}^{-1}\boldsymbol{\psi}_i = \sum_{i=1}^{p} \frac{c_i}{\lambda_i}\boldsymbol{\psi}_i.$$

Therefore

$$\text{Var}\,\{\hat{\beta}\} = \left(\sum_{i=1}^{p} \frac{c_i^2}{\lambda_i}\right)^{-1}.$$

On the other hand, the L.S.E. of β is

$$\tilde{\beta} = (\mathbf{X}'\mathbf{X})^{-1}\mathbf{X}'\mathbf{Y} = \mathbf{X}'\mathbf{Y} \tag{3.8.61}$$

with a variance of

$$\begin{aligned} \text{Var}\,\{\tilde{\boldsymbol{\beta}}\} &= (\mathbf{X}'\mathbf{R}\mathbf{X}) \\ &= \left(\sum_{i=1}^{p} \lambda_i c_i^2\right). \end{aligned} \tag{3.8.62}$$

Generally, $\text{Var}\,\{\hat{\beta}\} \leqslant \text{Var}\,\{\tilde{\beta}\}$. Under what condition does the L.S.E. have the same variance as the B.L.U.E.? Magness and McGuire prove that $\text{Var}\,\{\tilde{\beta}\} = \text{Var}\,\{\hat{\beta}\}$ if and only if $\mathbf{X}$ is an eigenvector of $\mathbf{R}$. If λ is the associated eigenvalue then the common variance is λ. Indeed, if $\mathbf{X}$ is an eigenvector of $\mathbf{R}$ with eigenvalue λ then $\mathbf{R}^{-1}\mathbf{X} = (1/\lambda)\mathbf{X}$ and

$$\hat{\beta} = \mathbf{X}'\mathbf{Y} = \tilde{\beta}. \tag{3.8.63}$$

Furthermore,

$$\text{Var}\,\{\hat{\beta}\} = \text{Var}\,\{\tilde{\beta}\} = \mathbf{X}'\mathbf{R}\mathbf{X} = \lambda\mathbf{X}'\mathbf{X} = \lambda. \tag{3.8.64}$$

On the other hand, assume that $\mathrm{Var}\{\hat{\beta}\} = \mathrm{Var}\{\tilde{\beta}\} = \lambda$. We want to prove that this implies that $\mathbf{X}$ is an eigenvector of $\mathbf{R}$ with eigenvalue λ. Let $\mathbf{X}^{\perp}$ be any unit vector perpendicular to $\mathbf{X}$. We can write

$$\mathbf{R}\mathbf{X} = \lambda\mathbf{X} + \mu\mathbf{X}^{\perp}, \tag{3.8.65}$$

where λ, μ are scalars. Multiply (3.8.65) by $\mathbf{X}'$; we have

$$\mathbf{X}'\mathbf{R}\mathbf{X} = \lambda > 0. \tag{3.8.66}$$

Multiply (3.8.65) from the left by $\mathbf{R}^{-1}$; then

$$\mathbf{X} = \lambda\mathbf{R}^{-1}\mathbf{X} + \mu\mathbf{R}^{-1}\mathbf{X}^{\perp}, \tag{3.8.67}$$

or

$$\mathbf{R}^{-1}\mathbf{X} = \frac{1}{\lambda}\mathbf{X} - \frac{\mu}{\lambda}\mathbf{R}^{-1}\mathbf{X}^{\perp}. \tag{3.8.68}$$

Multiplying again by $\mathbf{X}'$ yields

$$\mathbf{X}'\mathbf{R}^{-1}\mathbf{X} = \frac{1}{\lambda} - \frac{\mu}{\lambda}\mathbf{X}'\mathbf{R}^{-1}\mathbf{X}^{\perp}. \tag{3.8.69}$$

Hence

$$\lambda = \mathrm{Var}\{\hat{\boldsymbol{\beta}}\} = (\mathbf{X}'\mathbf{R}^{-1}\mathbf{X})^{-1} = \lambda(1 - \mu\mathbf{X}'\mathbf{R}^{-1}\mathbf{X}^{\perp})^{-1}. \tag{3.8.70}$$

The last equation can hold ($\lambda > 0$) if $\mu = 0$ or $\mathbf{X}'\mathbf{R}^{-1}\mathbf{X}^{\perp} = 0$. But $\mathbf{X}^{\perp}$ perpendicular to $\mathbf{R}^{-1}\mathbf{X}$ implies that $\mu = 0$. Hence $\mathbf{R}\mathbf{X} = \lambda\mathbf{X}$ and $\mathbf{R}^{-1}\mathbf{X} = (1/\lambda)\mathbf{X}$. This proves that $\mathbf{X}$ is an eigenvector of $\mathbf{R}$ with eigenvalue λ. This result is now given in a more general fashion by a theorem proven by Zyskind.

Theorem 3.8.4. (Zyskind [1].) *Consider a linear model* $\mathbf{Y} = \mathbf{X}\boldsymbol{\beta} + \boldsymbol{\epsilon}$ *where* $\mathbf{Y}$ *is n-dimensional,* $\mathbf{X}$ *of rank* $1 \leqslant r \leqslant n$, *and* $\boldsymbol{\epsilon}$ *a random vector satisfying* $E\{\boldsymbol{\epsilon}\} = \mathbf{0}$, $E\{\boldsymbol{\epsilon}\boldsymbol{\epsilon}'\} = \sigma^2\mathbf{V}$, *where* $\mathbf{V}$ *is a symmetric, non-negative definite matrix. Let* $\boldsymbol{\lambda}'\boldsymbol{\beta}$ *be an estimable linear function. A necessary and sufficient condition for the B.L.U.E. of* $\boldsymbol{\lambda}'\boldsymbol{\beta}$ *to coincide with the L.S.E. is that there exist a subset of r eigenvectors of* $\mathbf{V}$, *which constitutes a basis of the linear manifold generated by the column vectors of* $\mathbf{X}$.

Proof. (i) *Sufficiency*. Suppose that $\mathbf{V}$ has r orthogonal eigenvectors which form a basis for the linear manifold generated by the rows of $\mathbf{X}$. We denote by $\mathcal{M}(X)$ this column space of $\mathbf{X}$ and by $\mathbf{P}_0$ the $r \times n$ matrix whose rows consist of this r eigenvectors. The other $(n - r)$ eigenvectors of $\mathbf{V}$ form a basis of the orthogonal complement of $\mathcal{M}(X)$. Let $\mathbf{P}_1$ be the $(n - r) \times n$ matrix whose row vectors are these $(n - r)$ eigenvectors. We then have $\mathbf{P}_1\mathbf{X} = 0$. Let $\mathbf{P}$ be the matrix $\begin{bmatrix}\mathbf{P}_0 \\ \hline \mathbf{P}_1\end{bmatrix}$. The transformation $\mathbf{Z} = \mathbf{P}\mathbf{Y}$ reduces the linear model to a canonical form,

$$\mathbf{Z} = \begin{bmatrix} \mathbf{P}_0\mathbf{X}\boldsymbol{\beta} \\ \hdashline \mathbf{0} \end{bmatrix} + \boldsymbol{\epsilon}^*, \tag{3.8.71}$$

where $E\{\boldsymbol{\epsilon}^*\} = 0$ and $E\{\boldsymbol{\epsilon}^*\boldsymbol{\epsilon}^{*\prime}\} = \sigma^2\mathbf{PVP}' = \sigma^2\boldsymbol{\Lambda}$. The matrix $\boldsymbol{\Lambda}$ is a diagonal matrix whose diagonal elements are the eigenvalues of $\mathbf{V}$ (non-negative). If rank $(\mathbf{V}) < n$, then few of these diagonal elements are zero. This leads to the conclusion that the B.L.U.E. of any estimable function $\boldsymbol{\lambda}'\boldsymbol{\beta}$ should be a function only of $\mathbf{Z}_0 = \mathbf{P}_0\mathbf{Y}$. Indeed, let $\mathbf{a}'\mathbf{Y}$ be any unbiased estimator of $\boldsymbol{\lambda}'\boldsymbol{\beta}$. Let $\mathbf{a} = \mathbf{b} + \mathbf{b}^{\perp}$ where $\mathbf{b}$ is the orthogonal projection of $\mathbf{a}$ on the linear manifold generated by the column vectors of $\mathbf{X}$ and $\mathbf{b}^{\perp}$ is the orthogonal complement of $\mathbf{b}$. Now $\mathbf{b}'\mathbf{Y}$ is also an unbiased estimator of $\boldsymbol{\lambda}'\boldsymbol{\beta}$. To show this it is sufficient to notice that $\mathbf{a}'\mathbf{X}\boldsymbol{\beta} = \boldsymbol{\lambda}'\boldsymbol{\beta}$ and $(\mathbf{b}^{\perp})'\mathbf{X} = \mathbf{0}$. Hence $E\{\mathbf{b}'\mathbf{Y}\} = \boldsymbol{\lambda}'\boldsymbol{\beta}$.

The variance of $\mathbf{b}'\mathbf{Y}$ is smaller than or equal to that of $\mathbf{a}'\mathbf{Y}$. Indeed,

$$\begin{aligned} \text{Var}\,\{\mathbf{a}'\mathbf{Y}\} &= \sigma^2\mathbf{a}'\mathbf{V}\mathbf{a} \\ &= \sigma^2(\mathbf{b} + \mathbf{b}^{\perp})'\mathbf{P}'\boldsymbol{\Lambda}\mathbf{P}(\mathbf{b} + \mathbf{b}^{\perp}). \end{aligned} \tag{3.8.72}$$

Let

$$\boldsymbol{\Lambda} = \begin{bmatrix} \boldsymbol{\Lambda}_0 & \mathbf{0} \\ \mathbf{0} & \boldsymbol{\Lambda}_1 \end{bmatrix}.$$

Then since

$$\mathbf{Pb} = \begin{bmatrix} \mathbf{P}_0 \\ \hdashline \mathbf{P}_1 \end{bmatrix}\mathbf{b} = \begin{bmatrix} \mathbf{P}_0\mathbf{b} \\ \hdashline \mathbf{0} \end{bmatrix}, \tag{3.8.73}$$

and

$$\mathbf{Pb}^{\perp} = \begin{bmatrix} \mathbf{0} \\ \hdashline \mathbf{P}_1\mathbf{b}^{\perp} \end{bmatrix}, \tag{3.8.74}$$

we have

$$\begin{aligned} \text{Var}\,\{\mathbf{a}'\mathbf{Y}\} &= \sigma^2\{\mathbf{b}'\mathbf{P}_0'\boldsymbol{\Lambda}_0\mathbf{P}_0\mathbf{b} + (\mathbf{P}_1\mathbf{b}^{\perp})'\boldsymbol{\Lambda}_1\,\mathbf{P}_1\mathbf{b}^{\perp}\} \\ &\geqslant \sigma^2\mathbf{b}'\mathbf{P}_0'\boldsymbol{\Lambda}_0\,\mathbf{P}_0\mathbf{b} \\ &= \sigma^2\mathbf{b}'\mathbf{V}\mathbf{b} \\ &= \text{Var}\,\{\mathbf{b}'\mathbf{Y}\}. \end{aligned} \tag{3.8.75}$$

We infer from (3.8.75) that the B.L.U.E. of $\boldsymbol{\lambda}'\boldsymbol{\beta}$ is $\mathbf{b}'\mathbf{Y}$, in which $\mathbf{b} \in \mathcal{M}(\mathbf{X})$. Moreover, since $\mathbf{P}_0$ is a basis of $\mathcal{M}(\mathbf{X})$, $\mathbf{b}'\mathbf{Y} = \mathbf{C}'\mathbf{Z}_0$, where $\mathbf{C} = \mathbf{P}_0\mathbf{b}$. Finally, since $\mathcal{M}(\mathbf{X}) = \mathcal{M}(\mathbf{P}_0\mathbf{X})$, the B.L.U.E. is the L.S.E. of $\boldsymbol{\lambda}'\boldsymbol{\beta}$, which is the unique unbiased estimator of $\boldsymbol{\lambda}'\boldsymbol{\beta}$ for which $\boldsymbol{\lambda} \in \mathcal{M}(\mathbf{P}_0\mathbf{X})$.

(ii) *Necessity*. Suppose that the B.L.U.E. of an estimable function $\boldsymbol{\lambda}'\boldsymbol{\beta}$ coincided with the L.S.E. Let $\mathbf{P} = \begin{bmatrix} \mathbf{P}_0 \\ \hdashline \mathbf{P}_1 \end{bmatrix}$ be an orthogonal matrix such that

$\mathbf{P}_1\mathbf{X} = 0$. Make the transformation to the canonical form,

$$\mathbf{Z} = \mathbf{P}\mathbf{Y} = \left[\begin{array}{c} \mathbf{P}_0\mathbf{X}\boldsymbol{\beta} \\ \hdashline \mathbf{0} \end{array}\right] + \boldsymbol{\epsilon}^*. \tag{3.8.76}$$

Since $\mathbf{P}_0'\mathbf{P}_0 = \mathbf{I}_r$, $\mathbf{Z}_0 = \mathbf{P}_0\mathbf{Y}$ is the B.L.U.E. of $\boldsymbol{\zeta}_0 = \mathbf{P}_0\mathbf{X}\boldsymbol{\beta}$. This implies that the covariance matrix of $\boldsymbol{\epsilon}^*$ is of the form

$$\boldsymbol{\Sigma}(\boldsymbol{\epsilon}^*) = \sigma^2\left[\begin{array}{c:c} \mathbf{W}_0 & \mathbf{0} \\ \hdashline \mathbf{0} & \mathbf{W}_1 \end{array}\right]. \tag{3.8.77}$$

Otherwise, the L.S.E. $\mathbf{Z}_0$ of $\boldsymbol{\zeta}$ is not B.L.U.E. Furthermore, let $\mathbf{R}_0$ and $\mathbf{R}_1$ be orthogonal matrices such that $\mathbf{R}_0\mathbf{W}_0\mathbf{R}_0' = \mathbf{D}_0$ and $\mathbf{R}_1\mathbf{W}_1\mathbf{R}_1' = \mathbf{D}_0$ where $\mathbf{D}_i$ $(i = 0, 1)$ are diagonal. Define the orthogonal matrix

$$\mathfrak{S} = \left[\begin{array}{c:c} \mathbf{R}_0 & \mathbf{0} \\ \hdashline \mathbf{0} & \mathbf{R}_1 \end{array}\right]\left[\begin{array}{c} \mathbf{P}_0 \\ \hdashline \mathbf{P}_1 \end{array}\right] = \left[\begin{array}{c} \mathbf{R}_0\mathbf{P}_0 \\ \hdashline \mathbf{R}_1\mathbf{P}_1 \end{array}\right].$$

We have

$$\mathbf{S}\mathbf{V}\mathbf{S}' = \left[\begin{array}{c:c} \mathbf{D}_0 & \mathbf{0} \\ \hdashline \mathbf{0} & \mathbf{0}_1 \end{array}\right].$$

Thus the column vectors of $\mathbf{S}'$ form a set of n eigenvectors of $\mathbf{V}$. The row vectors of $\mathbf{R}_0\mathbf{P}_0$ are linear combinations of those of $\mathbf{P}_0$, and similarly those of $\mathbf{R}_1\mathbf{P}_1$. Hence $\mathbf{S}_0 = \mathbf{R}_0\mathbf{P}_0$ and $\mathbf{S}_1 = \mathbf{R}_1\mathbf{P}_1$ form respectively orthonormal bases for $\mathcal{M}(\mathbf{X})$ and $\mathcal{M}^\perp(\mathbf{X})$. (Q.E.D.)

It might be difficult in practice to apply this theorem directly. The following criterion for the equivalence of B.L.U.E. and L.S.E. may, however, be useful. This criterion is derived from the result of the previous theorem. The L.S.E. of $\boldsymbol{\beta}$, say $\hat{\boldsymbol{\beta}}$, satisfies the equation $\mathbf{X}\hat{\boldsymbol{\beta}} = \mathbf{T}\mathbf{Y}$, where the matrix $\mathbf{T}$ is the orthogonal projection of $\mathbf{Y}$ on $\mathcal{M}(\mathbf{X})$ given by $\mathbf{T} = \mathbf{X}(\mathbf{X}'\mathbf{X})^-\mathbf{X}'$, where $(\mathbf{X}'\mathbf{X})^-$ is any g-inverse of $\mathbf{X}'\mathbf{X}$ satisfying $(\mathbf{X}'\mathbf{X})(\mathbf{X}'\mathbf{X})^-(\mathbf{X}'\mathbf{X}) = (\mathbf{X}'\mathbf{X})$. The linear manifold generated by the column vectors of $\mathbf{T}$ is again $\mathcal{M}(\mathbf{X})$. The matrix $\mathbf{T}$ is a symmetric idempotent matrix and thus has r eigenvalues equal to 1 and $(n - r)$ eigenvalues equal to 0. Corresponding to the r eigenvalues 1, $\mathbf{T}$ has r eigenvectors in $\mathcal{M}(\mathbf{X})$, and corresponding to the $(n - r)$ eigenvalues 0 it has $(n - r)$ eigenvectors in $\mathcal{M}^\perp(\mathbf{X})$. Thus under the condition of Theorem 3.8.4, the row vectors of $\mathbf{P}$ are also eigenvectors of $\mathbf{T}$. Thus the matrices $\mathbf{V}$ and $\mathbf{T}$ are diagonalized by the same orthogonal matrix and therefore commute. Finally, since $\mathbf{V}$ and $\mathbf{T}$ are symmetric, $\mathbf{V}\mathbf{T} = \mathbf{T}\mathbf{V} = \mathbf{T}'\mathbf{V}' = (\mathbf{V}\mathbf{T})'$, we conclude that $\mathbf{V}\mathbf{T}$ is a *symmetric* matrix. It is relatively easy to compute $\mathbf{T}$ and to check whether $\mathbf{V}\mathbf{T}$ is a symmetric matrix. For further interesting results see Zyskind [1].

3.9. BEST LINEAR COMBINATIONS OF ORDER STATISTICS

Unbiased estimators which are linear combinations of order statistics are in many cases B.L.U.E. For example, the sample mean is a particular linear combination of order statistics. In the case of a rectangular distribution $R(\theta_1, \theta_2)$, $-\infty < \theta_1 < \theta_2 < \infty$, the B.L.U.E.'s of θ_1 and θ_2 are linear combinations of the extreme statistics $X_{(1)}$ and $X_{(n)}$. In many other cases, a linear combination of order statistics may not yield the best unbiased estimator. This is the case, for instance, when we are estimating the standard deviation σ of a normal distribution $\mathcal{N}(\mu, \sigma^2)$. The B.U.E. of σ, based on a sample $X_1, \ldots, X_n$ $(n \geqslant 2)$, is $\hat{\sigma}_n = C_n \cdot [\sum_{i=1}^n (X_1 - \bar{X})^2]^{1/2}$, where $C_n = \Gamma[(n-1)/2]/\sqrt{(2)}\,\Gamma(n/2)$. An unbiased estimator of σ in the normal case can be obtained faster as a simple function of the range, which is $\tilde{\sigma}_n = (X_{(n)} - X_{(1)})/d_n$ where d_n $(n \geqslant 2)$ is the expectation of the range of a random sample from the standard normal distribution $\mathcal{N}(0, 1)$. This estimator is in common use especially in the field of quality control. Tables of the d_n values can be found in every book on quality control (see *Biometrika Tables for Statisticians*, Vol. 1 [1]). The estimator $\tilde{\sigma}_n$ is less efficient (larger variance) than $\hat{\sigma}_n$, and its efficiency declines rapidly as the sample size n increases. This is due to the fact that $\tilde{\sigma}_n$ is not a function of the sufficient statistic $(\sum_{i=1}^n X_i, \sum_{i=1}^n X_i^2)$, and the amount of information in the sample that it discards is considerable. A better linear estimator of σ, based on the order statistics, can be obtained by a method which will be discussed presently. Although all estimators of σ which are linear combinations of order statistics are less efficient than the best unbiased estimator $\hat{\sigma}_n$, there are many advantages in the use of such linear estimators, especially in large sample situations. A good discussion of the usefulness of such less efficient estimators is provided by Mosteller [1]. The advantage of estimators based on order statistics is especially great in situations where trimming or censoring of observations in the extremes is part of the experimental model (e.g., in life testing experiments). The efficiency of such estimation procedures, when the distributions are normal, exponential, gamma, rectangular, extreme-value, etc., have been studied and discussed by Greenberg and Sarhan (see [1], Ch. 10C, 10D, 11B, 11C, 12A, 12C, 12D).

We present now the least-squares theory, as developed by E. H. Lloyd (see [1], Ch. 3, and [1]), for estimating scale and location parameters by linear combinations of order statistics.

Let $\mathcal{F}_0$ be a class of distributions depending on scale and location parameters only. That is, every distribution function in $\mathcal{F}_0$ is of the form $F[(x - \mu)/\sigma]$, $-\infty < \mu < \infty, 0 < \sigma < \infty$. The quantity μ designates the location parameter and σ the scale parameter. The distribution function of the standardized random variable $U = (X - \mu)/\sigma$ is $F(\mu)$, which is of a known functional form. Let $X_{(1)} \leqslant X_{(2)} \leqslant \cdots \leqslant X_{(n)}$ be the order statistic and $U_{(r)} = (X_{(r)} - \mu)/\sigma$. Basic quantities for the estimation of μ and σ are the parameters

$$\alpha_r = E\{U_{(r)}\}, \qquad r = 1, 2, \ldots, n,$$
(3.9.1)
$$v_{rs} = \text{cov}\,(U_{(r)}, U_{(s)}), \qquad r, s = 1, \ldots, n.$$

Let $\boldsymbol{\alpha}' = (\alpha_1, \ldots, \alpha_n)$, $\mathbf{V} = \|v_{rs}; r, s = 1, \ldots, n\|$ be the covariance matrix and $\mathbf{X}' = (X_{(1)}, \ldots, X_{(n)})$. We have the linear model

$$\mathbf{X} = (\mathbf{1}_n \mid \boldsymbol{\alpha})\begin{pmatrix}\mu \\ \sigma\end{pmatrix} + \boldsymbol{\epsilon}, \tag{3.9.2}$$

where $\mathbf{1}'_n = (1, 1, \ldots, 1)$; $\boldsymbol{\epsilon}' = (\epsilon_1, \ldots, \epsilon_n)$; $E\{\boldsymbol{\epsilon}\} = \mathbf{0}$; and $E\{\boldsymbol{\epsilon\epsilon}'\} = \sigma^2\mathbf{V}$. *Assume* that $\mathbf{V}$ is *positive* definite (it is always non-negative definite). Then according to the least-squares theory, the L.S.E. of (μ, σ) is

$$\begin{pmatrix}\hat{\mu} \\ \hat{\sigma}\end{pmatrix} = \begin{bmatrix}\mathbf{1}'_n\mathbf{V}^{-1}\mathbf{1}_n & \mathbf{1}'_n\mathbf{V}^{-1}\boldsymbol{\alpha} \\ \cdot & \boldsymbol{\alpha}'\mathbf{V}^{-1}\boldsymbol{\alpha}\end{bmatrix}^{-1}\begin{bmatrix}\mathbf{1}'_n\mathbf{V}^{-1}\mathbf{X} \\ \boldsymbol{\alpha}'\mathbf{V}^{-1}\mathbf{X}\end{bmatrix}. \tag{3.9.3}$$

If we let

$$\mathbf{C} = \frac{\mathbf{V}^{-1}(\mathbf{1}_n\boldsymbol{\alpha}' - \boldsymbol{\alpha}\mathbf{1}'_n)\mathbf{V}^{-1}}{\gamma}$$

where

$$\gamma = (\mathbf{1}'_n\mathbf{V}^{-1}\mathbf{1}_n)(\boldsymbol{\alpha}'\mathbf{V}^{-1}\boldsymbol{\alpha}) - (\mathbf{1}'_n\mathbf{V}^{-1}\boldsymbol{\alpha})^2,$$

then the L.S.E.'s of μ and σ are

(3.9.4)
$$\hat{\mu} = -\boldsymbol{\alpha}'\mathbf{C}\mathbf{X},$$
$$\hat{\sigma} = \mathbf{1}'_n\mathbf{C}\mathbf{X}.$$

According to the Gauss-Markov theorem, $\hat{\mu}$ and $\hat{\sigma}$ are the best linear combinations of the order statistics. Their variances are

$$V\{\hat{\mu}\} = \frac{\sigma^2(\boldsymbol{\alpha}'\mathbf{V}^{-1}\boldsymbol{\alpha})}{\gamma}$$
(3.9.5)
$$V\{\hat{\sigma}\} = \frac{\sigma^2(\mathbf{1}'_n\mathbf{V}^{-1}\mathbf{1}_n)}{\gamma},$$

and

$$\text{cov}\,(\hat{\mu}, \hat{\sigma}) = -\frac{\sigma^2(\mathbf{1}'_n\mathbf{V}^{-1}\boldsymbol{\alpha})}{\gamma}. \tag{3.9.6}$$

We now present a simple example of such a linear estimation for the case of a negative exponential distribution.

Example 3.16. Let $X_1, X_2, \ldots, X_n$ be i.i.d. random variables having a 2-parameter negative exponential distribution; that is,

$$F\left(\frac{x-\mu}{\sigma}\right) = \begin{cases} 0, & \text{if } x < \mu, \\ 1 - \exp\left\{-\dfrac{x-\mu}{\sigma}\right\}, & \text{if } x \geqslant \mu, \end{cases} \tag{3.9.7}$$

where $-\infty < \mu < \infty$ and $0 < \sigma < \infty$. Let $X_{(1)} \leqslant \cdots \leqslant X_{(n)}$ be the order statistics and $U_{(r)} = (X_{(r)} - \mu)/\sigma$.

We prove first that

$$\alpha_r = E\{U_{(r)}\} = \sum_{i=1}^{r} (n - i + 1)^{-1}, \qquad r = 1, 2, \ldots, n, \tag{3.9.8}$$

and then establish that, for all $1 \leqslant r < s \leqslant n$,

$$V\{U_{(r)}\} = \operatorname{cov}(U_{(r)}, U_{(s)}) = \sum_{i=1}^{r} (n - i + 1)^{-2}. \tag{3.9.9}$$

Consider the negative exponential distribution $F(x) = 1 - e^{-x}$ for $x \geqslant 0$. Suppose that the random variable Z, having a distribution $F(x)$, represents an observation on the lifetime of a certain apparatus. Let $Z_1, \ldots, Z_n$ be independent identically distributed random variables, representing the lifetime of n such apparatuses. Let $U_{(1)}, U_{(2)}, \ldots, U_{(n)}$ be the corresponding order statistics, representing the time of first failure, second failure, and so on. Since $Z_1, \ldots, Z_n$ are i.i.d., having a distribution $F(x)$, the distribution function of the first failure time $U_{(1)}$ is $F(nx)$, and hence $E\{U_{(1)}\} = 1/n$ and $\operatorname{Var}\{U_{(1)}\} = 1/n^2$.

Since the stochastic process governing the failure time points is a Poisson process with a mean of one failure per time unit, the distribution of $U_{(2)} - U_{(1)}$ is again negative exponential with an expectation of

$$E\{U_{(2)} - U_{(1)}\} = (n-1)^{-1},$$

and

$$\operatorname{Var}\{U_{(2)} - U_{(1)}\} = (n-1)^{-2}.$$

Furthermore, since $U_{(1)}$ and $U_{(2)} - U_{(1)}$ are independent we have

$$E\{U_{(2)}\} = n^{-1} + (n-1)^{-1},$$

and

$$\operatorname{Var}\{U_{(2)}\} = n^{-2} + (n-1)^{-2}.$$

In a similar manner (3.9.8) and the first part of (3.9.9) are proven for any $r = 1, \ldots, n$. Finally, to prove that $\operatorname{cov}(U_{(r)}, U_{(s)}) = \operatorname{Var}\{U_{(r)}\}$ for all $r = s$, we write $U_{(s)} = U_{(r)} + (U_{(s)} - U_{(r)})$. Since the increment $U_{(s)} - U_{(r)}$

is independent of $U_{(r)}$ the required result is proven. Thus

$$\boldsymbol{\alpha}' = \left(n^{-1}, n^{-1} + (n-1)^{-1}, \ldots, \sum_{i=1}^{r} (n-i+1)^{-1}, \ldots, \sum_{i=1}^{n} (n-i+1)^{-1}\right),$$

and

$$\mathbf{V} = \begin{bmatrix} v_{11} & v_{11} & \cdots & v_{11} \\ \cdot & v_{22} & \cdots & v_{22} \\ & & \ddots & \\ \cdot & \cdot & & v_{nn} \end{bmatrix},$$

where $v_{rs} = \sum_{r}^{i=1} (n-i+1)^{-2}$ for all $r \leqslant s$, $r = 1, \ldots, n$; and by the symmetry of $\mathbf{V}$, $v_{rs} = v_{sr}$. Greenberg and Sarhan [1] show that the inverse of $\mathbf{V}$ is

$$\mathbf{V}^{-1} = \begin{bmatrix} (n-1)^2 + n^2 & -(n-1)^2 & & & \\ -(n-1)^2 & (n-2)^2 + (n-1)^2 & -(n-2)^2 & & \mathbf{O} \\ & -(n-2)^2 & \ddots & \ddots & \\ & & \ddots & \ddots & -1 \\ \mathbf{O} & & & -1 & 1 \end{bmatrix}.$$

It is easy to verify that $\mathbf{1}_n'\mathbf{V}^{-1} = (n^2, \mathbf{0}_{n-1}')$ and $\boldsymbol{\alpha}'\mathbf{V}' = \mathbf{1}_n'$. From this we obtain that $\gamma = n^2(n-1)$ and

$$\hat{\mu} = \frac{nX_{(1)} - \bar{X}}{n-1}$$

$$\hat{\sigma} = \frac{n\bar{X} - X_{(1)}}{n-1}.$$

The covariance matrix of these estimators is:

$$\mathbf{\Sigma} = \sigma^2 \begin{bmatrix} n^{-1}(n-1)^{-1} & -n^{-1}(n-1)^{-1} \\ \cdot & (n-1)^{-1} \end{bmatrix}. \qquad \blacksquare$$

As shown in Example 2.3, $(X_{(1)}, \bar{X})$ is a minimal sufficient statistic for the 2-parameter negative exponential family. Hence by the Blackwell-Rao theorem, $\hat{\mu}$ and $\hat{\sigma}$ are not only B.L.U.E.'s but also B.U.E.'s. The derivation of $\hat{\mu}$ and $\hat{\sigma}$ on the basis of the Blackwell-Rao theorem is immediate and simple, while the use of the least-squares theory is very tedious (algebraically). Furthermore, in cases where the moments of $U_{(r)}$ have no simple formula (for instance in the normal case) the whole procedure is impractical, unless we are satisfied with numerical solutions only. To overcome this difficulty alternative estimation procedures were developed by Gupta, Jung, Blom, and others. The reader will find very interesting material on these alternative procedures in Sarhan and Greenberg's book [1]. Especially important are the procedures of estimation based on censored samples, in which r_1 of the smallest and r_2 of the largest values are trimmed. The reader will also find in that book extensive tables of coefficients, which enable actual application of the procedures.

PROBLEMS

Section 3.1

1. Show that if the loss function is quadratic, that is, $L(\mathbf{g}, \hat{\mathbf{g}}) = (\hat{\mathbf{g}} - \mathbf{g})'\mathbf{A}(\hat{\mathbf{g}} - \mathbf{g})$, and $\{F_\theta;\ \theta \in \Theta\}$ admits a complete sufficient statistic S, then the unbiased estimator $\hat{\mathbf{g}}(S)$ is the uniformly minimum risk estimator, among all estimators.

2. Is the uniformly minimum risk unbiased estimator $\hat{\mathbf{g}}(S)$ also a uniformly minimum risk estimator if $L(\mathbf{g}, \hat{\mathbf{g}})$ is convex but not quadratic?

3. Consider the statistical model specified in Problem 29, Section 2.11. Is the minimal sufficient statistic $(\bar{X}, \bar{Y}, s_X{}^2, s_Y{}^2)$ a complete one? Is there a uniformly minimum variance unbiased estimator of the common mean μ? (See Zacks [4].)

Section 3.2

4. Prove Ellison's [1] result: If $\beta[(n/2) - 1,\ (n/2) - 1]$ is independent of $\chi^2[n - 1]$ then $[2\beta((n/2) - 1, (n/2) - 1) - 1][\chi^2[n - 1]]^{1/2} \sim N(0, 1)$.

5. Derive the variance of the U.M.V.U. estimator of $\Phi(-\xi/\sigma)$ in the normal case. The estimator is given in (3.22). [The variance of this estimator is a complicated function of certain incomplete moments of noncentral t-statistics. (See Zacks and Milton [1].)]

6. Derive the U.M.V.U. estimator of $\Phi(-\xi/\sigma)$ in the normal case, when σ is known. What is the variance of this estimator? (See Zacks and Even [1].)

7. Find the variances and the covariance of the U.M.V.U. estimators of θ_1 and θ_2 of Example 3.3, case 2.

8. Determine the variance of the U.M.V.U. estimator, (3.2.20), of Poisson probabilities.

9. Determine the variance of the U.M.V.U. estimator, (3.2.32), of the reliability functions. (See Zacks and Even [1].)

10. Let $X_1, \ldots, X_n$ be i.i.d. discrete random variables having a negative binomial distribution N.B. (ψ, ν), $0 < \psi < 1$, ν known. Determine the U.M.V.U. estimator of ψ. (Hint: Consider $g(\psi) = 1 - P[X = 0]$.)

Section 3.3

11. Let X have a rectangular distribution $R(\theta, 2\theta)$, $0 < \theta < \infty$. Construct an unbiased estimator of $g\theta = \log \theta$, say $\hat{g}(X)$, such that $\text{Var}_{\theta=1}\{\hat{g}(X)\} = 0$.

12. Let $X_1, \ldots, X_n$ be i.i.d. random variables having a (truncated) location parameter exponential distribution, with a density function

$$f(x; \alpha) = \begin{cases} \dfrac{e^{-(x-\alpha)}}{1 - e^{-\alpha}}, & \alpha < x < 2\alpha, \\ 0, & \text{otherwise}, \end{cases}$$

$-\infty < \alpha < \infty$. Construct an unbiased estimator of $g(\alpha) = P_\alpha[X \geqslant (3/2)\alpha]$ having a zero variance at $\alpha = 0$.

13. Consider the function $A(\theta, \theta)$ defined in (3.3.34). Show that in the case of Example 3.7, $A(\theta, \theta)$ is not always finite.

Section 3.4

14. Let $X_1, \ldots, X_n$ be i.i.d. random variables distributed like $G(\theta, \nu)$, $0 < \theta < \infty$, ν known. Apply Corollary 3.4.2 to determine the U.M.V.U. estimator of $g(\theta) = \theta^k$.

15. As in Problem 14, what is the U.M.V.U. estimator of $e^{-\theta}$?

Section 3.5

16. (i) Prove that if $X_1, \ldots, X_n$ are i.i.d. having a common Type I truncation parameter density function, then $X_{(1)} = \min_{1 \leqslant i \leqslant n} \{X_i\}$ is a complete sufficient statistic.

(ii) Prove that the common density of $X_1, \ldots, X_n$ is of a Type II truncation parameter then $X_{(n)} = \max_{1 \leqslant i < n} \{X_i\}$ is a complete sufficient statistic.

17. Consider the Pareto family of distributions with density functions

$$f(x; \alpha, \sigma) = \begin{cases} \dfrac{\sigma}{\alpha} \cdot \left(\dfrac{\alpha}{x}\right)^{\sigma+1}, & \alpha \leqslant X \leqslant \infty, \\ 0, & \text{otherwise}, \end{cases}$$

where $0 < \sigma < \infty$; $0 < \alpha < \infty$. Let $X_1, \ldots, X_n$ be i.i.d. random variables having a common distribution from this family, with σ known. What is the U.M.V.U. estimator of the distribution function $F(x; \alpha, \sigma)$?

18. Let $X_1, \ldots, X_n$ be i.i.d. random variables distributed like $\alpha + G(1, 1)$; that is,

$$f(x; \alpha) = \begin{cases} \exp\{-(x - \alpha), & x \geqslant \alpha, \\ 0, & \text{otherwise} \end{cases} \qquad -\infty < \alpha < \infty.$$

Find the U.M.V.U. estimator of the distribution function $F(x; \alpha)$.

19. Let $X_1, \ldots, X_n$ be i.i.d. random variables having a common extreme value distribution (see Chapter 2, Problem 23). Apply Theorem 3.5.3 to derive the U.M.V.U. estimator of the common density function $f(x; \alpha)$. Provide also an unbiased estimator of α based on the sample minimum $X_{(1)}$.

Section 3.6

20. Consider the 2-parameter family of Pareto distributions with density functions $f(x; \alpha, \sigma)$ specified in Problem 17. Both the location and the scale parameters are unknown. On the basis of n i.i.d. random variables,

(i) derive the U.M.V.U. estimator of α^r;
(ii) determine the U.M.V.U. estimator of σ;
(iii) derive the U.M.V.U. estimator of the distribution function.

21. What parameters (functions of the location and scale parameters) of the family of Cauchy distributions with densities

$$f(x; \alpha, \sigma) = \frac{1}{\pi}\left[1 + \left(\frac{x - \alpha}{\sigma}\right)^2\right]^{-1}, \qquad -\infty < x < \infty,$$

$-\infty < \alpha < \infty, 0 < \sigma < \infty$, can you estimate unbiasedly, according to the theory provided in Section 3.6? Are these estimators U.M.V.U.?

Section 3.7

22. Let $X_1, \ldots, X_{n_1}$ and $Y_1, \ldots, Y_{n_2}$ be independent random variables. The X's have an identical absolutely continuous distribution $F(x)$, and the Y's have an identical absolutely continuous distribution $G(y)$. Assuming that the following variances exist, what is the U.M.V.U. estimator of Var $\{X\}$, of Var $\{Y\}$, and of Var $\{X + Y\}$? (See Fraser [5], p. 146.)

23. For the family of absolutely continuous distributions $F(x_1, x_2)$ over $\mathfrak{X}^{(2)}$, determine the U.M.V.U. estimators of Var $\{X_1\}$, Var $\{X_2\}$, and cov (X_1, X_2). (See Fraser [5], p. 158.)

24. Consider the family of all absolutely continuous distributions $\mathfrak{F}_x$ such that $E\{Y(x)\} = \alpha + \beta x$, $-\infty < \alpha < \infty$, $\beta \neq 0$, and Var $\{Y(x)\} = \sigma^2 < \infty$ for each fixed x in a prescribed interval D on the real line. Given observations on $\{Y(x_i); i = 1 \ldots n\}$, answer the following:

(i) What is the U.M.V.U. estimator of β?
(ii) What is the U.M.V.U. estimator of α?
(iii) What is the U.M.V.U. estimator of σ^2?

25. Prove that every U.M.V.U. estimator of (a functional) $g(F)$, $F \in \mathfrak{F}$, based on n i.i.d. random variables is a symmetric function of $(X_1, \ldots, X_n)$. Show examples in which $\mathfrak{F}$ is such that the U.M.V.U. estimator is not a U-statistic.

Section 3.8

26. Let $X_1, \ldots, X_n$ be random variables such that

$$X_n = \mu_n + \epsilon_n,$$
$$X_{n-1} = \mu_n + J_1 + \epsilon_{n-1},$$
$$\vdots$$
$$X_j = \mu_n + J_1 + \cdots + J_{n-j} + \epsilon_j, \qquad j = 1, \ldots, n-1,$$

where μ_n is a constant parameter, $-\infty < \mu_n < \infty$. The variables $J_1, \ldots, J_{n-1}$ are i.i.d. random variables

$$J_1 = \begin{cases} 1, & \text{w.p. } \theta \\ 0, & \text{w.p. } (1-\theta), \end{cases} \qquad 0 < \theta < 1,$$

and $\epsilon_1, \ldots, \epsilon_n$ are i.i.d. random variables $E\{\epsilon_1\} = 0$ Var $\{\epsilon_1\} = \sigma^2$, $0 < \sigma < \infty$. Find the B.L.U.E. of μ_n. (See Chernoff and Zacks [1].)

27. Consider a two-way layout experiment in which the random variables are independent and have Poisson distributions $X_{ij} \sim P(\lambda_{ij})$, $i = 1, \ldots, I, j = 1, \ldots, J$. The model is multiplicative. That is, $\lambda_{ij} = \zeta_i \beta_j$, where $\sum_{i=1}^{I} \zeta_i = 1$ and $0 < \zeta_i$, $\beta_j < \infty$. Determine the L.S.E.'s of ζ_i and of β_j. Are these estimators unbiased? (This model is applicable in system reliability studies.)

28. Consider a linear model $\mathbf{Y} = \mathbf{X}\boldsymbol{\beta} + \boldsymbol{\epsilon}$ where rank $(\mathbf{X}) = r$, $1 \leqslant r \leqslant p$. Determine a linear function $\boldsymbol{\lambda}'\boldsymbol{\beta}$ such that

(i) $\boldsymbol{\lambda}'\boldsymbol{\beta}$ is estimable;

(ii) Var $\{\boldsymbol{\lambda}'\hat{\boldsymbol{\beta}}\}/\boldsymbol{\lambda}'\boldsymbol{\lambda}$ is minimum, where $\hat{\boldsymbol{\beta}}$ is the (generalized) L.S.E. of $\boldsymbol{\beta}$. (See Rao [7], p. 249.)

29. Let $\mathbf{Y} = \mathbf{X}\boldsymbol{\beta} + \boldsymbol{\epsilon}$ where $E\{\boldsymbol{\epsilon}\} = 0$ and $E\{\boldsymbol{\epsilon}\boldsymbol{\epsilon}'\} = \sigma^2\mathbf{I}_n + \rho\sigma^2\mathbf{J}_n$, $\mathbf{J}_n = \mathbf{1}_n\mathbf{1}'_n$, $\mathbf{1}'_n = (1, 1, \ldots, 1)$. Make the (Helmert) orthogonal transformation of $\mathbf{Y} \to \mathbf{Z} = \mathbf{PY}$.

(i) What are the expectation and the covariance matrix of $(Z_2, \ldots, Z_n)'$?
(ii) What is the B.L.U.E. of β?
(iii) Do the B.L.U.E. and the L.S.E. coincide?

30. Consider a fractional weighing design (F.W.D.) problem in which four objects are under consideration. A Hadamard matrix is

$$\mathbf{C} = \begin{bmatrix} 1 & 1 & 1 & 1 \\ 1 & -1 & 1 & -1 \\ 1 & 1 & -1 & -1 \\ 1 & -1 & -1 & 1 \end{bmatrix}.$$

(i) What weighing operations should be performed in an F.W.D. of size $n = 3$ in order to estimate the linear function

$$\boldsymbol{\lambda}'\boldsymbol{\omega} = 6\omega_1 + 12\omega_2 - 4\omega_3 + 2\omega_4?$$

(ii) What is the generalized least-squares estimator of $\boldsymbol{\lambda}'\boldsymbol{\omega}$ associated with the F.W.D. of (i)?

(iii) What is the variance of the estimator determined in (ii) if $\sigma^2 = 1$?

31. Consider the linear model (3.8.47) with $r = 2$, $p = 8$, and four possible design matrices

$$\mathbf{X}_1^* = \left[\begin{array}{cc:cccccc} 1 & -1 & -1 & 1 & -1 & 1 & 1 & -1 \\ 1 & 1 & -1 & -1 & -1 & -1 & 1 & 1 \end{array}\right],$$

$$\mathbf{X}_2^* = \left[\begin{array}{cc:cccccc} 1 & -1 & 1 & -1 & -1 & 1 & -1 & 1 \\ 1 & 1 & 1 & 1 & -1 & -1 & -1 & -1 \end{array}\right],$$

$$\mathbf{X}_3^* = \left[\begin{array}{cc:cccccc} 1 & -1 & -1 & 1 & 1 & -1 & -1 & 1 \\ 1 & 1 & -1 & -1 & 1 & 1 & -1 & -1 \end{array}\right],$$

$$\mathbf{X}_4^* = \left[\begin{array}{cc:cccccc} 1 & -1 & 1 & -1 & 1 & -1 & 1 & -1 \\ 1 & 1 & 1 & 1 & 1 & 1 & 1 & 1 \end{array}\right].$$

The parameter vector is $\boldsymbol{\beta}_8' = (\boldsymbol{\alpha}_2' \mid \boldsymbol{\beta}_6^{*\prime})$.

(i) Show that the only design under which $\hat{\boldsymbol{\alpha}}_2$ is unbiased is the one in which $\mathbf{X}_v^*$ $(v = 1, \ldots, 4)$ is chosen from the above set at random, and with equal probabilities.

(ii) What is the variance of $\boldsymbol{\lambda}'\hat{\boldsymbol{\omega}}_2$ if $\sigma^2 = 1$. (See Zacks [1].)

32. Construct an example of a linear model $\mathbf{Y} = \mathbf{X}\boldsymbol{\beta} + \boldsymbol{\epsilon}$ where $E\{\boldsymbol{\epsilon}\} = 0$, $E\{\boldsymbol{\epsilon}\boldsymbol{\epsilon}'\} = \sigma^2\mathbf{V}$ where the covariance matrix is singular and $\mathbf{X}$ is a design matrix of rank r, $1 \leqslant r < n$, such that the B.L.U.E. of every estimable function $\boldsymbol{\lambda}'\boldsymbol{\beta}$ will coincide with its L.S.E.

Section 3.9

33. Let $X_1, \ldots, X_n$ be i.i.d. random variables having a rectangular distribution $\mathcal{R}(\theta_1, \theta_2)$, $-\infty < \theta_1 < \theta_2 < \infty$. Determine the best linear combinations of the order statistics for estimating θ_1 and θ_2.

REFERENCES

Anderson [1]; Banerjee [1], [2]; Barankin [1]; Batemen [1]; Basu [1]; Cramér [2]; Ellison [1]; Epstein and Sobel [1]; Fraser [3], [5]; Glasser [1]; Goldman and Zelen [1]; Graybill [2]; Greenberg and Sarhan [1]; Halmos [1]; Hotelling [1]; Jahnke and Emde [1]; Kitagawa [1]; Kolmogorov [1]; Lehmann [1]; Lehmann and Scheffé [1]; Liberman and Resnikoff [1]; Loeve [2]; Lloyd [1]; Magness and McGuire [1]; Morimoto and Sibuya [1]; Mosteller [1]; Olkin and Pratt [1]; Patil and Wani [1]; Pearson and Hartley [1]; Rao [4], [7]; Sarhan and Greenberg [1]; Scheffé [1]; Sethuraman [1]; Stein [3]; Takeuchi [1]; Tate [1]; U.S. Dept of Commerce, N.B.S. [1]; Washio, Morimoto, and Ikeda [1]; Widder [1]; Zacks [2], [4], [8]; Zyskind [1].

CHAPTER 4

The Efficiency of Estimators Under Quadratic Loss

4.1. THE CRAMÉR-RAO LOWER BOUND IN THE 1-PARAMETER REGULAR CASE

We derive the Cramér-Rao lower bound for the case of families of distribution functions depending on one real parameter, that is, $\mathcal{F} = \{P_\theta; \theta \in \Theta\}$, where Θ is a subset of the real line. We further assume that $\mathcal{F} \ll \mu$ so that for every $\theta \in \Theta$, $f(x; \theta)$ is the density function of P_θ with respect to μ. The random variable X is either real or vector valued. We impose the following regularity conditions:

(i) Θ is either the real line, or an interval on the real line.
(ii) $(\partial/\partial\theta)f(x; \theta)$ exists and is finite a.s. $[P_\theta]$ for every $\theta \in \Theta$.
(iii) $\int |(\partial^i/\partial\theta^i)f(x; \theta)| \mu(dx) < \infty$, for every $\theta \in \Theta$, and $i = 1, 2$.
(iv) $E_\theta[\partial/\partial\theta \log f(X; \theta)]^2 < \infty$, for every $\theta \in \Theta$.

If a family $\mathcal{F}$ satisfies these regularity conditions we call it C.R.-regular. We prefixed the initials C.R. to the term regular in order to emphasize that the family $\mathcal{F}$ satisfies the Cramér-Rao regularity conditions, rather than the weaker (Dynkin) regularity conditions required in Section 2.5.

Let $S(x; \theta) = (\partial/\partial\theta) \log f(x; \theta)$. This derivative of the log-likelihood function is called the *score* of the sample and is a necessary and sufficient statistic for $\mathcal{F}$. We now prove the following auxiliary lemmas.

Lemma 4.1.1. *Under the regularity conditions* (*i*)–(*iii*)

$$E_\theta\{S(X; \theta)\} = 0, \quad \textit{for all} \quad \theta \in \Theta. \tag{4.1.1}$$

Proof. Starting from the identity

$$\int f(x; \theta)\mu(dx) = 1 \quad \text{for all} \quad \theta \in \Theta,$$

we differentiate it with respect to θ to obtain

$$\frac{\partial}{\partial\theta}\int f(x;\theta)\mu(\mathrm{d}x) = 0, \quad \text{for all} \quad \theta \in \Theta. \tag{4.1.2}$$

Regularity conditions (ii) and (iii) permit differentiation of (4.1.2) under the integral sign. Hence

$$\int \frac{\partial}{\partial\theta} f(x;\theta)\mu(\mathrm{d}x) = 0, \quad \text{for all } \theta \in \Theta. \tag{4.1.3}$$

But,

$$E_\theta\{S(X;\theta)\} = \int\left(\frac{\partial}{\partial\theta}\log f(x;\theta)\right)f(x;\theta)\mu(\mathrm{d}x) \tag{4.1.4}$$

for all $\theta \in \Theta$

$$= \int \frac{\partial}{\partial\theta} f(x;\theta)\mu(\mathrm{d}x)$$

$$= 0, \quad \text{for all } \theta \in \Theta. \qquad \text{(Q.E.D.)}$$

Lemma 4.1.2. *Under the regularity conditions* (*i*)–(*iv*),

$$E_\theta\left\{\left[\frac{\partial}{\partial\theta}\log f(X;\theta)\right]^2\right\} = -E_\theta\left\{\frac{\partial^2}{\partial\theta^2}\log f(X;\theta)\right\}, \tag{4.1.5}$$

for all $\theta \in \Theta$

Proof. Obviously,

$$E_\theta\left\{\frac{\partial^2}{\partial\theta^2}\log f(x;\theta)\right\} \tag{4.1.6}$$

$$= \int \frac{(\partial^2/\partial\theta^2)f(x;\theta)\cdot f(x;\theta) - ((\partial/\partial\theta)f(x;\theta))^2}{f^2(x;\theta)} f(x;\theta)\mu(\mathrm{d}x)$$

$$= \int \frac{\partial^2}{\partial\theta^2} f(x;\theta)\mu(\mathrm{d}x) - E_\theta\{S^2(X;\theta)\}.$$

According to (iv), $E_\theta S^2(X;\theta) < \infty$ for each $\theta \in \Theta$. Furthermore, from (iii) we imply that

$$\frac{\partial^2}{\partial\theta^2}\int f(x;\theta)\mu(dx) = \int \frac{\partial^2}{\partial\theta^2} f(x;\theta)\mu(dx) \tag{4.1.7}$$

$$= 0, \quad \text{for every } \theta \in \Theta. \qquad \text{(Q.E.D.)}$$

The function

$$I(\theta) = E_\theta\{S^2(X;\theta)\}, \quad \theta \in \Theta,$$

is called the *Fisher information function*. This function measures in some sense the amount of information available in X about θ. We discuss the meaning and significance of the information function $I(\theta)$ after we prove the Cramér-Rao inequality. We proceed by establishing two important properties of the information function. In cases where several random variables are involved, X, Y, etc., we designate by $I^X(\theta), I^Y(\theta), \ldots$ the corresponding information functions, whereas $I^{X,Y}(\theta)$ designates the joint information function for X and Y, namely,

$$\begin{aligned} I^{X,Y}(\theta) &= E_\theta\{S^2(X, Y; \theta)\} \\ &= E_\theta\left\{\left[\frac{\partial}{\partial\theta}\log f(X, Y; \theta)\right]^2\right\} \end{aligned}$$

where $f(x, y; \theta)$ is the joint density of X, Y.

Lemma 4.1.3. *If X and Y are independent random variables having C.R.-regular distribution functions, then*

$$I^{X,Y}(\theta) = I^X(\theta) + I^Y(\theta), \qquad \theta \in \Theta. \tag{4.1.8}$$

The proof is very simple and is therefore omitted.

Lemma 4.1.4. *Let $(\mathfrak{X}, \mathfrak{B}, P_\theta)$ be the probability space corresponding to a random variable X where P_θ belongs to a C.R.-regular family $\mathfrak{F}$. Let $T(X)$ be a statistic $T: (\mathfrak{X}, \mathfrak{B}, P_\theta) \to (\mathcal{F}, \mathfrak{B}_T, P_\theta^T)$ such that $\mathfrak{F}^T = \{P_\theta^T; \theta \in \Theta\}$ is C.R.-regular. Then*

$$I^T(\theta) \leqslant I^X(\theta), \quad \textit{for all } \theta \in \Theta. \tag{4.1.9}$$

Equality holds if and only if T is a sufficient statistic for $\mathfrak{F}$.

Proof: Let μ^T be the induced sigma-finite measure on $(\mathcal{F}, \mathfrak{B}_T)$. Since $\mathfrak{F} \ll \mu$ we have that $\mathfrak{F}^T \ll \mu^T$. Thus let $g(t; \theta)$ be the density function of P_θ^T, that is, $g(t; \theta) = \mathrm{d}P_\theta^T(t)/\mathrm{d}\mu^T(t)$ a.s. Now let C be any Borel set in $\mathfrak{B}_T$. Then according to the definition of the conditional expectation and (iii),

$$\int_{T^{-1}(C)} S(x; \theta) P_\theta(\mathrm{d}x) = \int_C E\{S(X; \theta) \mid T = t\} P_\theta^T(\mathrm{d}t) \tag{4.1.10}$$

for every $\theta \in \Theta$. Moreover, since (i)–(iv) are satisfied by $\mathfrak{F}$ and by $\mathfrak{F}^T$ we have

$$\begin{aligned} &\int_{T^{-1}(C)} S(x; \theta) P_\theta(\mathrm{d}x) \qquad (4.1.11) \\ &= \int_{T^{-1}(C)} \frac{\partial}{\partial\theta} f(x; \theta)\mu(\mathrm{d}x) = \frac{\partial}{\partial\theta}\int_{T^{-1}(C)} f(x; \theta)\mu(\mathrm{d}x) \\ &= \frac{\partial}{\partial\theta}\int_C g(t; \theta)\mu^T(\mathrm{d}t) = \int_C \frac{\partial}{\partial\theta} g(t; \theta)\mu^T(\mathrm{d}t) = \int_C \frac{\partial}{\partial\theta}\log g(t; \theta) P_\theta^T(\mathrm{d}t), \end{aligned}$$

for every non-null $C \in \mathcal{B}_T$. Hence from (4.1.10) and (4.1.11) we deduce that

$$(4.1.12) \qquad \frac{\partial}{\partial\theta} \log g(t;\theta) = E\{S(X;\theta) \mid T = t\} \quad \text{a.s. } [P_\theta]$$

for all $\theta \in \Theta$. By definition $I^T(\theta) = E_\theta\{[(\partial/\partial\theta) \log g(T;\theta)]^2\}$. To prove that $I^T(\theta) \leqslant I^X(\theta)$, we write

$$(4.1.13) \qquad 0 \leqslant E_\theta\left\{\left[\frac{\partial}{\partial\theta}\log f(X;\theta) - \frac{\partial}{\partial\theta}\log g(T;\theta)\right]^2\right\}$$
$$= E_\theta\left\{\left[\frac{\partial}{\partial\theta}\log f(X;\theta)\right]^2\right\} + E_\theta\left\{\left[\frac{\partial}{\partial\theta}\log g(T;\theta)\right]^2\right\}$$
$$- 2E_\theta\left\{\frac{\partial}{\partial\theta}\log f(X;\theta)\cdot\frac{\partial}{\partial\theta}\log g(T;\theta)\right\}.$$

But according to (4.1.12),

$$(4.1.14) \quad E_\theta\left\{\frac{\partial}{\partial\theta}\log f(X;\theta)\cdot\frac{\partial}{\partial\theta}\log g(T;\theta)\right\}$$
$$= E_\theta\left\{\frac{\partial}{\partial\theta}\log g(T;\theta)\cdot E\left\{\frac{\partial}{\partial\theta}\log f(X;\theta) \mid T\right\}\right\}$$
$$= E_\theta\left\{\left[\frac{\partial}{\partial\theta}\log g(T;\theta)\right]^2\right\}.$$

Substituting (4.1.14) in (4.1.13), we obtain the inequality (4.1.9). It remains to prove that equality in (4.1.9) holds if and only if T is sufficient for $\mathcal{P}$.

If T is a sufficient statistic for $\mathcal{P}$, by the factorization theorem

$$f(x;\theta) = h(x)g^*(T(x);\theta), \quad \theta \in \Theta,$$

where $h \geqslant 0$ and $g^* \geqslant 0$; $h \in \mathcal{B}$ and $g^* \in T^{-1}(\mathcal{B}_T)$; and $h(x) = 0$ only on a $\mathcal{P}$-null set (independent of θ). Thus

$$\frac{\partial}{\partial\theta}\log f(x;\theta) = \frac{\partial}{\partial\theta}\log g^*(T(x);\theta) \quad \text{a.s. } [P_\theta] \quad \text{for all } \theta \in \Theta.$$

Hence

$$(4.1.15) \qquad I^X(\theta) = E_\theta\left\{\left[\frac{\partial}{\partial\theta}\log g^*(T(X);\theta)\right]^2\right\}, \quad \theta \in \Theta.$$

Since $g^*(T(x);\theta) \in T^{-1}(\mathcal{B}_T)$ for each $\theta \in \Theta$, there exists a statistic $Z(T)$ defined on $(\mathcal{F}, \mathcal{B}_T, P_\theta^T)$ whose density function is $J(t)g^*(t;\theta)$, where $J(t)$ does not depend on θ. Thus according to (4.1.9),

$$(4.1.16) \qquad E_\theta\left\{\left[\frac{\partial}{\partial\theta}\log g^*(T(X);\theta)\right]^2\right\} \leqslant I^T(\theta), \quad \theta \in \Theta.$$

Hence from (4.1.9), (4.1.15), and (4.1.16) we obtain that $I^X(\theta) = I^T(\theta)$ for all $\theta \in \Theta$. We now show that if $I^X(\theta) = I^T(\theta)$ for all $\theta \in \Theta$, then T is sufficient for $\mathfrak{F}$. Indeed, since $I^X(\theta)$ is the variance of $S(X; \theta)$ we write

$$I^X(\theta) = E_\theta\{\text{Var}\,\{S(X;\theta) \mid T\}\} + \text{Var}_\theta\,\{E\{S(X;\theta) \mid T\}\}. \tag{4.1.17}$$

As shown previously,

$$E_\theta\{S(X;\theta) \mid T\} = \frac{\partial}{\partial\theta} \log g(T;\theta) \quad \text{a.s. } [P_\theta].$$

Therefore, since $\mathfrak{F}^T$ is C.R.-regular,

$$I^T(\theta) = \text{Var}_\theta\,\{E_\theta\{S(X;\theta) \mid T\}\}, \quad \theta \in \Theta. \tag{4.1.18}$$

Thus since $I^X(\theta) = I^T(\theta)$ for all $\theta \in \Theta$, we obtain

$$E_\theta\{\text{Var}_\theta\,\{S(X;\theta) \mid T\}\} = 0 \quad \text{for all } \theta \in \Theta. \tag{4.1.19}$$

This implies that

$$\text{Var}_\theta\,\{S(X;\theta) \mid T\} = 0 \quad \text{a.s. } [P_\theta{}^T]. \tag{4.1.20}$$

In other words, $S(X;\theta) \in T^{-1}(\mathfrak{B}_T)$. Thus there exists a function $k(t;\theta)$ on $\mathcal{F}$ such that $S(X;\theta) = k(T(X);\theta)$ for all $\theta \in \Theta$. Finally,

$$f(X;\theta) = h(X)g^*(T(x);\theta), \quad \theta \in \Theta. \tag{4.1.21}$$

Indeed,

$$\log f(x;\theta) = \int_0^\theta k(T(x);\tau)\,dt + a(x), \quad \text{for all } \theta \in \Theta.$$

Thus $h \geqslant 0$ and $g^* \geqslant 0$, $h \in \mathfrak{B}$ and $g^* \in T^{-1}(\mathfrak{B}_T)$, where $h = 0$ only on a $\mathfrak{F}$-null set. Thus the conditions of the factorization theorem are satisfied, and T is a sufficient statistic for $\mathfrak{F}$. (Q.E.D.)

Inequality 4.1.9 shows that the only statistic which reduces the data in the sample without losing information is the sufficient statistic. If the family of probability measures admits a nontrivial sufficient statistic then the crudest partition possible of the sample space that does not lose information is the one obtained by the minimal sufficient statistic. We proceed now to the derivation of the Cramér-Rao lower bound.

Theorem. 4.1.1 *Let $(\mathfrak{X}, \mathfrak{B}, P_\theta)$ be a probability space, $P_\theta \in \mathfrak{F}$ where $\mathfrak{F}$ is a C.R.-regular family. Let g be a real valued function on Θ, which is estimable and differentiable. Let $\hat{g}(X)$ be an unbiased estimator of $g(\theta)$ having a finite variance and satisfying the following regularity condition:*

$$\int \left| \hat{g}(x) \cdot \frac{\partial}{\partial\theta} f(x;\theta) \right| \mu(dx) < \infty, \quad \text{for all } \theta \in \Theta.$$

Then

$$\text{Var}_\theta \{\hat{g}(X)\} \geqslant \frac{[g'(\theta)]^2}{I(\theta)}, \quad \textit{for all } \theta \in \Theta. \tag{4.1.22}$$

Equality in (4.1.22) *holds if and only if the density function of* $\hat{g}(X)$ *is of the exponential type; that is,*

$$f^{\hat{g}}(g; \theta) = \exp \{\psi_1(\theta)g + \psi_2(\theta) + k(g)\}, \quad g \in g(\Theta), \tag{4.1.23}$$

where $\psi_1'(\theta) \neq 0$.

Proof. Consider the covariance between $S(X; \theta)$ and $\hat{g}(X)$. This is given by

$$\begin{aligned} \text{cov}_\theta (S(X; \theta), \hat{g}(X)) &= E_\theta\{S(X; \theta)[\hat{g}(X) - g(\theta)]\} \\ &\leqslant [I(\theta) \cdot \text{Var}_\theta \{\hat{g}(X)\}]^{1/2}, \quad \theta \in \Theta. \end{aligned} \tag{4.1.24}$$

The inequality in (4.1.24) is verified by the Schwartz inequality. On the other hand, according to Lemma 4.1.1,

(4.1.25)

$$\begin{aligned} \int S(x; \theta)(\hat{g}(x) - g(\theta))f(x; \theta)\mu(\mathrm{d}x) &= \int \hat{g}(x)\left(\frac{\partial}{\partial\theta} \log f(x; \theta)\right)f(x; \theta)\mu(\mathrm{d}x) \\ &= \int \hat{g}(x) \cdot \frac{\partial}{\partial\theta} f(x; \theta)\mu(\mathrm{d}x). \end{aligned}$$

But according to (v) and the unbiasedness of $\hat{g}(X)$,

$$g'(\theta) = \int \frac{\partial}{\partial\theta} \hat{g}(x)f(x; \theta)\mu(\mathrm{d}x). \tag{4.1.26}$$

Hence from (4.1.24) and (4.1.26) we obtain the Cramér-Rao inequality (4.1.22). It remains to show that a necessary and sufficient condition for equality in (4.1.22) is (4.1.23). Indeed, equality in (4.1.25) holds if and only if $\hat{g}(X)$ is a linear function of $S(X; \theta)$, with a nonzero slope [since if the slope is zero we obtain that $\text{Var}_\theta \{\hat{g}(X)\} = 0$]. Thus

$$\hat{g}(X) = b(\theta)S(X; \theta) + a(\theta), \quad \text{for all } x \in \mathfrak{X}, \tag{4.1.27}$$

where $b(\theta) \neq 0$ for all $\theta \in \Theta$. From (4.1.27) we obtain that

$$\log f(x; \theta) = \hat{g}(x) \cdot \psi_1(\theta) + \psi_2(\theta) + h(x),$$

where $\psi_1(\theta)$ is a monotone function of θ, that is, $\psi_1'(\theta) = 0$. (Q.E.D.)

In many texts the Cramér-Rao inequality is given for the variances of unbiased estimators of the parameter θ itself. This inequality is obtained from (4.1.22), as a special case for $g(\theta) \equiv \theta$. In this case, if $\hat{\theta}$ is an unbiased estimator of θ then $\text{Var}\{\hat{\theta}\} \geqslant I^{-1}(\theta)$. We see thus that the Fisher information is inversely proportional to the lower bound for the variances of all unbiased estimators of θ. We can therefore say that the information value $I(\theta)$ is proportional to the precision of an unbiased estimator of θ. Inequality (4.1.22) also yields a lower bound for the mean square error of estimators of θ which are not necessarily unbiased. To derive this lower bound we substitute $g(\theta) = \theta + B(\theta)$, where $B(\theta)$ is the bias of an estimator of θ; that is,

$$B(\theta) = \int (\hat{\theta}(x) - \theta) P_\theta(\mathrm{d}x).$$

Assuming that the bias function $B(\theta)$ is differentiable we immediately obtain from (4.1.22),

$$E_\theta\{(\hat{\theta}(X) - \theta)^2\} \geqslant B^2(\theta) + \frac{(1 + B'(\theta))^2}{I(\theta)}. \tag{4.1.28}$$

This form of the Cramér-Rao inequality will be utilized in Chapter 8 for the study of the admissibility of estimators.

Example 4.1. (i) Let $X_1, X_2, \ldots, X_n$ be i.i.d. random variables having a Poisson distribution with mean $\theta, 0 < \theta < \infty$. The density of X is, with respect to the counting measure $N(x)$,

$$f(x; \theta) = \begin{cases} e^{-\theta} \dfrac{\theta^x}{x!}, & \text{for all } 0 \leqslant x < \infty, \\ 0, & \text{for all } x < 0. \end{cases}$$

A complete sufficient statistic is $T_n = \sum_{i=1}^n X_i$; T_n is distributed like a Poisson random variable with mean $E_\theta\{T_n\} = n\theta$. Since the density function of T_n belongs to a 1-parameter exponential family, and since T_n/n is an unbiased estimator of θ, we obtain from Theorem 4.1.1 that

$$I^{X_1,\ldots,X_n}(\theta) = \left[\text{Var}_\theta\left\{\frac{T_n}{n}\right\}\right]^{-1} = \frac{n}{\theta}.$$

Furthermore, according to Lemma 4.1.4, the Fisher information function associated with a Poisson random variable with mean θ is $I(\theta) = \theta^{-1}$. This can be checked by computing $I(\theta) = E_\theta\{[(\partial/\partial\theta) \log f(X; \theta)]^2\}$.

(ii) Consider now the problem of estimating the function $g(\theta) = e^{-\theta}$, which is the probability of $X = 0$. The U.M.V.U. estimator of $e^{-\theta}$ is, according to (3.2.20), $\hat{g}(T_n) = (1 - 1/n)^{T_n}$. The variance of this best unbiased estimator can be determined in the following manner. The generating function of a Poisson random variable, Y, with mean λ, is

$$\begin{aligned} M_\lambda(t) &= E_\lambda\{t^X\} \\ &= \exp\{-\lambda(1 - t)\}, \quad -\infty < t < \infty. \end{aligned} \tag{4.1.29}$$

Hence

$$\text{(4.1.30)}\quad \operatorname{Var}_\theta\left\{\left(1-\frac{1}{n}\right)^{T_n}\right\} = E_\theta\left\{\left(1-\frac{1}{n}\right)^{2T_n}\right\} - e^{-2\theta}$$

$$= \exp\left\{-n\theta\left[1-\left(1-\frac{1}{n}\right)^2\right]\right\} - e^{-2\theta}$$

$$= \exp\left\{-\theta\left(2-\frac{1}{n}\right)\right\} - e^{-2\theta} = e^{-2\theta}(e^{\theta/n}-1).$$

The Cramér-Rao lower bound is given, according to (4.1.22), by

$$\text{(4.1.31)}\qquad \frac{(g'(\theta))^2}{I(\theta)} = \frac{n}{\theta}e^{-2\theta}, \quad 0 < \theta < \infty.$$

We have thus obtained the expected result, that the U.M.V.U. estimator of $e^{-\theta}$ does not attain the Cramér-Rao lower bound for any value of θ. Indeed, the distribution of $(1 - 1/n)^{T_n}$ is not of the exponential type. The efficiency of the U.M.V.U. estimator, as a function of θ, is expressed as the ratio of the Cramér-Rao lower bound to the variance of the estimator. This efficiency function is

$$\text{(4.1.32)}\qquad \operatorname{eff}(\hat{g}(T_n) \mid \theta) = \frac{\theta}{n}(e^{\theta/n}-1)^{-1}, \qquad 0 < \theta < \infty.$$

This efficiency function converges to 1 for each fixed θ as $n \to \infty$; it converges to 0 for each fixed n as $\theta \to \infty$. ■

In the following section we present a system of more stringent lower bounds for the variance of unbiased estimators.

4.2. THE BHATTACHARYYA SYSTEM OF LOWER BOUNDS IN THE 1-PARAMETER REGULAR CASE

In this section we generalize the Cramér-Rao lower bound and provide a system of lower bounds in which the Cramér-Rao lower bound is a special case. These lower bounds were first given by Bhattacharyya in 1946 [1]. The Bhattacharyya lower bound of order k, $k \geqslant 2$, requires more stringent regularity conditions that the Cramér-Rao lower bound, which coincides with the Bhattacharyya lower bound of order $k = 1$. These more stringent regularity conditions are specified in the following theorem.

Theorem 4.2.1. *Let* $(\mathfrak{X}, \mathfrak{B}, P_\theta)$ *be a probability space and* $\mathfrak{F} = \{P_\theta; \theta \in \Theta\}$*a dominated family of probability measures satisfying regularity* (*i*) *and* (*ii*), *and* *in addition the following conditions:*

(*iii*)* $\int |(\partial^i/\partial\theta^i)f(x;\theta)|\,\mu(\mathrm{d}x) < \infty$, *for every* $i = 1, \ldots, k$, *and all* $\theta \in \Theta$;

(*iv*)* $\int (1/f(x;\theta))\,|(\partial^i/\partial\theta^i)f(x;\theta)\cdot(\partial^j/\partial\theta^j)f(x;\theta)|\,\mu(\mathrm{d}x) < \infty$, *for all* $i, j = 1, \ldots, k$ *and all* $\theta \in \Theta$.

Let $g(\theta)$ *be an estimable function which is k-times differentiable over* Θ. *Let* $\hat{g}(X)$ *be an unbiased estimator of* $g(\theta)$ *having a finite variance and satisfying the following condition:*

(v)* $\int |\hat{g}(x)\cdot(\partial^i/\partial\theta^i)f(x;\theta)|\,\mu(\mathrm{d}x)/f(x;\theta) < \infty$, *for all* $i = 1, \ldots, k$ *and all* $\theta \in \Theta$.

Furthermore, let $\mathbf{V}$ *be a* $k \times k$ *non-negative definite matrix, whose element is*

$$V_{ij} = E_\theta\left\{\frac{1}{f^2(X;\theta)}\cdot\frac{\partial^i}{\partial\theta^i}f(X;\theta)\cdot\frac{\partial^j}{\partial\theta^j}f(X;\theta)\right\}, \qquad i, j = 1, \ldots, k. \tag{4.2.1}$$

Then if $\mathbf{V}$ *is nonsingular at* θ,

$$\mathrm{Var}_\theta\{\hat{g}(X)\} \geqslant (g^{(1)}(\theta), \ldots, g^{(k)}(\theta))\mathbf{V}^{-1}(g^{(1)}(\theta), \ldots, g^{(k)}(\theta))', \tag{4.2.2}$$

where $g^{(i)}(\theta)$ *is the i-th order derivative of* $g(\theta)$.

Proof. Let $S_i(X;\theta) = [f(X;\theta)]^{-1}\cdot\partial^i/\partial\theta^i f(X;\theta)$, $i = 1, \ldots, k$. Consider the covariance matrix $\mathbf{\Sigma}$ of the random vector $(\hat{g}(X), S_1(X;\theta), \ldots, S_k(X;\theta))$. We show that $\mathbf{\Sigma}$ is a symmetric matrix.

$$\mathbf{\Sigma} = \begin{bmatrix} \mathrm{Var}\{\hat{g}(X)\} & g^{(1)}(\theta) & \cdots & g^{(k)}(\theta) \\ & V_{11} & \cdots & V_{1k} \\ & & \ddots & \vdots \\ \cdot & & & \\ & & & V_{kk} \end{bmatrix}. \tag{4.2.3}$$

Indeed, according to $(V)^*$,

$$\begin{aligned} \mathrm{cov}\,(\hat{g}(X), S_i(X;\theta)) &= \int \hat{g}(x)\frac{\partial^i}{\partial\theta^i}f(x,\theta)\mu(\mathrm{d}x) \\ &= \frac{\partial^i}{\partial\theta^i}\int \hat{g}(x)f(x;\theta)\mu(\mathrm{d}x) = g^{(i)}(\theta). \end{aligned} \tag{4.2.4}$$

Now, since $\mathbf{\Sigma}$ is non-negative definite,

$$\begin{aligned} \det\{\mathbf{\Sigma}\} = \det\{\mathbf{V}\}[\mathrm{Var}\,\{(\hat{g}X)\} \\ - (g^{(1)}(\theta), \ldots, g^{(k)}(\theta))\mathbf{V}^{-1}(g^{(1)}(\theta), \ldots, g^{(k)}(\theta))'] \geqslant 0. \end{aligned} \tag{4.2.5}$$

Moreover, since $\mathbf{V}$ is positive-definite at θ, $\det\{\mathbf{V}\} > 0$. Hence the Bhattacharyya inequality, (4.2.2), is immediately derived from (4.2.5). (Q.E.D.)

In the simple case of $g(\theta) = \theta$ we obtain from (4.2.2), since $g^{(1)}(\theta) = 1$ and $g^{(i)}(\theta) = 0$ for all $i \geqslant 2$, the k-th Bhattacharyya lower bound

$$\text{(4.2.6)}\quad \operatorname{Var}_\theta\{\hat{\theta}(X)\} \geqslant \begin{cases} \det\left\{\begin{bmatrix} V_{22} & \cdots & V_{2k} \\ & \cdot & \\ \cdot & \cdot & \\ & & \cdot \\ & & V_{kk} \end{bmatrix}\right\} \div \det\{\mathbf{V}\}, & k \geqslant 2 \\ I^{-1}(\theta), & k = 1. \end{cases}$$

We thus see that the Cramér-Rao lower bound is a special case of the Bhattacharyya lower bound for $k = 1$.

Example 4.2. Consider again the estimation problem discussed in Example 4.1, (ii). We have shown that the variance of the U.M.V.U. estimator of $e^{-\theta}$ in the Poisson case is $e^{-2\theta}(e^{\theta/n} - 1)$. This variance does not attain the Cramér-Rao lower bound $L_1 = (\theta/n)e^{-2\theta}$ for any value of θ. We now check whether it attains the Bhattacharyya lower bound for $k = 2$, say L_2.

As shown in Example 4.1,

$$V_{11} = I^{T_n}(\theta) = \frac{n}{\theta}.$$

Since

$$f^{T_n}(t;\theta) = e^{-n\theta}\frac{(n\theta)^t}{t!}, \qquad t = 0, 1, \ldots,$$

$$\frac{\partial}{\partial\theta} f^{T_n}(t;\theta) = f^{T_n}(t;\theta)\left(\frac{t}{\theta} - \theta_n\right)$$

and

$$\frac{\partial^2}{\partial\theta^2} f^{T_n}(t;\theta) = f^{T_n}(t;\theta)\left[\left(\frac{t}{\theta} - n\right)^2 - \frac{t}{\theta^2}\right].$$

Thus we obtain

$$V_{12} = E_\theta\left\{\left(\frac{T_n}{\theta} - n\right)\left[\left(\frac{T_n}{\theta} - n\right)^2 - \frac{T_n}{\theta}\right]\right\} = 0.$$

Furthermore,

$$V_{22} = \frac{1}{\theta^4} E_\theta\{[(T_n - n\theta)^2 - T_n]^2 = \frac{2n^2}{\theta^2}.$$

Hence

$$\mathbf{V} = \frac{n}{\theta}\begin{bmatrix} 1 & 0 \\ 0 & 2n/\theta \end{bmatrix},$$

and

$$L_2(\theta) = \frac{\theta}{n} e^{-2\theta}\left(1 + \frac{\theta}{2n}\right).$$

We have verified that the variance of the best unbiased estimator of $e^{-\theta}$ does not attain, at any value of θ, even the seccond Bhattacharyya bound. As a corollary of the next theorem we deduce that there is no value of k for which the variance of the U.M.V.U. estimator of $e^{-\theta}$ attains the corresponding Bhattacharyya bound. ∎

The following theorem from Fend [1] specifies the conditions under which an unbiased estimator attains the Bhattacharyya lower bound of the k-th order. For $k = 1$ we have show that a necessary and sufficient condition for the attainment of the Cramér-Rao lower bound is that the density of the unbiased estimator $\hat{g}(X)$ be of the exponential type. Fend's theorem states that if an unbiased estimator is a polynomial of degree k of the minimal sufficient statistic in an exponential family and if its variance does not attain the Bhattacharyya lower bounds for all $1 \leqslant j \leqslant k - 1$, then it attains the k-th bound, and conversely.

Theorem 4.2.2. (Fend [1].) *Let $(\mathfrak{X}, \mathfrak{B}, P_\theta)$ be a probability space, $P_\theta \in \mathfrak{F}$, where $\mathfrak{F} \ll \mu$ and satisfies the Bhattacharyya regularity conditions. Furthermore, let $f(x; \theta) = h(x) \exp\{t(x)\psi_1(\theta) + \psi_2(\theta)\}$ for all $\theta \in \Theta$, where $\psi_1'(\theta) \neq 0$. Let $g(\theta)$ be a k-times differentiable function over Θ having an unbiased estimator $\hat{g}(X)$, which satisfies condition* (v)*. *If at some θ, $\mathbf{V}$ is positive definite and the variance of $\hat{g}(X)$ attains the k-th Bhattacharyya bound but not the $(k - 1)$-th bound, then $\hat{g}(X)$ is a polynomial in $t(X)$ of degree k. Moreover, the variance of any polynomial in $t(X)$ of degree k achieves the k-th bound.*

Proof. According to the proof of Theorem 4.1.1 an unbiased estimator attains the k-th bound but does not attain the $(k - 1)$-th bound if and only if

$$\beta(X) = a_0(\theta) + \sum_{i=1}^{k} a_i(\theta) S_i(X; \theta) \quad \text{a.s. } [P_\theta] \tag{4.2.7}$$

for all θ, where $a_k(\theta) \neq 0$. It is simple to verify that if $f(X; \theta)$ is the above exponential type density, then

$$\begin{aligned} S_i(X; \theta) &= \frac{1}{f(X; \theta)} \frac{\partial^i}{\partial \theta^i} f(X; \theta) \\ &= [t(X)\psi_1'(\theta) + \psi_2(\theta)]^i + P_{i-1}(t(X); \theta), \qquad i = 1, \ldots, k, \end{aligned} \tag{4.2.8}$$

where $P_{i-1}(t(X); \theta)$ is a polynomial in $t(X)$ of degree at most $(i - 1)$. Since $\psi_1'(\theta) \neq 0$, $S_i(X; \theta)$ is a polynomial of degree i in $t(X)$. Thus according to (4.2.7), $\hat{g}(X)$ is a polynomial of degree k in $t(X)$. It remains to prove that the variance of any polynomial of degree k in $t(X)$ is equal to the k-th Bhattacharyya bound.

Write

$$P_{i-1}(t(X); \theta) = \sum_{j=0}^{i-1} t^j(X) U_{ij}(\theta),$$

and substitute (4.2.8) in (4.2.7). We then obtain

(4.2.9)

$$\begin{aligned}
\hat{g}(X) &= a_0(\theta) + \sum_{i=1}^{k} a_i(\theta)\left[(t(X)\psi_1'(\theta) + \psi_2'(\theta))^i + \sum_{j=0}^{i-1} t^j(X) U_{ij}(\theta)\right] \\
&= a_0(\theta) + \sum_{i=1}^{k} a_i(\theta)\left\{\sum_{j=0}^{i} \binom{i}{j} t^j(X)(\psi_1'(\theta))^j(\psi_2'(\theta))^{i-j} + \sum_{j=0}^{i-1} t^j(X) U_{ij}(\theta)\right\} \\
&= a_k(\theta)(\psi_1'(\theta))^k t^k(X) + \sum_{j=1}^{k-1} t^j(X)\left\{a_j(\theta)(\psi_1'(\theta))^j\right. \\
&\quad \left. + \sum_{\nu=j+1}^{k} a_\nu(\theta)\left[\binom{\nu}{j}(\psi_1'(\theta))^\nu(\psi_2'(\theta))^{\nu-j} + U_{\nu j}(\theta)\right]\right\} \\
&\quad + a_0(\theta) + \sum_{j=1}^{k} [(\psi_2'(\theta))^j + U_{j0}(\theta)] a_j(\theta).
\end{aligned}$$

Suppose $\hat{g}(X) = \sum_{i=0}^{k} c_i t^i(X)$, where $c_0, \ldots, c_k$ are arbitrary constants. Define

$$a_k(\theta) = c_k(\psi_1'(\theta))^{-k}$$

$$a_n(\theta) = \frac{c_n - \sum_{i=n+1}^{k} a_i(\theta)\left[\binom{i}{n}(\psi_2'(\theta))^{i-n} + U_{in}(\theta)\right]}{(\psi_1'(\theta))^n}, \qquad n = 1, \ldots, k-1$$

$$a_0(\theta) = c_0 - \sum_{i=1}^{k} a_i(\theta)[(\psi_2'(\theta))^i + U_{i0}(\theta)].$$

Then $\hat{g}(X)$ satisfies (4.2.7) and therefore its variance attains the k-th Bhattacharyya lower bound, but not the $(k-1)$-th. (Q.E.D.)

In Example 3.7 we studied the problem of estimating the parameter ξ^k in the case of n i.i.d. random variables distributed like $N(\xi, 1)$. The complete sufficient statistic is the sample mean $\bar{X}_n$. The joint density of $X, \ldots, X$ is

$$f(x_1, \ldots, x_n; \xi) = (2\pi)^{-n/2} \exp\left\{-\tfrac{1}{2}\sum_{i=1}^{n} x_i^2 + n\xi\bar{x}_n - \frac{n}{2}\xi^2\right\},$$

which is of the exponential type with $t(X_1, \ldots, X_n) = \bar{X}_n$. We have shown in that example that the U.M.V.U. estimator of $\xi^k (k \geqslant 1)$ is given by the polynomial, (3.4.29), of degree k. Hence we conclude from Theorem 4.2.2 that the variance of this U.M.V.U. estimator attains the k-th Bhattacharyya bound. We expressed this variance by (3.4.39), which is accordingly the formula of the k-th Bhattacharyya lower bound for a sample of random variables from $N(\xi, 1)$.

4.3. THE CASE OF REGULAR DISTRIBUTION FUNCTIONS DEPENDING ON VECTOR VALUED PARAMETER

We consider now the case where the family of probability measures $\mathfrak{F}$ is dominated and its density functions $f(x; \boldsymbol{\theta})$ depend on vectors of r real parameters. The regularity conditions of the Cramér-Rao theorem are generalized in the following manner:

(i)** The parameter space Θ is either the Euclidean r-space $E^{(r)}$ or a rectangle in $E^{(r)}$.

(ii)** $(\partial/\partial\theta_j)f(x; \theta_1, \ldots, \theta_r)$ exists for each $j = 1, \ldots, r$ and all $\boldsymbol{\theta}$.

(iii)** $\int |(\partial/\partial\theta_j)f(x; \theta_1, \ldots, \theta_r)| \, \mu(dx) < \infty$ for each $j = 1, \ldots, r$ and all $\boldsymbol{\theta}$.

(iv)** $I_{ij}(\boldsymbol{\theta}) = E_\theta\{(\partial/\partial\theta_i) \log f(X; \theta_1, \ldots, \theta_r) \cdot (\partial/\partial\theta_j) \log f(X; \theta_1, \ldots, \theta_r)\}$ exists and is finite for all $1 \leqslant i, j \leqslant r$ and all $\boldsymbol{\theta}$.

(v)** The matrix $\mathbf{I}(\boldsymbol{\theta}) = \|I_{ij}(\boldsymbol{\theta}): i, j = 1, \ldots, r\|$ is positive definite for all $\boldsymbol{\theta} \in \Theta$.

The matrix $\mathbf{I}(\boldsymbol{\theta})$ is called the *Fisher information matrix*. Let $g_1(\boldsymbol{\theta}), \ldots, g_k(\boldsymbol{\theta})$ be k estimable functions on Θ having the partial derivatives $D_{ij}(\boldsymbol{\theta}) = (\partial/\partial\theta_j)g_i(\boldsymbol{\theta})$, $i = 1, \ldots, k, j = 1, \ldots, r$, at all $\boldsymbol{\theta} \in \Theta$. Let

$$\mathbf{D}(\boldsymbol{\theta}) = \|D_{ij}(\boldsymbol{\theta}): i = 1, \ldots, k; j = 1, \ldots, r\|$$

be the $k \times r$ matrix of partial derivatives. Denote by $\mathbf{G}(\boldsymbol{\theta})$ the vector of the above k function of $\boldsymbol{\theta}$ and let $\hat{\mathbf{G}}(X) = (\hat{g}_1(X), \ldots, \hat{g}_k(X))'$ be an unbiased estimator of $\mathbf{G}(\boldsymbol{\theta})$. We further assume the following:

(vi)** $\int |\hat{g}_i(x)(\partial/\partial\theta_j)f(x; \theta_1, \ldots, \theta_r)| \, \mu(dx) < \infty$ for all $i = 1, \ldots, r$, $= 1, \ldots, k$, and all θ.

Theorem 4.3.1. (The generalized Cramér-Rao inequality.) *Let* $\{f(x; \theta);$ $\boldsymbol{\theta} \in \Theta\}$ *be a family of density functions satisfying regularity conditions* (i)**–(v)**. *Let* $\mathbf{G}(\boldsymbol{\theta}) = (g_1(\boldsymbol{\theta}), \ldots, g_k(\boldsymbol{\theta}))$ *be a vector valued function on* Θ *such that the matrix of partial derivatives* $\mathbf{D}(\boldsymbol{\theta})$ *exists for all* $\theta \in \Theta$. *Let* $\hat{\mathbf{G}}(X)$ *be an unbiased estimator of* $\mathbf{G}(\boldsymbol{\theta})$ *satisfying* (vi)**. *Let* $\boldsymbol{\Sigma}(\hat{\mathbf{G}}; \boldsymbol{\theta})$ *designate the covariance matrix of* $\hat{\mathbf{G}}(X)$. *Then* $\boldsymbol{\Sigma}(\hat{\mathbf{G}}; \boldsymbol{\theta}) - \mathbf{D}(\boldsymbol{\theta})\mathbf{I}^{-1}(\boldsymbol{\theta})\mathbf{D}'(\boldsymbol{\theta})$ *is a non-negative definite matrix for all* $\boldsymbol{\theta} \in \Theta$. *Moreover, if* $\mathbf{D}(\boldsymbol{\theta})$ *is nonsingular, then* $\hat{\mathbf{G}}(X)$ *attains the minimum possible generalized variance if and only if* $\hat{\mathbf{G}}(X) = (\hat{g}_1(X), \ldots, \hat{g}_k(X))'$ *is a function of the minimal sufficient statistic for* $\mathfrak{F}$, *and for each $i = 1, \ldots, r$,*

$$\text{(4.3.1)} \qquad \frac{\partial}{\partial\theta_i} \log f(X; \theta_1, \ldots, \theta_r) = \sum_{j=1}^{k} c_{ij}(\boldsymbol{\theta})[\hat{g}_j(X) - g_j(\boldsymbol{\theta})] \quad \text{a.s.},$$

where $c_{ij}(\boldsymbol{\theta})$ may depend on $\boldsymbol{\theta}$ but not on X. The minimal attainable generalized variance is

$$\text{(4.3.2)} \qquad \det\{\boldsymbol{\Sigma}(\hat{\mathbf{G}}; \boldsymbol{\theta})\} = \frac{[\det\{\mathbf{D}(\boldsymbol{\theta})\}]^2}{\det\{\mathbf{I}(\boldsymbol{\theta})\}}.$$

Proof. Consider the $(k + r) \times (k + r)$ covariance matrix $\mathbf{V}(\boldsymbol{\theta})$ of the random vector $((\hat{\mathbf{G}}(X))', S^{(1)}(X; \boldsymbol{\theta}), \ldots, S^{(r)}(X; \boldsymbol{\theta}))$. According to (ii)** and (vi)**,

$$\operatorname{cov}(\hat{g}_j(X), S^{(i)}(X; \boldsymbol{\theta})) = \int \hat{g}_j(x) \frac{\partial}{\partial\theta_i} f(x; \boldsymbol{\theta})\mu(dx)$$
$$= \frac{\partial}{\partial\theta_i} \int \hat{g}_j(x) f(x; \boldsymbol{\theta})\mu(dx) = \frac{\partial}{\partial\theta_i} g_j(\boldsymbol{\theta}).$$

Hence

$$\mathbf{V}(\boldsymbol{\theta}) = \left[\begin{array}{c|c} \boldsymbol{\Sigma}(\hat{\mathbf{G}}; \boldsymbol{\theta}) & \mathbf{D}(\boldsymbol{\theta}) \\ \hline \mathbf{D}'(\boldsymbol{\theta}) & \mathbf{I}(\boldsymbol{\theta}) \end{array}\right].$$

The matrix $\mathbf{V}(\boldsymbol{\theta})$ is non-negative definite. The matrix $\mathbf{I}(\boldsymbol{\theta})$ is positive definite [(v)**]. Hence

$$\text{(4.3.3)} \quad \det\{\mathbf{V}(\boldsymbol{\theta})\} = \det\{\mathbf{I}(\boldsymbol{\theta})\} \cdot \det\{\boldsymbol{\Sigma}(\hat{\mathbf{G}}; \boldsymbol{\theta}) - \mathbf{D}(\boldsymbol{\theta})\mathbf{I}^{-1}(\boldsymbol{\theta})\mathbf{D}'(\boldsymbol{\theta})\} \geqslant 0.$$

We conclude therefore that

$$\det\{\boldsymbol{\Sigma}(\hat{\mathbf{G}}; \boldsymbol{\theta}) - \mathbf{D}(\boldsymbol{\theta})\mathbf{I}^{-1}(\boldsymbol{\theta})\mathbf{D}'(\boldsymbol{\theta})\} \geqslant 0.$$

We now show that any determinant along the main diagonal of $\mathbf{C}(\hat{\mathbf{G}}; \boldsymbol{\theta}) = \boldsymbol{\Sigma}(\hat{\mathbf{G}}; \boldsymbol{\theta}) - \mathbf{D}(\boldsymbol{\theta})\mathbf{I}^{-1}(\boldsymbol{\theta})\mathbf{D}'(\boldsymbol{\theta})$ is non-negative. Let $\mathbf{C}_{(i_1,\ldots,i_s)}(\hat{\mathbf{G}}; \boldsymbol{\theta})$, where $1 \leqslant i_1 < \cdots < i_1 \leqslant k$ and $s = 1, 2, \ldots, k$, be an $s \times s$ matrix consisting of the elements in the $(i_1, \ldots, i_s)$ rows and $(i_1, \ldots, i_s)$ columns of $\mathbf{C}(\hat{\mathbf{G}}; \boldsymbol{\theta})$. Let $V_{(i_1,\ldots,i_s)}(\boldsymbol{\theta})$ designate the covariance matrix of the vector $(\hat{\mathbf{g}}_{i_1}(X), \ldots, \hat{\mathbf{g}}_{i_s}(X), S^{(1)}(X; \boldsymbol{\theta}), \ldots, S^{(r)}(X; \boldsymbol{\theta}))$. We have

$$\text{(4.3.4)} \qquad \mathbf{V}_{(i_1,\ldots,i_s)}(\boldsymbol{\theta}) = \left[\begin{array}{c|c} \boldsymbol{\Sigma}_{(i_1,\ldots,i_s)}(\hat{\mathbf{G}}; \boldsymbol{\theta}) & \mathbf{D}_{(i_1,\ldots,i_s)}(\boldsymbol{\theta}) \\ \hline \mathbf{D}'_{(i_1,\ldots,i_s)}(\boldsymbol{\theta}) & \mathbf{I}(\boldsymbol{\theta}) \end{array}\right],$$

where $\boldsymbol{\Sigma}_{(i_1,\ldots,i_s)}(\hat{\mathbf{G}}; \boldsymbol{\theta})$ consists of the $(i_1, \ldots, i_s)$ rows and columns of $\boldsymbol{\Sigma}(\hat{\mathbf{G}}; \boldsymbol{\theta})$ and $D_{(i_1,\ldots,i_s)}(\boldsymbol{\theta})$ consists of the $(i_1, \ldots, i_s)$ columns of $\mathbf{D}(\boldsymbol{\theta})$. The matrix $\mathbf{V}_{(i_1,\ldots,i_s)}(\boldsymbol{\theta})$ is non-negative definite. As before we obtain

$$\text{(4.3.5)} \qquad \det\{\boldsymbol{\Sigma}_{(i_1,\ldots,i_s)}(\hat{\mathbf{G}}; \boldsymbol{\theta}) - \mathbf{D}_{(i_1,\ldots,i_s)}(\boldsymbol{\theta})\mathbf{I}^{-1}(\boldsymbol{\theta})\mathbf{D}'_{(i_1,\ldots,i_s)}(\boldsymbol{\theta})\} \geqslant 0,$$

for all $(i_1, \ldots, i_s)$ such that $1 \leqslant i_1 < \cdots < i_s \leqslant k$; $s = 1, \ldots, k$. Finally, it is easy to verify that

$$\text{(4.3.6)} \quad \mathbf{C}_{(i_1,\ldots,i_s)}(\hat{\mathbf{G}}; \boldsymbol{\theta}) = \boldsymbol{\Sigma}_{(i_1,\ldots,i_s)}(\hat{\mathbf{G}}; \boldsymbol{\theta}) - \mathbf{D}_{(i_1,\ldots,i_s)}(\boldsymbol{\theta})\mathbf{I}^{-1}(\boldsymbol{\theta})\mathbf{D}'_{(i_1,\ldots,i_s)}(\boldsymbol{\theta}).$$

Thus we have proven that $\boldsymbol{\Sigma}(\mathbf{G}, \boldsymbol{\theta}) - \mathbf{D}(\boldsymbol{\theta})\mathbf{I}^{-1}(\boldsymbol{\theta})\mathbf{D}'(\boldsymbol{\theta})$ is non-negative definite. If $\mathbf{D}(\boldsymbol{\theta})$ is nonsingular then, according to (v)**, $\mathbf{D}(\boldsymbol{\theta})\mathbf{I}^{-1}(\boldsymbol{\theta})\mathbf{D}'(\boldsymbol{\theta})$ is symmetric and positive definite. Hence

$$\text{(4.3.7)} \qquad \det\{\boldsymbol{\Sigma}(\hat{\mathbf{G}}; \boldsymbol{\theta})\} \geqslant \frac{[\det\{\mathbf{D}(\boldsymbol{\theta})\}]^2}{\det\{\mathbf{I}(\boldsymbol{\theta})\}}$$

for all $\boldsymbol{\theta} \in \Theta$. If the generalized variance of, $\hat{\mathbf{G}}$ that is, $\det\{\boldsymbol{\Sigma}(\hat{\mathbf{G}}; \boldsymbol{\theta})\}$, attains the generalized Cramér-Rao lower bound, (4.3.2), $\hat{\mathbf{G}}$ must be a necessary statistic (a function of the minimal sufficient statistic). Indeed, if $\mathcal{B}_T$ is the sigma-subfield generated by the minimal sufficient statistic, and if $\hat{\mathbf{G}}$ is *not* $\mathcal{B}_T$-measurable, then the estimator $E\{\hat{\mathbf{G}}(X) \mid \mathcal{B}_T\}$ is also unbiased and has a generalized variance strictly smaller than that of $\hat{\mathbf{G}}$. (This is a generalized version of the Blackwell-Rao theorem.) But this contradicts the assumption that $\det\{\boldsymbol{\Sigma}(\hat{\mathbf{G}}; \boldsymbol{\theta})\}$ is minimal. Finally, the matrix $\mathbf{V}(\boldsymbol{\theta})$ is the covariance matrix of $S(X; \boldsymbol{\theta})$ and $\hat{\mathbf{G}}(X)$. Hence if the generalized variance of $\hat{\mathbf{G}}(X)$ attains the lower bound, (4.3.2), $\det\{\boldsymbol{\Sigma}(\hat{\mathbf{G}}; \boldsymbol{\theta}) - \mathbf{D}(\boldsymbol{\theta})\mathbf{I}^{-1}(\boldsymbol{\theta})\mathbf{D}'(\boldsymbol{\theta})\} = 0$. But this implies that the vector coefficient of correlation between $S(X; \boldsymbol{\theta})$ and $\hat{\mathbf{G}}(X)$ is one (see Anderson [1], pp. 244–245). Thus $\det\{\boldsymbol{\Sigma}(\hat{\mathbf{G}}; \boldsymbol{\theta}) - \mathbf{D}(\boldsymbol{\theta})\mathbf{I}^{-1}(\boldsymbol{\theta})\mathbf{D}'(\boldsymbol{\theta})\} = 0$ if and only if there exists a matrix of coefficients $\mathbf{C}(\boldsymbol{\theta}) = \|c_{i_j}(\theta); i, j = 1, \ldots, k\|$ independent of X such that

$$\mathbf{S}(X; \boldsymbol{\theta}) = \mathbf{C}(\boldsymbol{\theta})(\hat{\mathbf{G}}(X) - \mathbf{G}(\boldsymbol{\theta})) \quad \text{a.s.}$$

This proves (4.3.1). (Q.E.D.)

Example 4.3. Let $X_1, \ldots, X_n$, $n \geqslant 2$, be i.i.d. random variables having an $N(\mu, \sigma^2)$ distribution; $-\infty < \mu < \infty$, $0 < \sigma < \infty$. Both μ and σ are unknown. The complete sufficient statistic is $(\bar{X}, Q)$, where $\bar{X}$ is the sample mean and $Q = \sum_{i=1}^{n} (X_i - \bar{X})^2$. The family of normal densities is regular. It is simple to check that (i)**–(iv)** are satisfied. Since $(\bar{X}, Q)$ is a minimal sufficient statistic, we can determine the information matrix $\mathbf{I}(\boldsymbol{\theta})$, $\boldsymbol{\theta}' = (\mu, \sigma^2)$, from the distribution of $(\bar{X}, Q)$. Furthermore, $\bar{X}$ and Q are independent and $\bar{X} \sim N(\mu, (\sigma^2/n))$, $Q \sim \sigma^2\chi^2[n-1]$. Thus the joint density of $\bar{X}$ and Q is

$$\text{(4.3.8)} \quad f^{\bar{X},Q}(x, y; \mu, \sigma^2)$$
$$= \frac{\sqrt{n}}{\sqrt{(2\pi)}\, 2^{n-1/2}\sigma^n\Gamma((n-1/2))}\, y^{(n-3)/2} \exp\left\{-\frac{n}{2\sigma^2}(x-\mu)^2 - \frac{y}{2\sigma^2}\right\}.$$

We therefore obtain

$$\frac{\partial}{\partial \mu} \log f(\bar{X}, Q; \mu, \sigma^2) \sim \frac{\sqrt{n}}{\sigma} N(0, 1)$$

and

$$\frac{\partial}{\partial \sigma^2} \log f(\bar{X}, Q; \mu, \sigma^2) \sim -\frac{n}{2\sigma^2} + \frac{1}{2\sigma^2}(\chi_1^2[1] + \chi_2^2[n-1])$$

where $\chi_1^2[1]$ and $\chi_2^2[n-1]$ are independent and $\chi_1^2[1] \sim N^2(0, 1)$. Therefore the Fisher information matrix is

$$\mathbf{I}(\boldsymbol{\theta}) = \begin{bmatrix} \dfrac{n}{\sigma^2} & 0 \\ 0 & \dfrac{n}{2\sigma^4} \end{bmatrix}.$$

Let $g_1(\boldsymbol{\theta}) = \mu$ and $g_2(\boldsymbol{\theta}) = \sigma^2$. The U.M.V.U. estimator of $\mathbf{G}(\boldsymbol{\theta})$ is $\hat{\mathbf{G}}(\bar{X}, Q) = (\bar{X}, Q/(n-1))'$. The covariance matrix of $\hat{\mathbf{G}}(\bar{X}, Q)$ is

$$\boldsymbol{\Sigma}(\hat{\mathbf{G}}; \boldsymbol{\theta}) = \begin{bmatrix} \dfrac{\sigma^2}{n} & 0 \\ 0 & \dfrac{2\sigma^4}{n-1} \end{bmatrix}.$$

In the present example $D(\theta) = I$. Hence by the extended Cramér-Rao inequality, $\boldsymbol{\Sigma}(\mathbf{G}; \boldsymbol{\theta}) - \mathbf{I}^{-1}(\boldsymbol{\theta})$ is a non-negative definite matrix. Indeed,

$$\boldsymbol{\Sigma}(\hat{\mathbf{G}}; \boldsymbol{\theta}) - \mathbf{I}^{-1}(\boldsymbol{\theta}) = \begin{bmatrix} 0 & 0 \\ 0 & \dfrac{2\sigma^4}{n(n-1)} \end{bmatrix}.$$

We see here that the variance of the U.M.V.U. estimator of σ^2 does not attain the Cramér-Rao lower bound $2\sigma^4/n$. The difference becomes negligible when n grows. It should be noticed that $(\partial/\partial\sigma^2) \log f(\bar{X}, Q; \mu, \sigma^2)$ is *not* a linear function (for a given μ and σ^2) of $(\bar{X} - \mu)$ and $(Q - \sigma^2)$. ■

Example 4.4. Suppose that $X_1, \ldots, X_n$ are i.i.d. random variables having a log-normal distribution. That is, $X_i \sim \exp\{N(\mu, \sigma^2)\}$, $i = 1, \ldots, n$, where $-\infty < \mu < \infty$ and $0 < \sigma < \infty$ are unknown. It is easy to verify that the mean of a log-normal distribution is

$$\xi = E\{X\} = \exp\left\{\mu + \frac{\sigma^2}{2}\right\}.$$

The variance is

$$D^2 = \xi^2(e^{\sigma^2} - 1).$$

The Cramér-Rao lower bound to the variances of all unbiased estimators of $\xi = e^{\mu+\sigma^2/2}$ coincides with the Cramér-Rao lower bound in the normal case when one is estimating the function $g(\mu, \sigma^2) = \exp\{\mu + \sigma^2/2\}$. Therefore

$$L_1(\xi, \sigma^2) = \xi^2(1, \tfrac{1}{2})\begin{bmatrix} \dfrac{\sigma^2}{n} & 0 \\ 0 & \dfrac{2\sigma^4}{n} \end{bmatrix}\begin{pmatrix} 1 \\ \tfrac{1}{2} \end{pmatrix}$$

$$= \frac{\sigma^2}{n}\,\xi^2\left(1 + \frac{\sigma^2}{2}\right).$$

An unbiased estimator of ξ in common use is the sample mean $\bar{X}$. The variance of the sample mean is $\text{Var}\{\bar{X}\} = D^2/n = (\sigma^2/n)\xi^2(1 + \sigma^2/2 + \sigma^4/6 + \cdots)$. This is obviously greater than the lower bound $L_1(\xi, \sigma^2)$ for all (ξ, σ^2). But this result is anticipated, since $\bar{X}$ is not a function of the minimal sufficient statistic and is therefore not a U.M.V.U. estimator. It is interesting to check whether the U.M.V.U. estimator $\hat{\xi}$ has a variance which attains the lower bound. We shall now derive $\hat{\xi}$. The minimal sufficient statistic in the log-normal case is $(\bar{Y}, S)$, where $Y_i = \log X_i$ $(i = 1, \ldots,)$, $\bar{Y} = \sum_{i=1}^n Y_i/n$, and $S = \sum_{i=1}^n (Y_i - \bar{Y})^2$. The variables $Y_1, \ldots, Y_n$ are i.i.d. having a normal distribution $N(\mu, \sigma^2)$; $(\bar{Y}, S)$ is a *complete* sufficient statistic. We therefore have to find a function $\hat{\xi}(\bar{Y}, S)$, which is unbiased. This will be the U.M.V.U. estimator. Since $\bar{Y} \sim N(\mu, \sigma^2/n)$, $E\{e^{\bar{Y}}\} = \exp\{\mu + \sigma^2/2n\}$. Hence if we can find a function $\varphi(S)$ such that $E\{\varphi(S)\} = \exp\{(\sigma^2/2)(1 - 1/n)\}$, $\exp\{\bar{Y}\}\varphi(S)$ is the U.M.V.U. estimator of ξ. This follows from the independence of $\bar{Y}$ and S, which implies that $E\{\exp\{\bar{Y}\}\cdot \varphi(S)\} = \xi$ for all (μ, σ^2). Now expand $\exp\{\sigma^2/2(1 - 1/n)\}$ by a Taylor series,

$$\exp\left\{\frac{\sigma^2}{2}\left(1 - \frac{1}{n}\right)\right\} = \sum_{\nu=0}^{\infty} \frac{1}{\nu!}\left(\frac{\sigma^2}{2}\right)^{\nu}\left(1 - \frac{1}{n}\right)^{\nu}.$$

In order to estimate this function unbiasedly, we notice that

$$E\{S^{\nu}\} = 2^{\nu}\sigma^{2\nu}\frac{\Gamma(\nu + (n-1)/2)}{\Gamma((n-1)/2)}, \qquad \nu = 0, 1, \ldots.$$

Hence

$$\varphi(S) = \Gamma\left(\frac{n-1}{2}\right)\sum_{\nu=0}^{\infty}\frac{1}{\nu!\,\Gamma[((n-1)/2) + \nu]}\left(\frac{S}{4}\right)^{\nu}\left(1 - \frac{1}{n}\right)^{\nu} \tag{4.3.9}$$

is an unbiased estimator of $\exp\{\sigma^2/2(1 - 1/n)\}$. When $n = 2k + 1$ then $\varphi(S)$

can be expressed as a Bessel function of the first kind of order k. If $n = 2k$ we can express $\varphi(S)$ as a finite sum of sine and cosine functions.

The density function of $\hat{\xi}$ is obviously not of the exponential type, and its variance is therefore greater than the Cramér-Rao lower bound $L_1(\xi, \sigma^2)$. ■

A generalization of (4.1.28) to the multiparameter case is obtained by considering $g_i(\boldsymbol{\theta}) = \theta_i + B_i(\boldsymbol{\theta})$.

Theorem 4.3.2. *Let* $\{f(x; \boldsymbol{\theta}); \boldsymbol{\theta} \in \Theta\}$ *be a family of density functions satisfying regularity conditions* (i)**–(vi)**. *Let* $\mathbf{G}(\boldsymbol{\theta}) = (g_1(\boldsymbol{\theta}), \ldots, g_k(\boldsymbol{\theta})'$ *be such that* $g_i(\boldsymbol{\theta}) = \theta_i + B_i(\boldsymbol{\theta})$, $i = 1, \ldots, k$. *Let* $\tilde{\mathbf{D}}(\boldsymbol{\theta}) = \|(\partial/\partial\theta_j)B_j(\boldsymbol{\theta}); i = 1, \ldots, k; j = 1, \ldots, k\|$ *exist for all* $\boldsymbol{\theta} \in \Theta$. *Then for every* $\boldsymbol{\xi} = (\xi_1, \ldots, \xi_k)'$,

$$\text{(4.3.10)} \qquad \boldsymbol{\xi}'\boldsymbol{\Sigma}(\hat{\mathbf{G}}; \boldsymbol{\theta})\boldsymbol{\xi} \geqslant \boldsymbol{\xi}'\mathbf{I}^{-1}(\boldsymbol{\theta})\boldsymbol{\xi}\left[1 + \frac{\boldsymbol{\xi}'\tilde{\mathbf{D}}(\boldsymbol{\theta})\mathbf{I}^{-1}(\boldsymbol{\theta})\boldsymbol{\xi}}{\boldsymbol{\xi}'\mathbf{I}^{-1}(\boldsymbol{\theta})\boldsymbol{\xi}}\right].$$

Proof. Let $\mathbf{S}(X; \boldsymbol{\theta}) = \left\| \frac{\partial}{\partial\theta_i} \log f(X; \boldsymbol{\theta}); i = 1, \ldots, r \right\|$. Then by the Schwarz inequality,

$$\text{(4.3.11)} \quad E\{\boldsymbol{\xi}'(\hat{\mathbf{G}}(X) - \mathbf{G}(\boldsymbol{\theta})) \cdot \boldsymbol{\xi}'\mathbf{I}^{-1}(\boldsymbol{\theta})\mathbf{S}(X; \boldsymbol{\theta})\} \leqslant [\boldsymbol{\xi}'\boldsymbol{\Sigma}\boldsymbol{\xi} \cdot \boldsymbol{\xi}'\mathbf{I}^{-1}(\boldsymbol{\theta})\boldsymbol{\xi}]^{1/2}$$

On the other hand, since from (vi)**

$$\text{(4.3.12)} \qquad \int (\hat{g}_i(X) - g_i(\boldsymbol{\theta})) \frac{\partial}{\partial\theta_j} f(x; \boldsymbol{\theta})\mu(\mathrm{d}x) = \frac{\partial}{\partial\theta_j}(\theta_i + B_i(\boldsymbol{\theta})),$$

we obtain that

$$\text{(4.3.13)} \qquad E\{(\hat{\mathbf{G}}(X) - \mathbf{G}(\boldsymbol{\theta}))\mathbf{S}'(X; \boldsymbol{\theta})\} = \mathbf{I} + \tilde{\mathbf{D}}(\boldsymbol{\theta}), \quad \text{all } \theta \in \Theta.$$

Substituting (4.3.13) in the left-hand side of (4.3.11) we obtain

$$\text{(4.3.14)} \qquad [\boldsymbol{\xi}'(\mathbf{I} + \tilde{\mathbf{D}}(\boldsymbol{\theta}))\mathbf{I}^{-1}(\boldsymbol{\theta})\boldsymbol{\xi}]^2 \leqslant \boldsymbol{\xi}'\Sigma(\hat{\mathbf{G}}; \boldsymbol{\theta})\boldsymbol{\xi} \cdot \boldsymbol{\xi}'\mathbf{I}^{-1}(\boldsymbol{\theta})\boldsymbol{\xi},$$

which implies (4.3.10). (Q.E.D.)

We now prove a theorem of Hogg and Craig [1], which shows that if an unbiased estimator $\hat{\mathbf{G}}(X)$ of $\mathbf{G}(\boldsymbol{\theta})$ has a generalized variance which attains the Cramér-Rao lower bound (4.3.2) then it must have a multivariate normal distribution and vice-versa. This theorem is obviously valid also for the special univariate case and therefore has not been stated previously. For the following theorem we restrict attention to a family of the k-parameter exponential type. That is, the density function of P_θ with respect to $\mu(x)$ is

$$\text{(4.3.15)} \quad f(x; \boldsymbol{\theta}) = h(x) \exp\left\{\sum_{i=1}^{k} \theta_i U_i(x) + \psi(\theta_1, \ldots, \theta_r)\right\}, \qquad -\infty < x < \infty$$

and $\boldsymbol{\theta} \in \Theta$. We further assume that the six regularity conditions (i)**–(vi)** are satisfied. Given $X_1, \ldots, X_n$ which are i.i.d. random variables having an exponential type density (4.3.15), the minimal sufficient statistic is $\mathbf{T}(X) = (T_1(X), \ldots, T_k(X))$, where $T_i(X) = \sum_{j=1}^{n} U_i(X_j)$, $i = 1, \ldots, k$. As is well known, the distribution of $T(X)$ is also of the exponential type, with a density function of the form

$$f(t_1, \ldots, t_k; \boldsymbol{\theta}) = \exp\left\{\sum_{i=1}^{k} \theta_i t_i + n\psi(\theta_1, \ldots, \theta_k) + Q(t_1, \ldots, t_k)\right\}. \tag{4.3.16}$$

Let $\hat{\boldsymbol{\theta}}(\mathbf{T})$ be an unbiased estimator of $\boldsymbol{\theta}$ which attains the Cramér-Rao generalized lower bound (4.3.2).

Theorem 4.3.3. (Hogg and Craig [1].) *Let $X_1, \ldots, X_n$ be i.i.d. random variables having a k-parameter exponential type density* (4.3.15). *Let* $\mathbf{T}(X_1, \ldots, X_n)$ *be the vector of minimal sufficient statistic. Under the regularity conditions* (i)**–(vi)**, $\hat{\boldsymbol{\theta}}(\mathbf{T})$ *is an unbiased estimator of* $\boldsymbol{\theta}$ *attaining the generalized Cramér-Rao lower bound* (4.3.2) *if and only if the distribution function of* $T(X_1, \ldots, X_n)$ *is a nonsingular k-variate normal.*

Proof. (i) *Necessity.* Assume that $\hat{\boldsymbol{\theta}}(T)$ is an unbiased estimator attaining the Cramér-Rao lower bound for the generalized variance. According to (4.3.1) and (4.3.16) we have

$$\mathbf{T}(X) + n\frac{\partial}{\partial \theta}\psi(\boldsymbol{\theta}) = \mathbf{C}(\boldsymbol{\theta})[\hat{\boldsymbol{\theta}}(\mathbf{T}) - \boldsymbol{\theta}] \quad \text{a.s.} \tag{4.3.17}$$

The k linear equations on the left-hand side of (4.3.17) are linearly independent, since $f(t_1, \ldots, t_k; \theta)$ belongs to a regular family of rank $k + 1$ (see Section 2.5). Hence the matrix of coefficients $C(\theta)$ is nonsingular. We obtain from (4.3.17) that

$$\hat{\boldsymbol{\theta}}(\mathbf{T}) = \mathbf{C}^{-1}(\boldsymbol{\theta})\mathbf{T} + \mathbf{C}^{-1}(\boldsymbol{\theta})\left[n\frac{\partial}{\partial \theta}\psi(\boldsymbol{\theta}) + \mathbf{C}(\boldsymbol{\theta})\boldsymbol{\theta}\right] \quad \text{a.s.} \tag{4.3.18}$$

Since $\hat{\boldsymbol{\theta}}(\mathbf{T})$ is a statistic independent of $\boldsymbol{\theta}$, $\mathbf{C}^{-1}$ and hence $\mathbf{C}$ are independent of θ. Moreover,

$$\frac{\partial}{\partial \boldsymbol{\theta}}\psi(\boldsymbol{\theta}) = -\frac{1}{n}\mathbf{C}\boldsymbol{\theta}. \tag{4.3.19}$$

Integrating (4.3.19) we obtain that

$$\psi(\boldsymbol{\theta}) = -\frac{1}{2n}\boldsymbol{\theta}'\mathbf{C}\boldsymbol{\theta} + a, \tag{4.3.20}$$

all $\boldsymbol{\theta} \in \Theta$, where a is a scalar independent of θ. Finally, the characteristic function of **T** is

$$(4.3.21)\quad E\{\exp[i\boldsymbol{\lambda}'\mathbf{T}]\} = \exp\{\psi(\boldsymbol{\theta}) - (\psi\boldsymbol{\theta} + i\boldsymbol{\lambda})\}$$

$$= \exp\left\{-\frac{1}{2n}\boldsymbol{\theta}'\mathbf{C}\boldsymbol{\theta} + \frac{1}{2n}(\boldsymbol{\theta} + i\boldsymbol{\lambda})'\mathbf{C}(\boldsymbol{\theta} + i\boldsymbol{\lambda})\right\}$$

$$= \exp\left\{-\frac{1}{2n}\boldsymbol{\lambda}'\mathbf{C}\boldsymbol{\lambda} + \frac{i}{n}\boldsymbol{\lambda}'\mathbf{C}\boldsymbol{\theta}\right\}.$$

Thus the characteristic function of the distribution of **T** is that of a k-variate normal distribution, with a nonsingular covariance matrix $(1/n)\mathbf{C}$, and expectation vector $(1/n)\mathbf{C}\boldsymbol{\theta}$.

(ii) *Sufficiency.* Assume that **T** has a k-variate normal distribution $N(\boldsymbol{\theta}, \mathbf{V})$ where **V** is a positive definite covariance matrix independent of $\boldsymbol{\theta}$. The matrix **V** is *known*; otherwise, the distribution of **T** is not a k-parameter exponential one. Thus writing the density function of **T** in the form

$$(4.3.22)\quad f(t_1, \ldots, t_k; \boldsymbol{\theta}) = (2\pi)^{-n/2}|\mathbf{V}|^{-1/2}\exp\{-\tfrac{1}{2}\mathbf{t}'\mathbf{V}^{-1}\mathbf{t}\} \cdot \exp\{\boldsymbol{\theta}'\mathbf{V}^{-1}\mathbf{t} - \tfrac{1}{2}\boldsymbol{\theta}'\mathbf{V}^{-1}\boldsymbol{\theta}\},$$

we obtain by differentiating the integral of (4.3.22) with respect to θ,

$$(4.3.23)\quad E_\theta\{\mathbf{T}(X)\} = \boldsymbol{\theta}, \quad \text{all } \boldsymbol{\theta}.$$

Thus **T** is an unbiased estimator of $\boldsymbol{\theta}$ with a covariance matrix $\boldsymbol{\Sigma}(\mathbf{T}; \boldsymbol{\theta}) = \mathbf{V}$. The estimator **T** is a minimal sufficient statistic for this family of distributions. Hence the Fisher information matrix is

$$(4.3.24)\quad \mathbf{I}(\boldsymbol{\theta}) = \mathbf{V}^{-1}E_\theta\{(\mathbf{T} - \boldsymbol{\theta})(\mathbf{T} - \boldsymbol{\theta})'\}\mathbf{V}^{-1} = \mathbf{V}^{-1}.$$

Therefore $\det\{\boldsymbol{\Sigma}(\mathbf{T}; \boldsymbol{\theta})\} = 1/\det\{\mathbf{I}(\boldsymbol{\theta})\}$ and **T** attains the generalized Cramér-Rao lower bound. (Q.E.D.)

Example 4.5. Let $\mathbf{X} = (X_1, \ldots, X_n)$ be a random vector with an expectation $E\{\mathbf{X}\} = \mathbf{D}\boldsymbol{\alpha}$, where **D** is an $n \times p$ matrix and $\boldsymbol{\alpha}$ is a $p \times 1$ vector of unknown parameters. The matrix **D** is known and independent of $\boldsymbol{\alpha}$ and of a full rank $p(p \leqslant n)$. As shown in Section 3.8, if the covariance matrix of **X** is $\boldsymbol{\Sigma} = \sigma^2 I$, the minimum variance linear unbiased estimator of $\boldsymbol{\alpha}$ is the least-squares estimator $\hat{\boldsymbol{\alpha}} = (\mathbf{D}'\mathbf{D})^{-1}\mathbf{D}'\mathbf{X}$. The covariance matrix of $\hat{\boldsymbol{\alpha}}$ is $\sigma^2(\mathbf{D}'\mathbf{D})^{-1}$. As proven in Theorem 4.3.3, if $\mathbf{X} \sim N(\mathbf{D}\boldsymbol{\alpha}, \sigma^2\mathbf{I})$ and σ^2 is known, then the generalized variance of $\hat{\boldsymbol{\alpha}}$ attains the Cramér-Rao lower bound. Hence the Fisher information matrix in the normal case is $\sigma^{-2}\mathbf{D}'\mathbf{D}$. If by choosing between different experiments the statistician can vary **D** on a given compact set, the objective will be to maximize $|\mathbf{D}'\mathbf{D}|$. An experiment yielding a maximum determinant of $\mathbf{D}'\mathbf{D}$ is optimal according to the above criterion.

We conclude this by remarking that if the distribution of $\mathbf{X}$ is not a multivariate normal one then according to Theorem 4.3.3 the least-squares estimator $\hat{\boldsymbol{\alpha}}$ does not attain the generalized Cramér-Rao lower bound (if exists). ■

4.4. THE CRAMÉR-RAO INEQUALITY FOR SEQUENTIAL ESTIMATION IN THE REGULAR CASE

In this section we present an extension of the Cramér-Rao inequality established by Wolfowitz [1]. A sequential estimation procedure is a procedure defined by two sets of rules: stopping rules and estimation rules. In a sequential sampling procedure we evaluate the amount of information available after each observation and decide whether to terminate sampling or to take an additional observation. As described in Section 2.7, a stopping rule for a sequential sampling is a function of the observed random variable which maps the vector of the first n observations, namely, $(X_1, \ldots, X_n)$, into the set of integers $I_n = \{m; m = n, n + 1, \ldots\}$. Considering the sigma-fields $\mathcal{B}_1 \subset \mathcal{B}_2 \subset$ generated by X_1, $(X_1, X_2) \ldots$, a stopping rule R for a sequential procedure can be conveniently described by a sequence of sets $\{R_n; n = 1, 2, \ldots\}$, where $R_n \in \mathcal{B}_n$ for each $n = 1, 2, \ldots$. Sampling is continued as long as consecutive vectors $(X_1, \ldots, X_n)$, $n = 1, 2, \ldots$, do not enter one of the sets R_n. In other words, the sample size N (a random variable) is

$$N = \text{least integer } n, n \geqslant 1, \text{ such that } (X_1, \ldots, X_n) \in R_n.$$

Define the sets

$$\tilde{R}_n = \begin{cases} R_1, & \text{if } n = 1, \\ \bar{R}_1 \cap \cdots \cap \bar{R}_{n-1} \cap R_n, & \text{if } n \geqslant 2. \end{cases}$$

The set $\tilde{R}_n$ is the set of all sample points which lead to stopping at $N = n$.

The estimation rule for estimating a function $g(P_1, P_2, \ldots)$ is given by a sequence of functions $\hat{g}_1, \hat{g}_2, \ldots$ such that $\hat{g}_n \in \mathcal{B}_n$ for all $n = 1, 2, \ldots$; and if $N = n$ then the estimate of g is $\hat{g}_n$.

Lemma 4.4.1. (Wald [2].) *Let $X_1, X_2, \ldots$ be a sequence of i.i.d. random variables, distributed like X, satisfying $E\,|X| < \infty$. For any sequential stopping rule yielding $E\{N\} < \infty$,*

$$E\left\{\sum_{i=1}^{N} X_i\right\} = E\{X\} \cdot E\{N\}. \tag{4.4.1}$$

Proof. Let $\{R_1, R_2, \ldots\}$ be the sequence of stopping regions. Then

$$E\left\{\sum_{i=1}^{N} X_i\right\} = \sum_{n=1}^{\infty} \int_{\tilde{R}_n} \sum_{i=1}^{n} x_i \prod_{i=1}^{n} F(\mathrm{d}x_i), \tag{4.4.2}$$

where $F(x)$ is the common distribution of $X_1, X_2, \ldots$. The double summation on the right-hand side of (4.4.2) can be formally written as

$$\sum_{n=1}^{\infty} \int_{\tilde{R}_n} \sum_{i=1}^{n} x_i \prod_{i=1}^{n} F(\mathrm{d}x_i) = \sum_{n=1}^{\infty} \sum_{n=i}^{\infty} \int_{\tilde{R}_n} x_i \prod_{j=1}^{n} F(\mathrm{d}x_j). \tag{4.4.3}$$

Let

$$I(N \geqslant n) = \begin{cases} 1, & \text{if } N \geqslant n, \\ 0, & \text{otherwise.} \end{cases}$$

Then

$$\begin{aligned} \sum_{n=i}^{\infty} \int_{\tilde{R}_n} x_i \prod_{j=1}^{n} F(\mathrm{d}x_j) &= E\{X_i I(N \geqslant i)\} \\ &= P[N \geqslant i] E\{X_i \mid N \geqslant i\}, \qquad i = 1, 2, \ldots. \end{aligned} \tag{4.4.4}$$

The event $\{N \geqslant i\}$ is measurable $\mathcal{B}_{i-1}$ $(i = 1, 2, \ldots)$, where $\mathcal{B}_0 \equiv \mathcal{B}$. Therefore, since X_i is independent of $(X_1, \ldots, X_{i-1})$, we have

$$E\{X_i \mid N \geqslant i\} = E\{X_i\}, \quad \text{for all } i = 1, 2, \ldots. \tag{4.4.5}$$

Substituting (4.4.5) and (4.4.4) in (4.4.3) yields

$$\begin{aligned} E\left\{\sum_{i=1}^{N} X_i\right\} &= E\{X\} \sum_{i=1}^{\infty} P[N \geqslant i] \\ &= E\{X\} \cdot E\{N\}. \end{aligned} \tag{4.4.6}$$

The assumption that $E\,|X| < \infty$ is sufficient for the justification of the interchange in the summation order in (4.4.3). A rearrangement of the summation order in the right-hand side of (4.4.2) is justified if that series is absolutely convergent. But the above formal expression yields

$$\sum_{n=1}^{\infty} \int_{\tilde{R}_n} \sum_{i=1}^{n} |x_i| \prod_{i=1}^{n} F(\mathrm{d}x_i) = E\,|X| \cdot E\{N\} < \infty.$$

(Q.E.D.)

Lemma 4.4.2. *Let $X_1, X_2, \ldots$ be a sequence of i.i.d. random variables having a common distribution function $F(x)$ with zero mean and variance σ^2, $0 < \sigma^2 < \infty$. For any sequential stopping rule with $E\{N\} < \infty$, if $E\{(\sum_{i=1}^{n} |X_i|)^2\} < \infty$ then*

$$E\left\{\left(\sum_{i=1}^{N} X_i\right)^2\right\} = \sigma^2 E\{N\}. \tag{4.4.7}$$

Proof. As before,

(4.4.8)

$$\begin{aligned} E\left(\sum_{i=1}^{N} X_i\right)^2 &= \sum_{n=1}^{\infty} \int_{\tilde{R}_n} \left(\sum_{i=1}^{n} x_i\right)^2 \prod_{i=1}^{n} F(\mathrm{d}x_i) \\ &= \sum_{n=1}^{\infty} \int_{\tilde{R}_n} \sum_{i=1}^{n} x_i^2 \prod_{i=1}^{n} F(\mathrm{d}x_i) + 2 \sum_{n=1}^{\infty} \int_{\tilde{R}_n} \sum_{i=1}^{n-1} \sum_{j=i+1}^{n} x_i x_j \sum_{\nu=1}^{n} F(\mathrm{d}x_\nu). \end{aligned}$$

Since $\sigma^2 < \infty$ and $E\{N\} < \infty$ we can apply the previous lemma to obtain

$$\sum_{n=1}^{\infty} \int_{\tilde{R}_n} \sum_{i=1}^{n} x_i^2 \prod_{i=1}^{n} F(\mathrm{d}x_i) = \sigma^2 E\{N\}. \tag{4.4.9}$$

It remains to prove that the second term on the right-hand side of (4.4.8) is zero. Rearranging the summation order we obtain

$$\sum_{n=1}^{\infty} \int_{\tilde{R}_n} \sum_{i=1}^{n-1} \sum_{j=i+1}^{n} x_i x_j \prod_{\nu=1}^{n} F(\mathrm{d}x_\nu) = \sum_{i=2}^{\infty} \sum_{j=1}^{i-1} \sum_{n=i}^{\infty} \int_{\tilde{R}_n} x_i x_j \prod_{\nu=1}^{n} F(\mathrm{d}x_\nu). \tag{4.4.10}$$

As in the previous lemma, for all $j < i$, $i = 2, 3, \ldots,$

$$\sum_{n=i}^{\infty} \int_{\tilde{R}_n} x_i x_j \prod_{\nu=1}^{n} F(\mathrm{d}x_\nu) = P[N \geqslant i] E\{X_i X_j \mid N \geqslant i\}. \tag{4.4.11}$$

But since $\{N \geqslant i\} \in \mathcal{B}_{i-1}$,

$$\begin{aligned} E\{X_i X_j \mid N \geqslant i\} &= E\{X_i\} E\{X_j \mid N \geqslant i\} \\ &= 0, \quad \text{for all} \quad j < i,\ i = 2, 3, \ldots . \end{aligned} \tag{4.4.12}$$

Therefore either side of (4.4.10) is zero, provided the rearrangement of the summation order in (4.4.10) is justified. This is guaranteed, however, by the condition that $E\{(\sum_{i=1}^{n} |X_i|)^2\} < \infty$. Indeed,

$$\begin{aligned} E\left\{\left(\sum_{i=1}^{N} |X_i|\right)^2\right\} &= \sum_{n=1}^{\infty} \int_{\tilde{R}_n} \sum_{i=1}^{n} x_i^2 \prod_{\nu=1}^{n} F(\mathrm{d}x_\nu) + 2 \sum_{n=1}^{\infty} \int_{\tilde{R}_n} \sum_{i=1}^{n-1} \sum_{j=i+1}^{n} |x_i|\,|x_j| \prod_{\nu=1}^{n} F(\mathrm{d}x_\nu). \end{aligned} \tag{4.4.13}$$

But the second term on the right-hand side of (4.4.13) dominates the second term on the right-hand side of (4.4.8), which is a sufficient condition for the validity of (4.4.10). (Q.E.D.)

We now derive the extended Cramér-Rao inequality for sequential regular cases.

Theorem 4.4.1. (Wolfowitz [1].) *Let $X_1, X_2, \ldots$ be a sequence of i.i.d. random variables, whose common density function $f(x; \theta)$ with respect to μ belongs to a family $\psi = \{f(x; \theta); \theta \in \Theta\}$ on which the following regularity conditions are satisfied:*

(i) *Θ contains an interval in a Euclidean k-space.*
(ii) *$f(x; \theta)$ is differentiable with respect to θ on Θ.*
(iii) *$\int |(\partial/\partial\theta) f(x; \theta)|\, \mu(\mathrm{d}x) < \infty$, for all $\theta \in \Theta$.*
(iv) *$0 < \int ((\partial/\partial\theta) \log f(x; \theta))^2 f(x; \theta) \mu(\mathrm{d}x) < \infty$, all $\theta \in \Theta$.*

(v) *For each* $n = 1, 2, \ldots$ *and all* θ,

$$\int \left(\sum_{i=1}^{n} \frac{|(\partial/\partial\theta) f(x;\theta)|}{f(x_i;\theta)} \right)^2 \prod_{\nu=1}^{n} F(dx_\nu) < \infty.$$

Let $\{R_n; n = 1, 2, \ldots\}$ *be a sequence of stopping regions associated with a given sequential procedure. Let* $g(\theta)$ *be an estimable and differentiable function on* Θ. *Let* $\hat{g}(X_1, X_2, \ldots)$ *be unbiased estimator of* $g(\theta)$ *satisfying the following conditions:*

(vi) $\int |\hat{g}(x_1, \ldots, x_n) \cdot (\partial/\partial\theta) \prod_{\nu=1}^{n} f(x_\nu;\theta)| \prod_{\nu=1}^{n} \mu(dx_\nu) < \infty$, *for each* $n = 1, 2, \ldots$ *and all* θ.

(vii) $\sum_{n=1}^{\infty} dg_n(\theta)/d\theta$ *converges uniformly on* Θ, *where*

$$g_n(\theta) = \int_{\tilde{R}n} \hat{g}(x_1, \ldots, x_n) \prod_{\nu=1}^{n} F(dx_\nu).$$

Then

$$\text{(4.4.14)} \qquad \text{Var}_\theta\, \{\hat{g}(X_1, \ldots)\} \geqslant \frac{(g'(\theta))^2}{I(\theta)E\{N\}},$$

for all θ, *provided* $E\{N\} < \infty$.

Proof. Let N be the sample size associated with the given sequential procedure. Let $S(X_i;\theta) = (\partial/\partial\theta) \log f(X_i;\theta)$, $i = 1, 2, \ldots$. These are i.i.d. random variables, and (i)–(iv) guarantee that $E\{S(X_i;\theta)\} = 0$ and $I(\theta) = E\{S^2(X;\theta)\} < \infty$. Condition (iv) and the assumption that $E\{N\} < \infty$ imply by Lemma 4.4.1 that

$$\text{(4.4.15)} \qquad E\left\{ \sum_{i=1}^{N} S(X_i;\theta) \right\} = 0 \quad \text{for all} \quad \theta.$$

Furthermore, according to (v)

$$\text{(4.4.16)} \qquad E\left\{ \left(\sum_{i=1}^{N} |S(X_i;\theta)| \right)^2 \right\} < \infty.$$

Hence Lemma 4.4.2 implies that

$$\text{(4.4.17)} \qquad E\left\{ \left(\sum_{i=1}^{N} S(X_i;\theta) \right)^2 \right\} = I(\theta) \cdot E\{N\}.$$

Consider the expectation

$$E\left\{ \hat{g}(X_1, \ldots, X_N) \sum_{i=1}^{N} S(X_i;\theta) \right\}, \qquad \theta \in \Theta,$$

where $\hat{g}(X_1, \ldots)$ is an unbiased estimator of $g(\theta)$. According to (4.4.15) and the Schwartz inequality we have

$$\text{(4.4.18)} \quad E\left\{ \hat{g}(X_1, \ldots, X_N) \sum_{i=1}^{N} S(X_i;\theta) \right\}$$
$$\leqslant \left\{ E\{(\hat{g}(X_1, \ldots, X_N) - g(\theta))^2\} E\left\{ \left(\sum_{i=1}^{N} S(X_i;\theta) \right)^2 \right\} \right\}^{1/2}, \quad \text{for all} \quad \theta \in \Theta.$$

The quantity $E[\hat{g}(X_1, \ldots, X_N) - g(\theta)]^2$ is the variance of $\hat{g}(X_1, \ldots)$ under the sequential procedure. Furthermore, (vi) and (vii) allow the differentiation under the integral sign in

$$(4.4.19) \quad g'(\theta) = \frac{d}{d\theta} \sum_{n=1}^{\infty} \int_{\tilde{R}_n} \hat{g}(x_1, \ldots, x_n) \prod_{\nu=1}^{n} F(dX_\nu)$$
$$= \sum_{n=1}^{\infty} \int_{\tilde{R}_n} \hat{g}(x_1, \ldots, x_n) \left(\frac{\partial}{\partial \theta} \prod_{\nu=1}^{n} f(x_\nu; \theta) \right) \prod_{\nu=1}^{n} \mu(dx_\nu)$$
$$= \sum_{n=1}^{\infty} \int_{\tilde{R}_n} \hat{g}(x_1, \ldots, x_n) \sum_{\nu=1}^{n} S(x_i; \theta) \prod_{j=1}^{n} F(dx_j)$$
$$= E\left\{ \hat{g}(X_1, \ldots, X_N) \sum_{i=1}^{N} S(X_i; \theta) \right\}.$$

Equations 4.4.17–4.4.19 yield the extended Cramér-Rao inequality (4.4.17). (Q.E.D.)

In choosing a sequential estimation procedure we should adopt a criterion of optimality. Three commonly accepted criteria of optimality for choosing a sequential estimation procedure are the following:

(i) Subject to the condition $E_\theta\{N\} \leqslant m$ (m being a fixed integer bound) for all θ, minimize the variance of the best unbiased estimator, that is, $E_\theta\{(\hat{g}_N - g)^2\}$ uniformly in θ (if such an estimator exists).
(ii) Subject to the condition $E_\theta(\hat{g}_N - g)^2 \leqslant v < \infty$ (fixed finite positive value) for all θ, minimize the expected sample size $E_\theta\{N\}$.
(iii) Minimize the expected cost of sampling plus expected loss, that is,

$$cE_\theta\{N\} + E_\theta\{(\hat{g}_N - g)^2\}.$$

Generally there is no sequential estimation that can satisfy (iii) uniformly in θ, and we have to modify this criterion. Such modification is given by a Bayesian criterion of optimality which will be discussed in Chapter 6.

Concerning (i), we deduce immediately from the Cramér-Rao inequality (4.4.14) that a fixed sample of size m yields the optimal sequential procedure whenever the variance of its U.M.V.U. estimator attains the Cramér-Rao lower bound (4.1.22). This is the case, as we have proven, whenever the unbiased estimator $\hat{g}$ in a fixed sample size procedure has an exponential type density. If (ii) is adopted there is a justification for a strict sequential estimation procedure even in the case of exponential type distributions. This has been shown by DeGroot [1] and Wasan [1] for the binomial case. They have proven that a fixed sample size procedure in the binomial case does not minimize $E_\theta\{N\}$ with respect to all sequential procedures uniformly in θ, $0 < \theta < 1$, subject to the condition that $\sup_{0<\theta<1} \text{Var}_\theta \{\hat{g}\} \leqslant 1/4m$.

4.5. THE ASYMPTOTIC EFFICIENCY OF ESTIMATORS

When the sample size grows to infinity, estimators which are inefficient may converge in distribution to estimators with asymptotic variances which attain the Cramér-Rao lower bound or even become smaller than that. Examples of estimators with asymptotic variance which never exceed the Cramér-Rao lower bound and at some θ points are below it were provided by Hodges and reported by LeCam [1] in 1953. Such estimators are called *super-efficient*. As we illustrate in Example 4.6, super-efficient estimators which are not unbiased may considerably reduce the mean square error at some values of θ. The practice of restricting attention to consistent estimators which are asymptotically normal (C.A.N.) does not eliminate this phenomenon. The existence of super-efficient estimators among the C.A.N. estimators rejects an earlier conjecture of Fisher [3] that the asymptotic variance of C.A.N. estimators is at least equal to the Cramér-Rao lower bound. There are, however, some severe restrictions on this phenomenon. LeCam [1] has shown that the set of θ points on which a super-efficient estimator has an asymptotic variance smaller than the Cramér-Rao bound is of a Lebesgue measure zero. Stein proved that on intervals larger in order of magnitude than $1/\sqrt{n}$, the asymptotic variance of an estimator is not uniformly smaller than the inverse of the Fisher information.

Rao investigated the problems of asymptotic efficiency in a series of papers. See in particular [3] and [6]. He showed that super-efficient estimators among the C.A.N. ones can exist only if the approach to the asymptotic normal distribution is not uniform on compacts of θ. Such a result was established also by Bahadur [4]. Thus Rao suggested considering only C.A.N. estimators whose convergence in distribution to a normal estimator is uniform. Rao has shown that under certain regularity conditions the asymptotic variance of such estimators is always at least as large as the inverse of the Fisher information. Thus the uniformly best asymptotically normal (U.B.A.N.) estimators are the most efficient in this restricted class of estimators. Such U.B.A.N. estimators are called by Rao *first-order efficient* estimators. It was shown that many different kinds of estimators could be first-order efficient. When the sample size is not excessively large, however, these first-order efficient estimators may show different characteristics. It is important to characterize the rate of convergence of the distributions of these estimators to the asymptotic distribution. Thus Rao introduced the notion of *second-order efficiency* by which we can discriminate between various first-order efficient estimators. In this section we present the results of Bahadur [4]. We start with a few definitions and with an example of a super-efficient C.A.N. estimator.

A sequence of estimators $\{T_n; n = 1, 2, \ldots\}$ *is called consistent if* $T_n - g(\theta) = 0_p(1)$ *as* $n \to \infty$. *It is called consistent asymptotically normal (C.A.N.) for* θ *if*

(4.5.1)
$$\lim_{n\to\infty} P_\theta[\sqrt{n}\,(T_n - \theta) \leqslant x\sqrt{v(\theta)}] = \Phi(x), \qquad \text{all } -\infty < x < \infty,\ \theta \in \Theta.$$

As before, $\Phi(x)$ designates the standard normal integral. Following Pratt [1], we use the symbols $0_p(\cdot)$ and $O_p(\cdot)$ to denote the rate of convergence in probability. Thus $X_n = 0_p(Y_n)$ as $n \to \infty$ if $X_n/Y_n \xrightarrow{P} 0$; and $X_n = O_p(Y_n)$ if there exists a constant K, $0 < K < \infty$, such that $\lim_{n\to\infty} P[|X_n/Y_n| \leqslant K] = 1$. We call T_n a consistent estimator of $g(\theta)$ if $\{T_n\}$ is a consistent sequence. Furthermore, T_n is called a *best asymptotically normal* (B.A.N.) estimator of θ if T_n is a C.A.N. estimator and $v(\theta) = I^{-1}(\theta)$, where $I(\theta)$ is the Fisher information function (matrix). An important investigation of classes of B.A.N. estimators and a method of constructing such estimators by minimal separators was published by Barankin and Gurland [3].

We now present an example of a C.A.N. estimator which is super-efficient.

Example 4.6. Let X_1, X_2, be a sequence of i.i.d. random variable having an $N(\theta, 1)$ distribution, $-\infty < \theta < \infty$. The complete sufficient statistic for a sample of size n is the sample mean $\bar{X}_n$. Consider the estimator

$$T_n = \begin{cases} \frac{1}{2}\bar{X}_n, & \text{if } |\bar{X}_n| \leqslant \dfrac{(\log n)}{\sqrt{n}}, \\ \bar{X}_n, & \text{otherwise} \end{cases} \tag{4.5.2}$$

Obviously T_n is a biased estimator of θ. Moreover, since for every fixed n the distribution of T_n is not a 1-parameter exponential type in θ, the mean square error of T_n does not attain the Cramér-Rao lower bound. Let

$$\Lambda_n = \left\{(X_1, \ldots, X_n);\ |\bar{X}_n| \leqslant \frac{\log n}{\sqrt{n}}\right\}.$$

Since $(\log n)/\sqrt{n} \to 0$ as $n \to \infty$, $\Lambda_n \searrow \{0\}$, for every $\delta > 0$ we have

$$P_\theta[|T_n - \theta| \leqslant \delta] = P_\theta[|T_n - \theta| \leqslant \delta, \Lambda_n] + P_\theta[|T_n - \theta| \leqslant \delta, \bar{\Lambda}_n], \tag{4.5.3}$$

where $\bar{\Lambda}_n$ designates the complement of Λ_n in $\mathfrak{X}^{(n)}$. Since $X_1, X_2, \ldots$ are independent and $E\,|X| < \infty$, we obtain from the weak law of large numbers

$$P_\theta[|T_n - \theta| \leqslant \delta \mid (X_1, \ldots, X_n) \in \Lambda_n] \to \begin{cases} 1, & \text{if } |\theta| \leqslant \delta \\ 0, & \text{if } |\theta| > \delta \end{cases}, \text{ as } n \to \infty. \tag{4.5.4}$$

Furthermore,

$$(4.5.5)\qquad P_\theta[(X_1, \ldots, X_n) \in \Lambda_n] \to \begin{cases} 1, & \text{if } \theta = 0\,, \\ 0, & \text{if } \theta \neq 0 \end{cases} \qquad \text{as } n \to \infty.$$

Hence from (4.5.2)–(4.5.5) we obtain that

$$(4.5.6)\qquad P_\theta[|T_n - \theta| \leqslant \delta] \to 1,$$

for all θ in $(-\infty, \infty)$. This shows the asymptotic consistency of T_n. Proceeding with the example, the distribution law of $T_n = \frac{1}{2}\bar{X}_n I_{\Lambda_n}(\bar{X}_n) + \bar{X}_n(1 - I_{\Lambda_n}(\bar{X}_n))$, where $I_{\Lambda_n}(\bar{X}_n)$ is the indicator function of the set Λ_n, satisfies

$$(4.5.7)\qquad \lim_{n\to\infty} P_\theta[\sqrt{n}(T_n - \theta) \leqslant x] = \begin{cases} \Phi(x), & \text{if } \theta \neq 0, \\ \Phi(2x), & \text{if } \theta = 0, \end{cases}$$

where $\Phi(\cdot)$ is the standard normal integral. Thus the asymptotic variance of T_n is $1/n$ if $\theta \neq 0$ and $1/4n$ if $\theta = 0$. The Cramér-Rao lower bound for each n is n^{-1}. Thus T_n is asymptotically super-efficient at $\theta = 0$. Let us consider now the coverage probabilities of such a super-efficient estimator, T_n. Given a positive finite value δ, we define the coverage probability of T_n to be

$$(4.5.8)\qquad \varphi_n(\theta) = P_\theta[|T_n - \theta| < \delta], \qquad -\infty < \theta < \infty.$$

We have seen in (4.5.6) that the sequence of coverage probabilities $\{\varphi_n(\theta);\ n = 1, 2, \ldots\}$ converges to 1. The question is, How does it converge to this limit? It is instructive to recall that the B.A.N. estimator $\bar{X}_n$ of θ has a constant coverage probability $\varphi_n^*(\theta) = 2\Phi(\delta\sqrt{n}) - 1$ for each θ. Thus the convergence of φ_n^* to 1 is uniform in θ. This is, however, not the case with the sequence $\varphi_n(\theta)$. To verify this we write, according to (4.5.3),

$$(4.5.9)\qquad \varphi_n(\theta) = P_\theta\left[2\theta - 2\delta \leqslant \bar{X}_n \leqslant 2\theta + 2\delta,\ |\bar{X}_n| \leqslant \frac{1}{\sqrt{n}} \log n\right] + P_\theta\left[\theta - \delta \leqslant \bar{X}_n \leqslant \theta + \delta,\ |\bar{X}_n| > \frac{1}{\sqrt{n}} \log n\right].$$

Then we can easily show that, for $0 < \theta < \delta$,

(4.5.10)

$$\begin{aligned}
\varphi_n(\theta) &= P_\theta\left[\max\left(-\frac{\log n}{\sqrt{n}}, 2\theta - 2\delta\right) \leqslant \bar{X}_n \leqslant \min\left(2\theta + 2\delta, \frac{\log n}{\sqrt{n}}\right)\right] \\
&\quad + P_\theta\left[\frac{\log n}{\sqrt{n}} \leqslant \bar{X}_n \leqslant \theta + \delta\right] + P_\theta\left[\theta - \delta \leqslant \bar{X}_n \leqslant -\frac{\log n}{\sqrt{n}}\right] \\
&= \Phi(\min(2\delta\sqrt{n} + \theta\sqrt{n}, \log n - \theta\sqrt{n})) \\
&\quad + \Phi(\min(2\delta\sqrt{n} - \theta\sqrt{n}, \log n + \theta\sqrt{n})) - 1 \\
&\quad + [\Phi(\delta\sqrt{n}) - \Phi(\log n - \theta\sqrt{n})]^+ + [\Phi(\delta\sqrt{n}) - \Phi(\log n + \theta\sqrt{n})]^+,
\end{aligned}$$

where $a^+ = \max(0, a)$. When $\theta > \delta$ we have

$$\varphi_n(\theta) = \Phi(\min(\theta\sqrt{n} + 2\delta\sqrt{n}, \log n - \theta\sqrt{n})) - \Phi(\theta\sqrt{n} - 2\delta\sqrt{n}) + [\Phi(\delta\sqrt{n}) - \Phi(\log n - \theta\sqrt{n})]^+. \tag{4.5.11}$$

Similar expressions can be determined for negative values of θ. The important feature is that $\varphi_n(\theta)$ varies with θ. Moreover, since the dependence of $\varphi_n(\theta)$ on θ is not the same for $|\theta| \leqslant \delta$ or $|\theta| > \delta$, the convergence of $\varphi_n(\theta) \to 1$ is *not* uniform. ■

Bahadur's definition of efficiency in the large sample case is based on the coverage probabilities $\varphi_n(\theta)$ of consistent estimators. It provides an estimate of the rate of approach of the coverage probabilities to 1. We now present Bahadur's main results according to [4].

The effective standard deviation of a consistent sequence $\{T_n\}$ *of estimators of* $g(\theta)$ *is the root* $\tau \equiv \tau_g(T_n, \epsilon, \theta)$ *of the equation*

$$P_\theta[|T_n - g(\theta)| \geqslant \epsilon] = 2\left[1 - \Phi\left(\frac{\epsilon}{\tau}\right)\right], \tag{4.5.12}$$

where $\epsilon > 0$ is a specified value. The term on the left-hand side of (4.5.12) is the probability that θ is not covered by the interval $(T_n \pm \epsilon)$. As before, let $\varphi_n(\theta; T_n, \epsilon)$ designate the *coverage* probability of θ; then (4.5.12) has the unique solution

$$\tau_g(T_n, \epsilon, \theta) = \epsilon\,.\,[\Phi^{-1}(\tfrac{1}{2} + \tfrac{1}{2}\varphi_n(g(\theta); T_n, \epsilon))]^{-1} \tag{4.5.13}$$

where $\Phi^{-1}(\gamma)$ is the γ-fractile of $\Phi(\cdot)$. We thus see that $\tau_g(T_n, \epsilon, \theta)$ is a monotone decreasing function of the coverage probability $\varphi_n(g(\theta); T_n, \epsilon)$. Thus if $T_n - g(\theta) = 0_p(1)$ as $n \to \infty$, then $\tau_g(T_n, \epsilon, \theta) \to 0$, $n \to \infty$. Moreover, for the same n, ϵ and $g(\theta)$, if T_n and U_n are two consistent estimators of $g(\theta)$ and if $\varphi_n(g(\theta); T_n, \epsilon) < \varphi_n(g(\theta); U_n, \epsilon)$, then $\tau_g(T_n, \epsilon, \theta) > \tau_g(U_n, \epsilon, \theta)$. In the following example we compute the effective standard deviation for a log-normal case.

Example 4.7. We consider the problem of estimating $\xi = \exp\{\mu + \sigma^2/2\}$ in the log-normal case. If $Y_i = \log X_i$, $(i = 1, 2, \ldots)$ and $\bar{Y}_n$, S_n are the sample mean and the sample sum of squares of derivations then, a consistent estimator of $\xi = \exp\{\mu + \sigma^2/2\}$ is $\hat{\xi}_n = \exp\{\bar{Y}_n + (1/2n)S_n\}$. Indeed, $\bar{Y}_n - \mu = 0_p(1)$ and $S_n/n - \sigma^2 = 0_p(1)$. Thus

$$\begin{aligned}\varphi_n(\xi; \hat{\xi}_n, \epsilon) &= P\left\{\left|\exp\left\{\bar{Y}_n + \frac{1}{2n}S_n\right\} - \exp\{\mu + \tfrac{1}{2}\sigma^2\}\right| < \epsilon\right\}\\ &= P\left\{\exp\left\{\mu + \frac{\sigma^2}{2}\right\} - \epsilon \leqslant \exp\left\{\bar{Y}_n + \frac{1}{2n}S_n\right\} \leqslant \exp\left\{\mu + \frac{\sigma^2}{2}\right\} + \epsilon\right\}\\ &= P\left\{\ln(\xi - \epsilon)^+ \leqslant \bar{Y}_n + \frac{1}{2n}S_n \leqslant \ln(\xi + \epsilon)\right\}.\end{aligned} \tag{4.5.14}$$

Since $\bar{Y}_n$ and S_n are independent, we obtain

$$\begin{aligned}
&\varphi_n(\xi; \hat{\xi}_n, \epsilon) \\
&= P\left\{\ln (\xi - \epsilon)^+ - \frac{S_n}{2n} \leqslant \bar{Y}_n \leqslant \ln (\xi + \epsilon) - \frac{S_n}{2n}\right\} \\
&= E\left\{\Phi\left(\frac{\ln (\xi + \epsilon) - S_n/2n - \mu}{\sigma/\sqrt{n}}\right)\right\} - E\left\{\Phi\left(\frac{\ln (\xi - \epsilon)^+ - S_n/2n - \mu}{\sigma/\sqrt{n}}\right)\right\}.
\end{aligned}$$

The expectations on the right-hand side of (4.5.14) can be determined by a power-series expansion of $\Phi(\cdot)$ and integrating term by term with respect to the distribution of S, namely that of $\sigma^2\chi^2[n-1]$. Finally, the effective standard deviation of $\hat{\xi}_n$ is determined according to (4.5.13). ■

Let $\{T_n\}$ and $\{U_n\}$ be two consistent sequences of estimators of $g(\theta)$. The upper asymptotic efficiency of T_n relative to U_n is the upper limit

$$\bar{e}_g(T, U; \theta) = \overline{\lim_{\epsilon\to 0}}\ \overline{\lim_{n\to\infty}}\ \frac{\tau_g^2(U_n, \epsilon, \theta)}{\tau_g^2(T_n, \epsilon, \theta)}. \tag{4.5.15}$$

Similarly, the lower asymptotic relative efficiency $\underline{e}_g(T, U; \theta)$ is the lower limit of the ratios of $\tau_g^2(U_n, \epsilon, \theta)$ to $\tau_g^2(T_n, \epsilon, \theta)$.

A consistent estimator T_n of $g(\theta)$ is called asymptotically efficient if, for any other consistent estimator U_n,

$$\sup_{\theta\in\Theta} \bar{e}_g(U_n, T_n; \theta) \leqslant 1. \tag{4.5.16}$$

We prove that for any consistent estimator T_n of $g(\theta)$ and $\epsilon > 0$ the following holds:

(4.5.17)
$$\frac{2}{\epsilon^2} \log P_\theta[|T_n - g(\theta)| \geqslant \epsilon] = -\frac{1}{\tau_g^2(T_n, \epsilon, \theta)}(1 + 0(1)), \qquad \text{as } n \to \infty,$$

and, if $g(\theta)$ is differentiable at θ,

$$\underline{\lim_{\epsilon\to 0}}\ \underline{\lim_{n\to\infty}} \{n\tau_g^2(T_n, \epsilon, \theta)\} \geqslant \frac{(g'(\theta))^2}{I(\theta)}. \tag{4.5.18}$$

This will lead to the result that, if U_n is an asymptotically efficient estimator of $g(\theta)$,

$$P_\theta[|U_n - g(\theta)| \geqslant \epsilon] = \exp\left\{-\frac{n}{2}\epsilon^2 I_g(\theta)(1 + \delta_n(\epsilon, \theta))\right\}, \tag{4.5.19}$$

where

$$I_g(\theta) = \frac{I(\theta)}{(g'(\theta))^2},$$

and

$$\lim_{\epsilon\to 0}\ \overline{\lim_{n\to\infty}}\ \delta_n(\epsilon, \theta) = 0.$$

In the general framework for the following lemmas and theorems we make the following assumptions.

A.1 The family of probability measures on $(\mathfrak{X}^{(n)}, \mathfrak{B}^{(n)})$ for each $n \geqslant 1$ is dominated by a sigma-finite measure μ (the same for all n), and the corresponding density functions with respect to μ are $\{f(x_1, \ldots, x_n; \theta); \theta \in \Theta; n = 1, 2, \ldots\}$.

A.2 Θ is an interval in a Euclidean k-space and for each $\theta, \varphi \in \Theta$, if $\theta \neq \varphi$, then $\mu((x_1, \ldots, x_n): f(x_1, \ldots, x_n; \theta) \neq f(x_1, \ldots, x_n; \varphi)) > 0$ for all $n = 1, 2, \ldots$.

A.3 For each $n = 1, 2, \ldots$ and all $\theta, \varphi \in \Theta$, the carrier sets of $f(x_1, \ldots, x_n; \theta)$ and $f(x_1, \ldots, x_n; \varphi)$ are equivalent; that is if $\Lambda_n(\theta) = \{(x_1, \ldots, x_n); f(x_1, \ldots, x_n; \theta) > 0\}$, then

$$\int_{\Lambda_n} f(x_1, \ldots, x_n; \varphi)\mu(\mathrm{d}x_1, \ldots, \mathrm{d}x_n) = 1$$

for all $\varphi \in \Theta$.

Further assumptions concerning the family of densities will be imposed as we progress in the presentation.

Let $X_1, X_2, \ldots$ be a sequence of i.i.d. random variables. We then consider the family of the possible density functions of X_1 when $\theta \in \Theta$. Let $Z_i = \log(f(X_i; \varphi)/f(X_i; \theta))$ for a specified pair (φ, θ) in Θ. Let $M_\theta(t)$ and $M_\varphi(t)$ designate, respectively, the moment generating functions of Z_1 with respect to both $f(x; \theta)$ and $f(x; \varphi)$. We notice that

$$M_\theta(1) = E_\theta\{e^{\log(f(X;\varphi)/f(X;\theta))}\} = \int f(x; \varphi)\mu(\mathrm{d}x) = 1. \tag{4.5.20}$$

Similarly, $M_\varphi(-1) = 1$. Trivially, $M_\theta(0) = M_\varphi(0) = 1$. We further assume that

A.4

$$E_\omega\{Z_1^2 \exp\{tZ_1\}\} < \infty, \quad \text{for all} \begin{cases} t \in [0, 1], & \text{if } \omega = \theta, \\ t \in [-1, 0], & \text{if } \omega = \varphi. \end{cases}$$

This guarantees that the moment generating function $M_\theta(t)$ and $M_\varphi(t)$ are continuous and strictly convex in the closed intervals $[0, 1]$ and $[-1, 0]$, respectively. We also notice that wherever $M_\theta(t) < \infty$,

$$\begin{aligned} M_\theta(t) &= \int \left(\frac{f(x; \varphi)}{f(x; \theta)}\right)^t f(x; \theta)\mu(\mathrm{d}x) \\ &= \int \left(\frac{f(x; \varphi)}{f(x; \theta)}\right)^{t-1} f(x; \varphi)\mu(\mathrm{d}x) = M_\varphi(t-1). \end{aligned} \tag{4.5.21}$$

We now introduce another measure of information, called the *Kullback-Leibler information function* for the discrimination between $f(x; \theta)$ and $f(x; \varphi)$ (see [2]). This information measure is defined on $\Theta \times \Theta$. Its range is $(0, \infty)$ and is given by

$$I(\varphi, \theta) = E_\varphi\left\{\log \frac{f(X; \varphi)}{f(X; \theta)}\right\}. \tag{4.5.22}$$

Since the logarithmic function is concave, we immediately obtain that $I(\varphi, \theta) \geqslant 0$. Furthermore, according to A.1–A.4, $0 < I(\varphi, \theta) < \infty$ for all $(0, \varphi) \in \Theta \times \Theta$. Moreover, we can easily prove that the information function of independent random variables is additive. In other words, if X and Y are independent and $I^X(\varphi, \theta) = E_\varphi\{\log (f(X; \varphi)/f(X; \theta))\}$, $I_Y(\varphi, \theta) = E_\varphi\{\log(g(Y; \varphi)/g(Y; \theta))\}$, where $g(y; \theta)$, $\theta \in \Theta$, is the density function of Y, then $I^{X,Y}(\varphi, \theta) = I^X(\varphi, \theta) + I^Y(\varphi, \theta)$. The quantity $I^{X,Y}(\varphi, \theta)$ designates the joint information function; that is,

$$I^{X,Y}(\varphi, \theta) = E_\varphi\left\{\log \frac{f(X, Y; \varphi)}{f(X, Y; \theta)}\right\}.$$

Thus if $I_n(\varphi, \theta)$ designates the Kullback-Leibler joint information of a sample of n random variables and $X_1, \ldots, X_n$ are i.i.d. random variables, we obviously have $I_n(\varphi, \theta) = nI(\varphi, \theta)$. It is easy to show that if $T(X_1, \ldots, X_n)$ is any (measurable) statistic then $I_n(\varphi, \theta) \geqslant I^T(\varphi, \theta)$ for all (φ, θ). Equality holds for all (φ, θ) if and only if T is a sufficient statistic.

Lemma 4.5.1 (Chernoff [1].) *Let* X_1, X_2 *be a sequence of* i.i.d. *random variables with a common density in* $\{f(x; \theta); \theta \in \Theta\}$ *satisfying A.*1–*A.*3. *Let* $Z_i = \log (f(X_i; \varphi)/f(X_i; \theta))$, $\varphi, \theta \in \Theta$, $i = 1, 2, \ldots$. *Then*

$$P_\theta\left[\sum_{i=1}^n Z_i \geqslant nI(\varphi, \theta)\right] \leqslant \exp\{-nI(\varphi, \theta)\}. \tag{4.5.23}$$

Furthermore, for every $0 < \epsilon < e^{-I(\varphi,\theta)}$,

$$P_\theta\left[\sum_{i=1}^n Z_i \geqslant nI(\varphi, \theta)\right] \geqslant (e^{-I(\varphi,\theta)} - \epsilon)^n, \tag{4.5.24}$$

for an n sufficiently large.

Proof. Let $Z_i^* = Z_i - I(\varphi, \theta)$, $i = 1, 2, \ldots$; $P_\theta[\sum_{i=1}^u Z_i \geqslant nI(\varphi, \theta)] = P_\theta[\sum_{i=1}^n Z_i^* \geqslant 0]$. Consider the moment generating function

$$E_\theta\{\exp\{t \sum_{i=1}^n Z_i^*\}\}.$$

Since $X_1, X_2, \ldots$ are i.i.d.,

$$E_\theta\{\exp\{t \sum_{i=1}^n Z_i^*\}\} = (M_\theta(t))^n \exp\{-ntI(\varphi, \theta)\}.$$

We notice that $I(\varphi, \theta) = M_\theta'(t)\,|_{t=1}$ and is finite according to A.4. Write

$$\begin{aligned} E_\theta\left\{\exp\left\{t \sum_{i=1}^n Z_i^*\right\}\right\} &= P_\theta\left[\sum_{i=1}^n Z_i^* \geqslant 0\right] E_\theta\left\{e^{t\sum_{i=1}^n Z_j^*} \,\middle|\, \sum_{i=1}^n Z_i^* \geqslant 0\right\} \\ &\quad + P_\theta\left[\sum_{i=1}^n Z_i^* < 0\right] E_\theta\left\{e^{t\sum_{i=1}^u Z_i^*} \,\middle|\, \sum_{i=1}^n Z_i^* < 0\right\}. \end{aligned} \tag{4.5.25}$$

Thus

$$E_\theta\{e^{t\Sigma_{i=1}^n Z_i^*}\} \geqslant P_\theta\left[\sum_{i=1}^n Z_i^* \geqslant 0\right] E\left\{e^{t\Sigma_{i=1}^n Z_i^*} \,\middle|\, \sum_{i=1}^n Z_i^* \geqslant 0\right\}.$$

Furthermore, since $e^{t\Sigma Z_i^*} \geqslant 1$ on the set $\{\sum_{i=1}^n Z_i^* \geqslant 0\}$, for all $0 < t < \infty$, we obtain

$$(4.5.26) \qquad P_\theta\left[\sum_{i=1}^n Z_i^* \geqslant 0\right] \leqslant (e^{-tI(\varphi,\theta)} M_\theta(t))^n, \qquad \text{all } 0 < t < \infty.$$

It is easy to verify that $e^{-tI(\varphi,\theta)}M_\theta(t)$ is convex and attains its unique minimum at $t_0 = 1$. Hence $e^{-tI(\varphi,\theta)}M_\theta(t) > e^{-I(\varphi,\theta)}$ for all t. Thus (4.5.26) implies (4.5.23) for all $n \geqslant 1$.

The proof of (4.5.24) is considerably more delicate. Chernoff [1] proved this inequality by considering first the case of $X_1, X_2, \ldots$ being discrete random variables, and then approximating the probability $P_\theta[\sum_{i=1}^n Z_i^* \geqslant 0]$ by a sequence of probabilities corresponding to discrete random variables.

(Q.E.D.)

Corollary. *Under the conditions described by A.1–A.4,*

$$(4.5.27) \qquad \lim_{n\to\infty}\left\{\frac{1}{n}\log P_\theta\left[\sum_{i=1}^n Z_i \geqslant nI(\varphi,\theta)\right]\right\} = -I(\varphi,\theta)$$

for all $\varphi, \theta \in \Theta$.

Let $\{W_n; n = 1, 2, \ldots\}$ be a specified sequence of measurable sets in the corresponding measure spaces $(\mathfrak{X}^{(n)}, \mathcal{B}^{(n)})$. Let θ_1 and θ_2, $\theta_1 \neq \theta_2$, be two specified points in Θ and define

$$(4.5.28) \qquad \begin{aligned} \epsilon_n^{(1)} &= P_{\theta_1}[(X_1, \ldots, X_n) \in W_n] \\ \epsilon_n^{(2)} &= P_{\theta_2}[(X_1, \ldots, X_n) \in \overline{W}_n] \end{aligned}, \qquad n = 1, 2, \ldots,$$

where $\overline{W}_n$ is the complement of W_n in $\mathfrak{X}^{(n)}$. The quantities $\epsilon_n^{(1)}$ and $\epsilon_n^{(2)}$ correspond to the error probabilities of a sequence of tests of two simple hypotheses (see Lehmann [2], pp. 60–63).

Let $\mathcal{W}(\theta_2)$ be the class of all such sequences satisfying $\overline{\lim_{n\to\infty}}\, \epsilon_n^{(2)} < 1$. The lower slope of a sequence $\{W_n\}$ in $\mathcal{W}(\theta_2)$ is $\underline{c}(W_n) = \underline{\lim_{n\to\infty}}\{(2/n)\ln(1/\epsilon_n^{(1)})\}$. In a similar way we define the upper slope $\bar{c}(W_n)$ to be the limsup of $(2/n)\ln(1/\epsilon_n^{(1)})$. The following special sequence has an important role, namely,

$$W_n^* = \left\{(X_1, \ldots, X_n);\ \sum_{i=1}^n Z_i \geqslant \lambda_n\right\}, \qquad n = 1, 2, \ldots,$$

where $\{\lambda_n\}$ is a desirable sequence of positive constants. The fundamental theorem of testing hypotheses (the Neyman-Pearson lemma) is cited here without a proof. The proof of this important lemma can be found in Lehmann [2], p. 65. The theorem says: For a fixed n let $\Psi_\alpha^{(n)}$ be the class of all partitions of $\mathfrak{X}^n$, namely, $\{W_n, \overline{W}_n\}$, which satisfy

$$P_{\theta_1}[(X_1, \ldots, X_n) \in W_n] \leqslant \alpha, \qquad 0 < \alpha < 1.$$

Then

$$\epsilon_n^{(2)}(W_n^*) = \inf_{W_n \in \Psi_\alpha^{(n)}} \epsilon_n^{(2)}(W_n),$$

where $\{\lambda_n\}$ is chosen so that $W_n^* \in \Psi_\alpha^{(n)}$. Thus if for a given n, $\epsilon_n^{(2)}(W_n) < \epsilon_n^{(2)}(W_n^*)$, we imply that $\epsilon_n^{(1)}(W_n) > \epsilon_n^{(1)}(W_n^*)$.

Lemma 4.5.2. *Let $X_1, X_2, \ldots$ be i.i.d. random variables with a density function in a family satisfying A.1–A.4. Consider the sequence $\{W_n^*; n \geqslant 1\}$ with $\lambda_n = nI(\theta_2, \theta_1) + \sqrt{n}\, a \sqrt{v}$ where a is an arbitrary constant, $-\infty < a < \infty$, and $v = \mathrm{Var}_{\theta_2}\{Z_1\}$. Then*

(i) $W^* \in \mathcal{W}(\theta_2)$;
(ii) $\underline{c}(W^*) = \bar{c}(W^*) = 2I(\theta_2, \theta_1)$;
(iii) $\bar{c}(W) \leqslant 2I(\theta_2, \theta_1)$ *for all* $W \in \mathcal{W}(\theta_2)$.

Proof. (i) Since $Z_1, Z, \ldots$ are independent and $v_n < \infty$ (see A.4), we obtain from the central limit theorem that the asymptotic distribution of $(\sum_{i=1}^n Z_i - nI(\theta_2, \theta_1))/\sqrt{(nv)}$ is $\mathcal{N}(0, 1)$. Hence

$$\begin{aligned}\overline{\lim_{n\to 0}}\, \epsilon_n^{(2)}(W_n^*) &= \lim_{n\to\infty} P_{\theta_2}\left[\frac{\sum_{i=1}^n Z_i - nI(\theta_2, \theta_1)}{\sqrt{(nv)}} \leqslant a\right] \\ &= \Phi(a) < 1, \qquad \text{all } -\infty < a < \infty.\end{aligned}$$

This shows that $W_n^* \in \mathcal{W}(\theta_2)$.

$$\begin{aligned}\epsilon_n^{(1)}(W_n^*) &= P_{\theta_1}\left[\frac{\sum_{i=1}^n Z_i - nI(\theta_2, \theta_1)}{\sqrt{(nv)}} \geqslant a\right] \\ &= P_{\theta_1}\left[\sum_{i=1}^n Z_i \geqslant n\left(I(\theta_2, \theta_1) + a\left(\frac{v}{n}\right)^{1/2}\right)\right]\end{aligned}$$

Furthermore, since

$$\begin{aligned}&\lim_{n\to\infty}\left\{\frac{1}{n} \ln P_{\theta_1}\left[\sum_{i=1}^n Z_i \geqslant n\left(I(\theta_2, \theta_1) + a\left(\frac{v}{n}\right)^{1/2}\right)\right]\right\} \\ &\qquad = \lim_{n\to\infty}\left\{\frac{1}{n} \ln P_{\theta_1}\left[\sum_{i=1}^n Z_i \geqslant nI(\theta_2, \theta_1)\right]\right\},\end{aligned}$$

we obtain from the corollary to Lemma 4.5.1 that $\lim_{n\to\infty} \{(1/n) \ln \epsilon_n^{(1)}(W_n^*)\} = -I(\theta_2, \theta_1)$. Hence $\underline{c}(W_n^*) = \bar{c}(W_n^*) = 2I(\theta_2, \theta_1)$.

(iii) Let W_n be an arbitrary sequence in $\mathcal{W}(\theta_2)$. Then $\overline{\lim_{n\to\infty}}\, \epsilon_n^{(2)}(W_n) < 1$.

We can therefore choose real numbers a and $\epsilon > 0$ such that $\epsilon_n^{(2)}(W_n) \leqslant \Phi(a) - \epsilon$, for all n sufficiently large. Construct then the sequence W_n^* with this value of a, that is, $\lambda_n = nI(\theta_2, \theta_1) + a\sqrt{(nv)}$. Since $\epsilon_n^{(2)}(W_n^*) = P_{\theta_2}[(\sum_{i=1}^n Z_i - nI(\theta_2, \theta_1)/\sqrt{(nv)} < a] \to \Phi(a)$, we obtain that for all n sufficiently large, $\epsilon_n^{(2)}(W_n) < \epsilon_n^{(2)}(W_n^*)$. Hence from the Neyman-Pearson. lemma, $\epsilon_n^{(1)}(W_n) < \epsilon_n^{(2)}(W_n^*)$ for all n sufficiently large. Therefore $\bar{c}(W_n) \leqslant \bar{c}(W_n^*) = 2I(\theta_2, \theta_1)$. (Q.E.D.)

We are ready now to evaluate the asymptotic coverage probability of an interval of width 2ϵ centered at a consistent estimator T_n of $g(\theta)$. Since $g(\theta)$ is a general function, we have to introduce more assumptions which regulate $g(\theta)$.

A.5. The set

$$S_g(\epsilon, \theta_0) = \{\theta; \theta \text{ in } \Theta, |g(\theta) - g(\theta_0)| = \epsilon\}$$

is nonempty for all $\epsilon > 0$ sufficiently small. We define, for any $\epsilon > 0$ such that $S_g(\epsilon, \theta_0)$ is nonempty,

$$J_g(\epsilon; \theta_0) = \inf \{I(\theta, \theta_0); \theta \in S_g(\epsilon, \theta_0)\}. \tag{4.5.29}$$

Furthermore, for any real $r > 0$, let

$$K_g^{(r)}(\theta_0) = \overline{\lim_{\epsilon \to 0}} \left\{\frac{J_g(\epsilon; \theta_0)}{\epsilon^r}\right\}. \tag{4.5.30}$$

Assumptions A.1–A.5 guarantee that $J_g(\epsilon; \theta_0)$ is finite for every $\epsilon > 0$, but $K_g^{(r)}(\theta_0)$ may diverge to $+\infty$. For a given θ_0, $K_g^{(r)}(\theta_0) < \infty$ for at most one value of r.

Theorem 4.5.1. *If T_n is a consistent estimate of $g(\theta)$ based on a sequence of i.i.d. random variables, and if* A.1–A.5 *are satisfied, then for every θ in Θ and every $r > 0$,*

$$\underline{\lim_{\epsilon \to 0}}\ \underline{\lim_{n \to \infty}} \{(n\epsilon^r)^{-1} \ln P_\theta[|T_n - g(\theta)| \geqslant \epsilon]\} \geqslant -K_g^{(r)}(\theta). \tag{4.5.31}$$

Proof. Fix an $r > 0$ and $\theta_0 \in \Theta$. Let $\epsilon > 0$ be sufficiently small that $S_g(\epsilon; \theta_0)$ is nonempty and let $\theta_1 \in S_g(\epsilon; \theta_0)$. Let $0 < \lambda < 1$ and $\delta = \lambda\epsilon$. Consider the sequence $\{W_n; n \geqslant 1\}$ where $W_n = \{(X_1, \ldots, X_n); |T_n - g(\theta_0)| \geqslant \delta\}$. Then

$$\epsilon_n^{(2)}(W_n) = P_{\theta_2}[|T_n - g(\theta_0)| < \delta].$$

Since $|g(\theta_1) - g(\theta_0)| = \epsilon$, $|T_n - g(\theta_0)| = |T_n - g(\theta_1) + g(\theta_1) - g(\theta_0)| \geqslant |\epsilon - |T_n - g(\theta_1)||$. But under θ_1, $T_n - g(\theta_1) = o_p(1)$. Hence $\epsilon_n^{(2)}(W_n) = P_{\theta_1}[|T_n - g(\theta_0)| < \delta] \to 0$, as $n \to \infty$. Consequently, $W_n \in \mathcal{W}(\theta_1)$ for n

sufficiently large. Therefore, according to the definition of $\bar{c}(W_n)$ and Lemma 4.5.2,

$$(4.5.32)\quad \bar{c}(W_n) = \overline{\lim_{n\to\infty}}\left\{\frac{2}{n}\ln\,[P_{\theta_0}(|T_n - g(\theta_0)| \geqslant \delta)]^{-1}\right\} \leqslant 2I(\theta_1, \theta_0).$$

Hence

$$(4.5.33)\quad \underline{\lim}_{n\to\infty}\left\{\frac{1}{n}\ln P_{\theta_0}[|T_n - g(\theta_0)| \geqslant \delta]\right\} \geqslant -I(\theta_1, \theta_0).$$

Finally, since (4.5.33) holds for all $\theta \in S_g(\epsilon; \theta_0)$ we obtain that

$$(4.5.34)\quad \underline{\lim}_{n\to\infty}\left\{\frac{1}{n}\ln P_{\theta_0}[|T_n - g(\theta_0)| \geqslant \delta]\right\} \geqslant -J_g(\epsilon; \theta_0).$$

Divide both sides by δ^r; then

$$(4.5.35)\quad \underline{\lim}_{n\to\infty}\left\{\frac{1}{n\delta^r}\ln P_{\theta_0}[|T_n - g(\theta_0)| \geqslant \delta]\right\} \geqslant -\frac{J_g(\epsilon; \theta_0)}{\lambda^r\epsilon^r}.$$

Letting $\delta \searrow 0$ continuously through positive values, we obtain (4.5.31) from (4.5.35). (Q.E.D.)

In the remainder of the present section we assume that Θ is an interval on the real line and we impose the C.R.-regularity conditions (see Section 4.1). We then have, in a close neighborhood of θ_0,

$$(4.5.36)\quad I(\theta, \theta_0) = I(\theta_0, \theta_0) + (\theta - \theta_0)\frac{\partial}{\partial\theta}I(\theta, \theta_0)\bigg|_{\theta=\theta_0} + \tfrac{1}{2}(\theta - \theta_0)^2\frac{\partial^2}{\partial\theta^2}I(\theta, \theta_0)\bigg|_{\theta=\theta_0} + o(\theta - \theta_0)^2, \qquad \text{as } \theta \to \theta_0.$$

Obviously $I(\theta_0, \theta_0) = 0$. Furthermore, under the regularity conditions of Section 4.1,

$$\frac{\partial}{\partial\theta}I(\theta, \theta_0) = \int\left(\frac{\partial}{\partial\theta}\log f(x; \theta)\right)f(x; \theta)\mu(dx) + \int\log\frac{f(x; \theta)}{f(x; \theta_0)}\left(\frac{\partial}{\partial\theta}\log f(x; \theta)\right)f(x; \theta)\mu(dx).$$

Hence, $(\partial/\partial\theta)I(\theta, \theta_0)|_{\theta=\theta_0} = 0$ according to A.4 and the Lebesgue dominated convergence theorem. Similarly,

$$\frac{\partial^2}{\partial\theta^2}I(\theta, \theta_0) = \int\left(\frac{\partial}{\partial\theta}\log f(x; \theta)\right)^2 F_\theta(dx) + \int\log\frac{f(x; \theta)}{f(x; \theta_0)}\cdot\frac{\partial^2}{\partial\theta^2}\log f(x; \theta)\cdot F_\theta(dx).$$

Therefore $(\partial^2/\partial\theta^2)I(\theta, \theta_0)|_{\theta=\theta_0} = I(\theta_0)$. Combining with (4.5.36), we obtain

(4.5.37) $$I(\theta, \theta_0) = \tfrac{1}{2}(\theta - \theta_0)^2 I(\theta_0) + o(\theta - \theta_0)^2, \quad \text{as } \theta \to \theta_0.$$

Let $\epsilon > 0$ be sufficiently small and $S_g(\epsilon; \theta_0)$ nonempty. Let θ_ϵ be a point in $S_g(\epsilon; \theta_0)$, $\theta_\epsilon \to \theta$, as $\epsilon \to 0$. By definition, $J_g(\epsilon; \theta_0) \leqslant I(\theta_\epsilon, \theta_0)$ for all θ_ϵ in $S_g(\epsilon; \theta_0)$. According to the definition of $S_g(\epsilon; \theta_0)$ we obtain

(4.5.38) $$\frac{J_g(\epsilon; \theta_0)}{\epsilon^2} \leqslant \frac{I(\theta_\epsilon, \theta_0)}{[g(\theta_\epsilon) - g(\theta_0)]^2}, \qquad \text{all } \theta_\epsilon \text{ in } S_g(\epsilon; \theta_0).$$

Substituting (4.5.37) in (4.5.38) and letting $\epsilon \to 0$ we obtain

(4.5.39) $$K_g^{(2)}(\theta_0) = \overline{\lim_{\epsilon \to 0}} \frac{J_g(\epsilon; \theta_0)}{\epsilon^2} \leqslant \frac{I(\theta_0)}{2(g'(\theta_0))^2}.$$

According to the definition of the effective standard deviation,

$$P_\theta[|T_n - g(\theta)| \geqslant \epsilon] = 2\left[1 - \Phi\left(\frac{\epsilon}{\tau_g(T_n, \epsilon, \theta)}\right)\right].$$

Since $T_n - g(\theta) = o_p(1)$, $\tau_g(T_n, \epsilon, \theta) \to 0$ as $n \to \infty$. Hence $\epsilon/\tau_g(T_n, \epsilon, \theta) \to \infty$ as $n \to \infty$. It is well known that $\ln 2[1 - \Phi(x)] = -\tfrac{1}{2}x^2(1 + o(1))$ as $x \to \infty$. Hence we have proven (4.5.17), namely,

(4.5.40)
$$\ln P_{\theta_0}[|T_n - g(\theta)| \geqslant \epsilon] = -\frac{1}{2}\frac{\epsilon^2}{\tau_g^2(T_n, \epsilon, \theta)}(1 + o(1)), \qquad \text{as } n \to \infty.$$

Finally, Theorem 4.5.2 from (4.5.31), (4.5.39), and (4.5.40) we obtain, by setting $r = 2$.

Theorem 4.5.2. *If T_n is a consistent estimator of $g(\theta)$ and if* A.1–A.5 *and the C.R.-regularity conditions are satisfied,*

(4.5.41) $$\underline{\lim}_{\epsilon \to 0}\ \underline{\lim}_{n \to \infty} \{n\tau_g^2(T_n, \epsilon, \theta)\} \geqslant \frac{(g'(\theta))^2}{I(\theta)}, \qquad \textit{all } \theta \in \Theta.$$

In the next chapter we prove that if T_n is the maximum likelihood estimator, equality is attained in (4.5.41). Furthermore, from Theorem 4.5.2 and Definition 4.5.4 we infer that if a consistent estimator has an effective variance $\tau_g^2(T_n, \epsilon, \theta)$ which asymptotically attains the lower bound $(g'(\theta))^2/I(\theta)$ for all θ, then T_n is asymptotically efficient relative to the class of all estimators satisfying the conditions of Theorem 4.5.2. From Theorem 4.5.2 we also deduce that the phenomenon of super-efficiency discussed at the beginning of the section is due, as previously indicated, to a certain lack of uniformity in the approach of the distribution of $\sqrt{n}(T_n - g(\theta))$ to the asymptotic normal distribution. Indeed, let λ be a positive constant. Then if $v(\theta)$ is the asymptotic variance of $\sqrt{n}\, T_n$, and $\sqrt{n}\,(T_n - g(\theta))$ is asymptotically normal, then

$$\lim_{n\to\infty} P_\theta[\sqrt{n}\,|T_n - g(\theta)| \geqslant \lambda] = 2\left[1 - \Phi\left(\frac{\lambda}{\sqrt{(v(\theta))}}\right)\right].$$

Suppose that $v(\theta) \leqslant (g'(\theta))^2/I(\theta)$ for *all* θ. According to the definition of $\tau_g^2(T_n, \epsilon, \theta)$ and (4.5.41), we obtain

$$\lim_{n\to\infty}\left\{n\tau_g^2\left(T_n, \frac{\lambda}{\sqrt{n}}, \theta\right)\right\} = v(\theta)$$

uniformly in λ for each θ. If this limit holds uniformly in λ, (4.5.41) implies that $v(\theta)$ *cannot* be smaller than $(g'(\theta))^2/I(\theta)$.

For a generalization of Bahadur's results see Schmetterer [1]. See also Wolfowitz [4] for further comments on asymptotic efficiency and the phenomenon of super-efficiency.

PROBLEMS

Section 4.1

1. Let $X_1, \ldots, X_n$ be i.i.d. random variables, $X_1 \sim G(1/\theta, 1)$, $0 < \theta < \infty$. What is the Cramér-Rao lower bound for the variance of an unbiased estimator of $g(\theta) = e^{-1/\theta}$? Does the variance of the U.M.V.U. estimator of $g(\theta)$ attain this lower bound?

2. Determine the Cramér-Rao lower bound for the variance of an unbiased estimator of the tail probability of a normal distribution $\mathcal{N}(\xi, \sigma^2)$, that is, $g(\xi, \sigma^2) = \Phi(-\xi/\sigma)$, which is based on a sample of n i.i.d. random variables (a) when σ^2 is known and (b) when ξ is known.

3. Let $\hat{g}(x)$ be an unbiased estimator of $g(\theta)$ with a finite variance. Let $\{f(x; \theta;)\ \theta \in \Theta\}$ be the family of density functions corresponding to $\mathcal{F}$ and define

$$A(\phi, \theta) = \text{Var}_\theta\left\{\frac{f(X; \phi)}{f(X; \theta)}\right\}.$$

Then

$$\text{Var}_\theta\{\hat{g}(X)\} \geq \sup_{\phi\in\Theta} \frac{[g(\phi) - g(\theta)]^2}{A(\phi, \theta)}.$$

(See Chapman and Robbins [1].)

4. Let $X_1, \ldots, X_n$ be i.i.d. random variables having a common binomial distribution $\mathcal{B}(n; \theta)$, $0 < \theta < 1$.
   (i) What is the U.M.V.U. estimator of $\sigma^2(\theta) = \theta(1 - \theta)$?

(ii) What is the Cramér-Rao lower bound for the variance of an unbiased estimator of $\sigma^2(\theta)$?

(iii) What is the variance of the U.M.V.U. estimator of $\sigma^2(\theta)$?

5. Consider the family of extreme value distributions with density functions

$$f(x; \alpha) = \exp\{-(x - \alpha) - e^{-(x-\alpha)}\}, \qquad -\infty < x < \infty,$$

$-\infty < \alpha < \infty$. Make the reparametrization $\theta = e^{\alpha}$.

(i) What is the distribution function of $U(X) = e^{-X}$?

(ii) What is the Fisher information as a function of θ, $I(\theta)$?

(iii) What is the Fisher information function $I(\alpha)$?

Section 4.2

6. Consider the model of Problem 4. Does the variance of the U.M.V.U. estimator of $\sigma^2(\theta)$ attain the Bhattacharyya lower bound for some order k, $k \geqslant 1$?

7. Let $X_1, \ldots, X_n$ be i.i.d. random variables, $X_1 \sim \exp\{N(0, \sigma^2)\}$.

(i) Determine the U.M.V.U. estimator of $\xi = \exp\{\sigma^2/2\}$, say $\hat{\xi}_n$.

(ii) Determine the U.M.V.U. estimator of $D^2 = \xi^2(e^{\sigma^2} - 1)$, say $\hat{D}_n^2$.

(iii) What are the Bhattacharyya lower bounds of order k for the variance of unbiased estimators of ξ and of D^2. Do the variances of the above U.M.V.U. estimators attain the Bhattacharyya lower bounds for some k, $k \leqslant 1$?

Section 4.3

8. Consider Problem 2, estimating the tail probability of the normal distribution, that is, $g(\xi, \sigma^2) = \Phi(-\xi/\sigma)$. What is the Cramér-Rao lower bound for the variance of an unbiased estimator if both μ and σ^2 are unknown?

9. Consider the Model II of analysis of variance (see Chapter 2, Problem 18).

(i) What is the Fisher information matrix $\mathbf{I}(\boldsymbol{\theta})$, where $\theta = (\mu, \sigma^2, \tau^2)$?

(ii) What is the U.M.V.U. estimator of θ?

(iii) What is the covariance matrix of the U.M.V.U. estimator?

(iv) What is the lower bound to the covariance matrix derived in (iii)?

10. Consider the problem of estimating the common mean of two normal distributions, based on two random sampes of equal size n (see Chapter 2, Problem 29). What is the Cramér-Rao lower bound for the variance of an unbiased estimator of the common mean?

Section 4.4

11. Let $X_1, X_2, \ldots$ be a sequence of i.i.d. random variables having a common normal distribution $\mathcal{N}(\mu, \sigma^2)$, $-\infty < \mu < \infty$, $0 < \sigma < \infty$. Both μ and σ are unknown. The objective is to estimate μ unbiasedly with a prescribed degree of precision. That is, the variance of the estimator is not exceeding a prescribed value δ^2. If σ^2 is known, a sample of size $n(\sigma^2)$ = least integer n such that $n \geqslant \sigma^2/\delta^2$, with the sample mean as an unbiased estimator, attains the required precision. When the σ^2 is unknown we can follow the following sequential procedure.

Sample rule: Take initially $n_0 = 2$ observations and compute the sample variance $s_2^2 = (X_1 - X_2)^2$. If $s_2^2/\delta^2 > 2$, take an additional observation. Generally after n observations, $n \geqslant 2$, compute the sample variance $s_n^2 = (1/(n-1)) \sum_{i=1}^{n} (X_i - \bar{X}_n)^2$. If $s^2/\delta^2 > n$ take an additional observation. Stop at the smallest n, $n \geqslant 2$, such that $n > s_n^2/\delta^2$. Let N designate the number of observations at stopping (N is a random variable). After sampling terminates, estimate μ by the sample mean $\bar{X}_N$.

(i) Show that $P_\sigma\{N < \infty\} = 1$ for all σ, $0 < \sigma < \infty$.
(ii) Apply Robbins' theorem (see Theorem 10.9.3) to prove that $\bar{X}_N$ is an unbiased estimator of μ.
(iii) Prove that $E_\sigma\{N\} < \infty$ for all σ, $0 < \sigma < \infty$.
(iv) Imply from the inequality $N - 1 < s_{N-1}^2/\delta^2$ and the Wald theorem (Lemma 4.4.1) that

$$E_\sigma\{N\} \leqslant n(\sigma^2) + 2, \quad \text{all } \sigma, \quad 0 < \sigma < \infty.$$

(v) Use the result of (iv) and the extended Cramér-Rao inequality (4.4.14) to obtain a lower bound for $\text{Var}_\sigma\{\bar{X}_N\}$ for all σ, $0 < \sigma < \infty$. (In Section 10.9 we present the required theory by which an upper bound for $\text{Var}_\sigma\{\bar{X}_N\}$ can be determined.)

Section 4.5

12. Let $X_1, X_2, \ldots$ be a sequence of i.i.d. random variables $X_1 \sim N(\mu, \sigma^2)$, $-\infty < \mu < \infty$, $0 < \sigma < \infty$. Let $\bar{X}_n$ be the sample mean of $X_1, \ldots, X_n$ and M_n the sample median of $X_1, \ldots, X_n$.
(i) Find the asymptotic distribution of $\sqrt{n}\,(M_n - \mu)$, as $n \to \infty$ (see Fisz [1], p. 383).
(ii) What is the upper asymptotic efficiency of M_n relative to $\bar{X}_n$, as estimators of μ, that is, $\bar{e}_\mu(M_n, \bar{X}_n; \mu, \sigma)$?

13. What is the Kullback-Leibler information function for the following:
(i) the family of $\mathcal{N}(\mu, \sigma^2)$ distributions, $-\infty < \mu < \infty$, $0 < \sigma < \infty$?
(ii) the family of $\mathcal{G}(1/\theta, \nu)$ distributions, $0 < \theta < \infty$, ν fixed?
(iii) the family of binomial distributions $\mathcal{B}(n; \theta)$, $0 < \theta < 1$?

14. Let $X_1, \ldots, X_n$ be a sequence of i.i.d. random variables, $X_1 \sim G(1/\theta, \nu)$, $0 < \theta < \infty$, ν known. Let $T_n = \sum_{i=1}^{n} X_i$ and $\hat{\theta}_n = T_n/n\nu$, $n = 1, 2, \ldots$.
(i) Is $\hat{\theta}_n$ a consistent estimator of θ?
(ii) Is $\hat{\theta}_n$ efficient in the sense of Bahadur?

REFERENCES

Anderson [1]; Bahadur [4]; Barankin and Gurland [1]; Bhattacharyya [1]; Chernoff [1]; Cramér [1], [3]; DeGroot [1]; Doob [1]; Fend [1]; Fisher [3]; Gort [1]; Hogg and Craig [1]; LeCam [1]; Neyman [3]; Pratt [1]; Rao [1], [3], [6]; Stein [1]; Wald [2]; Wolfowitz [1], [4]; Wasan [1]; Kullback and Leibler [2]; Schmetterer [1].

CHAPTER 5

Maximum Likelihood Estimation

5.1. THE METHOD OF MAXIMUM LIKELIHOOD ESTIMATION

As in the previous chapters let the statistical model be represented by a parametric family of probability measures $\mathcal{F} = \{P_\theta; \theta \in \Theta\}$ on $(\mathfrak{X}, \mathcal{B})$, where $\mathcal{F} \ll \mu$. We designate the density function of P_θ with respect to μ by $f(x; \theta)$. The parameter space Θ is an interval in a k-dimensional Euclidean space $(k \geqslant 1)$. Thus $\theta = (\theta_1, \ldots, \theta_k)$ is vector valued. Since in this section we do not investigate asymptotic properties, we simplify the notation by letting the points of $\mathfrak{X}$ represent the whole sample. Thus the density function $f(x; \theta)$ designates the joint density of the whole sample.

The *likelihood function* of θ is a non-negative real valued function on $\Theta \times \mathfrak{X}$, say $\lambda(\theta; x)$, which is proportional to the density function $f(x; \theta)$; that is,

$$\lambda(\theta; x) = kf(x; \theta), \quad \theta \in \Theta, \quad x \in \mathfrak{X}, \tag{5.1.1}$$

where the proportionality factor k, $0 < k < \infty$, could depend on x but is independent of θ. The likelihood function is treated as a function of θ for a given sample value $X = x$.

Given a likelihood function $\lambda(\theta; x)$, the maximum likelihood estimator (M.L.E.) of θ is a function $\hat{\theta}: \mathfrak{X} \to \Theta$, $\mathcal{B}$-measurable, and satisfying

$$\lambda(\hat{\theta}(X); X) = \sup_{\theta \in \Theta} \lambda(\theta; X) \quad \text{a.s. } [\mu]. \tag{5.1.2}$$

We remark that $\lambda(\theta; X)$ does not have to be differentiable with respect to θ. We also remark that the maximum likelihood estimator is not necessarily unique. We demonstrate these points in a single example.

Example 5.1. (i) Let $X_1, X_2, \ldots, X_n$ be i.i.d. random variables having a uniform distribution over $[0, \theta]$, where $0 < \theta < \infty$. The likelihood function is

$$\lambda(\theta; X_1, \ldots, X_n) = \theta^{-n} J(\theta; X_1, \ldots, X_n)$$

where

$$J(\theta; X_1, \ldots, X_n) = \begin{cases} 1, & \text{if } \min_{1 \leqslant i \leqslant n} \{X_i\} \geqslant 0 \text{ and } \max_{1 \leqslant i \leqslant n} \{X_i\} \leqslant \theta, \\ 0, & \text{otherwise.} \end{cases}$$

Since θ^{-n} is a decreasing function of θ, to maximize $\lambda(\theta; X_1, \ldots, X_n)$ we should decrease θ as much as possible. Obviously the M.L.E. is $\hat{\theta} = \max_{1 \leqslant i \leqslant n} \{X_i\}$. In this case the M.L.E. is unique and is the complete sufficient statistic.

(ii) Suppose that $X_1, X_2, \ldots, X_n$ are i.i.d. random variables having a uniform distribution over $[\theta, \theta + 1]$, $-\infty < \theta < \infty$. The likelihood function is

$$\lambda(\theta; X_1, \ldots, X_n) = \begin{cases} 1, & \text{if } \theta \leqslant X_{(1)} \leqslant X_{(n)} \leqslant \theta + 1, \\ 0, & \text{otherwise,} \end{cases}$$

where $X_{(1)} \leqslant \cdots \leqslant X_{(n)}$ is the order statistic. In this case the M.L.E. is not unique. One M.L.E. is $\hat{\theta} = X_{(1)}$, another M.L.E. is $\tilde{\theta} = X_{(n)} - 1$. ∎

Now we introduce *the principle of invariance* for maximum likelihood estimation.

Theorem 5.1.1. (Zehna [1].) *Let $\mathcal{P} = \{P_\theta; \theta \in \Theta\}$ be a family of probability measures on $(\mathcal{X}, \mathcal{B})$. Let $g: \Theta \to \Omega$ be a function mapping Θ to an interval Ω in an r-dimensional Euclidean space $(1 \leqslant r \leqslant k)$. Then if $\hat{\theta}$ is an M.L.E. of θ, $g(\hat{\theta})$ is an M.L.E. of $g(\theta)$.*

Proof. For each $\omega \in \Omega$, let $G(\omega)$ be the coset of g in Θ; that is,

$$G(\omega) = \{\theta; \theta \in \Theta, g(\theta) = \omega\}.$$

Define

$$M(\omega; X) = \sup_{\theta \in G(\omega)} \lambda(\theta; X). \tag{5.1.3}$$

The function $M(\omega; X)$ on Ω is the induced likelihood function of $g(\theta)$. We remark that $\{G(\omega); \omega \in \Omega\}$ is a partition of Θ. Let $\hat{\theta}$ be any M.L.E. of θ; $\hat{\theta}$ belongs to one and only one set, say $G(\hat{\omega})$, of this partition. Morecver, since

$$\begin{aligned} \lambda(\hat{\theta}; X) &\leqslant \sup_{\theta \in G(\hat{\omega})} \lambda(\theta; X) = M(\hat{\omega}; X) \\ &\leqslant \sup_{\omega \in \Omega} M(\omega; X) = \sup_{\theta \in \Theta} \lambda(\theta; X) = \lambda(\hat{\theta}; X), \end{aligned} \tag{5.1.4}$$

$M(\hat{\omega}; X) = \sup M(\omega; X)$, and hence $\hat{\omega}$ is an M.L.E. of $g(\theta)$. Finally, since $\hat{\theta} \in G(\hat{\omega})$, $g(\hat{\theta}) = \hat{\omega}$. (Q.E.D.)

The principle of invariance for M.L.E.'s is cited in many books for the case of one-to-one mapping $g(\theta)$. As we have seen in Theorem 5.1.1, the restriction to one-to-one mapping is not necessary. We present now an example in which the usefulness of Theorem 5.1.1 is demonstrated.

Example 5.2. Let $X_1, X_2, \ldots, X_n$ be i.i.d. random variables having a common log-normal distribution, that is, $\log X_1 \sim N(\mu, \sigma^2)$, $-\infty < \mu < \infty$, $0 < \sigma < \infty$. As shown in Example 4.4, the expectation and variance of X are, respectively,

$$\xi = \exp\left\{\mu + \frac{\sigma^2}{2}\right\},$$

and

$$D^2 = \xi^2(e^{\sigma^2} - 1).$$

Consider the function $g(\mu, \sigma^2) = (\xi, D^2)$. We wish to find the M.L.E. $(\hat{\xi}, \hat{D}^2)$. It is easy to verify that the M.L.E. of (μ, σ^2) is $(\bar{Y}, \hat{\sigma}^2)$, where $Y_i = \log X_i$ $(i = 1, \ldots, n)$, $\bar{Y} = \sum_{i=1}^n Y_i/n$, and $\hat{\sigma}^2 = \sum_{i=1}^n (Y_i - \bar{Y})^2/n$. Invoking the invariance principle we immediately obtain

$$\hat{\xi} = \exp\left\{\bar{Y} + \frac{\hat{\sigma}^2}{2}\right\}$$

$$\hat{D}^2 = \hat{\xi}^2(e^{\hat{\sigma}^2} - 1). \qquad \blacksquare$$

We notice that the M.L.E. of θ is a function of the (minimal) sufficient statistic. Indeed, if $\mathfrak{F}$ admits a minimal sufficient statistic $T(X)$ then, according to the Neyman-Fisher factorization theorem, the likelihood function can be written in the form $\lambda(\theta; X) = g_\theta(T(X))$. That is, $\lambda(\theta; X)$ depends on X through $T(X)$. Hence the M.L.E. of θ is a function of $T(X)$ and a necessary statistic. An M.L.E. may, however, be an insufficient statistic for $\mathfrak{F}$. To illustrate such a case, let $X_1, X_2, \ldots, X_n$ be i.i.d. random variables having a common uniform distribution on $(\theta, 2\theta)$, $0 < \theta < \infty$. Let $X_{(1)} \leqslant \cdots \leqslant X_{(n)}$ be the order statistic. The minimal sufficient statistic is $(X_{(1)}, X_{(n)})$. However, an M.L.E. of θ is $\hat{\theta} = X_{(n)}/2$; $\hat{\theta}$ is *not* a sufficient statistic. In the following theorem we prove that if the density $f(x; \theta)$ is a k-parameter exponential in θ, then the M.L.E. of θ is unique and $\hat{\theta}$ is a minimal sufficient statistic.

Theorem 5.1.2. *Let $X_1, X_2, \ldots, X_n$ be i.i.d. random variables having a common k-parameter exponential density*

$$f(x; \theta) = h(x) \exp\left\{\sum_{i=1}^k \theta_i T_i(x) + \psi(\theta_1, \ldots, \theta_k)\right\}.$$

We further assume the following:

(i) *$\psi(\theta_1, \ldots, \theta_k)$ has continuous partial derivatives at all $(\theta_1, \ldots, \theta_k) \in \Theta$;*

(ii) $-\left\| \dfrac{\partial^2}{\partial\theta_i\, \partial\theta_j} \psi(\theta_1, \ldots, \theta_k);\ i, j = 1, \ldots, k \right\|$ *is positive definite for all $(\theta_1, \ldots, \theta_k) \in \Theta$;*

(iii) *$E_\theta\{|T_i(X)|\} < \infty$, all $i = 1, \ldots, k$; all $(\theta_1, \ldots, \theta_k)$ in Θ.*

Then the M.L.E. of $\theta = (\theta_1, \ldots, \theta_k)$ *is the root of the system of equations*

$$-\frac{1}{n}\sum_{j=1}^{n} T_i(X_j) = \frac{\partial}{\partial \theta_i}\psi(\hat{\theta}_1, \ldots, \hat{\theta}_k), \qquad i = 1, \ldots, k. \tag{5.1.5}$$

Furthermore, $\hat{\theta}$ *is unique and is a minimal sufficient statistic. If* $\nabla\psi(\theta_1, \ldots, \theta_k)$ *designates the gradient vector* $\|(\partial/\partial\theta_i)\psi(\theta_1, \ldots, \theta_k); i = 1, \ldots, k\|$ *then*

$$E_\theta\{\nabla\psi(\hat{\theta}_1, \ldots, \hat{\theta}_k)\} = \nabla\psi(\theta_1, \ldots, \theta_k), \quad \textit{all } \theta \textit{ in } \Theta. \tag{5.1.6}$$

Proof. The logarithm of the joint density function of $X_1, X_2, \ldots, X_n$ is

$$L(X_1, \ldots, X_n; \theta) = \sum_{j=1}^{n} \log h(X_j) + \sum_{i=1}^{k} \theta_i \sum_{j=1}^{n} T_i(X_j) + n\psi(\theta_1, \ldots, \theta_k).$$

Maximization of the likelihood function $\lambda(\theta; X)$ is equivalent to the maximization of $L(X_1, \ldots, X_n; \theta)$. Differentiating $L(X_1, \ldots, X_n; \theta)$ partially with respect to $\theta_1, \ldots, \theta_k$, we obtain that it is necessary for the M.L.E. $\hat{\theta}$ to be a root of (5.1.5). Since

$$\partial^2 L(X_1, \ldots, X_n; \theta)/\partial\theta_i\,\partial\theta_j = n\partial^2\psi(\theta_1, \ldots, \theta_k)/\partial\theta_i\,\partial\theta_j$$

(ii) implies that the root of (5.1.5) is unique and is a point of maximum of $L(X_1, \ldots, X_n; \theta)$. Moreover, since $(\sum_{j=1}^n T_1(X_j), \ldots, \sum_{j=1}^n T_k(X_j))$ is a minimal sufficient statistic, the uniqueness of $\hat{\theta}$ implies that $\hat{\theta} = (\hat{\theta}_1, \ldots, \hat{\theta}_k)$ is a sufficient statistic and therefore minimal sufficient. Finally, (iii) guarantees that $E_\theta\{\nabla L(X_1, \ldots, X_n; \theta)\} = 0$, for all $\theta \in \Theta$ (see Lemma 4.1.1). Hence

$$E_\theta\left\{\frac{1}{n}\sum_{j=1}^{n} T_i(X_j)\right\} = -\frac{\partial}{\partial\theta_i}\psi(\theta_1, \ldots, \theta_k), \qquad \text{all } i = 1, \ldots, k. \tag{5.1.7}$$

From (5.1.5) we have

$$E_\theta\left\{\frac{1}{n}\sum_{j=1}^{n} T_i(X_j)\right\} = -E_\theta\left\{\frac{\partial}{\partial\theta_i}\psi(\hat{\theta}_1, \ldots, \hat{\theta}_k)\right\} \qquad \text{all } i = 1, \ldots, k, \tag{5.1.8}$$

where $(\partial/\partial\theta_i)\psi(\hat{\theta}_1, \ldots, \hat{\theta}_k)$ is the partial derivative of $\psi(\theta_1, \ldots, \theta_k)$ at $(\hat{\theta}_1, \ldots, \hat{\theta}_k)$. Equations 5.1.7 and 5.1.8 imply (5.1.6). (Q.E.D.)

We remark that M.L.E.'s may exist even if (i) is not satisfied. Huzurbazar [1] indicated the following geometric property. Let θ be a real parameter and for a given X consider the log-likelihood function $L(X; \theta)$ for the exponential type density. The radius of curvature $\rho(\theta)$ of $L(X; \theta)$, at the point θ, satisfies the relationship

$$(5.1.9)\qquad \frac{1}{\rho(\theta)} = \frac{-\dfrac{\partial^2 L(X;\theta)}{\partial\theta^2}}{\left(1 + \left(\dfrac{\partial L(X;\theta)}{\partial\theta}\right)^2\right)^{3/2}}.$$

Since $(\partial/\partial\theta)L(X;\hat{\theta}) = 0$, we obtain that the radius of curvature at the M.L.E. $\hat{\theta}$ is

$$(5.1.10)\qquad \rho(\hat{\theta}) = -\left(\frac{\partial^2 L(X;\hat{\theta})}{\partial\theta^2}\right)^{-1}.$$

Moreover, since for exponential type densities, as considered in Theorem 5.1.2, the Fisher information function is $I(\theta) = -(\partial^2/\partial\theta^2)L(X;\theta)$, we obtain that the radius of curvature of $L(X;\theta)$ at $\theta = \hat{\theta}$ is the inverse of $I(\hat{\theta})$.

Notice that if the exponential density is of the form

$$(5.1.11)\quad f(X;\theta) = h(X)\exp\left\{\sum_{i=1}^{k}\varphi_i(\theta_1,\ldots,\theta_k)T_i(X) + \psi(\theta_1,\ldots,\theta_k)\right\}$$

we can find the M.L.E. of $\theta_1,\ldots,\theta_k$ by determining first the M.L.E. $\hat{\varphi}_1,\ldots,\hat{\varphi}_k$ according to Theorem 5.1.2. Then, by applying the invariance principle. we determine

$$(5.1.12)\qquad \hat{\theta}_i = \theta_i(\hat{\varphi}_1,\ldots,\hat{\varphi}_k),\quad i = 1,\ldots,k,$$

where $\theta_i(\varphi_1,\ldots,\varphi_k)$ are the inverse functions of $\varphi_i(\theta_1,\ldots,\theta_k)$. In order to apply this method we make a reparametrization of (5.1.11) according to (5.1.12). We then have

$$f(X;\theta) = h(X)\exp\left\{\sum_{i=1}^{k}\varphi_i T_i(X) + \xi(\varphi_1,\ldots,\varphi_k)\right\},$$

where $\xi(\varphi_1,\ldots,\varphi_k) = \psi(\theta_1(\varphi_1,\ldots,\varphi_k),\ldots,\theta_k(\varphi_1,\ldots,\varphi_k))$. A condition for the application of Theorem 5.1.2 is that $\xi(\varphi_1,\ldots,\varphi_k)$ has continuous partial derivatives with respect to $\varphi_i (i = 1,\ldots,k)$. This amounts to the requirement that $\psi(\theta_1,\ldots,\theta_k)$ will have continuous partial derivatives with respect to θ_i, and that $\theta_i(\varphi_1,\ldots,\varphi_k)$ will have continuous partial derivatives with respect to $\varphi_j (j = 1,\ldots,k)$. Consider the matrix of partial derivatives

$$\mathbf{H} = \left\|\frac{\partial\theta_i(\varphi)}{\partial\varphi_j};\ i,j = 1,\ldots,k\right\|$$

of the transformations (5.1.12). We assume that $\mathbf{H}$ is nonsingular, that is, $\theta_1,\ldots,\theta_k$ are linearly independent. We can express the gradient of $\xi(\boldsymbol{\varphi})$ with respect to $\varphi_1,\ldots,\varphi_k$ as a linear function of the gradient with respect to $(\theta_1, \ldots, \theta_k)$. Thus

$$\nabla \xi(\varphi_1, \ldots, \varphi_k) = \mathbf{H}'\nabla_\theta \psi(\boldsymbol{\theta}(\boldsymbol{\varphi})), \tag{5.1.13}$$

and the M.L.E. $\hat{\varphi}$ is the solution of the system of equations

$$\mathbf{M} = -\mathbf{H}'\nabla_\theta \psi(\boldsymbol{\theta}(\boldsymbol{\varphi})), \tag{5.1.14}$$

where $\mathbf{M} = \|M_i; i = 1, \ldots, k\|$ is the vector of sample means $M_i = \sum_{j=1}^n T_i(X_j)/n$. The quantity $\nabla_\theta \psi(\boldsymbol{\theta}(\boldsymbol{\varphi}))$ designates the gradient vector $\partial\psi(\theta)/\partial\theta$ evaluated at $\boldsymbol{\theta}(\boldsymbol{\varphi})$. The Fisher information function can be determined either with respect to $\boldsymbol{\varphi}$ or with respect to $\boldsymbol{\theta}$, according to the relationship

$$\begin{aligned} \mathbf{I}(\boldsymbol{\varphi}) &= \mathbf{H}'I(\boldsymbol{\theta}(\boldsymbol{\varphi}))\mathbf{H}; \\ \mathbf{I}(\boldsymbol{\theta}) &= (\mathbf{H}^{-1})'\mathbf{I}(\boldsymbol{\varphi}(\boldsymbol{\theta}))\mathbf{H}^{-1}. \end{aligned} \tag{5.1.15}$$

In the following example we illustrate this technique for estimating the parameters of the bivariate normal distribution.

Example 5.3. Let $(X_1, Y_1), \ldots, (X_n, Y_n)$ be independent vectors having an identical bivariate normal distribution, with zero means, and covariance matrix

$$\mathbf{V} = \sigma^2\begin{pmatrix} 1 & \rho \\ \rho & 1 \end{pmatrix}, \quad 0 < \sigma^2 < \infty, \quad -1 \leqslant \rho \leqslant 1.$$

We can write the density function in the exponential form

$$f(x, y) = (2\pi)^{-1} \exp\{\varphi_1 T_1(x, y) + \varphi_2 T_2(x, y) + \xi(\varphi_1, \varphi_2)\},$$

where

$$T_1(x, y) = x^2 + y^2, \quad T_2(x, y) = xy,$$

and

$$\varphi_1 = -\frac{1}{2\sigma^2(1-\rho^2)}$$

$$\varphi_2 = \frac{\rho}{\sigma^2(1-\rho^2)}$$

$$\xi(\varphi_1, \varphi_2) = -\tfrac{1}{2}\ln(4\varphi_1^2 - \varphi_2^2).$$

Thus we obtain

$$\frac{\partial}{\partial\varphi_1}\xi(\varphi_1, \varphi_2) = \frac{4\varphi_1}{4\varphi_1^2 - \varphi_2^2},$$

$$\frac{\partial}{\partial\varphi_2}\xi(\varphi_1, \varphi_2) = -\frac{\varphi_2}{4\varphi_1^2 - \varphi_2^2}.$$

Hence if $\hat{\varphi}_1$ and $\hat{\varphi}_2$ designate the M.L.E.'s of φ_1, φ_2, we have the equation

$$\frac{1}{n}\sum_{i=1}^{n}(X_i^2 + Y_i^2) = -\frac{4\hat{\varphi}_1}{4\hat{\varphi}_1^2 - \hat{\varphi}_2^2},$$

$$\frac{1}{n}\sum_{i=1}^{n}X_iY_i = \frac{\hat{\varphi}_2}{4\hat{\varphi}_1^2 - \hat{\varphi}_2^2}.$$

Since $\sigma^2 = -2\varphi_1/(4\varphi_1^2 - \varphi_2^2)$ and $\rho = -\varphi_2/2\varphi_1$, we obtain from the invariance principle

$$\hat{\sigma}^2 = \frac{1}{2n}\sum_{i=1}^{n}(X_i^2 + Y_i^2),$$

$$\hat{\rho} = \frac{2\sum_{i=1}^{n} X_iY_i}{\sum_{i=1}^{n}(X_i^2 + Y_i^2)}.$$

The Fisher information matrix for the parameters φ_1 and φ_2 can be obtained in the following manner:

$$\frac{\partial^2\xi(\varphi_1, \varphi_2)}{\partial\varphi_1^2} = -\frac{4(4\varphi_1^2 + \varphi_2^2)}{(4\varphi_1^2 - \varphi_2^2)^2},$$

$$\frac{\partial^2\xi(\varphi_1, \varphi_2)}{\partial\varphi_1\,\partial\varphi_2} = \frac{8\varphi_1\varphi_2}{(4\varphi_1^2 - \varphi_2^2)^2},$$

$$\frac{\partial^2\xi(\varphi_1, \varphi_2)}{\partial\varphi_2^2} = -\frac{4\varphi_1^2 + \varphi_2^2}{(4\varphi_1^2 - \varphi_2^2)^2}.$$

Thus

$$I(\varphi) = \frac{1}{(4\varphi_1^2 - \varphi_2^2)^2}\begin{bmatrix} 4(4\varphi_1^2 + \varphi_2^2) & -8\varphi_1\varphi_2 \\ -8\varphi_1\varphi_2 & 4\varphi_1^2 + \varphi_2^2 \end{bmatrix}.$$

To obtain the information matrix in terms of σ^2 and ρ we can use (5.1.15). We follow, however, another route and compute it directly. Letting $L(X, y; \sigma^2, \rho)$ designate the log-likelihood in terms of (σ^2, ρ), we have

$$\frac{\partial}{\partial\sigma^2}L(X, y; \sigma^2, \rho) = \frac{T_1(X, y) - 2\rho T_2(X, y)}{2\sigma^4(1 - \rho^2)} - \frac{1}{\sigma^2}.$$

Since the distribution laws of X and Y (marginals) are $\mathcal{N}(\theta, \rho^2)$, and the conditional distribution laws of X given Y and of Y given X are $\mathcal{N}(\rho Y, \sigma^2(1 - \rho^2))$ and $\mathcal{N}(\rho X, \sigma^2(1 - \rho^2))$, respectively, we obtain

$$EX^4 = EY^4 = 3\sigma^4,$$

$$EX^2Y^2 = \sigma^4(1 + 2\rho^2),$$

$$EX^3Y = EY^3X = 3\rho\sigma^4,$$

and

$$E\{(T_1(X, Y) - 2\rho T_2(X, Y))^2\} = 8\sigma^4(1 - \rho^2)^2.$$

Moreover,

$$E\{T_1(X, Y) - 2\rho T_2(X, Y)\} = 2\sigma^2(1 - \rho^2).$$

Hence

$$E\left\{\left(\frac{\partial}{\partial\sigma^2} L(X, Y; \sigma^2, \rho)\right)^2\right\} = \frac{1}{\sigma^4}.$$

Similarly, we find

$$E\left\{\frac{\partial}{\partial\sigma^2} L(X, Y; \sigma^2, \rho) \cdot \frac{\partial}{\partial\rho} L(X, Y, \sigma^2, \rho)\right\} = -\frac{\rho}{\sigma^2(1 - \rho^2)},$$

and

$$E\left\{\left[\frac{\partial}{\partial\rho} L(X, Y; \sigma^2, \rho)\right]^2\right\} = \frac{1 + \rho^2}{(1 - \rho^2)^2}.$$

Therefore the Fisher information matrix for (σ^2, ρ) is

$$I(\sigma^2, \rho) = \begin{bmatrix} \dfrac{1}{\sigma^4} & -\dfrac{\rho}{\sigma^2(1 - \rho^2)} \\ -\dfrac{\rho}{\sigma^2(1 - \rho^2)} & \dfrac{1 + \rho^2}{(1 - \rho^2)^2} \end{bmatrix}.$$ ■

To conclude this section we remark that M.L.E.'s are generally not unbiased ones. For example, as proven in Example 3.4 the best unbiased estimator of $e^{-\lambda}$ in the Poisson case is $(1 - 1/n)^{T_n}$, where T_n is the sample total. The M.L.E. in this case is $\exp\{-T_n/n\}$. According to the strong law of large numbers we obtain that both $(1 - 1/n)^{T_n}$ and $e^{-T_n/n}$ converge a.s. to $e^{-\lambda}$. Moreover, both estimators (properly normalized) have asymptotically the same distribution as $n \to \infty$. However, in small samples they reveal different characteristics at various λ intervals. For a comparison of the mean square errors and coverage probabilities see Zacks and Even [1]. In this estimation problem the M.L.E. is neither uniformly better nor uniformly worse than the U.M.V.U. estimator.

Some textbooks recommend converting the M.L.E. to an unbiased estimator (if it is not unbiased) in order to obtain a U.M.V.U. estimator. This recommendation *should not* be followed, unless one deals with special families of distributions like the exponential family. To show that the recommended procedure could be undesirable, consider the following example of Basu [1]. Let $X_1, \ldots, X_n$ be i.i.d. from a rectangular distribution over $(\theta, 2\theta)$, $0 < \theta < \infty$. The M.L.E. is $\hat{\theta} = X_{(n)}/2$ where $X_{(1)} \leqslant \cdots \leqslant X_{(n)}$. It is easy to show that $E_\theta\{\hat{\theta}\} = \theta[(2n + 1)/2(n + 1)]$. Hence the unbiased estimator based on $\hat{\theta}$ is $U(\hat{\theta}) = [(n + 1)/(2n + 1)]\hat{\theta}$. It can be shown that the unbiased estimator that is a linear combination of $X_{(1)}$ and $X_{(n)}$ has a uniformly smaller variance.

5.2. COMPUTATIONAL ROUTINES

It is often difficult to solve explicitly the likelihood equations [see (5.1.7) for the exponential case], even in cases where all the regularity conditions hold and a unique solution is known to exist. Equations 5.1.7 in the exponential case are very often nonlinear and difficult to solve. Moreover, if the family of distributions under consideration is not of the exponential type and multiple roots of the likelihood equations may exist, it may be difficult to locate the absolute maximum of the likelihood function. A number of papers were written on this subject; among them we mention in particular those of Barnett [1], Kale [1; 2], and Northan [1]. In Barnett's paper [1] we find an analysis of the various numerical techniques used to to approximate the roots of the likelihood equation. Kale [1; 2] investigated the large sample properties of iterative processes.

Fisher [3] was the first one to discuss and advocate the use of successive iterations to solve the likelihood equation. Fisher argued that in many "regular" cases it would be sufficient to execute only one cycle of iteration in order to arrive at a good approximation. Northan [1] has shown, however, that several cycles of iteration may be required to obtain a reasonable convergence. Barnett [1] illustrates the properties of several successive approximation techniques for small samples of sizes $n = 3, \ldots, 19$, when the random variables have a Cauchy distribution depending on a location parameter; that is,

$$f(x; \theta) = \frac{1}{\pi(1 + (x - \theta)^2)}, \qquad -\infty < x < \infty, \tag{5.2.1}$$

where $-\infty < \theta < \infty$. The derivative of the log-likelihood function, for n independent observations, is

$$\frac{\partial}{\partial \theta} L(\theta; X_1, \ldots, X_n) = \sum_{i=1}^{n} \frac{2(X_i - \theta)}{1 + (X_i - \theta)^2}. \tag{5.2.2}$$

The likelihood equation in this case is

$$\sum_{i=1}^{n} \frac{X_i - \theta}{1 + (X_i - \theta)^2} = 0. \tag{5.2.3}$$

An explicit solution of (5.2.3) is impossible to attain. The problem with likelihood functions for Cauchy distributions is that the log-likelihood

$$L(\theta; X_1, \ldots, X_n) = -n \log \pi - \sum_{i=1}^{n} \log [1 + (X_i - \theta)^2],$$

as a function of θ, may have several local maxima for a given sample $X_1, \ldots, X_n$. Indeed, $-\log[1 + (X_i - \theta)^2]$ has a maximum at $\theta = X_i$. Hence the sum $-\sum_i \log(1 + (X_i - \theta)^2)$ may have up to n different local maxima (it depends on the sample values). We now present several numerical methods for locating the relative maxima. These methods are variants of the celebrated Newton-Raphson method.

(i) The Newton-Raphson Method

The Newton-Raphson method is based on the expansion around $\hat{\theta}$ of the likelihood equation

$$(5.2.4)\quad 0 = \frac{\partial}{\partial\theta} L(\hat{\theta}; X) = \frac{\partial}{\partial\theta} L(\theta_1; X) + (\hat{\theta} - \theta_1)\frac{\partial^2}{\partial\theta^2} L(\theta_1 + \nu(\hat{\theta} - \theta_1), X),$$

for some $0 \leqslant \nu \leqslant 1$, where $\hat{\theta}$ is a root of the likelihood equation and θ_1 is an initial solution. If we take $\nu = 0$ in (5.2.4) we obtain an approximation for $\hat{\theta}$ in (5.2.4), namely,

$$(5.2.5)\qquad \theta_2 = \theta_1 - \frac{(\partial/\partial\theta)L(\theta_1; X)}{(\partial^2/\partial\theta^2)L(\theta_1; X)}.$$

The value of θ_2 can be substituted in (5.2.5) for θ_1 to obtain another value θ_3, and so on. Generally, starting from an initial solution θ_1 we generate a sequence $\{\theta_k; k \geqslant 1\}$, which is determined successively by the formula

$$(5.2.6)\qquad \theta_{k+1} = \theta_k - \frac{(\partial/\partial\theta)L(\theta_k; X)}{(\partial^2/\partial\theta^2)L(\theta_k; X)}, \qquad k = 1, 2, \ldots.$$

If the initial solution θ_1 was chosen close to the root of the likelihood equations $\hat{\theta}$, and if $(\partial^2/\partial\theta^2)L(\theta_k; X)$ for $k = 1, 2, \ldots$ is bounded away from zero, there is a good chance that the sequence generated by (5.2.6) will converge to the root $\hat{\theta}$. The sequence $\{\theta_k; k \geqslant 1\}$ generated by (5.2.6) depends actually on the sample values $X_1, \ldots, X_n$. As we show later, if the chosen initial solution θ_1 is a consistent estimator of θ (the true parameter value), a 1-cycle iteration provided by (5.2.5) yields a best asymptotically normal estimator of θ. However, in small sample situations the sequence $\{\theta_k; k \geqslant 1\}$ generated by (5.2.6) may reveal irregularities, due to the particular sample values obtained in the experiment. In order to avoid irregularities in the approximating sequence, two variants of this Newton-Raphson method were proposed. These variants are called the method of fixed-derivative and the method of scoring.

(ii) The Method of Fixed-Derivative Newton Approximation

In the fixed derivative Newton approximation the term $(\partial^2/\partial\theta^2)L(\theta_k; X)$ in (5.2.6) is replaced by $-n/a_k$, where $\{a_k; k \geqslant 1\}$ is a suitably chosen sequence of constants and n is the sample size. Thus we generate an approximating sequence by

$$\theta_{k+1} = \theta_k + \frac{a_k}{n} \cdot \frac{\partial}{\partial \theta} L(\theta_k; X), \qquad k = 1, 2, \ldots. \tag{5.2.7}$$

Such an approximating sequence for a well-chosen sequence $\{a_k\}$ might be more stable than (5.2.6) and coverage to the root $\hat{\theta}$ in a more regular fashion. In many cases, however, it was found that if the log-likelihood curve is steep in the neighborhood of a local maximum, (5.2.7) will very often fail to converge there. Barnett [1] illustrates how this approximation method fails to converge for a sample of size $n = 5$ from a Cauchy distribution. The special sequence computed by Barnett oscillated in cycles around the root of the equation. For the special case of $a_k = n$ for all $k = 1, 2, \ldots$, we obtain $\theta_{k+1} = \theta_k + (\partial/\partial\theta)L(\theta_k; X)$, $k = 1, 2, \ldots$. This iterative procedure has been used by many investigators.

(iii) The Method of Scoring

The method of scoring is a special case of the fixed-derivative Newton approximation (5.2.7), which has been introduced by Fisher [3]. The special $\{a_k\}$ sequence suggested by Fisher is $a_k = 1/I(\theta_k)$, where $I(\theta)$ is the Fisher information function (assumed to be positive for all θ) and θ_k is the value of the approximation after the k-th iteration. Thus Fisher's scoring method generates the sequence

$$\theta_{k+1} = \theta_k + \frac{1}{nI(\theta_k)} \cdot \frac{\partial}{\partial \theta} L(\theta_k; X). \tag{5.2.8}$$

For the k-th iteration ($k = 1, 2, \ldots$) we define

$$\zeta_k = \frac{\theta_{k-1}^{(1)} \cdot (\partial/\partial\theta)L(\theta_{k-1}^{(2)}; X) - \theta_{k-1}^{(2)}(\partial/\partial\theta)L(\theta_{k-1}^{(1)}; X)}{L(\theta_{k-1}^{(2)}; X) - L(\theta_{k-1}^{(1)}; X)}, \tag{5.2.9}$$

and

$$\left.\begin{aligned} \theta_k^{(1)} &= \zeta_k \\ \theta_k^{(2)} &= \theta_{k-1}^{(2)} \end{aligned}\right\} \quad \text{if } \frac{\partial}{\partial \theta} L(\zeta_k; X) > 0, \tag{5.2.10}$$

or

$$\left.\begin{aligned} \theta_k^{(1)} &= \theta_{k-1}^{(1)} \\ \theta_k^{(2)} &= \zeta_k \end{aligned}\right\} \quad \text{if } \frac{\partial}{\partial \theta} L(\zeta_k; X) < 0. \tag{5.2.11}$$

By this method of iteration continues to enclose and converge on a local maximum. [We have to assume that $(\partial/\partial\theta)L(\theta; X)$ is continuous in θ for each X.] It is not obvious, however, how to choose the two initial values $(\theta_1^{(1)}, \theta_1^{(2)})$. Barnett presented in his paper the results of an interesting Monte Carlo experiment for the estimation of the characteristics of local maxima in the Cauchy case, for small samples of sizes 3(2)19.

5.3. STRONG CONSISTENCY OF MAXIMUM LIKELIHOOD ESTIMATORS

In this section we prove that under certain regularity conditions the M.L.E.'s are strongly consistent. An outline of the proof of the strong consistency was first given by Wald [5] and Wolfowitz [2]. Later it was modified by LeCam [2] and presented also by Bahadur [4]. In 1956, Kiefer and Wolfowitz [3] proved the consistency of the M.L.E. for models with nuisance parameters.

As in Chapter 4, we assume that $\mathcal{F}$ is a family of probability measures on $(\mathcal{X}, \mathcal{B})$ dominated by μ. The corresponding class of density functions with respect to μ is $\{f(x; \theta); \theta \in \Theta\}$. We assume that Θ is an interval in a Euclidean space.

Let N_φ be a neighborhood of a point φ in Θ. The Kullback-Leibler information functions for discriminating between $f(X; \theta)$ and $f(X; \varphi')$, $\varphi' \in N_\varphi$, is defined as

$$I(\theta, N_\varphi) = E_\theta\left\{\inf_{\varphi' \in N_\varphi} \log \frac{f(X; \theta)}{f(X; \varphi')}\right\}. \tag{5.3.1}$$

We notice that although $I(\theta, \varphi) \geqslant 0$ for any two points θ and φ, $I(\theta, N_\varphi)$ is not necessarily non-negative. Furthermore, in order to assure the consistency of the M.L.E. in the general case, we have to impose a compactness condition on Θ. We therefore compactify Θ and include the point of infinity We designate by N_∞ a neighborhood of infinity (the complement of any closed sphere containing the origin). The following theorem establishes the strong consistency of the M.L.E. and gives an estimate to the rate of approach of the covering probabilities.

Theorem 5.3.1. (The consistency theorem.) *Let $\hat{\theta}_n$ designate an M.L.E. of θ based on n i.i.d. random variables $X_1, \ldots, X_n$. In addition to the above general conditions, assume the following:*

(i) $\mu(f(X; \theta) \neq f(X; \varphi)) > 0$ *for all* $\theta \neq \varphi$;
(ii) $f(X; \varphi') \to f(X; \varphi)$ *when* $\varphi' \to \varphi$;
(iii) $I(\theta, N_\varphi) > -\infty$ *for some neighborhood of* φ;
(iv) $f(X; \theta) \to 0$ *as* $|\theta| \to \infty$;
(v) $I(\theta, N_\infty) > -\infty$, *for some neighborhood of* ∞;

We further require that

(vi) $E_\theta\{Z^2(\theta, N_\varphi) \exp\{tZ(\theta, N_\varphi)\}\} < \infty$, *for all* $-1 \leqslant t \leqslant 0$, *and for some neighborhood of* φ, *for all* $\varphi \notin N_\theta$, *where*

$$Z(\theta, N_\varphi) = \inf_{\varphi' \in N_\varphi} \log \frac{f(X; \theta)}{f(X; \varphi')}.$$

Then $\hat{\theta}_n \to \theta$ a.s., and for some $K, a > 0$,

(5.3.2) $$P_\theta[|\hat{\theta}_n - \theta| \geqslant \epsilon] \leqslant Ke^{-na}, \quad \textit{for all } n.$$

Proof. Let $\eta_\varphi = \{N_\varphi^{(i)}; i = 1, 2, \ldots\}$ be a direction family of neighborhoods of φ; that is, $N_\varphi^{(i+1)} \subset N_\varphi^{(i)}$ for all $i = 1, 2, \ldots$. Then

$$Z(\theta; N_\varphi^{(i)}) \leqslant Z(\theta, N_\varphi^{(i+1)}) \quad \text{a.s.,} \qquad \text{all } i = 1, 2, \ldots.$$

We obtain from the Lebesgue monotone convergence theorem that

$$\lim_{i\to\infty} I(\theta, N_\varphi^{(i)}) = I(\theta, \varphi).$$

According to (iv), $I(\theta, \infty) = \infty$ and $I(\theta, N_\infty) \to \infty$ as $N_\infty \to \infty$. Hence there exists a neighborhood N_φ' such that $I(\theta, N_\varphi') > 0$ and a neighborhood N_∞ such that $I(\theta, N_\infty) > 0$, for all φ outside N_θ. Let $\bar{N}_\theta = \Theta - N_\theta$; $\bar{N}_\theta$ is covered by N_∞ and by a finite number of neighborhoods $N_{\varphi_1}, \ldots, N_{\varphi_k}$, where $\varphi_1, \ldots, \varphi_k$ are not in N_θ. This follows from the Heine-Borel theorem on the finite coverage of a compact set. In particular, for a given $\epsilon > 0$ (arbitrarily small) let $N_\theta = \{\theta'; |\theta - \theta'| < \epsilon\}$. We have

(5.3.3) $$\frac{\prod_{i=1}^n f(X_i;\theta)}{\sup_{\varphi\in\bar{N}_\theta} \prod_{i=1}^n f(X_i;\varphi)} > \prod_{i=1}^n \frac{f(X_i;\theta)}{\sup_{\varphi\in\bar{N}_\theta} f(X_i;\varphi)}.$$

Moreover, for each X_i $(i = 1, \ldots, n)$,

(5.3.4) $$Z_i(\theta, \bar{N}_\theta) = \inf_{\varphi\in\bar{N}_\theta} \log \frac{f(X_i;\theta)}{f(X_i;\varphi)} = \log \frac{f(X_i;\theta)}{\sup_{\varphi\in\bar{N}_\theta} f(X_i;\varphi)}.$$

To prove that $\hat{\theta}_n$ is strongly consistent we have to show that

(5.3.5) $$\psi_n(\theta) = P_\theta\left[\frac{\prod_{i=1}^n f(X_i;\theta)}{\sup_{\varphi\in\bar{N}_\theta} \prod_{i=1}^n f(X_i;\varphi)} > 1\right] > 1 - \epsilon \qquad \text{for all } n$$

sufficiently large. This means that for all n sufficiently large the value of θ which maximizes $\prod_{i=1}^n f(X_i;\theta)$, namely, the M.L.E. $\hat{\theta}_n$, belongs to N_θ with probability one.

Let $\varphi_{k+1} \equiv \infty$ and let $N_{\varphi_1}, \ldots, N_{\varphi_k}$ be open neighborhoods of $\varphi_1, \ldots, \varphi_k$ covering $\bar{N}_\theta$ and such that $I(\theta, N_{\varphi_v}) > 0$ for all $v = 1, \ldots, k+1$. As previously remarked, such neighborhoods exist. Then according to (5.3.3) and (5.3.4),

(5.3.6) $$\begin{aligned} P_\theta\left[\frac{\prod_{i=1}^n f(X_i;\theta)}{\sup_{\varphi\in\bar{N}_\theta} \prod_{i=1}^n f(X_i;\varphi)} > 1\right] &\geqslant P_\theta\left[\sum_{i=1}^n Z_i(\theta, \bar{N}_\theta) > 0\right] \\ &\geqslant P_\theta\left[\max_{1\leqslant v\leqslant k+1} \sum_{j=1}^n Z_j(\theta, N_{\varphi v}) > 0\right]. \end{aligned}$$

Thus from (5.3.6) we deduce

$$(5.3.7) \qquad 1 - \psi_n(\theta) \leqslant 1 - P_\theta\left[\sum_{i=1}^{n} Z_i(\theta, \bar{N}_\theta) > 0\right]$$
$$\leqslant P_\theta\left[\max_{1 \leqslant v \leqslant k+1} \sum_{i=1}^{n} Z(\theta, N_{\varphi_v}) \leqslant 0\right].$$

By the strong law of large numbers, the right-hand side of (5.3.7) is smaller than ϵ for all n sufficiently large (the required sample size may depend on θ). This proves the strong consistency of the M.L.E. To establish (5.3.2) we can bound the right-hand side of (5.3.7) in the following manner. Let $M_\theta^{(v)}(t)$ be the moment generating function of $Z_i(\theta, N_{\varphi_v})$. Assumption (vi) guarantees that $M_\theta^{(v)}(t)$ is strictly convex and

$$(5.3.8) \quad P_\theta\left[\sum_{i=1}^{n} Z_i(\theta, N_{\varphi_v}) \leqslant 0\right] \leqslant (M_\theta^{(v)}(t))^n \qquad \text{for all } -1 \leqslant t \leqslant 0.$$

Let $t^{(v)}$, $-1 < t^{(v)} < 0$, be the point of minimum of $M_\theta^{(v)}(t)$, and let $\rho_v = M_\theta^{(v)}(t^{(v)})$, $0 < \rho_v < 1$, for all $v = 1, \ldots, k+1$. Hence

$$(5.3.9) \qquad P_\theta[|\hat{\theta}_n - \theta| \geqslant \epsilon] \leqslant 1 - \psi_n(\theta) \leqslant (k+1)\left[\max_{1 \leqslant v \leqslant k+1} \rho_v\right]^n.$$

This proves (5.3.2). (Q.E.D.)

In the case of a finite number of points in the parameter space Θ, we can obtain the same results as in Theorem 5.3.1 with a weaker set of assumptions. The two necessary assumptions are the following:

(i) If θ, φ are points in Θ and $\theta \neq \varphi$, then $\mu(f(X; \theta) \neq f(X; \varphi)) > 0$.

(ii) If $Z(\theta, \varphi) = \log (f(X; \theta))/(f(X; \varphi))$ then

$$E_\theta\{Z^2(\theta, \varphi) \exp \{tZ(\theta, \varphi)\}\} < \infty,$$

all $-1 \leqslant t \leqslant 0$.

Condition (ii) is required for the strict convexity of the m.g.f. of $Z(\theta, \varphi)$ on $[-1, 0]$. It guarantees also that $E_\theta\{Z(\theta, \varphi)\} > 0$ and $E_\theta\{|Z(\theta, \varphi)|\} < \infty$. Condition (i) and the strong law of large numbers provide the strong consistency of the M.L.E.

(i) Examples of Inconsistent M.L.E.'s

There are in the literature many examples of inconsistent M.L.E.'s (see Basu [2], Bahadur [3], Kiefer and Wolfowitz [3], Hannan [1], and Neyman and Scott [1]). We present here two examples which show the necessity of some regularity conditions mentioned above. The important feature of consistent cases is that by increasing the sample size one can separate or distinguish between distributions having different parameter points by considering the likelihood function. This feature is absent from the following two examples.

Example 5.4. The following example (Basu [2]) is very artificial. It emphasizes, however, the requirement that small variations in the parameter value cause small variations in the density functions.

Let $X_1, X_2, \ldots$ be a sequence of i.i.d. Bernoulli random variables with a probability density

$$P_\theta[X_i = 1] = \begin{cases} \theta, & \text{if } \theta \text{ is rational,} \\ 1 - \theta, & \text{if } \theta \text{ is irrational,} \end{cases}$$

$0 < \theta < 1$; and $P_\theta[X_i = 0] = 1 - P_\theta[X_i = 1]$. The family of density of functions of X_i $(i = 1, 2, \ldots)$ with respect to the counting measure $\mu(x)$ is

$$f(X; \theta) = \begin{cases} \theta^X(1 - \theta)^{1-X}, & \text{if } \theta \text{ is rational,} \\ (1 - \theta)^X\theta^{1-X}, & \text{if } \theta \text{ is irrational,} \end{cases}$$

where $X = 0, 1$. The M.L.E. of θ based on the first n observations is $\hat{\theta}_n = \sum X_i/n$, since $\hat{\theta}_n$ is rational for all $n = 1, 2, \ldots$. But

$$\hat{\theta}_n \xrightarrow{\text{a.s.}} \begin{cases} \theta, & \text{if } \theta \text{ is rational,} \\ 1 - \theta, & \text{if } \theta \text{ is irrational.} \end{cases}$$

Hence $\hat{\theta}_n$ is an inconsistent estimator of θ.

If we analyze this example in terms of the conditions of Theorem 5.3.1 we realize that the continuity condition, (ii), is not satisfied. Indeed, any open neighborhood of θ_0 in (0, 1) contains rational and irrational points. Hence for both $x = 0, 1$ there exists no limit of $f(X; \theta)$ as $\theta \to \theta_0$. ■

Example 5.5. The following example is more meaningful statistically and is from Neyman and Scott [1].

Let

$$\begin{pmatrix} X_i \\ Y_i \end{pmatrix} \sim N\left(\begin{bmatrix} \mu_i \\ \mu_i \end{bmatrix}, \sigma^2 I\right), \quad i = 1, \ldots, n, \ldots$$

be independent vectors. Each of these vectors can represent two independent measurements of the same object. The two measurements have the same mean. The means may differ from one observed vector to another (e.g., weighing different objects) but the variance σ^2 remains the same. The objective is to estimate the unknown variance, σ^2, $0 < \sigma < \infty$. Since $Z_i = X_i - Y_i \sim N(0, 2\sigma^2)$ for all $i = 1, \ldots, n$, $\hat{\sigma}_n^2 = (1/2n)\sum_{i=1}^n Z_i^2$ is obviously an unbiased estimator of σ^2. Moreover, from the strong law of large numbers $\hat{\sigma}_n^2 \to \sigma^2$ a.s. for all $0 < \sigma < \infty$. The M.L.E. is, however, $\tilde{\sigma}_n^2 = (1/4n)\sum_{i=1}^n Z_i^2$, which is obviously inconsistent. To prove this, consider the joint density function of n independent vectors, that is,

$$f(x_1, y_1, \ldots, x_n, y_n; \mu_1, \ldots, \mu_n, \sigma^2)$$
$$= (2\pi)^{-n}\sigma^{-2n} \exp\left\{-\frac{1}{2\sigma^2}\sum_{i=1}^{n}[(x_i - \mu_i)^2 + (y_i - \mu_i)^2]\right\}.$$

The M.L.E. of μ_i is $\hat{\mu}_i = (x_i + y_i)/2$ $(i = 1, \ldots, n)$. Hence

$$f(x_1, y_1, \ldots, x_n, y_n; \hat{\mu}_1, \ldots, \hat{\mu}_n, \sigma^2) = (2\pi)^{-n}\sigma^{-2n} \exp\left\{-\frac{1}{4\sigma^2}\sum_{i=1}^{n} z_i^2\right\}$$

where $z_i = x_i - y_i$. It is very easy to verify that $\tilde{\sigma}_n^2 = (1/4n)\sum_{i=1}^{n} z_i^2$ maximizes $f(x_1, y_1, \ldots, x_n, y_n; \hat{\mu}_1, \ldots \hat{\mu}_n, \sigma^2)$. We can easily construct, as a variant of the present example a model in which the M.L.E. of σ^2 converges a.s. to an arbitrary fraction of σ^2.

In this example we have a case of estimation with nuisance parameters (unknown means). The number of nuisance parameters grows, to infinity as $n \to \infty$. Kiefer and Wolfowitz [3] gave sufficient conditions for the consistency of M.L.E.'s in the presence of nuisance parameters. ∎

(ii) Consistency of the M.L.E. in Cases of Discrete Distributions

Hannan [1] relaxed the conditions required for the consistency of an M.L.E. in cases of discrete distributions. The notion of M.L.E. has been generalized in that paper in the following manner.

Let $\mathcal{F}$ be a family of probability measures on $(\mathcal{X}, \mathcal{B})$ and $\mathcal{F} \ll \mu$. Let f be the density, with respect to μ, of P in $\mathcal{F}$, and $g(x) = \log f(x)$. Let $\{P_n\}$ be a sequence in $\mathcal{F}$. The sequence $\{P_n\}$ is called a strong maximum likelihood sequence if and only if

$$\text{(5.3.10)} \qquad \sup_{P\in\mathcal{F}}\left\{\frac{1}{n}\sum_{i=1}^{n} g(X_i)\right\} - \frac{1}{n}\sum_{i=1}^{n} g_n(X_i) \to 0 \quad \text{a.s. } [\mu].$$

Here g_n are the log-likelihood functions corresponding to $\{P_n\}$, that is, $\sum_{i=1}^{n} g_n(X_i) = \log \prod_{i=1}^{n} f_n(X_i)$. A sequence $\{P_n\}$ is called an *equi-likelihood sequence* if and only if, for each $P \in \mathcal{F}$,

$$\limsup_{n\to\infty}\left\{\frac{1}{n}\sum_{i=1}^{n} g(X_i) - \frac{1}{n}\sum_{i=1}^{n} g_n(X_i)\right\} \leqslant 0 \quad \text{a.s. } [\mu].$$

We notice that according to our previous notation,

$$\frac{1}{n}\sum_{i=1}^{n} g(X_i) - \frac{1}{n}\sum_{i=1}^{n} g_n(X_i) = \frac{1}{n}\sum_{i=1}^{n} Z(P, P_i).$$

We further notice that $E_P\{(1/n)\sum_{i=1}^{n} Z(P, P_i)\} = (1/n)\sum_{i=1}^{n} I(P, P_i)$, where $I(P, P_i)$ is the Kullback-Leibler information function to discriminate between

P and P_i. Hannan designates the function $(1/n)\sum_{i=1}^{n} Z(P, P_i)$ by $\Delta_n(P, \mathbf{P}_n)$. If, as assumed in Theorem 5.3.1. (see also Wald [5] and LeCam [2]), the family $\mathfrak{F}$ is compact in a given metric, then

$$\limsup_{n\to\infty}\left\{\frac{1}{n}\sum_{i=1}^{n} g(X_i) - \frac{1}{n}\sum_{i=1}^{n} g_n(X_i)\right\} \geqslant \limsup_{n\to\infty} \Delta_n(P, P_n).$$

Hannan considers two kinds of distance functions on $\mathfrak{F} \times \mathfrak{F}$, namely,

$$D(P, P') = \left[\int (\sqrt{f(x)} - \sqrt{(f'(x)})^2 \mu(dx)\right]^{1/2}, \tag{5.3.11}$$

and

$$V(P, P') = \int |f(x) - f'(x)|\, \mu(dx). \tag{5.3.12}$$

A maximum likelihood estimator of a *discrete* distribution function is the sample empiric distribution function $F_n(X)$, that is, $F_n(X) = 1/n$ (number of sample values $\leqslant X$). The Glivenko-Cantelli theorem implies that $F_n(X) \to F(X)$ a.s. This result is equivalent to

$$V(P_n^*, P) \to 0 \quad \text{a.s. } [\mu],$$

where P_n^* is the empiric probability measure corresponding to $F_n^*(x)$. Hannan [1] proves that if $E_P(-g) < \infty$, then $\Delta_n(P) \to 0$ a.s. $[\mu]$, where $\Delta_n(P)$ designates $\Delta_n(P, P_n^*)$. Moreover, he shows that for any P' and P in $\mathfrak{F}$, $I(P, P') \geqslant V(P, P')/4$. Hence $\Delta_n(P) \to 0$ a.s. implies that $V(P_n^*, P) \to 0$ a.s. Thus the only condition required for the strong consistency of the empiric distribution, as an M.L.E. of a discrete distribution P, is that

$$E_P\{|\log f_p(X)|\} < \infty.$$

Inconsistency of M.L.E.'s in discrete cases may happen when $E_P\{|\log f_p(X)|\} = \infty$. For such examples, see Hannan [1].

5.4. THE ASYMPTOTIC EFFICIENCY OF MAXIMUM LIKELIHOOD ESTIMATORS

In this section we prove that under the regularity conditions of Section 4.5, an M.L.E. is asymptotically efficient in the Bahadur sense. That is, the lower bound on the asymptotic effective variance, as given by (4.5.18), is attained by a sequence of M.L.E.'s. In our proof we follow Bahadur [4]. Assume that $\mathfrak{F}$ is a parametric family of probability measures, that is, $\mathfrak{F} = \{P_\theta;\ \theta \in \Theta\}$ where Θ is the real line and $\mathfrak{F}$ satisfies the Cramér-Rao conditions and conditions A.1–A.5, specified in Chapter 4. Furthermore, we assume that

$$\left|\frac{\partial^3}{\partial\theta^3} \log f(X;\theta)\right| \leqslant W(X;\theta_0) \tag{A.6}$$

for all θ in a neighborhood of θ_0, where $W(X; \theta_0)$ is an integrable function with respect to P_{θ_0}. In addition, if $S(X; \theta) = (\partial/\partial\theta)\log f(X; \theta)$, we assume that the moment generating functions of $S(X; \theta)$, $(\partial/\partial\theta)S(X; \theta_0)$, and $W(X; \theta_0)$ exist at $\theta = \theta_0$ for all θ_0 in a neighborhood of the origin.

Lemma 5.4.1. (Bahadur [4].) *Given a δ, $0 < \delta < I(\theta_0)$, there exists a $0 < \rho < 1$ (may depend on θ_0) such that, for every $\epsilon > 0$,*

$$(5.4.1)\quad P_{\theta_0}[|T_n - \theta_0| \geqslant \epsilon] \leqslant P_{\theta_0}\left[\left|\frac{1}{n}\sum_{i=1}^{n} S(X_i; \theta_0)\right| \geqslant \epsilon[I(\theta_0) - \delta]\right] + \rho^n$$

for all n sufficiently large, where T_n is an M.L.E. of θ.

Proof. Choose a neighborhood of θ_0 in Θ, that is, $N_{\theta_0} = \{\theta; |\theta - \theta_0| < h\}$, where $0 < h$ is as small as required to satisfy the following:

(i) $N_{\theta_0} \subset \Theta$;
(ii) A.6 is satisfied on N_{θ_0};
(iii) $E_\theta\{W(X; \theta_0)\} < h/2$ on Θ_{θ_0}.

According to Lemma 4.1.2, $E_{\theta_0}\{(\partial/\partial\theta)S(X; \theta_0)\} = -I(\theta_0)$. Since the m.g.f.'s of $S(X; \theta_0)$, $(\partial/\partial\theta)S(X; \theta_0)$ and $W(X; \theta_0)$ exist, we have, as in Lemma 4.5.1,

$$(5.4.2)\qquad P_{\theta_0}\left[I(\theta_0) + \frac{1}{n}\sum_{i=1}^{n}\frac{\partial}{\partial\theta} S(X_i; \theta_0) \geqslant \frac{\delta}{2}\right] \leqslant (e^{-(\delta/2)t}M_\theta(t))^n$$

for all t in a neighborhood of the origin, where $M_\theta(t)$ is the m.g.f. of $(\partial/\partial\theta)S(X; \theta_0)$. Hence there exists a number ρ_1, $0 < \rho_1 < 1$, such that

$$(5.4.3)\qquad P_{\theta_0}\left[I(\theta_0) + \frac{1}{n}\sum_{i=1}^{n}\frac{\partial}{\partial\theta} S(X_i; \theta_0) \geqslant \frac{\delta}{2}\right] \leqslant \rho_1{}^n.$$

Similarly,

$$(5.4.4)\qquad P_{\theta_0}\left[I(\theta_0) + \frac{1}{n}\sum_{i=1}^{n}\frac{\partial}{\partial\theta} S(X_i; \theta_0) \leqslant -\frac{\delta}{2}\right] \leqslant \rho_2{}^n, \qquad 0 < \rho_2 < 1.$$

Furthermore,

$$(5.4.5)\qquad P_{\theta_0}\left[\frac{1}{n}\sum_{i=1}^{n} W(X_i; \theta_0) > \frac{\delta}{h}\right] \leqslant \rho_3{}^n, \qquad 0 < \rho_3 < 1.$$

Equations 5.4.3–5.4.5 hold for every $n = 1, 2, \ldots$. Let T_n be an M.L.E. of θ, and for a given n, $|T_n - \theta_0| < h$; that is, $T_n \in N_{\theta_0}$. Suppose that in

N_{θ_0}, $\sum_{i=1}^{n} S(X_i; T_n) = 0$. Then there exists a point θ_* in N_{θ_0}, $\theta_* = \theta_0 + \vartheta(T_n - \theta_0)$, $0 \leqslant \vartheta \leqslant 1$, for which

(5.4.6)
$$\begin{aligned} 0 &= \sum_{i=1}^{n} S(X_i; T_n) \\ &= \sum_{i=1}^{n} S(X_i; \theta_0) + (T_n - \theta_0) \sum_{i=1}^{n} \frac{\partial}{\partial\theta} S(X_i; \theta_0) + \tfrac{1}{2}(T_n - \theta_0)^2 \sum_{i=1}^{n} \frac{\partial^2}{\partial\theta^2} S(X_i; \theta_*). \end{aligned}$$

From (5.4.6) we obtain

(5.4.7)
$$\frac{1}{n} \sum_{i=1}^{n} S(X_i; \theta_0) = (T_n - \theta_0)[I(\theta_0) + r_n],$$

where

$$r_n = -\left(I(\theta_0) + \frac{1}{n} \sum_{i=1}^{n} \frac{\partial}{\partial\theta} S(X_i; \theta_0)\right) - \frac{1}{2n}(T_n - \theta_0) \sum_{i=1}^{n} \frac{\partial^2}{\partial\theta^2} S(X_i; \theta_*).$$

Since $|(\partial^2/\partial\theta^2)S(X_i; \theta)| \leqslant W(X_i; \theta_0)$ for all θ in N_{θ_0}, we obtain that

(5.4.8)
$$|r_n| \leqslant \left| I(\theta_0) + \frac{1}{n} \sum_{i=1}^{n} \frac{\partial}{\partial\theta} S(X_i; \theta_0) \right| + \frac{1}{2n} h \sum_{i=1}^{n} W(X_i; \theta_0).$$

As proven in Theorem 5.3.1, $P_{\theta_0}[|T_n - \theta_0| \geqslant h] \leqslant \rho_0{}^n$ for some $0 < \rho_0 < 1$, for all n sufficiently large. Let $\eta_n = I(\theta_0) + (1/n) \sum_i (\partial/\partial\theta) S(X_i; \theta_0)$, $\xi_n = (1/n) \sum_i W(X_i; \theta_0)$. Since the event $\{|\eta_n| + \frac{1}{2}h\xi_n \geqslant \delta\}$ implies the union $\{|\eta_n| \geqslant \delta\} \cup \{\xi_n \geqslant \delta/h\}$, we obtain from (5.4.3)–(5.4.5) that there exists a ρ_4, $0 < \rho_4 < 1$, such that

$$P_{\theta_0}\{|r_n| \geqslant \delta\} \leqslant \rho_4{}^n.$$

Let A_n be the event $|T_n - \theta_0| \geqslant h_0$. Let B_n designate the event T_n does not maximize the likelihood function, and $C_n = \{|\eta_n| \geqslant \delta\} \cup \{\xi_n \geqslant \delta/h\}$. Let $E_n = A_n \cup B_n \cup C_n$. Then there exists a ρ, $0 < \rho < 1$, such that $P_{\theta_0}[E_n] \leqslant \rho^n$, for all n sufficiently large. Moreover,

(5.4.9)
$$\{|T_n - \theta_0| \geqslant \epsilon\} \subset \{E_n \cup [\bar{E}_n \cap \{|T_n - \theta_\theta| \geqslant \epsilon\}]\}.$$

Hence

(5.4.10)
$$P_{\theta_0}[|T_n - \theta_0| \geqslant \epsilon] \leqslant \rho^n + P_{\theta_0}[\bar{E}_n \cap \{|T_n - \theta_0| \geqslant \epsilon\}],$$

for all n sufficiently large. Finally,

(5.4.11)
$$\begin{aligned} P_{\theta_0}[\bar{E}_n \cap \{|T_n - \theta_0| \geqslant \epsilon\}] &= P_{\theta_0}[\{|T_n - \theta_\theta| \leqslant h\} \cap \bar{B}_n \cap \\ &\qquad \{|r_n| \leqslant \delta\} \cap \{|T_n - \theta_0| \geqslant \epsilon\}] \leqslant \\ &P_{\theta_0}\left[\left|\frac{1}{n} \sum_{i=1}^{n} S(X_i; \theta_0)\right| \geqslant \epsilon[I(\theta_0) - \delta]\right]. \end{aligned}$$

The inequality on the right-hand side of (5.4.11) follows from (5.4.7). The combination of (5.4.10) and (5.4.11) yields (5.4.1). (Q.E.D.)

We are ready now to prove that a sequence of M.L.E.'s is asymptotically efficient.

Theorem 5.4.2. (Bahadur.) *Under the regularity conditions of C.R., A.1–A.6 and the conditions for the consistency of an M.L.E., if $\hat{g}_n$ is an M.L.E. of $g(\theta)$, then*

$$\text{(5.4.12)} \qquad \overline{\lim_{\epsilon\to 0}}\ \overline{\lim_{n\to\infty}} \left\{ n\tau_g^{\,2}(\hat{g}_n, \epsilon, \theta) \right\} \leqslant \frac{(g'(\theta))^2}{I(\theta)},$$

at all points θ for which $g'(\theta)$ exists.

Remark. The quantity $\tau_g(\hat{g}_n, \epsilon, \theta)$ is the effective standard deviation, as defined in Section 4.5. We comment also that since $\hat{g}_n$ is a strongly consistent estimator of $g(\theta)$, Theorem 4.5.2 and Theorem 5.4.2 imply that an M.L.E. is an asymptotically efficient estimator, provided the regularity conditions hold.

Proof. Let δ be arbitrarly small, $0 < \delta < I(\theta_0)$. We start first with the case of $g(\theta) \equiv \theta$. According to the previous lemma,

(5.4.13)

$$\log P_{\theta_0}[|T_n - \theta_0| \geqslant \epsilon] \leqslant \log \left\{ P_{\theta_0}\left[\left| \frac{1}{n} \sum_{i=1}^{n} S(X_i; \theta_0) \right| \geqslant \epsilon[I(\theta_0) - \delta] \right] + \rho^n \right\}.$$

For any ρ, $0 < \rho < 1$, we can choose $\epsilon > 0$ sufficiently small so that

$$\text{(5.4.14)} \quad \lim_{n\to\infty} \left\{ \frac{1}{n} \log P_{\theta_0}\left[\left| \frac{1}{n} \sum_{i} S(X_i; \theta_0) \right| \geqslant \epsilon[I(\theta_0) - \delta] \right] \right\} > \log \rho.$$

Indeed, $(1/n)\sum_{i=1}^{n} S(X_i; \theta_0)$ converges a.s. to zero. Furthermore, by the central limit theorem, the effective variance of $(1/n)\sum_{i=1}^{n} S(X_i; \theta_0)$ is $(1/n)I(\theta_0)$. Hence according to (4.5.40),

$$\text{(5.4.15)} \quad \frac{1}{n} \ln P_{\theta_0}\left[\left| \frac{1}{n} \sum_{i=1}^{n} S(X_i; \theta_0) \right| \geqslant \epsilon(I(\theta_0) - \delta) \right] = -\frac{\epsilon^2[I(\theta_0) - \delta]^2}{2I(\theta_0)}(1 + o(1)), \qquad \text{as } n \to \infty.$$

The right-hand side of (5.4.15) can be made sufficiently close to zero by choosing ϵ sufficiently small. It follows that

$$\text{(5.4.16)} \quad \overline{\lim_{n\to\infty}} \left\{ \frac{1}{n} \ln P_{\theta_0}[|T_n - \theta_0| \geqslant \epsilon] \right\} \leqslant \overline{\lim_{n\to\infty}} \left\{ n^{-1} \ln P_{\theta_0}\left[\left| \frac{1}{n} \sum_{i=1}^{n} S(X_{ij}; \theta_0) \right| \geqslant \epsilon(I(\theta_0) - \delta) \right] \right\},$$

for all ϵ sufficiently small. Multiplying the terms in brackets by $1/\epsilon^2$ and letting $\epsilon \to 0$ we obtain from (5.4.15) and (5.4.16)

$$\overline{\lim_{\epsilon\to 0}}\ \overline{\lim_{n\to\infty}} \left\{\frac{1}{n\epsilon^2} \ln P_{\theta_0}[|T_n - \theta_0| \geqslant \epsilon]\right\} \geqslant \frac{(I(\theta_0) - \delta)^2}{2I(\theta_0)}. \tag{5.4.17}$$

Since δ is arbitrarily small, we obtain

$$\overline{\lim_{\epsilon\to 0}}\ \overline{\lim_{n\to\infty}} \left\{\frac{1}{n\epsilon^2} \ln P_\theta \{|T_n - \theta_0| \geqslant \epsilon\}\right\} \geqslant \tfrac{1}{2} I(\theta_0). \tag{5.4.18}$$

As in the proof of Theorem 4.5.2 we write

$$P_{\theta_0}[|T_n - \theta_0| \geqslant \epsilon] = 2\left[1 - \Phi\left(\frac{\epsilon}{\tau_\theta(T_n, \epsilon, \theta)}\right)\right],$$

and $\ln 2(1 - \Phi(X)) = -\frac{1}{2}X^2(1 + o(1))$, as $X \to \infty$, to obtain the required result. Finally, if we estimate $g(\theta)$ and the M.L.E. of θ is T_n, the invariance theorem implies that the M.L.E. of $g(\theta)$ is $U_n = g(T_n)$. Since $g(\theta)$ is differentiable at θ_0, it is continuous there. Therefore, for all $\epsilon > 0$ sufficiently small, if $|g(\theta) - g(\theta_0)| \geqslant \epsilon$ then $|\theta - \theta_0| > \delta$, where $\delta = \epsilon/(\lambda\, |g'(\theta_0)|)$, for a fixed $\lambda > 1$. Let $B_{n,\epsilon} = \{(X_1, \ldots, X_n); |U_n - g(\theta_0)| \geqslant \epsilon\}$; $B_{n,\epsilon} \subset \{|T_n - \theta_0| \geqslant \delta\}$. It follows that

$$\overline{\lim_{\epsilon\to 0}}\ \overline{\lim_{n\to\infty}} \left\{\frac{1}{n\epsilon^2} \ln P_{\theta_0}[B_{n,\epsilon}]\right\} \leqslant -\frac{1}{2} \cdot \frac{I(\theta_0)}{\lambda^2 (g'(\theta_0))^2}. \tag{5.4.19}$$

Since $\lambda > 1$ is arbitrary, we have

$$\overline{\lim_{\epsilon\to 0}}\ \overline{\lim_{n\to\infty}} \left\{\frac{1}{n\epsilon^2} \ln P_{\theta_0}[B_{n,\epsilon}]\right\} \leqslant -\frac{1}{2} \cdot \frac{I(\theta_0)}{(g'(\theta_0))^2}.$$

The end of the proof is as indicated above. (Q.E.D.)

Wolfowitz [4] presented in 1963 a very interesting paper concerning the asymptotic efficiency of the M.L.E. Wolfowitz considers a more general definition of asymptotic efficiency for classes of estimators which are uniformly consistent and whose (properly standardized) distributions converge to a limiting distribution $G(x; \theta)$ (not necessarily normal) uniformly in x and in θ. Wolfowitz's definition of asymptotic efficiency is based also on coverage probabilities. However, since he considers not only cases of asymptotically normal estimators, the coverage probability is not reflected by the effective standard deviation, as in Bahadur's theory. Thus Wolfowitz proves that the M.L.E. is asymptotically efficient not only among uniformly consistent asymptotically normal estimators but also in the wider class of all uniformly consistent estimators. These results of Wolfowitz were later generalized by Weiss and Wolfowitz [1] to nonregular cases.

Rao [3; 5; 6] investigated second-order properties of asymptotically efficient estimators. If $\{T_n\}$ is a sequence of estimators of θ, which are asymptotically efficient, there exist constants $\alpha(\theta)$, $\beta(\theta)$ such that

$$\frac{1}{\sqrt{n}}\sum_{i=1}^{n} S(X_{ij};\theta) - \alpha(\theta) - \sqrt{n}\,\beta(\theta)(T_n - \theta) \to 0 \tag{5.4.20}$$

in probability $[P_\theta]$. In order to investigate this convergence, Rao approximates $\sum_{i=1}^{n} S(X_i;\theta)$ by a polynomial of the second degree, namely: $\sqrt{n}\,\alpha(\theta) + n\beta(\theta)(T_n - \theta) + n\gamma(\theta)(T_n - \theta)^2$, where $\alpha(\theta)$ and $\beta(\theta)$ are as in (5.4.20). The coefficient $\gamma(\theta)$ is chosen so that $\{\sum_{i=1}^{n} S(X_i;\theta) - \sqrt{n}\,\alpha(\theta) - n\beta(\theta)(T_n - \theta) - n\gamma(\theta)(T_n - \theta)^2\}$ will have a minimum asymptotic variance. This minimal asymptotic variance is called the *index of the second-order efficiency of* T_n. An estimator is called second-order efficient if its index of second-order efficiency is minimal. In these terms, Rao [3] compared the second-order efficiency of various asymptotically efficient estimators of a parameter θ of a multinomial distribution $\{\pi_1(\theta), \ldots, \pi_k(\theta)\}$. The estimators compared by Rao are maximum likelihood, minimum-χ^2, minimum modified χ^2, minimum discrepancy, and minimum Kullback-Leibler separator. Rao's investigation established the second-order superiority of the M.L.E. over the other estimators mentioned here, in the above multinomial case. (See Problem 16.)

We conclude this section with some comments concerning the studies of Daniels [1] and Kallianpur [1]. Daniels discussed the consistency and asymptotic efficiency of M.L.E.'s with weaker regularity conditions. He motivated his study with the following example:

Suppose that $f(x;\theta)$ is the density function of a location parameter Laplace family, that is

$$f(x;\theta) = \tfrac{1}{2}\exp\{-|x - \theta|\}, \quad -\infty < x < \infty, \quad -\infty < \theta < \infty.$$

If we consider $(\partial/\partial\theta)\log f(x;\theta)$ we realize that this derivative does not exist for $\theta = x$. Moreover $(\partial^2/\partial\theta^2)\log f(x;\theta) = 0$ a.s. Thus this family of density functions does not satisfy the usual regularity conditions. Nevertheless, the M.L.E. of θ, the sample median, is an asymptotically normal and efficient estimator. Daniels's extended treatment of the problem also covers these cases.

Kallianpur [1] has shown that the M.L.E. can be, in regular cases, obtained as a special case of von Mises's functionals of the second order. These special kinds of functionals are defined and explained. Kallianpur has shown that a general treatment of the M.L.E. can be performed in terms of the von Mises functionals.

5.5. BEST ASYMPTOTICALLY NORMAL ESTIMATORS

The concept of best asymptotically normal (B.A.N.) estimator was introduced in Section 4.5. Here we prove that the asymptotic variance of any asymptotically normal estimator is at least $I^{-1}(\theta)$, where $I(\theta)$ is the Fisher information function. We then show that under the specified regularity conditions the M.L.E. is asymptotically normal, with asymptotic variance $I^{-1}(\theta)$. That is, the M.L.E. is B.A.N. At the end of the section we provide a method of constructing B.A.N. estimators.

Let $\mathcal{F} = \{P_\theta; \theta \in \Theta\}$ be a family of probability measures on $(\mathfrak{X}, \mathfrak{B})$. Let $f(x; \theta)$ be the generalized density of P_θ with respect to a sigma-finite measure μ on $(\mathfrak{X}, \mathfrak{B})$. We assume that P satisfies the Cramér-Rao regularity conditions, (i)*–(v)*, of Section 4.7. In the first theorem we treat the case of real parameters θ. The result will be extended then to the case of θ in a k-dimensional Euclidean space.

Theorem 5.5.1. (Walker [1].) *Under the regularity conditions,* (i)*–(v)*, *if* $\{T_n; n \geqslant 1\}$ *is a sequence of asymptotically normal estimators of* θ *with asymptotic variance* $\sigma^2(\theta)$, *that is,* $\sqrt{n}\,(T_n - \theta) \xrightarrow{d} N(0, \sigma^2(\theta))$ *as* $n \to \infty$; *and if, for each* $\theta \in \Theta$,

$$\text{(a)} \qquad E_\theta\, |\sqrt{n}\,(T_n - \theta)|^{2+\delta}\, \lambda_{2+\delta,n}(\theta) < \infty$$

for each n, *and the sequence* $\{\lambda_{2+\delta,n}(\theta)\}$ *is bounded for some positive* δ,

$$\text{(b)} \qquad b_n'(\theta) = \frac{\partial}{\partial\theta} E_\theta\{T_n - \theta\} \to 0, \qquad \text{as } n \to \infty,$$

then

$$\sigma^2(\theta) \geqslant I^{-1}(\theta).$$

Proof. Let $\psi_{n,\theta}(t) = E_\theta\,\{\exp\,\{it\sqrt{n}\,(T_n - \theta)\}\}$ be the characteristic function of $Y_n = \sqrt{n}\,(T_n - \theta)$. Assume that $0 \leqslant \delta \leqslant 1$. By the Taylor theorem, for any real y we have

$$(5.5.1) \qquad \begin{aligned} e^{ity} &= 1 + ity - t^2y^2 \int_0^1 e^{ityz}(1 - z)\, dz \\ &= 1 + ity - \tfrac{1}{2}t^2y^2 - t^2y^2 \int_0^1 (e^{ityz} - 1)(1 - z)\, dz \\ &= 1 + ity - \tfrac{1}{2}t^2y^2 + |ty|^2\, K_1(ty), \end{aligned}$$

where $|K_1(ty)| \leqslant 2^{1-\delta}/(1 + \delta)(2 + \delta) \leqslant 2$. Letting $b_n(\theta)$ denote the bias of T_n, that is, $b_n(\theta) = E_\theta(T_n - \theta)$, we obtain

$$(5.5.2) \quad \psi_{n,\theta}(t) = 1 + it\sqrt{n}\, b_n(\theta) - \tfrac{1}{2}t^2 n(v_n^{\,2}(\theta) + b_n^{\,2}(\theta)) + R_{n,\theta}(t)\, |t|^{2+\delta},$$

where $v_n(\theta) = \text{Var}\{T_n\}$ and $|R_{n,\theta}(t)| \leqslant 2\lambda_{2+\delta,n}(\theta)$. Furthermore, since $\sqrt{n}(T_n - \theta) \xrightarrow{d} N(0, \sigma^2(\theta))$ we have

$$(5.5.3) \quad \lim_{n\to\infty} \psi_{n,\theta}(t) = \exp\left\{-\frac{t^2}{2}\sigma^2(\theta)\right\} = 1 - \frac{t^2}{2}\sigma^2(\theta) + \tfrac{1}{8}t^4\sigma^4(\theta)K_2(t, \theta),$$

where $|K_2(t, \theta)| \leqslant 1$. Hence

$$(5.5.4) \quad \lim_{n\to\infty}\left\{it\sqrt{n}\, b_n(\theta) - \frac{t^2}{2}[nv_n^2(\theta) + nb_n^2(\theta) - \sigma^2(\theta)] + |t|^{2+\delta} R_{n,\theta}(t) - \tfrac{1}{8}t^4\sigma^4(\theta)K_2(t, \theta)\right\} = 0.$$

If $|t| < t_0$ the last two terms in the limit are bounded; that is, if

$$G_{n,\theta}(t) = |t|^{2+\delta} R_{n,\theta}(t) - \tfrac{1}{8}t^4\sigma^4(\theta)K_2(t, \theta)$$

then $|G_{n,\theta}(t)| \leqslant K(\theta)\,|t|^{2+\delta}$ for all t, $|t| < t_0$. Hence $\lim_{n\to\infty}\sqrt{n}\, b_n(\theta) = 0$ *and*

$$\lim_{n\to\infty} n(v_n^2(\theta) + b_n^2(\theta)) = \sigma^2(\theta);$$

otherwise (5.5.4) will be violated. Finally, since by the Cramér-Rao inequality

$$nv_n^2(\theta) \geqslant \frac{(1 + b_n'(\theta))^2}{I(\theta)}, \qquad \text{all } n = 1, 2, \ldots,$$

we obtain from (b),

$$\lim_{n\to\infty}\{nv_n^2(\theta)\} = \sigma^2(\theta) \geqslant \frac{1}{I(\theta)}. \qquad \text{(Q.E.D.)}$$

It should be remarked in this connection that the super-efficient estimator presented in Example 4.6 is asymptotically normal. However, it does not satisfy (b). We can show in that example that

$$\lim_{n\to\infty} b_n'(\theta) = \begin{cases} 0, & \text{if } \theta \neq 0, \\ -\frac{1}{2}, & \text{if } \theta = 0. \end{cases}$$

Since the distribution of any linear combination of normal random variables is normally distributed we immediately obtain, on the basis of Theorem 5.5.1, the following k-dimensional extension.

Theorem 5.5.2. (Walker [1].) *If the density functions* $f(x; \theta)$, $\theta \in \Theta \subset E^{(k)}$, *satisfy the Cramér-Rao regularity conditions of Section* 4.3; *if* T_n *is a sequence of asymptotically normal estimators, with an asymptotic covariance matrix* $n^{-1}\boldsymbol{\Sigma}(\theta)$; *and if, for each* $\theta \in \Theta$, (a) *the sequences* $\{E_\theta\,|\sqrt{n}(T_{i,n} - \theta_i)|^{2+\delta}\}$, $i = 1, \ldots, k$, *are all bounded, and* (b) $(\partial/\partial\theta_j)E_\theta\{T_{i,n} - \theta_i\} \to 0$ *for all* i,

$j = 1, \ldots, k$, *then* $\boldsymbol{\Sigma}(\theta) - \mathbf{I}^{-1}(\theta)$ *is non-negative definite, for each* $\theta \in \Theta$. (The matrix $\mathbf{I}(\theta)$ is the Fisher information matrix.)

We turn now to the asymptotic distribution of the M.L.E. Our treatment will be directly on the k-dimensional case, $\boldsymbol{\theta} = (\theta_1, \ldots, \theta_k)'$.

If we denote by $\mathbf{S}(X; \theta)$ the vector of partial derivatives $(\partial/\partial\theta_i) \log f(X; \boldsymbol{\theta})$, $i = 1, \ldots, k$, then an M.L.E. $\hat{\theta}_n$ is a solution of the system of equations $\sum_{i=1}^n \mathbf{S}(X_i; \boldsymbol{\theta}) = 0$. Let $\mathbf{g}_n(\boldsymbol{\theta})$ designate the log-likelihood gradient $\sum_{i=1}^n \mathbf{S}(X_i; \boldsymbol{\theta})$. Expanding $\mathbf{g}_n(\boldsymbol{\theta})$ in a neighborhood of $\mathbf{g}_n(\boldsymbol{\theta}_0)$, we obtain

$$(5.5.4) \qquad \mathbf{g}_n(\boldsymbol{\theta}) = \mathbf{g}_n(\boldsymbol{\theta}_0) + \left\| \frac{\partial}{\partial\theta_j} \mathbf{g}_n^{(i)}(\boldsymbol{\theta}'); i, j = 1, \ldots, k \right\| (\boldsymbol{\theta} - \boldsymbol{\theta}_0),$$

where $\boldsymbol{\theta}'$ is a point on the line segment connecting $\boldsymbol{\theta}$ and $\boldsymbol{\theta}_0$, $\mathbf{g}_n^{(i)}(\boldsymbol{\theta}')$ is the i-th component of $\mathbf{g}_n(\boldsymbol{\theta})$. $(\partial/\partial\theta_j)\mathbf{g}_n^{(i)}(\boldsymbol{\theta}) = (\partial^2/\partial\theta_j\,\partial\theta_i) \sum_{v=1}^n \log f(X_v; \boldsymbol{\theta})$. Define the k-dimensional vectors

$$(5.5.5) \qquad \mathbf{a}_n(\boldsymbol{\theta}) = \frac{1}{\sqrt{n}} \sum_{v=1}^{n} \mathbf{S}(X_v; \boldsymbol{\theta}), \qquad n = 1, 2, \ldots,$$

and the $k \times k$ matrices

(5.5.6)

$$\mathbf{B}_n(\boldsymbol{\theta}) = \frac{1}{n} \left\| \frac{\partial^2}{\partial\theta_j\,\partial\theta_i} \sum_{v=1}^{n} \log f(X_v; \boldsymbol{\theta}); i, j = 1, \ldots, k \right\|, \qquad n = 1, 2, \ldots.$$

According to (5.5.4), if $\hat{\boldsymbol{\theta}}_n$ is an M.L.E. of $\boldsymbol{\theta}$, then

$$(5.5.7) \qquad \mathbf{g}_n(\hat{\boldsymbol{\theta}}_n) = \mathbf{g}_n(\boldsymbol{\theta}) + n\mathbf{B}_n(\tilde{\boldsymbol{\theta}}_n)(\tilde{\boldsymbol{\theta}}_n - \boldsymbol{\theta}) = 0,$$

all $n = 1, 2, \ldots$, where $\tilde{\boldsymbol{\theta}}_n$ is a point on the line segment between $\tilde{\boldsymbol{\theta}}_n$ and $\boldsymbol{\theta}$. Thus we can write

$$(5.5.8) \qquad -\sqrt{n}\, \mathbf{a}_n(\boldsymbol{\theta}) = n\mathbf{B}_n(\tilde{\boldsymbol{\theta}}_n)(\tilde{\boldsymbol{\theta}}_n - \boldsymbol{\theta}).$$

Among the regularity assumptions we specify that the Fisher information matrix $\mathbf{I}(\boldsymbol{\theta})$ is positive definite for all $\boldsymbol{\theta}$. We show later that this implies the nonsingularity of $\mathbf{B}_n(\boldsymbol{\theta})$, with probability one, for sufficiently large n. Thus for sufficiently large n we can write

$$(5.5.9) \qquad \sqrt{n}(\tilde{\boldsymbol{\theta}}_n - \boldsymbol{\theta}) = -\mathbf{B}_n^{-1}(\tilde{\boldsymbol{\theta}}_n)\mathbf{a}_n(\boldsymbol{\theta}) \quad \text{a.s.}$$

From the regularity conditions we know that

$$E_\theta\left\{\frac{\partial^2}{\partial\theta_i\,\partial\theta_j} \log f(X; \boldsymbol{\theta})\right\} = -E_\theta\left\{\frac{\partial}{\partial\theta_i} \log f(X; \boldsymbol{\theta}) \cdot \frac{\partial}{\partial\theta_j} \log f(X; \boldsymbol{\theta})\right\}.$$

Hence for each $n = 1, 2, \ldots,$

$$E_\theta\{\mathbf{B}_n(\boldsymbol\theta)\} = -\mathbf{I}(\boldsymbol\theta). \tag{5.5.10}$$

Furthermore, the regularity conditions assure that the sequence of vectors $\mathbf{a}_n(\boldsymbol\theta)$ obeys the central limit theorem. Hence $\mathbf{a}_n(\boldsymbol\theta) \xrightarrow{d} N(\mathbf{0}, \mathbf{I}(\boldsymbol\theta))$. Moreover, assuming that $\hat{\boldsymbol\theta}_n$ is strongly consistent, since $\tilde{\boldsymbol\theta}_n = \nu\boldsymbol\theta_n + (1 - \nu)\boldsymbol\theta$ for some $0 \leqslant \nu \leqslant 1$, $\tilde{\boldsymbol\theta} \to 0$ a.s. By the strong law of large numbers

$$\mathbf{B}_n(\tilde{\boldsymbol\theta}_n) = -\mathbf{I}(\boldsymbol\theta) + o_p(1), \quad \text{as } n \to \infty. \tag{5.5.11}$$

Thus since $\mathbf{I}(\boldsymbol\theta)$ is positive definite, $\tilde{\mathbf{B}}_n^{-1}(\tilde{\boldsymbol\theta}_n)$ exists a.s. for n sufficiently large. Finally, since $\mathbf{a}_n(\boldsymbol\theta) = 0_p(1)$ as $n \to \infty$, we obtain

$$\sqrt{n}(\boldsymbol\theta_n - \boldsymbol\theta) = (\mathbf{I}(\boldsymbol\theta))^{-1}\mathbf{a}_n(\boldsymbol\theta) + o_p(1), \quad \text{as } n \to \infty. \tag{5.5.12}$$

The asymptotic distribution of $(\mathbf{I}(\boldsymbol\theta))^{-1}\mathbf{a}_n(\boldsymbol\theta)$ is $\mathcal{N}(\mathbf{0}, \mathbf{I}^{-1}(\boldsymbol\theta))$. Hence

$$\sqrt{n}(\hat{\boldsymbol\theta}_n - \boldsymbol\theta) \xrightarrow{d} N(\mathbf{0}, \mathbf{I}^{-1}(\boldsymbol\theta)) \quad \text{as } n \to \infty. \tag{5.5.13}$$

In the following theorem we summarize the required conditions for an M.L.E. to be a B.A.N. estimator.

Theorem 5.5.3. *If the following conditions hold;*

(i) $\boldsymbol\theta_n - \boldsymbol\theta = o_p(1)$ *as* $n \to \infty$;

(ii) $(\partial/\partial\theta_i) \log f(X; \boldsymbol\theta)$ *exists and is finite with probability* 1 *for all* $i = 1, \ldots, k$ *and all* $\boldsymbol\theta$ *in* Θ;

(iii)

$$\sup_{|\boldsymbol\theta - \boldsymbol\theta_0| \leqslant r} |(\partial^2/\partial\theta_i\,\partial\theta_j) \log f(X; \boldsymbol\theta) - (\partial^2/\partial\theta_i\,\partial\theta_j) \log f(X; \boldsymbol\theta_0)| \leqslant H(X; \boldsymbol\theta_0)o_p(1), \text{ as } r \to 0,$$

where $H(X; \boldsymbol\theta_0) \geqslant 0$ *and integrable* $P_{\boldsymbol\theta_0}$;

(iv) $\mathbf{I}(\boldsymbol\theta_0)$ *is positive definite;*

then an M.L.E. is B.A.N. (satisfying 5.5.13).

Example 5.6. We derive the asymptotic distribution of the M.L.E. of the mean of a log-normal distribution. Let $X_1, X_2, \ldots$ be a sequence of i.i.d. random variables distributed like $\exp\{N(\mu, \sigma^2)\}$, where $-\infty < \mu < \infty$, $0 < \sigma < \infty$. We have verified in Example 5.2 that the M.L.E. of $E\{X\} = \xi = \exp\{\mu + \sigma^2/2\}$ is

$$\hat\xi_n = \exp\left\{\bar{Y}_n + \frac{Q_n}{2n}\right\},$$

where $Y_i = \ln X_i$ $(i = 1, \ldots, n)$; $\bar{Y}_n = (1/n)\Sigma Y_i$, and $Q_n = \Sigma(Y_i - \bar{Y}_n)^2$.
Since Y_i $(i = 1, 2, \ldots)$ are i.i.d. distributed like $N(\mu, \sigma^2)$, $(\bar{Y}_n, Q_n/n)$ is

the M.L.E. of (μ, σ^2). It is easy to verify that the family of normal distributions satisfies the conditions of Theorem 5.5.3, and hence

$$\sqrt{n}\begin{bmatrix} \bar{Y}_n - \mu \\ Q_n/n - \sigma^2 \end{bmatrix} \xrightarrow{d} N\left(\begin{bmatrix} 0 \\ 0 \end{bmatrix}, \begin{bmatrix} \sigma^2 & 0 \\ 0 & 2\sigma^4 \end{bmatrix}\right).$$

We write, since $\bar{Y}_n$ and Q_n are independent,

$$\bar{Y}_n \sim \mu + \frac{\sigma}{\sqrt{n}} N_1(0, 1), \qquad \text{for all } n = 1, 2, \ldots,$$

$$\frac{Q_n}{n} \sim \sigma^2 + \frac{\sigma^2\sqrt{2}}{\sqrt{n}} N_2(0, 1) + o_p\left(\frac{1}{\sqrt{n}}\right), \qquad \text{as } n \to \infty,$$

where $N_1(0, 1)$ and $N_2(0, 1)$ are independent standard normal variables. Thus the M.L.E. $\hat{\xi}_n$ is distributed like

$$\exp\left\{\mu + \frac{\sigma^2}{2} + \left(\frac{\sigma}{\sqrt{n}} N_1(0, 1) + \frac{\sigma^2}{\sqrt{(2n)}} N_2(0, 1)\right) + o_p\left(\frac{1}{\sqrt{n}}\right)\right\}.$$

Furthermore,

$$\hat{\xi}_n - \xi \sim \xi\left[\exp\left\{\frac{\sigma}{\sqrt{n}} N_1(0, 1) + \frac{\sigma^2}{\sqrt{(2n)}} N_2(0, 1) + o_p\left(\frac{1}{\sqrt{n}}\right)\right\} - 1\right]$$

$$\sim \xi\left[\frac{\sigma}{\sqrt{n}} N\left(0, 1 + \frac{\sigma^2}{2}\right) + o_p\left(\frac{1}{\sqrt{n}}\right) + O_p\left(\frac{1}{n}\right)\right], \qquad \text{as } n \to \infty.$$

But $o_p(1/\sqrt{n}) + O_p(1/n) = o_p(1/\sqrt{n})$. Hence

$$\sqrt{n}\,(\hat{\xi}_n - \xi) \sim \xi\sigma N\left(0, 1 + \frac{\sigma^2}{2}\right) + o_p(1) \xrightarrow{d} N\left(0, \xi^2\sigma^2\left(1 + \frac{\sigma^2}{2}\right)\right).$$

As was shown in Example 4.4, $(1/n)\xi^2\sigma^2(1 + \sigma^2/2)$ is the Cramér-Rao lower bound for the variance of unbiased estimators of ξ; $\hat{\xi}_n$ is thus a B.A.N. estimator. This result can be obtained from the following more general consideration. ■

Suppose that $\hat{\boldsymbol{\theta}}^{(n)} = (\hat{\theta}_1^{(n)}, \ldots, \hat{\theta}_k^{(n)})$ is an M.L.E. of $\boldsymbol{\theta} = (\theta_1, \ldots, \theta_k)$ and that

$$\sqrt{n}(\hat{\boldsymbol{\theta}}^{(n)} - \boldsymbol{\theta}) \xrightarrow{d} N(\mathbf{0}, \mathbf{I}^{-1}(\boldsymbol{\theta})),$$

where $\mathbf{I}(\boldsymbol{\theta})$ is the Fisher information function.

Let $\mathbf{g}(\boldsymbol{\theta}) = (g_1(\boldsymbol{\theta}), \ldots, g_r(\boldsymbol{\theta}))'$ and assume that the partial derivatives

$(\partial/\partial\theta_j)g_i(\boldsymbol{\theta})$ $(i = 1, \ldots, r; j = 1, \ldots, k)$ exist. Let $\mathbf{D}_g(\boldsymbol{\theta}) = \|(\partial/\partial\theta_j)g_i(\boldsymbol{\theta}); i = 1, \ldots, r; j = 1, \ldots, k\|$ be the matrix of partial derivatives, of dimensionality $r \times k$. An M.L.E. of $\mathbf{g}(\boldsymbol{\theta}) = (g_1(\boldsymbol{\theta}), \ldots, g_r(\boldsymbol{\theta}))'$ is $\mathbf{g}(\boldsymbol{\theta}^{(n)})$. We now derive the asymptotic distribution of $\sqrt{n}(\mathbf{g}(\hat{\boldsymbol{\theta}}^{(n)}) - \mathbf{g}(\boldsymbol{\theta}))$. We assume that $\hat{\boldsymbol{\theta}}^{(n)}$ is a B.A.N. estimator. Expanding the vector $\mathbf{g}(\hat{\boldsymbol{\theta}}^{(n)})$ around $\mathbf{g}(\boldsymbol{\theta})$ we obtain

$$\mathbf{g}(\hat{\boldsymbol{\theta}}^{(n)}) = \mathbf{g}(\boldsymbol{\theta}) + \mathbf{D}_g(\tilde{\boldsymbol{\theta}}^{(n)})(\hat{\boldsymbol{\theta}}^{(n)} - \boldsymbol{\theta}), \tag{5.5.14}$$

where $\tilde{\boldsymbol{\theta}}^{(n)}$ is a point on the line segment connecting $\hat{\boldsymbol{\theta}}^{(n)}$ and $\boldsymbol{\theta}$. Since $\hat{\boldsymbol{\theta}}^{(n)} - \boldsymbol{\theta} = o_p(1)$ and $\sqrt{n}(\boldsymbol{\theta}^{(n)} - \boldsymbol{\theta}) = O_p(1)$, as $n \to \infty$, we obtain

$$\sqrt{n}(\mathbf{g}(\hat{\boldsymbol{\theta}}^{(n)}) - \mathbf{g}(\boldsymbol{\theta})) = \sqrt{n}\, \mathbf{D}_g(\boldsymbol{\theta})(\hat{\boldsymbol{\theta}}^{(n)} - \boldsymbol{\theta}) + o_p(1) \quad \text{as } n \to \infty. \tag{5.5.15}$$

Hence

$$\sqrt{n}(\mathbf{g}(\hat{\boldsymbol{\theta}}^{(n)}) - \mathbf{g}(\boldsymbol{\theta})) \xrightarrow{d} N(\mathbf{O}, \mathbf{D}_g(\boldsymbol{\theta})\mathbf{I}^{-1}(\boldsymbol{\theta})\mathbf{D}_g'(\boldsymbol{\theta})). \tag{5.5.16}$$

Comparing the asymptotic covariance matrix on the right-hand side of (5.5.16) to the Cramér-Rao lower bound given by Theorem 4.3.1, we immediately conclude that under the above differentiability condition on $\mathbf{g}(\boldsymbol{\theta})$, if $\hat{\boldsymbol{\theta}}^{(n)}$ is a B.A.N. estimator of $\boldsymbol{\theta}$, $\mathbf{g}(\hat{\boldsymbol{\theta}}^{(n)})$ is a B.A.N. estimator of $\mathbf{g}(\boldsymbol{\theta})$.

The following example is a continuation of Example 5.4 and uses the above results.

Example 5.7. As in Example 5.4, let $X_1, X_2, \ldots$ be a sequence of i.i.d. random variables having a log-normal distribution, that is, $X \sim \exp\{N(\mu, \sigma^2)\}$. In Example 5.4 we considered the M.L.E. estimator of $\xi = E\{X\} = \exp\{\mu + \sigma^2/2\}$. In this example we derive the asymptotic covariance matrix of the M.L.E. of EX and of $\text{Var}\{X\} = \tau^2 = \xi^2(e^{\sigma^2} - 1)$. Thus we have

$$g_1(\mu, \sigma^2) = \exp\left\{\mu + \frac{\sigma^2}{2}\right\},$$

$$g_2(\mu, \sigma^2) = \exp\{2\mu + \sigma^2\}(e^{\sigma^2} - 1).$$

The partial derivatives are

$$\frac{\partial}{\partial\mu} g_1(\mu, \sigma^2) = \exp\left\{\mu + \frac{\sigma^2}{2}\right\}; \qquad \frac{\partial}{\partial\sigma^2} g_1(\mu, \sigma^2) = \tfrac{1}{2}\exp\left\{\mu + \frac{\sigma^2}{2}\right\};$$

$$\frac{\partial}{\partial\mu} g_2(\mu, \sigma^2) = 2\exp\{2\mu + \sigma^2\}(e^{\sigma^2} - 1);$$

$$\frac{\partial}{\partial\sigma^2} g_2(\mu, \sigma^2) = \exp\{2\mu + \sigma^2\}(2e^{\sigma^2} - 1).$$

The asymptotic covariance matrix of $\sqrt{n}(g(\hat{\mu}_n, \hat{\sigma}_n^2))$ is

$$\text{(5.5.17)} \quad \mathbf{D}_g \mathbf{I}' \mathbf{D}_g' = \xi^2 \begin{bmatrix} 1 & \frac{1}{2} \\ \frac{2\tau^2}{\xi} & \frac{2\tau^2}{\xi} + \xi \end{bmatrix} \begin{bmatrix} \sigma^2 & 0 \\ 0 & 2\sigma^4 \end{bmatrix} \begin{bmatrix} 1 & 2\frac{\tau^2}{\xi} \\ \frac{1}{2} & \frac{2\tau^2}{\xi} + \xi \end{bmatrix}$$

$$= \xi^2 \begin{bmatrix} \sigma^2\left(1 + \frac{\sigma^2}{2}\right) & \frac{2\sigma^2\tau^2}{\xi}(1 + \sigma^2) + \sigma^4\xi \\ & 8\sigma^2\tau^2\left(\frac{\tau^2}{\xi^2} + \sigma^2\right) + 2\sigma^4\xi^2 \end{bmatrix},$$

as derived in Example 5.4. The M.L.E. of τ^2 is

$$\hat{\tau}_n^{\,2} = \exp\left\{2\,\bar{Y}_n + \frac{Q_n}{n}\right\}\left[\exp\left\{\frac{Q_n}{n}\right\} - 1\right],$$

where $\bar{Y}_n$ and Q_n are defined in Example 5.6. From (5.5.17) we find that the asymptotic variance of $\sqrt{n}\,\hat{\tau}_n^{\,2}$ is $2\sigma^2\xi^4[4(e^{\sigma^2} - 1)^2 + \sigma^2(4e^{\sigma^2} - 3)]$. We also find that the asymptotic covariance between $\sqrt{n}\,\hat{\xi}_n$ and $\sqrt{n}\,\hat{\tau}_n^{\,2}$ is $\sigma^2\xi^3[2(e^{\sigma^2} - 1)(1 + \sigma^2) + 1]$. ∎

In Section 5.2 we discussed the Newton-Raphson successive approximations to the roots of the likelihood equation. We prove now that a one-cycle approximation like that given by (5.2.5) yields a B.A.N. estimator, provided $\hat{\theta}_1$ is a consistent estimator of θ.

Theorem 5.5.4. *Under the conditions of Theorem 5.5.3, if* $\tilde{\boldsymbol{\theta}}_n^{(1)}$ *is a consistent estimator of* $\boldsymbol{\theta}$ *such that* $\tilde{\boldsymbol{\theta}}_n^{(1)} - \boldsymbol{\theta} = o_p(n^{-1/4})$ *as* $n \to \infty$, *then for sufficiently large* n,

$$\text{(5.5.18)} \qquad \tilde{\boldsymbol{\theta}}_n^{(2)} = \tilde{\boldsymbol{\theta}}_n^{(1)} - \frac{1}{\sqrt{n}}\,\mathbf{B}_n^{-1}(\tilde{\boldsymbol{\theta}}_n^{(1)})\mathbf{a}_n(\tilde{\boldsymbol{\theta}}_n^{(1)})$$

is a B.A.N. estimator.

Remark. The condition "for sufficiently large n" is required since $\mathbf{B}_n^{-1}(\tilde{\boldsymbol{\theta}}_n^{(1)})$ may not exist for values of n which are not large enough. However, if $\mathbf{B}_n^{-1}(\tilde{\boldsymbol{\theta}}_n^{(\nu)})$ does not exist, we can consider a definition of $\tilde{\boldsymbol{\theta}}_n^{(2)}$ in which a generalized inverse of $\mathbf{B}_n(\tilde{\boldsymbol{\theta}}_n^{(1)})$ is used. From the asymptotic distribution point of view it does not matter which particular generalized inverse is used. As we have previously shown, since $\tilde{\boldsymbol{\theta}}_n^{(1)} - \boldsymbol{\theta} = o_p(1)$, $\mathbf{B}_n^{-1}(\tilde{\boldsymbol{\theta}}_n^{(1)})$ exists with probability 1 as $n \to \infty$.

Proof. As $n \to \infty$ we can write, when $\tilde{\boldsymbol\theta}_n^{(1)} - \boldsymbol\theta_0 = o_p(1)$,

(5.5.19)

$$\begin{aligned}\sqrt{n}\,(\tilde{\boldsymbol\theta}_n^{(2)} - \boldsymbol\theta_0) &= \sqrt{n}\,(\tilde{\boldsymbol\theta}_n^{(1)} - \boldsymbol\theta_0) - [\mathbf{B}_n^{-1}(\boldsymbol\theta_0) + O_p(\tilde{\boldsymbol\theta}_n^{(1)} - \boldsymbol\theta_0)] \\ &\quad \times [\mathbf{a}_n(\boldsymbol\theta_0) + \sqrt{n}\,\mathbf{B}_n(\boldsymbol\theta_0)(\tilde{\boldsymbol\theta}_n^{(1)} - \boldsymbol\theta_0) + \sqrt{n}\,O_p(|\tilde{\boldsymbol\theta}_n^{(1)} - \boldsymbol\theta_0|^2)] \\ &= \mathbf{B}_n^{-1}(\boldsymbol\theta_0)\mathbf{a}_n(\boldsymbol\theta_0) + o_p(1) + O_p(\sqrt{n}\,|\tilde{\boldsymbol\theta}_n^{(1)} - \boldsymbol\theta_0|^2).\end{aligned}$$

As was shown in the proof of Theorem 5.5.1, $-\mathbf{B}_n^{-1}(\boldsymbol\theta_0)\mathbf{a}_n(\boldsymbol\theta_0) \xrightarrow{d} N(\mathbf{0}, \mathbf{I}^{-1}(\boldsymbol\theta_0))$. Finally, since $\tilde{\boldsymbol\theta}_n^{(1)} - \boldsymbol\theta_0 = o_p(n^{-1/4})$, $O_p(\sqrt{n}\,|\tilde{\boldsymbol\theta}_n^{(1)} - \boldsymbol\theta_0|^2) = o_p(1)$. Hence $\sqrt{n}\,(\tilde{\boldsymbol\theta}_n^{(2)} - \boldsymbol\theta_0) \xrightarrow{d} N(\mathbf{0}, \mathbf{I}^{-1}(\boldsymbol\theta_0))$. (Q.E.D.)

Example 5.8. We derive a B.A.N. estimator for the location parameter of a Cauchy distribution by applying Theorem 5.5.2. Thus let $X_1, X_2, \ldots$ be a sequence of i.i.d. random variables having a Cauchy distribution, with a density function

$$f(x;\theta) = \frac{1}{\pi}\cdot[1 + (x-\theta)^2]^{-1}, \qquad -\infty < x < \infty,$$

where the location parameter θ is any point on the real line, that is, $-\infty < \theta < \infty$. In Section 5.2 we have indicated the difficulty in deriving an M.L.E. of θ, which is due to the difficulty in solving the likelihood equation (5.2.3). As we show presently, a one-cycle Newton-Raphson iteration, according to (5.2.5), with the sample median as an initial estimator, leads to a B.A.N. estimator of θ.

Let m_n designate the median of a sample consisting of the first n observations. Letting $X_{(j),n}$, $j = 1, \ldots, n$, designate the j-th order statistic in a sample of n; the sample median is the statistic $m_n \equiv X_{([n/2+1]),n}$, where $[a]$ is the maximal integer not exceeding a. From Theorem 10.5.1 of Fisz [1], p. 380, we deduce that m_n is a C.A.N. estimator, that is,

$$\sqrt{n}\,(m_n - \theta) \xrightarrow{d} N\left(0, \frac{\pi^2}{4}\right) \quad \text{as} \quad n \to \infty. \tag{5.5.20}$$

Indeed, θ is the 0.5-fractile of the Cauchy distribution with density $f(x;\theta)$.

The Fisher information function (per one observation) is

$$I(\theta) = E_\theta\left\{\left(\frac{\partial}{\partial\theta}\log f(X;\theta)\right)^2\right\} = \frac{4}{\pi}\int_{-\infty}^{\infty}\frac{(x-\theta)^2}{[1+(x-\theta)^2]^3}\,dx. \tag{5.5.21}$$

Since (see Dwight [1], p. 31)

$$\int_{-\infty}^{\infty}\frac{y^2}{(1+y^2)^3}\,dy = \frac{\pi}{8}, \tag{5.5.22}$$

we obtain from (5.5.21) and (5.5.22) that $I(\theta) = \frac{1}{2}$ for all θ, $-\infty < \theta < \infty$.

Hence the asymptotic variance of $\sqrt{n}\, m_n$ is greater than $I^{-1}(\theta)$ for all θ, and m_n is *not* a B.A.N. estimator. From (5.5.20) we see that $m_n - \theta = O_p(n^{-1/2})$. Hence $m_n - \theta = o_p(n^{-1/4})$ and the conditions of Theorem 5.5.4 are satisfied. Substituting $\tilde{\theta}_n^{(1)} \equiv m_n$ in (5.5.18) we obtain that

$$\tilde{\theta}_n = m_n - \frac{\sum_{i=1}^{n} (X_i - m_n)/[1 + (X_i - m_n)^2]}{\sum_{i=1}^{n} [(X_i - m_n)^2 - 1]/[1 + (X_i - m_n)^2]^2} \tag{5.5.23}$$

is a B.A.N. estimator of θ. ■

Further results concerning the conditions which guarantee the existence of a B.A.N. estimator and the method of constructing such are given by Barankin and Gurland [1], Chiang [1], and Wijsman [1].

5.6. THE ASYMPTOTIC RISK OF MAXIMUM LIKELIHOOD ESTIMATORS

In this section we consider the efficiency problem of the M.L.E. from a risk function point of view. Suppose that, as before, θ is a parameter (real or vector valued) indexing a family $\mathcal{F}$ of probability measures P_θ. Let T_n be an estimator of θ based on a sample of size n. We designate by $L_n(\theta, t)$ the loss function associated with T_n and assume that $L_n(\theta, t)$ is bounded and that it can be expanded in a neighborhood of $t = \theta$ as

$$L_n(\theta, t) = c_{0n}(\theta) + c_{1n}(\theta)(t - \theta)^2 + o(t - \theta)^2, \tag{5.6.1}$$

where $\liminf_{n\to\infty} c_{1n}(\theta) > 0$, and $o(t - \theta)^2$ holds uniformly in n as $t \to 0$. We define by

$$L_n^*(\theta, t) = n\,\frac{L_n(\theta, t) - c_{0n}(\theta)}{c_{1n}(\theta)} \tag{5.6.2}$$

the normalized loss function, having the following properties:

(i) $L_n^*(\theta, t) = n(t - \theta)^2(1 + o(1))$ uniformly in n, as $t \to \theta$;
(ii) $L_n^*(\theta, t) = nM$ for all $|t - \theta| > \delta$, where M, $0 < M < \infty$, is independent of n and of t.

In this section we investigate the conditions under which, if T_n is an M.L.E., then $\lim_{n\to\infty} E_\theta\{L_n^*(\theta, T_n)\} = I^{-1}(\theta)$, where $I(\theta)$ is the Fisher information function. In our presentation we follow Chernoff [5] and Commins [1].

We notice first that if T_n is any consistent estimator of θ and $\sigma^2(\theta)$ is its asymptotic variance, that is, $\sigma^2(\theta) = \lim_{k\to\infty} \lim_{n\to\infty} E_\theta\{\min\{n(T_n - \theta)^2, k^2\}\}$, then

$$\sigma^2(\theta) \leqslant \varliminf_{n\to\infty} E_\theta\{L_n^*(\theta, T_n)\}, \qquad \text{all } \theta. \tag{5.6.3}$$

Indeed, fix $k > 0$ and ϵ, $0 < \epsilon < 1$; then for all n sufficiently large

$$L_n^*(\theta, T_n) \geqslant (1 - \epsilon) \min [n(T_n - \theta)^2, k^2]. \tag{5.6.4}$$

Hence

$$\lim_{k\to\infty} \liminf_{n\to\infty} \frac{E_\theta\{L_n^*(\theta, T_n)\}}{E_\theta\{\min [n(T_n - \theta)^2, k^2]\}} \geqslant 1 - \epsilon, \qquad \text{all } \theta. \tag{5.6.5}$$

The denominator of (5.6.5) is the asymptotic variance of T_n. Furthermore, since $L_n^*(\theta, T_n)/(1 + \epsilon) < n(T_n - \theta)^2$ when $|T_n - \theta| < \delta$, we obtain

(5.6.6)

$$\frac{1}{1+\epsilon} E_\theta\{L_n^*(\theta, T_n)\} \leqslant n \int_{|T_n-\theta|<\delta} (t - \theta)^2 \, dP_n^\theta(t) + \frac{nM}{1+\epsilon} P_n^\theta[|T_n - \theta| \geqslant \delta].$$

Hence if $T_n - \theta = o_p(n^{-1})$, then $\lim_{n\to\infty} \{nP_n^\theta[|T_n - \theta| \geqslant \delta]\} = 0$. Hence whenever $T_n - \theta = o_p(n^{-1})$,

$$\frac{1}{1+\epsilon} \liminf_{n\to\infty} E_\theta\{L_n^*(\theta, T_n)\} \leqslant \liminf_{n\to\infty} \{nE_\theta\{(T_n - \theta)^2\}\}. \tag{5.6.7}$$

Thus the normalized risk $E_\theta\{L_n^*(\theta, T_n)\}$ is asymptotically sandwiched between the asymptotic variance $\sigma^2(\theta)$ and the expected (normalized) mean square error.

In the following example we illustrate a case in which the normalized risk approaches infinity, while the asymptotic variance is one.

Example 5.9. This example is from Commins [1]. Let $X_1, X_2, \ldots$ be a sequence of i.i.d. random variables distributed like $N(\theta, 1)$, $-\infty < \theta < \infty$. Let $\bar{X}_n = (1/n) \sum X_i$, and consider the estimator T_{2n}, which is defined without loss of generality for an even sample size $2n$ as

$$T_{2n} = \begin{cases} \bar{X}_{2n}, & \text{if } |\bar{X}_{2n} - \bar{X}_n| < \dfrac{\sqrt{(\log n)}}{\sqrt{(2n)}}, \\ \bar{X}_n + 1, & \text{otherwise.} \end{cases} \tag{5.6.8}$$

We notice that $\bar{X}_{2n} - \bar{X}_n \sim N(0, 1/2n)$. Hence

$$P_\theta\left[|\bar{X}_{2n} - \bar{X}_n| \geqslant \left(\frac{\log n}{2n}\right)^{1/2}\right] = 2[1 - \Phi(\sqrt{(\log n)})] \approx \frac{2}{\sqrt{(2\pi n \log n)}} \tag{5.6.9}$$

as $n \to \infty$. Thus the asymptotic distribution of $\sqrt{(2n)}(T_{2n} - \theta)$ is like that of $\sqrt{(2n)}(\bar{X}_{2n} - \theta)$. That is, T_{2n} is a B.A.N. estimator of θ with an asymptotic variance $\sigma^2(\theta) \equiv 1$. Consider the normalized loss function

$$L_n^*(\theta, T_n) = \begin{cases} 2n(T_{2n} - \theta)^2, & \text{if } (T_n - \theta)^2 \leqslant 3, \\ 6n, & \text{otherwise.} \end{cases} \tag{5.6.10}$$

Let $A_n = \{|\bar{X}_{2n} - \bar{X}_n| \geqslant ((\log n)/2n)^{1/2}\}$, and $B_n = \{(\bar{X}_n - \theta + 1)^2 \leqslant 3\}$; then

(5.6.11)

$$E_\theta\{L_n^*(\theta, T_{2n})\} \geqslant \int_{A_n \cap B_n} L_n^*(\theta, t)\, dP^\theta[T_{2n} \leqslant t]$$

$$= 2n \int_{A_n \cap B_n} (\bar{X}_n - \theta + 1)^2\, dP$$

$$\geqslant 2n \int_{A_n} (\bar{X}_n - \theta + 1)^2\, dP - 2n \int_{\bar{B}_n} (\bar{X}_n - \theta + 1)^2\, dP,$$

where $\bar{B}_n$ is the complement of B_n. Furthermore,

(5.6.12) $$\int_{B_n} (\bar{X}_n - \theta)\, dP$$

$$= \frac{1}{\sqrt{n}} \int_{-\infty}^{\infty} y\varphi(y)[1 - \Phi(2q_n\sqrt{n} + y) + \Phi(-2q_n\sqrt{n} + y)]\, dy,$$

where $\varphi(y)$ and $\Phi(y)$ are the density function and the distribution function of $N(0, 1)$, and $q_n = (\log n)^{1/2}/\sqrt{(2n)}$. It is easy to verify from (5.6.12) that $\int_{A_n}(\bar{X}_n - \theta)\, dP = 0$. Hence according to (5.6.12) and (5.6.9),

(5.6.13) $$2n \int_{A_n} (\bar{X}_n - \theta + 1)^2\, dP \geqslant 2nP[A_n] \approx \frac{4n}{(2\pi n \log n)^{1/2}},$$

as $n \to \infty$. On the other hand, since the event $\bar{B}_n$ implies that $(\bar{X}_n - \theta)^2 > \frac{1}{2}$, we have

(5.6.14) $$-2n \int_{\bar{B}_n} (\bar{X}_n - \theta + 1)^2\, dP \geqslant -2n \int_{\{(\bar{X}_n - \theta)^2 > \frac{1}{2}\}} (\bar{X}_n - \theta + 1)^2\, dP.$$

It is easy to check that the right-hand side of (5.6.14) is of order of magnitude $o(n^{-1})$ as $n \to \infty$. Finally, from (5.6.11),

(5.6.15) $$\liminf_{n\to\infty} E_\theta\{L_n^*(\theta, T_{2n})\} \geqslant 2 \liminf_{n\to\infty} \{nP_\theta[A_n]\} = \infty.$$

This completes the example. ■

Turning now to the normalized risk of the maximum likelihood estimator we notice from (5.6.6) that if an M.L.E., T_n, is not only consistent but $T_n - \theta = o_p(n^{-1})$ and

$$\lim_{n\to\infty} n \int_{\{|t-\theta|<\delta\}} (t - \theta)^2\, dP_\theta[T_n \leqslant t] = \frac{1}{I(\theta)},$$

then $\limsup_{n\to\infty} E_\theta\{L_n^*(\theta, T_n)\} \leqslant 1/I(\theta)$. This combined with (5.6.5) will assure that the asymptotic normalized risk to the M.L.E. coincides with its asymptotic variance, $1/I(\theta)$. To attain this result we must impose more stringent conditions on the family of distribution functions.

The proof of Commins [1] that $P[|T_n - \theta| \geqslant \delta] = o(n^{-1})$ is a modification of Wald's proof of the consistency of the M.L.E. (see Section 5.3). Whereas Wald's proof requires the existence of certain first moments and the law of large numbers, Commins's proof requires the existence of corresponding third moments. The analysis of the convergence of $n \int_{|t-\theta|<\delta} (t - \theta)^2 \, dP[T_n \leqslant t]$ involves a modification of the standard proof that the distribution of $\sqrt{n}(T_n - \theta)$ converges to a normal distribution. Thus if $\mathfrak{F} = \{P_\theta; \theta \in \Theta\}$ is a family of probability measures on $(\mathfrak{X}, \mathfrak{B})$, and each P_θ of $\mathfrak{F}$ has a generalized density $f(x; \theta)$ with respect to a sigma-finite measure μ, Commins's study imposes on these density functions, in addition to the usual Cramér-Rao conditions, the following regularity conditions. We assume here (Commins's results are more general) that Θ is an interval on the real line. As previously assumed, if $\{\theta_k\} \subset \Theta$ is any sequence such that $\lim_{k\to\infty} |\theta_k| = \infty$, then $\lim_{k\to\infty} f(x; \theta_k) = 0$. We further define the following functions:

$$f(X, \theta, \rho) = \sup_{|\theta'-\theta|\leqslant\rho} f(X; \theta');$$

$$\varphi(X; r) = \sup_{|\theta|>r} f(X; \theta);$$

$$f^*(X, \theta, \rho) = \max\{f(X, \theta, \rho), 1\};$$

$$\varphi^*(X; r) = \max\{\varphi(X; r), 1\};$$

$$\psi(X, \theta_0, \epsilon) = \sup_{|\theta'-\theta_0|\leqslant\epsilon} \left\{-\frac{\partial^2}{\partial\theta^2} \log f(X; \theta')\right\};$$

$$\psi^*(X, \theta_0, \epsilon) = \inf_{|\theta'-\theta_0|\leqslant\epsilon} \left\{-\frac{\partial^2}{\partial\theta^2} \log f(X; \theta')\right\}.$$

Then the regularity conditions require the following:

(i) $E_\theta[\log f^*(X, \theta, \rho)]^3 < \infty$, for each θ and all $0 < \rho \leqslant \rho(\theta)$;
(ii) there exists an r, $0 < r < \infty$, such that $E[\log \varphi^*(X; r)]^3 < \infty$;
(iii) $E_\theta |\log f(X; \theta_0)|^3 < \infty$;
(iv) $(\partial^2/\partial\theta^2) \log f(X; \theta)$ exists for all θ and all X and is continuous in some neighborhood of θ_0;
(v) there exists an ϵ, $\epsilon > 0$, such that

$$E_{\theta_0} |\psi(X, \theta_0, \epsilon)|^3 < \infty, \qquad E_{\theta_0} |\psi^*(X, \theta_0, \epsilon)|^3 < \infty.$$

The proof of the consistency of the M.L.E. under these regularity conditions is similar to that in Section 5.3. It is shown that under these conditions the probability that the M.L.E. does not lie in a neighborhood of the true value θ_0 is of order $0(n^{-3/2})$. Thus $T_n - \theta = o_p(n^{-1})$. Similarly, Commins proved that under the above regularity conditions, for any given $\delta > 0$, there exists an $\epsilon > 0$ such that

$$\limsup_{n\to\infty} n \int_{|\hat{\theta}_n - \theta| \leqslant \epsilon} (\hat{\theta}_n - \theta)^2 \, dP \leqslant \frac{1+\delta}{I(\theta)},$$

where $\hat{\theta}_n$ is the M.L.E. of θ. Thus the above regularity conditions secure the convergence of $E_\theta\{L_n^*(\theta, \hat{\theta}_n)\}$ to its asymptotic variance $I^{-1}(\theta)$.

5.7. MAXIMUM LIKELIHOOD ESTIMATION BASED ON GROUPED, TRUNCATED, AND CENSORED DATA

In many experimental or data collecting processes we do not obtain the values of the random variable under consideration (which is unobserved) but of a less informative random variable. For example, suppose that M electronic devices operate simultaneously and independently. Suppose also that the lifetime of each of these devices is exponentially distributed with the same mean. An observer visits the plant at preassigned epochs, say $\xi_1, \xi_2, \ldots, \xi_k$, where $0 < \xi_1 < \xi_2 < \cdots < \xi_k < \infty$; he records at each visit how many devices have failed in the time interval between the last and the present visit. Failed devices are not replaced. The original random variables $X_1, X_2, \ldots, X_n$, which represent the lifetime of the devices, are unobserved. If these values were known the M.L.E. of the mean lifetime θ would be $\hat{\theta}_n = 1/n \sum_{i=1}^n X_i$. However, the observed random variables are the non-negative integers $N_1, N_2, \ldots, N_{k+1}$ such that $\sum_{i=1}^{k+1} N_i = n$. The quantity N_i is the number of devices failed in the interval $(\xi_{i-1}, \xi_i]$; $\xi_0 \equiv 0$ and $\xi_{k+1} \equiv \infty$. The likelihood function is

$$(5.7.1) \qquad f(N_1, \ldots, N_k; \theta) = \prod_{i=1}^{k+1} [e^{-\xi_{i-1}/\theta} - e^{-\xi_i/\theta}]^{N_i}.$$

Letting $P_i(\theta) = e^{-\xi_{i-1}/\theta} - e^{-\xi_i/\theta}$, we obtain that the M.L.E. $\tilde{\theta}_n$ is the root of the likelihood equation

$$(5.7.2) \qquad \sum_{i=1}^{k+1} N_i \frac{1}{P_i(\theta)} \cdot \frac{\partial}{\partial \theta} P_i(\theta) = 0.$$

The solution of (5.7.2) is not simple. We can use the Newton-Raphson iterative technique, according to (5.2.5), with the initial solution $\tilde{\theta}_1 = 1/n \sum N_i t_i$, where $t_i = \frac{1}{2}(\xi_{i-1} + \xi_i)$ $i = 1, \ldots, k$; and $t_{k+1} = \xi_{k+1} + 1$. Rao [2] investigated the properties of maximum likelihood estimates from grouped data, as in the above example. Conditions are given there for the asymptotic consistency of the M.L.E. $\tilde{\theta}_n$. Properties of the M.L.E. from grouped data were also studied by Kulldorf [1]. It is intuitively clear that the grouping of observations, as in the above example, leads to a certain loss of information. Kullback [1], pp. 15–16, shows this fact with respect to the Kullback-Leibler information function $I(\theta_1, \theta_2) = E_{\theta_1}\{\log (f(X; \theta_1)/f(X; \theta_2))\}$. Kale [3] investigated the question of whether there exists a grouping of observations with an associated measure of information, say $I_g(\theta_1, \theta_2)$, which is close to $I(\theta_1, \theta_2)$ to within a given $\epsilon > 0$. He gave conditions for the existence of such a grouping, and later he showed [4] that there exists a sequence of groupings $\{G_k; k = 1, 2, \ldots\}$ which yield M.L.E.'s $\tilde{\theta}_n^{(k)}$ approaching in distribution, as $n \to \infty$, the M.L.E. $\hat{\theta}_n$ of ungrouped data.

Ehrenfeld [1] studied the above grouping problem from the design of experiments point of view. He considered the question of determining the optimal partition (grouping) $\{E_1, \ldots, E_{k+1}\}$ of the positive part of the real line $E_i = (\xi_{i-1}, \xi_i]$, $i = 1, \ldots, k + 1$, so that the loss of information would be minimized. The solution is not simple since the amount of information lost by a given partition is a function of the unknown parameter θ.

Swamy [1] discussed the question of how much information is lost if, in addition to grouping, there is a truncation or censoring of observations. Truncated grouped data are a set of grouped data in which the frequencies of the first r intervals and those of the last s intervals are missing. In censored grouped data, the total frequencies of cases falling in the first r intervals and in the last s intervals are given. Swamy compared the determinants of the corresponding Fisher information matrices, employing interesting results about the determinants of Gram matrices. His results are applicable, in particular, to cases of truncated and censored ungrouped samples. We do not reproduce the formulae here. The interested reader is advised to consider the cited papers.

PROBLEMS

Section 5.1

1. Let $X_1, \ldots, X_n$ be i.i.d. random variables,

$$X_1 \sim \mathcal{R}(\theta, \theta + |\theta|), \qquad \theta \in \Theta.$$

(i) What is the M.L.E. of θ when $\Theta = (-\infty, 0)$?
(ii) Compare the mean square error of the M.L.E. of θ in (i) to the variance of the U.M.V.U. estimator of that case.
(iii) Show that when $\Theta = (0, \infty)$ there exists no U.M.V.U. estimator of θ.

(iv) Show that $(n + 1)[(X_{(1)}/2(n + 2)) + (X_{(n)}/2(2n + 1))]$ is an unbiased estimator of θ for the statistical model of (iii). What is the M.L.E.?

(v) Compare the mean square errors of the unbiased estimator and the M.L.E. of (iv).

2. Compare the variance and the mean square error of the U.M.V.U. and the M.L.E. of λ^2 in the Poisson case.

3. (i) What is the M.L.E. of the components of variance σ^2 and τ^2 in Model II of analysis of variance, given in Problem 18 of Chapter 2?

(ii) Compute the mean square errors of these M.L.E.'s.

(iii) Show that the mean square error of the M.L.E. of τ^2 is uniformly smaller than the variance of its U.M.V.U. estimator. (See Klotz, Milton, and Zacks [1].)

(iv) What is the Fisher information function for this model?

4. Let $X_1, \ldots, X_n$ be i.i.d., $X_1 \sim G(1/\theta, 1)$.

(i) What is the M.L.E. of $e^{-1/\theta}$, say $g(\hat{\theta})$?

(ii) What is the mean square error of this M.L.E.? (See Zacks and Even [1].)

(iii) What is the coverage probability

$$C(\theta; \delta) = P_\theta\{|g(\hat{\theta}) - e^{-1/\theta}| < \delta e^{-1/\theta}\}, \qquad 0 < \delta < 1?$$

5. (i) What is the M.L.E. of $\theta = (\mu, \sigma^2, \rho)$ in the common mean model of Problem 29, Chapter 2?

(ii) What is the Fisher information function?

6. Determine the M.L.E. of the location parameter, of the translation type, in the extreme value distribution with standard density

$$f(x) = \exp\{-x - e^{-x}\}, \qquad -\infty < x < \infty.$$

7. Let $X_1, \ldots, X_n$ be i.i.d. random variables $(n \geqslant 2)$, $X_1 \sim N(\mu, \sigma^2)$. Let $Q = \sum (X_i - \bar{X})^2$, $\bar{X} = \sum X_i/n$. Show that $Q/(n + 1)$ is an estimator of σ^2 with a mean square error which is uniformly smaller than that of the M.L.E.

8. Let $X_1, \ldots, X_n$ be i.i.d. random variables having a common multinomial distribution. What is the M.L.E. of the parameters of this distribution? What is the Fisher information matrix?

9. Let $X_1, \ldots, X_k$ be certain doses of a drug that is administered to k groups, each of n patients. Let K_j be the number of patients which respond positively to the drug at dose X_j $(j = 1, \ldots, k)$. It is assumed that $K_1, \ldots, K_j, \ldots, K_k$ are independent and that K_j has a binomial distribution $\mathcal{B}(n; \theta_j)$, $j = 1, \ldots, k$. Furthermore, the model assumes that $\Phi^{-1}(\theta_j) = \alpha + \beta X_j$, $j = 1, \ldots, k$. The transformation $\Phi^{-1}(K_j/n)$ is called the normit transformation of K_j/n.

(i) Determine the M.L.E. of (α, β).

(ii) What is the Fisher information matrix as a function of α and β?

10. Read the paper of Teicher [1] on the maximum likelihood characterization of distribution.

Section 5.2

11. Let $X_1, \ldots, X_n$ be i.i.d. random variables whose common distribution belongs to the family of truncated normal distributions, with densities

$$f(x; \mu, \sigma) = \frac{1}{\sigma\Phi(\mu/\sigma)}\,\varphi\left(\frac{x-\mu}{\sigma}\right), \qquad x \geqslant 0,$$

$-\infty < \mu < \infty$, $0 < \sigma < \infty$. Which of the methods described in Section 5.2 will you use to approximate the M.L.E. of (μ, σ)? Which initial values will you use for an iterative process? Can you determine the Fisher information matrix? (For similar methods of estimating the parameters of truncated distributions, see Cohen [1; 2; 3]).

Section 5.3

12. Let $f(x; \theta) = \varphi(x - \theta)$, $-\infty < \theta < \infty$, be the family of location parameter normal densities. The standard normal density is denoted by $\varphi(x)$.

(i) For $\theta \neq \varphi$ compute the Kullback-Leibler information

$$I(\theta, \varphi) = E_\theta\left\{\log \frac{f(X; \theta)}{f(X; \varphi)}\right\}.$$

(ii) Let $N_\varphi = \{\varphi'; |\varphi' - \varphi| < (10)^{-1}\}$. Determine

$$I(\theta, N_\varphi) = E_\theta\left\{\inf_{\varphi' \in N_\varphi} \log \frac{f(X; \theta)}{f(X; \varphi')}\right\}$$

for $\theta = 1$ and $\varphi = 0$.

13. Check whether the conditions of Theorem 5.3.1 are satisfied by the following:

(i) the family of binomial distributions $\mathcal{B}(n; \theta)$, $0 < \theta < 1$;
(ii) the general family of a 1-parameter exponential type;
(iii) the family of a k-parameter exponential type.

14. Compare the variance of the unbiased estimator of σ^2 to the mean square error of its M.L.E. in Example 5.5. (This is a striking example of a case in which the M.L.E. is very inefficient compared to an unbiased estimator, when the sample is large.)

Section 5.4

15. Consider the problem of estimating $\Phi(-(\mu/\sigma))$ in the $\mathcal{N}(\mu, \sigma^2)$ case, based on a sample of n i.i.d. random variables.

(i) What is the M.L.E., say $\hat{p}_n$, of $\Phi(-(\mu/\sigma))$?
(ii) What is the asymptotic distribution of $\sqrt{n}(\hat{p}_n - \Phi(-(\mu/\sigma)))$?

16. Let X have a k-nomial distribution function; that is,

$$P_\theta[K = x] = \begin{cases} \pi_j(\theta), & x = j, \quad j = 1, 2, \ldots, k. \\ 0, & \text{otherwise.} \end{cases}$$

Consider the problem of estimating θ on the basis of n i.i.d. random variables. We consider the following estimators:

(i) M.L.E. = roots of the equation

$$\sum_{j=1}^{k} p_j \frac{\pi_j'(\theta)}{\pi_j(\theta)} = 0.$$

(ii) Minimum-χ^2 = root of the equation

$$\sum_{j=1}^{k} p_j^2 \frac{\pi_j'(\theta)}{\pi_j^2(\theta)} = 0,$$

where $(p_1, \ldots, p_k)$ is a random vector representing the proportion of sample values, out of n, in each one of the k values of X.

Determine the index of second-order efficiency for these two estimators (see Rao [3]).

Section 5.5

17. Consider the problem of estimating the common mean, μ, of two normal distributions $\mathcal{N}(\mu, \sigma_1^2)$, $\mathcal{N}(\mu, \sigma_2^2)$, $-\infty < \mu < \infty$, $0 < \sigma_1, \sigma_2 < \infty$. Let $\bar{X}$, Q_1, $\bar{Y}$, Q_2 be the mean and sum of squares of deviations about the means of two random samples of equal size. Show that the estimator

$$\hat{\mu} = \frac{\bar{X}Q_2 + \bar{Y}Q_1}{Q_1 + Q_2}$$

is B.A.N., although it is not an M.L.E.

18. Consider the problem of estimating the parameters of Model II of A.O.V., as in Problem 18 of Chapter 2. Is the M.L.E. of (μ, σ^2, τ^2) a B.A.N. estimator?

19. Let $X_1, \ldots, X_n$ be a sequence of i.i.d. random variables which have a common extreme value distribution that belongs to a scale parameter family $f(x; \sigma) = (1/\sigma)f(x/\sigma)$, $0 < \sigma < \infty$, where $f(x) = \exp\{-x - e^{-x}\}$, $-\infty < x < \infty$

(i) Provide a consistent estimator of σ.
(ii) Construct a B.A.N. estimator of σ.

Section 5.6

20. Let $X_1, \ldots, X_n$ be a sequence of i.i.d. random variables $X_1 \sim \exp\{N(\mu, \sigma^2)\}$, $-\infty < \mu < \infty$, $0 < \sigma < \infty$, that is, a sequence of log-normal r.v.'s.

(i) Consider the M.L.E. estimator of $\xi = E\{X\}$, say $\hat{\xi}_n$. What is the normalized loss function $L_n^*(\xi, \hat{\xi}_n)$?
(ii) What is the asymptotic variance of $\sqrt{n}(\hat{\xi}_n - \xi)/\xi^2\sigma^2(1 + (\sigma^2/2))$?
(iii) What is the $\lim_{n\to\infty} E_{(\xi,\sigma)}\{L_n^*(\xi, \hat{\xi}_n)\}$?

REFERENCES

Bahadur [3], [4]; Barankin and Gurland [1]; Barnett [1]; Basu [2]; Chernoff [2], [5]; Chiang [1]; Commins [1]; Daniels [1]; Dwight [1]; Ehrenfeld [1]; Fisher [2], [3]; Fisz [1]; Hannan [1]; Huzurbazar [1]; Kale [1], [2], [3], [4]; Kallianpur [1]; Kiefer and Wolfowitz [3]; Kraft and LeGam [1]; Kullback [1]; Kulldorf [1]; LeCam [2]; Neyman and Scott [1] Northan [1]; Rao [2], [3], [5], [6]; Swamy [1], [2]; Wald [5]; Walker [1]; Weiss and Wolfowitz [1]; Wijsman [1]; Wolfowitz [2], [4]; Zacks and Even [1]; Zehna [1].

CHAPTER 6

Bayes and Minimax Estimation

6.1. THE STRUCTURE OF BAYES ESTIMATORS

The following are the four essential elements in the Bayesian decision framework:

1. The statistical model represented by the probability space $(\mathfrak{X}, \mathcal{B}, P)$, where the probability measure P belongs to a specified family of probability measures $\mathcal{P} = \{P_\theta; \theta \in \Theta\}$. We assume that the parameter θ of P is real or vector valued and Θ is a (k-dimensional) interval in a Euclidean k-space, $E^{(k)}$.
2. A family $\mathcal{H}$ of *prior probability* measures $\xi(\theta)$ defined on the measurable space $(\Theta, \mathcal{F})$, where $\mathcal{F}$ is the smallest Borel sigma-field on Θ.
3. A set of possible decisions $\mathfrak{D}$ such that every element d of $\mathfrak{D}$ is a $\mathcal{B}$-measurable function on $\mathfrak{X}$. In this chapter on estimation the set of decision functions $\mathfrak{D}$ contains all estimators of θ, that is, d: $\mathfrak{X} \to \Theta$.
4. A loss (regret) function $L(\theta, d)$ defined on $\Theta \times \mathfrak{D}$. This function represents the regret due to an erroneous estimation. In many studies we find that the loss function specified is of the general form $L(\theta, d) = \lambda(\theta)W(|d - \theta|)$, where $W(0) = 0$ and $W(t)$ are monotone increasing, $t > 0$. Furthermore, $\lambda(\theta)$ is positive, finite, and $\mathcal{F}$-measurable. In order to avoid difficulties it is often assumed that $L(\theta, d)$ is a bounded function of θ for each d. Sometimes it is also assumed that $L(\theta, d)$ is continuous in both θ and d and uniformly bounded in (θ, d). In this section we impose the boundedness condition on $L(\theta, d)$. In the next section we consider unbounded loss functions, that are convex in d for each θ. In this case we assume certain integrability assumptions.

Let $f(x; \theta)$ designate the density function of $F_\theta(x)$ with respect to a sigma-finite measure $\mu(x)$; $f(x; \theta)$ is $\mathcal{F}$-measurable for each $x \in \mathfrak{X}$. Let $H(\theta)$ designate a prior (probability) distribution function over Θ and $h(\theta)$ the prior density of $H(\theta)$ with respect to a sigma-finite measure $\zeta(\theta)$ on Θ. Let $H(\theta \mid X)$ designate the posterior distribution of θ, given X. We discussed the notion of posterior distributions in Section 2.10. We have shown, using the Bayes formula, that the posterior density function of θ, given X, is given by

$$h(\theta \mid X) = \frac{f(X;\theta)h(\theta)}{f_H(X)} \quad \text{a.s. } [\mu],$$

where

$$f_H(X) = \int H(\mathrm{d}\theta)f(X;\theta) \quad \text{a.s. } [\mu].$$

Since $H(\theta)$ is a distribution function, $f_H(X) < \infty$ a.s. $[\mu]$.

In Section 2.10 we discussed the notion of Bayesian sufficiency and showed that if $\mathcal{H}$ is a family of prior distributions on Θ, then T is a sufficient statistic for $\mathcal{F} = \{F_\theta;\, \theta \in \Theta\}$ if and only if T is Bayesian sufficient for $\mathcal{H}$. Hence if T is sufficient for $\mathcal{F}$, then $H(\theta \mid X) = H(\theta \mid T(X))$ a.s. $[\mu]$. It is generally more convenient to derive the posterior distribution on the basis of the induced distribution of the sufficient statistic T.

In this section we derive various formulae, assuming that the loss function is of the general form $L(\theta, d) = \lambda(\theta)W(|d - \theta|)$. Suppose that $d \in \mathcal{D}$ is an estimator of θ. The *risk function* of d is

$$R(\theta, d) = \lambda(\theta)\int W(|d(x) - \theta|)f(x;\theta)\mu(\mathrm{d}x). \tag{6.1.1}$$

For a given estimator d, $R(\theta, d)$ is considered a function on Θ, which is $\mathcal{F}$-measurable. The prior risk of d with respect to H is the prior expectation of the risk function

$$\mathrm{R}(H, d) = \int H(\mathrm{d}\theta)R(\theta, d). \tag{6.1.2}$$

Since $L(\theta, d)$ is bounded in θ for each d, $R(\theta, d)$ is bounded for each d. Hence since $W(t) \geqslant 0$, we can write

$$\begin{aligned} R(H, d) &= \int_\Theta \lambda(\theta)H(\mathrm{d}\theta)\int_{\mathcal{X}} W(|d(x) - \theta|)f(x;\theta)\mu(\mathrm{d}x) \\ &= \int_{\mathcal{X}} f_H(x)\mu(\mathrm{d}x)\int_\Theta \lambda(\theta)W(|d(x) - \theta|)H(\mathrm{d}\theta \mid x). \end{aligned} \tag{6.1.3}$$

The integral $\int \lambda(\theta)W(|d(x) - \theta|)H(\mathrm{d}\theta \mid x)$ is called the *posterior risk* of d, given $X = x$. We thus see that the prior risk $R(H, d)$ is the expectation of the posterior risk, with respect to the marginal distribution of X under H, $F_H(x)$.

From the Bayesian point of view, once X has been observed, the relevant function to consider is the posterior risk rather than the prior risk. Thus a

Bayes estimator of θ with respect to the prior distribution H is the estimator in $\mathfrak{D}$ which minimizes the posterior risk, given X. Let $\hat{\theta}_H(X)$ designate the Bayes estimator; then

$$(6.1.4) \quad \int \lambda(\theta)W(|\hat{\theta}_H(X)-\theta|)H(\mathrm{d}\theta \mid X) = \inf_{d\in\mathfrak{D}} \int \lambda(\theta)W(|d(X)-\theta|)H(\mathrm{d}\theta \mid X)$$

a.s. $[\mu]$. The Bayes estimator for a given H is *not* necessarily unique. As is proven in Section 6.2, if the loss function $L(\theta, d)$ is strictly convex in d for each θ, then $\hat{\theta}_H(X)$ is essentially unique. We notice also that, according to (6.1.3) and the Fatou lemma,

$$\begin{aligned}(6.1.5) \quad \inf_{d\in\mathfrak{D}} R(H, d) &= \inf_{d\in\mathfrak{D}} \int f_H(x)\mu(\mathrm{d}x) \int \lambda(\theta)W(|d(x)-\theta|)H(\mathrm{d}\theta \mid x) \\ &\geqslant \int f_H(x)\mu(\mathrm{d}x) \inf_{d\in\mathfrak{D}} \int \lambda(\theta)W(|d(x)-\theta|)H(\mathrm{d}\theta \mid x). \\ &= \int f_H(x)\mu(\mathrm{d}x) \int \lambda(\theta)W(|\hat{\theta}_H(x)-\theta|)H(\mathrm{d}\theta \mid x) \\ &= R(H, \hat{\theta}_H) \geqslant \inf_{d\in\mathfrak{D}} R(H, d).\end{aligned}$$

Hence a Bayes estimator $\hat{\theta}_H$ also minimizes the prior risk. In many textbooks and papers (see, for example, Girshick and Savage [1], DeGroot and Rao [1], and others) a Bayes estimator is defined as an element of $\mathfrak{D}$ which minimizes the *prior* risk. As shown in (6.1.5), the two definitions yield the same estimator $\hat{\theta}_H$.

Raiffa and Schlaifer [1] discuss at length the question of which prior distribution to choose for various models, dealing in particular with the notion of conjugate prior distributions. We do not discuss these topics here, but we show some of the properties of conjugate prior distributions in examples.

Example 6.1. This example is based on the paper of Wolfowitz [3]. Let $X_1, X_2, \ldots, X_n$ be i.i.d. random variables having a common normal distribution $\mathcal{N}(\mu, 1)$, with an unknown mean μ, $-\infty < \mu < \infty$. A complete sufficient statistic is $T_n = \sum_{1=i}^n X_i$. Assuming a prior normal distribution for μ with mean zero and variance τ^2, we wish to derive a Bayes estimator of μ with respect to the loss function $L(\theta, d) = W(|d-\theta|)$, where

$$W(|d-\mu|) = \begin{cases} 0, & \text{if } |d-\mu| < \delta, \\ 1, & \text{otherwise.} \end{cases}$$

The quantity d is an estimator of μ. First we remark that, given T_n, the posterior distribution of μ is the normal distribution,

$$\mathcal{N}\left(\frac{T_n}{n + 1/\tau^2}, \left(n + \frac{1}{\tau^2}\right)^{-1}\right).$$

This can be easily verified by the theory of the normal distribution. Hence the posterior risk of an estimator $d(T_n)$ is

$$\begin{aligned} &R(\tau^2, d(T_n)) \\ &= 1 - P[d(T_n) - \delta < \mu < d(T_n) + \delta \mid T_n] \\ &= 1 - \Phi\left(\frac{d(T_n) + \delta - T_n/(n + \tau^{-2})}{(n + \tau^{-2})^{-1/2}}\right) + \Phi\left(\frac{d(T_n) - \delta - T_n/(n + \tau^{-2})}{(n + \tau^{-2})^{-1/2}}\right). \end{aligned}$$

To minimize $R(\tau^2, d(T_n))$ we have to choose $d(T_n)$ so as to maximize

$$\Phi\left(\frac{d(T_n) + \delta - T_n/(n + \tau^{-2})}{(n + \tau^{-2})^{-1/2}}\right) - \Phi\left(\frac{d(T_n) - \delta - T_n/(n + \tau^{-2})}{(n + \tau^{-2})^{-1/2}}\right).$$

If we consider the general expression $\Phi(X + \delta - \xi) - \Phi(X - \delta - \xi)$, we obtain, by differentiating with respect to X, that the value X^0 which maximizes this expression should satisfy the equation

$$\varphi(X^0 + \delta - \xi) = \varphi(X^0 - \delta - \xi),$$

where $\varphi(\ )$ is the $\mathcal{N}(0, 1)$ density function. Since $\varphi(\delta) = \varphi(-\delta)$, we immediately obtain that $X^0 = \xi$. It is easy to verify that $X^0 = \xi$ is the unique root. The second-order derivative of $\Phi(X + \delta - \xi) - \Phi(X - \delta - \xi)$, at the point $X^0 = \xi$, is $-2\delta\varphi(\delta)$. Hence X^0 maximizes $\Phi(X + \delta - \xi) - \Phi(X - \delta - \xi)$. Setting $X = d(T_n)/(n + \tau^{-2})^{-1/2}$, $\xi = T_n/(n + \tau^{-2})^{1/2}$, we obtain that the unique Bayes estimator of μ for the above function is $\hat{\mu}_\tau(T_n) = T_n/(n + \tau^{-2})$. We notice, moreover, that $\lim_{\tau\to\infty} \hat{\mu}_\tau(T_n) = \bar{X}_n$, which is the sample mean. ■

We conclude this section with the following comment. As will be shown later, there are optimality criteria and cases in which one can improve his decision by considering a wider class of estimators containing, in addition to the point estimators, randomized estimators. Randomized estimators, $\pi(\theta \mid X)$, are conditional distribution functions defined on the parameter space Θ. Given an observed value $X = x$, the statistician generates a random value of θ according to $\pi(\theta \mid x)$ and uses the outcome as an estimate of θ. The nonrandomized point estimators of θ could be considered as special randomized estimators which assign probability 1 to the point $\hat{\theta}(X)$. We now show that if the loss function is convex in d for each θ, then, given any randomized estimator $\pi(\theta \mid X)$, there exists a nonrandomized estimator $\hat{\theta}_\pi(X)$ whose risk function does not exceed that of $\pi(\theta \mid X)$ at every θ. Indeed, the risk function of $\pi(\theta \mid X)$ is

$$R(\theta, \pi) = \int_{\mathfrak{X}} f(x; \theta)\mu(\mathrm{d}x) \int_{\Theta} \pi(\mathrm{d}\hat{\theta} \mid x) L(\theta, \hat{\theta}). \tag{6.1.6}$$

But since $L(\theta, \hat{\theta})$ is convex in $\hat{\theta}$ for each θ, then by the Jensen inequality

$$\int_{\Theta} \pi(\mathrm{d}\hat{\theta} \mid X) L(\theta, \hat{\theta}) \geqslant L(\theta, E_\pi\{\hat{\theta} \mid X\}) \quad \text{a.s. } [\mu]. \tag{6.1.7}$$

Let $\hat{\theta}_\pi(X) = E_\pi\{\hat{\theta} \mid X\}$ a.s. $[\mu]$ be a nonrandomized estimator corresponding to $\pi(\hat{\theta} \mid X)$. Then from (6.1.6) and (6.1.7),

$$R(\theta, \pi) \geqslant \int_{\mathfrak{X}} f(x; \theta)\mu(\mathrm{d}x) L(\theta, \hat{\theta}_\pi(x)) = R(\theta, \hat{\theta}_\pi). \tag{6.1.8}$$

Thus in cases of convex loss functions the Bayes estimators, if exist, can be found within the class of nonrandomized estimators. Hodges and Lehmann [1] constructed an example in which the loss function is not convex and the optimal estimator is randomized. This example is discussed in Section 6.5.

6.2. BAYES ESTIMATORS FOR QUADRATIC AND CONVEX LOSS FUNCTIONS

The quadratic loss function, and in particular the squared-error loss, are the most commonly applied ones. In large sample situations the squared-error loss yields a good approximation to any loss function $\lambda(\theta)W(|d - \theta|)$, which is smooth in the neighborhood of the origin. A convex loss function of the type $W(|d - \theta|) = |d - \theta|^k$, $k \geqslant 1$, is widely used too. We notice that these loss functions are not bounded. The unboundedness of the loss function may cause some difficulty. The first difficulty is that the estimator which minimizes the posterior risk may have an infinite prior risk. We demonstrate such a case in the following example, provided by Girshick and Savage [1].

Example 6.2. Consider a random variable X with a family of rectangular distributions on $(0, |\theta|^{-1})$, where θ is any real number such that $1 \leqslant |\theta| < \infty$; that is, $f(X; \theta) = |\theta|$ if $0 \leqslant X \leqslant |\theta|^{-1}$.

Let the prior distribution $H(\theta)$ be absolutely continuous on the real line, with a density

$$h(\theta) = \begin{cases} \frac{1}{2}|\theta|^{-2}, & \text{if } 1 \leqslant |\theta| < \infty \\ 0, & \text{otherwise.} \end{cases}$$

The posterior density function of θ, given $X = x$, is

$$h(\theta \mid X) = \begin{cases} \dfrac{\frac{1}{2}|\theta|^{-1}}{-\log X}, & \text{if } X \leqslant |\theta|^{-1} \leqslant 1, \\ 0, & \text{otherwise.} \end{cases}$$

Thus if the loss function is squared error, that is, $W(|d - \theta|) = (d - \theta)^2$, the Bayes estimator, as defined in Section 6.1, should minimize the posterior risk

$$R(d, x) = -\frac{1}{2} \int_{\{1 \geqslant |\theta|^{-1} \geqslant X\}} (d - \theta)^2 \frac{|\theta|^{-1}}{\log X} \, d\theta.$$

The unique $d(X)$ minimizing $R(d, X)$ is the posterior expectation of θ given $X = x$, $E\{\theta \mid X = x\}$. Since $h(\theta \mid x)$ is symmetric about $x = 0$, we immediately obtain that the Bayes estimator is $\hat{\theta}(X) = 0$ a.s. The posterior risk associated with the Bayes estimator is

$$R(0, X) = \frac{1}{2} \int_{\{1 \geqslant |\theta|^{-1} \geqslant X\}} \frac{|\theta|}{-\log X} \, d\theta = \frac{1}{2} \frac{1 - X^2}{X^2(-\log X)} < \infty \quad \text{a.s. } [\mu].$$

The prior risk of $\hat{\theta}(X)$ is, however,

$$R(0, H) = \int_1^\infty d\theta = \infty.$$

We further remark that the Bayes estimator $\hat{\theta}(X) = 0$ a.s. is of no value since it does not depend on the observation. This is due to the special structure of the problem. The variable X can give information only on $|\theta|$, and we cannot infer from the observed value of X whether θ is positive or negative. The example has been designed, however, to show the possible anomality. ■

Although the logic of the Bayesian inference leads us to consider estimators which minimize the posterior risk, it is desirable for various theoretical purposes (which are discussed later) that the Bayes estimators also minimize the prior risk. This can be guaranteed if we impose the condition that the posterior risk, given $X = x$, is an integrable function with respect to the marginal distribution $F_H(x)$, that is,

$$\int_{\mathfrak{X}} F_H(dx) \int_\Theta \lambda(\theta) W(|\hat{\theta}(x) - \theta|) H(d\theta \mid x) < \infty. \tag{6.2.1}$$

The quantity $\hat{\theta}(X)$ is the estimator that minimizes the posterior risk. If (6.2.1) is satisfied, then by Fubini's theorem (6.1.3) holds. In this case, as shown by (6.1.5), the estimator that minimizes the posterior risk also minimizes the prior risk. Thus, assuming (6.2.1), we can define the Bayes estimator as an estimator that minimizes the prior risk.

We start the discussion with the case of a real parameter θ and consider the question of existence of a Bayes estimator for a quadratic loss function $L(\theta, d) = \lambda(\theta)W(|d - \theta|)$, where $W(|d - \theta|) = (d - \theta)^2$, $0 < \lambda(\theta) < \infty$ for all θ. We notice that, given $X = x$, the posterior risk function is

$$(6.2.2)\quad E_H\{\lambda(\theta)[d(X) - \theta]^2 \mid X = x\} = d^2(x)E_H\{\lambda(\theta) \mid X = x\} - 2d(x)E_H\{\lambda(\theta)\cdot\theta \mid X = x\} + E_H\{\lambda(\theta)\cdot\theta^2 \mid X = x\},$$

where $E_H\{\cdot \mid X = x\}$ designates the posterior expectation of the term in brackets. This posterior risk is a quadratic function of $d(X)$; $E_H\{\lambda(\theta) \mid X = x\} > 0$ a.s. Thus if the coefficients of this quadratic function are finite, there exists a unique point $\hat{\theta}(X)$ that minimizes (6.2.2) a.s. The following theorem gives a criterion for the finiteness of the posterior risk.

Theorem 6.2.1. (Girshick and Savage [1].) *Consider a quadratic loss function $\lambda(\theta)(d - \theta)^2$. Given a prior distribution $H(\theta)$, let $R_H(d, x)$ designate the posterior risk (6.2.2) against $H(\theta)$. Then for each x, $R_H(d, x) < \infty$ either for no d, for exactly one d, or for every d. If $R(d, x) < \infty$ for exactly one d then $E_H\{\lambda(\theta) \mid X = x\} = \infty$. If $R(d, x) < \infty$ for all d then $E_H\{\lambda(\theta) \mid X = x\} < \infty$.*

Proof. The first part of the proof is designed to verify that if $R(d, x) < \infty$ for any two real values, say $d_1(x)$ and $d_2(x)$, then $R(d, x) < \infty$ for all d. Suppose that $d_1(x) < d_2(x)$ and $R(d_1, x) < \infty$, $R(d_2, x) < \infty$. Since $\lambda(\theta)(h - \theta)^2$ is a convex function of h for each θ, we obtain for all h in the interval $[d_1(x), d_2(x)]$ that

$$(6.2.3)\quad \lambda(\theta)(h - \theta)^2 \leqslant \frac{d_2(x) - h}{d_2(x) - d_1(x)}\cdot\lambda(\theta)[d_1(x) - \theta]^2 + \frac{h - d_1(x)}{d_2(x) - d_1(x)}\cdot\lambda(\theta)[d_2(x) - \theta]^2.$$

Hence the posterior risk is finite for each h such that $h(x) \in [d_1(x), d_2(x)]$. Indeed,

$$(6.2.4)\quad R(h, x) = \int \lambda(\theta)(h(x) - \theta)^2 H(d\theta \mid x) \leqslant \frac{d_2(x) - h(x)}{d_2(x) - d_1(x)} R(d_1, x) + \frac{h(x) - d_1(x)}{d_2(x) - d_1(x)} R(d_2, x) < \infty$$

We now show that $E_H\{\lambda(\theta) \mid X = x\} < \infty$, $E_H\{\theta\lambda(\theta) \mid X = x\} < \infty$, and $E_H\{\lambda(\theta)\cdot\theta^2 \mid X = x\} < \infty$. This result combined with (6.2.2) implies that $R(d, x) < \infty$ for an arbitrary d. For this purpose, choose $h_1(x)$ in $[d_1(x), d_2(x)]$, $h_1(x) \neq 0$, and $d(x)$ arbitrary, $d(x) \neq h_1(x)$. It is easy to verify that, for each θ,

$$(6.2.5)\quad (h_1(x) - d(x))^2 + 2(h_1(x) - d(x))(d(x) - \theta) \leqslant (h_1(x) - \theta)^2.$$

Therefore, since $0 < \lambda(\theta) < \infty$,

$$(6.2.6)\quad [h_1(x) - d(x)][\lambda(\theta)(h_1(x) - d(x)) + 2\lambda(\theta)(d(x) - \theta)] \leqslant \lambda(\theta)(h_1(x) - \theta)^2.$$

Hence

$$(6.2.7)\quad (h_1(x) - d(x))\{E_H\{\lambda(\theta) \mid X = x\}(h_1(x) - d(x)) + 2d(x)E_H\{\lambda(\theta) \mid X = x\} - 2E_H\{\lambda(\theta)\cdot\theta \mid X = x\}\} \leqslant R(H, x) < \infty.$$

By setting $d(x) = 0$ we obtain from (6.2.7) that

$$(6.2.8)\quad h_1^2(x)E_H\{\lambda(\theta) \mid X = x\} - 2h_1(x)E_H\{\theta\lambda(\theta) \mid X = x\} < \infty.$$

Since on the left-hand side of (6.2.8) we have a finite quadratic (convex) form in $h_1(x)$, both coefficients are finite. Hence

$$(6.2.9)\quad \begin{aligned} 0 < E_H\{\lambda(\theta) \mid X = x\} < \infty; \\ |E_H\{\theta\lambda(\theta) \mid X = x\}| < \infty. \end{aligned}$$

Let d be an arbitrary estimator. Since $R(h_1, x) < \infty$ and Inequalities 6.2.9 imply that $E_H\{\theta^2\lambda(\theta) \mid X = x\} < \infty$, we obtain from (6.2.2) that $R(d, x) < \infty$. Finally, it remains to prove that if $d_0(x)$ is the only estimator with a finite posterior risk then $E_H\{\lambda(\theta) \mid X = x\} = \infty$. Let $d(x) \neq d_0(x)$; then $R(d, x) = \infty$. Write

$$(6.2.10)\quad \begin{aligned} R(d, x) &= E_H\{\lambda(\theta)[d(x) - d_0(x) + d_0(x) - \theta]^2\} \\ &= (d(x) - d_0(x))^2 E_H\{\lambda(\theta) \mid X = x\} \\ &\quad + 2(d(x) - d_0(x))E_H\{\lambda(\theta)(d_0(x) - \theta) \mid X = x\} \\ &\quad + R(d_0, x). \end{aligned}$$

Since $R(d_0, x) < \infty$, $E_H\{\lambda(\theta)(d_0(x) - \theta) \mid X = x\}$ is finite, and hence $R(d, x) = \infty$ if and only if $E_H\{\lambda(\theta) \mid X = x\} = \infty$. (Q.E.D.)

We notice that if the prior distribution $H(\theta)$ is such that the posterior risk is finite a.s. for all $d \in \mathfrak{D}$, then the Bayes estimator for the quadratic loss function $\lambda(\theta)(d - \theta)^2$ is

$$(6.2.11)\quad \hat{\theta}(X) = \frac{E_H\{\theta\lambda(\theta) \mid X\}}{E_H\{\lambda(\theta) \mid X\}} \quad \text{a.s. } [\mu].$$

If $R(d, X) < \infty$ a.s. only for $d = d_0$ then $d_0(X)$ is the Bayes estimator of θ. In any case the Bayes estimator is essentially unique.

Example. 6.3 Let $X_1, X_2, \ldots, X_n$ be i.i.d. random variables having a negative exponential distribution $\mathcal{G}(\theta, 1)$, representing the lifetime of a certain equipment. The reliability of the equipment is defined as $g(t_0; \theta) = e^{-t_0\theta}$ for some $0 < t_0 < \infty$. The objective is to estimate $g(t_0; \theta)$ with a squared-error loss function $W(|\hat{g} - g|) = (\hat{g} - g)^2$. We assume that θ has a gamma prior distribution $\mathcal{G}(1/\tau, \nu)$; $0 < \tau < \infty$, $0 < \nu < \infty$ are specified.

The minimal sufficient statistic $T_n = \sum_{i=1}^{n} X_i$ has the gamma distribution $\mathcal{G}(\theta, n)$. It is easy to verify that the posterior distribution of θ, given T_n, is the gamma distribution $\mathcal{G}(T_n + (1/\tau), n + \nu)$. Thus the Bayes estimator of $e^{-t_0\theta}$ for the squared-error loss is its posterior expectation

$$\text{(6.2.12)} \qquad \hat{g}(t_0; T_n) = \frac{(T_n + (1/\tau))^{n+\nu}}{\Gamma(n + \nu)} \int_0^\infty e^{-t_0\theta}\theta^{n+\nu-1}e^{-\theta(T_n+(1/\tau))}\, d\theta$$
$$= \left(1 + \frac{t_0}{T_n + (1/\tau)}\right)^{-(n+\nu)}.$$

It is easy to check that for this class of gamma prior distributions, $0 < \tau$, $\nu < \infty$, (6.2.1) is satisfied. We remark that this Bayes estimator is a consistent estimator of $e^{-t_0\theta}$, since $\bar{X}_n = T_n/n \xrightarrow{p} \theta^{-1}$. Indeed,

$$\text{p}\cdot\lim_{n\to\infty} \hat{g}(t_0; T_n) = \text{p}\cdot\lim_{n\to\infty} \left(1 + \frac{t_0}{n\bar{X}_n(1 + (1/\tau n))}\right)^{-n(1+(\nu/n))} = e^{-t_0\theta}.$$

It is interesting to compare the Bayes estimator (6.2.12) to the uniformly minimum variance unbiased estimator of $e^{-t_0\theta}$, given by (3.2.32). While the U.M.V.U. estimator is warranted only if $T_n > t_0$, the above Bayes estimator can be used also for extrapolation purposes when $T_n \leqslant t_0$. A numerical investigation of the mean square error function of $\hat{g}(t_0; T_n)$ can be found in the article by Gaver and Hoel [1]. ■

The following example, taken from Zacks [5], presents the method of Bayes estimation for sampling from finite populations.

Example 6.4. Given a finite population with k strata of known sizes $N_1, \ldots, N_k$, let $M_1, \ldots, M_k$ be the number of units in these strata having a specified attribute. The objective is to estimate the parameter $\theta = \sum_{i=1}^{k} \lambda_i P_i$, where $P_i = M_i/N_i$ and λ_i are given coefficients ($i = 1, \ldots, k$). Simple random samples without replacement are drawn from the k strata independently. Let $(n_1, \ldots, n_k)$ be the sample sizes and X_i ($i = 1, \ldots, k$) the number of units in the sample having the attribute under consideration. The density function of X_i given (N_i, M_i, n_i), $i = 1, \ldots, k$, is the hypergeometric

$$f(x \mid N_i, M_i, n_i) = \begin{cases} \dfrac{\dbinom{M_i}{x}\dbinom{N_i - M_i}{n_i - x}}{\dbinom{N_i}{n_i}}, & \text{if } x = 0, \ldots, n_i, \\[2ex] 0, & \text{otherwise.} \end{cases}$$

Consider a prior distribution for $M = (M_1, \ldots, M_k)$ according to which $M_1, \ldots, M_k$ are priorly independent and have uniform prior densities

$$h_i(m) = \begin{cases} (N_i + 1)^{-1}, & \text{if } m = 0, 1, \ldots, N_i, \\ 0, & \text{otherwise.} \end{cases}$$

We now derive the Bayes estimator of θ, against this joint prior distribution, and a squared-error loss function.

Mixing binomial densities with a beta distribution we can easily establish the identity

$$\sum_{r=0}^{N-n} \binom{r + x}{r} \binom{N - x - r}{n - x} = \binom{N + 1}{N - n}. \tag{6.2.13}$$

Thus the prior marginal densities of X_i under the above prior assumptions are

$$\begin{aligned} f(x \mid h_i, N_i, n_i) &= \frac{1}{N_i + 1} \sum_{m=0}^{N_i} \frac{\binom{m}{x}\binom{N_i - m}{n_i - x}}{\binom{N_i}{n_i}} \\ &= \frac{1}{N_i + 1} \sum_{m=x}^{N_i - n_i + x} \frac{\binom{m}{x}\binom{N_i - m}{n_i - x}}{\binom{N_i}{n_i}} \\ &= \frac{1}{n_i + 1}, \qquad x = 0, 1, \ldots, n_i. \end{aligned} \tag{6.2.14}$$

Thus the prior marginal densities of $X_1, \ldots, X_k$ are uniform on $\{0, 1, \ldots, n_i\}$, $i = 1, \ldots, k$.

It is easy to verify that the posterior density of M_i given (N_i, n_i, X_i) is, for $X_i = x$,

$$h_i(m \mid N_i, n_i, x) = \frac{\binom{m}{x}\binom{N_i - m}{n_i - x}}{\binom{N_i + 1}{N_i - n_i}}, \qquad m = x, \ldots, N_i - n_i + x \tag{6.2.15}$$

Hence the Bayes estimator of $P_i = M_i/N_i$ is the posterior expectation

$$\hat{P}_i = \frac{1}{N_i\binom{N_i + 1}{N_i - n_i}} \sum_{m=x}^{N_i - n_i + x} m \binom{m}{x}\binom{N_i - m}{n_i - x} = \frac{x + 1}{n_i + 2} \cdot \left(1 + \frac{2}{N_i}\right) - \frac{1}{N_i}.$$

Since $M_1, \ldots, M_k$ are priorly independent and $X_1, \ldots, X_k$, given $M_1, \ldots, M_k$, are independent, the posterior joint density of $M_1, \ldots, M_k$ is the product $\prod_{i=1}^k h_i(m_i \mid N_i, n_i, X_i)$. Hence the Bayes estimator of $\theta = \sum_{i=1}^k \lambda_i M_i/N_i$ is

$$\hat{\theta} = \sum_{i=1}^{k} \lambda_i \left[\frac{X_i + 1}{n_i + 2} \cdot \frac{N_i + 2}{N_i} + \frac{1}{N_i}\right]. \tag{6.2.16}$$

The associated Bayes posterior risk is the posterior variance of θ, given $(X_1, \ldots, X_k)$. Again, due to the posterior independence of $M_1, \ldots, M_k$ and since

$$\frac{1}{\binom{N+1}{N-n}} \sum_{m=x}^{N-n+x} m^2 \binom{m}{x}\binom{N-m}{n-x} \tag{6.2.17}$$

$$= (N+2)(N+3)\frac{(x+1)(x+2)}{(n+2)(n+3)} - 3(N+2)\frac{x+1}{n+2} + 1,$$

the posterior risk of $\hat{\theta}$ is

$$R_H(\hat{\theta}, x) = \sum_{i=1}^{k} \lambda_i^2 \frac{(x_i + 1)(n_i + 1 - x_i)}{(n_i + 2)^2(n_i + 3)} \cdot \frac{N_i + 2}{N_i}\left(1 - \frac{n_i}{N_i}\right). \tag{6.2.18}$$

Taking the expectation of (6.2.18) with respect to the marginal distribution (6.2.14) and noticing that

$$\frac{1}{n+1}\sum_{y=1}^{n+1} \frac{y(n+2-y)}{(n+2)^2(n+3)} = \frac{1}{6(n+2)}, \tag{6.2.19}$$

we obtain that the prior risk associated with the Bayes estimator $\hat{\theta}$ is

$$R(H, \hat{\theta}) = \frac{1}{6}\sum_{i=1}^{k} \lambda_i^2 \frac{N_i + 2}{N_i(n_i + 2)}\left(1 - \frac{n_i}{N_i}\right). \tag{6.2.20}$$

■

In the rest of the section we reproduce the main theorem of DeGroot and Rao [1] concerning Bayes estimators under convex loss. DeGroot and Rao studied the structure of Bayes estimators when the loss function is $L(\theta, d) = |\theta - d|^k$, $k = 1, 2, \ldots$. We know that when $k = 1$ the Bayes estimator is the median of the posterior distribution of θ. When $k = 2$ it is the mean of the posterior distribution. The general theory developed by DeGroot and Rao is restricted to the symmetric case, that is, $W(t) = W(-t)$ for all real t. The function $W(t)$ does not have to be strictly convex.

Let $W'(t)$ designate the derivative of $W(t)$ at $t \neq 0$. If $W(t)$ is not differentiable at t, take the left derivative if $t < 0$ and the right derivative if $t > 0$. Then DeGroot and Rao prove the following theorem.

Theorem 6.2.2. (DeGroot and Rao [1].) *An estimator $\tilde{\theta} \in \mathfrak{D}$ is a Bayes estimator if and only if it satisfies a.s. the following inequalities:*

$$(6.2.21) \qquad \int_{\{\theta \geqslant \hat{\theta}(X)\}} W'(\theta - \hat{\theta}(X))H(d\theta \mid X) \geqslant \int_{\{\theta < \hat{\theta}(X)\}} W'(\hat{\theta}(X) - \theta)H(d\theta \mid X),$$

and

$$(6.2.22) \qquad \int_{\{\theta > \hat{\theta}(X)\}} W'(\theta - \hat{\theta}(X))H(d\theta \mid X) < \int_{\{\theta \leqslant \hat{\theta}(X)\}} W'(\hat{\theta}(X) - \theta)H(d\theta \mid X).$$

Moreover, if $W(t)$ is strictly convex there is, for each $x \in \mathfrak{X}$, a unique Bayes estimator $\hat{\theta}(X)$ satisfying (6.2.21) and (6.2.22). If $W'(0) = 0$ the inequalities (6.2.21) and (6.2.22) reduce to the equation

$$(6.2.23) \qquad \int_{\{\theta > \hat{\theta}(X)\}} W'(\theta - \hat{\theta}(X))H(d\theta \mid X) = \int_{\{\theta < \hat{\theta}(X)\}} W'(\hat{\theta}(X) - \theta)H(d\theta \mid X).$$

We do not provide here the proof of this theorem, which is long and technical, but rather we illustrate it with several examples.

Example 6.5. (i) We start with the case of a loss function $W(t) = |t|$, $-\infty < t < \infty$. Here

$$W'(t) = \begin{cases} 1, & \text{if } t > 0 \\ -1, & \text{if } t < 0. \end{cases}$$

Hence inequalities (6.2.21) and (6.2.22) become

$$\int_{\theta \geqslant \hat{\theta}(X)} H(d\theta \mid X) \geqslant \int_{\theta < \hat{\theta}(X)} H(d\theta \mid X),$$

and

$$\int_{\theta > \hat{\theta}(X)} H(d\theta \mid X) < \int_{\theta \leqslant \hat{\theta}(X)} H(d\theta \mid X).$$

The value $\hat{\theta}(X)$ for which these inequalities are satisfied simultaneously is defined as the median of $H(\theta \mid X)$. The Bayes estimator $\hat{\theta}(X)$ is unique.

(ii) Let $W(t) = t^2$. Here $W'(0) = 0$ and $W(t) = 2t$. Thus, according to (6.2.23), the Bayes estimator (unique!) should satisfy the equation

$$\int_{\{\theta > \hat{\theta}(X)\}} (\theta - \hat{\theta}(X))H(d\theta \mid X) = \int_{\theta \leqslant \hat{\theta}(X)\}} (\hat{\theta}(X) - \theta)H(d\theta \mid X) \quad \text{a.s.}$$

From this equation we immediately obtain the special case

$$\hat{\theta}(X) = \int \theta H(\mathrm{d}\theta \mid X) \quad \text{a.s. } [\mu].$$

(iii) Consider the following special case: $X_1, X_2, \ldots, X_n$ are i.i.d. having a normal distribution $\mathcal{N}(\theta, 1)$, where $-\infty < \theta < \infty$. Let θ have a prior $\mathcal{N}(0, \tau^2)$ distribution. It is easy to verify that the posterior distribution of θ, given $(X_1, \ldots, X_n)$, is $\mathcal{N}(\bar{X}_n(1 + (1/n\tau^2))^{-1}, (1/n)(1 + (1/n\tau^2))^{-1})$, where $\bar{X}_n$ is the sample mean. Let $W(t) = |t|^{k+1}$, $k = 1, 2, \ldots$. Then

$$W'(t) = \begin{cases} (k+1)t^k, & t \geqslant 0, \\ -(k+1)t^k, & t < 0, \end{cases}$$

and $W'(0) = 0$. Then the essentially unique Bayes estimator of θ should satisfy (6.2.23). We notice that if $\xi_n = \bar{X}_n(1 + (1/n\tau^2)^{-1}$ and $\omega_n = (1/n)(1 + (1/n\tau^2))^{-1}$ then, since the above posterior normal distribution is symmetric about ξ_n,

$$\text{(6.2.24)} \quad \int_{\xi_n}^{\infty} (\theta - \xi_n)^k H(\mathrm{d}\theta \mid \bar{X}_n)$$

$$= \omega_n^{k/2} \int_0^{\infty} u^k \varphi(u)\, \mathrm{d}u$$

$$= \omega_n^{k/2} \int_{-\infty}^{0} (-u)^k \varphi(u)\, \mathrm{d}u = \int_{-\infty}^{\xi_n} (\xi_n - \theta)^k H(\mathrm{d}\theta \mid \bar{X}_n) \quad \text{a.s.}$$

Hence (6.2.23) is satisfied and ξ_n is the Bayes estimator. ■

6.3. GENERALIZED BAYES ESTIMATORS

Generalized Bayes estimators are regular limits of sequences of Bayes estimators which can be formally expressed as Bayes estimators. For example, consider the case where X has an $\mathcal{N}(\theta, 1)$ distribution. As shown in Example 6.4, a sequence of Bayes estimators of θ for any loss function $W(t) = |t|^k$, $k \geqslant 1$, is $\hat{\theta}_\tau(X) = X(1 + (1/\tau^2))^{-1}$, $0 < \tau < \infty$; X is the regular limit of $\hat{\theta}_\tau(X)$ as $\tau \to \infty$. However, X itself is not a Bayes estimator. Can X be expressed formally as a Bayes estimator? If $W(t) = t^2$, the Bayes estimators are the posterior means of θ. Since

$$\frac{1}{\sqrt{(2\pi)}} \int_{-\infty}^{\infty} \exp\{-\tfrac{1}{2}(\theta - X)^2\}\, \mathrm{d}\theta = 1, \qquad \text{all } -\infty < X < \infty,$$

$$\frac{1}{\sqrt{(2\pi)}} \int_{-\infty}^{\infty} \theta \exp\{-\tfrac{1}{2}(\theta - X)^2\}\, \mathrm{d}\theta = X, \qquad \text{all } -\infty < X < \infty.$$

The variable X can therefore be expressed *formally* as a Bayes estimator for the (prior) Lebesgue measure $\mathrm{d}\theta$.

Generally, let $f(X;\theta)$ be the density function of an observed random variable X, $\theta \in \Theta$. Let $\xi_n(\theta)$ be a sequence of prior distribution functions on Θ. Let $\hat{\theta}_n$ be the corresponding sequence of Bayes estimators, with respect to the loss function $L(\theta, d) = (\theta - d)^2$. Let $\theta^*(X) = \lim_{n\to\infty} \hat{\theta}_n(X)$ a.s. Can we determine a sigma-finite measure $G(d\theta)$ on $(\Theta, \mathfrak{J})$ such that

$$\theta^*(X) = \frac{\int \theta f(X;\theta)G(d\theta)}{\int f(X;\theta)G(d\theta)} \qquad ? \tag{6.3.1}$$

If there exists such a $G(d\theta)$, θ^* is a generalized Bayes estimator.

Sacks [1] established sufficient conditions under which the above question has an affirmative answer. We present here his main result.

Let $f(X;\theta)$ be a 1-parameter exponential density with respect to $\mu(dx)$. That is, Θ is an interval on the real line. We assume that

$$\int e^{\theta x}\mu(dx) < \infty, \qquad \text{all } \theta \in \Theta, \tag{6.3.2}$$

and the density function of X with respect to $\mu(dx)$ is of the general form

$$f(x;\theta) = \rho(\theta)e^{\theta x}; \qquad \rho^{-1}(\theta) = \int e^{\theta X}\mu(dx). \tag{6.3.3}$$

We further assume that $\mu(dx)$ is such that Θ can be considered as the whole real line, that is, $\Theta = (-\infty, \infty)$. There is no loss of generality due to this assumption; if (6.3.2) does not hold for every real θ, modify $\mu(dx)$ by multiplying it by a proper non-negative function of X. For example, if $\mu^*(dx) = dx$ set $\mu(dx) = e^{-\frac{1}{2}x^2}\,dx$.

Let $\xi_n(\theta)$ be a prior probability measure on $(\Theta, \mathcal{F})$ such that

$$\int \theta^2 e^{\theta X}\rho(\theta)\xi_n(d\theta) < \infty \quad \text{a.s. } [\mu].$$

We notice that in this case the Bayes estimator of θ for a squared-error loss can be obtained as the derivative, with respect to X, of the logarithm of the Fourier-Stieltjes transform

$$\mathfrak{L}_n(X) = \int_{-\infty}^{\infty} e^{\theta X}\rho(\theta)\xi_n(d\theta). \tag{6.3.4}$$

Indeed,

$$\frac{\partial}{\partial X}\log \mathfrak{L}_n(X) = \frac{\partial/\partial X \int_{-\infty}^{\infty} e^{\theta X}\rho(\theta)\xi_n(d\theta)}{\int_{-\infty}^{\infty} e^{\theta X}\rho(\theta)\xi_n(d\theta)} = \frac{\int_{-\infty}^{\infty} \theta e^{\theta X}\rho(\theta)\xi_n(d\theta)}{\int_{-\infty}^{\infty} e^{\theta X}\rho(\theta)\xi_n(d\theta)}. \tag{6.3.5}$$

Differentiation under the integral sign in (6.3.5) is allowed, since $\int_{-\infty}^{\infty} |\theta|\, e^{\theta X}\rho(\theta)\xi_n(d\theta) < \infty$ a.s. $[\mu]$.

To illustrate the method used by Sacks in the proof of his main results we follow the example given in the first section of Sacks [1]. Later we formulate the more general assumptions and state the theorems, deleting the highly technical proofs.

Consider the $\mathcal{N}(\theta, 1)$ case, in which we can write $f(X; \theta) = e^{X\theta} \cdot e^{-\frac{1}{2}\theta^2}$. Suppose that θ is restricted to the interval $[0, \infty)$ and that $W(|d - \theta|) = (d - \theta)^2$. Let $\{\xi_n\}$ be a sequence of prior distributions on $[0, \infty)$ and $\{\hat{\theta}_n(X)\}$ the corresponding sequence of Bayes estimators (all belong to $(0, \infty)$), that is,

$$\hat{\theta}_n(X) = \frac{\int_0^\infty \theta e^{X\theta - \frac{1}{2}\theta^2} \xi_n(d\theta)}{\int_0^\infty e^{X\theta - \frac{1}{2}\theta^2} \xi_n(d\theta)}. \tag{6.3.6}$$

Suppose that $\hat{\theta}_n(X) \to \hat{\theta}(X)$ a.s. and $\hat{\theta}(X) < \infty$. The assumption that $\hat{\theta}(X) < \infty$ leads to the conclusion that there exists a finite positive value A such that for *all* $v > 0$,

$$\limsup_{n\to\infty} \frac{\xi_n(v)}{\xi_n(A)} < \infty. \tag{6.3.7}$$

This is proven by negation in the following manner. If (6.3.7) is not true, then for each $0 < A < \infty$ there exists a v, $A < v < \infty$, and a subsequence $\{\xi_{n_j}\}$ such that $\lim_{j\to\infty} (\xi_{n_j}(v))/(\xi_{n_j}(A)) = \infty$. Let $\gamma(A \mid X) = \inf\{f(X; \theta);\ \theta \in [A, v]\}$. For any X,

$$\begin{aligned}\lim_{j\to\infty} \int_0^\infty f(X; \theta) \frac{\xi_{n_j}(d\theta)}{\xi_{n_j}(A)} &\geqslant \lim_{j\to\infty} \int_A^v f(X; \theta) \frac{\xi_{n_j}(d\theta)}{\xi_{n_j}(A)} \\ &\geqslant \gamma(A \mid X) \lim_{j\to\infty} \frac{\xi_{n_j}(v) - \xi_{n_j}(A)}{\xi_{n_j}(A)} = \infty, \quad \text{a.s.}\end{aligned} \tag{6.3.8}$$

Hence

$$\hat{\theta}(X) = \lim_{j\to\infty} \hat{\theta}_n(X) = \lim_{j\to\infty} \frac{\left\{\int_0^A \theta f(X; \theta) \frac{\xi_{n_j}(d\theta)}{\xi_{n_j}(A)} + \int_A^\infty \theta f(X; \theta) \frac{\xi_{n_j}(d\theta)}{\xi_{n_j}(A)}\right\}}{\left(\int_0^A + \int_A^\infty\right) f(X; \theta) \frac{\xi_{n_j}(d\theta)}{\xi_{n_j}(A)}}. \tag{6.3.9}$$

But according to (6.3.8), since

$$\int_0^A f(X; \theta) \frac{\xi_{n_j}(d\theta)}{\xi_{n_j}(A)} \leqslant \sup_{0 \leqslant \theta \leqslant A} f(X; \theta) < \infty,$$

$$\hat{\theta}(X) = \lim_{j\to\infty} \frac{\int_A^\infty \theta f(X; \theta) \frac{\xi_{n_j}(d\theta)}{\xi_{n_j}(A)}}{\int_A^\infty f(X; \theta) \frac{\xi_{n_j}(d\theta)}{\xi_{n_j}(A)}} \geqslant A \quad \text{a.s. } [\mu], \tag{6.3.10}$$

for *all* $A > 0$. Hence, $\hat{\theta}(X) = \infty$ a.s. $[\mu]$, which contradicts the previous assumption.

Define the function $F_n(\theta) = \xi_n(\theta)/\xi_n(A)$. Since $\xi_n(\theta)$ is a prior distribution, $F_n(\theta)$ has the properties of a distribution function but $F_n(+\infty) = 1/\xi_n(A) > 1$. According to the first theorem of Helly-Bray there exists a subsequence $\{F_{n_j}(\theta)\}$ and a nondecreasing function $F(\theta)$ such that $F_{n_j}(\theta) \to F(\theta)$ at all the continuity points of F. To simplify the notation, assume that $F_n \to F$ (weakly). The next step is to verify that

$$\limsup_{n\to\infty} \int_0^\infty f(X;\theta)F(d\theta) < \infty \quad \text{a.s. } [\mu]. \tag{6.3.11}$$

This is also proven by negation. We can show that if (6.3.11) is not true then $\hat{\theta}(y) = \infty$ for all $y \geqslant X$. This contradicts the assumption that $\hat{\theta}(X) < \infty$ a.s. $[\mu]$. Finally, we have to prove that

$$\lim_{n\to\infty} \int_0^\infty \theta f(X;\theta)F_n(d\theta) = \int_0^\infty \theta f(X;\theta)F(d\theta) \quad \text{a.s. } [\mu]. \tag{6.3.12}$$

Since $\theta f(X;\theta) = \theta e^{X\theta - \frac{1}{2}\theta^2}$ is a continuous function of θ for each X, we need only to establish the uniform integrability, that is,

$$\lim_{A\to\infty} \overline{\lim_{n\to\infty}} \int_A^\infty \theta f(X;\theta)F_n(d\theta) = 0 \quad \text{a.s.} \tag{6.3.13}$$

To prove (6.3.13) we notice that for $Y > X$,

$$\begin{aligned}\int_A^\infty \theta f(X;\theta)F_n(d\theta) &= \int_A^\infty \theta e^{(X-Y)\theta} f(Y;\theta)F_n(d\theta) \\ &\leqslant R(A)\int_A^\infty f(Y;\theta)F_n(d\theta),\end{aligned} \tag{6.3.14}$$

where

$$R(A) = \sup_{\theta \geqslant A} \{\theta e^{(X-Y)\theta}\}. \tag{6.3.15}$$

Since $R(A) \to 0$ as $A \to \infty$, (6.3.13) is violated and we have

$$\overline{\lim_{n\to\infty}} \int_A^\infty f(Y;\theta)F_n(d\theta) = \infty$$

for all $Y > X$. But this contradicts (6.3.11).

In the above example of normal densities we have shown how the sigma-finite measure $F(d\theta)$ can be constructed so that if $\hat{\theta}_n(X) \to \hat{\theta}(X)$ a.s., $\hat{\theta}(X)$ is a formal Bayes estimator against F.

The following three assumptions are required for the general discussion:

A.1 The loss function $L(\theta, d)$ is non-negative, finite, and continuous in both variables. Moreover,

$$\lim_{|d|\to\infty} L(\theta, d) = \infty$$

uniformly on compacts of θ.

A.2 If $\theta > d > d'$ then $L(\theta, d) < L(\theta, d')$ and $\inf_{d\leqslant\theta} \{L(\theta, d') - L(\theta, d)\} > 0$. Conversely, if $\theta < d < d'$ then $L(\theta, d) < L(\theta, d')$ and $\inf_{\theta\leqslant d} \{L(\theta, d') - L(\theta, d)\} > 0$.

A.3 For each d in $(-\infty, \infty)$ and every $\epsilon > 0$,

$$\sup_{\theta\leqslant 0} \{e^{\theta\epsilon} L(\theta, d)\} < \infty,$$

$$\sup_{\theta\geqslant 0} \{e^{-\theta\epsilon} L(\theta, d)\} < \infty,$$

and

$$\lim_{A\to\infty} \sup_{\theta\leqslant A} \{e^{-\theta\epsilon} L(\theta, d)\} = \lim_{A\to\infty} \sup_{\theta\leqslant -A} \{e^{\theta\epsilon} L(\theta, d)\}.$$

The loss functions $L(\theta, d) = |d - \theta|^p$, $p > 0$, satisfy these assumptions. In parallel to the above illustration the following lemmas and theorem are proven.

Lemma 6.3.1. *There exists a subsequence* $\{\xi_{n_k}\}$ *and a number* A, $-\infty < A < \infty$ *such that for all* $v < \infty$,

$$\overline{\lim_{k\to\infty}} \frac{\xi_{n_k}(-v, v)}{\xi_{n_k}(-A, A)} < \infty. \tag{6.3.16}$$

The quantity $\xi_{n_k}(A, B)$ is the prior probability of the interval (A, B). Equation 6.3.16 is a generalization of (6.3.7). Define $F_{n_k}(d\theta) = \xi_{n_k}(d\theta)/\xi_{n_k}(-A, A)$; $F_{n_k} \xrightarrow{W} F$.

Lemma 6.3.2. *For almost all* $X[\mu]$,

$$\overline{\lim_{k\to\infty}} \int_{-\infty}^{\infty} f(X; \theta) F_n(d\theta) < \infty. \tag{6.3.17}$$

This is a generalization of (6.3.11).

Lemma 6.3.3. *For almost all* $X[\mu]$,

$$\lim_{n\to\infty} \int_{-\infty}^{\infty} f(X; \theta) F_n(d\theta) = \int_{-\infty}^{\infty} f(X; \theta) F(d\theta) < \infty, \tag{6.3.18}$$

and

$$\lim_{n\to\infty} \int_{-\infty}^{\infty} L(\theta, d) f(X; \theta) F_n(d\theta) = \int_{-\infty}^{\infty} L(\theta, d) f(X; \theta) F(d\theta) < \infty, \tag{6.3.19}$$

uniformly on compacts of d. The right-hand side of (6.3.19) *is continuous in d.*

Finally, Sacks proved the following.

Theorem 6.3.4. *Under A.1–A.3, if* $\{\hat{\theta}_n\}$ *is a sequence of Bayes estimators and* $\hat{\theta}_n \to \hat{\theta}$ *a.s., then there exists a sigma-finite measure* $F(\mathrm{d}\theta)$ *on* $(-\infty, \infty)$ *such that, for almost all* $X[\mu]$, $\int L(\theta, d(X))f(X; \theta)F(\mathrm{d}\theta)$ *is minimized by* $\hat{\theta}(X)$.

We illustrate the above results by the case of Example 6.3. The random variable there, T_n, has a 1-parameter exponential distribution. The parameter space Θ of this exponential distribution is $(0, \infty)$. Set $\mathfrak{D} = (-\infty, \infty)$ and the loss function $L(\theta, d) = (d - e^{-t_0\theta})^2$. It is easy to verify that A.1–A.3 are satisfied. In Example 6.3 we have shown that the Bayes estimators against the prior measures $\xi_{\tau,\nu}(\mathrm{d}\theta) = (1/\tau^\nu\Gamma(\nu))\theta^{\nu-1}e^{-\theta/\tau}\,\mathrm{d}\theta$ are $\hat{g}_{\tau,\nu}(T_n) = (1 + [t_0/T_n + (1/\tau)])^{-(n+\nu)}$. Set $\nu = 1$, and for each $\tau > 1$ let $a_\tau \in (0, \infty)$ be such that $\xi_{\tau,1}(0, a_\tau) = 1/\tau$. Obviously, $a_\tau = -\tau \ln (1 - 1/\tau)$. Consider the sequence of sigma-finite measures

$$\xi_\tau^*(\mathrm{d}\theta) = e^{-\theta/\tau}\,\mathrm{d}\theta = \frac{\xi_{\tau,1}(\mathrm{d}\theta)}{\xi_{\tau,1}(0, a_\tau)}\,.$$

As $\tau \to \infty$, $\xi_\tau^*(\mathrm{d}\theta)$ converges weakly to the Lebesgue measure $\mathrm{d}\theta$. Thus, $F(\mathrm{d}\theta) = \mathrm{d}\theta$. Furthermore,

$$\lim_{\tau\to\infty} \hat{g}_{\tau,1}(T_n) = \left(1 + \frac{t_0}{T_n}\right)^{-(n+1)}.$$

Thus according to Theorem 6.3.4, this limit is a generalized Bayes estimator against $F(\mathrm{d}\theta) = \mathrm{d}\theta$. Indeed, the minimization of $\int_0^\infty (d - e^{-t_0\theta})^2 e^{-T_n\theta}\theta^n\,\mathrm{d}\theta$ yields the estimator

$$\hat{g}(T_n) = \frac{\int_0^\infty e^{-t_0\theta}e^{-T_n\theta}\theta^n\,\mathrm{d}\theta}{\int_0^\infty e^{-T_n\theta}\theta^n\,\mathrm{d}\theta} = \left(1 + \frac{t_0}{T_n}\right)^{-(n+1)}.$$

Farrel [2] generalized the results of Sacks and showed that a proper compactification of the parameter space Θ is essential for the problem studied here.

6.4. ASYMPTOTIC BEHAVIOR OF BAYES ESTIMATORS

Suppose that one derives a Bayes estimator of θ with respect to a given loss function $L(\theta, d)$ and a prior distribution $\xi(\theta)$. It is interesting to investigate the large sample properties of such an estimator in the framework of a non-Bayesian large sample theory. We remark that this kind of investigation has no place in a Bayesian theory since, if one believes that the parameter θ is indeed a random variable having a nondegenerate prior distribution $\xi(\theta)$, there is no sense in asking what the asymptotic properties of Bayes estimators are (as the sample size increases) when actually $\theta = \theta_0$. On the other hand, many statisticians apply the Bayes method without believing that the prior distributions which they have chosen are the correct ones, or even without believing that θ is a random variable. They use the Bayes method to derive estimators having certain properties, and then they study the properties of these estimators in a non-Bayesian framework. In many cases Bayes estimators are asymptotically consistent and often converge to the maximum likelihood estimator. The following example illustrates such a situation.

Example 6.6. Let $X_1, \ldots, X_n, \ldots$ be a sequence of i.i.d. random variables having a Poisson distribution $P(\lambda)$ with mean λ, $0 < \lambda < \infty$. The objective is to estimate the Poisson density

$$p(i; \lambda) = e^{-\lambda} \cdot \frac{\lambda^i}{i!}, \qquad i = 0, 1, \ldots.$$

Given the values of the first n observations, let $T_n = \sum_{i=1}^n X_i$; $T_n \sim P(n\lambda)$. We now derive the Bayes estimator of $p(i, \lambda)$ for a squared-error loss function and the $\mathcal{G}(1, 1)$ prior distribution of λ. This Bayes estimator, say $\hat{p}(i; T_n)$, is the posterior expectation of $p(i; \lambda)$ given T_n. It is easy to verify that

$$\begin{aligned}\hat{p}(i; T_n) &= \frac{1}{i!} \cdot \frac{\int_0^\infty \lambda^i e^{-\lambda(n+2)} \lambda^{T_n}\, d\lambda}{\int_0^\infty e^{-\lambda(n+1)} \lambda^{T_n}\, d\lambda} \\ &= \frac{1}{i!} \cdot \frac{(T_n + i)!}{T_n!} \left(\frac{1}{n+2}\right)^i \left(1 - \frac{1}{n+2}\right)^{T_n+1}, \qquad i = 0, 1, \ldots.\end{aligned}$$

The Bayes estimator of the Poisson density $p(i; \lambda)$ is the negative binomial (Pascal) density $\hat{p}(i; T_n)$. It is easy to verify that $\lim_{n\to\infty} \hat{p}(i; T_n) = p(i; \lambda)$ a.s., for each $i = 0, 1, \ldots$. Indeed, $T_n/n \to \lambda$ a.s. Hence

$$\frac{(T_n + i)!}{T_n!} \frac{1}{(n+2)^i} \approx \frac{n^i}{(n+2)^i}\left(\lambda + \frac{1}{n}\right) \cdots \left(\lambda + \frac{i}{n}\right) \to \lambda^i \qquad \text{as } n \to \infty.$$

Furthermore, $(1 - 1/(n+2))^{T_n+1} \to e^{-\lambda}$ a.s. Thus the Bayes estimator $\hat{p}(i; T_n)$ is a strongly consistent estimator of $p(i; \lambda)$. The maximum likelihood estimator of $p(i; \lambda)$ is $(1/i!)(\bar{X}_n)^i e^{-\bar{X}_n}$, where $\bar{X}_n$ is the sample mean T_n/n. From the results of Chapter 5 we know that this M.L.E. is best asymptotically normal. Since $\hat{p}(i; T_n) - p(i; \lambda) \to 0$ a.s., the asymptotic distribution of $\sqrt{n}(\hat{p}(i; T_n) - p(i; \lambda))$ is $\mathcal{N}(0, \lambda[p(i; \lambda) - p(i-1; \lambda)]^2)$, $i = 0, 1, \ldots$, where $p(-1; \lambda) \equiv 0$. ■

The above example illustrates some general results of Freedman [1] and Fabius [1] which will be discussed later. We first present some heuristic results of Lindley [1]. We assume in this exposition that θ is real. Lindley displays the results for vector valued parameters.

Let $L(\theta, d)$ be a loss function, continuous in both d and θ. Let $\hat{\theta}_n$ be a maximum likelihood estimator of θ. We assume that the family of distribution functions $\mathfrak{F} = \{F_\theta; \theta \in \Theta\}$ satisfies the regularity conditions specified in Chapter 5 so that $\hat{\theta}_n$ is B.A.N. The quantity $\hat{\theta}_n$ is a function of n i.i.d. random variables $X_1, \ldots, X_n$ having a common distribution F_θ in $\mathfrak{F}$ with a density $f(x; \theta)$ with respect to $\mu(dx)$. Thus if $\mathbf{X}_n = (X_1, \ldots, X_n)'$ and

$$\Lambda(\mathbf{X}_n; \theta) = \sum_{i=1}^{n} \log f(X_i; \theta),$$

the posterior risk function is

$$R(d; \mathbf{X}_n) = \frac{\int L(\theta, d(\mathbf{X}_n)) \exp\{\Lambda(\mathbf{X}_n; \theta)\}\xi(d\theta)}{\int \exp\{\Lambda(\mathbf{X}_n; \theta)\}\xi(d\theta)}. \tag{6.4.1}$$

To obtain the Bayes estimator of θ we minimize (6.4.1). This is equivalent to finding a $d(\mathbf{X}_n)$ which minimizes

$$K(d; \mathbf{X}_n) = \int \xi(d\theta) L(\theta, d(\mathbf{X}_n)) \exp\{\Lambda(\mathbf{X}_n; \theta)\}. \tag{6.4.2}$$

Let $C(\mathbf{X}_n; \theta) = -1/n \cdot (\partial^2/\partial\theta^2)\Lambda(\mathbf{X}_n; \theta)$. Under the required regularity conditions, $C(\mathbf{X}_n; \theta)$ converges strongly to the Fisher information function $\mathbf{I}(\theta)$. Furthermore, the M.L.E. $\hat{\theta}_n$ is B.A.N. and we can write

$$\Lambda(\mathbf{X}_n; \theta) = \Lambda(\mathbf{X}_n; \hat{\theta}_n) - \frac{n}{2}(\hat{\theta}_n - \theta)^2 C(\mathbf{X}_n; \theta) + o_p(1) \qquad \text{as } n \to \infty.$$

Let $h(\theta)$ designate the prior density corresponding to $\xi(\theta)$, and let $\omega(\theta, d) = L(\theta, d)h(\theta)$. We assume that, for each d in $\mathfrak{D}$, $\omega(\theta, d)$ is a continuously differentiable function of θ, which is bounded away from zero at $\theta = \hat{\theta}_n$. Expanding $\omega(\theta, d)$ around $\theta = \hat{\theta}_n$ and considering only the first-order term, we represent $K(d; \mathbf{X}_n)$ asymptotically, as $n \to \infty$, by

$$\begin{aligned} &(6.4.3) \quad K(d; \mathbf{X}_n) \\ &\approx \omega(\hat{\theta}_n; d(\mathbf{X}_n)) \exp\{\Lambda(\mathbf{X}_n; \hat{\theta}_n)\} \cdot \int_{-\infty}^{\infty} \exp\left\{-\frac{n}{2} C(\mathbf{X}_n; \theta) \cdot (\hat{\theta}_n - \theta)^2\right\} d\theta \\ &= \sqrt{(2\pi)}\, \omega(\hat{\theta}_n; d(\mathbf{X}_n)) \prod_{i=1}^{n} f(X_i; \hat{\theta}_n)(C(\mathbf{X}_n; \hat{\theta}_n))^{-1/2}. \end{aligned}$$

Thus the problem of minimizing $K(d; \mathbf{X}_n)$ is asymptotically equivalent to the problem of minimizing $\omega(\hat{\theta}_n; d)$. But $\omega(\hat{\theta}_n; d) = 0$ when $d = \hat{\theta}_n$. Hence the Bayes estimator and the M.L.E. estimator are asymptotically equivalent. We can therefore call the M.L.E. an *asymptotically Bayes* estimator.

The exact form of the prior distribution is irrelevant, since in very large samples we can substitute the M.L.E. $\hat{\theta}_n$ for the unknown value of θ. If

$L(\theta, d)$ is a monotone increasing function of $|\theta - d|$, the above large sample rule of Lindley leads to the use of the M.L.E., irrespective of the prior distribution. As implied from (6.4.3), this theory may be appropriate for *very* large samples. Not much is known about its medium sample size properties. Bickel and Yahav [2] provide a more rigorous proof of a similar result for the case of a 1-parameter exponential family and a squared-error loss. Their result is discussed in Section 6.10.

Freedman [1] studied the asymptotic properties of Bayes estimators of discrete distribution functions. The family of distribution functions which he considered consists of all distributions concentrated on the set I of all positive integers. Let Λ be the family of all probability density functions on I together with all defective probability density functions; that is, $\Lambda = \{\lambda \mid 0 \leqslant \lambda(i) \leqslant 1, \text{all } i \in I, \sum_{i \in I} \lambda(i) \leqslant 1\}$. On the set Λ we generate a Borel sigma-field $\mathcal{F}$ and choose a prior probability measure $\xi(\mathrm{d}\lambda)$ on $(\Lambda, \mathcal{F})$. In Example 6.5 we considered a more restricted family of density functions, namely, all the Poisson distributions $P(\lambda)$, where $0 < \lambda < \infty$. Thus the set $\Lambda = \{\lambda; 0 < \lambda < \infty\}$ and $\mathcal{F}$ is the common Borel sigma-field on Λ. The prior probability measure in Example 6.5 is the one induced by the exponential distribution on $(0, \infty)$.

The *topological carrier* $\mathcal{C}(\xi)$ of the prior probability measure ξ is the smallest compact set in Λ for which the ξ-probability is 1. Thus $\lambda \in \mathcal{C}(\xi)$ if and only if $\lambda \in \Lambda$ and every Λ-neighborhood of λ has a positive ξ-probability. In Example 6.5 the topological carrier of ξ is $(0, \infty)$.

Let $X_1, X_2, \ldots$ be a sequence of i.i.d. (discrete) random variables having a distribution λ belonging to Λ. Given the values $X_1 = i_1, \ldots, X_n = i_n$ of n observations, we determine the *posterior* probability measure on $(\Lambda, \mathcal{F})$,

$$\xi(A \mid X_n) = \frac{\int_A (\prod_{i=1}^{n} \lambda(X_i))\xi(\mathrm{d}\lambda)}{\int (\prod_{v=1}^{n} \lambda(X_v))\xi(\mathrm{d}\lambda)}, \qquad A \in \mathcal{F}. \tag{6.4.4}$$

Let $\theta(i)$ be the "true" distribution of $X_1, X_2, \ldots; \theta \in \Lambda$. For a squared-error loss function the Bayes estimator of θ, given $(X_1, \ldots, X_n)$, is

$$\hat{\theta}(i \mid X_n) = \frac{\int \lambda(i)(\prod_{v=1}^{n} \lambda(X_v))\xi(\mathrm{d}\lambda)}{\int (\prod_{v=1}^{n} \lambda(X_v))\xi(\mathrm{d}\lambda)}. \tag{6.4.5}$$

We now present the main results of Freedman without proofs. The proofs are very technical and tedious. We start with the definition of (θ, ξ) consistency.

If $\xi(\mathrm{d}\lambda)$ is a prior probability measure on $(\Lambda, \mathcal{F})$, and $\theta \in \Lambda$, then (θ, ξ) is called consistent if and only if the posterior distribution $\xi(\mathrm{d}\lambda \mid X_n) \to \xi_\theta$, a.s. $[P_\theta]$, where ξ_θ *designates the prior distribution concentrated on* θ.

Theorem 6.4.1. *Let $\theta \in \Lambda$ and $\{i \mid \theta(i) > 0\}$ be finite. Let ξ be a prior probability measure on $(\Lambda, \mathcal{F})$. Then, (θ, ξ) is consistent if and only if $\theta \in \mathcal{C}(\xi)$.*

Although the proof of this theorem is very delicate, the result itself is intuitively clear. We can illustrate this very simply. Suppose that $\lambda(i)$ is a binomial distribution $i = 0, \ldots, k$, $B(k; \lambda), 0 \leqslant \lambda \leqslant 1$. Suppose that $\theta = \frac{1}{3}$ but $\xi\,(d\theta)$ is a uniform prior distribution on the interval $[\frac{1}{2}, 1]$; that is, $\mathcal{C}(\xi) = [\frac{1}{2}, 1]$. Given $X_1, \ldots, X_n$, the complete sufficient statistic is $T_n = \sum_{i=1}^n X_i$ and $T_n \sim B(nk, \lambda)$. The posterior density of λ, given T_n, is

$$\xi(\mathrm{d}\lambda \mid T_n) = \frac{\lambda^{T_n}(1 - \lambda)^{nk-T_n}\,\mathrm{d}\lambda}{\int_{1/2}^{1} \lambda^{T_n}(1 - \lambda)^{nk-T_n}\,\mathrm{d}\lambda}, \qquad \text{if } \tfrac{1}{2} \leqslant \lambda \leqslant 1.$$

Thus for each n and all T_n, $\xi(\mathrm{d}\lambda \mid T_n)$ is positive only on λ values in the interval $[\frac{1}{2}, 1]$. Thus the posterior distribution $\xi(\mathrm{d}\lambda \mid T_n)$ *cannot* converge to $\xi_{1/3}$.

Let $N_n(i) = \sum_{v=1}^n I_{\{X_v = i\}}$, which is the number of independent observations equal to i among the n observed ones. Obviously, $\sum_{i \in I} N_n(i) = n$. It is easy to verify that when I is finite, $N_n(i)/n$ is the M.L.E. of $\lambda(i)$. If θ is the true parameter value, $\lim_{n\to\infty} N_n(i)/n = \theta(i)$ a.s. $[P_\theta]$. The following theorem states that asymptotically the Bayes estimator $\hat{\theta}(i \mid X_n)$ and the M.L.E. $N_n(i)/n$ converge a.s.

Theorem 6.4.2. *If I is finite then*

$$\lim_{n\to\infty} [\hat{\theta}(i \mid \mathbf{X}_n) - n^{-1}N_n(i)] = 0 \quad \text{a.s. } [P_\theta]. \tag{6.4.6}$$

Freedman proved also that when I is finite the posterior distribution of λ, given $\mathbf{X}_n$, when centered at its mean and rescaled by $\sqrt{n}$, converges a.s. $[P_\theta]$ to a normal distribution. This normal distribution is the asymptotic joint distribution of the properly standardized maximum likelihood estimators of $\lambda(1), \ldots, \lambda(k)$.

Theorem 6.4.1 is false without the assumption that $\{i \mid \theta(i) > 0\}$ if finite. Freedman [1] gives a counter-example and proves (see [1], Theorem 5) that if $\{i \mid \theta(i) > 0\}$ is infinite, there exists a prior distribution $\xi(\lambda)$ such that $\xi(\mathrm{d}\lambda \mid \mathbf{X}_n) \to \xi_q$ a.s. $[P_\theta]$, where $q \neq \theta$. Thus a condition must be imposed on the prior distribution $\xi(\lambda)$ which will insure the consistency of (θ, ξ) when $\{i \mid \theta(i) > 0\}$ is infinite. Freedman defines the notion of *tail-free* prior distributions and shows that this property is sufficient for consistency. The idea beyond this property is to be able to reduce the problem for the given prior probability measures to one of essentially a finite set $\{i \mid \theta(i) > 0\}$. Thus we have the following:

A prior probability measure $\xi(\mathrm{d}\lambda)$ *on* $(\Lambda, \mathcal{F})$ *is called tail-free if there exists a natural number* K *such that,* $\{\lambda(i); i = 1, \ldots, K\}$ *and* $\lambda(K+\nu)/[1 - \sum_{i=1}^{K+\nu-1} \lambda(i)]$ *are independent under* ξ, *for each* $\nu = 1, 2, \ldots$. The natural number K which specifies the tail may depend on ξ. We notice that

$$\frac{\lambda(K+\nu)}{1 - \sum_{i=1}^{K+\nu-1} \lambda(i)}, \qquad \nu = 1, 2, \ldots,$$

is the probability density restricted to the tail. Theorem 6.4.1 is then generalized to infinite sets I by adding the condition that $\xi(\mathrm{d}\lambda)$ is a tail-free prior measure.

Fabius [1] generalized some results of Freedman [1]. He provided a theorem for discrete distributions with a tail-free prior ξ for which $K = 0$. Its conditions guarantee that the posterior distribution of $\sqrt{n}\{\sum_{i=1}^{\infty} a(i)[\lambda(i) - \hat{\theta}(i \mid X_n)]\}$ converges a.s. $[P_\theta]$ to a normal distribution with zero mean and variance $\sum_{i=1}^{\infty} a^2(i)\theta(i) - \{\sum_{i=1}^{\infty} a(i)\theta(i)\}^2$, where $\sup_{1 \leqslant i < \infty} |a(i)| < \infty$. Fabius generalized the theory also to the case of random variables with continuous distributions.

In Freedman's paper [1] the reader can find several other relevant references to studies in this field.

Farrel [1] studied the asymptotic properties of generalized Bayes estimators when the family of distribution functions depends on a single real location parameter, θ. Thus let $X_1, \ldots, X_n, \ldots$ be a sequence of i.i.d. random variables having a common density function (with respect to the Lebesgue measure) $f(X - \theta)$, $-\infty < \theta < \infty$. Let $\lambda(\mathrm{d}\theta)$ be a sigma-finite measure on $(\Theta, \mathcal{C})$, and for a given set of n observations $X_1, \ldots, X_n$, let $\delta_n(X_1, \ldots, X_n)$ be an estimator of θ. Let $\{\hat{\theta}_n(X_1, \ldots X_n); n \geqslant 1\}$ be a sequence of generalized Bayes estimators of θ with respect to $\lambda(\mathrm{d}\theta)$ and a loss function $L(\theta, d)$. Such a generalized Bayes estimator, $\hat{\theta}_n(X_1, \ldots, X_n)$, is defined as

$$(6.4.7) \quad \int \lambda(\mathrm{d}\theta) L(\theta, \hat{\theta}_n(X_1, \ldots, X_n)) \prod_{i=1}^{n} f(X_i - \theta) = \inf_c \int \lambda(\mathrm{d}\theta) L(\theta, c) \prod_{i=1}^{n} f(X_i - \theta) \quad \text{a.s.}$$

We assume that the sequence $\{\hat{\theta}_n; n \geqslant 1\}$ is not vacuous. That is, there exists an integer $M_1 \geqslant 1$ such that if $n \geqslant M_1$, then $\hat{\theta}_n(X_1, \ldots, X_n)$ exists. Farrel considered loss functions $L(\theta, d)$ of the form $L(\theta, d) = W(\theta - d)$, which have the following monotonicity property

$$(6.4.8) \qquad \text{If } |t_1| > |t_2|, \quad \text{then} \quad W(t_1) \geqslant W(t_2).$$

Moreover, $W(t)$ satisfies the following conditions:

(i) if $\epsilon \neq 0$ then $W(\epsilon) > 0$;

(ii) $W(t)$ is either bounded or convex, and there exists a constant $K > 0$ such that

(6.4.9)
$$W(x + y) \leqslant K(W(x) + W(y)),$$
$$W(-x) \leqslant KW(x),$$

for all $-\infty < x, y < \infty$;

(iii) if $W(t)$ is bounded, it is continuous or $\lambda(d\theta)$ is nonatomic. If $W(t)$ is bounded, $\lim_{t \to 0} W(t) = 0$.

Let $\mathcal{E}_\infty$ be the set of all real valued functions defined on the non-negative integers, $n \geqslant 0$. Let e designate an element of $\mathcal{E}_\infty$, and let $\mathcal{F}$ be the smallest Borel sigma-field on $\mathcal{E}_\infty$. Let μ be a sigma-finite measure on $(\mathcal{E}_\infty, \mathcal{F})$ satisfying the following condition: If $n \geqslant 0$, and $A_0, \ldots, A_n \in \mathcal{B}$, and if

$$E = \{e; e(i) \in A_i, 0 \leqslant i \leqslant n\}$$

then

(6.4.10)
$$\mu(E) = \int_{A_0} \lambda(d\theta) \int_{A_1} \cdots \int_{A_n} \prod_{i=1}^{n} f(x_i - \theta)\, dx_i,$$

where $\mathcal{B}$ is the Borel sigma-field on the real line. Let X_n, $n \geqslant 1$, be a function on $\mathcal{E}_\infty$ such that $X_n(e) = e(n)$ for all $e \in \mathcal{E}_\infty$. Let $\mathcal{F}_n$ be the smallest Borel sigma-field generated by $\{X_1, \ldots, X_n\}$; $\mathcal{F}_n \subset \mathcal{F}$. We assume the following additional property.

(iv) There exists an integer $M_2 \geqslant 1$ such that if $n \geqslant M_2$, then μ restricted to $\mathcal{F}_n$ is a sigma-finite measure. Furthermore, there exists in $\mathcal{F}_{M_2}$ a sequence of sets $\{A_\nu; \nu \geqslant 1\}$ satisfying $A_\nu \subset A_{\nu+1}$, $\mu(A_\nu) < \infty$, $\cup_{\nu=1}^{\infty} A_\nu = \mathcal{E}_\infty$, and $\int_{A_\nu} W(e(0))\mu(de) < \infty$.

Define the function $\theta(e) = e(0)$ for all $e \in \mathcal{E}_\infty$. Farrel [1] has proven the following strong consistency property of generalized Bayes estimators.

Theorem 6.4.3. (Farrel [1].) *Under (i)–(iv), if $\{\hat{\theta}_n; n \geqslant 1\}$ is a sequence of estimators satisfying (6.4.7), for almost all $e \in \mathcal{E}_\infty[\mu]$,*

(6.4.11)
$$\lim_{n \to \infty} \hat{\theta}_n(X_1(e), \ldots, X_n(e)) = \theta(e).$$

For a proof of this theorem, see Farrel [1].

6.5. MINIMAX ESTIMATION

Much has been written on minimax procedures in statistical decision theory. In this chapter we consider minimax procedures for estimation only, and therefore the treatment of the subject here and in the following sections will be somewhat limited in scope. We return to minimax procedures later in the context of testing of hypotheses.

The minimax principle for estimation problems can be stated as follows: Let $L(\theta, d)$ be a specified loss function, where $\theta \in \Theta$ and d: $\mathfrak{X} \to \Theta$ is an estimator. Let $R(\theta, d)$ be the corresponding risk function, that is,

$$R(\theta, d) = \int L(\theta, d(x))F_\theta(dx).$$

Without loss of generality, assume that $R(\theta, d) \leqslant M < \infty$ for each (θ, d). *The minimax principle* is to choose d^* from the set $\mathfrak{D}$ of possible estimators so that

$$\sup_{\theta \in \Theta} R(\theta, d^*) = \inf_{d \in \mathfrak{D}} \sup_{\theta \in \Theta} R(\theta, d). \tag{6.5.1}$$

Accordingly, the minimax approach is to choose an estimator which protects against the largest risk possible when θ varies over Θ. We can also say that a minimax estimator is a Bayes estimator against a prior distribution on Θ, which is least favorable for the estimation problem. We explain this point as we proceed with the material. We exhibit examples in which the minimax principle leads to commonly used estimators and, on the other hand, cases in which the minimax estimator is not useful.

The following inequality is easily verified:

$$\sup_\theta \inf_d R(\theta, d) \leqslant \inf_d \sup_\theta R(\theta, d). \tag{6.5.2}$$

A minimax estimator exists if equality holds in (6.5.2). There is a considerable amount of published results on the existence of minimax estimators. The reader is especially referred to Chapter 2 of Wald [6]. We provide in this section two theorems which give sufficient conditions for an estimator to be minimax. A general minimax theorem for multiple testing is given in Section 9.1. We proceed with some comments on an extension of (6.5.1).

Let $\mathcal{H}$ be the class of *all* prior distributions on Θ. Each point θ_0 of Θ can be represented by a degenerate prior distribution ξ_{θ_0} of $\mathcal{H}$. Furthermore, for each d of $\mathfrak{D}$ and every ξ of $\mathcal{H}$,

$$R(\xi, d) = \int R(\theta, d)\xi(\mathrm{d}\theta) \leqslant \sup_\theta R(\theta, d).$$

Hence

$$\sup_{\xi \in \mathcal{H}} R(\xi, d) \leqslant \sup_\theta R(\theta, d).$$

Let $\mathcal{H}^* = \{\xi_\theta; \theta \in \Theta\}$, $\mathcal{H}^* \subset \mathcal{H}$. Hence $\sup\limits_\theta R(\theta, d) = \sup\limits_{\xi \in \mathcal{H}} R(\xi, d)$ for each $d \in \mathfrak{D}$, since $\sup\limits_\theta R(\theta, d) = \sup\limits_{\xi \in \mathcal{H}^*} R(\xi, d) \leqslant \sup\limits_{\xi \in \mathcal{H}} R(\xi, d)$. The class of estimators $\mathfrak{D}$ can be extended to consist of all randomized estimators $\pi(\theta \mid X)$. Let Ψ

be the class of all randomized estimators on Θ, $\mathfrak{D} \subset \Psi$. Therefore, for each $\xi \in \mathcal{H}$,

$$\inf_{\pi \in \Psi} R(\xi, \pi) \leqslant \inf_{d \in \mathfrak{D}} R(\xi, d).$$

Hence

$$\sup_{\xi \in \mathcal{H}} \inf_{\pi \in \Psi} R(\xi, \pi) \leqslant \sup_{\xi \in \mathcal{H}} \inf_{d \in \mathfrak{D}} R(\xi, d).$$

This means that the minimax risk can be reduced in general by considering the extended class Ψ of randomized estimators. In a general context, therefore, we define a minimax estimator π^* of Ψ to be a randomized estimator for which

$$\sup_{\xi \in \mathcal{H}} R(\xi, \pi^*) = \inf_{\pi \in \Psi} \sup_{\xi \in \mathcal{H}} R(\xi, \pi). \tag{6.5.3}$$

We have shown in Section 6.1 that if the loss function $L(\theta, d)$ is convex in d for each θ, we can restrict attention to nonrandomized estimators. In this case a minimax estimator will also be nonrandomized. Thus if the loss function is convex we define a minimax estimator d^* to be one in $\mathfrak{D}$ satisfying

$$\sup_{\xi \in \mathcal{H}} R(\xi, d^*) = \inf_{d \in \mathfrak{D}} \sup_{\xi \in \mathcal{H}} R(\xi, d). \tag{6.5.4}$$

Equations (6.5.1) and (6.5.4) are equivalent.

If the loss function is not convex, we can construct examples in which we can reduce the minimax risk by considering the class of all randomized estimators. One such example is provided by Hodges and Lehmann [1]. Their example consists of a family $\mathcal{F}$ of binomial distributions $B(n; \theta)$ with unknown θ, $0 \leqslant \theta \leqslant 1$. Let $\hat{\theta} \in [0, 1]$ be an estimator of θ. They consider a loss function $L(\theta, \hat{\theta}) = |\theta - \hat{\theta}|^\delta$ with $0 < \delta < 1$. This is a concave function of $|\theta - \hat{\theta}|$. Let $h(X)$ be a non-randomized minimax estimator with $\sup_{0 \leqslant \theta \leqslant 1} R(\theta, h) = M$. Define the randomized estimator

$$T_X = \begin{cases} h(X), & \text{if} \quad 0 < X < n, \\ h(X) + \alpha Y, & \text{if} \quad X = 0 \quad \text{or} \quad X = n, \end{cases}$$

where Y is a random variable independent of X, $Y = \pm 1$ with probability $\frac{1}{2}$ and α is a small positive number. They prove that for a suitable choice of α, $\sup_{0 \leqslant \theta \leqslant 1} R(\theta, T_X) < M$. Hence if the class of estimators contains all the randomized estimators as well, the minimax estimator is not in the subclass of nonrandomized estimators.

Now we present several theorems which establish sufficient conditions under which a specified nonrandomized estimator is minimax. The usefulness of these theorems will be shown in examples. All these theorems can be found in the papers of Hodges and Lehmann [1], Blyth [1], and Girshick and Savage [1].

Theorem 6.5.1. *Let $\mathfrak{F} = \{F_\theta; \theta \in \Theta\}$ be a family of distribution functions and $\mathfrak{D}$ a class of estimators of Θ. Suppose that d^* in $\mathfrak{D}$ is Bayes against a prior distribution $\xi^*(\theta)$ on Θ, and $R(\theta, d^*)$ = constant on Θ; then d^* is a minimax estimator of θ.*

Proof. Since $R(\theta, d^*) = \rho^*$ for all θ and d^* is Bayes against ξ^*, we have

$$\rho^* = \int R(\theta, d^*)\xi^*(d\theta) = \inf_{d \in \mathcal{H}} \int R(\theta, d)\xi^*(d\theta).$$

We implicitly assume here that the loss function $L(\theta, d)$ is such that (6.1.3) holds, and the Bayes estimator minimizes the prior risk. Moreover,

$$\rho^* = \inf_{d \in \mathfrak{D}} \int R(\theta, d)\xi^*(d\theta) \leqslant \sup_{\xi \in \mathcal{H}} \inf_{d \in \mathfrak{D}} \int R(\theta, d)\xi(d\theta)$$

$$\leqslant \inf_{d \in \mathfrak{D}} \sup_{\xi \in \mathcal{H}} \int R(\theta, d)\xi(d\theta) \leqslant \inf_{d \in \mathfrak{D}} \sup_{\epsilon \in \Theta} R(\theta, d).$$

On the other hand, since $\rho^* = R(\theta, d^*)$ for all $\theta \in \Theta$,

$$\rho^* = \sup_{\theta \in \Theta} R(\theta, d^*) \geqslant \inf_{d \in \mathfrak{D}} \sup_{\theta \in \Theta} R(\theta, d).$$

Thus

$$\sup_{\theta \in \Theta} R(\theta, d^*) = \inf_{d \in \mathfrak{D}} \sup_{\theta \in \Theta} R(\theta, d).$$

Hence d^* is minimax. (Q.E.D.)

In the following two examples we illustrate an application of Theorem 6.5.1.

Example 6.6. The following example was given by Hodges and Lehmann [12]. Let X be a random variable having a binomial distribution $B(n; \theta)$, $0 < \theta < 1$; θ is unknown. We show that

$$d^*(X) = \frac{X}{n} \cdot \frac{\sqrt{n}}{1 + \sqrt{n}} + \frac{1}{2(1 + \sqrt{n})} \tag{6.5.5}$$

is a minimax estimator of θ for the squared-error loss function. To verify it, consider the linear estimator $d_{\alpha,\beta}(X) = (\alpha/n)X + \beta$, $\alpha \neq 0$. The risk function of such an estimator is

$$\begin{aligned} R(\theta, d_{\alpha\beta}) &= E_\theta\left\{\left[\alpha \frac{X}{n} + \beta - \theta\right]^2\right\} \\ &= \beta^2 + \frac{\theta}{n}[1 - 2(1 - \alpha) + (1 - \alpha)^2 - 2n\beta(1 - \alpha)] \\ &\quad - \frac{\theta^2}{n}[1 - 2(1 - \alpha) + (1 - \alpha)^2(1 - n)]. \end{aligned} \tag{6.5.6}$$

Choose α^*, β^* so that $R(\theta, d_{\alpha^*\beta^*}) = \beta^{*2}$, all $0 < \theta < 1$. For this purpose we solve the system of equations

$$\begin{cases} 1 - 2(1-\alpha) + (1-\alpha)^2 - 2n\beta(1-\alpha) = 0, \\ 1 - 2(1-\alpha) + (1-\alpha)^2(1-n) = 0. \end{cases} \tag{6.5.7}$$

The two roots of (6.5.7) are

$$\alpha^* = \frac{\sqrt{n}}{1+\sqrt{n}}, \qquad \beta^* = \tfrac{1}{2}(1+\sqrt{n})^{-1}. \tag{6.5.8}$$

Hence the estimator $d^*(X)$ given by (6.5.5) has a constant risk $\frac{1}{4}(1+\sqrt{n})^{-2}$ for all $0 \leqslant \theta \leqslant 1$. It remains to show that $d^*(X)$ is Bayes against some prior distribution $\xi^*(\theta)$.

If the prior distribution of θ is a beta distribution, say $\beta(p, q)$, $0 < p$, $q < \infty$, then the posterior distribution of θ, given X, is $\beta(X+p, n-X+q)$. Hence the posterior expectation of θ is $E_\xi\{\theta \mid X\} = (X+p)/(n+p+q)$. This is the Bayes estimator of θ for a squared-error loss. It follows immediately that if $p = q = \sqrt{n}/2$, the Bayes estimator is

$$\frac{X + (\sqrt{n}/2)}{n+\sqrt{n}} = \frac{X}{n}\cdot\frac{\sqrt{n}}{1+\sqrt{n}} + \tfrac{1}{2}(1+\sqrt{n})^{-1} = d^*(X).$$

This completes the verification that (6.5.5) is a minimax estimator for a squared-error loss. It is interesting to compare the mean square error function of $d^*(X)$ to that of the M.V.U.E. $\hat{\theta} = X/n$, which is also the M.L.E. The mean square error coincides here with the risk function and is therefore the constant β^{*2} for $d^*(X)$. The risk function of $\hat{\theta}$ is, however, $\theta(1-\theta)/n$. It is easy to show that the mean square error of d^* is smaller than that of $\hat{\theta}$ if and only if $|\theta - \frac{1}{2}| < (1+2\sqrt{n})^{1/2}/2(1+\sqrt{n})$. The same method is used by Hodges and Lehmann [12] to establish the minimaxity of

$$d^*(X) = \frac{N}{n + [n(N-n)/(N-1)]^{1/2}} X + \frac{N}{2}\left\{1 - \frac{n}{n+[n(N-n)/(N-1)]^{1/2}}\right\},$$

as an estimator of the parameter D of the hypergeometric distribution, with density

$$f(X \mid N, D, n) = \frac{\binom{D}{X}\binom{N-D}{n-X}}{\binom{N}{n}}, \qquad X = 0, 1, \ldots, n. \qquad \blacksquare$$

The above result has been applied to estimation problems in statistical quality control, in sampling surveys, and so on. We now present another commonly applied theorem.

Theorem 6.5.2. *Let $\{\xi_k; k = 1, 2, \ldots\}$ be a sequence of prior distributions on Θ, and let $\{d_k; k = 1, 2, \ldots\}$ and $\{R(\xi_k, d_k); k = 1, 2, \ldots\}$ be the corresponding sequences of Bayes estimators and prior risks. If d^* is an estimator of θ whose risk function satisfies*

$$\sup_{\theta\in\Theta} R(\theta, d^*) \leqslant \limsup_{k\to\infty} R(\xi_k, d_k), \tag{6.5.9}$$

then d^* is minimax.

Proof. If d^* is not a minimax estimator, there exists an estimator $\tilde{d}$ such that

$$\sup_{\theta\in\Theta} R(\theta, \tilde{d}) < \sup_{\theta\in\Theta} R(\theta, d^*). \tag{6.5.10}$$

Moreover, since d_k are Bayes against ξ_k, $k \geqslant 1$,

$$R(\xi_k, d_k) \leqslant \int R(\theta, \tilde{d})\xi_k(d\theta) \leqslant \sup_{\theta} R(\theta, \tilde{d}) \tag{6.5.11}$$

for all $k \geqslant 1$. Hence from (6.5.10) and (6.5.11),

$$\limsup_{k\to\infty} R(\xi_k, d_k) < \sup_{\theta\in\Theta} R(\theta, d^*). \tag{6.5.12}$$

This contradicts (6.5.9). Hence d^* is minimax. (Q.E.D.)

Corollary 6.5.1. *If d^* is an estimator having a constant risk function, that is, $R(\theta, d^*) = \rho^*$ for all $\theta \in \Theta$, and if there exists a sequence of prior distributions $\{\xi_k\}$ such that $\lim_{k\to\infty} R(\xi_k, d_k) = \rho^*$, then d^* is minimax.*

We illustrate several applications of Theorem 6.5.2 and its corollary.

Example 6.7. In Example 6.1 we considered an estimation of the mean of a normal distribution with a 0–1 loss function. We have shown that the estimators $\hat{\theta}_\tau(\bar{X}_n) = \bar{X}_n(1 + (1/n\tau^2))^{-1}$, $\tau = \tau_1, \tau_2, \ldots$, are Bayes against the prior $\mathcal{N}(0, \tau^2)$ distributions. We show now that the sample mean $\bar{X}_n$ is minimax. The risk function of the sample mean is

$$R(\theta, \bar{X}_n) = P_\theta[|\bar{X}_n - \theta| > \delta] = 2[1 - \Phi(\delta\sqrt{n})] = \rho^* \tag{6.5.13}$$

for all θ, $-\infty < \theta < \infty$. We now show that the prior risks $R(\xi_\tau, \hat{\theta}_\tau) \to \rho^*$ as $\tau \to \infty$. It is easy to verify that the risk function of $\hat{\theta}_\tau(\bar{X}_n)$ is

(6.5.14)

$$R(\theta, \hat{\theta}_\tau) = 2 - \left\{\Phi\left(\sqrt{n}\,\frac{\delta + \theta(1 + n\tau^2)^{-1}}{(1 + (1/n\tau^2))^{-1}}\right) + \Phi\left(\frac{\delta - \theta(1 + n\tau^2)^{-1}}{(1 + (1/n\tau^2))^{-1}}\sqrt{n}\right)\right\}.$$

The Bayes prior risk is

$$R(\xi_\tau, \hat{\theta}_\tau) = E_{\xi_\tau}\{R(\theta, \hat{\theta}_\tau)\}. \tag{6.5.15}$$

Since the function under the expectation on the right-hand side of (6.5.15) is uniformly bounded by 2, we obtain from the Lebesgue dominated convergence theorem

$$(6.5.16)\quad \lim_{\tau\to\infty} R(\xi_\tau, \hat{\theta}_\tau)$$
$$= 2 - E\left\{\lim_{\tau\to\infty}\left[\Phi\left(\sqrt{n}\,\frac{\delta + \theta(1 + n\tau^2)^{-1}}{(1 + (1/n\tau^2))^{-1}}\right) + \Phi\left(\sqrt{n}\,\frac{\delta - \theta(1 + n\tau^2)^{-1}}{(1 + (1/n\tau^2))^{-1}}\right)\right]\right\}$$
$$= 2[1 - \Phi(\delta\sqrt{n})] = \rho^*.$$

Hence $\bar{X}_n$ is minimax. ■

Example 6.8. Let $X_1, \ldots, X_n$ and $Y_1, \ldots, Y_n$ be two sets (samples) of i.i.d. random variables, $X_i \sim N(\mu, \sigma^2)$, $i = 1, \ldots, n$, and $Y_i \sim N(\mu, \rho\sigma^2)$, $i = 1, \ldots, n$. The two distributions have a common mean μ (unknown), $-\infty < \mu < \infty$, and unknown variances. The ratio of variances, ρ, is unknown; $0 < \rho < \infty$. The objective is to estimate the common mean μ. We show that if the loss function is $L_1(\mu, \hat{\mu}) = (\hat{\mu} - \mu)^2/\sigma^2$, a minimax estimator is $\mu^* = \bar{X}_n$, the mean of the first sample. This is an example of a minimax estimator that generally will not be used. If, on the other hand, we consider a loss function symmetric in the arguments

$$L_2(\mu, \hat{\mu}) = \frac{(\hat{\mu} - \mu)^2}{\sigma^2 \max(1, \rho)},$$

then a minimax estimator is $\frac{1}{2}(\bar{X} + \bar{Y})$.

(i) Consider the sequence of prior distributions with densities

$$\xi_k(d\mu, d\sigma^2, d\rho) = \begin{cases} \dfrac{1}{\sqrt{(2\pi k)}}\, e^{-(1/2k)\mu^2}\, d\mu, & \text{if } \rho = \rho_k,\ \sigma^2 = \sigma_k^2, \\ 0, & \text{otherwise.} \end{cases}$$

It is easy to verify that the corresponding Bayes estimator of μ for $L_1(\mu, \hat{\mu})$ is

$$\hat{\mu}_k = \frac{n\bar{X}_n/\sigma_k^2 + n\bar{Y}_n/\rho_k\sigma_k^2}{n/\sigma_k^2 + n/\rho_k\sigma_k^2 + 1/k} = \frac{\rho_k\bar{X}_n + \bar{Y}_n}{(1 + \rho_k + \rho_k\sigma_k^2/nk)}.$$

Assume further that $\lim_{k\to\infty} \rho_k = \infty$ but, $\overline{\lim}_{k\to\infty} \rho_k\sigma_k^2/k = 0$. We notice that $\bar{Y}_n = 0_p(1)$ as $k \to \infty$. Therefore $\lim_{k\to\infty} \hat{\mu}_k = \bar{X}_n$. Furthermore, let $\theta = (\mu, \sigma^2, \rho)$; then $R(\theta, \mu^*) = 1/n$ for all θ. We have to show then that $\lim_{k\to\infty} R(\xi_k, \hat{\mu}_k) = 1/n$. The risk function of $\hat{\mu}_k$ is

$$R(\theta, \hat{\mu}_k) = \frac{\rho_k^2 + \rho}{n(1 + \rho_k + \rho_k\sigma_k^2/nk)^2} + \mu^2\,\frac{\rho_k^2\sigma_k^4}{\sigma^2 n^3 k^2(1 + \rho_k + \rho_k\sigma_k^2/nk)^2}.$$

The corresponding prior risks are

$$R(\xi_k, \hat{\mu}_k) = \frac{\rho_k(1+\rho_k)}{n(1+\rho_k+\rho_k\sigma_k^2/nk)^2} + \frac{\rho_k^2\sigma_k^2}{n^3k(1+\rho_k+\rho_k\sigma_k^2/nk)^2}.$$

Hence

$$\lim_{k\to\infty} R(\xi_k, \hat{\mu}_k) = \frac{1}{n}\lim_{k\to\infty}\frac{\rho_k}{1+\rho_k} + \frac{1}{n^3}\lim_{k\to\infty}\frac{\rho_k\sigma_k^2}{k}\cdot\lim_{k\to\infty}\frac{\rho_k}{(1+\rho_k)^2} = \frac{1}{n}.$$

This completes the proof that $\bar{X}_n$ is minimax.

(ii) We show now that $\tilde{\mu} = (\bar{X}_n + \bar{Y}_n)/2$ is minimax under the loss function $L_2(\hat{\mu}, \mu)$. Indeed, the risk function of $\tilde{\mu}$ under $L_2(\hat{\mu}, \mu)$ is

$$R(\theta, \hat{\mu}) = \begin{cases} \dfrac{1}{4n}(1+\rho), & \text{if } 0 < \rho \leqslant 1 \\ \dfrac{1}{4n}\left(1+\dfrac{1}{\rho}\right), & \text{if } 1 \leqslant \rho < \infty. \end{cases} \qquad \text{all } (\mu, \sigma^2)$$

Hence $\sup_\theta R(\theta, \hat{\mu}) = 1/2n$. On the other hand, consider the sequence of Bayes estimators against priors ξ_k, which assign $\sigma^2 = 1$ and $\rho = 1$ prior probability 1, independently of μ, and μ has a prior $\mathcal{N}(0, k)$ distribution. The corresponding Bayes estimators are

$$\hat{\mu}_k = \frac{\bar{X} + \bar{Y}}{2 + 1/nk}.$$

These Bayes estimators converge to $\tilde{\mu}$ as $k \to \infty$. The risk functions of $\hat{\mu}_k$ are

$$R(\theta, \hat{\mu}_k) = \begin{cases} \dfrac{(1+\rho)}{[n(2+1/nk)^2]} + \mu^2(1+2nk)^{-2}, & \text{if } 0 < \rho \leqslant 1, \\ \dfrac{(1+\rho)}{[n\rho(2+1/nk)^2]} + \mu^2(1+2nk)^{-2}, & \text{if } 1 \leqslant \rho < \infty. \end{cases}$$

The prior risks are

$$R(\xi_k, \hat{\mu}_k) = \frac{2}{n(2+1/nk)^2} + \frac{k}{(1+2nk)^2}, \qquad k = 1, 2, \ldots.$$

Hence $\lim_{k\to\infty} R(\xi_k, \hat{\mu}_k) = 1/2n$. This proves that $\tilde{\mu} = (\bar{X}_n + \bar{Y}_n)/2$ is minimax.

■

6.6. SUGGESTED PROCEDURES OF ESTIMATION IN CASES OF PARTIAL PRIOR INFORMATION

In practical situations a statistician may often find it very difficult to assign an appropriate prior distribution $\xi(\theta)$ over the parameter space Θ. In many studies we find that prior distributions are assigned to the model under consideration according to some pragmatic criteria of convenience or mathematical tractability. Almost always, however, the scientist has a considerable partial prior information at his disposal. Although he cannot guarantee that a certain prior distribution $\xi(\theta)$ is the most appropriate one to use, he may nevertheless say that the appropriate prior distribution belongs to a certain family $\mathcal{H}$ of prior distributions. The question is what procedure of estimation is optimal under such a partial information, and in what sense is it optimal.

Blum and Rosenblatt [3] studied minimax procedures which are restricted to certain families of prior distributions and compared them to general unrestricted minimax procedures. We show later on some of their results. Hodges and Lehmann [3] argued that although the statistician may find it too difficult to assign the appropriate prior distribution, he can often choose a prior distribution $\xi_0(\theta)$ and say that with a confidence (subjective) probability p, $0 \leqslant p \leqslant 1$, the prior distribution chosen is the appropriate one. In this case they set up a criterion of minimizing

$$p \int R(\theta, d)\xi_0(d\theta) + (1 - p) \sup_{\theta \in \Theta} R(\theta, d), \tag{6.6.1}$$

with respect to all estimators d in $\mathfrak{D}$. This criterion is a mixture of the Bayes and the minimax criteria. Kudō [2] provided a general theory of estimation under partial information that contains as special cases the Bayes, minimax, restricted minimax, and other methods. In this section we present the main results of Blum and Rosenblatt and of Hodges and Lehmann. Kudō's theory is discussed in the next section.

We start with two examples. The first example illustrates cases in which the restricted minimax estimators yield minimax risks smaller than those of the unrestricted minimax estimators. In these cases the partial prior information that θ belongs to a certain subset of Θ leads to a reduction of the minimax risk. We also illustrate cases in which such partial prior information does not reduce the minimax risk. In the second example we show similar cases when the partial prior information is formulated in terms of families of prior distributions.

Example 6.9. Let X be a random variable such that $X = \alpha + \mathbf{H}'\boldsymbol{\beta} + \epsilon$, where $\mathbf{H}' = (H_1, H_2)$, $H_1 + H_2 = 0$, $H_1^2 + H_2^2 = 1$, and $\boldsymbol{\beta}' = (\beta_1, \beta_2)$. Furthermore ϵ is a random variable such that $E\{\epsilon\} = 0$ and $E\{\epsilon^2\} = 1$. The parameters α, β_1, and β_2 are unknown; the vector $\mathbf{H}$ is specified. The objective is to estimate α, whereas β is a nuisance parameter. The loss function considered is $L(\alpha, \hat{\alpha}) = (\hat{\alpha} - \alpha)^2$.

Consider the class of linear estimators

$$\hat{\alpha}(X; \boldsymbol{\gamma}) = X - \mathbf{H}'\boldsymbol{\gamma}, \qquad \boldsymbol{\gamma}' = (\gamma_1, \gamma_2).$$

A choice of an estimator is equivalent to a choice of a vector $\boldsymbol{\gamma}$. The risk function associated with $\hat{\alpha}(X; \boldsymbol{\gamma})$ is

$$R(\boldsymbol{\beta}, \boldsymbol{\gamma}) = 1 + |\boldsymbol{\beta} - \boldsymbol{\gamma}|^2. \tag{6.6.2}$$

It is simple to show that if a prior distribution $\xi(\boldsymbol{\beta})$ is available such that $E_\xi\{\boldsymbol{\beta}'\boldsymbol{\beta}\} < \infty$, the Bayes estimator of α, against, ξ, is

$$\hat{\alpha}(X; \xi) = X - \mathbf{H}'E_\xi\{\boldsymbol{\beta}\}. \tag{6.6.3}$$

Suppose now that only a partial information is available, according to which β belongs to $\Theta_1^* = \{\boldsymbol{\beta}; |\boldsymbol{\beta}|^2 \leqslant D^2\}$; Θ_1^* is a prescribed circle around the origin. It is easy to show then that a minimax estimator of α against Θ_1^* is $\alpha_1^*(X) = X$, with a minimax risk of $R_1^* = 1 + D^2$. If the partial prior information is that $\beta \in \Theta_2^*$, where $\Theta_2^* = \{\boldsymbol{\beta}; \beta_1 \geqslant 0, \beta_2 \geqslant 0, |\boldsymbol{\beta}|^2 \leqslant D^2\}$, then the minimax estimator is

$$\alpha_2^*(X) = X - \tfrac{1}{2}D\mathbf{H}'(1, 1)'. \tag{6.6.4}$$

The minimax risk associated with $\alpha_2^*(X)$ is $1 + D^2/2$. We see that the set Θ_2^* contains more relevant prior information about β than Θ_1^*. We now show another set contained in Θ_1^*, which induces the minimax estimator $\alpha_3^*(X)$ and has the same minimax risk as Θ_1^*. This is the set

$$\Theta_3^* = \{\boldsymbol{\beta}; \beta_1 \leqslant 0, \beta_2 \leqslant 0, |\boldsymbol{\beta}|^2 \leqslant D^2\} \cup \Theta_2^*.$$

We notice that if the sets Θ_2^* and $\Theta_4^* = \{\boldsymbol{\beta}; \beta_1 \leqslant 0, \beta_2 \leqslant 0, |\boldsymbol{\beta}|^2 \leqslant D^2\}$ have equal prior probabilities, and if we consider the estimator $\alpha_4^*(X) = X + (D/2)\mathbf{H}'(1, 1)'$, which is the minimax estimator for Θ_4^*, then $\alpha_3^*(X) = \frac{1}{2}(\alpha_2^*(X) + \alpha_4^*(X))$. But if the set Θ_2^* has a prior probability p, and Θ_4^* has a prior probability $(1 - p)$, the minimax estimator $\tilde{\alpha}(X; \boldsymbol{\gamma})$ should apply γ such that

$$p \sup_{\beta \in \Theta_2^*} |\boldsymbol{\beta} - \boldsymbol{\gamma}|^2 + (1 - p) \sup_{\beta \in \Theta_4^*} |\boldsymbol{\beta} - \boldsymbol{\gamma}|^2 \tag{6.6.5}$$

is minimized. It is a straightforward matter to verify that $\gamma = (D/2)(2p - 1, 2p - 1)$.

The criterion of minimizing (6.6.4) can be extended to general finite partitions of the parameter space Θ, as will be shown later. ■

Consider now the case in which it is known that the prior distribution belongs to a family of distribution functions $\mathcal{F}$. Blum and Rosenblatt [3] define an $\mathcal{F}$-minimax estimator as one which minimizes the supremum of the prior risks over $\mathcal{F}$. In other words, let

$$G(\mathcal{F}, d) = \sup_{\xi \in \mathcal{F}} \int R(\theta, d)\xi(d\theta), \qquad d \in \mathfrak{D}; \tag{6.6.6}$$

then d^* is called $\mathcal{F}$-minimax if

$$G(\mathcal{F}, d^*) = \inf_{d \in \mathfrak{D}} G(\mathcal{F}, d). \tag{6.6.7}$$

In the following example we illustrate two cases. In one case $\mathcal{F}$ is such that the $\mathcal{F}$-minimax estimator coincides with the nonrestricted minimax estimator. In the other case the $\mathcal{F}$-minimax estimator is different from the nonrestricted minimax estimator. This example is a modification of Theorems 1 and 3 of Blum and Rosenblatt [3].

Example 6.10. Let X be a random variable having an $\mathcal{N}(\theta, 1)$ distribution. The quantity θ is unknown, and $\Theta = (-\infty, \infty)$. Consider the loss function $L(\theta, d) = (\theta - d)^2$; $\mathfrak{D} = \Theta$. An application of Theorem 6.5.2 with the sequence of prior distributions specified in Example 6.7 yields that $\delta_0(X) = X$ is an unrestricted minimax estimator. We also have $R(\theta, \delta_0) = 1$ for all $\theta \in \Theta$.

Case 1. Let $\mathcal{F}$ be the family of all prior normal distributions for θ. Let F be a member of $\mathcal{F}$, according to which $\theta \sim N(\mu_0, \tau^2)$. Let $\hat{\theta}$ be the Bayes estimator of θ against F. We have $\hat{\theta} = (X + \mu_0/\tau^2)(1 + 1/\tau^2)^{-1}$. The corresponding prior risk is $R(F, \hat{\theta}) = (1 + 1/\tau^2)^{-1}$. For any estimator d we have

$$G(\mathcal{F}, d) \geqslant R(F, d) \geqslant R(F, d_F), \quad \text{all} \quad F \in \mathcal{F} \tag{6.6.8}$$

and all $d \in \mathfrak{D}$; d_F designates the Bayes estimator against F. For any $\epsilon > 0$ arbitrarily small, choose F^ϵ in $\mathcal{F}$ so that $R(F^\epsilon, d_{F^\epsilon}) \geqslant 1 - \epsilon$. Then

$$\inf_{d \in \mathfrak{D}} G(\mathcal{F}, d) \geqslant \inf_{d \in \mathfrak{D}} R(F^\epsilon, d) \geqslant 1 - \epsilon. \tag{6.6.9}$$

Finally, $G(\mathcal{F}, \delta_0) = \sup_{F \in \mathcal{F}} \int R(\theta, \delta_0)F(d\theta) = 1$. Hence we obtain from (6.6.9), since ϵ is arbitrary, that $\delta_0(X) = X$ is $\mathcal{F}$-minimax.

Case 2. Consider the same model as in Case 1 but with another family of prior distributions

$$\mathcal{F}_p = \{\xi_H;\ \xi_H = p\xi_0 + (1 - p)H\}, \qquad 0 < p < 1,$$

where p is a given number; ξ_0 is a degenerate distribution assigning the value $\theta = 0$ prior probability 1; and H is an arbitrary prior distribution on the real line. The prior risk function for a given H and any estimator d is

$$R(\xi_H, d) = pE_0\{d^2(X)\} + (1-p)\int R(\theta, d)H(d\theta). \tag{6.6.10}$$

Hence

$$G(\mathcal{F}_p, d) = pE_0\{d^2(X)\} + (1-p)G(\mathcal{H}, d), \tag{6.6.11}$$

where $\mathcal{H}$ is the family of all prior distributions H on the real line. According to (6.6.8),

$$G(\mathcal{H}, d) \geqslant G(\mathcal{F}, d) \geqslant 1, \quad \text{for all} \quad d \in \mathfrak{D}. \tag{6.6.12}$$

Suppose that d_1 is an estimator satisfying

$$E_0\{d_1^2(X)\} < 1 \quad \text{and} \quad G(\mathcal{H}, d_1) < \infty. \tag{6.6.13}$$

Then, by choosing p so that

$$1 \geqslant p > \frac{G(\mathcal{H}, d_1) - 1}{G(\mathcal{H}, d_1) - E_0\{d_1^2(X)\}}, \tag{6.6.14}$$

we obtain from (6.6.11) that

$$\begin{aligned} G(\mathcal{F}_p, d_1) &< E_0\{d_1^2(X)\} + \frac{1 - E_0\{d_1^2(X)\}}{G(\mathcal{H}, d_1) - E_0\{d_1^2(X)\}}\, G(\mathcal{H}, d_1) \\ &\leqslant E_0\{d_1^2(X)\} + 1 - E_0\{d_1^2(X)\} = 1. \end{aligned} \tag{6.6.15}$$

Indeed, $G(\mathcal{H}, d_1) \geqslant 1$ and therefore

$$1 \leqslant \frac{G(\mathcal{H}, d_1)}{G(\mathcal{H}, d_1) - E_0\{d_1^2(X)\}} < \infty. \tag{6.6.16}$$

Finally, $\inf_{d \in \mathfrak{D}} G(\mathcal{F}_p, d) \leqslant G(\mathcal{F}_p, d_1) < 1$. Therefore, if there exists an estimator d_1 satisfying (6.6.13), then the $\mathcal{F}_p$-minimax risk for p satisfying (6.6.14) is smaller than the unrestricted minimax risk. We now show that the estimator

$$d_1(X) = \begin{cases} 0, & \text{if } |X| \leqslant 1, \\ X, & \text{if } |X| > 1, \end{cases} \tag{6.6.17}$$

satisfies (6.6.13). First we observe that

$$\begin{aligned} E_0\{d_1^2(X)\} &= \frac{1}{\sqrt{(2\pi)}}\left(\int_{-\infty}^{\infty} - \int_{-1}^{1}\right) x^2 e^{-\frac{1}{2}x^2}\, dx \\ &= 1 - \frac{1}{\sqrt{(2\pi)}}\int_{-1}^{1} x^2 e^{-\frac{1}{2}x^2}\, dx < 1. \end{aligned} \tag{6.6.18}$$

It remains to show that $G(\mathcal{H}, d_1) < \infty$. For any prior distribution H the prior risk of d_1 is

(6.6.19) $R(H, d_1)$

$$= \frac{1}{\sqrt{(2\pi)}} \int H(\mathrm{d}\theta) \left\{ \int_{-1}^{1} \theta^2 e^{-\frac{1}{2}(x-\theta)^2} \,\mathrm{d}x + \int_{\{|x|>1\}} (x-\theta)^2 e^{-\frac{1}{2}(x-\theta)^2} \,\mathrm{d}x \right\}.$$

But

$$\frac{1}{\sqrt{(2\pi)}} \int_{|x|>1} (x-\theta)^2 e^{-\frac{1}{2}(x-\theta)^2} \,\mathrm{d}x \leqslant \frac{1}{\sqrt{(2\pi)}} \int_{-\infty}^{\infty} (x-\theta)^2 e^{-\frac{1}{2}(x-\theta)^2} \,\mathrm{d}x = 1$$

for *all* θ. Thus

$$R(H, d)_1 \leqslant 1 + \int \theta^2 H(\mathrm{d}\theta) \left\{ \int_{-1}^{1} \frac{1}{\sqrt{(2\pi)}} e^{-\frac{1}{2}(x-\theta)^2} \,\mathrm{d}x \right\}. \tag{6.6.20}$$

Hence $G(\mathcal{H}, d_1) \leqslant 1 + M < \infty$, where $M = \sup_{-\infty<\theta<\infty} \{\theta^2[\Phi(1+\theta) + \Phi(1-\theta) - 1]\}$. This completes the verification that d_1 satisfies (6.6.13), and therefore the $\mathcal{F}_p$-minimax risk is smaller than 1. ■

We notice in the above example that with the family $\mathcal{F}_p$, for any value of p, $0 < p < 1$, if $G(\mathcal{H}, d) = \infty$ for all d then $G(\mathcal{F}_p, d) = \infty$ for all d, and therefore $\liminf_{p\to 1}\, \inf_{d\in\mathfrak{D}} G(\mathcal{F}_p, d) = \infty$. The criterion of applying an $\mathcal{F}_p$-minimax estimator may in many cases lead to an infinite minimax risk despite the fact that the knowledge of the true prior distribution $\xi_0(\theta)$ is almost complete, $p \to 1$. It is desirable that as $p \to 1$, $\inf_d G(\mathcal{F}_p, d) \to R(\xi_0, d_{\xi_0})$, where $\mathcal{F}_p = \{\xi;\ \xi = p\xi_0 + (1-p)H,\ H \in \mathcal{H}\}$ and $R(\xi_0, d_{\xi_0})$ is the minimal prior risk associated with $\xi_0(\theta)$. Blum and Rosenblatt [3] furnished sufficient conditions for the convergence of the $\mathcal{F}_p$-minimax risk to the Bayes prior risk of $\xi_0(\theta)$ as $p \to 1$. We present an adjusted version of their Theorem 4.

Theorem 6.6.1. *Let* $\mathcal{F}_p = \{\xi;\ \xi = p\xi_0 + (1-p)H,\ H \in \mathcal{H}\}$, *where* $\mathcal{H}$ *is the class of all prior distributions on* Θ. *Let* d^0 *be a Bayes estimator against* ξ_0, *satisfying*

$$0 < G(\mathcal{H}, d^0) < \infty. \tag{6.6.21}$$

Then

$$\liminf_{p\to 1}\, \inf_{d\in\mathfrak{D}} G(\mathcal{F}_p, d) = R(\xi_0, d^0). \tag{6.6.22}$$

Proof. Since $G(\mathcal{H}, d^0) < \infty$ we have, for each p, $0 < p < 1$,

$$G(\mathcal{F}_p, d^0) = pR(\xi_0, d^0) + (1-p)G(\mathcal{H}, d^0) < \infty. \tag{6.6.23}$$

Choose a sequence $\{\epsilon_k\}$, $\epsilon_k \searrow 0$, and let p_k, for k sufficiently large, be defined by

$$0 < 1 - p_k \leqslant \frac{\epsilon_k}{G(\mathcal{H}, d^0)} < 1. \tag{6.6.24}$$

Then for all k sufficiently large,

$$\inf_d G(\mathcal{F}_p, d) \leqslant G(\mathcal{F}_{p_k}, d^0) \leqslant R(\xi_0, d^0) + \epsilon_k. \tag{6.6.25}$$

This implies that

$$\overline{\lim_{k\to\infty}} \inf_d G(\mathcal{F}_{p_k}, d) \leqslant R(\xi_0, d^0). \tag{6.6.26}$$

On the other hand, for each $d \in \mathfrak{D}$, since $\xi_0 \in \mathcal{H}$,

$$\begin{aligned} G(\mathcal{F}_{p_k}, d) &= p_k R(\xi_0, d) + (1 - p_k) G(\mathcal{H}, d) \\ &\geqslant p_k R(\xi_0, d) + (1 - p_k) R(\xi_0, d) = R(\xi_0, d). \end{aligned} \tag{6.6.27}$$

Therefore

$$\underline{\lim_{k\to\infty}} \inf_d G(\mathcal{F}_p, d) \geqslant \inf_d R(\xi_0, d) = R(\xi_0, d^0). \tag{6.6.28}$$

Equations 6.6.26 and 6.6.28 imply (6.6.22). (Q.E.D.)

We remark that if for each $d \in \mathfrak{D}$ the loss function $L(\theta, d)$ is bounded and $R(\theta, d)$ is positive on some interval of θ then d^0 satisfies (6.6.21) and Theorem 6.6.1 holds.

As shown by Hodges and Lehmann [3], if $\mathcal{F}_p$ is as defined in Theorem 6.6.1, and if d_p is an $\mathcal{F}_p$-minimax estimator such that $G(\mathcal{H}, d_p) = C_0 < \infty$, then d_p is a restricted Bayes estimator of θ against ξ_0. That is, if $\mathfrak{D}^0 = \{d; \sup_{\theta\in\Theta} R(\theta, d) \leqslant C_0\}$ then $R(\xi_0, d_p) = \inf_{d\in\mathfrak{D}^0} R(\xi_0, d)$. Restricted Bayes estimators have the merit that even if ξ_0 is not the true (or appropriate) prior distribution, the resulting risk function is bounded by C_0. The restricted Bayes estimator may be uniformly better than the unrestricted minimax estimator.

6.7. PARTIAL PRIOR INFORMATION AND PARAMETRIC SUFFICIENCY

Consider an estimation problem with a risk function $R(\theta, d)$, $\theta \in \Theta$ and $d \in \mathfrak{D}$. Let $\mathcal{C}$ be the smallest Borel sigma-field on Θ. The function $R(\theta, d)$ is assumed to be $\mathcal{C}$-measurable, for each d in $\mathfrak{D}$ and ξ-integrable, where ξ is a prior probability measure on $(\Theta, \mathcal{C})$.

There are problems in which $R(\theta, d)$ depends on θ only through a function $u(\theta)$, that is, $R(\theta, d) = R(u(\theta), d)$ for all (θ, d). If we designate by $\mathcal{U}$ the sigma-subfield generated by u on Θ, then $R(\theta, d)$ is $\mathcal{U}$-measurable for each d. Moreover, the prior risk $R(\xi, d)$ can be determined by integrating $R(u(\theta), d)$ with respect to the prior distribution ξ^u induced by $u(\theta)$. That is,

$$R(\xi, d) = \int R(u, d)\xi^u(du). \tag{6.7.1}$$

Since (6.7.1) holds for every prior distribution ξ, we can say that the subfield $\mathcal{U}$ is parametric sufficient for the decision problem. This concept was introduced by Barankin [2]. We present in this section the results of Kudō [2] on parametric sufficiency. We remark that if a sigma-subfield $\mathcal{U}$ is parametric sufficient, then, due to (6.7.1), we do not need the complete prior information given by $\xi(\theta)$. The partial prior information $\xi^u(u)$ is sufficient for all decision purposes. For examples, if $u(\theta) = I_\omega(\theta)$ is the indicator function of the subset $\omega \subset \Theta$, and $\mathcal{U}$ is the corresponding sigma-subfield, then $\xi^u(u(\theta)) = E_\xi\{I_\omega(\theta)\}$ is the prior probability of ω, if $u(\theta) = 1$, and the prior probability of $\bar{\omega} = \Theta - \omega$, if $u(\theta) = 0$.

We consider here the problem of partial prior information and parametric sufficiency from the other end. We start with a given partial prior information, given by a prior probability measure $\xi^u(\cdot)$ defined on $(u, \mathcal{U})$, where $\mathcal{U}$ is a sigma-subfield of $\mathcal{T}$. We then ask, Under what condition is $\mathcal{U}$ a parametric sufficient subfield? In particular we are interested in subfields, $\mathcal{U}$, which are generated by given finite partitions $\{T_1, \ldots, T_k\}$ of Θ. The partial prior information available in these cases is just the prior probabilities $\pi_j = \xi(T_j)$, $j = 1, \ldots, k$. The prior probability measure $\xi(d\theta)$ on $(\Theta, \mathcal{T})$ is unknown. Kudō [2] introduced the mean-max criterion according to which, if the partial prior information available is $\pi_j = \xi(T_j), j = 1, \ldots, k, T_i \cap T_j = \phi$ and $\bigcup_{j=1}^k T_i = \Theta$, then the optimal estimator of θ (if exists) is the one which minimizes

$$\sum_{j=1}^{k} \xi(T_j) \sup_{\theta \in T_j} R(\theta, d).$$

The function $\bar{R}(T_j, d) = \sup_{\theta \in T_j} R(\theta, d)$ is measurable with respect to the subfield $\mathcal{U}$ generated by the finite partition $\{T_1, \ldots, T_k\}$. The principle of estimating by an estimator which minimizes the mean-max risk is a generalization of both the Bayes and minimax principles. An example of an estimator which minimizes the mean-max risk is given in Example 6.9. If the partial prior information in that example is given by $p = \xi(\Theta_2^*)$ and $1 - p = \xi(\Theta_4^*)$, $0 < p < 1$, whereas $\xi(\Theta - \Theta_2^* - \Theta_4^*) = 0$, then the estimator which minimizes the mean-max risk is $\hat{\alpha}(\gamma_p) = X - (D/2)H'(2p - 1, 2p - 1)'$.

The above definition of a mean-max risk is with respect to a given partition of Θ. For the purpose of studying the conditions necessary or sufficient for parametric sufficiency of a subfield $\mathcal{U}$, Kudō generalized the above definition and defined the mean-max risk of d with respect to $(\mathcal{U}, \xi)$ as

$$R(\mathcal{U}, \xi, d) = \inf_{\mathcal{W} \subset \mathcal{U}} \sum_{i=1}^{k} \xi(W_i) \sup_{\theta \in W_i} R(\theta, d), \tag{6.7.2}$$

where $\mathcal{W}$ is a subfield of $\mathcal{U}$ generated by a partition $\{W_1, \ldots, W_k\}$ of Θ, $\{W_1, \ldots, W_k\} \in \mathcal{U}$. The number k of disjoint sets may vary from one partition to another. We notice that if $R(\theta, d)$ is $\mathcal{U}$-measurable, $R(\mathcal{U}, \xi, d) = \int R(\theta, d)\xi(d\theta)$. Generally, from the definition of $R(\mathcal{U}, \xi, d)$, we obtain that

$$R(\mathcal{U}, \xi, d) \geqslant R(\xi, d) = \int R(\theta, d)\xi(d\theta). \tag{6.7.3}$$

Obviously $R(\mathcal{T}, \xi, d) = R(\xi, d)$ for all d in $\mathfrak{D}$ and any ξ-integrable $R(\theta, d)$. The quantity $R(\mathcal{U}, \xi, d)$ may be greater than the prior risk $R(\xi, d)$ for some d. *A sigma-subfield $\mathcal{U}$ of $\mathcal{T}$ is called ξ-sufficient if*

$$\inf_d R(\mathcal{U}, \xi, d) = \inf_d R(\xi, d). \tag{6.7.4}$$

If (6.7.4) *holds for all ξ on* $(\Theta, \mathcal{T})$ *then $\mathcal{U}$ is called a sufficient subfield for the estimation problem.*

According to the above criterion of mean-max optimality, given a partial information $(\mathcal{U}, \xi)$, we select from $\mathfrak{D}$ an estimator d^* which minimizes $R(\mathcal{U}, \xi, d)$; d^* is called $(\mathcal{U}, \xi)$-optimal. We notice that if d^* minimizes the mean-max risk $R(\mathcal{U}, \xi, d)$ and if $\mathcal{U}$ is ξ-sufficient, d^* is Bayes against ξ. On the other hand, if $\mathcal{O} = \{\phi, \Theta\}$ then $R(\mathcal{O}, \xi, d) = \sup_{\theta \in \Theta} R(\theta, d)$ for all ξ on Θ. Hence if d^* minimizes the mean-max risk $R(\mathcal{O}, \xi, d)$, then d^* is minimax. The trivial subfield $\mathcal{O} = \{\phi, \Theta\}$ represents a state of complete ignorance about the true θ.

Let $\rho(\mathcal{U}, \xi) = \inf_d R(\mathcal{U}, \xi, d)$. This is the minimal mean-max risk attainable under the partial information $(\mathcal{U}, \xi)$. The relationships between the various attainable prior risks can be schematically presented by the following diagram:

$$\begin{array}{ccc} R(\mathcal{T}, \xi, d) & \geqslant & \rho(\mathcal{T}, \xi) \\ \wedge\!\!| & & \wedge\!\!| \\ R(\mathcal{U}, \xi, d) & \geqslant & \rho(\mathcal{U}, \xi) \end{array} \quad \text{for all} \quad \mathcal{U} \subset \mathcal{T} \text{ and all } \quad d \in \mathfrak{D}. \tag{6.7.5}$$

Let $E_\xi\{R(\theta, d) \mid \mathcal{U}\}$ designate the conditional expectation of the risk function, according to the prior probability measure ξ. If $R(\theta, d) \in \mathcal{U}$ then $R(\theta, d) = E_\xi\{R(\theta, d) \mid \mathcal{U}\}$ a.s. $[\xi]$. Otherwise $\xi(\theta; R(\theta, d) \neq E_\xi\{R(\theta, d) \mid \mathcal{U}\}) > 0$.

Theorem 6.7.2. (Kudō [2].) *For each $d \in \mathfrak{D}$ and $U \subset \mathcal{C}$, the following inequality holds:*

$$\frac{1}{2}\int |R(\theta, d) - E_\xi\{R(\theta, d) \mid \mathfrak{U}\}| \, \xi(d\theta) \leqslant R(\mathfrak{U}, \xi, d) - R(\xi, d), \tag{6.7.6}$$

where $R(\xi, d) = \int R(\theta, d)\xi(d\theta)$ is the prior risk of d.

Proof. Let B be any Borel set of $\mathfrak{U}$; then by the definition of conditional expectation

$$\int_B R(\theta, d)\xi(d\theta) = \int_B E_\xi\{R(\theta, d) \mid \mathfrak{U}\}\xi(d\theta). \tag{6.7.7}$$

Define the sets

$$B_+ = \{\theta;\, R(\theta, d) \geqslant E_\xi\{R(\theta, d) \mid \mathfrak{U}\}\} \cap B,$$
$$B_- = \{\theta;\, R(\theta, d) < E_\xi\{R(\theta, d) \mid \mathfrak{U}\}\} \cap B.$$

Then from (6.7.7) we obtain, since $B = B_+ \cup B_-$ and $B_+ \cap B_- = \phi$,

$$\begin{aligned} 0 &= \int_B (R(\theta, d) - E_\xi\{R(\theta, d) \mid \mathfrak{U}\})\xi(d\theta) \\ &= \int_{B+} [R(\theta, d) - E_\xi\{R(\theta, d) \mid \mathfrak{U}\}]\xi(d\theta) \\ &\quad + \int_{B-} [R(\theta, d) - E_\xi\{R(\theta, d) \mid \mathfrak{U}\}]\xi(d\theta) \end{aligned} \tag{6.7.8}$$

Hence for every $B \in \mathfrak{U}$,

$$\begin{aligned} &\int_{B+} [R(\theta, d) - E_\xi\{R(\theta, d) \mid \mathfrak{U}\}]\xi(d\theta) \\ &\qquad = -\int_{B-} [R(\theta, d) - E_\xi\{R(\theta, d) \mid \mathfrak{U}\}\xi(d\theta). \end{aligned} \tag{6.7.9}$$

Thus

$$\begin{aligned} &\frac{1}{2}\int_B |R(\theta, d) - E_\xi\{R(\theta, d) \mid \mathfrak{U}\}| \, \xi(d\theta) \\ &\qquad = \int_{B+} [R(\theta, d) - E_\xi\{R(\theta, d) \mid \mathfrak{U}\}]\xi(d\theta). \end{aligned} \tag{6.7.10}$$

Moreover, $E_\xi\{R(\theta, d) \mid \mathfrak{U}\} \leqslant \sup_{\theta \in B} \{R(\theta, d)\}$ a.s. $[\xi]$ on B. Thus for every $B \in \mathfrak{U}$,

$$\begin{aligned} &\int_{B+} [R(\theta, d) - E_\xi\{R(\theta, d) \mid \mathfrak{U}\}]\xi(d\theta) \\ &\leqslant \int_{B+} \left[\sup_{\theta \in B+} R(\theta, d) - E_\xi\{R(\theta, d) \mid \mathfrak{U}\}\right]\xi(d\theta) \\ &\leqslant \int_B \left[\sup_{\theta \in B} R(\theta, d) - E_\xi\{R(\theta, d) \mid \mathfrak{U}\}\right]\xi(d\theta). \end{aligned} \tag{6.7.11}$$

Combining (6.7.10) and (6.7.11) we obtain

$$\text{(6.7.12)} \qquad \frac{1}{2}\int_B |R(\theta, d) - E_\xi\{R(\theta, d) \mid \mathcal{U}\}|\, \xi(d\theta)$$

$$\leqslant \sup_{\theta \in B} R(\theta, d)\xi(B) - \int_B R(\theta, d)\xi(d\theta).$$

Let $\mathcal{W} = \{B_1, \ldots, B_k\}$ be any partition of Θ which is $\mathcal{U}$-measurable. Then since (6.7.12) holds for each B_j in $\mathcal{W}$ $(j = 1, \ldots, k)$,

$$\text{(6.7.13)} \qquad \frac{1}{2}\sum_{j=1}^{k}\int_{B_j} |R(\theta, d) - E_\xi\{R(\theta, d) \mid \mathcal{U}\}|\, \xi(d\theta)$$

$$= \frac{1}{2}\int_\Theta |R(\theta, d) - E_\xi\{R(\theta, d) \mid \mathcal{U}\}|\, \xi(d\theta)$$

$$\leqslant \sum_{j=1}^{k} \sup_{\theta \in B_j} R(\theta, d) \cdot \xi(B_j) - \int_\Theta R(\theta, d)\xi(d\theta).$$

Furthermore, since (6.7.13) holds for any partition of Θ we obtain (6.7.6) by taking the infimum of both sides of (6.7.13). (Q.E.D.)

Equation 6.7.6 gives a lower bound to the amount of missing partial information, as measured by the difference between the mean-max and the prior risks; that is, $R(\mathcal{U}, \xi, d) - R(\xi, d)$. Thus, if for some d' in $\mathfrak{D}$, $R(\mathcal{U}, \xi, d') = R(\xi, d')$, then $E_\xi\{|R(\theta, d') - E_\xi\{R(\theta, d') \mid \mathcal{U}\}|\} = 0$, which implies that $R(\theta, d')$ is $\mathcal{U}$-measurable, except perhaps on a set Λ of ξ-measure zero. From (6.7.5) we also infer that if $\mathcal{U}$ is ξ-sufficient, that is, $\rho(\mathcal{U}, \xi) = \rho(\mathcal{T}, \xi)$, and if d^* is $(\mathcal{U}, \xi)$-optimal, then d^* is Bayes against ξ. Hence if $\mathcal{U}$ is ξ-sufficient we obtain from (6.7.6) that $R(\theta, d^*)$ is $\mathcal{U}$-measurable, where d^* is Bayes against ξ.

Similarly, if $R(\theta, d^*)$ is $\mathcal{U}$-measurable and d^* is Bayes against ξ, then $\mathcal{U}$ is ξ-sufficient and d^* is $(\mathcal{U}, \xi)$-optimal.

In the following example we illustrate these results.

Example 6.11. Let $\{e_1, \ldots, e_N\}$ be the elements of a finite population (a lot of N products). Let $\{\theta_1, \ldots, \theta_N\}$ designate the (unknown) values of the elements. The vector $\boldsymbol{\theta} = (\theta_1, \ldots, \theta_N)'$ is considered in the modern approach to sampling theory (see Godambe [1]) as a parametric point in an N-dimensional Euclidean space $E^{(N)}$. Let $\zeta_n = \langle e_{i_1}, \ldots, e_{i_n}\rangle$ be a sample (an ordered set) of n elements chosen at random and without replacement from the given population. The objective is to estimate the population total $\omega = \sum_{i=1}^{N} \theta_i$. Suppose that $\theta_i = 0, 1$ $(i = 1, \ldots, N)$. Then a sufficient statistic for ω is the sample total T_n. The conditional distribution of T_n, given ω, is hypergeometric. Let $\hat{\omega}(T_n)$ be an estimator of ω, and consider the loss function $L(\omega, \hat{\omega}) = (\omega - \hat{\omega})^2$. The risk function is

$$R(\omega, \hat{\omega}) = \sum_{t=0}^{n} (\hat{\omega}(t) - \omega)^2 \frac{\binom{\omega}{t}\binom{N-\omega}{n-t}}{\binom{N}{n}}.$$

We see that the risk function of any estimator depends on θ only through $\omega = \sum_{i=1}^{N} \theta_i$. Let $\mathcal{U}$ be the sigma-subfield of $\mathcal{C}$ on Θ, which is generated by ω. Since $R(\omega, \hat{\omega}) \in \mathcal{U}$ for each $\hat{\omega}$, $\mathcal{U}$ is sufficient for $\mathcal{C}$. It is enough to specify a prior distribution η of ω and consider the Bayes estimator against η (such as in Example 6.4).

A specification of a prior distribution ξ on Θ will not lead to a reduction of the minimal prior risk. The ξ-Bayes estimator of ω is not different from the η-Bayes estimator. ■

6.8. SOME MINIMAX SEQUENTIAL ESTIMATION PROCEDURES

Minimax sequential estimation procedures were studied by Stein and Wald [1], Wolfowitz [3], Blyth [1], Kiefer [1], and others. Models of sequential estimation were previously discussed in Chapter 2, Section 2.7 and in Chapter 4, Section 4.4. Here we outline only the essential elements of the model.

Let $X_1, X_2, \ldots$ be a sequence of i.i.d. random variables having a common distribution function $F_\theta(x)$, where $\theta \in \Theta$. A sequential estimation procedure is prescribed with a rule by which, given m observations $X_1, \ldots, X_m$ ($m = 0, 1, \ldots$), we can make one of the following decisions:

(i) Take an additional observation on X_{m+1}.
(ii) Terminate sampling and estimate θ by an estimator $d(X_1, \ldots, X_m)$.

The sequential rule consists thus of two elements, a stopping rule and an estimation rule. We designate a sequential rule generically by $\zeta = (S, d)$. In a Bayesian framework, the estimation rule $d(X_1, \ldots, X_m)$, $m = 0, 1, \ldots,$ is independent of the stopping rule $S(X_1 \ldots, X_m)$, $m = 0, 1, \ldots,$ since once sampling is terminated the posterior distribution of θ does not depend on the stopping rule. For the sake of determining an optimal stopping rule, as well as an optimal estimator, we set up two loss functions $L_1(\theta; \zeta)$ and $L_2(\theta; \zeta)$. The loss function $L_2(\theta; \zeta)$ assigns a certain cost due to termination according to the rule S. The function $L_1(\theta; \zeta)$ is the loss due to erroneous estimation. Let ξ be a prior probability measure on $(\Theta, \mathcal{C})$. We designate by $R_1(\xi, \zeta)$ and by $R_2(\xi, \zeta)$ the prior risks associated with the particular sequential rule ζ. In many studies the loss function $L_2(\theta, \zeta)$ is simply the cost of $N(\xi, \zeta)$ observations. In this case $L_2(\theta, \zeta) = cN(\xi, \zeta)$, where c,

$0 < c < \infty$, is the cost of one observation. The corresponding risk is

$$R_2(\xi, \zeta) = c \int \xi(d\theta) \sum_{n=0}^{\infty} n \int_{\{N(\xi,\varphi)=n\}} \prod_{\nu=1}^{n} F_\theta(dx_\nu). \tag{6.8.1}$$

Similarly,

$$R_1(\xi, \zeta) = \int \xi(d\theta) \sum_{n=0}^{\infty} \int_{\{N(\xi,\varphi)=n\}} L_1(\theta, d(x_1, \ldots, x_n)) \prod_{\nu=1}^{n} F_\theta(dx_\nu). \tag{6.8.2}$$

Minimax sequential rules were determined in the literature according to one of the following three criteria:

(i) minimize $\sup_\theta R_1(\theta, \zeta)$, subject to the condition $\sup_\theta R_2(\theta, \zeta, c) \leqslant L_2$;

(ii) minimize $\sup_\theta R_2(\theta, \zeta, c)$, subject to the condition $\sup_\theta R_1(\theta, \zeta) \leqslant L_1$;

(iii) minimize $\sup_\theta \{R_1(\theta, \zeta) + R_2(\theta, \zeta, c)\}$.

The risk functions $R_2(\theta, \zeta, c)$ and $R_1(\theta, \zeta)$ are determined as in (6.8.1) and (6.8.2). The three criteria of minimax sequential rules can be formulated in a more general manner, using supremum over the prior risks $R_1(\xi, \zeta)$ and $R_2(\xi, \zeta, c)$.

Stein and Wald [1] considered minimax sequential procedures for a fixed-width interval estimation of the mean of a normal distribution with a known variance. This is an estimation problem with a loss function as in Example 6.1. They followed (i) and (ii). Wolfowitz [3] considered the same problem, following (iii). Blyth [1] gave an example in which he provided the minimax sequential rule for estimating θ, when X has a rectangular distribution over $(\theta - \frac{1}{2}, \theta + \frac{1}{2})$, $\Theta = (-\infty, \infty)$. He solved the problem under (iii). Kiefer [1] considered the rectangular distribution over $(0, \theta)$, $\theta \in (0, \infty)$.

The main result of Wolfowitz [3] can be followed from Example 6.7. We have shown there that for each sample of size $n \geqslant 1$, the minimax estimator of the mean θ of an $\mathcal{N}(\theta, 1)$ distribution is the sample mean $\bar{X}_n$. If $n = 0$ we estimate θ by $\hat{\theta}_0 \equiv 0$. The loss function is a zero-one function specified in Example 6.1. If $R_2(\theta, \zeta, c) = cE_\theta\{N(\zeta)\}$, we can find the minimax stopping rule in the following manner. Given n observations ($n \geqslant 0$), if an additional observation is taken, the minimax risk is reduced from $2[1 - \Phi(\delta\sqrt{n})]$ to $2[1 - \Phi(\delta\sqrt{(n+1)})]$. On the other hand, the cost of sampling will be increased by c. Let $\Delta(n) = 2[\Phi(\delta\sqrt{(n+1)}) - \Phi(\delta\sqrt{n})]$. It is easy to verify that $\Delta(n)$ is a decreasing function of n, $n = 0, 1, \ldots$. Hence the minimax rule is to stop sampling at the least integer n, $n \geqslant 0$, for which $\Delta(n) \leqslant c$. Since $\Delta(n)$ is independent of the observed values, the sample size can be determined before observations commence; that is, the minimax estimation procedure is a fixed sample size procedure.

Kiefer [1] studied minimax sequential estimation of the range θ of a rectangular distribution over $(0, \theta)$, $0 < \theta < \infty$. He considered the quadratic loss function $L(\theta, \hat{\theta}) = (\theta - \hat{\theta})^2/\theta^2$. The minimal sufficient statistic, for each sample of size $n \geqslant 1$, is $T_n = \max_{1 \leqslant i \leqslant n} \{X_i\}$. Kiefer proved that the minimax sequential procedure is a procedure with a fixed sample size n_0. The minimax estimator $\hat{\theta}(T_n) = T_{n_0} \cdot (n_0 + 2)/(n_0 + 1)$. The sample size n_0 is an integer which minimizes the function $r(m) = cm + (m + 1)^{-2}$, where c, $0 < c < \infty$, is the fixed cost of observation. He showed that there are at most two integers, n_0 and $n_0 + 1$, over which $r(m)$ is minimized for each c. The method of proof used by Kiefer is the following. Consider a sequence of prior distributions for θ with density $h_\tau(\theta)$ given by

$$h_\tau(\theta) = \begin{cases} \dfrac{1}{\log 1/\tau} \cdot \dfrac{1}{\theta}, & \tau < \theta < 1, \\ 0, & \text{otherwise}, \end{cases} \tag{6.8.3}$$

where $0 < \tau < 1$. The posterior density of θ, given T_n, is then

$$h_\tau(\theta \mid T_n) = \begin{cases} \dfrac{nz^n}{1 - z^n} \cdot \dfrac{1}{\theta^{n+1}}, & z < \theta < 1, \\ 0, & \text{otherwise}, \end{cases} \tag{6.8.4}$$

where $z = \max(\tau, T_n)$. The Bayes estimator against $h_\tau(\theta)$ is then

$$\hat{\theta}_\tau(T_n) = \frac{n+2}{n+1} \cdot \frac{1 - z^{n+1}}{1 - z^{n+2}} z, \tag{6.8.5}$$

with a posterior risk

$$R(\tau, T_n) = 1 - \frac{n(n+2)}{(n+1)^2} \cdot \frac{(1 - z^{n+1})^2}{(1 - z^n)(1 - z^{n+2})}, \tag{6.8.6}$$

provided $n \geqslant 1$. For $n = 0$ we use the prior distribution and obtain the prior Bayes risk of

$$R(\tau) = 1 - \frac{2(1 - \tau)}{(1 + \tau) \log 1/\tau}. \tag{6.8.7}$$

The next step is to find, for each $n \geqslant 1$, the prior risk of the Bayes procedure, which is the expectation of the posterior risk (6.8.6) with respect to the marginal distribution of T_n. Kiefer proves that a lower bound for this Bayes prior risk approaches, as $\tau \to 0$, the function $r(m) = cm + (m + 1)^{-2}$. Applying Theorem 6.5.2 we obtain that the procedure specified above is minimax.

6.9. EMPIRICAL BAYES PROCEDURES

Consider an estimation problem which repeats itself independently at an infinite sequence of epochs. At each epoch a random variable X is observed.

This variable X has a distribution function $F_\theta(x)$. The distribution function F_θ belongs to a family whose elements are indexed by θ, $\theta \in \Theta$. As in the Bayesian framework, we assume that there exists a prior distribution $H(\theta)$ on Θ and that at each epoch, θ is generated at random according to $H(\theta)$; then the observation on X is generated according to $F_\theta(x)$. We consider here the problem of estimating the current value of θ.

In contrast to the purely Bayesian framework, we do not have to specify the prior distribution $H(\theta)$, according to which θ is generated, but only the family of prior distributions to which H belongs. The objective is to provide a sequence of estimators, that will approach (in probability) the Bayes estimator of θ, say $\hat{\theta}_H$, for a specified loss function, so that the corresponding sequence of prior risks will converge to the Bayes risk. Such an approach was first suggested by Robbins [2], who called it an *empirical Bayes* procedure. Further research in this area was performed by Johns [1], Miyasawa [1], Neyman [4], Robbins [2; 4], Rutherford and Krutchkoff [1], and others. In this section we provide the general framework for an empirical Bayes estimation as well as a theorem of Robbins [4]. Relevant results concerning testing procedures are given in Chapter 9. An example of an empirical Bayes determination of an asymptotically optimal control of stock levels is also given.

As in the previous sections, let $(\mathcal{X}, \mathcal{B})$ be the measure space associated with each observation, and let $\{P_\theta;\ \theta \in \Theta\}$ be a family of probability measures on $(\mathcal{X}, \mathcal{B})$. For each $\theta \in \Theta$, let $f(x; \theta)$ designate the density function of P_θ with respect to a sigma-finite measure $\mu(dx)$ on $(\mathcal{X}, \mathcal{B})$. Let $(\Theta, \mathcal{C})$ be the measure space associated with the parameter space Θ and $H(\theta)$ a fixed prior distribution of θ. The distribution $H(\theta)$ is unspecified, but is assumed to belong to a family $\mathcal{H}$ of prior distributions. Let $\hat{\theta}_H(X)$ designate the Bayes estimator of θ against H, with respect to a specified loss function $L(\theta, \hat{\theta})$, and let $\rho(H)$ designate the Bayes (prior) risk associated with $\hat{\theta}_H(X)$. The functional $\rho(H)$, $H \in \mathcal{H}$, is called the *Bayes envelope functional.*

Let (X_1, θ_1), (X_2, θ_2), ... be a sequence of independent random vectors; $X_1, X_2, \ldots$ are the observed random variables, whereas $\theta_1, \theta_2, \ldots$ are unobserved. After the n-th observation, X_n, we have to estimate the associated random parameter, θ_n. (Without loss of generality, assume that $n \geqslant 2$.) Since $\theta_1, \ldots, \theta_{n-1}$ are generated by the same prior distribution as θ_n, the observations $X_1, \ldots, X_{n-1}$ contain some information on θ_n also. We are looking, therefore, for an estimator $\hat{\theta}_n(X_n; X_1, \ldots, X_{n-1})$ which will be asymptotically "close" to the Bayes estimator $\hat{\theta}_H(X)$.

The variables $X_1, X_2, \ldots$ are independent random variables with a common marginal density

$$f_H(x) = \int f(x; \theta) H(d\theta). \tag{6.9.1}$$

Let

$$R_H(\hat{\theta}; X) = \int L(\theta; \hat{\theta}(X)) f(X; \theta) H(\mathrm{d}\theta). \tag{6.9.2}$$

If we use the estimator $\hat{\theta}_n = \hat{\theta}_n(X_n; X_1, \ldots, X_{n-1})$, $R_H(\hat{\theta}_n; X)$ depends on $X_1, \ldots, X_{n-1}$ as well. Thus let

$$R_H^*(\hat{\theta}_n; X) = \int_{\mathfrak{X}^{(n-1)}} \prod_{i=1}^{n-1} f_H(x_i)\mu(\mathrm{d}x_i) \int_{\Theta} L(\theta; \hat{\theta}_n(X; \mathbf{x}_{n-1})) f(X; \theta) H(\mathrm{d}\theta). \tag{6.9.3}$$

The prior risk associated with an estimator $\hat{\theta}_n(X_n; \mathbf{X}_{n-1})$ is then

$$R_n(\hat{\theta}_n; H) = \int \mu(\mathrm{d}x) R_H^*(\hat{\theta}_n; x). \tag{6.9.4}$$

A sequence of estimators $\{\hat{\theta}_n\}$ is called *asymptotically optimal* (A.O.) relative to H if

$$\lim_{n \to \infty} R_n(\hat{\theta}_n; H) = \rho(H). \tag{6.9.5}$$

The question is whether there exists a sequence $\{\theta_n\}$ which is A.O. relative to each H in $\mathcal{H}$. The method of empirical Bayes procedures provides such sequences of A.O. estimators. In the following theorem we give some sufficient conditions for the A.O. of a sequence $\{\theta_n\}$. We proceed with some notation.

Let $\mathfrak{D}$ be the class of all estimators of θ and

$$0 \leqslant \bar{L}(\theta, x) = \sup_{\theta \in \mathfrak{D}} L(\theta, \hat{\theta}(x)) \leqslant \infty \tag{6.9.6}$$

be integrable with respect to $H(\theta)$, and let

$$\int_{\Theta} \bar{L}(\theta, x) H(\mathrm{d}\theta) < \infty \quad \text{a.s. } [\mu]. \tag{6.9.7}$$

For a fixed estimator $\hat{\theta}_0$, define

$$\Delta_H(\hat{\theta}; x) = \int_{\Theta} [L(\theta, \hat{\theta}(x)) - L(\theta, \hat{\theta}_0(x))] f(x; \theta) H(\mathrm{d}\theta), \tag{6.9.8}$$

and

$$R_H{}^0(x) = \int L(\theta, \hat{\theta}_0(x)) f(x; \theta) H(\mathrm{d}\theta). \tag{6.9.9}$$

Obviously,

$$R_H(\hat{\theta}; x) = R_H{}^0(x) + \Delta_H(\hat{\theta}; x) \quad \text{a.s. } [\mu]. \tag{6.9.10}$$

Let $\Delta_n(\hat{\theta}, x; X_1, \ldots, X_{n-1})$, $n = 2, 3, \ldots$ be a sequence of random functions such that

$$\text{(6.9.11)} \qquad \underset{n\to\infty}{\text{p. lim}} \sup_{\theta \in \mathfrak{D}} |\Delta_n(\hat{\theta}, x; \mathbf{X}_{n-1}) - \Delta_H(\hat{\theta}; x)| = 0 \quad \text{a.s. } [\mu].$$

Let $\{\epsilon_n\}$ be a sequence of constants converging to zero. Define the sequence of estimators $\{\hat{\theta}_n\}$, $\hat{\theta}_n(X) \equiv \hat{\theta}_n(X; \mathbf{X}_{n-1})$, which satisfy

$$\text{(6.9.12)} \qquad \Delta_n(\hat{\theta}_n, x; \mathbf{X}_{n-1}) \leqslant \inf_{\theta \in \mathfrak{D}} \Delta_n(\hat{\theta}, x; \mathbf{X}_{n-1}) + \epsilon_n.$$

for all n sufficiently large.

Theorem 6.9.1. (Robbins [4].) *If H is a prior distribution for which (6.9.7) holds and $\{\hat{\theta}_n\}$ a sequence of estimators satisfying (6.9.12), where the functions $\Delta_n(\hat{\theta}, x; \mathbf{X}_{n-1})$ satisfy (6.9.11), then $\{\hat{\theta}_n\}$ is A.O. relative to H, for all H in $\mathcal{H}$.*

Proof. According to the definition of the Bayes estimator $\hat{\theta}_H$,

$$\text{(6.9.13)} \qquad \Delta_H(\hat{\theta}_n, x) - \Delta_H(\hat{\theta}_H, x) \geqslant 0 \quad \text{a.s. } [\mu],$$

for all $n \geqslant 1$. Furthermore,

$$\text{(6.9.14)} \qquad \begin{aligned} &|\Delta_H(\hat{\theta}_n, x) - \Delta_H(\hat{\theta}_H, x)| \\ &\quad \leqslant |\Delta_H(\hat{\theta}_n, x) - \Delta_n(\hat{\theta}_n, x)| + |\Delta_n(\hat{\theta}_n, x) - \Delta_n(\hat{\theta}_H, x)| \\ &\quad\quad + |\Delta_n(\hat{\theta}_H, x) - \Delta_H(\hat{\theta}_H, x)|. \end{aligned}$$

According to (6.9.11), given any $\epsilon > 0$, the first and the third terms of the right-hand side of (6.9.14) are smaller than ϵ, with probability close to 1, if n is sufficiently large. Moreover, for n sufficiently large (6.9.12) implies that the second term on the right-hand side is smaller than ϵ_n. Hence

$$\text{(6.9.15)} \qquad \underset{n\to\infty}{\text{p. lim}}\, \Delta_H(\hat{\theta}_n(x; \mathbf{X}_{n-1}), x) = \Delta_H(\hat{\theta}_H, x) \quad \text{a.s. } [\mu].$$

Define

$$\text{(6.9.16)} \qquad K(x) = \int \bar{L}(\theta, x) f(x; \theta) H(\mathrm{d}\theta) \quad \text{a.s. } [\mu].$$

The quantity $K(x) \geqslant 0$ and, according to Fatou's lemma,

$$\text{(6.9.17)} \qquad R_H(\hat{\theta}, x) \leqslant K(x) \quad \text{a.s. } [\mu].$$

According to (6.9.7) and (6.9.16)

$$\text{(6.9.18)} \qquad \int K(x) \mu(\mathrm{d}x) < \infty.$$

Hence by the Lebesgue dominated convergence theorem,

$$\text{(6.9.19)} \qquad \lim_{n\to\infty} R(\hat{\theta}_n, H) = \int \underset{n\to\infty}{\text{p. lim}}\, R_H^*(\hat{\theta}_n, x) \mu(\mathrm{d}x).$$

Furthermore, according to (6.9.10) and (6.9.15),

$$\text{(6.9.20)} \qquad \operatorname*{p.\,lim}_{n\to\infty} R_H^*(\hat\theta_n, x) = R_H(\hat\theta_H, x) \quad \text{a.s. } [\mu].$$

From (6.9.19) and (6.9.20) we obtain that $\{\theta_n\}$ is A.O. with respect to H for all H in $\mathcal{H}$. (Q.E.D.)

In certain cases we can provide an A.O. sequence of estimators without the need to check all the conditions of Theorem 6.9.1. As previously mentioned, the conditions of this theorem are sufficient but not necessary. To illustrate it we provide the following example.

Example 6.12. In this example we provide an empirical Bayes procedure for an adaptive control of a stock level, which has been derived by Zacks [8].

The monthly demand for a certain commodity is a random variable X having a Poisson distribution $P(\lambda)$, $0 < \lambda < \infty$. The value of λ is determined at the beginning of each month according to some prior gamma distribution $\mathcal{G}(1/\tau, \nu)$, $0 < \tau, \nu < \infty$. The true values of ν and τ are unknown. The family $\mathcal{H}$ is the family of all gamma distributions $\mathcal{G}(1/\tau, \nu)$; $0 < \nu$, $\tau < \infty$. As above, we assume that the random vectors $(X_1, \lambda_1), (X_2, \lambda_2), \ldots$ are independent. The marginal distribution of X_1, X_2 for a specified pair of τ and ν is a negative binomial N.B. (ν, ψ), with density

$$\text{(6.9.21)} \quad g(x \mid \nu, \psi) = \frac{\Gamma(x+\nu)}{\Gamma(\nu)\Gamma(x+1)}\, \psi^x (1-\psi)^\nu, \qquad x = 0, 1, \ldots,$$

where

$$\text{(6.9.22)} \qquad \psi = \frac{\tau}{(1+\tau)}.$$

The optimal stock level at the beginning of each month, if the expected demand λ is known, is the function $k^0(\lambda)$ which minimizes the expected loss for the loss function

$$\text{(6.9.23)} \qquad L(k(\lambda), X) = c(k(\lambda) - X)^+ + p(k(\lambda) - X)^-,$$

where $k(\lambda)$ is the stock level; c is the cost per month of carrying a unit not in demand; p is the penalty cost for shortage of one unit; and $a^+ = \max(0, a)$, $a^- = -\min(0, a)$. When λ is unknown, the Bayes stock level for a prior distribution $\mathcal{G}(1/\tau, \nu)$ is the $p/(c+p)$-th fractile of the N.B. (ν, ψ) distribution, which is (see Zacks [8])

$$\text{(6.9.24)} \quad k^0(\nu, \psi) = \text{least integer } k,\ k \geqslant 0, \quad \text{such that } I_{1-\psi}(\nu, k+1) \geqslant p/(c+p);$$

and $I_a(p, q)$ designates the incomplete beta function ratio.

After the demand values of the first n months have been observed, the posterior distribution of λ, given $S_n = \sum_{i=1}^n X_i$, is the gamma distribution $\mathcal{G}(1/\tau + n, \nu + S_n)$. Thus if we define

$$\nu_{n+1} = \nu + S_n, \qquad n = 0, 1, \ldots,$$

(6.9.25)
$$\psi_{n+1} = \frac{\tau}{1 + (n+1)\tau}, \qquad n = 0, 1, \ldots,$$

with $S_0 \equiv 0$, the Bayes stock level of the beginning of the $(n+1)$-th month is $k^0(\nu_{n+1}, \psi_{n+1})$. This is determined by substituting ν_{n+1} and ψ_{n+1} in (6.9.24). If ν and τ are unknown, we can provide strongly consistent estimators of ν_n and τ_n. Indeed, since all the moments of N.B. (ν, ψ) exist, for each (ν, ψ), the sequences $\{\sum_{i=1}^n X_i/n; n = 1, 2, \ldots\}$ and $\{\sum_{i=1}^n X_i^2/n; n = 1, 2, \ldots\}$ obey the strong law of large numbers. Hence

(6.9.26)
$$\bar{X}_n \xrightarrow{\text{a.s.}} E\{X\} = \frac{\nu\psi}{1-\psi},$$

and

(6.9.27)
$$\hat{\sigma}_n^2 \xrightarrow{\text{a.s.}} \text{Var}\{X\} = \frac{\nu\psi}{(1-\psi)^2},$$

where $\bar{X}_n = \sum_{i=1}^n X_i/n$ and $\hat{\sigma}_n^2 = \sum_{i=1}^n X_i^2/n - \bar{X}_n^2$. It follows that a strongly consistent estimator of $\psi = \tau/(1+\tau)$ is

(6.9.28)
$$\hat{\psi} = \left(1 - \frac{\bar{X}_n}{\hat{\sigma}_n^2}\right)^+,$$

Similarly, a strongly consistent estimator of ν_{n+1} is

(6.9.30)
$$\bar{\nu}_{n+1} = \bar{X}_n \frac{1-\hat{\psi}}{\hat{\psi}} + S_n.$$

If we substitute $\hat{\psi}_{n+1}$ and $\hat{\nu}_{n+1}$ in (6.9.24) we obtain a strongly consistent estimator of the Bayes stock level $k^0(\nu_{n+1}, \psi_{n+1})$. That is,

$$\lim_{n\to\infty} \frac{k^0(\bar{\nu}_{n+1}, \hat{\psi}_{n+1})}{k^0(\nu_{n+1}, \psi_{n+1})} = 1 \quad \text{a.s.}$$

As shown by Zacks [8], the Bayes posterior risk corresponding to

$$k^0(\nu_{n+1}, \psi_{n+1})$$

is

(6.9.31)
$$R(\nu_{n+1}, \psi_{n+1}) \cong (c+p)\frac{\Gamma(k^0(\nu_{n+1}, \psi_{n+1}) + \nu_{n+1} + 1)}{\Gamma(\nu_{n+1})\Gamma(k^0(\nu_{n+1}, \psi_{n+1}) + 1)} \times \psi_{n+1}^{k^0(\nu_{n+1}, \psi_{n+1})+1}(1-\psi_{n+1})^{\nu_{n+1}-1}.$$

It follows that

(6.9.32) $$\operatorname{p.lim}_{n\to\infty} \{R(\bar{\nu}_{n+1}, \hat{\psi}_{n+1}) - R(\nu_{n+1}, \psi_{n+1})\} = 0.$$

Finally, since

(6.9.33) $$\sup_{k\geqslant 0} \frac{\Gamma(k+\nu)}{\Gamma(\nu)\Gamma(k+1)} \psi^k(1-\psi)^\nu \equiv K(\nu, \psi) < \infty$$

for all ψ, ν, we obtain from the Lebesgue dominated convergence theorem and (6.9.32) that

(6.9.34) $$\lim_{n\to\infty} E\{R(\bar{\nu}_{n+1}, \hat{\psi}_{n+1}) - R(\nu_{n+1}, \psi_{n+1})\} = 0,$$

for all ν, ψ. This proves that $\{k^0(\hat{\nu}_n, \hat{\psi}_n)\}$ is an A.O. sequence for all ν and τ. ■

6.10. ASYMPTOTICALLY OPTIMAL BAYES SEQUENTIAL ESTIMATION

In Section 6.8 we outlined the general framework of sequential estimation. As explained there, every sequential procedure ζ prescribes a stopping rule s and an estimator d. Moreover, the posterior distribution of the parameter θ, given $\{S = n\}$, is independent of the stopping rule. Thus the Bayes estimator for any specified prior distribution H and loss function $L(\theta, d)$ is independent of the stopping rule s. We have designated by $L_1(\theta, \zeta)$ the loss due to erroneous estimation and by $L_2(\theta, \zeta)$ the loss (cost) due to stopping according to the rule s. The prior risk functions for a prior probability measure ξ on $(\Theta, \mathscr{T})$ were designated by $R_1(\xi, \zeta)$ and $R_2(\xi, \zeta)$ and given by (6.8.1) and (6.8.2) for the case where $L_2(\theta, \zeta) = cN(\xi, \zeta)$. It is generally very difficult to determine the optimal Bayes sequential stopping rule. We can determine, however, stopping rules which are asymptotically as good as the Bayes stopping rules. We notice that if the loss $L_2(\theta, \zeta)$ is a function $cK(N(\xi, \zeta))$, where $N(\xi, \zeta)$ is the random sample size obtained by ζ under ξ, then if $R_2(\xi, \zeta) < \infty$, $R_2(\xi, \zeta) \to 0$ as $c \to 0$. Thus as $c \to 0$, the optimal stopping rule will generally yield very large samples. Let $\mathscr{S}$ be the class of all stopping rules.

A stopping rule s^0 in $\mathscr{S}$ is *asymptotically optimal* if

(6.10.1) $$\limsup_{c\to 0} \left\{R(\xi, s^0)\left[\inf_{s\in\mathscr{S}} R(\xi, s)\right]^{-1}\right\} \leqslant 1,$$

where

(6.10.2) $$R(\xi, s^0) = R_1(\xi, (s^0, \hat{\theta}_\xi)) + R_2(\xi, s^0).$$

Bickel and Yahav [1; 2] also investigated stopping rules which are asymptotically pointwise optimal in the Bayesian sense. Let $R(\xi, X_n)$ designate the Bayes (posterior) risk associated with the Bayes estimator when the sample size is n and $L_2(\theta, \zeta) = cK(N(\xi, \zeta))$. A stopping rule t is called *asymptotically pointwise optimal* if

$$\text{(6.10.3)}\quad \limsup_{c\to 0}\left\{[R(\xi, X_t) + cK(t)]\left[\inf_{s\in S}\{R(\xi, X_s) + cK(s)\}\right]^{-1}\right\} \leqslant 1, \quad \text{a.s.}$$

The property of asymptotically pointwise optimality (A.P.O.) implies generally the weaker property of asymptotical optimality (A.O.). We devote the rest of this section to some theorems proven by Bickel and Yahav [2] concerning A.P.O. and A.O. stopping rules. We proceed by remarking that the concept of asymptotic optimality which we discussed in the previous section, in the context of empirical Bayes estimation, differs from the present concept since in empirical Bayes estimation there is no role for a stopping rule. The empirical Bayes problem repeats itself indefinitely. Otherwise there is an analogy, in the sense that both concepts pertain to properties of estimators whose prior risks approach the Bayes risks.

For the following discussion we designate the Bayes risk, when the sample is of size n, by Y_n. Obviously, $Y_n > 0$ a.s. Assume also that $Y_n \to 0$ a.s. The conditions for the validity of this assumption will be discussed later. Let $K(n)$ be a strictly increasing function of n, $\lim_{n\to\infty} K(n) = \infty$. Define, for each c, $0 < c < \infty$,

$$\text{(6.10.4)}\qquad X(n, c) = Y_n + cK(n).$$

If $\mathcal{B}_n$ designates the Borel sigma-subfield generated by the first n observations, $Y_n \in \mathcal{B}_n$, all $n \geqslant 1$. Let $\mathcal{S}$ be the class of all stopping rules $t(c)$ defined on the sequence $\{\mathcal{B}_n; n \geqslant 1\}$. We have $\{t(c) = n\} \in \mathcal{B}_n$ for all $n \geqslant 1$ and all c. Let β be a positive real and suppose the following:

(i) $\eta^\beta Y_n \to V$ a.s., where $P\,[0 < V < \infty] = 1$;

(ii) For any c, $0 < c < \infty$, and X, $0 < X < \infty$, let $N(X, c)$ be the least positive integer n which minimizes the function $xn^{-\beta} - cK(n)$. We assume that $N(X, c) < \infty$.

(iii) Let $\underset{n}{\Delta} K(n) = K(n + 1) - K(n)$. Then $n^{-(1+\beta)} = o\left(\underset{n}{\Delta} K(n)\right)$, as $n \to \infty$.

Assumption (iii) guarantees that it will be optimal to stop sampling. Indeed, according to (i) the Bayes risk is asymptotically of the order $O(n^{-\beta})$. Hence if $\underset{n}{\Delta} K(n) = O(n^{-(1+\beta)})$, it is worthwhile to continue sampling, provided the sample is sufficiently large.

Consider the following stopping rule:

(6.10.5)

$$\bar{t}(c) = \text{least positive integer } n, \text{ such that } Y_n\left(1 - \left(\frac{n}{n + 1}\right)^\beta\right) \leqslant c \underset{n}{\Delta} K(n).$$

We prove now that $\tilde{t}(c)$ is an A.P.O. stopping rule. That is,

$$\lim_{c\to 0}\left\{X(\tilde{t}(c), c)\left[\inf_{S} X(s, c)\right]^{-1}\right\} \leqslant 1 \quad \text{a.s.} \tag{6.10.6}$$

Theorem 6.10.1. (Bickel and Yahav [2].) *Under (i)-(iii), the stopping rule $\tilde{t}(c)$ is A.P.O.*

Proof. We notice first that $n(1 - (n/(n+1))^\beta) \to \beta$ as $n \to \infty$. Hence according to (i), $Y_n(1 - (n/(n+1))^\beta) = O(n^{-(1+\beta)})$. Combined with (iii) we imply that $\tilde{t}(c) < \infty$ a.s.

The function $Xn^{-\beta}$ is strictly decreasing with n. Thus $\Delta(xn^{-\beta} + cK(n)) < 0$ for all $n < N(X, c)$, or $N(X, c) = 1$. Moreover,

$$\begin{aligned} \underset{n}{\Delta}(xn^{-\beta} + cK(n)) &= -x\,\frac{(n+1)^\beta - n^\beta}{n^\beta(n+1)^\beta} + c\,\underset{n}{\Delta}\,K(n) \\ &= -\frac{x}{n^\beta}\left(1 - \left(\frac{n}{n+1}\right)^\beta\right) + c\,\underset{n}{\Delta}\,K(n). \end{aligned} \tag{6.10.7}$$

Thus $\tilde{t}(c)$ is the least integer n minimizing $Y_n + cK(n) = X(n, c)$. If we let $h(x, c, n) = xn^{-\beta} + cK(n)$, then

$$X(\tilde{t}(c), c) = \min_{n} h((\tilde{t}(c))^\beta Y_{\tilde{t}(c)}, c, n). \tag{6.10.8}$$

Let $n_0(c)$ be the least $n \geqslant 1$ which minimizes $X(n, c)$. Then

$$X(n_0(c), c) \leqslant X(\tilde{t}(c), c) \leqslant (\tilde{t}(c))^\beta Y_{\tilde{t}(c)}(n_0(c))^{-\beta} + cK(n_0(c)). \tag{6.10.9}$$

According to the definition of $n_0(c)$ and the assumption that $Y_n \to 0$ a.s., $K(n) \nearrow \infty$ as $n \to \infty$; $n_0(c) \nearrow \infty$ a.s., as $c \to 0$. Hence, according to (i),

$$(\tilde{t}(c))^\beta Y_{\tilde{t}(c)} \to V \quad \text{a.s.,} \qquad \text{as } c \to 0, \tag{6.10.10}$$

and

$$(n_0(c))^\beta Y_{n_0(c)} \to V \quad \text{a.s.,} \qquad \text{as } c \to 0. \tag{6.10.11}$$

Dividing $X(\tilde{t}(c), c)$ by the right-hand side of (6.10.9) we obtain

(6.10.12)
$$\limsup_{c\to 0} \frac{X(\tilde{t}(c), c)}{(\tilde{t}(c))^\beta Y_{\tilde{t}(c)}(n_0(c))^\beta + cK(n_0(c))} = \limsup_{c\to 0} \frac{X(\tilde{t}(c), c)}{X(n_0(c), c)} \leqslant 1 \quad \text{a.s.}$$

On the other hand, the left-hand side of (6.10.9) yields

$$\liminf_{c\to 0} \frac{X(\tilde{t}(c), c)}{X(n_0(c), c)} \geqslant 1 \quad \text{a.s.} \tag{6.10.13}$$

Equations 6.10.12 and 6.10.13 imply that $\tilde{t}(c)$ is A.P.O. (Q.E.D.)

As a corollary of Theorem 6.10.1 we can prove (see Bickel and Yahav [2]) that, under the assumptions of Theorem 6.10.1,

$$\limsup_{c\to 0}\left\{X(\bar{t}(c), c)\left[\inf_n h(V, c, n)\right]^{-1}\right\} = 1. \tag{6.10.14}$$

A variant of $\bar{t}(c)$, which is also A.P.O., is the following stopping rule:

(6.10.15) $t'(c) =$ least positive integer n such that $Y_n\beta(n + 1)^{-1} \leqslant c$.

Bickel and Yahav [1] prove the following theorem.

Theorem 6.10.2. *If a sequence* $\{Y_n\}$ *obeys the conditions of Theorem 6.10.1 and, with* $K(n) = n$,

$$\sup_{n\geqslant 1} n^\beta E\{Y_n\} < \infty, \tag{6.10.16}$$

then the stopping rules $\bar{t}(c)$ *and* $t'(c)$ *are A.O.*

We apply these results now to the problem of sequential Bayes estimation. Consider first the case where $L_1(\theta, d) = (d - \theta)^2$, θ real, and the distribution of X is regular in the Cramér-Rao sense (see Section 4.1). We can prove then that the posterior risk of the Bayes estimator is asymptotically like the variance of the M.L.E.

Theorem 6.10.3. (Bickel and Yahav [1].) *If the density function of* X *is* $f(X; \theta)$, $\theta \in \Theta$, *which is an interval on the real line, and the following conditions hold,*

(i) $(\partial^2/\partial\theta^2)\log f(X; \theta)$ is continuous a.s. $[P_\theta]$;

(ii) $E_\theta\left\{\sup_{|\varphi-\theta|\geqslant\epsilon}[\log f(X; \varphi) - \log f(X; \theta)]\right\} < 0$ *for almost all* θ *and all* $\epsilon > 0$;

(iii) $E_\theta\left\{\sup_{|\varphi-\theta|<\epsilon}[(\partial^2/\partial\theta^2)\log f(X; \varphi)]^\dagger\right\} < \infty$ *for some* $\epsilon > 0$ *and almost all* θ;

(iv) *the Fisher information function* $I(\theta)$ *exists and is positive;*

(v) *the M.L.E.'s* $\hat{\theta}_n$ *of* θ *exist and* $\hat{\theta}_n \to \theta$ a.s. $[P_\theta]$.

(vi) *the prior distribution* $H(\theta)$ *satisfies* $\int \theta^2 H(d\theta) < \infty$;

then $nY_n \to I^{-1}(\theta)$ *a.s.* $[P_\theta]$, *where* $Y_n = \operatorname{Var}\{\theta \mid X_1, \ldots, X_n\}$.

For a proof see [24]. We notice that if $\beta = 1$ then Theorem 6.10.1 is satisfied and the stopping rules $\bar{t}(c)$ and $t'(c)$ are A.P.O. Conditions are given in [24] for the A.P.O. of $\bar{t}(c)$ and $t'(c)$ when the loss function is not necessarily quadratic, but $L(\theta, d) = W(|d - \theta|)$. The parameter θ is a k-dimensional vector. The cost of observations is $cK(n)$, and β is a positive real.

PROBLEMS

Section 6.1

1. Let $X_1, \ldots, X_n$ be i.i.d. discrete random variables having a negative binomial distribution, N.B. (ψ, ν), $0 < \psi < 1$, ν known. Suppose that the prior distribution of ψ is a beta (a, b), $0 < a, b < \infty$.

   (i) What is the posterior distribution of ψ, given $\mathbf{X}_n = (X_1, \ldots, X_n)$?
   (ii) What is the Bayes estimator of ψ for a bilinear convex loss function $L(\psi, \hat{\psi}) = |\psi - \hat{\psi}|$ on $[0, 1]$?

2. Let X have a density function $f(x; \theta)$, $\theta \in \Theta$. Suppose that Θ is a finite set of possible parameter points. Let $L(\theta, d)$ be any non-negative bounded loss function on $\Theta \times \Theta$. Let $H(\theta)$ be any discrete prior distribution function concentrated on Θ. What is the Bayes estimator of θ? Is it randomized or nonrandomized?

3. Let X have a binomial distribution $\mathscr{B}(n; \theta)$, $0 < \theta < 1$, n known. Let θ have a prior beta (a, b) distribution, $0 < a, b < \infty$. An interval $(\hat{\theta}(X) - \delta, \hat{\theta}(X) + \delta)$ for a specified value of δ, $0 < \delta < \frac{1}{2}$, is called a modal interval (see Blackwell and Girshick [1], p. 305) if the posterior probability $P_H[|\theta - \hat{\theta}(X)| < \delta \mid X]$ is maximal. Determine the modal interval for θ.

Section 6.2

4. Compute the Bayes estimator of the current mean, for the model given in Problem 36, Chapter 2, with a squared-error loss function. What is the associated Bayes risk (see Chernoff and Zacks [1]).

5. Suppose that X has a noncentral χ^2-distribution, that is, $X \sim \chi^2[\nu; \lambda]$, $0 < \lambda < \infty$. The number of degrees of freedom, ν, is known. For a gamma prior distribution for λ and squared-error loss, determine the Bayes estimator of λ. (Hint: $\chi^2[\nu; \lambda] \sim \chi^2[\nu + 2J]$ where J is a Poisson r.v. with mean λ.)

6. Determine the Bayes estimator, for squared-error loss, of the tail probability $\Phi(-\xi/\sigma)$ of a normal distribution $\mathscr{N}(\xi, \sigma^2)$ under the prior distribution

   (i) $P_H[\sigma^2 = 1] = 1$, $\mu \sim N(0, \tau^2)$
   (ii) The conditional prior distribution of μ, given σ^2, is $\mathscr{N}(0, \sigma^2)$ and $1/2\sigma^2 \sim G(\zeta, \nu)$.

Section 6.3

7. Consider a sequence of Bayes estimators of $\Phi(-\xi/\sigma)$ derived in Problem 6(i). What is the regular limit of the sequence as $\tau \to \infty$? Is it a generalized Bayes estimator, against an improper prior distribution?

8. Consider the problem of estimating the common mean μ of two normal distributions $\mathscr{N}(\mu, \sigma^2)$, $\mathscr{N}(\mu, \rho\sigma^2)$, $-\infty < \mu < \infty$, $0 < \sigma < \infty$, $0 < \rho < \infty$. The variance ratio ρ is unknown. Given a minimal sufficient statistic $(\bar{X}, Q_1, \bar{Y}, Q_2)$ based on two samples of equal size n, for these distributions, derive a sequence of

Bayes estimators for the quadratic loss function $L(\hat{\mu}, \mu) = (\hat{\mu} - \mu)^2/\sigma^2$ for the following sequence of prior distributions:

$$\mu \sim N(0, k), \quad k = 1, 2, \ldots, \quad P_H[\sigma^2 = \sigma_0^2] = 1,$$

$P_H[\rho = \rho_0] = 1$. Show that the regular limit of this sequence, as $k \to \infty$, is an L.M.V.U. estimator of μ.

Section 6.4

9. Let $X_1, \ldots, X_n$ be i.i.d. binomial random variables, $X_1 \sim \mathcal{B}(k; \theta), 0 < \theta < 1$.
   (i) Show that (θ, ξ) are consistent for any prior probability measure $\xi(d\theta)$, which is equivalent to the rectangular distribution on (0, 1).
   (ii) What is the Bayes estimator of $P_\theta[X = j]$, $j = 0, \ldots, k$, for a squared-error loss function and a prior beta (1, 1) distribution?
   (iii) Does the Bayes estimator derived in (ii) converge to the probability density function of $\mathcal{B}(k; \theta)$ as $n \to \infty$?

10. Let $X_1, \ldots, X_n$ be a sequence of i.i.d. random variables having a common location parameter exponential distribution (of the truncation type), with a density function

$$f(x; \alpha) = \exp\{-(x - \alpha)\}, \quad x \geqslant \alpha, \quad -\infty < \alpha < \infty.$$

Let $\xi(\alpha)$ be any prior probability measure whose topological carrier is the real line. Is (α, ξ) consistent?

Section 6.5

11. What is a minimax estimator of the current mean of the normal sequence specified in Problem 36 of Chapter 2?

12. Compare the mean square error functions of the M.L.E. and minimax estimator of the parameter θ in $\mathcal{B}(n; \theta)$ (see Example 6.6).

13. Let $X_1, \ldots, X_n$ be i.i.d. random variables having a common normal distribution $\mathcal{N}(\mu, \sigma^2)$. Both μ and σ are unknown; $-\infty < \mu < \infty$, $0 < \sigma < \infty$. What is a minimax estimator of σ^2 for the loss function $L(\sigma^2, \hat{\sigma}^2) = (\sigma^2 - \hat{\sigma}^2)^2/\sigma^4$?

Section 6.6

14. Suppose that $X \sim N(\mu, 1)$. Let $\mathcal{H}_0$ be the family of prior normal distributions $\mathcal{N}(0, \tau^2)$, $0 < \tau^2 < \infty$, for μ. Let ξ_0 be the prior distribution which is concentrated at $\{\mu = 0\}$. For a specified p, $0 < p < 1$, let $\mathcal{F}_p^{\,0} = \{\xi(\mu);\ \xi(\mu) = p\xi_0 + (1 - p)H,\ H \in \mathcal{H}_0\}$. What is the $\mathcal{F}_p^{\,0}$-minimax estimator of μ, based on a sample of n observations; and what is the associated minimax risk when the loss function is $L(\hat{\mu}, \mu) = (\hat{\mu} - \mu)^2$?

15. Let $X_1, \ldots, X_n$ be i.i.d. random variables having a binomial distribution $\mathcal{B}(k; \theta)$, $0 < \theta < 1$; k known. Let $\mathcal{H}$ be the family of all prior beta (a, b) distributions. If, with (confidence) probability p, $0 < p < 1$, we let $\theta = \frac{1}{2}$, what is the $\mathcal{F}_p$-minimax estimator of θ; and what is the associated minimax risk? For a loss function consider
   (i) $L(\theta, \hat{\theta}) = (\theta - \hat{\theta})^2$;
   (ii) $L(\theta, \hat{\theta}) = (\theta - \hat{\theta})^2/\theta(1 - \theta)$.

Section 6.7

16. Consider the problem of estimating the variance σ^2 of a normal distribution $\mathcal{N}(\mu, \sigma^2)$, with a squared-error loss. Let $\theta = (\mu, \sigma)$ and let $(\Theta, \mathcal{C})$ be the Borel-measurable space on the parameter space Θ. Let $\mathcal{U}$ be the smallest Borel sigma-field generated by $\mathcal{F} = \{B_{\sigma_0} = \{\theta; -\infty < \mu < \infty, \sigma = \sigma_0\}, 0 < \sigma_0 < \infty\}$. Show that $\mathcal{U}$ is sufficient for $\mathcal{F}$.

17. Consider the problem of estimating the mean of a normal distribution $\mathcal{N}(\mu, 1)$, $-\infty < \mu < \infty$, with a squared-error loss. Suppose that only a partial prior information is available. This is given by

$$\pi_j = P_H[j - 1 \leqslant |\mu| < j] = \frac{1}{2^j}, \qquad j = 1, 2, \ldots.$$

(i) What is the mean-max risk of the sample mean $\bar{X}_n$ with respect to the specified partial prior?

(ii) Is the estimator

$$\hat{\mu}_K(\bar{X}_n) = \begin{cases} -K, & \text{if } \bar{X}_n \leqslant -K \\ \bar{X}_n, & \text{if } |\bar{X}_n| \leqslant K \\ K, & \text{if } \bar{X}_n \geqslant K \end{cases}$$

for some K, $0 < K < \infty$, has a smaller mean-max risk than $\bar{X}_n$?

(iii) A lower-bound for the mean-max risk of $\hat{\mu}_K(\bar{X}_n)$ can be obtained by considering the Bayes risk associated with the prior distribution for μ, according to which the conditional distribution law of μ, given $\{J = j\}$, is

$$\mathcal{L}(\mu \mid J = j) = \begin{cases} \mathcal{R}(j - 1, j), & j = 1, 2, \ldots, \\ \mathcal{R}(j, j + 1), & j = -1, -2, \ldots, \end{cases}$$

and where $P[J = j] = (2^{|j|+1})^{-1}$, $j = \pm 1, \pm 2, \ldots$. Approximate the Bayes risk for this prior distribution by the Bayes of a normal prior distribution with mean zero and the same prior variance. How does this Bayes risk compare with the mean-max risk of (i)?

Section 6.9

18. Let $X_1, X_2, \ldots$ be a sequence of independent Poisson random variables. The means $\lambda_1, \lambda_2, \ldots$ of the Poisson distributions are i.i.d. and have a common prior distribution $H(\lambda)$, which is absolutely continuous on the positive part of the real line. What is an empirical Bayes estimator of λ_n for a squared-error loss? (See Robbins [2].)

REFERENCES

Barankin [2]; Bickel and Yahav [1], [2]; Blum and Rosenblatt [3]; Blyth [1]; Chernoff and Moses [1]; DeGroot and Rao [1]; Fabius [1]; Farrel [2]; Freedman [1]; Gaver and Hoel [1]; Girshick and Savage [1]; Godambe [1]; Hodges and Lehmann [1], [3]; Johns [1]; Kiefer [1]; Kudō [2]; Lindley [1]; Miyasawa [1]; Neyman [4]; Raiffa and Schlaiffer [1]; Rutherford [1]; Rutherford and Krutchkoff [1]; Sacks [1]; Stein and Wald [1]; von Neumann and Morgenstern [1]; Wald [6]; Wolfowitz [3]; Zacks [5], [6].

CHAPTER 7

Equivariant Estimators

7.1. THE STRUCTURE OF EQUIVARIANT ESTIMATORS

Let $\mathcal{F} = \{P_\theta;\ \theta \in \Theta\}$ be a family of probability measures on $(\mathcal{X}, \mathcal{B})$ and Θ a parameter space. In accordance with our previous framework, we consider Θ to be an interval in a Euclidean k-space.

Let $\mathcal{G}$ be a group of one-to-one transformations on $\mathcal{X}$ such that $g\mathcal{X} = \mathcal{X}$ and $g\mathcal{B} = \mathcal{B}$ for all $g \in \mathcal{G}$. Let $\bar{\mathcal{G}}$ be the induced group of transformations on Θ, defined by the identity

$$P_\theta[B] = P_{\bar{g}\theta}[gB], \tag{7.1.1}$$

every $B \in \mathcal{B}$ and each $g \in \mathcal{G}$. We assume that $\bar{\mathcal{G}}$ is one-to-one and $\bar{g}\Theta = \Theta$ for all $\bar{g} \in \bar{\mathcal{G}}$. Examples illustrating groups of transformations and (7.1.1) were given in Section 2.7. We also discussed in that section the notions of maximal invariant statistics and of orbits of $\mathcal{G}$ and $\bar{\mathcal{G}}$ in $\mathcal{X}$ and in Θ, respectively. In the exposition of the theory in this section, we designate the whole vector of sample values by x and the corresponding random variable by X. In the following examples we may give more details, according to the requirement of the derivations.

Consider now the problem of estimating the parameter θ of the distribution of X, with a loss function $L(\theta, d)$. An estimator (nonrandomized) $d(X)$ of θ preserves the structure of the estimation problem if and only if

$$d(gX) = \bar{g}d(X) \quad \text{for all} \quad g \in \mathcal{G}. \tag{7.1.2}$$

Indeed, a transformation g on $\mathcal{X}$ induces the transformation $\bar{g}$ on Θ. If the probability measure corresponding to the distribution of X is P_θ, then the one corresponding to gX is $P_{\bar{g}\theta}$. Thus if $d(X)$ estimates θ, it is desirable that $d(gX)$ estimate $\bar{g}\theta$. This is not always the case. As we illustrate in examples, (7.1.2) implies a certain structure of the estimator, which is specifically related to the group of transformations $\mathcal{G}$. In nontrivial cases, the structural preserving relationship, (7.1.2), imposes generally a restriction on the class of estimators and thus confines our attention to a subclass of estimators.

In such a subclass of estimators there may often exist an estimator which minimizes the risk function uniformly in θ, whereas a uniformly minimum estimator may not exist in the general class of estimators. For a reason that will be discussed later on, all the estimators which satisfy (7.1.2) for all x are called *equivariant* estimators. Those which satisfy (7.1.2) for almost all x (the null set over which (7.1.2) is violated may depend on the transformation g) are called *essentially equivariant*. Obviously, the class of all essentially equivariant estimators contains all the equivariant estimators. Another important assumption in the theory of equivariant estimators is that the loss function remains invariant under the group $\mathcal{G}$. That is,

$$L(\theta, d) = L(\bar{g}\theta, \bar{g}d), \quad \text{for all} \quad \bar{g} \in \bar{\mathcal{G}}, \tag{7.1.3}$$

all θ in Θ and almost all X. We present several examples.

Example 7.1. Let $X_1, X_2, \ldots, X_n$ be i.i.d. random variables having a common $\mathcal{N}(\mu, \sigma^2)$ distribution, $-\infty < \mu < \infty$, $0 < \sigma < \infty$. Both μ and σ are unknown. The minimal sufficient statistic is $[\bar{X}, \sum_{i=1}^{n} (X_i - \bar{X})^2]$. Consider the group of real translations

$$\mathcal{G} = \{g; gX_i = X_i + g, \text{ all } i = 1, \ldots, n, -\infty < g < \infty\}.$$

The induced group of transformations on the space $\mathcal{T}$ of sufficient statistics is

$$\mathcal{G}^* = \{g^*; g^*(\bar{X}, \sum (X_i - \bar{X})^2) = (\bar{X} + g, \sum (X_i - \bar{X})^2), -\infty < g < \infty\}.$$

Furthermore, the maximal invariant statistic on the space $\mathcal{T}$ is, according to $\mathcal{G}^*$, $U(\bar{X}, \sum (X_i - \bar{X})^2) = \sum (X_i - \bar{X})^2$. Thus every invariant function of $(\bar{X}, \sum (X_i - \bar{X})^2)$, with respect to the group $\mathcal{G}^*$, is of the general form $\psi(\sum (X_i - \bar{X})^2)$.

Suppose that we wish to estimate the mean μ of the distribution. Then the loss function to consider should be invariant with respect to the group $\bar{\mathcal{G}}$ on Θ, namely,

$$\bar{\mathcal{G}} = \{\bar{g}; \bar{g}(\mu, \sigma^2) = (\mu + g, \sigma^2), -\infty < g < \infty\}.$$

Thus for each estimator $\hat{\mu}$ of μ, $L(\mu, \hat{\mu})$ should satisfy the invariance condition $L(\mu + g, \hat{\mu} + g) = L(\mu, \hat{\mu})$ for all g, $-\infty < g < \infty$. According to (7.1.2) an estimator of μ is equivariant with respect to g if

$$\hat{\mu}(g^*(\bar{X}, \sum (X_i - \bar{X})^2) = \hat{\mu}(\bar{X} + g, \sum (X_i - \bar{X})^2) = \hat{\mu}(\bar{X}, \sum (X_i - \bar{X})^2) + g,$$

for all g, $-\infty < g < \infty$. In particular, setting $g = \bar{X}$ we obtain that any equivariant estimator of μ should be of the general form

$$\hat{\mu}_\psi(\bar{X}, \sum (X_i - \bar{X})^2) = \bar{X} + \psi(\sum (X_i - \bar{X})^2).$$

Consider now the risk function of an equivariant estimator $\hat{\mu}_\psi$ for the loss function $L(\mu, \hat{\mu}) = (\hat{\mu} - \mu)^2$. This is

$$R(\mu, \sigma^2; \psi) = E_{\mu,\sigma^2}\{[\bar{X} - \mu + \psi(\sum (X_i - \bar{X})^2)]^2\}$$
$$= \frac{\sigma^2}{n} + E\{\psi^2(\sigma^2\chi^2[n-1])\}.$$

We observe first that $R(\mu, \sigma^2; \psi)$ does not depend on μ. Moreover, since $\chi^2[n-1] > 0$ a.s., all σ^2, there exists a uniformly minimum risk equivariant estimator of μ, namely, $\hat{\mu}_0 = \bar{X}$. This is a special case of a more general result established by Pitman [1], which will be discussed later on.

Suppose now that we wish to estimate the parameter σ^2, using equivariant estimators under the above group of translations. If $\hat{\sigma}^2(\bar{X}, \sum (X_i - \bar{X})^2)$ is an equivariant estimator of σ^2 it should satisfy, according to (7.1.2),

$$\hat{\sigma}^2(\bar{X}, \sum (X_i - \bar{X})^2) = \hat{\sigma}^2(\bar{X} + g, \sum (X_i - \bar{X})^2),$$

for all g, $-\infty < g < \infty$. In particular, by setting $g = -\bar{X}$ we obtain that

$$\hat{\sigma}_\psi{}^2(\bar{X}, \sum (X_i - \bar{X})^2) = \psi(\sum (X_i - \bar{X})^2).$$

Thus every equivariant estimator of σ^2 is translation invariant. The risk function of $\hat{\sigma}_\psi{}^2$ is

$$R(\mu, \sigma^2; \psi) = E\{[\psi(\sigma^2\chi^2[n-1]) - \sigma^2]^2\}.$$

This is a function of σ^2 only. We notice that since any measurable function of $\sum (X_i - \bar{X})^2$ is an equivariant estimator of σ^2, with respect to the group of translations, we do not have any guide as to which estimator is optimal. If we require an additional property from the estimator, like the property of unbiasedness, then we may find an estimator with uniformly minimum risk. Essentially, there is only one estimator which is unbiased. This is the estimator $\hat{\psi}(\sum (X_i - \bar{X})^2) = \sum (X_i - \bar{X})^2/(n-1)$. Other types of estimators may be Bayes equivariant against some prior distribution of σ^2. We conclude the present example with the following remark. The orbits of the group $\bar{\mathcal{G}}$ in Θ are the straight lines in the (μ, σ^2)-plane, which are parallel to the μ-axis. We see that in both cases (estimating μ or estimating σ^2) the risk function associated with an equivariant estimator depends only on σ^2 and is therefore a constant on each of the orbits of $\bar{\mathcal{G}}$. ■

This property is proven in the following theorem.

Theorem 7.1.1. *Let $\mathcal{G}$ be a group of one-to-one transformations on $\mathcal{X}$, and let $\bar{\mathcal{G}}$ be the corresponding group of transformations on Θ. Let $L(\theta, d)$ be an invariant loss function with respect to $\mathcal{G}$. Let $\Omega = \{\Omega(s); s \in S\}$ be the orbits of*

$\bar{\mathcal{G}}$ *in* Θ, *where* S *is some index set. Then for each equivariant estimator of* θ, *and each* $s \in S$,

$$R(\theta, \hat{\theta}) = R_s(\hat{\theta}) \quad \textit{for all} \quad \theta \in \Omega(s). \tag{7.1.4}$$

Proof. Let $\bar{g}$ be an arbitrary transformation in $\bar{\mathcal{G}}$ and let g be the corresponding transformation in $\mathcal{G}$. Choose a point θ_0 in Θ. Consider the risk function of an equivariant estimator $\hat{\theta}$ at $\bar{g}\theta_0$. This is given by

$$R(\bar{g}\theta_0, \hat{\theta}) = \int L(\bar{g}\theta_0, \hat{\theta}(x)) \, \mathrm{d}P_{\bar{g}\theta_0}(X \leqslant x). \tag{7.1.5}$$

Let $y = g^{-1}x$. Then since $\hat{\theta}$ is equivariant, $\hat{\theta}(gy) = \bar{g}\hat{\theta}(y)$ and $P_{\bar{g}\theta_0}[X \leqslant gy] = P_{\theta_0}[X \leqslant y]$.

$$\begin{aligned} R(\bar{g}\theta_0, \hat{\theta}) &= \int L(\bar{g}\theta_0, \bar{g}\hat{\theta}(y)) \, \mathrm{d}P_{\theta_0}(X \leqslant y) \\ &= \int L(\theta_0, \hat{\theta}(y)) \, \mathrm{d}P_{\theta_0}(X \leqslant y) = R(\theta_0, \hat{\theta}). \end{aligned} \tag{7.1.6}$$

Since (7.1.6) holds for any $\bar{g} \in \bar{\mathcal{G}}$, $R(\theta, \hat{\theta})$ is constant on the orbit of $\bar{\mathcal{G}}$ which contains θ_0. The risk function may change its value only as θ changes from one orbit to another. (Q.E.D.)

The significance of the term "equivariant" lies in the property of having equal risk on orbits of $\bar{\mathcal{G}}$. The nonrandomized estimators which satisfy (7.1.2) are not invariant under $\mathcal{G}$; their values may change by the transformations $X \to gX$. The term "equivariant" is therefore more appropriate than the widely used term "invariant."

As shown in Section 6.2, when the loss function $L(\theta, d)$ is convex in d for each θ, the statistician may confine attention only to nonrandomized estimators. Bayes and minimax estimators (if exist) are nonrandomized. The question is whether these nonrandomized Bayes or minimax estimators are also equivariant. Generally these estimators are *not* equivariant. In Section 7.5 we establish sufficient conditions for the equivariance of minimax estimators.

We comment now about randomized invariant estimators. As in Section 6.1, let $\pi(\theta \mid X = x)$ be a distribution function on Θ which, given $X = x$, estimates θ according to $\pi(\theta \mid x)$; $\pi(\theta \mid X)$ is a randomized estimator. We assume that

$$\int_{\mathcal{X}} \left\{ \int_{\Theta} L(\theta, \hat{\theta}) \pi(\mathrm{d}\hat{\theta} \mid X = x) \right\} f(x; \theta) \mu(\mathrm{d}x) < \infty \tag{7.1.7}$$

for all $\theta \in \Theta$. A randomized estimator $\pi(\theta \mid X)$ is called *invariant* under the group of transformations $\mathcal{G}$ if and only if

$$\pi(\bar{g}\theta \mid gX) = \pi(\theta \mid X), \quad \text{all} \quad g \in \mathcal{G}, \quad \text{and all} \quad x \in \mathcal{X}. \tag{7.1.8}$$

If (7.1.8) holds for almost all $x[\mu]$ then $\pi(\theta \mid X)$ is called on *almost invariant* randomized estimator.

Let $\tilde{g}\pi(\theta \mid X) = \pi(\bar{g}\theta \mid gX)$. According to (7.1.8), if $\pi(\theta \mid X)$ is almost invariant, then $\tilde{g}\pi(\theta \mid X) = \pi(\theta \mid X)$ a.s. $[\mu]$. Moreover, for every $g \in \mathcal{G}$,

$$
\begin{aligned}
(7.1.9) \qquad R(\theta, \tilde{g}\pi) &= \int \mathrm{d}P_\theta[X \leqslant x] \int L(\theta, \hat{\theta})\tilde{g}\pi(\mathrm{d}\hat{\theta} \mid x) \\
&= \int \mathrm{d}P_\theta[X \leqslant x] \int L(\bar{g}\theta, \bar{g}\hat{\theta})\pi(d\bar{g}\hat{\theta} \mid x) \\
&= \int \mathrm{d}P_{\bar{g}\theta}[X \leqslant y] \int L(\bar{g}\theta, \theta')\pi(\mathrm{d}\theta' \mid y) \\
&= R(\bar{g}\theta, \pi).
\end{aligned}
$$

Hence if $\pi(\theta \mid X)$ is almost invariant, then $R(\theta, \tilde{g}\pi) = R(\bar{g}\theta, \pi) = R(\theta, \pi)$ for all $\bar{g} \in \bar{\mathcal{G}}$. This is a generalization of Theorem 7.1.1 to the case of a randomized estimator. Almost invariant randomized estimators have risk functions which assume constant values on the orbits of $\bar{\mathcal{G}}$.

Kiefer [2] proved that if $\mathcal{G}$ is a transitive group, we can restrict attention to nonrandomized estimators.

7.2. EQUIVARIANT ESTIMATORS OF LOCATION PARAMETERS

We start the discussion by deriving the Pitman estimator for the location parameter. Let $X_1, \ldots, X_n$ be i.i.d. random variables having a common density function $f(x - \theta)$, $-\infty < \theta < \infty$, such that $E_\theta\{X^2\} < \infty$ for each θ. Let $\mathcal{G} = \{g; gX_i = X_i + g$, all $i = 1, \ldots, n;\ -\infty < g < \infty\}$ be the group of real translations, and let $\hat{\theta}(X)$ be an equivariant estimator with respect to $\mathcal{G}$. The loss function is the squared-error loss, $L(\theta, \hat{\theta}) = (\hat{\theta} - \theta)^2$; $L(\theta, \hat{\theta})$ is invariant with respect to $\bar{\mathcal{G}}$.

Consider the n-dimensional interval $A = \{X; -\infty < X_1 \leqslant a_1, \ldots, -\infty < X_n \leqslant a_n\}$. The probability of A under θ is

$$P_\theta(A) = \int_{-\infty}^{a} \cdots \int \prod_{i=1}^{n} f(x_i - \theta)\mu(\mathrm{d}x_i).$$

Let $y_i = x_i + g$, $-\infty < g < \infty$, and $gA = \{x; -\infty < x_i \leqslant a_i + g$, $i = 1, \ldots, n\}$. Then

$$
\begin{aligned}
(7.2.1) \qquad P_\theta(A) &= \int_{-\infty}^{a_1+g} \cdots \int_{-\infty}^{a_n+g} \prod_{i=1}^{n} f(y_i - (\theta + g))\mu(\mathrm{d}y_i) \\
&= P_{\bar{g}\theta}(gA).
\end{aligned}
$$

Thus the problem of estimating the location parameter, θ, with the squared-error loss function is invariant (preserves its structure) under the group of real translations $\mathcal{G}$. The risk function of an equivariant (nonrandomized) estimator is

$$(7.2.2)\quad R(\theta, \hat{\theta}) = \int_{-\infty}^{\infty} \cdots \int [\hat{\theta}(x_1, \ldots, x_n) - \theta]^2 \prod_{i=1}^{n} f(x_i - \theta)\mu(\mathrm{d}x_i).$$

A maximal invariant statistic with respect to $\mathcal{G}$ based on the sufficient statistic $X_{(1)} \leqslant \cdots \leqslant X_{(n)}$ is $U(X_{(1)}, \ldots, X_{(n)}) = (X_{(2)} - X_{(1)}, \ldots, X_{(n)} - X_{(1)})$. Consider the transformation

$$(7.2.3)\quad \begin{aligned} Y_1 &= X_{(1)} \\ Y_2 &= X_{(2)} - X_{(1)} \\ &\ \vdots \\ Y_n &= X_{(n)} - X_{(1)}. \end{aligned}$$

We consider only equivariant estimators which depend on $X_1, X_2, \ldots, X_n$ through the order statistic $X_{(1)} \leqslant \cdots \leqslant X_{(n)}$. Thus the risk function of such estimators is

$$(7.2.4)\quad R(\theta, \hat{\theta}) = n! \int_0^{\infty} \cdots \int_0^{\infty} \prod_{i=2}^{n} \mu(\mathrm{d}y_i) \int_{-\infty}^{\infty} \mu(\mathrm{d}y_1)$$
$$\times\ [\hat{\theta}(y_1, y_1 + y_2, \ldots, y_1 + y_n) - \theta]^2 f(y_1 - \theta) \prod_{i=2}^{n} f(y_1 + y_i - \theta).$$

Let $z = y_1 - \theta$; then according to the equivariance property of $\hat{\theta}$,

$$(7.2.5)\quad R(\theta, \hat{\theta}) = n! \int_0^{\infty} \cdots \int \prod_{i=2}^{n} \mu(\mathrm{d}y_i) \int_{-\infty}^{\infty} [\hat{\theta}(z, y_2 + z, \ldots, y_n + z)]^2$$
$$\times\ f(z) \prod_{i=2}^{n} f(y_i + z)\mu(\mathrm{d}z).$$

Furthermore, $\hat{\theta}(z, y_2 + z, \ldots, y_n + z) = z + \psi(y_2, \ldots, y_n)$ for all z, due to the equivariance of $\hat{\theta}$. We also notice that the joint distribution function of the maximal-invariant statistic $u = (y_2, \ldots, y_n)$ does not depend on θ. Hence the minimum mean square estimator of the location parameter is

$$(7.2.6)\quad \hat{\theta}_0(X_{(1)}, \ldots, X_{(n)}) = X_{(1)} + \psi_0(X_{(2)} - X_{(1)}, \ldots, X_{(n)} - X_{(1)}),$$

where $\psi_0(y_2, \ldots, y_n)$ minimizes almost surely the conditional risk function

$$(7.2.7)\quad R(y_2, \ldots, y_n; \psi) = \int_{-\infty}^{\infty} [z + \psi(y_2, \ldots, y_n)]^2 f(z) \prod_{i=2}^{n} f(z + y_i)\mu(\mathrm{d}z).$$

It is immediately obtained that

$$\text{(7.2.8)} \qquad \psi_0(y_2, \ldots, y_n) = -\frac{\int_{-\infty}^{\infty} zf(z)\prod_{i=2}^{n} f(z+y_i)\mu(\mathrm{d}z)}{\int_{-\infty}^{\infty} f(z)\prod_{i=2}^{n} f(z+y_i)\mu(\mathrm{d}z)}.$$

Substituting (7.2.8) in (7.2.6), we obtain the Pitman estimator

$$\text{(7.2.9)} \qquad \hat{\theta}_0(X_1, \ldots, X_n) = X_{(1)} - \frac{\int_{-\infty}^{\infty} zf(z)\prod_{i=2}^{n} f(z+Y_i)\mu(\mathrm{d}z)}{\int_{-\infty}^{\infty} f(z)\prod_{i=2}^{n} f(z+Y_i)\mu(\mathrm{d}z)}.$$

We remark that the assumption about the independence of $X_1, X_2, \ldots, X_n$ is not necessary. We can consider any family of joint densities

$$f(X_1 - \theta, \ldots, X_n - \theta)$$

such that $E_\theta\{X_i^2\} < \infty$ for all θ and all $i = 1, \ldots, n$ and express the Pitman estimator in the form

$$\text{(7.2.10)} \quad \hat{\theta}_0(X_1, \ldots, X_n) = X_{(1)} - \frac{\int_{-\infty}^{\infty} zf(z, Y_2 + z, \ldots, Y_n + z)\mu(\mathrm{d}z)}{\int_{-\infty}^{\infty} f(z, Y_2 + z, \ldots, Y_n + z)\mu(\mathrm{d}z)}.$$

We further remark that

$$\text{(7.2.11)} \quad \frac{\int_{-\infty}^{\infty} zf(z, Y_2 + z, \ldots, Y_n + z)\mu(\mathrm{d}z)}{\int_{-\infty}^{\infty} f(z, Y_2 + z, \ldots, Y_n + z)\mu(\mathrm{d}z)} = E\{X_{(1)} \mid X_{(2)} - X_{(1)}, \ldots, X_{(n)} - X_{(1)}\}.$$

Since the Pitman estimator $\hat{\theta}_0(X_1, \ldots, X_n)$ is equivariant, it is an *unbiased* estimator of θ. Indeed, $\hat{\theta}_0(X_1, \ldots, X_n) - \theta = \hat{\theta}_0(X_1 - \theta, \ldots, X_n - \theta)$ for all θ, $-\infty < \theta < \infty$. Hence

$$\text{(7.2.12)} \quad E_\theta\{\hat{\theta}_0(X_1, \ldots, X_n) - \theta\} = E_\theta\{X_{(1)} - \theta - E\{X_{(1)} - \theta \mid X_{(2)} - X_{(1)}, \ldots, X_{(n)} - X_{(1)}\}\} = 0,$$

for all θ. This proves that $\hat{\theta}_0(X_1, \ldots, X_n)$ is an unbiased estimator of θ.

In the following we prove a result of Girshick and Savage [1], showing that the Pitman estimator $\hat{\theta}_0(X_1, \ldots, X_n)$ is a minimax estimator (in the general class of estimators). We proceed with an example.

Example 7.2. Let $X_1, \ldots, X_n$ be i.i.d. random variables having a location parameter exponential distribution, with density function

$$f(x - \theta) = \exp\{-(x - \theta)\}, \quad -\infty < \theta < \infty, \ \theta \leqslant x < \infty.$$

The Pitman estimator of θ is then

$$\hat{\theta}_0(X_1, \ldots, X_n) = X_{(1)} - \frac{\int_{-\infty}^{\infty} ze^{-z}\prod_{i=2}^{n} e^{-(z+y_i)}\,\mathrm{d}z}{\int_{-\infty}^{\infty} e^{-z}\prod_{i=2}^{n} e^{-(z+y_i)}\,\mathrm{d}z} = X_{(1)} - \frac{1}{n}.$$

Since $X_{(1)} - \theta \sim G(n, 1)$, $X_{(1)} - (1/n)$ is indeed an unbiased estimator of θ. Furthermore, the variance of the Pitman estimator is $1/n^2$.

We conclude this example by deriving the Pitman estimator in another manner. The minimal sufficient statistic in the present case is $T(X_1, \ldots, X_n) = X_{(1)}$. Thus every translation equivariant function based on $X_{(1)}$ is of the form $\varphi(X_{(1)}) = X_{(1)} + \varphi$, where φ is a real constant. The risk function associated with $\hat{\theta}_\varphi(X_{(1)}) = X_{(1)} + \varphi$ is

$$R(\theta, \hat{\theta}_\varphi) = E\{(X_{(1)} - \theta + \varphi)^2\} = \frac{1}{n^2} + \left(\frac{1}{n} + \varphi\right)^2.$$

Obviously, the minimum risk equivariant estimator is $\hat{\theta}_0(X_{(1)}) = X_{(1)} - (1/n)$. This is the Pitman estimator. ■

It is easy to verify also that the Pitman estimator of θ, in the $\mathcal{N}(\theta, 1)$ case, is $\hat{\theta}_0(X_1, \ldots, X_n) = \bar{X}$.

Girshick and Savage [1] considered a more general representation of the minimum mean square error equivariant estimator with respect to the group of translations. This representation is obtained in the following fashion. Let $t(X)$ be any measurable function of X satisfying (7.1.2); that is, $t(X + g) = t(X) + g$ for all g, $-\infty < g < \infty$. We remark that if $X = (X_1, \ldots, X_n)'$ is an n-dimensional vector, then $X + g$ designates the vector $X + g\mathbf{1} = (X_1 + g, \ldots, X_n + g)'$, for any real g, $-\infty < g < \infty$. Let $d(X)$ be any translation equivariant estimator of the translation parameter θ. Then the risk function of $d(X)$ is

$$\begin{aligned} R(\theta, d) &= E_\theta\{[d(X) - \theta]^2\} \\ &= E_\theta\{[d(X - t(X)) + t(X) - \theta]^2\} \\ &= E_\theta\{[d(X - t(X)) + t(X - \theta)]^2\} \\ &= E_0\{[d(Z) + t(X)]^2\}, \end{aligned} \tag{7.2.13}$$

where $Z = X - t(X)$. We notice that Z is translation invariant. Furthermore, if $d^*(X)$ designates the minimum mean square error translation equivariant estimator then, from (7.2.13),

$$d^*(X) = t(X) - E_0\{t(X) \mid X - t(X)\}. \tag{7.2.14}$$

The Pitman estimator $\hat{\theta}_0(X_1, \ldots, X_n)$ is a special case of (7.2.14), with $t(X) = X_{(1)}$. Thus we can start from an arbitrary translation equivariant estimator $t(X)$, having a finite risk at each θ, and reduce it to $d^*(X)$. The estimator $d^*(X)$ is an unbiased estimator of θ having the same variance as $\hat{\theta}_0(X)$, independently of $t(X)$.

We prove now that the Pitman estimator of the location parameter is a minimax estimator, for the mean square error risk function. We follow in our proof the method of Girshick and Savage [1]. The method consists of constructing sequences of similar problems with bounded loss functions and compact carriers of the distribution $F_0(x)$ and then approaching the Pitman estimator by a proper limiting process.

Let

$$R(\theta, d) = \int W(d(X) - \theta) F_\theta(\mathrm{d}X),$$

where $W(d - \theta) = L(\theta, d)$ is the risk function of $d(X)$ under θ.

Let $\mathfrak{D}$ be the class of *all* (nonrandomized) estimators of θ and let $\mathfrak{D}_I$ designate the class of translation equivariant estimators. Let $s(X) = d(X) - \bar{X}$, where $\bar{X}$ is the sample mean. As before, X designates the observed sample vector $(X_1, \ldots, X_n)$. We notice then that

$$R(\theta, d) = \int W(d(x) - \theta) f(x - \theta) \mu(\mathrm{d}x) \tag{7.2.15}$$

$$= \int W(s(x + \theta) + \bar{x}) f(x) \mu(\mathrm{d}x), \qquad \text{all } \theta, -\infty < \theta < \infty.$$

If $d(x)$ is translation equivariant then $s(x)$ is invariant; that is, $s(x + \theta) = s(x)$ for all θ, $-\infty < \theta < \infty$. Hence, if $d \in \mathfrak{D}_I$, then $R(\theta, d) = \rho(d)$ for all θ, $-\infty < \theta < \infty$. Let

$$\rho = \inf_{d \in \mathfrak{D}_I} \rho(d), \tag{7.2.16}$$

$$\bar{\rho}^* = \inf_{d \in \mathfrak{D}} \sup_{\xi} R(\xi, d), \tag{7.2.17}$$

and

$$\underline{\rho}^* = \sup_{\xi} \inf_{d \in \mathfrak{D}} R(\xi, d), \tag{7.2.18}$$

where ξ is any prior distribution of θ. The following relationship holds:

$$\rho \geqslant \bar{\rho}^* \geqslant \underline{\rho}^*, \quad \text{all } \theta. \tag{7.2.19}$$

Thus if $\rho = \underline{\rho}^* < \infty$, the minimum mean square error translation equivariance estimator is a minimax one. The invariance minimax theorem gives the condition for $\rho = \underline{\rho}^* < \infty$. We start with two lemmas.

Lemma 7.2.1. *Let F and W be such that for every $\epsilon > 0$ there exists a d.f. F' and a loss function W' for which*

(i) $F'(s) \leqslant F(s)$, *for all* $-\infty \leqslant s \leqslant \infty$;
(ii) $W'(\theta) \leqslant W(\theta)$, *for all* $-\infty \leqslant \theta \leqslant \infty$;
(iii) $\rho' \geqslant \rho - \epsilon$, *where* $\rho' = \inf_{d \in \mathfrak{D}_I} \int W'(d(x) - \theta) F'(dx)$;
(iv) $\rho' = \underline{\rho}'^*$;

Then $\rho = \underline{\rho}^*$.

Proof.

$$\begin{aligned} \underline{\rho}^* &= \inf_s \sup_\xi \iint W(s(x + \theta) + \bar{x}) F(dx) \xi(d\theta) \\ &\geqslant \inf_s \sup_\xi \iint W'(s(x + \theta) + \bar{x}) F'(dx) \xi(d\theta) \\ &= \underline{\rho}'^*. \end{aligned} \tag{7.2.20}$$

Hence from (iii), (iv), (7.2.19), and (7.2.20),

$$\rho \geqslant \underline{\rho}^* \geqslant \rho - \epsilon. \tag{7.2.21}$$

Equality holds since ϵ is arbitrary. (Q.E.D.)

Sequences of distribution functions and of loss functions which satisfy the conditions of Lemma 7.2.1 will be constructed in the proof of the main theorem.

Lemma 7.2.2. *Suppose that the loss function is bounded and let F be such that for some* $m > 0$, $P_F\{X; |\bar{X}| > m\} = 0$. *Then* $\rho = \underline{\rho}^*$.

Proof. Let $T > m$. Then

$$\begin{aligned} \sup_\xi \inf_{d \in \mathfrak{D}} \int R(\theta, d) \xi(d\theta) &= \sup_\xi \inf_{d \in \mathfrak{D}} \iint W(s(x + \theta) + \bar{x}) F(dx) \xi(d\theta) \\ &\geqslant \frac{1}{2T} \int_{-T}^{T} d\theta \cdot \inf_{d \in \mathfrak{D}} \int W(s(x + \theta) + \bar{x}) F(dx) \end{aligned} \tag{7.2.22}$$

Hence there exists $d^0 \in \mathfrak{D}$ such that

$$\begin{aligned} \rho^* + \epsilon &\geqslant \frac{1}{2T} \int_{-T}^{T} d\theta \int W(s^0(x + \theta) + \bar{x}) F(dx) \\ &= \frac{1}{2T} \int F(dx) \int_{-T}^{T} W(s^0(x + \theta) + \bar{x})\, d\theta. \end{aligned} \tag{7.2.23}$$

The equality on the right-hand side of (7.2.23) is due to the non-negativity of $W(\theta)$ and Fubini's theorem. Write

$$(7.2.24)\quad \int_{-T}^{T} W(s^0(x+\theta)+\bar{x})\,d\theta$$
$$= \int_{-T+\bar{X}}^{T+\bar{X}} W(s^0(x+\varphi-\bar{x})+\bar{x})\,d\varphi$$
$$= \int_{-T}^{T} W(s^0(x+\varphi-\bar{x})+\bar{x})\,d\varphi - \int_{-T}^{-T+\bar{X}} W(s^0(x+\varphi-\bar{x})+\bar{x})\,d\varphi$$
$$+ \int_{T}^{T+\bar{X}} W(s^0(x+\varphi-\bar{x})+\bar{x})\,d\varphi.$$

Let W^0 be an upper bound for $W(\theta)$. Then

$$(7.2.25)\quad \frac{1}{2T}\int F(dx)\left|\int_{-T}^{-T-\bar{X}} W(s^0(x+\varphi-\bar{x})+\bar{x})\,d\varphi\right.$$
$$\left. - \int_{T}^{T+\bar{X}} W(s^0(x+\varphi-\bar{x})+\bar{x})\,d\varphi\right| \leqslant \frac{W^0}{T}\int_{-m}^{m} |\bar{x}|\,F(dx)$$
$$\leqslant \frac{W^0 m}{T}\,[F(m)-F(-m)].$$

By choosing T large enough, we can make the right-hand side of (7.2.25) as small as we wish, and we obtain from (7.2.23)

$$(7.2.26)\quad \rho^* + \tfrac{3}{2}\epsilon \geqslant \frac{1}{2T}\int F(dx)\int_{-T}^{T} W(s^0(x+\varphi-\bar{x})+\bar{x})\,d\varphi.$$

The quantity $x-\bar{x}+\varphi$ is translation invariant. Hence $d_\varphi{}^0(x) = \bar{x} + s^0(x+\varphi-\bar{x})$ belongs to $\mathfrak{D}_I$. Therefore

$$(7.2.27)\quad \frac{1}{2T}\int F(dx)\int_{-T}^{T} W(s^0(x+\varphi-\bar{x})+\bar{x})\,d\varphi$$
$$= \frac{1}{2T}\int_{-T}^{T} d\varphi \int F(dx) W(s^0(x+\varphi-\bar{x})+\bar{x})\,d\varphi$$
$$\geqslant \frac{1}{2T}\int_{-T}^{T} d\varphi \inf_{d\in\mathfrak{D}_I} \rho(d) = \rho.$$

From (7.2.26) and (7.2.27) we obtain $\rho = \rho^*$. (Q.E.D.)

Theorem 7.2.3. *If W is bounded then $\rho = \rho^*$ for every θ.*

Proof. Let W^0 be an upper bound of $W(\theta)$. For any $\epsilon > 0$ there exists an M_ϵ such that,

$$P_F[|\bar{X}| > M_\epsilon] \leqslant \frac{\epsilon}{W^0}.$$

Define the defective d.f. $F'(x)$ by

$$F'(dx) = \begin{cases} F(dx), & \text{if } |\bar{X}| \leqslant M_\epsilon, \\ 0, & \text{otherwise.} \end{cases} \tag{7.2.28}$$

By Lemma 7.2.2, $\rho' = \underline{\rho}'^*$. Moreover,

(7.2.29)

$$\begin{aligned} \rho &= \inf_{d \in \mathfrak{D}_I} \int W(s(x) + \bar{x})F(dx) \\ &= \inf_{d \in \mathfrak{D}_I} \left\{ \int_{\{|\bar{X}| \leqslant M_\epsilon\}} W(s(x) + \bar{x})F(dx) + \int_{\{|\bar{X}| > M_\epsilon\}} W(s(x) + \bar{x})F(dx) \right\} \\ &\leqslant \inf_{d \in \mathfrak{D}_I} \int W(s(x) + \bar{x})F'(dx) + \epsilon \\ &= \underline{\rho}'^* + \epsilon = \rho' + \epsilon. \end{aligned}$$

Hence according to Lemma 7.2.1, $\rho = \underline{\rho}^*$. (Q.E.D.)

Lemma 7.2.4. *Let F be a d.f. of X and F_m the defective distribution*

$$F_m(dx) = \begin{cases} F(dx), & \text{if } |\bar{x}| \leqslant m, \\ 0, & \text{otherwise.} \end{cases} \tag{7.2.30}$$

Let the loss function be $W(\theta) = \theta^2$. Then

$$\rho_m = \inf_{d_I \in \mathfrak{D}} \int W(d(x) - \theta)F_{m,\theta}(dx) \nearrow \rho \text{ as } m \to \infty. \tag{7.2.31}$$

Proof. Since $W(\theta) = \theta^2$ we obtain

$$\rho_m = \inf_{d \in \mathfrak{D}_I} \int_{-m}^{m} (d(x) + \bar{x})^2 F(dx). \tag{7.2.32}$$

Hence

$$\rho_m = v_2(m) - \frac{v_1^2(m)}{v_0(m)}, \tag{7.2.33}$$

where

$$v_i(m) = \int_{-m}^{m} X^i F(\mathrm{d}x), \qquad i = 0, 1, 2,$$

provided m is sufficiently large so that $v_0(m) > 0$.

Case 1. $\rho < \infty$. By the Lebesgue monotone convergence theorem $v_i(m) \to v_i = \int x^i F(\mathrm{d}x)$ ($i = 0, 1, 2$). Hence $\rho_m \nearrow \rho$.

Case 2. $\rho = \infty$. In this case we can show that $\rho_m \nearrow \infty$. (Q.E.D.)

We now prove the main theorem.

Theorem 7.2.5. (Girshick and Savage [1].) *The estimator $d^*(X)$ given by (7.2.14) is minimax for the squared-error loss function if all equivariant estimators have finite risk functions. If there are some estimators with infinite risk then the minimax risk is infinity.*

Proof. As we have established previously, $d^*(X)$ can be obtained from any translation equivariant estimator $t(x)$. Thus let $d^*(X) = \bar{X} - E_0\{\bar{X} \mid Z\}$, where $Z = X - \bar{X}$.

Let m be a positive number and define

$$F_m(\mathrm{d}x) = \begin{cases} F(\mathrm{d}x), & \text{if } |\bar{x}| \leqslant m, \\ 0, & \text{otherwise.} \end{cases}$$

Consider the bounded loss function.

$$W_m(d(x) - \theta) = I(x, m)[d(x) - \theta]^2, \tag{7.2.34}$$

where $I(x, m)$ is the indicator function of the set $\{x;\ |\bar{x}| \leqslant m\}$. If $d(X)$ is a translation equivariant estimator and $s(X) = d(X) - \bar{X}$,

$$\rho_m = E\{E\{I(X, m)[\bar{X} - f_m(Z)]^2 \mid Z\}\}, \tag{7.2.35}$$

where

$$f_m(Z) = \frac{E\{I(X, m)\bar{X} \mid Z\}}{E\{I(X, m) \mid Z\}}. \tag{7.2.36}$$

Thus since F_m and W_m satisfy the conditions of Lemma 7.2.2, $\rho_m = \rho_m^*$. Furthermore, by Lemma 7.2.4 $\rho_m \nearrow \rho$. If $E_0\{\bar{X}^2 \mid Z\} < \infty$ a.s. then $\rho < \infty$, and for each $\epsilon > 0$ we can assign m_ϵ so large that $\rho_{m_\epsilon} \geqslant \rho - \epsilon$. It is easy to check that all the four conditions of Lemma 7.2.1 are satisfied and therefore $\rho = \underline{\rho}^*$. We also notice that $\lim_{m\to\infty} f_m(Z) = E\{\bar{X} \mid Z\}$. Hence the minimum mean square error translation equivariant estimator $d^*(X)$ is minimax.

(Q.E.D.)

7.3. THE GROUP-THEORETIC STRUCTURAL MODEL AND INVARIANT HAAR INTEGRALS

The theory of the Haar integral and invariant Haar measure plays an important role in the development of optimal invariant decision procedures. We present here the fundamental ideas of this theory. The reader is referred to the book by Nachbin [1] on the Haar integral and is advised to read Fraser's book [9] on the structural model.

We present the main ideas in Fraser's structural model by the following special case. Let $\mathbf{X}_1, \ldots, \mathbf{X}_n$ be i.i.d. p-variate normal random vectors with mean zero and an unknown covariance matrix $\boldsymbol{\Sigma}$, i.e., $\mathbf{X}_i \sim N(\mathbf{0}, \boldsymbol{\Sigma})$ all $i = 1, \ldots, n$. We consider the problem of estimating $\boldsymbol{\Sigma}$ equivariantly with respect to the group $\mathcal{G}$ of nonsingular linear transformations. Thus each element $\mathbf{G}$ of $\mathcal{G}$ is a $p \times p$ nonsingular matrix. We notice that if $\mathbf{Y} = \mathbf{GX}$ then $\mathbf{Y} \sim N(\mathbf{0}, \mathbf{G}\boldsymbol{\Sigma}\mathbf{G}')$. Thus the group $\mathcal{G}$ induces an isomorphic group $\bar{\mathcal{G}}$, whose elements $\bar{G}$ operate on the group of positive definite symmetric matrices $\boldsymbol{\Theta}$ according to $\bar{\mathbf{G}}\boldsymbol{\Sigma} = \mathbf{G}\boldsymbol{\Sigma}\mathbf{G}'$, for all $\mathbf{G} \in \mathcal{G}$. The space of all matrix parameters $\boldsymbol{\Sigma}$ will be designated by Θ.

Given the set of n vectors $\{\mathbf{X}_1, \ldots, \mathbf{X}_n\}$ we can construct an equivariant estimator $\mathbf{T}_X$ of $\boldsymbol{\Sigma}$, where $\mathbf{T}_X = \sum_{i=1}^n \mathbf{X}_i\mathbf{X}_i'$. Indeed, $\sum_{i=1}^n (\mathbf{GX}_i)(\mathbf{GX}_i)' = \mathbf{GT}_X\mathbf{G}'$ for all $\mathbf{G} \in \mathcal{G}$. Moreover, $\mathbf{T}_X$ is symmetric and positive definite with probability one. We consider $\mathbf{T}_X$, henceforth, as a positive definite matrix; $\mathbf{T}_X \in \Theta$.

Consider the $N = n \times p$-dimensional vector $\mathbf{X} = (\mathbf{X}_1', \ldots, \mathbf{X}_n')'$. Let $\mathcal{X}$ be the sample space of $\mathbf{X}$, which is an open interval in the Euclidean N-space. An orbit in $\mathcal{X}$ containing the point $\mathbf{X}$ is the set $\{\underset{\sim}{\mathbf{y}}; \underset{\sim}{\mathbf{y}} = ((\mathbf{GX}_1)', \ldots, (\mathbf{GX}_n)')'$, $\mathbf{G} \in \mathcal{G}\}$. On each such orbit we define a *reference-point* with respect to the estimator $\mathbf{T}_X$, namely, $\mathbf{R}(\underset{\sim}{\mathbf{X}}) = \mathbf{T}_X^{-1}\underset{\sim}{\mathbf{X}}$, where $\mathbf{G}\underset{\sim}{\mathbf{X}} = ((\mathbf{GX}_1)', \ldots, (\mathbf{GX}_n)')'$. The reference points $\mathbf{R}(\underset{\sim}{\mathbf{X}})$ can be used as indices of orbits. These points play a central role in the theory of invariant differentials.

Let $\mathbf{e}_1, \ldots, \mathbf{e}_n$ be i.i.d. p-variate standard normal vectors, that is, $\mathbf{e}_i \sim N(\mathbf{0}, \mathbf{I})$. Let $\mathbf{e} = (\mathbf{e}_1', \ldots, \mathbf{e}_n')'$. Let D be a nonsingular matrix such that $\boldsymbol{\Sigma} = \mathbf{DD}'$. Then $\mathbf{X} \sim \mathbf{De}$. We denote by $\mathbf{T}_e$ the matrix $\sum_{i=1}^n \mathbf{e}_i\mathbf{e}_i'$; notice that $\mathbf{T}_X = \mathbf{DT}_e\mathbf{D}'$ whenever $\mathbf{X} = \mathbf{De}$. Thus we can illustrate the relationships between the points on orbits of $\mathcal{X}$ and elements of the parameter-space Θ by the following scheme:

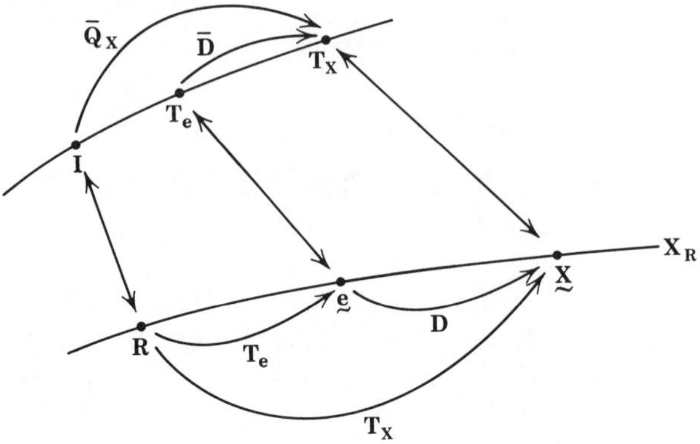

$T_X = Q_X Q'_X$
$\Sigma = DD'$

We construct now left and right invariant differentials on $\mathcal{X}$ and on $\mathcal{G}$. We notice that the parameter space Θ is a subgroup of $\mathcal{G}$, and the invariant differentials which will be defined on $\mathcal{G}$ yield an invariant distribution function on Θ, called by Fraser the *structural distribution*. This distribution coincides in special cases with the Fisher *fiducial distribution*, which is discussed in the next section.

The volume element or differential of a vector $\mathbf{X} = (\mathbf{X}_1, \ldots, \mathbf{X}_n)'$ is $d\mathbf{X} = d\mathbf{X}_1 \cdots d\mathbf{X}_N$. If we are given a matrix G of order $p \times q$ we convert it to a $p \times q$-dimensional vector $\mathbf{G}^*$, consisting of the q column vectors of G. The quantity dG is then defined as the differential element of $\mathbf{G}^*$.

When a nonsingular linear transformation A is applied to $\mathbf{X}$, that is, $\mathbf{Y} = \mathbf{A}\mathbf{X}$, the differential element of $\mathbf{X}$ changes to $d\mathbf{Y} = |\partial \mathbf{Y}/\partial \mathbf{X}|\, d\mathbf{X}$, where $|\partial \mathbf{Y}/\partial \mathbf{X}| = |\det\{\mathbf{A}\}|$ is the positive Jacobian factor.

Suppose that $\mathbf{X}$ is a $p \times p$ matrix, and let $\mathbf{A}$ be a $p \times p$ nonsingular matrix. We investigate now how the differential element $d\mathbf{X}$ changes when $\mathbf{A}$ multiplies $\mathbf{X}$ from the right or from the left. Formally, we write it as before:

$$d\mathbf{Y} = \begin{cases} J^*(\mathbf{A}; \mathbf{X})\, d\mathbf{X}, & \text{if} \quad \mathbf{Y} = \mathbf{X}\mathbf{A}, \\ J(\mathbf{A}; \mathbf{X})\, d\mathbf{X}, & \text{if} \quad \mathbf{Y} = \mathbf{A}\mathbf{X}, \end{cases}$$

where $J^*(\mathbf{A}; \mathbf{X})$ and $J(\mathbf{A}; \mathbf{X})$ are the right and the left positive Jacobian factors. More specifically, consider the right transformation

$$\mathbf{Y} = \mathbf{X}\mathbf{A} = \left\| \sum_{k=1}^{p} X_{ik} A_{kj};\ i, j = 1, \ldots, p \right\|.$$

Since the right transformation **XA** transforms the *row* vectors of **X** into the *row* vectors of **Y**, we convert **Y** into a p^2-dimensional vector $\mathbf{Y}^*$ by going along the rows of **Y**, that is,

$$\mathbf{Y}^* = \left(\sum_{k=1}^{k} X_{1k}A_{k1}, \ldots, \sum_{k=1}^{p} X_{1k}A_{kp} \,\middle|\, \cdots \,\middle|\, \sum_{k=1}^{p} X_{pk}A_{k1}, \ldots, \sum_{k=1}^{} X_{pk}A_{kp}\right).$$

In a similar manner we set

$$\mathbf{X}^* = (X_{11}, \ldots, X_{1p} \mid \cdots \mid X_{p1}, \ldots, X_{pp}),$$

and then

$$\left|\frac{\partial \mathbf{Y}}{\partial \mathbf{X}}\right| \equiv \left|\frac{\partial \mathbf{Y}^*}{\partial \mathbf{X}^*}\right|.$$

It is easy to verify that

$$\frac{\partial \mathbf{Y}^*}{\partial \mathbf{X}^*} = \begin{bmatrix} \mathbf{A} & \cdots & 0 \\ \cdot & \cdot & \cdot \\ 0 & \cdot & \cdot \\ \cdot & \cdot & \cdot \\ 0 & \cdots & \mathbf{A} \end{bmatrix},$$

and hence

$$\left|\frac{\partial \mathbf{Y}}{\partial \mathbf{X}}\right| = |\det\{\mathbf{A}\}|^p.$$

If

$$\mathbf{Y} = \mathbf{AX} = \left\| \sum_{k=1}^{p} A_{ik}X_{kj};\ i, j = 1, \ldots, p \right\|,$$

we form the vector Y^* from the *column* vectors of Y; that is,

$$\mathbf{Y}^* = \left(\sum_{k=1}^{p} A_{1k}X_{k1}, \ldots, \sum_{k=1}^{p} A_{pk}X_{k1} \,\middle|\, \cdots \,\middle|\, \sum_{k=1}^{p} A_{1k}X_{kp}, \ldots, \sum_{k=1}^{p} A_{pk}X_{kp}\right).$$

Similarly,

$$\mathbf{X}^* = (X_{11}, \ldots, X_{p1} \mid \cdots \mid X_{1p}, \ldots, X_{p1}).$$

Then

$$\left|\frac{\partial \mathbf{Y}}{\partial \mathbf{X}}\right| \equiv \left|\frac{\partial \mathbf{Y}^*}{\partial \mathbf{X}^*}\right| = |\det \cdot \{\mathbf{A}\}|^p.$$

Thus in the case of nonsingular linear transformations,

$$J^*(\mathbf{A}; \mathbf{X}) = J(\mathbf{A}; \mathbf{X}) = |\det\{\mathbf{A}\}|^p. \tag{7.3.1}$$

Left and right invariant differentials on the orbits of $\mathfrak{X}$ are defined so that along each orbit the differential elements of the points **X** are equal to that of the reference point $R(\mathbf{X})$. Since $\mathbf{X} = (\mathbf{X}_1', \ldots, \mathbf{X}_n')'$ can be written as a vector of order $N \times 1$, where $N = np$, we define left invariant differential

$$m(\mathrm{d}\mathbf{X}) = \frac{\mathrm{d}\mathbf{X}}{J(\mathbf{T}_X;\, R(\mathbf{X}))} \tag{7.3.2}$$

and right invariant differential

$$m^*(\mathrm{d}\mathbf{X}) = \frac{\mathrm{d}\mathbf{X}}{J^*(\mathbf{T}_X;\, R(\mathbf{X}))}.$$

It is easy to verify that, as in (7.3.1),

$$J(\mathbf{T}_X;\, R(\mathbf{X})) = J^*(\mathbf{T}_X;\, R(\mathbf{X})) = |\det\{\mathbf{T}_X\}|^{np} \tag{7.3.3}$$

Thus in this case of nonsingular linear transformations,

$$m(\mathrm{d}\mathbf{X}) = m^*(\mathrm{d}\mathbf{X}) = \frac{\mathrm{d}\mathbf{X}}{|\det \cdot \{\mathbf{T}_X\}|^{np}}. \tag{7.3.4}$$

In a similar manner we define left and right invariant differentials on $\mathcal{G}$, respectively, as

$$\mu(\mathrm{d}\mathbf{G}) = \frac{\mathrm{d}\mathbf{G}}{J(\mathbf{G};\, \mathbf{I})}, \tag{7.3.5}$$

and

$$\nu(\mathrm{d}\mathbf{G}) = \frac{\mathrm{d}\mathbf{G}}{J^*(\mathbf{G};\, \mathbf{I})}. \tag{7.3.6}$$

In this case $J(\mathbf{G};\mathbf{I}) = J^*(\mathbf{G};\mathbf{I}) = |\det\{\mathbf{G}\}|^p$.

Generally the left and right invariant differentials are not necessarily equal. For example, consider the group of affine transformations, which is presented by the lower triangular matrices

$$[a, c] = \begin{pmatrix} 1 & 0 \\ a & c \end{pmatrix}, \quad -\infty < a < \infty, \quad 0 < c < \infty.$$

The differential element of $[a, c]$ is $\mathrm{d}[a, c] = \mathrm{d}a\,\mathrm{d}c$. Now a right transformation of $[a, c]$ by $[b, d]$ yields

$$[a, c][b, d] = \begin{pmatrix} 1 & 0 \\ a + cb & cd \end{pmatrix} = [a + cb, cd].$$

A left transformation yields

$$[b, d][a, c] = [b + da, cd].$$

Thus

$$J^*([b, d];\, [a, c]) = \left| \det \left\{ \begin{pmatrix} 1 & 0 \\ b & d \end{pmatrix} \right\} \right| = d. \tag{7.3.7}$$

On the other hand,

$$J([b, d];\, [a, c]) = \left| \det \left\{ \begin{pmatrix} d & 0 \\ 0 & d \end{pmatrix} \right\} \right| = d^2. \tag{7.3.8}$$

The unit transformation on this affine group is [0, 1]. According to (7.3.8), the left invariant differential is

$$\mu(\mathrm{d}[a, c]) = \frac{\mathrm{d}a\,\mathrm{d}c}{J([a, c]; [0, 1])} = \frac{\mathrm{d}a\,\mathrm{d}c}{c^2}. \tag{7.3.9}$$

The right invariant differential is

$$\nu(\mathrm{d}[a, c]) = \frac{\mathrm{d}a\,\mathrm{d}c}{J^*([a, c]; [0, 1])} = \frac{\mathrm{d}a\,\mathrm{d}c}{c}. \tag{7.3.10}$$

We notice that

$$\mu(\mathrm{d}[a, c]) = \frac{J^*([a, c]; [0, 1])}{J([a, c]; [0, 1])}\,\nu(\mathrm{d}[a, c]). \tag{7.3.11}$$

The ratio $J^*([a, c]; [0, 1])/J([a, c]; [0, 1])$ is called the *modular function.* The *modular function* of the group $\mathcal{G}$ of nonsingular linear transformations is, as previously shown, the constant 1 for all $\mathcal{G}$. We present now the corresponding generalization when $\mathcal{G}$ is an arbitrary locally compact group. The theory becomes quite trivial when the group $\mathcal{G}$ is compact (which is the case, for example, when $\mathcal{G}$ is finite). A locally compact group G is one in which every element G of $\mathcal{G}$ has a neighborhood in $\mathcal{G}$ that is compact. In many cases we deal with groups $\mathcal{G}$ which are topological Hausdorff groups. These are groups for which every two elements G and G' have disjoint open neighborhoods and the mapping $(\mathbf{G}_1, \mathbf{G}_2) \to \mathbf{G}_1\mathbf{G}_2^{-1}$ is continuous.

The Haar integral of functions $f(\cdot)$ on $\mathcal{G}$ is an invariant operator, which will be introduced now.

Let f be a function defined on $\mathcal{G}$, mapping each element G of $\mathcal{G}$ into a range A. The range A of f could be real, vector valued, or some abstract set (it could be $\mathcal{G}$ itself). In the context of estimation theory $\mathcal{G}$ is often isomorphic to $\mathcal{X}$, A is the parameter space Θ, and f an estimator. Let $\mathcal{A} = \{f; f(\mathbf{G}) \in A$, for all $\mathbf{G} \in \mathcal{G}\}$. We define a homomorphism $\mathcal{G} \to \bar{\mathcal{G}}$ by introducing a 1–1 mapping $\bar{\mathcal{G}}$: $\mathcal{A} \to \mathcal{A}$ such that for each $\mathbf{G} \in \mathcal{G}$ and any $\mathbf{H} \in \mathcal{G}$

$$\bar{\mathbf{H}}f(\mathbf{G}) = f(\mathbf{H}^{-1}\mathbf{G}), \qquad \bar{\mathbf{H}}f \in \mathcal{A}. \tag{7.3.12}$$

Similarly, we define the right transformations

$$f\bar{\mathbf{H}}(\mathbf{G}) = f(\mathbf{G}\mathbf{H}^{-1}), \qquad f\bar{\mathbf{H}} \in \mathcal{A}. \tag{7.3.13}$$

It is immediately implied that

$$\bar{\mathbf{K}}(\bar{\mathbf{G}}f) = \overline{\mathbf{KG}}f \quad \text{for all } f \in \mathcal{A}, \quad \mathbf{K}, \mathbf{G} \in \mathcal{G}, \tag{7.3.14}$$

and

$$(f\bar{\mathbf{G}})\bar{\mathbf{K}} = f\overline{\mathbf{GK}} \quad \text{for all } f \in \mathcal{A}, \quad \mathbf{K}, \mathbf{G} \in \mathcal{G}.$$

Let $\mathfrak{C}(f) \subset \mathfrak{G}$ be a compact support of the function f in $\mathcal{A}$. We notice that $\mathfrak{C}(f)$ is left invariant with respect to $\bar{\mathfrak{G}}$. Indeed, $\mathfrak{C}(\bar{\mathfrak{s}}f) = \mathfrak{C}(f)$ for all $\bar{\mathfrak{s}} \in \bar{\mathfrak{G}}$.

Suppose that $\mu(\mathrm{d}\mathbf{G})$ is a left invariant measure on $\mathfrak{G}$ and let $E_\mu\{f\}$ denote a positive integral of f with respect to μ, that is,

$$E_\mu\{f\} = \int_{\mathfrak{G}} f(\mathbf{G})\mu(\mathrm{d}\mathbf{G}). \tag{7.3.15}$$

If f is integrable with respect to μ, that is, $E_\mu\{f\}$ exists, then $\bar{\mathbf{G}}f$ is integrable with respect to μ for every $\mathbf{G} \in \mathfrak{G}$. Furthermore,

$$\begin{aligned} E_\mu\{\bar{\mathbf{G}}f\} &= \int \bar{\mathbf{G}}f(\mathbf{H})\mu(\mathrm{d}\mathbf{H}) \\ &= \int f(\mathbf{G}^{-1}\mathbf{H})\mu(\mathrm{d}\mathbf{H}). \end{aligned}$$

If $E_\mu\{\bar{\mathbf{G}}f\} = E_\mu\{f\}$ for every $\mathbf{G} \in \mathfrak{G}$ we say that $E_\mu\{f\}$ is a left invariant Haar integral. Thus a left invariant Haar integral satisfies

$$\int f(\mathbf{G}^{-1}\mathbf{H})\mu(\mathrm{d}\mathbf{H}) = \int f(\mathbf{H})\mu(\mathrm{d}\mathbf{H}), \qquad \text{all } \mathbf{G} \in \mathfrak{G}. \tag{7.3.16}$$

Haar proved (see Nachbin [1], pp. 65) that on every locally compact group $\mathfrak{G}$ there exists at least one left invariant integral $E_\mu\{f\} \neq 0$. Such an integral is *unique*, except for a strictly positive factor of proportionality. In case $\mathfrak{G}$ is a group of nonsingular linear transformations on an N-dimensional vector space we have shown that $\mu(\mathrm{d}\mathbf{H}) = \mathrm{d}\mathbf{H}/|\det\{\mathbf{H}\}|^N$. We give another example which presents a left invariant Haar integral. This example is taken from Nachbin [1], pp. 67.

Example 7.3. Let $\mathfrak{C}$ be a compact multiplicative group of complex numbers on the unit circle; that is, if $V \in \mathfrak{C}$ then $V = e^{2\pi i X}$; $\mathfrak{C}$ is isomorphic with the locally compact additive group $\mathfrak{R}$ of real numbers. Let

$$V_n = \{(V_1, \ldots, V_n);\ V_j = e^{2\pi i X_j};\ -\infty < X_j < \infty, j = 1, \ldots, n\}.$$

Then V_n is isomorphic to $\mathfrak{R}^n = \{(X_1, \ldots, X_n); -\infty < X_i < \infty, i = 1, \ldots, n\}$. Let r: $\mathfrak{R}^n \to V_n$ be the covering mapping

$$r(X_1, \ldots, X_n) = (e^{2\pi i X_1}, \ldots, e^{2\pi i X_n}).$$

Let fr be a function on $\mathfrak{R}^n$; fr is periodic, with period 1 in each variable. Let $I = \{0 \leqslant X_i \leqslant 1, \text{all } i = 1, \ldots, n\}$. We show that the Haar left invariant integral is

$$E\{f\} = \int_I f(r(x_1, \ldots, x_n))\, \mathrm{d}x_1 \cdots \mathrm{d}x_n,$$

which is the Lebesgue integral of fr over I, *provided* fr is Lebesgue-integrable. Let $s \in \mathcal{R}^n$ and $I + s = \{s_i \leqslant x_i \leqslant 1 + s_i, \text{ all } i = 1, \ldots, n\}$. Since $f(r(x_1, \ldots, x_n))$ is periodic, with period 1 for each variable, it follows that

(7.3.17)

$$\int_{I+s} f(r(x_1, \ldots, x_n))\, \mathrm{d}x_1 \cdots \mathrm{d}x_n = \int_I f(r(x_1, \ldots, x_n))\, \mathrm{d}x_1 \cdots \mathrm{d}x_n,$$

for *all* $s \in \mathcal{R}^n$. Thus (7.3.17) is a special case of (7.3.16) for the translation group on $\mathcal{R}^n$. According to Haar's theorem

$$E\{f\} = \int_I f(r(x_1, \ldots, x_n))\, \mathrm{d}x_1 \cdots \mathrm{d}x_n$$

is essentially unique. Since the group $\mathcal{C}$ is compact, we can normalize this integral by the condition that $\mu(\mathcal{C}) = 1$. ■

We remark that a left invariant Haar integral $E\{f\}$ can assign a group $\mathcal{G}$ a finite measure if and only if $\mathcal{G}$ is compact. Furthermore the left invariant measure $\mu(\cdot)$ gives a positive measure to the elements of $\mathcal{G}$ if and only if $\mathcal{G}$ is discrete. (Consider for example the group $\mathcal{G}$ of all $n!$ permutations of the coordinates of $x = (x_1, \ldots, x_n)$). We introduce now the definition of a right invariant Haar integral.

Let $\bar{\mathcal{G}}$ be the induced group of transformations of f, and let $\bar{T} \in \bar{\mathcal{G}}$. A right invariant Haar integral of f could be defined as

$$E^*\{f\} = E\{f\bar{T}\}, \qquad \bar{T} \in \bar{\mathcal{G}}, \tag{7.3.18}$$

where $E\{\cdot\}$ is the left invariant Haar integral. Since the Haar integral is determined up to a positive constant, we obtain from (7.3.18)

$$\begin{aligned} E^*\{f\} &= \int f(G)\nu(\mathrm{d}G) \\ &= \int f\bar{T}(G)\mu(\mathrm{d}G) \\ &= \int f(GT^{-1})\mu(\mathrm{d}G) \\ &= \Delta_R(T) \int f(G)\mu(\mathrm{d}G). \end{aligned} \tag{7.3.19}$$

The function $\Delta_R(T)$ is the right-hand *modular* function of $\mathcal{G}$. The transformation $\Delta_R\colon T \to \Delta_R(T)$ is a homomorphism of $\mathcal{G}$ into the multiplicative group of positive reals R_+^*. This transformation is continuous and satisfies

(i) $$\Delta_R(I) = 1;$$

(ii) $$\Delta_R(T_1T_2) = \Delta_R(T_1)\,\Delta_R(T_2), \quad \text{all} \quad T_1, T_2 \in \mathcal{G};$$

(iii) $$\Delta_R(T^{-1}) = \Delta_R^{-1}(T).$$

If we start from a given right invariant Haar integral $E^*\{f\} = \int f(G)\nu(\mathrm{d}G)$ and define a left invariant Haar integral $E\{f\}$ to satisfy

$$E^*\{f\} = E\{\bar{T}f\}, \quad \text{all} \quad \bar{T} \in \bar{\mathcal{G}},$$

we obtain

(7.3.20) $$\int f(G)\nu(\mathrm{d}G) = \Delta_L(T)\int f(G)\mu(\mathrm{d}G).$$

Thus if $E^*\{f\} \neq 0$, then (7.3.19) and (7.3.20) imply

(7.3.21) $$\Delta_R(T)\,\Delta_L(T) = 1, \quad \text{all} \quad T \in \mathcal{G}.$$

If $\Delta_R(T) = \Delta_L(T) = 1$ for all $T \in \mathcal{G}$, the group $\mathcal{G}$ is called *unimodular*. Every commutative group is unimodular. Previously we saw a case of a non-commutative group which was unimodular.

7.4. FIDUCIAL THEORY AND MINIMUM RISK EQUIVARIANT ESTIMATORS

The method of fiducial inference was introduced by Fisher [4] in 1930 and studied further by Fisher in a series of papers [5; 6], Pitman [1] showed in 1939 how Fisher's fiducial argument can be used to obtain minimum mean square error equivariant estimators of location and of scale parameters. Pitman's estimator for location parameters was discussed in Section 7.2. Fisher did not provide a general theory of fiducial inference but only some special cases. As noted by Barnard [1], a general framework for fiducial inference can be provided by the theory of invariant decision procedures. Fraser [6] gave a precise mathematical framework to fiducial theory by a group-theoretic approach, using invariant Haar measures. The definition of fiducial distributions in terms of the Haar invariant measures coincides in the special case of location and scale parameters with that of Fisher. Hora and Buehler [1] showed the usefulness of fiducial theory in deriving minimum risk equivariant estimators. Their result can be considered as a generalization of Pitman's representation. We start with the following example.

Example 7.4. Let $X_1, \ldots, X_n$ be i.i.d. random variables such that $X_i = \mu + \sigma U_i$, $i = 1, \ldots, n$, where U_1 is a negative exponential random variable with density $f(x) = e^{-x}$, when $x \geqslant 0$. A minimal sufficient statistic is

$T_x = (X_{(1)}, \sum_{i=2}^{n} (X_{(i)} - X_{(1)})/(n-1))$, where $X_{(1)} \leqslant \cdots \leqslant X_{(n)}$ is the other statistic. We notice that T_x is an equivariant estimator of $\theta = (\mu, \sigma)$ with respect to the group of affine transformations

$$x_i \to a + cx_i, \quad \text{all} \quad i = 1, \ldots, n.$$

In Example 2.3 we have proven that $X_{(1)}$ and $s_x = \sum_{i=2}^{n} (X_{(i)} - X_{(1)})/(n-1)$ are independent and that

$$X_{(1)} \sim \mu + \sigma G(n, 1),$$
$$s_x \sim \sigma G(n-1, n-1).$$

Let $E_i = (1/\sigma)(X_i - \mu)$, and let $T_e = (E_{(1)}, s_e)$. The probability element of T_e is

$$f(E_{(1)}, s_e)\, dE_{(1)}\, ds_e,$$

where

$$f(E_{(1)}, s_e) = \frac{(n-1)^{n-1}n}{(n-2)!} s_e^{n-2} \exp\{-nE_{(1)} - (n-1)s_e\}.$$

In the previous section we defined the group of operations $[a, c][x, s] = [a + cx, cs]$, which represent the affine transformations. The structural model is

$$T_x = [\mu, \sigma]T_e.$$

Making the transformation

$$T_x[E_{(1)}, s_e]^{-1} = T_x\left[-\frac{E_{(1)}}{s_e}, \frac{1}{s_e}\right] = \left(X_{(1)} - E_{(1)}\frac{s_x}{s_e}, \frac{s_x}{s_e}\right),$$

we obtain from the structural model

$$[\mu, \sigma] = \left(X_{(1)} - E_{(1)}\frac{s_x}{s_e}, \frac{s_x}{s_e}\right).$$

The Jacobian of this transformation is

$$J = \left|\frac{\partial[\mu, \sigma]}{\partial[E_{(1)}, s_e]}\right| = \left|\det\left\{\begin{pmatrix} -s_x/s_e & 0 \\ E_{(1)}s_x/s_e & -s_x/s_e^{\,2} \end{pmatrix}\right\}\right|$$
$$= \frac{s_x^{\,2}}{s_e^{\,3}} = \frac{\sigma^3}{s_x}.$$

Thus the probability element of T_e can be written as

$$f(E_{(1)}, s_e)\, dE_{(1)}\, ds_e = f([\mu, \sigma]^{-1}T_x)\frac{s_x}{\sigma^3}\, d\mu\, d\sigma. \tag{7.4.1}$$

Thus $f([\mu, \sigma]^{-1}T_x)(s_x/\sigma^3)$ could be interpreted as the probability element at the point (μ, σ) given the observed value of T_x. If we let

$$f([\mu, \sigma]^{-1}T_x)\frac{s_x}{\sigma^3}\,\mathrm{d}\mu\,\mathrm{d}\sigma \equiv g(\mu, \sigma \mid T_x)\,\mathrm{d}\mu\,\mathrm{d}\sigma$$

then

$$g(\mu, \sigma \mid T_x)$$

is the *fiducial density* of (μ, σ) given T_x. Fraser [9] calls the distribution of (μ, σ), represented by the density $g(\mu, \sigma \mid T_x)$, the *structural distribution.* We showed in the previous section that the left invariant Haar measure of $[\theta] = [\mu, \sigma]$ for the present group of affine transformations is $\mu(\mathrm{d}[a, c]) = \mathrm{d}a\,\mathrm{d}c/c^2$. The right invariant Haar measure is $\nu(\mathrm{d}[a, c]) = \mathrm{d}a\,\mathrm{d}c/c$, and the modular function is $\Delta([a, c]) = 1/c$. Hence we can write the fiducial probability element as

$$g(\mu, \sigma \mid T_x) = f([\mu, \sigma]^{-1}T_x)\,\Delta([\mu, \sigma]^{-1}T_x)\mu(\mathrm{d}[\mu, \sigma]). \tag{7.4.2}$$

According to the group properties of $\Delta(\cdot)$ we have

$$\begin{aligned}\Delta([\mu, \sigma]^{-1}T_x)\mu(\mathrm{d}[\mu, \sigma])\\ &= \Delta(T_x)\,\Delta([\mu, \sigma]^{-1})\mu[\mathrm{d}[\mu, \sigma]) = \Delta(T_x)\nu(\mathrm{d}[\mu, \sigma]).\end{aligned} \tag{7.4.3}$$

Thus we can write

$$g(\mu, \sigma \mid T_x)\,\mathrm{d}\mu\,\mathrm{d}\sigma = f([\mu, \sigma]^{-1}T_x)\,\Delta(T_x)\nu(\mathrm{d}[\mu, \sigma]). \tag{7.4.4}$$

A generalization of (7.4.4) is provided by Hora and Buehler [1], who give the following general definition of fiducial distributions.

Let $\mathcal{F} = \{F_\theta;\ \theta \in \Theta\}$ be a family of distribution functions of a random variable X; Θ is real or vector valued. Let $\mathcal{G}$ be a group of 1:1 transformations on $\mathcal{X}$; $\mathcal{G}$ is locally compact and $\Theta \leftrightarrow \mathcal{G}$ so that for each $\theta \in \Theta$ there is a unique $[\theta] \in \mathcal{G}$. The random variable X is represented as a pair (T, V), where V is invariant with respect to $\mathcal{G}$. The variable V could be the maximal-invariant reduction of X with respect to $\mathcal{G}$. The distribution of V does not depend on θ. They assume that for each $\theta \in \Theta$ the probability element of F_θ can be expressed in the form

$$F(\mathrm{d}x) = f([\theta]^{-1}t \mid v)\lambda(\mathrm{d}v)m(\mathrm{d}t), \tag{7.4.5}$$

where $m(\mathrm{d}t)$ is left invariant. It is essential to this framework that the conditional density function of T given V will depend on θ according to $f([\theta]^{-1}t \mid V)$. This connects the family of distribution functions with the group structure and can serve as a guide in choosing the appropriate group of transformations. Thus the location parameter family of distributions corresponds to the group of real translations. The family of location and scale parameters is connected with the real group of affine transformations, etc.

In analogy to (7.4.4) we can follow Hora and Buehler and define the probability element of the fiducial distribution as

$$g(\theta \mid (T, V))\,\mathrm{d}\theta = f([\theta]^{-1}T \mid V)\,\Delta([T])\nu(\mathrm{d}[\theta]), \tag{7.4.6}$$

where $[T]$ is the transformation in $\mathcal{G}$ that carries $V \to X = (T, V)$; that is, $[T]V = (T, V)$. Furthermore, $\nu(\mathrm{d}G)$ and $\Delta(G)$ designate, respectively, the right invariant Haar measure and the modular function of $\mathcal{G}$.

We remark in this connection that, since the group $\mathcal{G}$ is isomorphic with the parameter space Θ, many writers, and Hora and Buehler in particular, simplify the notation by letting θ designate $[\theta]$, and so on. We distinguish above between the parametric point θ and the transformation $[\theta]$ because, as shown already by examples, the operations associated with the group multiplication $[\theta_1][\theta_2]$ or those of $\theta_1\theta_2$ might be entirely different. Before proving the first lemma we remark that generally the left invariant measure $m(\mathrm{d}t)$ and the left invariant Haar measure $\mu(\mathrm{d}t)$ coincide (see Fraser [9], p. 58) since

$$m(\mathrm{d}(t, v)) = \lambda(\mathrm{d}v)\mu(\mathrm{d}[t]). \tag{7.4.7}$$

Hence $m(\mathrm{d}t) = m(\mathrm{d}(t, v))/\lambda(\mathrm{d}v)$ implies that $m(\mathrm{d}t) = \mu(\mathrm{d}[t])$.

In the following lemma we prove the basic *expectation identity*.

Lemma 7.4.1. (Hora and Buehler [1].) *Given a group* $\mathcal{G}$ *and a family of density functions of the form* (7.4.5), *let* $H(x; \theta)$ *be an integrable* $[\mathcal{F}]$ *function, which is invariant with respect to* $\mathcal{G}$. *Then*

$$E^{(f)}_{\theta|X}\{H(X; \theta)\} = E^{(\theta)}_{X|V}\{H(X; \theta)\}. \tag{7.4.8}$$

Remark. The notation in (7.4.8) has the following meaning: $E^{(f)}_{\theta|X}\{\cdot\}$ is the expectation according to the fiducial distribution of θ given X; $E^{(\theta)}_{X|V}\{\cdot\}$ is the expectation operator with respect to the conditional distribution of X, given V, under F_θ.

Proof.

$$E_{X|V}\{H(X; \theta)\} = \int H((t, V); \theta) f([\theta]^{-1}t \mid V) m(\mathrm{d}t). \tag{7.4.9}$$

Since $H(X; \theta)$ is invariant we have $H((t, v); \theta) = H(([\theta]^{-1}t, v); e)$, where e is the identity of Θ. Thus

$$\begin{aligned} E^{(\theta)}_{X|V}\{H(X; \theta)\} &= \int H(([\theta]^{-1}t; V); e) f([\theta]^{-1}t \mid V)\mu(\mathrm{d}[t]) \\ &= \int H((s, V); e) f(s \mid V)\mu(\mathrm{d}[\theta][s]). \end{aligned} \tag{7.4.10}$$

But according to the properties of the invariant Haar measures,

$$\begin{aligned}(7.4.11)\qquad \mu(d[\theta][s]) &= \Delta([s])\mu(d[\theta]) \\ &= \Delta([\theta]^{-1}[t])\mu(d[\theta]) \\ &= \Delta([t])\,\Delta^{-1}([\theta])\mu(d[\theta]) \\ &= \Delta([t])\nu(d[\theta]).\end{aligned}$$

Substituting (7.4.11) in (7.4.10) we obtain

$$\begin{aligned}(7.4.12)\qquad E^{(\theta)}_{X\mid V}\{H(X;\theta)\} &= \int H(([\theta]^{-1}t, V); e) f([\theta]^{-1}t \mid V)\,\Delta([t])\nu(\mathrm{d}[\theta]) \\ &= \int H((t, V); \theta) f([\theta]^{-1}t \mid V)\,\Delta([t])\nu(\mathrm{d}[\theta]) \\ &= E^{(f)}_{\theta\mid X}\{H(X;\theta)\}. \qquad \text{(Q.E.D.)}\end{aligned}$$

On the basis of the previous theorem we prove now the main theorem.

Theorem 7.4.2. (Hora and Buehler [1].) *For a loss function $L(\theta, \hat{\theta}) = W(\theta^{-1}\hat{\theta})$, the minimum risk equivariant estimator of θ is the one minimizing the fiducial risk*

$$\rho^{(f)}(X) = E^{(f)}_{\theta\mid X}\{W(\theta^{-1}\hat{\theta}(X))\}.$$

Moreover, if the loss function $L(\theta, \hat{\theta}) = \varphi(\theta)[\hat{\theta}(X) - \theta]^2$ is invariant then the minimum risk equivariant estimator is

$$(7.4.13)\qquad \hat{\theta}_0(X) = \frac{E^{(f)}_{\theta\mid X}\{\varphi(\theta)\cdot\theta\}}{E^{(f)}_{\theta\mid X}\{\varphi(\theta)\}}.$$

Proof. (i) Since $W(\theta^{-1}\hat{\theta})$ is invariant with respect to $\mathcal{G}$ whenever $\hat{\theta}$ is equivariant, we obtain from Lemma 7.4.1 that

$$(7.4.14)\qquad E^{(f)}_{\theta\mid X}\{W(\theta^{-1}\hat{\theta}(X))\} = E^{(\theta)}_{X\mid V}\{W(\theta^{-1}\hat{\theta}(X))\}.$$

The risk function $R(\theta, \hat{\theta}) = E^{(\theta)}_V\{E^{(\theta)}_{X\mid V}\{W(\theta^{-1}\hat{\theta}(X))\}\}$ and $E_{X\mid V}\{W(\theta^{-1}\hat{\theta}(X))\}$ is independent of θ. It depends only on $\hat{\theta}$. Thus if $\hat{\theta}_0(X)$ minimizes the fiducial risk it minimizes a.s. the non-negative random variables $E_{X\mid V}\{W(\theta^{-1}\hat{\theta}(X))\}$. Hence the minimum fiducial risk estimator is a minimum risk equivariant estimator.

(ii) In the case of the quadratic loss function $\varphi(\theta)(\hat{\theta} - \theta)^2$, which is invariant, the proof is as follows. As given by (7.4.13), $\hat{\theta}_0(X)$ minimizes a.s. the fiducial risk $E^{(f)}_{\theta\mid X}\{\varphi(\theta)[\hat{\theta}(X) - \theta]^2\}$. Now, according to Lemma 7.4.1,

$$\begin{aligned}(7.4.15)\qquad \varphi(\theta)E^{\theta}_{X\mid V}\{[\hat{\theta}_0(X) - \theta]^2\} &= E^{(f)}_{\theta\mid X}\{\varphi(\theta)[\hat{\theta}_0(X) - \theta]^2\} \\ &\leqslant E^{(f)}_{\theta\mid X}\{\varphi(\theta)[\hat{\theta}(X) - \theta]^2\} \\ &= \varphi(\theta)E^{\theta}_{X\mid V}\{(\hat{\theta}(X) - \theta)^2\},\end{aligned}$$

for all equivariant estimators $\hat{\theta}(X)$. The rest of the argument is like that of (i).

(Q.E.D.)

Example 7.5. In continuation of Example 7.4, we derive now the minimum risk equivariant estimator of the scale parameter σ. We derive this estimator in two ways. The first way is to derive the minimum fiducial risk estimator, (7.4.13), for the loss function $L(\theta, \hat{\theta}) = (\hat{\theta} - \theta)^2/\theta^2$. We have shown that the fiducial density of $\theta = (\mu, \sigma)$, for the real group of affine transformations, given $T_x = (X_{(1)}, s_x)$, is

$$f([\mu, \sigma]^{-1}T_x)\frac{s_x}{\sigma^3},$$

where

$$f(E_{(1)}, s_e) = \frac{n(n-1)^{n-1}}{(n-2)!} s_e^{n-2} \exp\{-n(E_{(1)} - (n-1)s_e\}.$$

Hence the fiducial density of (μ, σ), given $(X_{(1)}, s_x)$, is

$$\text{(7.4.16)}\quad g(\mu, \sigma \mid X_{(1)}, s_x) = \frac{n(n-1)^{n-1}}{(n-2)!} s_x^{n-1}\sigma^{-(n-1)} \exp\left\{-n\frac{X_{(1)} - \mu}{\sigma} - (n-1)\frac{s_x}{\sigma}\right\}.$$

Thus, according to (7.4.13), with $\varphi(\sigma) = \sigma^{-2}$, we obtain that the minimum risk equivariant estimator of σ is

(7.4.17)

$$\hat{\sigma}_0(T_x) = \frac{\displaystyle\int_0^\infty d\sigma\cdot\sigma^{-(n+2)}\exp\left\{-(n-1)\frac{s_x}{\sigma}\right\}\int_{-\infty}^{X_{(1)}}\exp\left\{-n\frac{X_{(1)}-\mu}{\sigma}\right\}d\mu}{\displaystyle\int_0^\infty d\sigma\cdot\sigma^{-(n+3)}\exp\left\{-(n-1)\frac{s_x}{\sigma}\right\}\int_{-\infty}^{X_{(1)}}\exp\left\{-n\frac{X_{(1)}-\mu}{\sigma}\right\}d\mu}.$$

Letting $u = X_{(1)} - \mu/\sigma$, $d\mu = -\sigma\, du$, we immediately obtain that

$$\int_{-\infty}^{X_{(1)}} \exp\{-n(X_{(1)} - \mu/\sigma)\}\, d\mu = \sigma/n.$$

Substituting this in (7.4.17) we obtain

$$\text{(7.4.18)}\qquad \hat{\sigma}_0(T_x) = \frac{\displaystyle\int_0^\infty d\sigma\cdot\sigma^{-n(+1)}\exp\left\{-(n-1)\frac{s_x}{\sigma}\right\}}{\displaystyle\int_0^\infty d\sigma\cdot\sigma^{-(n+2)}\exp\left\{-(n-1)\frac{s_x}{\sigma}\right\}} = \frac{(n-1)s_x}{n} = \frac{1}{n}\sum_{i=2}^{n}(X_{(i)} - X_{(1)}).$$

The minimum risk estimator $\hat{\sigma}_0(T_x)$, given by (7.4.18), can be derived by using Stein's theorem in the following manner. Let $T_{n-1}^* = (n-1)s_x = \sum_{l=z}^{m}(X_{(i)} - X_{(1)})$; $(X_{(1)}, T_{n-1}^*)$ is a minimal sufficient statistic. Let $\hat{\sigma}(X_{(1)}, T_{n-1}^*)$ be an equivariant estimator of σ. Thus

$$\hat{\sigma}(X_{(1)}, T_{n-1}^*) = \hat{\sigma}(0, T_{n-1}^*) = cT_{n-1}^*,$$

where c is some positive constant; $T_{n-1}^* \sim \sigma G(1, n-1)$. Hence the risk of cT_{n-1}^* (which is independent of σ) is given by

$$\text{(7.4.19)} \qquad R(c) = E\{[cG(1, n-1) - 1]^2\} = c^2 n(n-1) - 2c(n-1) + 1$$

independently of σ. Hence the unique value of c for which (7.4.19) is minimized is $c_0 = 1/n$. The estimator $(1/n)T_{n-1}^*$ coincides with (7.4.18). ■

To conclude this section we remark that, as expressed by (7.4.13), the minimum risk equivariant estimator is a formal (or generalized) Bayes estimator of θ, with a quasi-prior distribution $(d\theta)$, whenever $\mathcal{G}$ can be identified with Θ. The quantity $\nu(d\theta)$ is a σ-finite measure. If Θ is not compact, $\nu(d\theta)$ is not finite. In other words,

$$\text{(7.4.20)} \qquad \xi(d\theta \mid X) = \frac{f(\theta^{-1}T \mid V)\,\Delta(T)\nu(d\theta)}{\int_\Theta f(\theta^{-1}T \mid V)\,\Delta(T)\nu(d\theta)}$$

is a posterior probability measure if and only if Θ is compact. In the compact case the fiducial distribution of θ is a special posterior distribution, with an invariant prior. This Bayesian nature of fiducial distributions was discussed by Fisher [7], Ch. 3; Jeffreys [1], Ch. 3; Hora and Buehler [1]; and others.

Stone [1] investigated the following problem. He considered a class Q of *relatively invariant* quasi-prior densities $q(\theta)$ such that

(i) $\int q(\bar{G}\theta)\nu(d\bar{G}) = \infty$, for all $\theta \in \Theta$;
(ii) $q(\bar{G}_1\bar{G}_2\theta) = q(\bar{G}_1\theta)q(\bar{G}_2\theta)$, for all $\theta \in \Theta$.

Here $\bar{\mathcal{G}}$ is the induced group operating on Θ and $\nu(d\bar{G})$ is a right invariant Haar measure. The trivial function $q(\theta) \equiv 1$ belongs to Q. A quasi-posterior distribution is the function $q(\theta \mid X)$ obtained by substituting $q(\theta)$ of Q in the Bayes formula.

Construct a sequence of compact θ sets $\Theta_1, \Theta_2, \ldots$ in Θ, and let $q(\theta)$ be a relatively invariant quasi-prior density. A sequence of *proper* prior densities is defined as the truncation of $q(\theta)$ on the compact sets. That is,

$$\text{(7.4.21)} \qquad \xi_i(\bar{G}\theta_0)\nu(d\bar{G}) = \begin{cases} q(\bar{G}\theta_0)\nu(d\bar{G}) \Big/ \int_{\Theta_i} q(\bar{G}\theta_0)\nu(d\bar{G}), & \text{if } \bar{G}\theta_0 \in \Theta_i, \\ 0, & \text{otherwise.} \end{cases}$$

The corresponding sequence of posterior probability measures is

$$\text{(7.4.22)} \qquad \xi_i(\bar{G}\theta_0 \mid (T, V)) = \frac{f(G^{-1}T, V)\xi_i(\bar{G}\theta_0)\nu(d\bar{G})}{\int_{\Theta_i} f(H^{-1}T, V)\xi_i(\bar{H}\theta_0)\nu(d\bar{H})},$$

$i = 1, 2, \ldots$. Stone [1] proved that the sequence $\xi_i(\bar{G}\theta_0 \mid (T, V))$ converges in probability to the *quasi*-posterior density

$$\xi^*(\bar{G}\theta_0 \mid (T, V)) = \frac{f(G^{-1}T, V)q(\bar{G}\theta_0)\nu(d\bar{G})}{\int f(G^{-1}T, V)q(\bar{G}\theta_0)\nu(d\bar{G})} \tag{7.4.23}$$

if and only if $q(\theta) \equiv 1$. The quasi-posterior probability measure, when $q(\theta) \equiv 1$, is the fiducial probabilty measure (7.4.6). Other properties of invariant prior distributions were investigated by Hartigan [1].

7.5. THE MODIFIED MINIMAX PRINCIPLE AND THE GENERALIZED HUNT-STEIN THEOREM

In Section 7.2 we have proven that the Pitman estimator is minimax. This is an example of a case where a minimax estimator exists among the equivariant estimators. As we have previously verified, an equivariant estimator has a constant risk function along the orbits of $\bar{\mathcal{G}}$, which partition the parameter space Θ. Thus it may generally be easier to find minimax estimators if one restricts attention to the equivariant estimators. The question is whether minimax equivariant estimators are minimax at large. The first answer to this question was given in 1945 by Hunt and Stein [1] in an unpublished paper. Their theorem is valid for problems of testing hypotheses or related estimation problems with a zero-one loss function. This result of Hunt and Stein for testing hypotheses is described in Lehmann's book [2] and the reader can find a proof of this theorem also in Wesler's paper [2]. Kiefer [2] investigated the problem in the context of a fixed sample estimation and of sequential estimation. Kiefer's results will be discussed in the following section. This section is devoted to the results of Wesler [1], who generalized the Hunt-Stein theorem to general decision problems with invariant loss functions.

Let $(\mathcal{X}, \mathcal{B})$ be the basic sample space and the Borel sigma-field generated by X. Let $\mathcal{F} = \{P_\theta; \theta \in \Theta\}$ be the family of probability measures of $(\mathcal{X}, \mathcal{B})$. We can then make the following basic assumptions.

A.1 There exists a sigma-finite measure μ on $(\mathcal{X}, \mathcal{B})$ which is equivalent to $\mathcal{F}$.

Let $\mathcal{D}$ be the set of all estimators of θ; $\mathcal{D}^*$ the set of the essentially equivariant estimators; and $\mathcal{D}^{**}$ the set of equivariant estimators, with respect to a group $\mathcal{G}$, for which the estimation problem preserves its structure.

A.2 There is a topology on $\mathcal{D}$ such that $\mathcal{D}$ is a *separable metric space* and such that $\mathcal{B}_D$ is the Borel sigma-field generated by the *compact* subsets of $\mathcal{D}$. The loss function $L(\theta, d)$ is, for each $\theta \in \Theta$, a non-negative continuous function in d and, for each real τ, the set $\{d; L(\theta, d) \leqslant \tau\}$ is compact.

A.3 Let $\mathcal{C}$ be a Borel sigma-field of subsets of the group of transformations $\mathcal{G}$, on which there is an asymptotically right invariant sequence of measures, that is, a sequence $\{\nu_n\}$ on $(\mathcal{G}, \mathcal{C})$ such that for every $G \in \mathcal{G}$ and $C \in \mathcal{C}$,

$$\lim_{n \to \infty} \{\nu_n(CG) - \nu_n(C)\} = 0. \tag{7.5.1}$$

Assumption 3 is fulfilled whenever $\mathcal{G}$ is locally compact, and we can take the right invariant Haar measure $\nu(C)$ for each element of the sequence.

The *modified minimax* principle proposed by Wesler is to compare different estimators by the maximal risk they assume on each orbit. Thus let $\{\Theta_s; s \in \mathcal{S}\}$ be the orbits of $\bar{\mathcal{G}}$ in Θ, indexed by $s \in \mathcal{S}$. Given two estimators d_1 and d_2 we say that d_1 is *at least as good as* d_2, in the modified minimax sense, if

$$\sup_{\theta \in \Theta_s} R(\theta, d_1) \leqslant \sup_{\theta \in \Theta_s} R(\theta, d_2), \qquad \text{for } all\ s \in \mathcal{S}. \tag{7.5.2}$$

A subclass of estimators $\mathfrak{D}^*$ is called *essentially complete* in the *modified minimax* sense if, given any estimator d, there exists an estimator d^* in $\mathfrak{D}^*$ that is at least as good as in the modified minimax sense.

Theorem 7.5.1. (the generalized Hunt-Stein Theorem.) (a) *Under A1.–A.3 the class* $\mathfrak{D}^*$ *of essentially equivariant estimators is essentially complete*, and (b) *if instead of A.*3 *we assume the following*,

A.4 *The group of transformations* $\mathcal{G}$ *is a locally compact topological group and* $\mathcal{C}$ *is generated by compact subsets of* $\mathcal{G}$,

then $\mathfrak{D}^{**}$ *is essentially complete in the modified minimax sense.*

Before we prove this important theorem let us consider it's implication. If an estimator is minimax it is either equivariant or it is equivalent, in the modified minimax sense, to an estimator which is equivariant. This equivariant estimator should also be minimax. Thus if Theorem 7.5.1 holds, a minimax equivaraint estimator is minimax at large.

Proof. *When* $\mathcal{G}$ *is compact*. The proof of the theorem when $\mathcal{G}$ is compact is significantly simpler than the general proof. The reason is that the right invariant Haar measure $\nu(G)$ can be normalized so that $\nu(\mathcal{G}) = 1$. In this case every essentially equivariant procedure has the same risk function as some equivariant procedure. The proof that we present here for the case of compact groups is similar to that of Blackwell and Girshick [1], p. 226, for the case of finite groups.

Case 1. The loss function $L(\theta, d)$ is convex in d for each θ.

Given an estimator d which is not in $\mathfrak{D}^{**}$, we should construct an estimator d^* in $\mathfrak{D}^{**}$ which is at least as good as d, in the modified minimax sense.

Since $\mathcal{G}$ is compact, there exists a right invariant Haar measure $\nu(G)$ on $(\mathcal{G}, \mathcal{C})$. construct the estimator

$$d^*(X) = \int \bar{G}^{-1} d(GX)\nu(\mathrm{d}G). \tag{7.5.3}$$

We first prove that $d^*(X)$ is equivariant. Indeed, let H be any element of $\mathcal{G}$. Then

$$\begin{aligned}\bar{H}^{-1} d^*(HX) &= \bar{H}^{-1} \int \bar{G}^{-1} d(GHX)\nu(\mathrm{d}G)\\ &= \int (\bar{G}\bar{H})^{-1} d(GHX)\nu(\mathrm{d}G)\\ &= \int (\bar{G}')^{-1} d(G'X)\nu(\mathrm{d}G'H^{-1})\\ &= \int (\bar{G}')^{-1} d(G'X)\nu(\mathrm{d}G')\\ &= d^*(X).\end{aligned} \tag{7.5.4}$$

Hence d^* is equivariant. We show now that d^* is at least as good as d, in the modified minimax sense. Since $L(\theta, d)$ is convex in d for each θ, Jensen's inequality yields

$$\begin{aligned}R(\theta, d^*) &= \int \mathrm{d}P_\theta(X \leqslant x) L\left(\theta, \int \bar{G}^{-1} d(Gx)\nu(\mathrm{d}G)\right)\\ &\leqslant \int \nu(\mathrm{d}G) \int \mathrm{d}P_\theta(X \leqslant x) L(\theta, \bar{G}^{-1} d(Gx)).\end{aligned} \tag{7.5.5}$$

Furthermore, from the invariance of $L(\theta, d)$ and since $P_\theta[B] = P_{\bar{G}\theta}[GB]$ for all G, θ and $B \in \mathcal{B}$, we obtain

(7.5.6)

$$\begin{aligned}\int \nu(\mathrm{d}G) \int L(\theta, \bar{G}^{-1} d(Gx))\, \mathrm{d}P_\theta(X \leqslant x) &= \int \nu(\mathrm{d}G) \int L(\bar{G}\theta, d(y))\, \mathrm{d}P_{\bar{G}\theta}(X \leqslant y)\\ &\leqslant \sup_{\bar{G} \in \bar{\mathcal{G}}} R(\bar{G}\theta, d).\end{aligned}$$

Since d^* is equivariant, $R(\theta_0, d^*) = R(\theta, d^*)$ for all θ that belong to the orbit containing θ_0. Hence for each θ_0 in Θ,

$$\sup_{\theta \in \Omega(\theta_0)} R(\theta, d^*) \leqslant \sup_{\theta \in \Omega(\theta_0)} R(\theta, d), \tag{7.5.7}$$

where $\Omega(\theta_0)$ is an orbit containing θ_0. This proves that d^* is at least as good as d.

Case 2. $L(\theta, d)$ is *not* convex in d for each θ. Consider a randomized estimator $\pi(\theta \mid X)$, and construct the randomized estimator

$$\pi^*(\theta \mid X) = \int \nu(\mathrm{d}G)\pi(\bar{G}\theta \mid GX). \tag{7.4.8}$$

It is easy to verify that $\pi^*(\theta \mid X)$ is equivariant and that π^* is at least as good as π. The details are similar to those of Case 2 and will be omitted.

General proof. According to A.3, let $\{\nu_\alpha\}$ be a net of σ-finite measures on $(\mathcal{G}, \mathcal{C})$, converging to a right invariant measure ν. Let π_α designate the randomized estimator

$$\pi_\alpha(\theta \mid X) = \int \nu_\alpha(\mathrm{d}G)\pi(\bar{G}\theta \mid GX)$$

Let $\mathcal{K}$ be the set of all *compact* subsets K of $\mathfrak{D}$, where $\mathfrak{D}$ here is the class of all randomized estimators. Fix a K in $\mathcal{K}$. The following topological arguments imply that every net $\{\pi_\alpha(K \mid X)\}$ has at least one cluster point $\bar{\pi}(K \mid X)$, where $\pi_\alpha(K \mid X)$ is the probability of $\{\theta;\ \theta = G\theta_0,\ G \in K\}$ for some θ_0. Every randomized estimator $\pi(\theta \mid X)$ is a $\mathcal{B}$-measurable function on $\Theta \to [0, 1]$. The Banach space $L^\infty(\mu)$ of all bounded $\mathcal{B}$-measurable functions is the adjoint of the Banach space $L'(\mu)$ of all extended real valued functions, integrable with respect to μ. Alaoglu's theorem tells us that the solid unit sphere of $L'(\mu)$ is compact in the weak topology. Let $\psi_\alpha(K \mid x)$ be the class of all randomized estimators $\pi_\alpha(K \mid x)$; $\psi_\alpha(K \mid x)$ is the intersection of a closed set with the solid unit sphere and is therefore compact in the weak* topology on $L^\infty(\mu)$. By Tychonoff's compactness theorem the cartesian product of $\psi(K \mid x)$ for all K in $\mathcal{K}$ is a compact set in the product topology. Therefore, for each x, $\{\pi_\alpha(K \mid x)\}$ has a cluster point $\bar{\pi}(K \mid x)$. Hence for any finite number of compact sets K, any finite number of μ-integrable f functions, and any $\epsilon > 0$, there exists an $\alpha(\epsilon)$ such that for all $\alpha \geqslant \alpha(\epsilon)$

$$\left| \int \pi_\alpha(K \mid x) f(x)\mu(\mathrm{d}x) - \int \bar{\pi}(K \mid x) f(x)\mu(\mathrm{d}x) \right| < \epsilon. \tag{7.5.9}$$

Since $\bar{\pi}(K \mid x)$ is a cluster point, there exists a subnet $\{\pi_{\alpha_j}(K \mid x)\}$, $\pi_{\alpha_j}(K \mid x) \xrightarrow{W^*} \bar{\pi}(K \mid x)$.

The quantity $\bar{\pi}(K \mid x)$ is generally not a randomized estimator. We notice that since every randomized estimator $\pi(\theta \mid x)$ has the properties of a probability measure, $\bar{\pi}(K \mid x)$ also satisfies the following:

(i) if $K_1 \subset K_2$, then $\bar{\pi}(K_1 \mid x) \leqslant \bar{\pi}(K_2 \mid x)$;
(ii) if $K_1 \cap K_2 = \phi$, then $\bar{\pi}(K_1 \cup K_2 \mid x) = \bar{\pi}(K_1 \mid x) + \bar{\pi}(K_2 \mid x)$;
(iii) $\bar{\pi}(K \mid x) \leqslant 1$ for all K.

We leave it to the reader to check that $\bar{\pi}(K \mid x)$ satisfies these three properties. We notice that (i)–(iii) hold almost everywhere $[\mu]$ on $\mathfrak{X}$. The exceptional null sets may depend on K. In order that there will be only a denumerable number of exceptional null sets we restrict $\bar{\pi}(\cdot \mid x)$ in the following manner.

Let $\mathfrak{R}$ be a countable subset of $\mathfrak{K}$ such that every finite union and finite intersection of sets in $\mathfrak{R}$ is also in $\mathfrak{K}$, and such that every open set U of $\mathfrak{D}$ is a countable union of interior elements of $\mathfrak{R}$, which are themselves subsets of U. Assumption 2 guarantees the existence of such a collection. Restricting $\bar{\pi}$ to elements of $\mathfrak{R}$, it satisfies (i)–(iii) with only a countable number of exceptional null sets. Moreover, since a countable union of null sets is a null set, $\bar{\pi}$, which is restricted to $\mathfrak{R}$, satisfies (i)–(iii) almost everywhere $[\mu]$. We define now an outer measure:

$$\varphi^*(U \mid x) = \sup_{K \subset U} \bar{\pi}(K \mid x), \quad \text{for each} \quad x \in \mathfrak{X}, \tag{7.5.10}$$

where $\varphi^*(U \mid x)$ is $\mathfrak{B}$-measurable. Moreover, we extend φ^* by defining for every subset of $\mathfrak{D}$,

$$\varphi^*(K \mid x) = \inf_{K \subset U} \varphi^*(U \mid x), \quad \text{for each} \quad x \in \mathfrak{X}. \tag{7.5.11}$$

The quantity φ^* is thus a measure on $(\mathfrak{D}, \mathfrak{B}_D)$ for each x. We show now that $\varphi^*(\mathfrak{D} \mid x) = 1$ a.s. $[\mu]$. Then, φ^* is a randomized estimator of θ.

Let K be a compact set in $\mathfrak{R}$, and let U be an open set, $U \supset K$. Let $U = \bigcup_{R_i \subset U}$ (interior R_i). Then U is an open covering of the compact set K. Since K is compact there exists a finite n such that $K \subset \bigcup_{i=1}^{n} R_i = R \in \mathfrak{R}$. Hence $\bar{\pi}(K \mid x) \leqslant \bar{\pi}(R \mid x)$ a.s. $[\mu]$. Since $R \subset U$, $\varphi^*(U \mid x) \geqslant \bar{\pi}(K \mid x)$ a.s. $[\mu]$. Let $K = \bigcap_j U_j$; then $\varphi^*(K \mid x) = \lim_{j \to \infty} \varphi^*(U_j \mid x)$ everywhere. Combining the inequalities, we get $\varphi^*(K \mid x) \geqslant \bar{\pi}(K \mid x)$ a.s. $[\mu]$.

On the other hand, for any $\theta_0 \in \Theta$ we have for each α, $R(\theta_0, \varphi_\alpha) \leqslant m(\theta_0)$, where $m(\theta_0)$ is a finite upper bound of the risk values of $R(\theta, \varphi_\alpha)$, $\theta \in \Omega(\theta_0)$. Define the compact set $K = \{d; L(\theta_0, d) \leqslant (m(\theta_0))/\epsilon\}$, where $\epsilon > 0$ is arbitrary. Then for every α,

$$\int \pi_\alpha(K \mid x)\, dP_{\theta_0}(X \leqslant x) > 1 - \epsilon. \tag{7.5.12}$$

Indeed,

$$\begin{aligned} m(\theta_0) \geqslant R(\theta_0, \pi_\alpha) &\geqslant \int_{\mathfrak{X}} \int_{\bar{K}} L(\theta_0, \hat{\theta}) \pi_\alpha(d\hat{\theta} \mid x)\, dP_\theta (X \leqslant x) \\ &> \frac{m(\theta_0)}{\epsilon} \int_{\mathfrak{X}} \pi_\alpha(\bar{K} \mid x)\, dP_{\theta_0}(X \leqslant x), \end{aligned} \tag{7.5.13}$$

where $\bar{K}$ is the complement of K. Equation 7.5.13 implies (7.5.12). Now for (7.5.12) we obtain that

$$\int \pi_\alpha(K \mid x)\, \mathrm{d}P_{\theta_0}(X \leqslant x) \geqslant 1 - \epsilon, \tag{7.5.14}$$

and therefore

$$\int \varphi^*(\mathfrak{D} \mid x)\, \mathrm{d}P_{\theta_0}(X \leqslant x) \geqslant 1 - \epsilon. \tag{7.5.15}$$

But $\mathfrak{D}$ is open, ϵ is arbitrary, and $\varphi^*(\mathfrak{D} \mid x) \leqslant 1$ everywhere. Hence $\varphi^*(\mathfrak{D} \mid x) = 1$ a.s. $[\mu]$.

After we have established that $\varphi^*(\mathfrak{D} \mid x) = 1$ a.s., that is, φ^* is a randomized estimator for almost all x, the proof is essentially reduced to the proof of the theorem for the compact case. It remains to show that the procedure is at least as good as π and that φ^* is almost equivariant $[\mu]$. The reader is referred to Wesler [1] for the remaining details. (Q.E.D.)

It is instructive to compare the main points in the proof of Theorem 7.2.5 and its preceding lemmas to those of the generalized Hunt-Stein Theorem. In conclusion we remark that whenever a locally compact group $\mathcal{G}$ is given, and the loss function $L(\theta, d)$ is invariant with respect to $\mathcal{G}$ and satisfies A.2, the minimum fiducial risk estimator discussed in Theorem 7.4.2 is minimax. This is a generalization of Theorem 7.2.5.

7.6. EQUIVARIANT SEQUENTIAL ESTIMATION

In Section 6.8 we discussed the problem of sequential minimax estimation and provided few cases in which the minimax sequential estimation is a fixed size sample minimax estimation. In this section we provide a theorem according to which, if the estimation problem is invariant with respect to a certain group of transformations $\mathcal{G}$, then under the union of the conditions of the generalized Hunt-Stein theorem and three additional conditions, the minimax equivariant sequential estimation is a fixed sample size procedure with a minimax equivariant (nonrandomized) estimator. This result has been proven by Kiefer [2] and our discussion in this section provides the main results of Kiefer. The first part of Kiefer's paper is devoted to the problem of equivariant minimax estimation in the fixed sample case. We discussed this problem in the previous section in a slightly different manner. For this reason the formulation of Kiefer's results varies somewhat (but not significantly) from the original formulation.

We start with a general discussion of the theoretical framework appropriate to the sequential minimax equivariant estimation. In the presentation of this part we follow Kiefer quite closely.

Let $X_1, X_2, \ldots$ be a sequence of i.i.d. random variables having a common distribution function F_θ; $\theta \in \Theta$. A sequential sampling design is a rule that is represented by a sequence of integers $\mathbf{e} = (e_1, e_2, \ldots)$. The meaning of $\mathbf{e}$ is the following. Sample e_1 of the above random variables, then e_2 additional variables, etc. Let $\mathbf{e}^{(k)} = (e_1, \ldots, e_k)$; $\mathbf{e}^{(k)}$ represent a k-stage sampling design. The fixed sample designs are represented by $\mathbf{e}^{(1)} = (s)$, where s is the sample size. If $\mathcal{B}^{(m)}$ designates the sigma-subfield generated by $(X_1, \ldots, X_k)$, then generally in k-stage sampling design $e_j \in \mathcal{B}^{(e_1+\cdots+e_{j-1})}$, $j = 2, \ldots$. That is, the size of the j-th stage is a function of the observed random variables in the previous stages.

Let $\mathfrak{D}$ represent the class of all estimators (randomized); and let $\mathcal{E}$ designate the class of all ordered k-tuples $\mathbf{e}^{(k)}$, with *positive* integers $k = 0, 1, \ldots$. The case of *no* sampling is represented by $\mathbf{e}^{(0)}$. A couple $(d, \mathbf{e})$ will be called a *strategy* of sequential estimation.

Let $\mathcal{G}$ be a group of transformations on $\mathfrak{X}$. If $G \in \mathcal{G}$, then $G(X_1, X_2, \ldots) = (GX_1, GX_2, \ldots)$. If $\bar{\mathcal{G}}$ is the induced group of transformations on Θ, $\bar{G}(d, e) = (\bar{G}d, e)$. In other words, the group $\bar{\mathcal{G}}$ acts trivially on $\mathcal{E}$.

As in Section 6.8, the loss function is the sum of two additive components,

$$L(\theta, (d, \mathbf{e})) = L_1(\theta, d) + L_2(\mathbf{e}). \tag{7.6.1}$$

The quantity $L_1(\theta, d)$ is the component of the loss function due to erroneous terminal estimation, while $L_2(\mathbf{e})$ is the cost of observing the sample $\mathbf{e}$. We assume that $0 < L_2(\mathbf{e}^{(k)}) < \infty$ for each finite $k \geqslant 1$, and $L_2(\mathbf{e}^{(\infty)}) = \infty$. Furthermore, it is necessary to impose the following restrictions on $L_2(\mathbf{e})$. These restrictions are formulated in the following assumption.

A.5 There exists a finite number q and a real nondecreasing function h, $h(x) \nearrow \infty$ as $x \to \infty$, such that for all $k < \infty$,

$$L_2((e_1, \ldots, e_k, 1)) - L_2((e_1, \ldots, e_k)) < q,$$

$$L_2(\mathbf{e}^{(1)}) > h(e_1), \quad \text{where} \quad e^{(1)} = (e_1), \qquad L_2(\mathbf{e}^{(k+1)}) \geqslant L_2(\mathbf{e}^{(k)}). \tag{7.6.2}$$

In many studies we find the loss function $L_2(\mathbf{e}^{(k)}) = c\sum_{i=1}^{k} e_i$, where $0 < c < \infty$. The sixth assumption required imposes a restriction on the family of distribution functions.

A.6 Let $X_1, \ldots, X_r$ be a sequence of i.i.d. random variables having a common distribution function which belongs to the family $\mathcal{F} = \{F_\theta; \theta \in \Theta\}$. Let $\{T(X_1, \ldots, X_r); r \geqslant 1\}$ be a sequence of sufficient statistics for $\mathcal{F}$. For each $r \geqslant 1$, let $U(X_1, \ldots, X_r)$ be a maximal invariant statistic with respect to a group $\mathcal{G}$. Then for each $r \geqslant 1, \ldots, U(X_1, \ldots, X_r)$ and $T(X_1, \ldots, X_r)$ are independent.

Example 7.6. The following is an example of a family of distributions that satisfies A.6. Let $\mathcal{F}$ be the family of negative exponential distributions with a scale parameter θ, $0 < \theta < \infty$; that is, the density of X is $f(x;\theta) = (1/\theta)e^{-x/\theta}$, $0 \leqslant x \leqslant \infty$. For each $n \geqslant 1$, consider the sample $\{X_1, \ldots, X_n\}$. A minimal sufficient statistic for $\mathcal{F}$ is $T_n = \sum_{i=1}^n X_i$. The group of transformations that should be considered for estimating θ is the group $\mathcal{G}$ of real linear transformations $\mathcal{G} = \{g; g(x) = gx, g > 0\}$. The maximal invariant statistic is $U(X_1, \ldots, X_n) = (X_2/X_1, \ldots, X_n/X_1)$. It is easy to prove that the conditional joint distribution of $(u_2, \ldots, u_n)$ given T_n, where $u_i = X_i/X_1$ $(i = 2, \ldots, n)$, has the density

$$f(\mu_2, \ldots, \mu_n \mid T_n) = \begin{cases} (n-1)!\,[1 + u_2 + \cdots + u_n]^{-n}, & \text{if } 0 < u_2, \ldots, u_n < \infty, \\ 0, & \text{otherwise.} \end{cases}$$

Hence $(u_2, \ldots, u_n)$ is independent of T_n for each $n \geqslant 2$. If $n = 1$ we have $u_1 = 1$, which is obviously independent of $T_1 = X_1$. ■

The seventh assumption required is that if for any $m \geqslant 1$, we consider the class of all sequential sampling strategies which require at least m observations, and if the minimax risk associated with this class is finite, then there exists a fixed sample strategy with a sample size $m' \geqslant m$ for which the maximal risk over Θ is finite. To formulate this assumption more precisely, let $\mathcal{E}_m$ be the class of all sequential (and fixed sample) samples which require at least m observations, $m \geqslant 1$. Let $\Xi_m = \mathfrak{D} \times \mathcal{E}_m$ be the class of all the associated stategies.

A.7 Given a class of stategies Ξ_m, $m \geqslant 1$, either $\sup_{\theta\in\Theta} R(\theta; (\pi, \mathbf{e})) = \infty$ for *all* $(\pi, \mathbf{e}) \in \Xi_m$, or there exists a fixed sample strategy $(\pi, (m'))$ of size $m' \geqslant m$ for which $\rho^{(1)}(m') < \infty$, where

$$\rho^{(1)}(m') = \inf_{\pi\in\mathfrak{D}} \sup_{\theta\in\Theta} R(\theta, (\pi, (m'))). \tag{7.6.3}$$

Kiefer [2] proved that A.5–A.7 imply that, for each $m \geqslant 1$ and any $\epsilon > 0$, there exists a fixed sample strategy with sample size $m' \geqslant m$ such that

$$\rho^{(1)}(m') \leqslant \inf_{\pi\in\Xi_m} \sup_{\theta\in\Theta} R(\theta; (\pi, \mathbf{e})) + \epsilon. \tag{7.6.4}$$

Since ϵ is arbitrary, this fixed sample strategy is minimax. Furthermore, if the loss function $L_1(\theta, d)$ is invariant with respect to $\mathcal{G}$, and $\mathcal{G}$ is transitive, the above fixed sample minimax strategy employs a nonrandomized estimator. If in addition we also suppose A.1–A.3 of the previous section, the estimator in that minimax fixed-sample strategy is almost equivariant. If A.4 replaces A.3, a fixed sample strategy with sample size m' and an equivariant estimator which is the minimum fiducial risk estimator is a minimax strategy. The sample size m' is determined so that the associated total risk will be minimized.

We conclude by remarking that if we allow estimators which are not based on observations, the above results may not hold. An illustration of such cases was given by Kiefer [2].

7.7. BAYES EQUIVARIANT AND FIDUCIAL ESTIMATORS FOR CASES WITH NUISANCE PARAMETERS

Consider a statistical model in which the prescribed family of distribution functions depends on a vector $\theta = (\theta_1, \theta_2)$ of parameters. The dimension of θ is $k \geqslant 2$; the dimension of θ_1 is r, $1 \leqslant r \leqslant k - 1$. Suppose, furthermore, that if θ_2 is known there exists a uniformly minimum risk equivariant estimator of θ_1 with respect to a group of transformations $\mathcal{G}$. It is often the case that when θ_2 is unknown, there is no uniformly minimum risk equivariant estimator of θ_1 with respect to $\mathcal{G}$. The risk function of every equivariant estimator is not a constant on each of the orbits of $\bar{\mathcal{G}}$ but varies on each orbit as a function of θ_2. In such a case θ_2 is called a *nuisance parameter* for the equivariance estimation.

Given a sample of n i.i.d. random variables $X_1, \ldots, X_n$ having a common distribution $F_{(\theta_1,\theta_2)}$ and a group $\mathcal{G}$ of transformations on $\mathcal{X}$, we assume that the induced group $\bar{\mathcal{G}}$ transforms (θ_1, θ_2) to $(\bar{G}\theta_1, \theta_2)$; that is, θ_2 is invariant under $\bar{\mathcal{G}}$. Let $U(X_1, \ldots, X_n)$ be a maximal invariant statistic with respect to $\mathcal{G}$. The distribution of $U(X_1, \ldots, X_n)$ does not depend on θ_1, but since θ_2 is invariant with respect to $\bar{\mathcal{G}}$, the distribution of $U(X_1, \ldots, X_n)$ will generally depend on θ_2. We designate this distribution by $K_{\theta_2}(u)$. A general theory concerning the existence of such distributions was given by Wijsman [3].

Let $(X_1, \ldots, X_n) = (T(X_1, \ldots, X_n), U(X_1, \ldots, X_n))$, and let $P_{\theta_1,\theta_2}(t \mid u)$ designate the conditional distribution function of $T(X_1, \ldots, X_n)$ given $U(X_1, \ldots, X_n)$. This conditional distribution function depends generally on both θ_1 and θ_2.

Write $d(\mathbf{X}) = \psi(T(\mathbf{X}), U(\mathbf{X}))$. If $d(\mathbf{X})$ is an equivariant estimator with respect to $\mathcal{G}$, it should have the structure

$$\psi(t, U) = \bar{G}_t\phi(U), \tag{7.7.1}$$

where $\bar{G}_t$ is a transformation in $\mathcal{G}$ such that the corresponding transformation G_t in $\mathcal{G}$ satisfies, for each $\mathbf{X} = (T(\mathbf{X}), U(\mathbf{X}))$,

$$G^{-1}_{T(\mathbf{X})}(\mathbf{X}) = U(\mathbf{X}). \tag{7.7.2}$$

We show now that on each orbit of θ_1 values, the risk function of an equivariant estimator $d_\phi(X) = \bar{G}_T\phi(U)$ depends only on θ_2 and on the particular orbit on which θ_1 lies. To verify it, we notice that since $L(\theta, d(\mathbf{X}))$ is invariant we have, for any equivariant estimator of θ_1,

$$(7.7.3) \qquad R(\theta_1, \theta_2; \phi) = \iint L(\theta, \bar{G}_t\phi(u))P_{\theta_1,\theta_2}(\mathrm{d}t \mid u)K_{\theta_2}(\mathrm{d}u)$$
$$= \iint L(\bar{H}\theta, \bar{H}\bar{G}_t\phi(u))P_{\theta_1,\theta_2}(\mathrm{d}t \mid u)K_{\theta_2}(\mathrm{d}u)$$
$$= \iint L(\bar{H}\theta, \bar{G}_t\phi(u))P_{\bar{H}\theta_1,\theta_2}(\mathrm{d}t \mid u)K_{\theta_2}(\mathrm{d}u),$$

for all transformations H of $\mathcal{G}$. In particular, choose H in $\mathcal{G}$ so that $\bar{H}\theta_1 = \tau(\theta_1)$, where $\tau(\theta_1)$ is a maximal invariant transformation of θ_1 according to $\bar{\mathcal{G}}$. The transformation $\tau(\theta_1)$ is the index of the orbit in Θ_1 on which θ_1 lies. Thus since $\bar{\mathcal{G}}$ is transitive, $R(\theta_1, \theta_2; \phi) = R_\tau(\theta_2; \phi)$ for all θ_1 which lie on the orbit indexed by τ.

We now introduce the notion of the *Bayes equivariant* estimator. Let $\xi(\theta_2)$ be any prior distribution function on the space of θ_2, Θ_2. Then the prior risk associated with an equivariant estimator $d_\phi(\mathbf{X})$ and an orbit τ is

$$(7.7.4) \qquad \rho_\tau(\xi; \phi) = \int \xi(\mathrm{d}\theta_2)R_\tau(\theta_2; \phi)$$
$$= \int \xi(\mathrm{d}\theta_2) \iint L(\tau, \bar{G}_t\phi(u))P_{\tau,\theta_2}(\mathrm{d}t \mid u)K_{\theta_2}(\mathrm{d}u).$$

We see that a choice of an equivariant estimator is equivalent to a choice of a function ϕ of U. A Bayes choice of an invariant function $\phi(u)$ is one which minimizes (7.7.4) We designate such a function by $\phi_\xi(u)$, and the corresponding Bayes equivariant estimator is $d_\xi(X) = \bar{G}_T\phi_\xi(U)$. Let

$$(7.7.5) \qquad W_\tau(\theta_2, U; \phi) = \int L(\tau, \bar{G}_t\phi(U))P_{\tau,\theta_2}(\mathrm{d}t \mid U) \quad \text{a.s.},$$

and define the marginal prior distribution of U,

$$(7.7.6) \qquad K_\xi(u) = \int \xi(\mathrm{d}\theta_2)K_{\theta_2}(u).$$

Since $W_\tau(\theta_2, U; \phi) > 0$ a.s., a trivial application of Fubini's theorem yields

$$(7.7.7) \qquad \rho_\tau(\xi; \phi) = \int K_\xi(\mathrm{d}u) \int \xi(\mathrm{d}\theta_2 \mid u)W_\tau(\theta_2, u; \phi),$$

where

$$(7.7.8) \qquad \xi(\mathrm{d}\theta_2 \mid u) = \frac{k_{\theta_2}(u)\xi(\mathrm{d}\theta_2)}{\int k_\tau(u)\xi(\mathrm{d}\tau)}, \quad \text{a.s.}$$

The function $k_{\theta_2}(u)$ is the density function of $K_{\theta_2}(u)$ with respect to some sigma-finite measure $\lambda(\mathrm{d}u)$. An equivariant estimator is called *Bayes equivariant* against ξ if it minimizes

$$R_\tau(U, \phi; \zeta) = \int \xi(\mathrm{d}\theta_2 \mid U) W_\tau(\theta_2, U; \phi). \tag{7.7.9}$$

Later on we present two examples in which Bayes equivariant estimators are derived. Before we start with these examples we would like to discuss the connection between Bayes equivariant estimators and fiducial estimators when nuisance parameters exist.

Consider the same structure as before, with the additional condition that the density of the conditional distributions of $T(X)$ given $U(X)$, with respect to a left invariant measure $m(\mathrm{d}t)$, is of the form

$$f_{\theta_1,\theta_2}(t \mid U) = f_{\theta_2}([\theta_1]^{-1}t \mid U) \quad \text{a.s.} \tag{7.7.10}$$

Let $\xi(\mathrm{d}\theta_2 \mid U)$ be the conditional probability measure of θ_2, given U, as defined in (7.7.8); then

$$f_\xi([\theta_1]^{-1}t \mid U) = \int \xi(\mathrm{d}\theta_2 \mid U) f_{\theta_2}([\theta_1]^{-1}t \mid U) \quad \text{a.s.} \tag{7.7.11}$$

is a conditional density function, with respect to the left invariant measure $m(\mathrm{d}t)$. Let $L(\theta_1, \bar{G}_T\phi(u))$ be an invariant loss function. As proven in Theorem 7.4.2, the minimum fiducial risk estimator, that is, the one which minimizes the fiducial risk

$$\rho^{(f)}(T, U) = \int L(\theta_1, \bar{G}_T\phi(U)) f_\xi([\theta_1]^{-1}T \mid U)\, \Delta(G_T)\nu(\mathrm{d}[\theta_1]), \tag{7.7.12}$$

is a minimum risk equivariant estimator of θ_1 for the family of densities $f_\xi([\theta_1]t \mid U)$. Applying the expectation identity, proven in Lemma 7.4.1, we obtain

$$\rho^{(f)}(T, U) = \int L(\theta_1, \bar{G}_t\phi(U)) P_{\theta_1}^{(\xi)}(\mathrm{d}t \mid U) \quad \text{a.s.,} \tag{7.7.13}$$

where

$$\begin{aligned} P_{\theta_1}^{(\xi)}(\mathrm{d}t \mid U) &= f_\xi([\theta_1]^{-1}t \mid U) m(\mathrm{d}t) \\ &= \int \xi(\mathrm{d}\theta_2 \mid U) P_{\theta_1,\theta_2}(\mathrm{d}t \mid U) \quad \text{a.s.} \end{aligned} \tag{7.7.14}$$

Substituting the right-hand side of (7.7.14) in (7.7.13) and interchanging the order of integration, we easily prove that (7.7.12) is equal a.s. to (7.7.9). Hence the best fiducial estimator of θ_1 for densities of the form $f_\xi([\theta_1]^{-1}t \mid U)$ coincides with the Bayes equivariant estimator.

Example 7.7. In this example we present the results of Zacks [6] concerning Bayes equivariant estimators of variance components.

Let Y_{ij} $(i = 1, \ldots, I; j = 1, \ldots, J)$ be random variables following Model II of the analysis of variance, namely,

$$Y_{ij} = \mu + a_i + e_{ij}, \qquad i = 1, \ldots, I, j = 1, \ldots, J,$$

where e_{ij} are mutually independent, having a common $\mathcal{N}(0, \sigma_e^2)$ distribution; and a_i are normally distributed random variables, independent of each other and of the e_{ij}, having a common $\mathcal{N}(0, \sigma_a^2)$ distribution. The problem is to estimate the variance components σ_e^2 and σ_a^2.

Let $Y_{..} = \sum_{i=1}^{I} \sum_{j=1}^{J} Y_{ij}/IJ$, $S_e^2 = \sum_{i=1}^{I} \sum_{j=1}^{J} (Y_{ij} - \bar{Y}_i)^2$, $\bar{Y}_i = (1/J) \times \sum_{j=1}^{J} Y_{ij}$ $(i = 1, \ldots, I)$, and $S_a^2 = J \sum_{i=1}^{I} (\bar{Y}_i - Y_{..})^2$. A minimal sufficient statistic is $(S_e^2, S_a^2, Y_{..})$. Consider the real affine group

$$\mathcal{G} = \{g_{\alpha\beta};\, g_{\alpha\beta} Y_{ij} = \alpha Y_{ij} + \beta,\, \alpha > 0,\, -\infty < \beta < \infty;$$
$$\text{for all } i = 1, \ldots, I; j = 1, \ldots, J\}$$

and quadratic loss functions $L(\sigma_e^2, \hat{\sigma}_e^2) = ((\hat{\sigma}_e^2/\sigma_e^2) - 1)^2$ and $L(\sigma_a^2, \hat{\sigma}_a^2) = ((\hat{\sigma}_a^2/\sigma_a^2) - 1)^2$. The estimation problem preserves its structure under this group and these loss functions.

Estimation of σ_e^2. A function $\psi(S_e^2, S_a^2, Y_{..})$ is equivariant if it is of the form

$$\psi(S_e^2, S_a^2, Y_{..}) = S_e^2 \phi\left(\frac{S_a^2}{S_e^2}\right). \tag{7.7.15}$$

The statistic $U = S_a^2/S_e^2$ is a maximal invariant statistic based on the minimal sufficient statistic $(S_e^2, S_a^2, Y_{..})$. Thus in terms of the general theory which has been developed in this section, $X = (S_e^2, S_a^2, Y_{..})$, $T(X) = (Y_{..}, S_e^2)$, $U(X) = S_a^2/S_e^2$. Furthermore, $\bar{G}_T = S_e^2$, and every equivariant estimator $d_\phi(X)$ is of the form $\bar{G}_T \phi(U)$. We present now some distribution theory relevant to these statistics. Let $\rho = \sigma_a^2/\sigma_e^2$; then

(i) $Y_{..}, S_e^2, S_a^2$ are independent,

$$S_e^2 \sim \sigma_e^2 \chi^2[I(J - 1)]$$
$$S_a^2 \sim \sigma_e^2 (1 + J\rho)\chi^2[I - 1]$$
$$IJY_{..}^2 \sim \sigma_e^2(1 + J\rho)\chi^2\left[1; \frac{IJ\mu^2}{2\sigma_e^2(1 + J\rho)}\right],$$

where $\chi^2[\nu]$ designates a chi-square r.v. with ν degrees of freedom; $\chi^2[\nu; \lambda]$ designates a noncentral χ^2 with ν degrees of freedom and a parameter of noncentrality λ, $0 < \lambda < \infty$.

(ii) As proven by Zacks [6],

$$S_a^2 \mid U \sim \frac{\sigma_e^2 U(1 + J\rho)}{1 + U + J\rho} \chi^2[IJ - 1],$$

$$S_e S_a \mid U \sim \frac{\sigma_e^2 U^{1/2}(1 + J\rho)}{1 + U + J\rho} \chi^2[IJ - 1],$$

where $X \mid Y \sim Z$ designates that the conditional distribution of X given Y is like that of Z. Hence

$$E\{S_a^2 \mid U\} = \frac{\sigma_e^2 U(1 + J\rho)}{1 + U + J\rho}(IJ - 1), \tag{7.7.16}$$

$$E\{S_a^4 \mid U\} = \frac{\sigma_e^4 U^2(1 + J\rho)^2}{(1 + U + J\rho)^2}(I^2J^2 - 1), \tag{7.7.17}$$

and

$$E\{S_e^2 S_a^2 \mid U\} = \frac{\sigma_e^4 U(1 + J\rho)^2}{(1 + U + J\rho)^2}(I^2J^2 - 1). \tag{7.7.18}$$

We write the equivariant estimator of σ_e^2 in the form

$$\hat{\sigma}_e^2 = S_e^2 \phi(U) = \frac{S_e^2}{IJ + 1}[1 + Uf(U)]. \tag{7.7.19}$$

The result will be that if $\xi(\rho)$ is any prior distribution of $\rho = \sigma_a^2/\sigma_e^2$, then the Bayes equivariant estimator of σ_e^2 is given by the right-hand side of (7.7.19) with

$$f_\xi(U) = \frac{E_{\rho \mid U, \xi}\{(1 + J\rho)/(1 + U + J\rho)^2\}}{E_{\rho \mid U, \xi}\{(1 + J\rho)^2/(1 + U + J\rho)^2\}}, \tag{7.7.20}$$

where $E_{\rho \mid U, \xi}\{\cdot\}$ designates the posterior expectation of the function in the brackets, given U and the prior distribution ξ. To verify it one notices first that the risk function of an equivariant estimator of the form (7.7.19) is

$$\begin{aligned} &R(\sigma_e^2, \rho; f) \\ &= \frac{2}{IJ + 1} + \frac{IJ - 1}{IJ + 1} \cdot E_{U \mid \rho}\left\{U^2 \frac{(1 + J\rho)^2}{(1 + U + J\rho)^2}\left[f(U) - \frac{1}{1 + J}\right]^2\right\}. \end{aligned} \tag{7.7.21}$$

We notice, as in (7.7.3), that $R(\sigma_e^2, \rho; f)$ is independent of σ_e^2. Since σ_e^2 is real, there is only one orbit of (μ, σ_e^2) values for each ρ. Thus the function $f_\xi(U)$ is the one which minimizes

$$Q_f(\xi) = E_{\rho \mid \xi}\left\{E_{U \mid \rho}\left\{U^2 \frac{(1 + J\rho)^2}{(1 + U + J\rho)^2}\left[f(U) - \frac{1}{1 + J\rho}\right]^2\right\}\right\}. \tag{7.7.22}$$

We can verify immediately that (7.7.20) minimizes (7.7.22).

In a similar fashion we show that the Bayes equivariant estimator of σ_a^2 is

$$\hat{\sigma}_a^{\,2} = \frac{S_e^{\,2}}{IJ+1} \cdot \frac{E_{\rho|U,\xi}\left\{\dfrac{\rho(1+J\rho)}{1+U+J\rho}\right\}}{E_{\rho|U,\xi}\left\{\dfrac{(1+J\rho)^2}{(1+U+J\rho)^2}\right\}}. \qquad \blacksquare$$

Example 7.8. In this example we derive the Bayes equivariant estimator of the common mean of two normal distributions $\mathcal{N}(\mu, \sigma^2)$ and $\mathcal{N}(\mu, \rho\sigma^2)$, based on two samples of equal sizes.

As explained in previous examples, if $\bar{X}$, $\bar{Y}$, S_1, and S_2 designate the sample means and sample sum of squares of deviations about their respective means, an equivariant estimator of the common mean μ, with respect to the group $\mathcal{G}$ of real affine transformations, is of the form

$$\text{(7.7.23)} \quad \hat{\mu}_\phi(\bar{X}, \bar{Y}, S_1, S_2) = \bar{X} + (\bar{Y} - \bar{X})\phi\left(\frac{S_1}{(\bar{Y}-\bar{X})^2}, \frac{S_2}{(\bar{Y}-\bar{X})^2}\right).$$

A maximal invariant reduction of the sufficient statistic is $U = (S_1/(\bar{Y}-\bar{X})^2, S_2/(\bar{Y}-\bar{X})^2)$. We notice also that $T(\bar{X}, \bar{Y}, S_1, S_2) = (\bar{X}, \bar{Y}-\bar{X})$ and $\bar{G}_T\phi = \bar{X} + (\bar{Y}-\bar{X})\phi$, for every ϕ. As an invariant loss function for estimating μ we consider $L(\hat{\mu}, \mu) = (\hat{\mu}-\mu)^2/\sigma^2$.

It is easy to verify that every equivariant estimator $\hat{\mu}_\phi$ is unbiased. Furthermore, if $V_i = S_i/(\bar{X}-\bar{Y})^2$, $i = 1, 2$, we show that the risk function of an equivariant estimator $\hat{\mu}_\phi$ is

$$\text{(7.7.24)} \quad R(\rho; \phi) = \frac{1}{n} + 2\left(1 - \frac{1}{2n}\right) \times E\left\{\left[\phi^2(V_1, V_2) - \frac{2}{1+\rho}\,\phi(V_1, V_2)\right]\frac{1+\rho}{1 + (1+\rho)\dfrac{V_1}{n} + \dfrac{1+\rho}{\rho}\dfrac{V_2}{n}}\right\},$$

where n designates the size of each sample. To prove (7.7.24) we notice that

$$\text{(7.7.25)} \quad \operatorname{Var}\{\bar{X} + (\bar{Y}-\bar{X})\phi(V_1, V_2)\} = \frac{\sigma^2}{n} + 2E\{(\bar{X}-\mu)(\bar{Y}-\bar{X})\phi(V_1, V_2)\} + E\{(\bar{Y}-\bar{X})^2\phi^2(V_1, V_2)\}.$$

Furthermore,

$$\begin{pmatrix} \bar{X}-\mu \\ \bar{Y}-\bar{X} \end{pmatrix} \sim N\left(O, \frac{\sigma^2}{n}\begin{bmatrix} 1 & -1 \\ -1 & 1+\rho \end{bmatrix}\right).$$

Hence the conditional distribution of $\bar{X} - \mu$, given $\bar{Y} - \bar{X}$, is

$$\mathcal{N}\left(-\frac{1}{1+\rho}(\bar{Y} - \bar{X}), \frac{\sigma^2}{n} \cdot \frac{\rho}{1+\rho}\right).$$

Thus

$$E\{(\bar{X} - \mu)(\bar{Y} - \bar{X})\phi(V_1, V_2) \mid (\bar{Y} - \bar{X}), S_1, S_2\} = -\frac{1}{1+\rho}(\bar{Y} - \bar{X})^2\phi(V_1, V_2) \quad \text{a.s.} \tag{7.7.26}$$

It is not difficult to verify (see Zacks [7]) that the conditional distribution of $W = (\bar{Y} - \bar{X})^2$, given (V_1, V_2), is the gamma distribution

$$\mathcal{G}\left(\frac{1}{2\sigma^2}\left(\frac{n}{1+\rho} + V_1 + \frac{V_2}{\rho}\right), n - \tfrac{1}{2}\right).$$

Hence

$$E(W \mid V_1, V_2) = \frac{2(n - \frac{1}{2})\sigma^2}{\dfrac{n}{1+\rho} + V_1 + \dfrac{V_2}{\rho}}, \quad \text{a.s.} \tag{7.7.27}$$

From (7.7.25), (7.7.26), and (7.7.27) we obtain (7.7.24). If $H(\rho)$ designates a prior distribution of ρ, the prior risk associated with an equivariant estimator $\hat{\mu}_\phi$ is

$$R(H; \phi) = \int H(d\rho)R(\rho; \phi).$$

Thus according to (7.7.24), a Bayes equivariant estimator of μ, against H, is one using the function $\phi_H(V_1, V_2)$, which is obtained by minimizing

$$Q(H; \phi) = \int_0^\infty H(d\rho)E\left\{\left[\phi^2(V_1, V_2) - \frac{2}{1+\rho}\phi(V_1, V_2)\right] \cdot \left[\frac{1}{1+\rho} + \frac{V_1}{n} + \frac{V_2}{n\rho}\right]^{-1}\right\}. \tag{7.7.28}$$

Thus we obtain

(7.7.29)

$$\phi_H(V_1, V_2) = \frac{E_{\rho \mid V_1, V_2}\left\{\left[1 + (1+\rho)\dfrac{V_1}{n} + \dfrac{1+\rho}{\rho} \cdot \dfrac{V_2}{n}\right]^{-1}\right\}}{E_{\rho \mid V_1, V_2}\left\{(1+\rho)\left[1 + (1+\rho)\dfrac{V_1}{n} + \dfrac{1+\rho}{\rho} \cdot \dfrac{V_2}{n}\right]^{-1}\right\}} \quad \text{a.s.}$$

where $E_{\rho \mid V_1, V_2}\{\cdot\}$ designates the posterior expectation of $\{\cdot\}$, given (V_1, V_2). ∎

PROBLEMS

Section 7.1

1. Let $X_1, \ldots, X_n$ be i.i.d. random variables having a common extreme value distribution that depends on a translation parameter; that is,

$$f(x; \alpha) = \exp\{-(x - \alpha) - e^{-(x-\alpha)}\}, \quad -\infty < x < \infty; \quad -\infty < \alpha < \infty.$$

Let $\mathcal{G}$ be the group of real translations and $L(\hat{\alpha}, \alpha) = (\hat{\alpha} - \alpha)^2$. Let $\hat{\alpha}(X_1, \ldots, X_n)$ be a translation equivariant estimator of α based on the minimal sufficient statistic T_n. Determine the minimum risk equivariant estimator $\hat{\alpha}(T_n)$ of α.

2. Consider the Pareto distribution with a location parameter of the translation type. The family of density functions is

$$f(x; \alpha) = \frac{3}{\alpha} \cdot \left(\frac{\alpha}{x}\right)^2, \qquad \alpha \leqslant x < \infty,$$

where $0 < \alpha < \infty$. Find the best translation equivariant, for a squared-error loss, of the location parameter α.

3. Consider the problem of estimating the common mean of two normal distributions. In continuation to Problem 29, Chapter 2, determine the following:

(i) The general form of an equivariant estimator of μ, based on the sufficient statistic $(\bar{X}, \bar{Y}, S_X^2, S_Y^2)$ with respect to the group $\mathcal{G}$ of affine transformations.

(ii) Show that every equivariant estimator, with respect to $\mathcal{G}$, is unbiased.

(iii) On which parameter does the parameter of the variance of an equivariant estimator depend?

(iv) Is there a uniformly minimum risk equivariant estimator of μ?

Section 7.2

4. Let $X_1, \ldots, X_n$ be i.i.d. random variables having a common rectangular distribution $\mathcal{R}(\theta, \theta + 1)$, $-\infty < \theta < \infty$.

(i) Determine the Pitman estimator of the location parameter θ with respect to the group of real translations.

(ii) What is the mean square error of this estimator?

5. Let $\{F_\theta; -\infty < \theta < \infty\}$ be a family of translation parameter Cauchy distributions. Let $X_1, \ldots, X_n$ be i.i.d. random variables having one of these Cauchy distributions.

(i) Does a Pitman estimator of θ exist?

(ii) If the above Cauchy distributions are truncated in the following manner,

$$f^{(K)}(x; \theta) = \begin{cases} \dfrac{c}{\pi}\,[1 + (x - \theta)^2]^{-1}, & \text{if } |X - \theta| \leqslant K, \\ 0, & \text{otherwise,} \end{cases}$$

where c is a known positive number, what is the Pitman estimator of θ for squared-error loss?

Section 7.3

6. Consider n i.i.d. random vectors $\{(X_i, Y_i); i = 1, \ldots, n\}$. The common density of $(\mathbf{X}, \mathbf{Y})$ depends on a location vector $\boldsymbol{\theta} = (\theta_1, \theta_2)$ and a positive definite covariance matrix $\boldsymbol{\Sigma}$; that is,

$$f(\mathbf{X}, \mathbf{y}; \boldsymbol{\theta}, \boldsymbol{\Sigma}) = |\boldsymbol{\Sigma}|^{-1/2} f((\boldsymbol{Z} - \boldsymbol{\theta})'\boldsymbol{\Sigma}^{-1}(\boldsymbol{Z} - \boldsymbol{\theta})),$$

where $\boldsymbol{Z} = (X, Y)'$. Consider the group $\mathcal{G}$ of affine transformations

$$\mathbf{G}(1, X, Y)' = C(X, Y)' + (g_1, g_2)',$$

where $-\infty < g_1, g_2 < \infty$, and C is positive definite.

(i) Determine all the components of the structural model.

(ii) Letting $\mathbf{G} = \begin{bmatrix} 1 & 0 & 0 \\ g_1 & & \\ g_2 & C & \end{bmatrix}$ show that

the invariant Haar measures are,

$$\text{left:}\quad \mu(\mathrm{d}\mathbf{G}) = \frac{\mathrm{d}g_1\mathrm{d}g_2\mathrm{d}c_{11}\mathrm{d}c_{12}\mathrm{d}c_{21}\mathrm{d}c_{22}}{|\mathbf{G}|^3},$$

$$\text{right:}\quad \nu(\mathrm{d}\mathbf{G}) = \frac{\mathrm{d}g_1\mathrm{d}g_2\mathrm{d}c_{11}\mathrm{d}c_{12}\mathrm{d}c_{21}\mathrm{d}c_{22}}{|\mathbf{G}|^2},$$

that the modular functions $\Delta(\mathbf{G}) = |\mathbf{G}|^{-1}$, and that the invariant differential is

$$m(\mathrm{d}\mathbf{X}) = \frac{\prod_{i=1}^{n} \mathrm{d}x_i \mathrm{d}y_i}{|\boldsymbol{Z}\boldsymbol{Z}'|^{n/2}}, \qquad \boldsymbol{Z} = \begin{pmatrix} x_1 \cdots x_n \\ y_1 \cdots y_n \end{pmatrix}.$$

(iii) What is the fiducial distribution of θ given T_x? (See Fraser [9], pp. 80–81.)

7. Let V be a real vector space of dimension n. Let $\mathcal{G}$ be the locally compact group of nonsingular linear transformations. Let x be points of V, and designate by T_g the matrix corresponding to $g \in \mathcal{G}$. Let $f(T)$ be a real valued functional on the group of nonsingular matrices. Show that the Haar integral of f is

$$\mu(f) = \int_{\mathcal{G}} \frac{f(T_g)}{|T_g|^n}\, \mathrm{d}T_g.$$

Section 7.4

8. Let $(\bar{X}, S_X)$ be the sample mean and sample standard deviation based on n i.i.d. random variables having an $\mathcal{N}(\mu, \sigma^2)$ distribution. Let $(\bar{Y}, S_Y)$ be the corresponding statistics of n i.i.d. random variables having an $\mathcal{N}(\mu, \rho\sigma^2)$ distribution; $-\infty < \mu < \infty$, $0 < \sigma < \infty$, $0 < \rho < \infty$. For a specified value of $\rho = \rho_0$, derive the fiducial probability element of (μ, σ), given $\{(\bar{X}, S_X), (\bar{Y}, S_Y)\}$, with respect to the group of affine transformations. Specify all the components of the structural model.

9. Show that for the group of real translations and squared-error loss the Pitman estimator of the location parameter coincides with the expected fiducial value of θ, given X.

Section 7.5

10. (i) Let $X_1, \ldots, X_n$ be i.i.d. random variables having a common $\mathcal{N}(\theta, 1)$ distribution, where $-\infty < \theta < \infty$. Use the Hunt-Stein theorem to prove that the sample mean $\bar{X}$ is a minimax estimator for a squared-error loss.
(ii) Use the same model as in (i), but $\Theta = [\theta_0, \infty)$. What is the minimax estimator?
(iii) If $\mathbf{X} = (X_1, \ldots, X_n)'$ has a multivariate normal distribution $\mathcal{N}(\boldsymbol{\xi}, \boldsymbol{\Sigma})$, what is a minimax estimator of $\boldsymbol{\xi}$ for a quadratic loss function $L(\boldsymbol{\xi}, \hat{\boldsymbol{\xi}}) = (\hat{\boldsymbol{\xi}} - \boldsymbol{\xi})'\boldsymbol{\Sigma}^{-1}(\hat{\boldsymbol{\xi}} - \boldsymbol{\xi})$?

11. Let $X_1, \ldots, X_n$ be i.i.d. random variables having a common 2-parameter negative exponential distribution; that is, the family of density functions is

$$f(x; \theta, \sigma) = \frac{1}{\sigma} \exp\left\{-\frac{x - \theta}{\sigma}\right\}, \quad x \geqslant \theta, \quad -\infty < \theta < \infty, \quad 0 < \sigma < \infty.$$

Determine minimax estimators to θ and of σ for squared-error loss functions.

Section 7.6

12. Consider the family of rectangular distributions $\mathcal{R}(\theta, \theta + \sigma)$, $-\infty < \theta < \infty$, $0 < \sigma < \infty$. Let $X_1, X_2, \ldots$ be a sequence of i.i.d. random variables having a common distribution from this family. Suppose that the cost of each observation is fixed, c. Derive a minimax sequential estimation procedure for θ, σ with a squared-error loss.

Section 7.7

13. Let F be an absolutely continuous d.f. on the real line. Consider the problem of estimating $F(X)$ with a loss function

$$L(F, \hat{F}) = \int |F(X) - \hat{F}(X)|^r \, dX;$$

where r, $r = 1, 2, \ldots$, is a positive integer. Let $\mathcal{G}$ be the group of all one-to-one monotone transformations of $\mathcal{X}$. Prove that if $X_{(1)} \leqslant \cdots \leqslant X_{(n)}$ is the sample order statistic, then the minimax invariant estimator of F is

$$\hat{F}(X) = \frac{j+1}{n+2}, \quad \text{if} \quad X_{(j)} \leqslant x < X_{(j+1)},$$

where $X_{(0)} \equiv -\infty$, $X_{(n+1)} \equiv \infty$. (See Aggarwal [1].)

REFERENCES

Barnard [1]; Blackwell and Girshick [1]; Fisher [4], [5], [6], [7]; Fraser [6], [8], [9]; Girshick and Savage [1]; Hartigan [1]; Hora and Buehler [1], [2]; Hunt and Stein [1]; Jeffreys [1]; Kiefer [2]; Kudō [1]; Lehmann [2]; Nachbin [1]; Pitman [1]; Stone [1]; Wesler [1]; Wijsman [3]; Zacks [6], [7].

CHAPTER 8

Admissibility of Estimators

8.1. BASIC THEORY OF ADMISSIBILITY AND COMPLETE CLASSES

Let $(\mathfrak{X}, \mathfrak{B})$ be the measure space related to an observed random variable X and let $\mathfrak{F} = \{P_\theta; \theta \in \Theta\}$ be a family of probability measures on $(\mathfrak{X}, \mathfrak{B})$. As in the previous chapters we assume that Θ is an interval in Euclidean k-space. Consider the problem of estimating the parameter θ with a loss function $L(\theta, d)$. Let $\mathfrak{D}$ be a class of (possibly randomized) estimators of θ. Let $R(\theta, d)$ designate the risk function corresponding to $L(\theta, d)$, $\theta \in \Theta$ and $d \in \mathfrak{D}$. We introduce now the following partial ordering on $\mathfrak{D}$. An estimator d_1 in $\mathfrak{D}$ is said to be *at least as good as* d_2 in $\mathfrak{D}$ if $R(\theta, d_1) \leqslant R(\theta, d_2)$ for *all* $\theta \in \Theta$. An estimator d_1 is said to be *better* than d_2 (or d_1 *strictly dominates* d_2) if $R(\theta, d_1) \leqslant R(\theta, d_2)$ for *all* $\theta \in \Theta$, with strict inequality for *at least one point*, say θ_0, of Θ. If $R(\theta, d_1) = R(\theta, d_2)$ for *all* $\theta \in \Theta$, then d_1 is said to be *equivalent* to d_2.

The ordering of estimators relative to a specified loss function is only partial, since generally certain estimators are uncomparable. For example, consider the problem of estimating the probability of success θ in a family of binomial distributions $\{B(\theta, n), 0 < \theta < 1\}$, n known. We have seen in Example 6.6 that for the squared-error loss function neither is the minimax estimator of θ better than the U.M.V.U. one nor vice versa. The two estimators are uncomparable. On the other hand, the reader can easily check that, in case $X_1, \ldots, X_n$ are i.i.d. such as $N(\mu, \sigma^2)$, where $-\infty < \mu < \infty$ and $0 < \sigma < \infty$, then the estimator $\sum_{i=1}^n (X_i - \bar{X})^2/(n+1)$ where $\bar{X} = (1/n) \sum_{i=1}^n X_i$, dominates all the equivariant estimators of σ^2 for the squared-error loss.

Every estimator which is not dominated by any other estimator is called *admissible* (in the sense of Wald). An estimator d_1 is called *almost admissible* if, for any estimator d_2 satisfying $R(\theta, d_2) \leqslant R(\theta, d_1)$, strict inequality holds on a set of θ values of a Lebesgue measure zero. From the point of view of selecting estimators which minimize the risk function (if such exist), it is obvious that inadmissible estimators should be discarded. However, we should make the following observation. There may be cases in which the minimum risk equivariant estimator is minimax but inadmissible. On the other hand, many admissible estimators are generally not recommendable. Such a case is illustrated in the following example.

Example 8.1. Let $X_1, \ldots, X_n$ be i.i.d. random variables having the common distribution $\mathcal{N}(\mu, \sigma^2)$ when the mean μ and the variance σ^2 are unknown. The objective is to estimate σ^2. For the quadratic loss function $L(\sigma^2, \hat{\sigma}^2) = (\hat{\sigma}^2 - \sigma^2)^2/\sigma^4$, the best equivariant estimator with respect to the group of real affine transformations is, as mentioned previously, $\hat{\sigma}_n^{\,2} = 1/(n+1) \times \sum_{i=1}^n (X_i - \bar{X})^2$. It is easy to check that the risk of $\hat{\sigma}_n^{\,2}$ is the constant $2/(n+1)$. The estimator $\hat{\sigma}_n^{\,2}$ is minimax, according to the generalized Hunt-Stein theorem (Theorem 7.5.1). However, as will be shown in Section 8.5, $\hat{\sigma}_n^{\,2}$ is inadmissible. It is dominated by the nonequivariant estimator $D_n^{\,2} = \min\{\sum_{i=1}^n (X_i - \bar{X})^2/(n+1), \sum_{i=1}^n X_i^2/(n+2)\}$.

We see thus that the principle of invariance is not always compatible with the principle of using only admissible estimators. Moreover, many of the admissible estimators are not uniformly good ones. For example, we prove in Section 8.2 that the estimator of σ^2, $\tilde{\sigma}_n^{\,2} = \sum_{i=1}^n X_i^2/(n+2)$, is admissible. If $\mu = 0$ the risk function of $\tilde{\sigma}_n^{\,2}$ yields the value $2/(n+2)$, which is very close to $2/(n+1)$ whenever n is moderately large. On the other hand, since the risk function of $\tilde{\sigma}_n^{\,2}$ is, for $\lambda = n\mu^2/2\sigma^2$,

$$R(\lambda, \tilde{\sigma}^2) = \frac{1}{(n+2)^2}\,[(2\lambda - 1)^2 + 2n + 3 + 4\lambda], \tag{8.1.1}$$

$R(\lambda, \tilde{\sigma}^2) \to \infty$ when $\lambda \to \infty$. Thus the risk values of $\tilde{\sigma}^2$ are smaller than those of $\hat{\sigma}^2$ only if $0 \leqslant \lambda \leqslant [\frac{1}{2}(n+2)(n+1)]^{1/2}$. Thus although $\tilde{\sigma}^2$ is admissible it is unadvisable to apply it, unless there is a strong indication that the true value of μ is smaller in absolute value than $(\sigma/\sqrt{n})(2(n+2)/(n+1))^{1/4}$. On the other hand, no one will use the admissible estimator $\hat{\sigma}_n^{\,2} \equiv 5$. ■

We prove now several basic theorems.

Theorem 8.1.1.

(i) *If d_0 is the unique minimax estimator for a loss function $L(\theta, d)$, then d_0 is admissible.*

(ii) *If d_0 is admissible and has a constant risk over Θ, then d_0 is minimax.*

Proof. (i) Suppose that d_0 is not an admissible estimator. Then there exists an estimator d_1 such that

$$R(\theta, d_1) \leqslant R(\theta, d_0) \quad \text{for all } \theta \in \Theta, \tag{8.1.2}$$

with a strict inequality for some θ in Θ. Hence $\sup_{\theta\in\Theta} R(\theta, d_1) \leqslant \sup_{\theta\in\Theta} R(\theta, d_0)$. Therefore d_1 is also a minimax estimator, which contradicts the assumption that d_0 is the unique minimax one.

(ii) If d_0 is not minimax then there exists an estimator d_1 such that

$$\sup_{\theta \in \Theta} R(\theta, d_1) < \sup_{\theta \in \Theta} R(\theta, d_0). \tag{8.1.3}$$

But since $R(\theta, d_0) = \text{constant}$ for all $\theta \in \Theta$, $\sup_\theta R(\theta, d_1) < \sup_\theta R(\theta, d_0)$ implies that

$$R(\theta, d_1) < R(\theta, d_0) \quad \text{for } all\ \theta. \tag{8.1.4}$$

Thus d_1 dominates d_0, and this contradicts the assumption that d_0 is admissible. Hence d_0 is minimax. (Q.E.D.)

In Section 8.2 we prove that if $X \sim N(\mu, 1)$ and X is 1-dimensional (real), then X is an admissible estimator of μ, for a quadratic loss function. Since the risk function of X is the constant 1, X is minimax. The minimaxity of X has been established in Chapter 6. Although Theorem 8.1.1 seems to be a trivial one, there are many occasions on which it can be readily applied. Another commonly applied theorem is the following.

Theorem 8.1.2. (Blyth [1].) *If the risk function $R(\theta, d)$ is continuous in θ for each d, and if $\xi(\theta)$ is a prior distribution over Θ, whose carrier is Θ, then the Bayes estimator against ξ, d_ξ, is admissible.*

Proof. If d_ξ is not admissible, there exists an estimator d^* satisfying

$$R(\theta, d^*) \leqslant R(\theta, d_\xi) \quad \text{for all } \theta \in \Theta, \tag{8.1.5}$$

and

$$R(\theta_1, d^*) < R(\theta_1, d_\xi) \quad \text{for some } \theta_1 \in \Theta. \tag{8.1.6}$$

There exists a positive real ϵ_1, $\epsilon_1 > 0$, such that

$$R(\theta_1, d^*) \leqslant R(\theta_1, d_\xi) - \epsilon_1. \tag{8.1.7}$$

Since $R(\theta, d^*)$ is continuous in θ, there exists a θ-neighborhood of θ_1, say $S_{\epsilon_1}^{(1)}$, such that

$$R(\theta, d^*) \leqslant R(\theta, d_\xi) - \epsilon_1 \qquad \text{for all } \theta \in S_{\epsilon_1}^{(1)}.$$

Finally,

$$\begin{aligned} R(\xi, d^*) &= \int_{S_{\epsilon_1}^{(1)}} R(\theta, d^*)\xi(d\theta) + \int_{\bar{S}_{\epsilon_1}^{(1)}} R(\theta, d^*)\xi(d\theta) \\ &\leqslant \int_{S_{\epsilon_1}^{(1)}} [R(\theta, d_\xi) - \epsilon_1]\xi(d\theta) + \int_{\bar{S}_{\epsilon_1}^{(1)}} R(\theta, d_\xi)\xi(d\theta) \\ &= R(\xi, d_\xi) - \epsilon_1 \xi(S_{\epsilon_1}^{(1)}). \end{aligned} \tag{8.1.8}$$

But since $\xi(\mathrm{d}\theta) > 0$ for all $\theta \in \Theta$, $\xi(S_{\epsilon_2}^{(1)}) > 0$. Hence $R(\xi, d^*) < R(\xi, d_\xi)$. This contradicts the hypothesis that d_ξ is Bayes against ξ. Hence d_ξ is admissible. (Q.E.D.)

This theorem is of great practical value, since it provides us with a method of generating a subclass of estimators which are all admissible.

Example 8.2. As a continuation of Example 8.1, consider the problem of estimating the variance of $\mathcal{N}(\mu, \sigma^2)$, $-\infty < \mu < \infty$, $0 < \sigma < \infty$. Given a sample of i.i.d. random variables $X_1, \ldots, X_n$, the minimal sufficient statistic is $(\bar{X}, Q)$, where $Q = \sum_{i=1}^n (X_i - \bar{X})^2$. The variables $\bar{X} \sim N(\mu, \sigma^2/n)$ and $Q \sim \sigma^2\chi^2[n-1]$ are independent. Choose a prior distribution $\xi(\mu, \sigma^2)$ such that (a) μ and σ^2 are priorly independent and (b) $\mu \sim N(0, \tau^2)$, $1/(2\sigma^2) \sim G(\lambda, \nu)$, $0 < \lambda, \nu < \infty$. We obtain the following Bayes estimator for σ^2, with respect to a squared-error loss

$$(8.1.9)\quad \hat{\sigma}^2_{\tau,\lambda,\nu}(\bar{X}, Q) = \frac{\displaystyle\int_0^\infty \theta^{n-5/2+\nu}(1 + 2\theta n\tau^2)^{-1/2} \exp\left\{-\frac{\theta n\bar{X}^2}{1 + 2n\theta\tau^2}\right\} \exp\{-\theta(Q+\lambda)\}\, \mathrm{d}\theta}{\displaystyle\int_0^\infty \theta^{n-3/2+\nu}(1 + 2n\theta\tau^2)^{-1/2} \exp\left\{-\frac{\theta n\bar{X}^2}{1 + 2n\theta\tau^2} - \theta(Q+\lambda)\right\} \mathrm{d}\theta}.$$

This estimator is admissible since $\xi(\mathrm{d}\mu, \mathrm{d}\sigma^2) > 0$ for all (μ, σ^2), and the risk function

$$(8.1.10)\quad R(\sigma^2, \mu; d) = \frac{\sqrt{n}}{\sigma\sqrt{(2\pi)}} \cdot \frac{1}{2^{(n-1)/2}\sigma^{n-1}} \int_{-\infty}^{\infty} \mathrm{d}x \int_0^\infty \mathrm{d}y \times (d - \sigma^2)^2 y^{(n-3)/2} \exp\left\{-\frac{n}{2\sigma^2}(x-\mu)^2 - \frac{1}{2\sigma^2}y\right\}$$

is a continuous function of (μ, σ^2) over $\Theta = \{(\mu, \sigma^2); -\infty < \mu < \infty, 0 < \sigma < \infty\}$. We notice that $\hat{\sigma}^2_{\tau,\lambda,\nu}(\bar{X}, Q)$ is a function not only of Q but also of $\bar{X}^2$. All the equivariant estimators (for affine transformations) are functions of Q only and therefore (as will be verified later) are inadmissible. The quantity $\bar{X}^2 \sim (\sigma^2/n)\chi^2[1; (n\mu^2)/(2\sigma^2)]$ and therefore has some information on σ^2, too. We notice that the best equivariant estimator of σ^2, that is, $Q/(n+1)$, is obtained from $\hat{\sigma}^2_{\tau,\lambda,\nu}(\bar{X}, Q)$ by letting $\tau^2 \to \infty$, $\lambda \to 0$, and $\nu = 3$. A regular limit of a sequence of admissible estimators may thus yield an inadmissible estimator. This point will be discussed further later on. ■

We now provide Stein's characterization of admissible estimators [4]. This characterization gives a sufficient condition for the admissibility of an estimator. We first introduce the notion of an ϵ-Bayes estimator.

Let $\xi(\theta)$ be a prior distribution on Θ and $R(\theta, d)$ a risk function associated with an estimation problem. An estimator $d_1 \in \mathfrak{D}$ is called ϵ-Bayes against

ξ if, for any $\epsilon > 0$,

$$R(\xi, d_1) \leqslant \inf_{d \in \mathfrak{D}} R(\xi, d) + \epsilon. \tag{8.1.11}$$

As previously, we designate by ξ_θ a prior distribution on Θ which assigns probability 1 to θ.

Theorem 8.1.3. (Stein [4].) *If d_0 is an estimator in $\mathfrak{D}$ such that for any θ_1 in Θ and $\epsilon > 0$ there exists a prior distribution $\xi^{(1)}$ on Θ, and $0 < p < 1$ such that d_0 is ϵp-Bayes against $(1 - p)\xi^{(1)} + p\xi_{\theta_1}$, then d_0 is admissible.*

Proof. If d_0 is inadmissible, there exists an estimator d_1 in $\mathfrak{D}$ such that $R(\theta, d_1) \leqslant R(\theta, d_0)$ for all $\theta \in \Theta$, and $R(\theta_1, d_1) < R(\theta_1, d_0)$ for some $\theta_1 \in \Theta$. According to the hypothesis of the theorem, d_0 is ϵp-Bayes against $(1 - p)\xi^{(1)} + p\xi_{\theta_1}$ for some $\xi^{(1)}$ on Θ and p, $0 < p < 1$; that is,

$$\begin{aligned} R((1 - p)\xi^{(1)} + p\xi_{\theta_1}, d_0) &\leqslant \inf_{d \in \mathfrak{D}} R((1 - p)\xi^{(1)} + p\xi_{\theta_1}, d) + \epsilon p \\ &\leqslant R((1 - p)\xi^{(1)} + p\xi_{\theta_1}, d_1) + \epsilon p. \end{aligned} \tag{8.1.12}$$

Since $R(\xi, d)$ is a linear function of ξ for each d, we obtain from (8.1.12)

$$\begin{aligned} (1 - p)R(\xi^{(1)}, d_0) + pR(\theta_1, d_0) &\leqslant (1 - p)R(\xi^{(1)}, d_1) + \\ pR(\theta_1, d_1) + \epsilon p &\leqslant (1 - p)R(\xi^{(1)}, d_0) + pR(\theta_1, d_1) + \epsilon p. \end{aligned} \tag{8.1.13}$$

Hence

$$R(\theta_1, d_0) \leqslant R(\theta_1, d_1) + \epsilon. \tag{8.1.14}$$

Since ϵ is arbitrary, $R(\theta_1, d_0) \leqslant R(\theta_1, d_1)$. This contradicts the assumption that $R(\theta_1, d_1) < R(\theta_1, d_0)$. Hence d_0 is admissible. (Q.E.D.)

Example 8.3. In Example 6.7 we have shown that the sample mean $\bar{X}$ is a minimax estimator of the mean μ of a normal distribution $\mathcal{N}(\mu, 1)$, with the loss function

$$L(\mu, \hat{\mu}) = \begin{cases} 0, & \text{if } |\mu - \hat{\mu}| < \delta, \\ 1, & \text{otherwise.} \end{cases}$$

We also have shown that the sequence of normal prior distributions ξ_k; $\mathcal{N}(0, k)$ induces the Bayes estimators $\hat{\mu}_k(\bar{X}) = \bar{X}(1 + 1/nk)^{-1}$ and a sequence of prior risks $R(\xi_k, \hat{\mu}_k)$ which converges, as $k \to \infty$, to the constant risk value of $\bar{X}$; that is, $\rho^* = P[|\bar{X} - \mu| > \delta] = 2[1 - \Phi(\delta\sqrt{n})]$. Thus for every $\epsilon^* > 0$, there exists a $k^*(\epsilon^*)$ such that, for all $k \geqslant k^*(\epsilon^*)$, $|R(\xi_k, \hat{\mu}_k) - \rho^*| < \epsilon^*$. Furthermore, for each k and μ_1

$$\inf_{-\infty < \hat{\mu} < \infty} R((1 - p)\xi_k + p\xi_{\mu_1}, \hat{\mu}) \geqslant (1 - p)R(\xi_k, \hat{\mu}_k). \tag{8.1.15}$$

Choose p close enough to 1 so that for each k, $k \geqslant k^*(\epsilon^*)$,

$$(1 - p)R(\xi_k, \hat{\mu}_k) \geqslant \rho^* - \epsilon^*. \tag{8.1.16}$$

Hence from (8.1.16) and (8.1.17),

$$\begin{aligned} \rho^* &\leqslant (1 - p)R(\xi_k, \hat{\mu}_k) + \epsilon^* \\ &\leqslant \inf_{-\infty < \hat{\mu} < \infty} R((1 - p)\xi_k + p\xi_{\mu_1}, \hat{\mu}) + \epsilon^*. \end{aligned} \tag{8.1.17}$$

Let $\epsilon^* = \epsilon/2$. Since $p > \frac{1}{2}$, $\epsilon^* < \epsilon p$. Hence from (8.1.17), $\bar{X}$ is an ϵp-Bayes estimator against $(1 - p)\xi_k + p\xi_{\mu_1}$ for any given μ_1 and ϵ. *Thus* $\bar{X}$ *is admissible.* ■

Admissible estimators are those which are not strictly dominated by other estimators. Among the admissible estimators we may find some with a stronger property, called strict admissibility. For every point $\theta_0 \in \Theta$ and a specified $\epsilon > 0$, let $\mathfrak{D}_\epsilon(\theta_0)$ designate the set of all estimators in $\mathfrak{D}$ such that $R(\theta_0, d) - R(\theta, d_0) \leqslant -\epsilon$, for some fixed $d_0 \in \mathfrak{D}$. The estimator d_0 is called *strictly admissible* if for any $\epsilon > 0$ and $\theta_0 \in \Theta$ there exists a $\delta > 0$ and $\theta_1 \in \Theta$ (which may depend on θ_0 and ϵ) such that $R(\theta_1, d) - R(\theta_1, d_0) \geqslant \delta$ for *all* $d \in \mathfrak{D}_\epsilon(\theta_0)$.

Strict admissibility obviously implies admissibility but not vice versa. The notion of strict admissibility can be defined also in the following manner. If for any sequence of estimators $\{d_i\} \subset \mathfrak{D}$, and any θ_0 and $\epsilon > 0$, $R(\theta_0, d_i) - R(\theta_0, d_0) \leqslant -\epsilon$ for all $i = 1, 2, \ldots$ implies the existence of $\delta > 0$ and some $\theta_1 \neq \theta_0$ such that $R(\theta_1, d_i) - R(\theta_1, d_0) \geqslant \delta$ for all $i = 1, 2, \ldots$, then d_0 is strictly admissible. Hence if d_0 is not strictly admissible, there exists a sequence $\{d_i\} \subset \mathfrak{D}$ such that, for some $\theta_0 \in \Theta$ and $\epsilon > 0$, $R(\theta_0, d_i) - R(\theta_0, d_0) \leqslant -\epsilon$ for all $i = 1, 2, \ldots$ and

$$\varliminf_{i\to\infty} \sup_{\theta\in\Theta} \{R(\theta, d_i) - R(\theta, d_0)\} \leqslant 0. \tag{8.1.18}$$

We introduce now the notion of weak-compactness, in the sense of Wald.

The class $\mathfrak{D}$ of estimators is called *weakly compact* with respect to $R(\theta, d)$ if for every sequence $\{d_i\} \subset \mathfrak{D}$ there exists a d^* in $\mathfrak{D}$ such that

$$\varliminf_{i\to\infty} R(\theta, d_i) \geqslant R(\theta, d^*), \qquad \text{all } \theta \in \Theta.$$

Weak-compactness plays an important role in the following lemma.

Lemma 8.1.4. (Stein [4].) *If* d_0 *is admissible and* $\mathfrak{D}$ *is weakly compact in the sense of Wald, then* d_0 *is strictly admissible.*

Proof. If d_0 is not strictly admissible there exists a sequence $\{d_i\} \subset \mathfrak{D}$, a point θ_0 in Θ and $\epsilon > 0$ such that

(i) $R(\theta_0, d_i) - R(\theta_0, d_0) \leqslant -\epsilon$, all $i = 1, 2, \ldots$;

(ii) $\overline{\lim_{i\to\infty}} \sup_{\theta\in\Theta} \{R(\theta, d_i) - R(\theta, d_0)\} \leqslant 0$.

Since $\mathfrak{D}$ is weakly compact, there exists a subsequence $\{d_{i_j}\} \subset \{d_i\}$ and an estimator d^* in $\mathfrak{D}$ such that $\overline{\lim_{j\to\infty}} R(\theta, d_{i_j}) \geqslant R(\theta, d^*)$ for *all* θ in Θ. Hence from the weak-compactness and (ii),

$$\sup_{\theta\in\Theta} \{R(\theta, d^*) - R(\theta, d_0)\} \leqslant \sup_{\theta\in\Theta} \left\{ \underline{\lim}_{j\to\infty} R(\theta, d_{i_j}) - R(\theta, d_0)\right\} \tag{8.1.19}$$

$$\leqslant \underline{\lim}_{j\to\infty} \sup_{\theta\in\Theta} \{R(\theta, d_{i_j}) - R(\theta, d_0)\} \leqslant 0.$$

Moreover, from (i) above we obtain that at $\theta = \theta_0$,

$$R(\theta_0, d^*) - R(\theta_0, d_0) \leqslant \underline{\lim}_{j\to\infty} R(\theta_0, d_{i_j}) - R(\theta_0, d_0) \leqslant -\epsilon. \tag{8.1.20}$$

Equations 8.1.19 and 8.1.20 imply that d_0 is inadmissible. This contradicts the hypothesis. (Q.E.D.)

Theorem 8.1.5. (Stein [4].) *Suppose that the class of estimators $\mathfrak{D}$ is weakly compact in the sense of Wald. Then $d_0 \in \mathfrak{D}$ is admissible if and only if, for every $\theta_0 \in \Theta$,*

$$\overline{\lim_{\gamma\to\infty}} \inf_{d\in\mathfrak{D}} \sup_{\theta\in\Theta} \{R(\theta_0, d) - R(\theta_0, d_0) + \gamma[R(\theta, d) - R(\theta, d_0)]\} \geqslant 0. \tag{8.1.21}$$

Proof. According to Lemma 8.1.4, every admissible estimator is strictly admissible. We prove now that (8.1.21) is a necessary and sufficient condition of strict admissibility. Define, for a given d_0,

$$M(\theta, d) = R(\theta, d) - R(\theta, d_0), \qquad \theta \in \Theta. \tag{8.1.22}$$

(i) *Necessity.* Let $\rho(\theta) = \inf_{d\in\mathfrak{D}} M(\theta, d)$. Since $R(\theta, d) \geqslant 0$ for all (θ, d) and $\sup_{\theta\in\Theta} R(\theta, d_0) < \infty$, $\rho(\theta) > -\infty$ for all $\theta \in \Theta$. Now

$$\overline{\lim_{\gamma\to\infty}} \inf_{d\in\mathfrak{D}} \sup_{\theta\in\Theta} \{M(\theta_0, d) + \gamma M(\theta, d)\} \tag{8.1.23}$$

$$= \overline{\lim_{\gamma\to\infty}} \min\left\{ \inf_{d\in\mathfrak{D}_\epsilon(\theta_0)} \sup_{\theta\in\Theta} [M(\theta_0, d) + \gamma M(\theta, d)], \right.$$

$$\left. \inf_{d\notin\mathfrak{D}_\epsilon(\theta_0)} \sup_{\theta\in\Theta} [M(\theta_0, d) + \gamma M(\theta, d)]\right\}.$$

Since d_0 is strictly admissible, there exists $\delta_\epsilon > 0$ such that

(8.1.24) $$\sup_{\theta\in\Theta} M(\theta, d) \geqslant \delta_\epsilon, \quad \text{all} \quad d \in \mathfrak{D}_\epsilon(\theta_0).$$

Hence

(8.1.25)
$$\inf_{d\in\mathfrak{D}_\epsilon(\theta_0)} \sup_{\theta\in\Theta} \{M(\theta_0, d) + \gamma M(\theta, d)\} \geqslant \inf_{d\in\mathfrak{D}_\epsilon(\theta_0)} \{M(\theta_0, d) + \gamma\delta_\epsilon\} \geqslant \rho(\theta_0) + \gamma\delta_\epsilon.$$

Furthermore, the admissibility of d_0 implies that $\sup_{\theta\in\Theta} M(\theta, d) \geqslant 0$ for all $d \in \mathfrak{D}_\epsilon(\theta_0)$. Hence

(8.1.26) $$\inf_{d\in\mathfrak{D}_\epsilon(\theta_0)} \sup_{\theta\in\Theta} \{M(\theta_0, d) + \gamma M(\theta, d)\} \geqslant \inf_{d\in\mathfrak{D}_\epsilon(\theta_0)} M(\theta_0, d) > -\epsilon.$$

Substituting (8.1.25) and (8.1.26) in (8.1.23), we obtain

(8.1.27) $$\overline{\lim_{\gamma\to\infty}} \inf_{d\in\mathfrak{D}} \sup_{\theta\in\Theta} \{M(\theta_0, d) + \gamma M(\theta, d)\} \geqslant \lim_{\gamma\to\infty} \min \{\rho(\theta_0) + \gamma\delta_\epsilon, -\epsilon\}.$$

Finally, since $\rho(\theta_0) + \gamma\delta_\epsilon$ is an increasing function of γ, and since $\rho(\theta_0) > -\infty$,

(8.1.28) $$\lim_{\gamma\to\infty} \min \{\rho(\theta_0) + \gamma\delta_\epsilon, -\epsilon\} = -\epsilon.$$

Equations 8.1.27 and 8.1.28 imply (8.1.21), since ϵ is arbitrary.

(ii) *Sufficiency*. Suppose that (8.1.21) holds; then

(8.1.29)
$$\begin{aligned} 0 &\leqslant \overline{\lim_{\gamma\to 0}} \inf_{d\in\mathfrak{D}} \sup_{\theta\in\Theta} \{M(\theta_0, d) + \gamma M(\theta, d)\} \\ &\leqslant \overline{\lim_{\gamma\to\infty}} \inf_{d\in\mathfrak{D}_\epsilon(\theta_0)} \sup_{\theta\in\Theta} \{M(\theta_0, d) + \gamma M(\theta, d)\} \\ &\leqslant \overline{\lim_{\gamma\to\infty}} \inf_{d\in\mathfrak{D}_\epsilon(\theta_0)} \sup_{\theta\in\Theta} \{-\epsilon + \gamma M(\theta, d)\} \\ &= -\epsilon + \overline{\lim_{\gamma\to\infty}} \inf_{d\in\mathfrak{D}_\epsilon(\theta_0)} \sup_{\theta\in\Theta} \gamma M(\theta, d). \end{aligned}$$

Hence there exists a $\gamma_\epsilon > 0$ such that

(8.1.30) $$\inf_{d\in\mathfrak{D}_\epsilon(\theta_0)} \sup_{\theta\in\Theta} M(\theta, d) \geqslant \frac{\epsilon}{2\gamma_\epsilon}.$$

Let $\delta_\epsilon = \epsilon/2\gamma_\epsilon$, $\delta_\epsilon > 0$. From (8.1.30) we deduce that there exists a θ_1 in Θ such that $M(\theta_1, d) \geqslant \delta_\epsilon$ for *all* $d \in \mathfrak{D}_\epsilon(\theta_0)$. Hence d_0 is strictly admissible.

(Q.E.D.)

We introduce the notion of complete classes with respect to a specified risk function. Let $\mathcal{C}$ be a subclass of estimators, that is, $\mathcal{C} \subset \mathfrak{D}$. The subclass $\mathcal{C}$ is said to be *essentially complete* with respect to $R(\theta, d)$ if, for every $d \in \mathfrak{D} - \mathcal{C}$, there exists an estimator d_* in $\mathcal{C}$ which is at least as good as d. The subclass $\mathcal{C}$ is called *complete* if for every $d \in \mathcal{D} - \mathcal{C}$ there exists a $d^* \in \mathcal{C}$ which *strictly dominates d.* A complete subclass $\mathcal{C}^*$ is called *minimal complete* if no proper subclass of $\mathcal{C}^*$ is complete.

Let $\mathcal{A}$ designate the subclass of all admissible estimators in $\mathcal{D}$ (if nonempty) and let $\mathcal{C}$ be a complete subclass of $\mathcal{D}$. Then $\mathcal{A} \subset \mathcal{C}$. Indeed, if d_0 is admissible then $d_0 \in \mathcal{C}$; otherwise there exists in $\mathcal{C}$ an estimator which strictly dominates d_0. Furthermore, *if* the subclass of admissible estimators $\mathcal{A}$ is complete, it is *minimal* complete. In cases of *finite* parameter spaces Θ, that is, $\Theta = \{\theta_1, \ldots, \theta_k\}$ for some $k < \infty$, let $\mathcal{R} = \{R(\theta_1, d), \ldots, R(\theta_k, d);\ d \in \mathcal{D}\}$ be the set of all risk points and $\mathcal{R}^*$ its convex hull. If $\mathcal{R}^*$ is closed from below (it is obviously bounded from below since $R(\theta_i, d) \geqslant 0$ for all $d \in \mathcal{D}$, $i = 1, \ldots, k$), then every admissible estimator is Bayes, and the subclass of all Bayes estimators $\mathcal{B}$ is complete. Moreover, the subclass of all the admissible estimators is minimal complete (see proof in Ferguson [1], Sec. 2.10 and Wald [6], Ch. 3). In order to prove completeness theorems in the case of an infinite parameter space Θ Wald [6] introduced the notion of weak semiseparability. Thus Θ is said to be *weakly semiseparable* in the sense of Wald if and only if there exists a sequence of parametric points $\{\theta_i\} \subset \Theta$ such that

$$\overline{\lim_{\gamma \to \infty}} R(\theta_i, d) \geqslant R(\theta, d), \qquad \text{for all } (\theta, d). \tag{8.1.31}$$

Wald proved ([6], Theorems 2.22 and 2.25) the following theorem.

Theorem 8.1.6. (Wald.) *Let $L(\theta, d)$ be bounded in θ for each $d \in \mathcal{D}$. If $\mathcal{D}$ is weakly compact and Θ is weakly semiseparable, the class $\mathcal{A}$ of all admissible estimators is complete; and the subclass of all proper and generalized Bayes estimators is essentially complete.*

The proof of Theorem 8.1.6 is omitted. We remark only that if the loss function $L(\theta, d)$ is not bounded in θ for each d it should satisfy assumptions A.2 and A.3 of Section 6.3.

Ferguson proves a theorem which establishes the conditions for the class of all ϵ-Bayes estimators to be essentially complete. The theorem (Ferguson [1], Theorem 2.10.3) is the following.

Theorem 8.1.7. *If the risk function $R(\theta, d)$ is lower semicontinuous in $d \in \mathcal{D}$ for each $\theta \in \Theta$, and if $\mathcal{D}$ is compact (in a proper metric topology), then the class of all ϵ-Bayes estimators is essentially complete.*

8.2. ADMISSIBILITY UNDER QUADRATIC LOSS

The problem of characterizing admissible estimators when the loss function is quadratic was studied in several papers. We present here the main results of Hodges and Lehmann [2], Girshick and Savage [1], and Karlin [4].

Hodges and Lehmann [2] used the Cramér-Rao lower bound to the mean square error of estimators to show that certain estimators which are minimax are also admissible. They confined their investigation to the case of a real parameter. Thus we assume that the family of probability measures $\mathfrak{T}$ is represented by a family of density functions $\{f(x; \theta); \theta \in \Theta\}$ with respect to a sigma-finite measure $\mu(dx)$ on $(\mathfrak{X}, \mathfrak{B})$, which satisfies the Cramér-Rao regularity conditions of Section 4.1. Let $\mathfrak{D}$ be the class of all (nonrandomized) regular estimators of θ. With each d in $\mathfrak{D}$ there is associated a bias function $B_d(\theta)$ on Θ.

The Cramér-Rao lower bound (see (4.1.28)) is

$$C_d(\theta) = B_d^2(\theta) + \frac{(1 + B_d'(\theta))^2}{I(\theta)}. \tag{8.2.1}$$

In the following theorem a sufficient condition for admissibility is established.

Theorem 8.2.1. (Hodges and Lehmann [2].) *Assume that the mean square error of an estimator d_0 in $\mathfrak{D}$ attains the Cramér-Rao lower bound for all θ in Θ. If for any estimator d_1 in $\mathfrak{D}$ the relationship $C_{d_1}(\theta) \leqslant C_{d_0}(\theta)$ for all θ implies that $B_{d_1}(\theta) = B_{d_0}(\theta)$ for all θ, then d_0 is admissible with respect to the squared-error loss.*

Proof. If d_0 is inadmissible there exists an estimator d_1 in $\mathfrak{D}$ whose mean square error strictly dominates that of d_0; that is,

$$R(\theta, d_1) \leqslant R(\theta, d_0) \qquad \text{for all } \theta \in \Theta, \tag{8.2.2}$$

with strict inequality for at least one θ_1 in Θ. Since $R(\theta, d_0) = C_{d_0}(\theta)$ for all θ, we have from (8.2.2),

$$C_{d_1}(\theta) \leqslant R(\theta, d_1) \leqslant R(\theta, d_0) = C_{d_0}(\theta), \tag{8.2.3}$$

for all $\theta \in \Theta$. According to the hypothesis, $B_{d_1}(\theta) = B_{d_0}(\theta)$ for all θ. Therefore $B'_{d_1}(\theta) = B'_{d_0}(\theta)$ for all θ, and $C_{d_1}(\theta) = C_{d_0}(\theta)$ for all θ. But this contradicts the assumption that d_1 strictly dominates d_0. Hence d_0 is admissible. (Q.E.D.)

We remark that the theorem still holds if the loss function is quadratic, that is, $L(\theta, d) = \lambda(\theta)(d - \theta)^2$, with any $0 < \lambda(\theta) < \infty$. Furthermore, if d_0 is a constant risk estimator and satisfies the condition of the theorem, it is also minimax.

Example 8.4. Let $X_1, \ldots, X_n$ be i.i.d. random variables distributed like $N(0, \sigma^2)$, $0 < \sigma^2 < \infty$. The minimal sufficient statistic is $S_n = \sum_{i=1}^n X_i^2$; S_n is distributed like $\sigma^2\chi^2[n]$. It is easy to verify that the mean square error of the estimator $\hat{\sigma}_n^2 = S_n/(n + 2)$ attains the Cramér-Rao lower bound $2\sigma^4/(n + 2)$. The bias function of $\hat{\sigma}_n^2$ is $B^{(0)}(\sigma^2) = -(2/(n + 2))\sigma^2$. Thus

$\hat{\sigma}_n^2$ is admissible with respect to the loss function $L(\sigma^2, \hat{\sigma}^2) = (\hat{\sigma}^2 - \sigma^2)^2/\sigma^4$ if, for any estimator d, the inequality

$$\text{(8.2.4)} \qquad \frac{1}{\sigma^4} B_d^2(\sigma^2) + \frac{2[1 + B_d'(\sigma^2)]^2}{n} \leqslant \frac{2}{n+2}, \qquad \text{for all } \sigma^2,$$

implies that $B_d(\sigma^2) = -2\sigma^2/(n+2)$ for all σ^2. Since $2/n > 2/(n+2)$ for all $n \geqslant 1$, and $B_d^2(\sigma^2)/\sigma^4 \geqslant 0$, $B_d'(\sigma^2) < 0$ for all σ^2, *and* $|B_d(\sigma^2)| < \sigma^2 \times (2/(n+2))^{1/2}$. It follows that $\lim_{\sigma^2 \searrow 0} B_d(\sigma^2) = 0$ and $B_d(\sigma^2) < 0$ for all $0 < \sigma^2 < \infty$. If $B_d'(\sigma^2) = B_d(\sigma^2)/\sigma^2$, then $B_d'(\sigma_2)$ must be $-2(n+2)$, since $X^2 + (2/n)(1 + X^2)$ has a minimum of $2/(n+2)$ at $X = -2/(n+2)$. We observe next that $B_d'(\sigma^2) \leqslant B_d(\sigma^2)/\sigma^2$, since otherwise we obtain

$$\text{(8.2.5)} \quad \left(\frac{B_d(\sigma^2)}{\sigma^2}\right)^2 + \frac{2}{n}[1 + B_d'(\sigma^2)] > \left(\frac{B_d(\sigma^2)}{\sigma^2}\right)^2 + \frac{2}{n}\left[1 + \frac{B_d(\sigma^2)}{\sigma^2}\right] \geqslant \frac{2}{n+2};$$

but (8.2.5) contradicts (8.2.4). Moreover, since $\sigma^4(d/d\sigma^2)[B_d(\sigma^2)/\sigma^2] = \sigma^2 B_d'(\sigma^2) - B_d(\sigma^2)$, we conclude that $(d/d\sigma^2)[B_d(\sigma^2)/\sigma^2] \leqslant 0$; that is, $B_d(\sigma^2)/\sigma^2$ is a nonincreasing function of σ^2. Since $\lim_{\sigma^2 \searrow 0} B_d(\sigma^2) = 0$, we obtain that

$$\text{(8.2.6)} \qquad \lim_{\sigma^2 \searrow 0} \frac{B_d(\sigma^2)}{\sigma^2} = \lim_{\sigma^2 \searrow 0} B_d'(\sigma^2).$$

Furthermore, writing (8.2.4) in the form

$$\text{(8.2.7)} \quad \left\{\left(\frac{B_d(\sigma^2)}{\sigma^2}\right)^2 + \frac{2}{n}\left[1 + \frac{B_d(\sigma^2)}{\sigma^2}\right]^2\right\} + \frac{2}{n}\left\{\left[B_d'(\sigma^2) - \frac{B_d(\sigma^2)}{\sigma^2}\right]^2 - 2\left[B_d'(\sigma^2) - \frac{B_d(\sigma^2)}{\sigma^2}\right]\left[1 + \frac{B_d(\sigma^2)}{\sigma^2}\right]\right\} \leqslant \frac{2}{n+2}.$$

we obtain that

$$\text{(8.2.8)} \qquad \overline{\lim_{\sigma^2 \searrow 0}} \left\{\left(\frac{B_d(\sigma^2)}{\sigma^2}\right)^2 + \frac{2}{n}\left[1 + \frac{B_d(\sigma^2)}{\sigma^2}\right]^2\right\} \leqslant \frac{2}{n+2}.$$

On the other hand,

$$\text{(8.2.9)} \qquad \left(\frac{B_d(\sigma^2)}{\sigma^2}\right)^2 + \frac{2}{n}\left[1 + \frac{B_d(\sigma^2)}{\sigma^2}\right]^2 \geqslant \frac{2}{n+2}.$$

Hence

$$\text{(8.2.10)} \qquad \lim_{\sigma^2 \searrow 0} \left\{\left(\frac{B_d(\sigma^2)}{\sigma^2}\right)^2 + \frac{2}{n}\left[1 + \frac{B_d(\sigma^2)}{\sigma^2}\right]^2\right\} = \frac{2}{n+2},$$

and

$$\text{(8.2.11)} \qquad \lim_{\sigma^2 \searrow 0} \frac{B_d(\sigma^2)}{\sigma^2} = -\frac{2}{n+2}.$$

We now prove that $\lim_{\sigma^2\to\infty} B_d(\sigma^2)/\sigma^2 = \lim_{\sigma^2\to\infty} B_d'(\sigma^2) = -2/(n+2)$. Indeed, if for some $\epsilon > 0$ there exists a σ_0^2 such that

$$B_d'(\sigma^2) \leqslant \frac{B_d(\sigma^2)}{\sigma^2} - \epsilon, \qquad \text{all } \sigma^2 \geqslant \sigma_0^2, \tag{8.2.12}$$

we can construct a function $G(\sigma^2)$ having the following properties:

(i) $G(\sigma_0^2) = B_d(\sigma_0^2)$
(ii) $G'(\sigma^2) = G(\sigma^2)/\sigma^2$, all $\sigma^2 \geqslant \sigma_0^2$.

The solution of (ii) under the condition (i) yields, for all $\sigma^2 \geqslant \sigma_0^2$.

$$G(\sigma^2) = \epsilon[\sigma^2 \log \sigma^2 - \sigma_0^2 \log \sigma_0^2] + \sigma^2 \frac{B_d(\sigma_0^2)}{\sigma_0^2}. \tag{8.2.13}$$

Moreover, $B(\sigma^2) \leqslant G(\sigma^2)$ for all $\sigma^2 \geqslant \sigma_0^2$. From (8.2.13) we obtain

$$\frac{B_d(\sigma^2)}{\sigma^2} \leqslant \epsilon \log \sigma^2 - \epsilon\sigma^2\sigma_0^2 \log \sigma_0^2 + \frac{B_d(\sigma_0^2)}{\sigma_0^2}. \tag{8.2.14}$$

Hence, since $B_d(\sigma_0^2)/\sigma_0^2 < (2/(n+2))^{1/2}$,

$$\overline{\lim_{\sigma^2\to\infty}} \frac{B_d(\sigma^2)}{\sigma^2} = -\infty. \tag{8.2.15}$$

But (8.2.15) contradicts the previous result that $|B_d(\sigma^2)/\sigma^2| < (2/(n+2))^{1/2}$. Hence

$$\lim_{\sigma^2\to\infty} \frac{B_d'(\sigma^2)}{B_d(\sigma^2)/\sigma^2} = 1. \tag{8.2.16}$$

Finally, letting $\sigma^2 \to \infty$ in (8.2.17), we obtain that

$$\lim_{\sigma^2\to\infty} \frac{B_d(\sigma^2)}{\sigma^2} = -\frac{2}{n+2}. \tag{8.2.17}$$

From (8.2.11) and (8.2.17) and the fact that $B_d(\sigma^2)/\sigma^2$ is nonincreasing, we obtain that

$$\frac{B_d(\sigma^2)}{\sigma^2} = -\frac{2}{n+2}, \qquad \text{all } 0 < \sigma^2 < \infty. \tag{8.2.18}$$

This completes the proof that $\hat{\sigma}_n^2 = S_n/(n+2)$ is admissible. The estimator $\hat{\sigma}_n^2$ is also minimax. Indeed, for every d in $\mathfrak{D}$,

$$\sup_{0<\sigma^2<\infty} R(\sigma^2, d) \geqslant \sup_{0<\sigma^2<\infty} \left\{\left(\frac{B_d(\sigma^2)}{\sigma^2}\right)^2 + \frac{2}{n}[1 + B_d'(\sigma^2)]^2\right\} \tag{8.2.19}$$
$$\geqslant \frac{2}{n+2},$$

where the right-hand side of (8.2.19) is attained by the risk function of $\hat{\sigma}_n^2$.

We notice that if the parameter space Θ of σ^2 is truncated, that is, $\Theta = \{\sigma^2; \sigma_0^2 \leqslant \sigma^2 < \infty\}$ where $0 < \sigma_0^2 < \infty$, then $\hat{\sigma}_n^2$ is still minimax but no longer admissible. This is due to the fact that $P_{\sigma^2}\{\hat{\sigma}_n^2 < \sigma_0^2\} > 0$ for all σ^2 in Θ. Therefore the estimator $\tilde{\sigma}_n^2 = \max\{\sigma_0^2, S_n/(n+2)\}$ assumes risk values smaller than those of $\hat{\sigma}_n^2 = S_n/(n+2)$ for all σ^2 in Θ. The verification that $\tilde{\sigma}_n^2$ has mean square error values smaller than those of $\hat{\sigma}_n^2$, for all $\sigma^2 \in \Theta$, is straightforward and left to the reader. ■

In Theorem 4.1.1 we proved that if the risk function of an estimator d_0 for quadratic loss attains the Cramér-Rao lower bound $C_{d_0}(\theta)$ for all θ then the distribution of d_0 must be of the 1-parameter exponential type. In the case of Example 8.4 the admissible estimator $\hat{\sigma}_n^2 = S_n/(n+2)$ is distributed like $G((n+2)/2\sigma^2, n/2)$, $0 < \sigma^2 < \infty$, which has an exponential type density. However, all the estimators of σ^2 of the form $d = \lambda S_n$ have exponential type densities. This shows that having a density of the exponential family does not guarantee admissibility. In this connection Girshick and Savage [1] proved the following theorem.

Theorem 8.2.2. (Girshick and Savage.) *Let U be a random variable with a distribution function of the 1-parameter exponential family, that is,*

$$F(du) = \beta(\tau)e^{u\tau}\psi(du), \qquad -\infty < \mu < \infty, \tag{8.2.20}$$

$\tau \in \Omega$. Let $\theta(\tau) = E_\tau\{U\}$ and $\sigma^2(\tau) = \underset{\tau}{\text{Var}}\{U\}$. If Ω is the entire real line, then U is an admissible minimax estimator of $\theta(\tau)$, with respect to the quadratic loss function $L(\theta, d) = (d - \theta(\tau))^2/\sigma^2(\tau)$.

Proof. Since the distribution of U is of the exponential type, its variance attains the Cramér-Rao lower bound; that is,

$$\sigma^2(\tau) = I^{-1}(\tau)(\theta'(\tau))^2. \tag{8.2.21}$$

Let $\omega(\tau) = \beta^{-1}(\tau) = \int_{-\infty}^{\infty} e^{u\tau}\psi(du) < \infty$. It is easy to verify that

$$\theta(\tau) = \frac{\omega'(\tau)}{\omega(\tau)}, \tag{8.2.22}$$

and

$$\sigma^2(\tau) = \frac{\omega''(\tau)\omega(\tau) - (\omega'(\tau))^2}{\omega^2(\tau)} = \theta'(\tau). \tag{8.2.23}$$

Hence from (8.2.21) and (8.2.23) we obtain that $I^{-1}(\tau) = 1/\sigma^2(\tau)$. Let $d(U)$ be a regular estimator of $\theta(\tau)$. Let $B_d(\tau) = E_\tau\{d(U)\} - \theta(\tau)$. Then according to the Cramér-Rao lower bound,

$$E_\tau\{(d(U) - \theta(\tau))^2\} \geqslant B_d^2(\tau) + \frac{(B_d'(\tau) + \sigma^2(\tau))^2}{\sigma^2(\tau)}. \tag{8.2.24}$$

If d is any estimator of $\theta(\tau)$ which strictly dominates U, then

$$B_d^2(\tau) + \frac{(B_d'(\tau) + \sigma^2(\tau))^2}{\sigma^2(\tau)} \leqslant \sigma^2(\tau) \qquad \text{for all } \tau \in \Omega, \tag{8.2.25}$$

with strict inequality for some τ_0 in Ω. Equivalently,

$$\sigma^2(\tau)B_d^2(\tau) + [B_d'(\tau) + \sigma^2(\tau)]^2 \leqslant \sigma^4(\tau), \qquad \text{for all } \tau.$$

Hence $B_d'(\tau) \leqslant 0$ for all τ. We also obtain from (8.2.25) that

$$B_d^2(\tau) + 2B_d'(\tau) \leqslant 0, \tag{8.2.26}$$

or, for all τ such that $B_d(\tau) \neq 0$,

$$(\mathrm{d}/\mathrm{d}\tau)\{B_d^{-1}(\tau)\} \geqslant \tfrac{1}{2}. \tag{8.2.27}$$

Since $B_d'(\tau) \leqslant 0$, then either $B_d(\tau) = 0$ for all τ sufficiently large or there exists a τ_0 such that for all $\tau \geqslant \tau_0$, $B_d(\tau) \neq 0$. In the latter case we obtain from (8.2.27) that $B_d^{-1}(\tau) \geqslant G(\tau)$ for all $\tau \geqslant \tau_0$, where $B_d^{-1}(\tau_0) = G(\tau_0)$ and $G'(\tau) = \frac{1}{2}$ for all $\tau \geqslant \tau_0$. Thus

$$G(\tau) = \tfrac{1}{2}\tau - \tfrac{1}{2}(\tau_0 - 2B_d'(\tau_0)), \qquad \text{all } \tau \geqslant \tau_0. \tag{8.2.28}$$

It follows that

$$\lim_{\tau\to+\infty} B_d^{-1}(\tau) = \infty, \quad \text{or} \quad \lim_{\tau\to\infty} B_d(\tau) = 0. \tag{8.2.29}$$

Similarly, we can show that

$$\lim_{\tau\to-\infty} B_d(\tau) = 0. \tag{8.2.30}$$

Since $B_d(\tau)$ is nonincreasing, we obtain from (8.2.29) and (8.2.30) that

$$B_d(\tau) = 0, \quad \text{for all } \tau \in \Omega. \tag{8.2.31}$$

Thus every estimator d which satisfies (8.2.25) is unbiased. Finally, since the exponential family with densities of the form (8.2.20) is complete (in the sense of Section 2.6), every unbiased estimator d of $\theta(\tau)$ is essentially equal to U, and its risk function is $R(\theta, d) = 1$ for all θ. Thus there is no estimator d which strictly dominates U. This completes the proof of the admissibility of U. The estimator U is minimax since it is admissible and has a constant risk function $R(\theta, U) = 1$ for all θ. (Q.E.D.)

The present theorem can be applied to prove the admissibility and minimaxity under quadratic loss of the sample mean when $X_1, \ldots, X_n$ are i.i.d. like $\mathcal{N}(\mu, 1)$.

Karlin [4] extended the results of Girshick and Savage for the exponential family and investigated admissibility when the family of distributions has densities of the form

(8.2.32)
$$f(x;\theta) = \begin{cases} q(\theta)r(x), & 0 \leqslant x \leqslant \theta, \\ 0, & \text{otherwise,} \end{cases}$$

and $\mu(\mathrm{d}x) = \mathrm{d}x$. These densities are called extremal densities, since for $r(x) = x^{n-1}$ and $q(\theta) = \theta^{-n}$, $f(x;\theta)$ is the density of the maximum of n i.i.d. random variables from a rectangular distribution over $[0, \theta]$. Karlin's paper also contains results concerning the admissibility of estimators of the location parameter when $f(x;\theta) = p(x - \theta)$. These results are discussed in Section 8.4.

The first extension in Karlin's results, concerning the exponential family of the form (8.2.20), is that all the estimators $d_\gamma(U) = \gamma U$, with $0 < \gamma \leqslant 1$, are admissible, provided Ω is the entire real line. Girshick and Savage's theorem (Theorem 8.2.2) pertains only to the case of $\gamma = 1$, in which we have proven admissibility. If Ω is not the entire real line this result is invalid, as shown in Example 8.3. In the case of restricted parameter space Karlin proved that if $\beta^{-\lambda}(\tau)$ is *not* integrable in the neighborhood of both boundaries of Ω, where $\beta^{-1}(\tau) = \int_{-\infty}^{\infty} e^{u\tau}\psi(\mathrm{d}u)$, then $u/(\lambda + 1)$ is an almost admissible estimator of $\theta(\tau)$. The method used by Karlin to prove almost admissibility under squared-error loss is the following. Given an estimator $d_0(X)$ of $g(\theta)$, $\theta \in \Theta$, its risk function is $R(\theta, d_0) = \int (d_0(x) - g(\theta))^2 F_\theta(\mathrm{d}x)$. We assume that $R(\theta, d) < \infty$ for all θ. Suppose that another estimator, say $d^*(X)$, yields a risk function $R(\theta, d^*)$ such that

(8.2.33)
$$R(\theta, d^*) \leqslant R(\theta, d_0) \quad \text{for all } \theta.$$

This inequality is equivalent to

(8.2.34)
$$\int [d^*(x) - d_0(x)]^2 F_\theta(\mathrm{d}x) \leqslant 2\int [d_0(x) - d^*(x)][d_0(x) - g(\theta)]F_\theta(\mathrm{d}x),$$

for all θ. Suppose that $\mathcal{F} = \{P_\theta;\, \theta \in \Theta\}$ is dominated by a sigma-finite measure $\mu(\mathrm{d}x)$ so that $F_\theta(\mathrm{d}x) = f(x;\theta)\mu(\mathrm{d}x)$, all $\theta \in \Theta$. If there exists a sigma-finite measure $\xi(\mathrm{d}\theta)$ on $(\Theta, \mathcal{C})$ such that

(i) $\xi(\mathrm{d}\theta) = h(\theta)\,\mathrm{d}\theta$, where $h(\theta) > 0$ for all θ,
(ii) $f_\xi(x) = \int f(x;\theta)\xi(\mathrm{d}\theta) < \infty$ a.s. $[\mu]$,
(iii) $\int |g(\theta)| f(x;\theta)\xi(\mathrm{d}\theta) < \infty$ a.s. $[\mu]$,
(iv) $d_0(x)f_\xi(x) = \int g(\theta)f(x;\theta)\xi(\mathrm{d}\theta)$ a.s. $[\mu]$,

then d_0 is almost admissible. Indeed, integrating both sides of (8.2.34) with respect to $\xi(d\theta)$ and interchanging integrals, we obtain

$$(8.2.35)\quad 0 \leqslant \int (d^*(x) - d_0(x))^2 f_\xi(x)\mu(dx)$$
$$\leqslant 2\int (d_0(x) - d^*(x))\left\{\int (d_0(x) - g(\theta))f(x;\theta)\xi(d\theta)\right\}\mu(dx)$$

According to Fubini's theorem, the interchange of integrals in (8.2.35) is legitimate if

$$\int |d_0(x) - d^*(x)|\left\{\int |d_0(x) - g(\theta)|\, f(x;\theta)\xi(d\theta)\right\}\mu(dx) < \infty.$$

This condition is satisfied according to (iii) and the finiteness of the risk function. Thus if (iv) holds, the right-hand side of (8.2.35) is zero and the equality sign must hold in (8.2.35). According to (i), a strict inequality can hold in (8.2.34) at most on a set of θ values of a Lebesgue measure zero. Thus d_0 is almost *admissible*. We show now how this method can be used to prove the following theorem.

Theorem 8.2.3. (Karlin [4].) *Let* $F_\tau(dx) = \beta(\tau)e^{x\tau}\,\psi(dx)$, *where* $\tau \in \Omega$; $\Omega = (\underline{\omega}, \bar{\omega})$ *is the interval of all* τ *values for which* $\int_{-\infty}^{\infty} e^{x\tau}\psi(dx) < \infty$. *Let* $\underline{\omega} < c < \bar{\omega}$. If

$$(8.2.36)\qquad \lim_{b\to\bar{\omega}} \int_c^b \beta^{-\lambda}(\tau)\,d\tau = +\infty$$

and

$$(8.2.37)\qquad \lim_{a\to\underline{\omega}} \int_a^c \beta^{-\lambda}(\tau)\,d\tau = +\infty,$$

then $(\lambda + 1)^{-1}X$ *is an almost admissible estimator of* $\theta(\tau) = E_\tau\{X\}$, *with a squared-error loss.*

Proof. Consider an estimator $d_\gamma(X) = \gamma X$. If $d(X)$ is any estimator of $\theta(\tau)$ whose risk does not exceed that of $d_\gamma(X)$ we should have, according to (8.2.34),

$$(8.2.38)\quad \int_{-\infty}^{\infty} [d(x) - \gamma x]^2\beta(\tau)e^{\tau x}\psi(dx)$$
$$\leqslant 2\int_{-\infty}^{\infty} [\gamma x - d(x)][\gamma x\beta(\tau) + \beta'(\tau)]e^{x\tau}\psi(dx).$$

Indeed, $\theta(\tau) = \omega'(\tau)/\omega(\tau) = \beta(\tau)\cdot(d/d\tau)\{\beta^{-1}(\tau)\} = -\beta'(\tau)/\beta(\tau)$. Let $\xi(d\tau) = \beta^\lambda(\tau)\,d\tau$, $\lambda \neq 1$, and let $a < b$ in Ω. Then, integrating both sides of (8.2.38), we obtain

$$\int_a^b \beta^\lambda(\tau)\cdot\left\{\int_{-\infty}^{\infty}[\gamma x-d(x)]^2\beta(\tau)e^{x\tau}\psi(\mathrm{d}x)\right\}\mathrm{d}\tau \tag{8.2.39}$$
$$\leqslant 2\int_a^b \beta^\lambda(\tau)\left\{\int_{-\infty}^{\infty}(\gamma x-d(x))(\gamma x\beta(\tau)+\beta'(\tau))e^{x\tau}\psi(\mathrm{d}x)\right\}\mathrm{d}\tau.$$

We notice that

$$\frac{d}{\mathrm{d}\tau}\left\{\frac{1}{\lambda+1}\beta^{\lambda+1}(\tau)e^{x\tau}\right\}=\beta^\lambda(\tau)\beta'(\tau)e^{x\tau}+\frac{x}{\lambda+1}\beta^{\lambda+1}(\tau)e^{x\tau}.$$

Hence

$$\int_a^b \beta^\lambda(\tau)\left\{\int_{-\infty}^{\infty}(\gamma x-d(x))(\gamma x\beta(\tau)+\beta'(\tau))e^{x\tau}\psi(\mathrm{d}x)\right\}\mathrm{d}\tau \tag{8.2.40}$$
$$=\int_c^b \mathrm{d}\tau\int_{-\infty}^{\infty}(\gamma x-d(x))\left[\frac{1}{\lambda+1}\cdot\frac{\mathrm{d}}{\mathrm{d}\tau}e^{\tau x}\beta^{\lambda+1}(\tau)\right.$$
$$\left.+\,x\left(\gamma-\frac{1}{\lambda+1}\right)e^{\tau x}\beta^{\lambda+1}(\tau)\right]e^{x\tau}\,\psi(\mathrm{d}x).$$

It is easy to verify that Fubini's theorem holds, and therefore

(8.2.41)

$$\int_a^b \beta^\lambda(\tau)D(\tau)\,\mathrm{d}\tau\leqslant 2\int_{-\infty}^{\infty}(\gamma x-d(x))\left[\frac{e^{bx}\beta^{\lambda+1}(b)}{\lambda+1}-\frac{e^{ax}\beta^{\lambda+1}(a)}{\lambda+1}\right]\psi(\mathrm{d}x)$$
$$+\,2\int_{-\infty}^{\infty}(\gamma x-\mathrm{d}(x))\left(\gamma x-\frac{x}{\lambda+1}\right)\left(\int_a^b\beta^{\lambda+1}(\tau)e^{\tau x}\,\mathrm{d}\tau\right)\psi(\mathrm{d}x),$$

where

$$D(\tau)=\int_{-\infty}^{\infty}(\gamma x-d(x))^2e^{\tau x}\beta(\tau)\psi(\mathrm{d}x). \tag{8.2.42}$$

In particular, for $\gamma=1/(\lambda+1)$ the second term on the right-hand side of (8.2.41) vanishes. The Schwartz inequality yields

$$\int_{-\infty}^{\infty}(\gamma x-d(x))\left[\frac{e^{bx}\beta^{\lambda+1}(b)}{\lambda+1}-\frac{e^{ax}\beta^{\lambda+1}(a)}{\lambda+1}\right]\psi(\mathrm{d}x) \tag{8.2.43}$$
$$\leqslant\frac{1}{\lambda+1}\sqrt{\beta^\lambda(b)}\cdot\sqrt{D(b)\beta^\lambda(b)}+\frac{1}{\lambda+1}\sqrt{\beta^\lambda(a)}\cdot\sqrt{D(a)\beta^\lambda(a)}.$$

Hence from (8.2.42) and (8.2.43) we obtain

$$\int_a^b\beta^\lambda(\tau)D(\tau)\,\mathrm{d}\tau \tag{8.2.44}$$
$$\leqslant\frac{2}{\lambda+1}\sqrt{\beta^\lambda(b)}\cdot\sqrt{D(b)\beta^\lambda(b)}+\frac{2}{\lambda+1}\sqrt{\beta^\lambda(a)}\sqrt{D(a)\beta^\lambda(a)}.$$

We now investigate what happens if (8.2.36) and (8.2.37) hold.

Case 1.

$$\Delta = \varlimsup_{b\to\bar{\omega}} \beta^{\lambda}(b)\sqrt{D(b)} > 0.$$

Fix 'a' and let $H(b) = \int_a^b \beta^{\lambda}(\tau)D(\tau)\,\mathrm{d}\tau$; $H'(b) = \beta^{\lambda}(b)D(b)$. According to (8.2.44) there exists an appropriate constant $C > 0$ such that

$$H(b) \leqslant C\sqrt{\beta^{\lambda}(b)}\,\sqrt{H'(b)}, \tag{8.2.45}$$

or

$$C^2 \frac{H'(b)}{H^2(b)} \geqslant \beta^{-\lambda}(b).$$

which after integration yields

$$C^2\left[\frac{1}{H(b_1)} - \frac{1}{H(b_2)}\right] \geqslant \int_{b_1}^{b_2} \beta^{-\lambda}(b)\,\mathrm{d}b, \tag{8.2.46}$$

where $b_1 < b_2$ and $H(b_1) > 0$. As $b_2 \to \bar{\omega}_1$, $\int_{b_1}^{b_2} \beta^{-\lambda}(b)\,\mathrm{d}b \to \infty$, but the left-hand side of (8.2.46) remains bounded, which is impossible. Thus Case 1 cannot happen.

Case 2.

$$\varlimsup_{b\to\bar{\omega}} \beta^{\lambda}(b)\sqrt{D(b)} = 0.$$

Let

$$\rho(a) = \int_a^{\bar{\omega}} \beta^{\lambda}(\tau)D(\tau)\,\mathrm{d}\tau. \tag{8.2.47}$$

According to (8.2.44) and the assumption of Case 2,

$$\rho(a) = \int_a^{\bar{\omega}} \beta^{\lambda}(\tau)D(\tau)\,\mathrm{d}\tau \leqslant \frac{2}{\lambda+1}\sqrt{\beta^{\lambda}(a)}\,\sqrt{-\rho'(a)}. \tag{8.2.48}$$

From this inequality we obtain

$$-\frac{\rho'(a)}{\rho(a)} = \frac{d}{\mathrm{d}a}\left\{\frac{1}{\rho(a)}\right\} \geqslant \left(\frac{\lambda+1}{2}\right)^2 \beta^{-\lambda}(a). \tag{8.2.49}$$

Integration of (8.2.49) from a_1 to a_0, $a_1 < a_0$ and $\rho(a_0) > 0$, yields

$$\left(\frac{2}{\lambda+1}\right)^2\left[\frac{1}{\rho(a_0)} - \frac{1}{\rho(a_1)}\right] \geqslant \int_{a_1}^{a_0} \beta^{-\lambda}(a)\,\mathrm{d}a. \tag{8.2.50}$$

The right-hand side of (8.2.50) approaches ∞ as $a_1 \to \underline{\omega}$ and the left-hand side is bounded, which is impossible. Since $\beta^{\lambda}(b)\sqrt{D(b)} \geqslant 0$, the assumption

$\lim_{b \to \bar{\omega}} \beta^{\lambda}(b)\sqrt{D(b)} = 0$ must hold. Hence there exists no a_0 for which $\rho(a_0) > 0$. This proves that $d(X) = (\lambda + 1)^{-1}X$ a.s. for almost all θ; Or that $X/(\lambda + 1)$ is an almost admissible estimator. (Q.E.D.)

Example 8.5. (i) If $\Omega = (-\infty, \infty)$ and $\psi(x)$ possesses a positive measure in each of the intervals $(-\infty, 0)$ and $(0, \infty)$, then

$$\beta(\tau) = \left[\int_{-\infty}^{\infty} e^{\tau x}\psi(dx)\right]^{-1} \to 0, \qquad \text{as } |\tau| \to \infty.$$

Consequently, for each $\lambda \geqslant 0$

$$\int_c^b \beta^{-\lambda}(\tau)\, d\tau \to \infty \qquad \text{as } b \to \infty$$

$$\int_a^c \beta^{-\lambda}(\tau)\, d\tau \to \infty \qquad \text{as } a \to -\infty.$$

Let $\gamma = (1 + \lambda)^{-1}$. If $\lambda \geqslant 0$ then $0 < \gamma \leqslant 1$. Hence γX is an almost admissible estimator of $\theta(\tau)$ whenever $0 < \gamma \leqslant 1$.

(ii) If $\Omega = (-\infty, \infty)$ but no further properties of $\psi(x)$ are given, then at least X is almost admissible, since for $\lambda = 0$ both (8.2.36) and (8.2.37) are satisfied.

(iii) Consider the family of gamma distributions with densities

$$f(x; \theta) = \frac{1}{\theta^{\nu}\Gamma(\nu)}\, x^{\nu-1}e^{-x/\theta}, \qquad 0 \leqslant x \leqslant \infty, 0 < \theta < \infty.$$

Letting $\omega = -1/\theta$, we obtain the 1-parameter exponential family

$$f(x; \omega)\, dx = (-\omega)^{\nu}e^{x\omega}\psi(dx), \qquad 0 \leqslant x \leqslant \infty, -\infty < \omega < 0,$$

where $\psi(x) = x^{\nu-1}/\Gamma(\nu)$. The estimator X/ν is an unbiased estimator of $\theta(\omega) = -1/\omega$.

Let $0 < a < c < \infty$; then since $\beta(\omega) = (-\omega)^{\nu}$ we obtain

$$\int_{-a}^{-c} (-\omega)^{\nu\lambda}\, d\omega = \int_a^c \omega^{-\lambda\nu}\, d\omega.$$

The only value of λ for which both (8.2.36) and (8.2.37) are satisfied as $c \to \infty$, $a \to 0$, is $\lambda = 1/\nu$. Hence an almost admissible estimator of θ of the form λX is $(1 + \nu^{-1})^{-1}X/\nu = (1 + \nu)^{-1}X$. In the case of $Y_1, \ldots, Y_n$ i.i.d. like $N(0, \sigma^2)$ we proved that the unique admissible estimator of σ^2 is $(n + 2)^{-1}\sum_{i=1}^n Y_i^2$. Since $\sum_{i=1}^n Y_i^2 \sim \sigma^2\chi^2[n]$, we have to substitute above $\theta = 2\sigma^2$, $\nu = n/2$, and $X = \frac{1}{2}\sum_{i=1}^n Y_i^2$. ■

8.3. ADMISSIBILITY AND MINIMAXITY IN TRUNCATED SPACES UNDER QUADRATIC LOSS

In the previous section we demonstrated a case in which the unique admissible estimator of θ becomes inadmissible when the parameter space Θ is truncated. In this section we extend the previous results on admissibility in exponential families and quadratic loss to cases with truncated parameter spaces. This extension was provided by Katz [1].

Consider the exponential family of density functions $f(x; \theta) = \beta(\theta)e^{x\theta}$, $-\infty < x < \infty$, $\theta \in \Theta$, with respect to the sigma-finite measure $\mu(\mathrm{d}x)$. The natural parameter space is Ω, that is,

$$\Omega = \left\{\theta;\ \int_{-\infty}^{\infty} e^{x\theta}\mu(\mathrm{d}x) < \infty\right\}.$$

The actual parameter space Θ is a subinterval of Ω. We assume here that

$$\Theta = \{\theta;\ \theta \geqslant \theta_0,\ \theta \in \Omega\}.$$

We are concerned with the estimation of $g(\theta) = -\beta'(\theta)/\beta(\theta)$ under quadratic loss; $g(\theta)$ is the expectation of X. The variance of X is $g'(\theta)$. Thus $g'(\theta) > 0$ for all θ, and therefore $g(\theta)$ is a monotone increasing function of θ. If $\theta > \theta_0$, then $g(\theta) > g(\theta_0)$. We define now, for each real x,

$$H(x) = \int_0^{\infty} \beta(\theta)e^{x\theta}\, \mathrm{d}\theta. \tag{8.3.1}$$

The function $H(x)$ may assume the value ∞. Since

$$H''(x) = \int_0^{\infty} \beta(\theta)\theta^2 e^{\theta x}\, \mathrm{d}\theta > 0,$$

$H(x)$ is a nondecreasing convex function, and $I = \{x;\ H(x) < \infty\}$ is an interval with left-hand limit $-\infty$ and right-hand limit b; $H(b)$ may be ∞.

We assume now without loss of generality that $\theta_0 = 0$; that is, $\Theta = [0, \infty)$. Let $\xi_\sigma(\theta)$ be a prior distribution on Θ such that

$$\xi_\sigma(\mathrm{d}\theta) = \begin{cases} 1/\sigma \exp\{-\theta/\sigma\}, & \theta \geqslant 0 \\ 0, & \text{otherwise.} \end{cases}$$

The Bayes estimator of $g(\theta)$, for squared-error loss, is

$$\hat{g}_\sigma(X) = \frac{\int_0^{\infty} g(\theta)\beta(\theta)\exp\{\theta X - \theta/\sigma\}\, \mathrm{d}\theta}{\int_0^{\infty} \beta(\theta)\exp\{\theta X - \theta/\sigma\}\, \mathrm{d}\theta}. \tag{8.3.2}$$

But since $g(\theta) = -\beta'(\theta)/\beta(\theta)$ we obtain

$$\hat{g}_\sigma(X) = -\int_0^\infty \beta'(\theta) \exp\left\{\theta\left(X - \frac{1}{\sigma}\right)\right\} d\theta \Big/ H\left(X - \frac{1}{\sigma}\right). \tag{8.3.3}$$

We further notice that, for each $x < \bar{b}$, where $\bar{b} = \sup\{x; H(x) < \infty\}$, $\bar{b} \leqslant b$,

$$\lim_{\theta \to \infty} \beta(\theta) e^{x\theta} = 0. \tag{8.3.4}$$

Indeed, if $\overline{\lim_{\theta \to \infty}} \beta(\theta) e^{x\theta} > 0$ there exists a positive number δ and $\theta_0(\delta)$ such that $\beta(\theta) e^{x\theta} \geqslant \delta$ for all $\theta \geqslant \theta_0(\delta)$. Moreover, since $x < b$, there exists an $\epsilon > 0$ so that $x + \epsilon \leqslant b$. Then

$$\infty > H(x + \epsilon) = \int_0^\infty \beta(\theta) e^{\theta(x+\epsilon)}\, d\theta \geqslant \delta \int_0^\infty e^{\theta\epsilon}\, d\theta = \infty. \tag{8.3.5}$$

This is a contradiction. Hence (8.3.4) holds. Integration in parts and (8.3.4) reduce (8.3.3) to

$$\hat{g}_\sigma(X) = X - \frac{1}{\sigma} + \frac{\beta(0)}{H(X - 1/\sigma)} \tag{8.3.6}$$

Let

$$\hat{g}(X) = \lim_{\sigma \to \infty} \hat{g}_\sigma(X) = X + \frac{\beta(0)}{H(X)}. \tag{8.3.7}$$

We now show that $\hat{g}(X)$ is an admissible estimator of $g(\theta)$.

The risk function of the Bayes estimator $\hat{g}_\sigma(X)$ is

(8.3.8)

$$\begin{aligned} R(\theta, \hat{g}_\sigma) &= E_\theta\{[\hat{g}_\sigma(X) - g(\theta)]^2\} \\ &= g'(\theta) + 2\beta(0) E_\theta\left\{\frac{X - g(\theta)}{H(X - 1/\sigma)}\right\} + E_\theta\left\{\left[\frac{\beta(0)}{H(X - 1/\sigma)} - \frac{1}{\sigma}\right]^2\right\}. \end{aligned}$$

The prior risk of $\hat{g}_\sigma$ is

$$\begin{aligned} \rho(\sigma, \hat{g}_\sigma) &= \frac{1}{\sigma} \int_0^\infty e^{-\theta/\sigma} R(\theta, \hat{g}_\sigma)\, d\theta \\ &= \frac{1}{\sigma} \int_0^\infty g'(\theta) e^{-\theta/\sigma}\, d\theta - \frac{\beta^2(0)}{\sigma} \\ &\quad \times \int_{-\infty}^\infty \frac{1}{H(x - 1/\sigma)}\, \mu(dx) + \frac{1}{\sigma^2}. \end{aligned} \tag{8.3.9}$$

The prior risk of $\hat{g}(X)$ is

$$\text{(8.3.10)}\quad \rho(\sigma, \hat{g}) + \frac{1}{\sigma}\int_0^\infty g'(\theta)e^{-\theta/\sigma}\,d\theta + \frac{\beta^2(0)}{\sigma}\int_{-\infty}^{\infty}\frac{H(x-1/\sigma)}{H^2(x)}\mu(dx) - \frac{2\beta^2(0)}{\sigma}\int_{-\infty}^{\infty}\frac{1}{H(x)}\mu(dx) + 2\frac{\beta(0)}{\sigma}\int_{-\infty}^{\infty}\frac{H(x-1/\sigma)}{H(x)}\mu(dx).$$

If $\hat{g}(X)$ is inadmissible there exists another estimator $g^*(x)$, for which

$$\text{(8.3.11)}\qquad R(\theta, g^*) \leqslant R(\theta, \hat{g}) \quad \text{for all } \theta \in \Theta,$$

with strict inequality for some $\theta \in \Theta$. The function $R(\theta, g)$ is continuous in θ for each g. Let (a, b) be an interval in Θ over which

$$\text{(8.3.12)}\qquad R(\theta, g^*) \leqslant R(\theta, \hat{g}) - \epsilon \quad \text{for some } \epsilon > 0.$$

Consider the ratio

$$\text{(8.3.13)}\qquad K(\sigma; \hat{g}, g^*) = \frac{\rho(\sigma, \hat{g}) - \rho(\sigma, g^*)}{\rho(\sigma, \hat{g}) - \rho(\sigma, \hat{g}_\sigma)},$$

where $\rho(\sigma, \hat{g})$ and $\rho(\sigma, g^*)$ are the prior risks of $\hat{g}$ and g^* with respect to $\xi_\sigma(d\theta)$. Since $\hat{g}_\sigma$ is Bayes against $\xi_\sigma(d\theta)$, $\rho(\sigma, \hat{g}) > \rho(\hat{\sigma}, \hat{g}_\sigma)$. We verify now that for sufficiently large σ, $K(\sigma; \hat{g}, g^*) > 1$. This implies that $\rho(\sigma, \hat{g}_\sigma) > \rho(\sigma, g^*)$, a result which contradicts the assumption that $\hat{g}_\sigma$ is Bayes. The function $H(x - 1/\sigma)$ is an increasing function of σ. Hence

(8.3.14)

$$\rho(\sigma, \hat{g}) - \rho(\sigma, \hat{g}_\sigma) \leqslant \frac{\beta^2(0)}{\sigma}\int_{-\infty}^{\infty}\left[\frac{1}{H(x-1/\sigma)} - \frac{1}{H(x)}\right]\mu(dx) + \frac{1}{\sigma^2}.$$

By the Lebesgue monotone convergence theorem,

$$\text{(8.3.15)}\qquad \int_{-\infty}^{\infty}\left[\frac{1}{H(x-1/\sigma)} - \frac{1}{H(x)}\right]\mu(dx) \to 0 \qquad \text{as } \sigma \to \infty.$$

Since we assumed that $R(\theta, g^*) \leqslant R(\theta, \hat{g}) - \epsilon$ on (a, b), we have

$$\text{(8.3.16)}\qquad \rho(\sigma, \hat{g}) - \rho(\sigma, g^*) > \frac{\epsilon}{\sigma}\int_a^b e^{-\theta/\sigma}\,d\theta = \epsilon(e^{-a/\sigma} - e^{-b/\sigma}).$$

Hence from (8.3.15) and (8.3.16),

$$K(\sigma; \hat{g}, g^*) > \frac{\epsilon(b - a + O(\sigma^{-1}))}{\int_{-\infty}^{\infty}\left[\frac{1}{H(x-1/\sigma)} - \frac{1}{H(x)}\right]\mu(dx)} \to \infty \qquad \text{as } \sigma \to \infty.$$

(Q.E.D.)

We notice that in the general case, when $\theta_0 \neq 0$, $H(x)$ is defined as $\int_{\theta_0}^{\infty} \beta(\theta)e^{x\theta}\, d\theta$, and we have the following theorem.

Theorem 8.3.1. (Katz [1].) *If $\Theta = \{\theta; \theta \geqslant \theta_0\}$, then an admissible estimator of $g(\theta) = E_\theta\{X\}$ is*

$$\hat{g}(X) = X + \frac{\beta(\theta_0)e^{\theta_0 X}}{\int_\theta^\infty \beta(\tau)e^{\tau X}\, d\tau}. \tag{8.3.17}$$

Example 8.6. Let X have a binomial distribution $B(p, n)$, $p \geqslant p_0$. Let $\theta = \ln p/(1-p)$. The function θ is an increasing function of p, and we have the truncated space $\Theta = \{\theta; \theta \geqslant \theta_0\}$, where $\theta_0 = \ln p_0/(1-p_0)$. Write $\beta(\theta) = e^{-n\theta}/(1+e^{-\theta})^n$. Then the density function of X can be written as

$$f(x; \theta) = \binom{n}{x}\beta(\theta)e^{\theta x}, \qquad x = 0, 1, \ldots, n.$$

Then according to (8.3.17), we compute

$$\begin{aligned}\int_{\theta_0}^{\infty} \beta(\theta)e^{\theta x}\, d\theta &= \int_{\theta_0}^{\infty} \left(\frac{e^{-\theta}}{1+e^{-\theta}}\right)^{n-x}\left(\frac{1}{1+e^{-\theta}}\right)^{x} d\theta \\ &= \int_{p_0}^{1} p^{x-1}(1-p)^{n-x-1}\, dp.\end{aligned} \tag{8.3.18}$$

An admissible estimator of $g(\theta) = p$ is

$$\hat{p}(X) = \frac{X}{n} + \frac{p_0^{X}(1-p_0)^{n-X}}{n\int_{p_0}^{1} p^{X-1}(1-p)^{n-X-1}\, dp}. \tag{8.3.19}$$

The M.L.E., X/n, is admissible when $p_0 = 0$.

Another theorem of Katz extends the theorem of Karlin (Theorem 8.2.3) to the case of a truncated sample space.

Theorem 8.3.2. (Katz [1].) *Let $f(x; \theta)$ be a density function with respect to $\mu(dx)$. Let Θ be the interval $[a, \bar{\theta}]$ and let $\hat{\theta}$ be an estimator of $g(\theta)$ with a bounded risk function. Let $Q(\theta)$ be a positive measurable function on Θ, satisfying*

$$\infty > \int_a^b \frac{1}{Q(\theta)}\, d\theta \to \infty \qquad \text{as } b \to \bar{\theta}. \tag{8.3.20}$$

If there exists a $K > 0$ such that

$$\left|\int_a^b [\hat{\theta}(x) - g(\theta)]Q(\theta)f(x; \theta)\, d\theta\right| \leqslant KQ(b)f(x; b), \tag{8.3.21}$$

for all $b \in (a, \bar{\theta})$ and all x, then $\hat{\theta}$ is almost admissible.

Proof. Let $\theta^*(X)$ be an arbitrary estimator which is at least as good as $\hat{\theta}(X)$. Then as in (8.2.35),

$$(8.3.22)\quad \begin{aligned} \int_a^b D(\theta)Q(\theta)\,\mathrm{d}\theta &\leqslant 2\int_a^b Q(\theta)\int_{-\infty}^{\infty}[\hat{\theta}(x)-\theta^*(x)][\hat{\theta}(x)-g(\theta)]f(x;\theta)\mu(\mathrm{d}x)\,\mathrm{d}\theta \\ &= 2\int_{-\infty}^{\infty}(\hat{\theta}(x)-\theta^*(x))\left\{\int_a^b Q(\theta)[\hat{\theta}(x)-g(\theta)]f(x;\theta)\,\mathrm{d}\theta\right\}\mu(\mathrm{d}x) \\ &\leqslant 2KQ(b)\sqrt{D(b)}, \end{aligned}$$

where $D(\theta)=\int_a^b [\hat{\theta}(x)-\theta^*(x)]^2 f(x;\theta)\mu(\mathrm{d}x)$. Let $\rho(b)=\int_a^b D(\theta)Q(\theta)\,\mathrm{d}\theta$. We show now that $\rho(b)=0$ for *all* $b\in(a,\bar{\theta})$. This will prove the theorem. By negation, if there exists a $c\in(a,\bar{\theta})$ such that $\rho(c)>0$ then, for every b in $[c,\bar{\theta})$, we have, from (8.3.22),

$$(8.3.23)\qquad \frac{1}{4K^2Q(b)}\leqslant\frac{1}{\rho^2(b)}Q(b)D(b),\qquad \text{all } c\leqslant b\leqslant\bar{\theta}.$$

Integrating both sides of the inequality we obtain

$$(8.3.24)\qquad \frac{1}{4K^2}\int_c^B\frac{\mathrm{d}b}{Q(b)}\leqslant\int_c^B\frac{Q(b)D(b)}{\rho^2(b)}\,\mathrm{d}b,\qquad c\leqslant B\leqslant\bar{\theta}.$$

According to (8.3.20), the left-hand side of (8.3.24) approaches to ∞ as $B\to\bar{\theta}$. On the other hand, the right-hand side of (8.3.24) is bounded. This leads to a contradiction. Indeed, if

$$(8.3.25)\qquad M(B)=\int_c^B\frac{Q(b)D(b)}{\rho^2(b)}\,\mathrm{d}b+\frac{1}{\rho(B)},\qquad \text{for } c\leqslant B\leqslant\bar{\theta},$$

we obtain by differentiating $M(B)$ with respect to B that, for almost all $B\in[c,\bar{\theta})$,

$$(8.3.26)\qquad M'(B)=\frac{Q(B)D(B)}{\rho^2(B)}-\frac{\rho'(B)}{\rho^2(B)}.$$

But

$$\rho'(B)=\frac{d}{\mathrm{d}B}\int_a^B Q(b)D(b)\,\mathrm{d}b=Q(B)D(B).$$

Hence $M'(B)=0$ for almost all $B\in[c,\bar{\theta})$. Also $M(B)$ is absolutely continuous on each interval $[c,d]$, $c<d<\bar{\theta}$. Hence $M(B)$ is a constant on $[c,\bar{\theta})$. This implies that the right-hand side of (8.3.24) is

$$\int_c^B\frac{Q(b)D(b)}{\rho^2(b)}\,\mathrm{d}b=\frac{1}{\rho(c)}-\frac{1}{\rho(B)},$$

which remains bounded as $B\to\bar{\theta}$, since we assumed that $\rho(B)>0$ for all $B\geqslant c$. (Q.E.D.)

Example 8.7. Let $X \sim N(\theta, 1)$. It is known that $\theta \geqslant \theta_0$. We write $f(x; \theta) = \beta(\theta)e^{x\theta}$, with $\beta(\theta) = e^{-\frac{1}{2}\theta^2}$; $f(x; \theta)$ is the density of $N(\theta, 1)$ with respect to the standard normal density $\varphi(x) = (2\pi)^{-\frac{1}{2}}e^{-\frac{1}{2}x^2}$. Let $Q(\theta) = e^{-\frac{1}{2}\lambda\theta^2}$ for an arbitrary positive λ.

$$\int_{\theta_0}^{b} \frac{1}{Q(\theta)}\, \mathrm{d}\theta = \int_{\theta_0}^{b} e^{\frac{1}{2}\lambda\theta^2}\, \mathrm{d}\theta \to \infty \qquad \text{as } b \to \infty.$$

Hence (8.3.20) is satisfied. We can then show that the estimator

$$(8.3.27) \quad \hat{\theta}(X) = \theta_0 + \frac{X - \theta_0}{1 + \lambda} + \frac{1}{(1 + \lambda)^{\frac{1}{2}}} \cdot \frac{\varphi((X - \theta_0)/(1 + \lambda)^{\frac{1}{2}})}{\Phi((X - \theta_0)/(1 + \lambda)^{\frac{1}{2}})}$$

is almost admissible. ■

8.4. THE ADMISSIBILITY OF THE PITMAN ESTIMATOR OF A SINGLE LOCATION PARAMETER

In Section 7.4 we showed that for the squared-error loss the Pitman estimator of the location parameter is the (essentially unique) minimum risk translation equivariant estimator. We proved also that the Pitman estimator is minimax. We investigate here whether the Pitman estimator is admissible for quadratic loss. The results are somewhat surprising. If the distribution of X depends on at most two location parameters (when θ is a real or two-dimensional vector), the Pitman estimator is admissible. If the dimensionality of the observed vector X is *at least* three, the Pitman estimator is generally inadmissible. In this section we prove the admissibility under mild conditions in the case of a single location parameter. The proof which is presented here follows that of Stein [7]. The reader is referred to Stein and James's article [1] for a proof of the admissibility of Pitman's estimator in the 2-dimensional case. We remark also that Karlin [4] furnished another proof of the admissibility of the Pitman estimator in the case of a single location parameter.

Let $X_1, \ldots, X_n$ be i.i.d. random variables having a common distribution $F(x - \theta)$, where θ is a real location parameter. We assume here that Θ is the whole real line. Let $f(x - \theta)$ designate the density function of $F(x - \theta)$, with respect to a sigma-finite measure $\mu(\mathrm{d}x)$ on $(\mathfrak{X}, \mathcal{B})$. We have shown that among all estimators that are translation equivariant, the Pitman estimator

$$\hat{\theta}(X_1, \ldots, X_n) = X_1 - E\{X_1 \mid X_2 - X_1, \ldots, X_n - X_1\}$$

minimizes the risk function for the squared-error loss uniformly in θ. Make the transformation

$$(8.4.1) \quad \begin{aligned} X &= X_1 - E\{X_1 \mid Y_1, \ldots, Y_{n-1}\}, \\ Y_i &= X_{i+1} - X_1, \qquad i = 1, \ldots, n - 1. \end{aligned}$$

The conditional distribution of X, given $Y = (Y_1, \ldots, Y_{n-1})$, is assumed to be absolutely continuous with density

$$p(x - \theta \mid \mathbf{y}) = \frac{1}{q(\mathbf{y})} f(x + e(\mathbf{y}) - \theta) \prod_{i=1}^{n-1} f(y_i + e(\mathbf{y}) + x - \theta) \tag{8.4.2}$$

where

$$q(\mathbf{y}) = \int_{-\infty}^{\infty} f(x_1 - \theta) \prod_{i=1}^{n-1} f(y_i + x_1 - \theta)\, dx_1.$$

Thus if $(X, \mathbf{Y})$ is a vector of random variables such that the conditional distribution of $X - \theta$, given $\mathbf{Y}$, is $p(x \mid \mathbf{y})$ then $E_\theta\{X\} = \theta$ for all θ; X is the Pitman estimator, and

$$\begin{cases} \displaystyle\int_{-\infty}^{\infty} p(x \mid \mathbf{y})\, dx = 1 \quad \text{a.s.} \\ \displaystyle\int_{-\infty}^{\infty} x p(x \mid \mathbf{y})\, dx = 0 \quad \text{a.s.} \end{cases} \tag{8.4.3}$$

We then have the following theorem.

Theorem 8.4.1. (Stein [7].) *If the conditional distribution of $X - \theta$ given $\mathbf{Y}$ satisfies* (8.4.3), *and if in addition*

$$\int q(\mathbf{y})\nu(d\mathbf{y}) \left[\int x^2 p(x \mid \mathbf{y})\, dx \right]^{3/2} < \infty, \tag{8.4.4}$$

then X is an admissible estimator of θ for the squared-error loss.

Proof. Suppose that $\phi(X, \mathbf{Y})$ is any estimator satisfying, for all θ,

$$\begin{aligned} \int Q(d\mathbf{y}) \int [\phi(x, \mathbf{y}) - \theta]^2 p(x - \theta \mid \mathbf{y})\, dx \\ \leqslant \int Q(d\mathbf{y}) \int (x - \theta)^2 p(x - \theta \mid \mathbf{y})\, dx \\ = \int Q(d\mathbf{y}) \int x^2 p(x \mid \mathbf{y})\, dx, \end{aligned} \tag{8.4.5}$$

where $Q(d\mathbf{y}) = q(\mathbf{y})\nu(d\mathbf{y})$. We prove first that the two sides of (8.4.5) are equal for almost all θ. This will establish the almost admissibility of X. From this we will be able to deduce the admissibility of X.

Suppose that X is not almost admissible. Then strict inequality holds in (8.4.5) on a set of θ values having a positive Lebesgue measure. The risk function $R(\theta, \phi) = \int Q(d\mathbf{y}) \int [\phi(x, \mathbf{y}) - \theta]^2 p(x - \theta \mid \mathbf{y})\, dx$ is a continuous function of θ. Hence if θ_1 is a point at which $R(\theta_1, \phi) < R(\theta_1, \hat{\theta})$, where $\hat{\theta} = X$ a.s., then for any $\epsilon > 0$ there exists an interval containing θ_1, say S_ϵ, on which $R(\theta, \phi) \leqslant R(\theta, \hat{\theta}) - \epsilon$. The Lebesgue measure of S_ϵ is positive. Assign θ a prior distribution with density $(1/\sigma)h(\theta/\sigma)$, where $h(\theta)$ is the Cauchy density, that is, $h(\theta) = (1/\pi)(1 + \theta^2)^{-1}$, $-\infty < \theta < \infty$. Hence the prior risk of ϕ satisfies, for any σ,

$$\text{(8.4.6)}\quad \rho(\sigma, \phi) = \frac{1}{\sigma}\int h\left(\frac{\theta}{\sigma}\right)\left\{\int Q(\mathrm{d}\mathbf{y})\int [\phi(x, \mathbf{y}) - \theta]^2 p(x - \theta \mid \mathbf{y})\, \mathrm{d}x\right\} \mathrm{d}\theta$$
$$\leqslant \int Q(\mathrm{d}\mathbf{y})\int x^2 p(x \mid \mathbf{y})\, \mathrm{d}x - \epsilon\xi_\sigma(S_\epsilon).$$

We will establish that

$$\text{(8.4.7)}\quad \inf_{\phi} \rho(\sigma, \phi) \geqslant \int Q(\mathrm{d}\mathbf{y})\int x^2 p(x \mid \mathbf{y})\, \mathrm{d}x - \frac{f(\sigma)}{\sigma},$$

where $\lim_{\sigma\to\infty} f(\sigma) = 0$. We can show for the Cauchy prior distribution with density $(1/\sigma)h(\theta/\sigma)$ that $1/\sigma \int_{S_\epsilon} h(\theta/\sigma)\, \mathrm{d}\theta = O(\sigma^{-1})$ as $\sigma \to \infty$. But since $(f(\sigma))/\sigma = o(\sigma^{-1})$ as $\sigma \to \infty$, there exists a σ_0, sufficiently large, for which (8.4.7) contradicts (8.4.6). This will prove the almost admissibility of $\hat{\theta} = X$ a.s.

To establish (8.4.7), we derive first the formula

$$\text{(8.4.8)}\quad \int Q(\mathrm{d}\mathbf{y})\int x^2 p(x \mid \mathbf{y})\, \mathrm{d}x - \inf_{\phi} E_\sigma\{R(\theta, \phi)\}$$
$$= \frac{1}{\sigma}\int Q(\mathrm{d}\mathbf{y})\int \frac{[\int \eta h((x - \eta)/\sigma)p(\eta \mid \mathbf{y})\, \mathrm{d}\eta]^2}{\int h((x - \eta)/\sigma)p(\eta \mid \mathbf{y})\, \mathrm{d}\eta}\, \mathrm{d}x.$$

Indeed,

$$\text{(8.4.9)}\quad \inf_{\phi} E_\sigma\{R(\theta, \phi)\} = \inf_{\phi} E_\sigma E_\theta\{[\phi(X, \mathbf{Y}) - \theta]^2\}$$
$$= \inf_{\phi} E\{E_\sigma\{[\phi(X, \mathbf{Y}) - \theta]^2 \mid X, \mathbf{Y}\}\}$$
$$= E\{[E\{\theta \mid X, \mathbf{Y}\} - \theta]^2\}.$$

Hence

$$\text{(8.4.10)}\quad \int Q(\mathrm{d}\mathbf{y})\int x^2 p(x \mid \mathbf{y})\, \mathrm{d}x - \inf_{\phi} E_\sigma\{(\phi(X, \mathbf{Y}) - \theta)^2\}$$
$$= E_\sigma\{(X - \theta)^2\} - E_\sigma\{(E\{\theta \mid X, \mathbf{Y}\} - \theta)^2\}$$
$$= E_\sigma\{(X - E(\theta \mid X, \mathbf{Y}))^2\}.$$

Moreover,

$$\text{(8.4.11)}\quad E_\sigma\{(X - E\{\theta \mid X, \mathbf{Y}\})^2\} = \frac{1}{\sigma}\int h\left(\frac{\theta}{\sigma}\right) \mathrm{d}\theta \int Q(\mathrm{d}\mathbf{y})$$
$$\times \int \left[x - \frac{\int \tau h(\tau/\sigma)p(x - \tau \mid \mathbf{y})\, \mathrm{d}\tau}{\int h(\tau/\sigma)p(x - \tau \mid \mathbf{y})\, \mathrm{d}\tau}\right]^2 \cdot p(x - \theta \mid \mathbf{y})\, \mathrm{d}x.$$

But

$$x - \frac{\int \tau h(\tau/\sigma)p(x-\tau \mid \mathbf{y})\,d\tau}{\int h(\tau/\sigma)p(x-\tau \mid \mathbf{y})\,d\tau} = \frac{\int (x-\tau)h(\tau/\sigma)p(x-\tau \mid \mathbf{y})\,d\tau}{\int h(\tau/\sigma)p(x-\tau \mid \mathbf{y})\,d\tau} = \frac{\int \eta h((x-\eta)/\sigma)p(\eta \mid \mathbf{y})\,d\eta}{\int h((x-\eta)/\sigma)p(\eta \mid \mathbf{y})\,d\eta}. \tag{8.4.12}$$

Substituting (8.4.12) in (8.4.11) we obtain, after further manipulations, (8.4.8). Let

$$f(\sigma) = \int Q(d\mathbf{y}) \int dx \frac{[\int \eta h((x-\eta)/\sigma)p(\eta \mid \mathbf{y})\,d\eta]^2}{\int h((x-\eta)/\sigma)p(\eta \mid \mathbf{y})\,d\eta}. \tag{8.4.13}$$

Equation 8.4.8 can then be written as (8.4.7), and it remains to show that $\lim_{\sigma\to\infty} f(\sigma) = 0$. Let

$$I(P, \sigma) = \int dx \frac{[\int \eta h((x-\eta)/\sigma)P(d\eta)]^2}{\int h((x-\eta)/\sigma)P(d\eta)}, \tag{8.4.14}$$

where P is any distribution function on the real line. We notice that

$$I(P, \sigma) = \sigma^3 \int dx \frac{[\int \eta' h(x-\eta')\,dP(\sigma\eta')]^2}{\int h(x-\eta')\,dP(\sigma\eta')}. \tag{8.4.15}$$

Let Ψ_λ, $0 < \lambda < \infty$, be the family of all distribution functions satisfying

$$\begin{cases} \int \eta\, dP(\eta) = 0, \\ \int \eta^2\, dP(\eta) = \lambda. \end{cases} \tag{8.4.16}$$

Let $S(\lambda, \sigma) = \sup_{P\in\Psi_\lambda} I(P, \sigma)$. Then according to (8.4.15),

$$S(\lambda, \sigma) = \sigma^3 S\left(\frac{\lambda}{\sigma^2}, 1\right) = \sigma^3 S^*\left(\frac{\lambda}{\sigma^2}\right). \tag{8.4.17}$$

From (8.4.15) we immediately obtain that $I(P, 1) \leqslant \lambda$ for all $P \in \Psi_\lambda$. Hence $S^*(\lambda) \leqslant \lambda$. Furthermore, for all $P \in \Psi_\lambda$,

$$\int h(x-\eta)\,dP(\eta) \geqslant P[-1, 1] \inf_{\eta\in[-1,1]} h(x-\eta), \tag{8.4.18}$$

By the Tchebytchev inequality, $P[-1, 1] \geqslant 1 - \lambda^2$. Hence if $\lambda \leqslant \frac{1}{2}$ we obtain

$$\int h(x-\eta)\,dP(\eta) \geqslant \frac{3}{4}\cdot\frac{2}{5}\cdot\frac{1}{\pi(1+x^2)} = \frac{3}{10\pi(1+x^2)}. \tag{8.4.19}$$

Also since $\int \eta \, \mathrm{d}P(\eta) = 0$ for all $P \in \Psi_\lambda$, we obtain from the Schwartz inequality

$$(8.4.20)\quad \left(\int \eta h(x-\eta)\,\mathrm{d}P(\eta)\right)^2 = \left(\int \eta(h(x-\eta) - h(x))\,\mathrm{d}P(\eta)\right)^2 \leqslant \int \eta^2\,\mathrm{d}P(\eta)\cdot\int (h(x-\eta)-h(x))^2\,\mathrm{d}P(\eta),$$

Thus from (8.4.19) and (8.4.20) we derive the inequality, for each $P \in \Psi_\lambda$,

$$(8.4.21)\quad I(P, I) \geqslant \tfrac{10}{3}\lambda \int \mathrm{d}x(1+x^2)\int \left[\frac{1}{1+(x-\eta)^2} - \frac{1}{1+x^2}\right]^2 \mathrm{d}P(\eta) = \tfrac{10}{3}\lambda \int \mathrm{d}P(\eta)\int \mathrm{d}x\left[\frac{1}{1+(x-\eta)^2} - \frac{1}{1+x^2}\right]^2 (1+x^2).$$

Moreover,

$$(8.4.22)\quad \int \mathrm{d}x\left[\frac{1}{1+(x-\eta)^2} - \frac{1}{1+x^2}\right]^2 (1+x^2) = \eta^2 \int \frac{\mathrm{d}x}{(1+x^2)^2}.$$

Hence, substituting (8.4.22) in (8.4.21), we obtain

$$(8.4.23)\quad I(P, 1) \geqslant \tfrac{10}{3}\lambda^2 \int_{-\infty}^{\infty} \frac{\mathrm{d}x}{(1+x^2)^2} = c\lambda^2, \qquad \text{all } P \in \Psi_\lambda.$$

Hence

$$(8.4.24)\quad S^*(\lambda) \leqslant c\lambda^2 \quad \text{for} \quad \lambda \leqslant \tfrac{1}{2}.$$

From (8.4.24) and since $S^*(\lambda) \leqslant \lambda$ for all λ, we obtain

$$S(\lambda, \sigma) = \sigma^3 S^*\left(\frac{\lambda}{\sigma^2}\right) \leqslant \begin{cases} \dfrac{c}{\sigma}\lambda^2, & \text{if } \lambda \leqslant \tfrac{1}{2} \\ \sigma\lambda, & \text{if } \lambda > \tfrac{1}{2}. \end{cases}$$

We return now to (8.4.13). Let $G(\lambda)$ be the distribution function of $\int \eta^2\,\mathrm{d}P(\eta \mid \mathbf{y})$. Then from (8.4.24),

$$(8.4.25)\quad f(\sigma) \leqslant \frac{c}{\sigma}\int_0^{\frac{1}{2}\sigma^2} \lambda^2\,\mathrm{d}G(\lambda) + \sigma \int_{\frac{1}{2}\sigma^2}^{\infty} \lambda\,\mathrm{d}G(\lambda).$$

For any ϵ in (0, 1) choose σ_0 so large that

$$(8.4.26)\quad \int_{(\epsilon/2)\sigma_0^2}^{\infty} \lambda^{3/2}\,\mathrm{d}G(\lambda) < \epsilon.$$

The existence of such a σ_0 is guaranteed by (8.4.4). Then for all $\sigma \geqslant \sigma_0$,

$$\text{(8.4.27)} \quad \frac{1}{\sigma}\int_0^{\frac{1}{2}\sigma^2} \lambda^2\, dG(\lambda)$$

$$= \frac{1}{\sigma}\int_0^{(\epsilon/2)\sigma^2} \lambda^2\, dG(\lambda) + \frac{1}{\sigma}\int_{(\epsilon/2)\sigma^2}^{\sigma^2/2} \lambda^2\, dG(\lambda)$$

$$\leqslant (\epsilon/2)^{\frac{1}{2}} \int_0^{(\epsilon/2)\sigma^2} \lambda^{\frac{3}{2}}\, dG(\lambda) + \frac{1}{\sqrt{2}} \int_{(\epsilon/2)\sigma^2}^{\sigma^2/2} \lambda^{\frac{3}{2}}\, dG(\lambda) \leqslant \left(\frac{\epsilon}{2}\right)^{\frac{1}{2}} \int_0^{\infty} \lambda^{\frac{3}{2}}\, dG(\lambda) + \frac{\epsilon}{\sqrt{2}}.$$

Similarly,

$$\text{(8.4.28)} \qquad \sigma \int_{\sigma^2/2}^{\infty} \lambda\, dG(\lambda) \leqslant \sqrt{2} \int_{\sigma^2/2}^{\infty} \lambda^{\frac{3}{2}}\, dG(\lambda) \leqslant \sqrt{2}\,\epsilon$$

for all $\sigma > \sigma_0/\epsilon$. Hence from (8.4.25), (8.4.27), and (8.4.28), we obtain that, for all σ sufficiently large,

$$\text{(8.4.29)} \qquad f(\sigma) \geqslant \sqrt{\epsilon}\left[\frac{1}{\sqrt{2}} \int_0^{\infty} \lambda^{\frac{3}{2}}\, dG(\lambda) + \frac{\sqrt{\epsilon}}{\sqrt{2}} + \sqrt{(2\epsilon)}\right].$$

Since ϵ is arbitrary we obtain that $\lim_{\sigma\to\infty} f(\sigma) = 0$. This completes the proof that X is almost admissible.

We prove now that X is admissible. If X is inadmissible, there exists an estimator $\phi(X, Y)$ which satisfies (8.4.5) for all θ, with strict inequality for some θ. Whenever $R(\theta, \phi) < R(\theta, \hat{\theta})$ the probability of the set $S_{\mathbf{y}} = \{(x, \mathbf{Y});\ \phi(x, \mathbf{Y}) \neq x\}$ is positive. Let T be the set of all θ values for which

$$\text{(8.4.30)} \qquad \int Q(\mathrm{d}\mathbf{y}) \int_{S_{\mathbf{y}}} p(x - \theta \mid \mathbf{y})\, \mathrm{d}x > 0.$$

That is, T is the set of θ values on which $R(\theta, \phi) < R(\theta, \hat{\theta})$. Inequality 8.4.30 implies that

(8.4.31)

$$\int_{-\infty}^{\infty} \mathrm{d}\theta \int Q(\mathrm{d}\mathbf{y}) \int_{S_y} p(x - \theta \mid \mathbf{y})\, \mathrm{d}x = \int Q(\mathrm{d}\mathbf{y}) \int_{S_{\mathbf{y}}} \mathrm{d}x \int_{-\infty}^{\infty} p(x - \theta \mid \mathbf{y})\, \mathrm{d}\theta$$

$$= \int Q(\mathrm{d}\mathbf{y}) \int_{S_{\mathbf{y}}} \mathrm{d}x > 0.$$

Indeed, if $\int Q(\mathrm{dy}) \int_{S_\mathbf{y}} \mathrm{d}x = 0$ then $\int Q(\mathrm{dy}) \int_{S_\mathbf{y}} p(X - \theta \mid \mathbf{y})\, \mathrm{d}x = 0$. Furthermore,

$$\int_T \mathrm{d}\theta \int Q(\mathrm{dy}) \int_{S_\mathbf{y}} p(x - \theta \mid \mathbf{y})\, \mathrm{d}x = 0. \tag{8.4.32}$$

Hence

$$\int_T \mathrm{d}\theta \int Q(\mathrm{dy}) \int_{S_\mathbf{y}} p(x - \theta \mid \mathbf{y})\, \mathrm{d}x > 0, \tag{8.4.33}$$

which implies that the Lebesgue measure of T is positive. This contradicts the result that X is almost admissible. (Q.E.D.)

We notice that an important feature in the proof of admissibility is that the conditional distribution of $X - \theta$, given $\mathbf{Y}$, is absolutely continuous. This accounts for the property that $\int_{-\infty}^{\infty} p(x - \theta \mid \mathbf{y})\, \mathrm{d}\theta = 1$ a.s. for all x. If the distribution of X is discrete then X may not be admissible (see Blackwell [2]). Formula 8.4.4 adjusted to the discrete case secures only the *almost* admissibility of X.

An important question arises concerning (8.4.4). Can admissibility be generally secured if we require a weaker condition than (8.4.4)? Roughly speaking, (8.4.4) requires one moment more than needed for $R(\theta, \hat{\theta}) < \infty$ for all θ. Can we guarantee the admissibility of X when (8.4.3) is satisfied, but instead of (8.4.4) we require that $E\{|x|^{3-\alpha}\} < \infty$ for some $0 < \alpha < 1$? The answer is *negative*. Formula 8.4.4 can be made weaker for some families of distribution functions but not for all families satisfying (8.4.3). This was shown by Perng [1], who proved the following theorem.

Theorem 8.4.2. (Perng [1].) *In the fixed sample size case, if the loss function is* $L(\theta, d) = |\theta - d|^k$, $k \geqslant 1$, *then for every* α, $0 \leqslant \alpha < 1$, *there exists a family of distribution functions for which* $E_\theta\{|X|^\alpha\, |X - \theta|^k\} < \infty$ *for all* θ, *but the best equivariant estimator* X *is inadmissible.*

The family of distributions for which Theorem 8.4.2 holds is, for a sample of size $n = 2$, the one for which

$$Q(\mathrm{d}y) = \begin{cases} c(y^{k+2-\eta})^{-1}\, \mathrm{d}y, & y > 1, \\ 0, & \text{otherwise}, \end{cases} \tag{8.4.34}$$

and

$$p(x - \theta \mid y) = \begin{cases} \dfrac{1}{2by}, & \text{if } \left|\dfrac{x - \theta}{y}\right| \leqslant b \\ 0, & \text{otherwise}. \end{cases} \tag{8.4.35}$$

The parameters η, b, and c are all positive. Perng has proven that given any α in (0, 1) there exists a distribution in this family for which $E_0\, |X|^{k+\alpha} < \infty$ but X is inadmissible. The proof is very long and tedious and will be omitted.

Brown [2] proved that if the sample consists of one observation, the weaker condition $\int_{-\infty}^{\infty} x^2 p(x)\, dx < \infty$ is sufficient for the admissibility of X under the squared-error loss. In Stein's theorem (Theorem 8.4.1) and in Perng's counter-example (Theorem 8.4.2) the sample size n is at least 2.

Fox and Rubin [1] considered the problem of estimating a single location parameter when the loss function is a bilinear convex function

$$L(\theta, d) = \begin{cases} a(\theta - d), & \text{if } d \leqslant \theta, \\ b(d - \theta), & \text{if } d \geqslant \theta, \end{cases} \tag{8.4.36}$$

$0 < a,\ b < \infty$. It is easy to prove that, given θ, if X has a distribution $F_\theta(x)$, the value of d which minimizes $E_\theta\{L(\theta, d)\}$ is the $b/(a + b)$ (upper) fractile of $F_\theta(x)$. As before, let $X_1, \ldots, X_n$ be i.i.d. random variables having a common distribution function which depends on a single location parameter; $Y_i = X_{i+1} - X_1$ $(i = 1, \ldots, n - 1)$. Let $f(x - \theta)$ designate the density function of $F_\theta(x)$, with respect to $\mu(dx)$. Consider the fiducial density of θ, given $(X_1, Y_1, \ldots, Y_{n-1})$, namely,

$$h(\theta \mid X_1, \mathbf{Y}) = \frac{f(X_1 - \theta) \prod_{i=1}^{n-1} f(y_i + X_1 - \theta)}{\int_{-\infty}^{\infty} f(X_1 - \theta) \prod_{i=1}^{n-1} f(y_i + X_1 - \theta)\, d\theta}. \tag{8.4.37}$$

The γ-fractile of the fiducial distribution of θ, given $(X_1, \mathbf{Y})$, is

$$\theta_\gamma(X_1, \mathbf{Y}) = \text{least value of } \tau \text{ such that } \int_{-\infty}^{\tau} h(\theta \mid X_1, \mathbf{Y})\, d\tau \geqslant \gamma. \tag{8.4.38}$$

The estimator $\theta_\gamma(X_1, \mathbf{Y})$ is the minimum risk equivariant estimator of θ with respect to the real translations group, under the loss function (8.4.36), $\gamma = b/(a + b)$. We show now that $\theta_\gamma(X_1, \mathbf{Y})$ is indeed translation equivariant. Make the transformation $\zeta = X_1 - \theta$. Then according to (8.4.38), we can write

(8.4.39)

$$\theta_\gamma(X_1, \mathbf{Y}) = \text{least value of } \tau \text{ such that } \frac{\int_{X_1-\tau}^{\infty} f(\zeta) \prod_{i=1}^{n-1} f(y_i + \zeta)\, d\zeta}{\int_{-\infty}^{\infty} f(\zeta) \prod_{i=1}^{n-1} f(y_i + \zeta)\, d\zeta} \geqslant \gamma.$$

Define the translation invariant function

(8.4.40)

$$\psi_\gamma(Y) = \text{least value of } \tau \text{ such that } \frac{\int_{-\tau}^{\infty} f(\zeta) \prod_{i=1}^{n-1} f(y_i + \zeta)\, d\zeta}{\int_{-\infty}^{\infty} f(\zeta) \prod_{i=1}^{n-1} f(y_i + \zeta)\, d\zeta} \geqslant \gamma$$

Then

$$\theta_\gamma(X_1, \mathbf{Y}) = X_1 - \psi_\gamma(\mathbf{Y}), \tag{8.4.41}$$

which is translation equivariant. Thus the minimum risk translation equivariant estimator $\theta_\gamma(X_1; \mathbf{Y})$ is analogous to the Pitman estimator.

Consider the transformation $X = X_1 - \psi_\gamma(\mathbf{Y})$, and let $p(x - \theta \mid \mathbf{Y})$ designate the conditional density, under θ, of X given $\mathbf{Y}$. The conditional density of $X - \theta$ given $\mathbf{Y}$ is $p(x \mid \mathbf{Y})$, and we notice that the $a/(a + b)$-fractile of this conditional density is zero for all $\mathbf{Y}$. Indeed,

(8.4.42)

$$\begin{aligned} P_0[X \leqslant 0 \mid \mathbf{Y}] &= P_0[X_1 \leqslant \psi_\gamma(\mathbf{Y}) \mid \mathbf{Y}] \\ &= 1 - P_0[X_1 \geqslant \psi_\gamma(\mathbf{Y}) + 0 \mid \mathbf{Y}] \geqslant 1 - \gamma = \frac{a}{a + b}. \end{aligned}$$

Thus the following theorem of Fox and Rubin proves the admissibility of $\theta_\gamma(X_1, \mathbf{Y})$ whenever the conditional distribution of $X - \theta$ given Y is absolutely continuous.

Theorem 8.4.3. (Fox and Rubin [1].) *Let $p(x \mid \mathbf{Y})$ be the conditional density of $X - \theta$ given $\mathbf{Y}$ satisfying the following conditions:*

(i) *The $a/(a + b)$-fractile of $p(x \mid \mathbf{Y})$ is 0 for each $\mathbf{Y}$;*
(ii) $\int Q(\mathrm{d}\mathbf{y}) \int x^2 p(X \mid \mathbf{y})\, \mathrm{d}x < \infty$.

Then X is an admissible estimator of θ.

The proof of this theorem is very similar to the proof of Theorem 8.4.1 and therefore will be omitted. We remark also that if we do not assume that the conditional distribution of $X - \theta$, given $\mathbf{Y}$, is absolutely continuous or has its points of increase on a fixed lattice for all $\mathbf{Y}$, then we can only prove the almost admissibility of X.

8.5. THE INADMISSIBILITY OF SOME COMMONLY USED ESTIMATORS

Two examples of commonly used equivariant estimators which are inadmissible are presented in this section. These two examples are from Stein [5; 8] and are considered by now the classical counter-examples for admissibility of best equivariant estimators.

Example 8.8. *The inadmissibility of the sample mean in the p-variate normal case* ($p \geqslant 3$). Consider the problem of estimating the mean of a p-variate normal distribution ($p \geqslant 3$) when the covariance matrix is known. Given $X \sim N(\theta, I)$, the Pitman estimator of $\theta = (\theta_1, \ldots, \theta_p)'$ is $\hat{\theta} = X$. This estimator is minimax for the squared-error loss function $L(\theta, d) = \|d - \theta\|^2$ Stein [5] has shown, however, that X is inadmissible. It is strictly dominated, uniformly in θ, by the estimator

(8.5.1) $$\theta^*(X) = \left(1 - \frac{p - 2}{X'X}\right)X.$$

We could raise the question, What is the physical meaning of the estimator $\theta^*(X)$, when the components of X are not pure numbers? Actually $\theta^*(X)$ is a special case of the more general estimator $\theta_v^*(X) = (1 - (p-2)/X'V^{-1}X)X, p \geqslant 3$, where $X \sim N(\theta, V)$. The number $X'V^{-1}X$ is real (regardless of the physical dimensions of the components of X). Thus $X'X$ in $\theta^*(X)$ should be regarded as a pure number. We show now that

$$R(\theta, \hat{\theta}) > R(\theta, \theta^*) \quad \text{for all } \theta. \tag{8.5.2}$$

Indeed,

$$R(\theta, \hat{\theta}) - R(\theta, \theta^*) = E_\theta\left\{\left(2 - \frac{p-2}{X'X}\right)(p-2) + 2(p-2)\frac{X'\theta}{X'X}\right\}. \tag{8.5.3}$$

We notice first that

$$E_\theta\{X'\theta \mid X'X\} = 0 \quad \text{a.s.,} \quad \text{for all } \theta. \tag{8.5.4}$$

This is due to the symmetry of the distribution of X. Hence, given $X'X$, the conditional distribution of X is the symmetric distribution on the sphere with radius $X'X$. Thus it remains to prove that

$$E_\theta\left\{\frac{p-2}{X'X}\right\} < 2, \qquad \text{for all } \theta. \tag{8.5.5}$$

Since $X'X \sim \chi^2[p; \lambda]$ where $\lambda = \theta'\theta/2$, we have

$$X'X \sim G\left(\frac{1}{2}, \frac{p}{2} + J\right), \tag{8.5.6}$$

where J is a random variable having a Poisson distribution with mean λ. Hence for all θ,

$$E_\theta\left\{\frac{p-2}{X'X}\right\} = (p-2)E_\lambda\left\{G^{-1}\left(\frac{1}{2}, \frac{p}{2} + J\right)\right\} = E_\lambda\left\{\frac{p-2}{p-2+2J}\right\}. \tag{8.5.7}$$

Formula 8.5.7 reflects nicely why we have to require that $p \geqslant 3$. For every $J = 0, 1, \ldots, (p-2)/(p-2+2J) \leqslant 1$. Hence (8.5.5) holds for all θ. ■

Example 8.9. *The inadmissibility of the usual estimator of the variance of a normal distribution with unknown mean.* Let $X_1, \ldots, X_n$ be i.i.d. random variables, having a common $\mathcal{N}(\mu, \sigma^2)$ distribution. Both μ and σ are unknown; $-\infty < \mu < \infty, 0 < \sigma < \infty$. The minimal sufficient statistic is $(\bar{X}, Q)$ where $\bar{X} = \sum_{i=1}^n X_i/n$, $Q = \sum_{i=1}^n (X_i - \bar{X})^2$. The quantity $\bar{X}$ is independent of Q, and

(i) $Q \sim \sigma^2\chi_1^2[n-1]$;
(ii) $n\bar{X}^2 \sim \sigma^2\chi_2^2[1; \lambda]$, $\lambda = n\mu^2/2\sigma^2$. Both $\chi_1^2[n-1]$ and $\chi_2^2[1; \lambda]$ are independent. Hence if $S = Q + n\bar{X}^2$,
(iii) $S \sim \sigma^2\chi^2[n + 2J]$, where $J \sim P(\lambda)$.

Define the family of estimators

$$\phi(S, Q) = S\varphi\left(\frac{Q}{S}\right). \tag{8.5.8}$$

The estimator $\hat{\sigma}^2 = Q/(n + 1)$ is the special case of $\varphi_1(Q/S) = Q/((n + 1)S)$. Consider the estimator D^2 obtained by taking $\varphi_2(Q/S) = \min\{Q/(n + 1)S, 1/(n + 2)\}$. Now the risk function of (8.5.8) under $\theta = (\mu, \sigma^2)$ is

$$\begin{aligned}(8.5.9)\quad R(\theta, \varphi) &= E_\theta\left\{\left[S\varphi\left(\frac{Q}{S}\right) - \sigma^2\right]^2\right\} \\ &= \sigma^4 E_\theta\left\{E\left\{\left[\chi^2[n + 2J]\varphi\left(\frac{\chi_1^2[n - 1]}{\chi_1^2[n - 1] + \chi_2^2[1 + 2J]}\right) - 1\right]^2 \,\middle|\, \frac{Q}{S}, J\right\}\right\}.\end{aligned}$$

Given J, S is independent of Q/S. Hence for every φ and all θ,

$$\begin{aligned}(8.5.10)\quad R(\theta, \varphi) &= \sigma^4 E_\theta\left\{\varphi^2\left(\frac{Q}{S}\right)(n + 2J)(n + 2 + 2J) - 2\varphi\left(\frac{Q}{S}\right)(n + 2J) + 1\right\} \\ &= \sigma^4 E_\theta\left\{\frac{1}{(n + 2J)(n + 2 + 2J)}\left[\varphi\left(\frac{Q}{S}\right) - \frac{1}{n + 2 + 2J}\right]^2 + \frac{2}{n + 2 + 2J}\right\}.\end{aligned}$$

Since

$$\varphi_1\left(\frac{Q}{S}\right) = \frac{Q}{(n + 1)S} \geqslant \min\left\{\frac{Q}{(n + 1)S}, \frac{1}{n + 2}\right\} = \varphi_2\left(\frac{Q}{S}\right) \tag{8.5.11}$$

where equality holds only if $Q/(n + 1)S \leqslant 1/(n + 2)$,

$$\left[S\varphi_1\left(\frac{Q}{S}\right) - \frac{1}{n + 2 + 2J}\right]^2 \geqslant \left[S\varphi_2\left(\frac{Q}{S}\right) - \frac{1}{n + 2 + 2J}\right]^2 \tag{8.5.12}$$

for each $J = 0, 1, \ldots$. We notice also that

$$P_\theta\left\{(S, Q); \frac{Q}{(n + 1)S} > \frac{1}{n + 2}\right\} > 0, \qquad \text{for all } \theta. \tag{8.5.13}$$

Hence from (8.5.10), (8.5.12), and (8.5.13),

$$R(\theta, \varphi_2) < R(\theta, \varphi_1), \quad \text{for all } \theta. \tag{8.5.14}$$

■

8.6. MORE GENERAL THEORY OF ADMISSIBILITY OF EQUIVARIANT ESTIMATORS OF A SINGLE LOCATION PARAMETER

The results presented in the previous sections have been generalized by Farrel [1] for cases of location parameter distributions in which the conditional distribution of $x - \theta$, given $\mathbf{Y}$, is absolutely continuous. We treat the problem still in the framework of fixed-sample procedures. As in the previous sections, we designate by $p(x \mid \mathbf{Y})$ the density function of $x - \theta$, given $\mathbf{Y}$, and by $Q(\mathrm{d}\mathbf{y})$ the probability measure on the space of $\mathbf{Y}$. The loss function $L(\theta, d)$ is expressed in the form $L(\theta, d) = W(d - \theta)$. We assume that if $t_1 \geqslant t_2 \geqslant 0$, then $W(t_1) \geqslant W(t_2)$, and if $0 \geqslant t_2 \geqslant t_1$, then $W(t_1) \geqslant W(t_2)$. We also assume, naturally, that $W(0) = 0$ and $W(t) \geqslant 0$ for all t, $-\infty < t < \infty$.

Let $\lambda(\mathrm{d}\theta)$ be a sigma-finite measure on $(\Theta, \mathcal{C})$. Here Θ is the whole real line and $\mathcal{C}$ the smallest Borel sigma-field on the real line. An estimator $\hat{\theta}_\gamma(X, \mathbf{Y})$ is called *generalized Bayes* against λ with respect to the loss function $W(t)$ if

(8.6.1)
$$\int_{-\infty}^{\infty} \lambda(\mathrm{d}\theta) W(\hat{\theta}_\lambda(X, \mathbf{Y}) - \theta) p(X - \theta \mid \mathbf{Y}) = \inf_d \int_{-\infty}^{\infty} \lambda(\mathrm{d}\theta) W(d - \theta) p(X - \theta \mid \mathbf{Y}), \quad \text{a.s. } [\mu].$$

In Section 6.3 we defined the notion of the generalized Bayes estimator as a regular limit of a sequence of proper Bayes estimators. We showed there that if the family of distributions is of the exponential type then, subject to assumptions A.1–A.3, every estimator which is a regular limit of a sequence of proper Bayes estimators satisfies (8.6.1) for *some* sigma-finite measure $\lambda(\mathrm{d}\theta)$. Thus under A.1–A.3 of Section 6.3, every generalized Bayes estimator in that sense is a generalized Bayes estimator according to the present definition. The present definition is thus more general.

A generalized Bayes estimator $\hat{\theta}_\lambda(X, \mathbf{Y})$, as defined in (8.6.1), is not necessarily unique. Farrel therefore defines the maximal generalized Bayes estimator as the largest real value d^0 satisfying

(8.6.2)
$$\int_{-\infty}^{\infty} \lambda(\mathrm{d}\theta) W(d^0 - \theta) p(X - \theta \mid \mathbf{Y}) = \inf_d \int_{-\infty}^{\infty} \lambda(\mathrm{d}\theta) W(d - \theta) p(x - \theta \mid \mathbf{Y}).$$

Similarly, the minimal real value d^0 satisfying (8.6.2) is called the minimal generalized Bayes estimator. The following lemma gives sufficient conditions for the existence of these estimators.

Lemma 8.6.1. *If $W(t)$ has the above monotonicity property and there exists a null set $N \in R^{(1)} \times \mathcal{R}^{(n-1)}$ such that if $(X, \mathbf{Y}) \notin N$ then*

$$\inf_d \int_{-\infty}^{\infty} \lambda(\mathrm{d}\theta) W(d - \theta) p(x - \theta \mid \mathbf{Y}) < \infty;$$

and if either one of the following conditions holds,

(i) *$W(t)$ is continuous, $\lambda(\mathrm{d}\theta)$ sigma-finite,*
(ii) *$\lambda(\mathrm{d}\theta)$ is a nonatomic sigma-finite measure,*

then there exists a maximal and a minimal generalized Bayes estimator. Furthermore, if $W(t)$ is convex and $\lim_{|b|\to\infty} W(t) = \infty$ then every generalized Bayes estimator satisfies

$$\int_{-\infty}^{\infty} W'(\hat{\theta}_\lambda(X, \mathbf{Y}) - \theta)p(X - \theta \mid \mathbf{Y})\lambda(\mathrm{d}\theta) = 0. \tag{8.6.3}$$

The proof of the lemma is left to the reader. We discuss now the problem of the admissibility of equivariant estimators of the location parameter θ. Given $(X, \mathbf{Y})$, every equivariant estimator of θ with respect to the group of real translations is $\phi(X, \mathbf{Y}) = X + \psi(\mathbf{Y})$. The risk function of a translation equivariant estimator is thus

$$\begin{aligned} R(\theta, \psi) &= \int Q(\mathrm{d}\mathbf{y}) \int W(x + \psi(\mathbf{y}) - \theta)p(x - \theta \mid \mathbf{y})\, \mathrm{d}x \\ &= \int Q(\mathrm{d}\mathbf{y}) \int W(x + \psi(\mathbf{y}))p(x \mid \mathbf{y})\, \mathrm{d}x \\ &= R(0, \psi), \qquad \text{all } \theta \quad -\infty < \theta < \infty. \end{aligned} \tag{8.6.4}$$

Indeed, there is only one orbit of θ values, with respect to the group of real translations, which is the whole real line. It follows that an equivariant estimator $X + \psi^0(\mathbf{Y})$ is minimax in the class of equivariant estimators with respect to a loss function $W(t)$ if it is a generalized Bayes estimator for the Lebesgue linear measure $\lambda(\mathrm{d}\theta) = \mathrm{d}\theta$. That is,

$$\int_{-\infty}^{\infty} W(\theta + \psi^0(\mathbf{y}))p(\theta \mid \mathbf{y})\, \mathrm{d}\theta = \inf_{\psi} \int_{-\infty}^{\infty} W(\theta + \psi(\mathbf{y}))p(\theta \mid \mathbf{y})\, \mathrm{d}\theta \quad \text{a.s.} \tag{8.6.5}$$

The Pitman estimator is such an estimator for the squared-error loss function $W(t) = t^2$. In this case $\psi^0(\mathbf{y})$ is *unique* and, as we have previously established, the Pitman estimator is minimax at large and admissible. As will be proven in the next theorem, the uniqueness of $\psi^0(\mathbf{y})$ is a necessary condition for the admissibility of any translation equivariant estimator. Before stating the theorem we remind the reader that, as proven in Theorem 8.1.1, if d^0 is the (essentially) unique minimax estimator, it is admissible. The unique minimax estimator may not however be, equivariant. What we prove in the following theorem is that if within the class of translation equivariant estimators there are two minimax estimators which are not equivalent, then all equivariant estimators of the location parameter are inadmissible.

Theorem 8.6.2. (Farrel [1].) *Let $\mathfrak{D}_I$ designate the class of estimators of the location parameter θ, which are translation equivariant. Let the loss function $L(\theta, d) = W(d - \theta)$ satisfy the strict monotonicity condition*

$$W(t_2) > W(t_1) \quad \textit{for all} \quad |t_2| > |t_1| \geqslant 0. \tag{8.6.6}$$

Suppose, furthermore, that $d_1(X, \mathbf{Y}) = X + \psi_1(\mathbf{Y})$ *and* $d_2(X, \mathbf{Y}) = X + \psi_2(\mathbf{Y})$ *are two nonequivalent minimax estimators for* $\mathfrak{D}_I$; that is,

$$\int_A Q(\mathrm{d}\mathbf{y}) > 0, \tag{8.6.7}$$

where

$$A = \{\mathbf{y};\, \psi_1(\mathbf{y}) \neq \psi_2(\mathbf{y})\}.$$

Then all the equivariant estimators in $\mathfrak{D}_I$ *are inadmissible.*

Proof. To prove the theorem we show that, for every real η, the non-equivariant estimator

$$d_\eta(X, \mathbf{Y}) = \begin{cases} X + \min\{\psi_1(\mathbf{Y}), \psi_2(\mathbf{Y})\}, & \text{if } X > \eta, \\ X + \max\{\psi_1(\mathbf{Y}), \psi_2(\mathbf{Y})\}, & \text{if } X \leqslant \eta, \end{cases} \tag{8.6.8}$$

strictly dominates every equivariant estimator. We have seen that the risk function of every equivariant estimator $X + \psi(\mathbf{Y})$ does not depend on θ. Hence if $X + \psi_1(\mathbf{Y})$ and $X + \psi_2(\mathbf{Y})$ are two minimax estimators within $\mathfrak{D}_I$, we have

$$\int_{-\infty}^{\infty} W(x + \psi_i(\mathbf{y}))p(x \mid \mathbf{y})\, \mathrm{d}x = \inf_{\psi} \int_{-\infty}^{\infty} W(x + \psi(\mathbf{y}))p(x \mid \mathbf{y})\, \mathrm{d}x \quad \text{a.s.} \tag{8.6.9}$$

for $i = 1, 2$. Let N_i, $i = 1, 2$, be a null set in the space of $\mathbf{Y}$, with respect to $Q(\mathrm{d}\mathbf{y})$; that is,

$$\int_{N_i} Q(\mathrm{d}\mathbf{y}) = 0, \qquad i = 1, 2,$$

such that if $\mathbf{Y} \notin N_i$, $\psi_i(\mathbf{y})$ satisfies (8.6.9). Define the sets

$$\begin{aligned} A_+ &= \{\mathbf{y};\, \psi_1(\mathbf{y}) > \psi_2(\mathbf{y})\} \\ A_- &= \{\mathbf{y};\, \psi_1(\mathbf{y}) \leqslant \psi_2(\mathbf{y})\}. \end{aligned} \tag{8.6.10}$$

Since $N_1 \cup N_2 = N$ is a null set with respect to $Q(\mathrm{d}\mathbf{y})$ we have for $m(\mathbf{y}) = \min\{\psi_1(\mathbf{y}), \psi_2(\mathbf{y})\}$,

$$\begin{aligned} (8.6.11)\quad R(\theta, m) &= R(0, m) = \left(\int_{A_+ \cap \bar{N}} + \int_{A_- \cap \bar{N}}\right) Q(\mathrm{d}\mathbf{y}) . \int_{-\infty}^{\infty} W(x + m(\mathbf{y}))p(x \mid \mathbf{y})\, \mathrm{d}x \\ &= \int_{A_+ \cap \bar{N}} Q(\mathrm{d}\mathbf{y}) \int_{-\infty}^{\infty} W(x + \psi_2(\mathbf{y}))p(x \mid \mathbf{y})\, \mathrm{d}x \\ &\quad + \int_{A_- \cap \bar{N}} Q(\mathrm{d}\mathbf{y}) \int_{-\infty}^{\infty} W(x + \psi_1(\mathbf{y}))p(x \mid \mathbf{y})\, \mathrm{d}x. \end{aligned}$$

Furthermore, according to (8.6.9), if $\mathbf{Y} \in \bar{N}$ then

$$\int_{-\infty}^{\infty} W(x + \psi_2(\mathbf{y}))p(x \mid \mathbf{y})\, dx = \int_{-\infty}^{\infty} W(x + \psi_1(\mathbf{y}))p(x \mid \mathbf{y})\, dx. \tag{8.6.12}$$

Substituting (8.6.12) in the first integral on the right-hand side of (8.6.11), we obtain that

$$\begin{aligned} R(0, m) = & \int_{A_+ \cap \bar{N}} Q(d\mathbf{y}) \int_{-\infty}^{\infty} W(x + \psi_1(\mathbf{y}))p(x \mid \mathbf{y})\, dx \\ & + \int_{A_- \cap \bar{N}} Q(d\mathbf{y}) \int_{-\infty}^{\infty} W(x + \psi_1(\mathbf{y}))p(x \mid \mathbf{y})\, dx = R(0, \psi_1). \end{aligned} \tag{8.6.13}$$

This implies that $X + m(\mathbf{Y})$ is a minimax estimator within $\mathfrak{D}_I$. Similarly, we show that $X + M(\mathbf{Y})$, where $M(\mathbf{Y}) = \max\{\psi_1(\mathbf{Y}), \psi_2(\mathbf{Y})\}$ is a minimax estimator within $\mathfrak{D}_I$. We assume now, without loss of generality, that $\psi_1(\mathbf{y}) \leqslant \psi_2(\mathbf{y})$ a.s. Let $R(\theta, d_\eta; \mathbf{Y})$ designate the conditional expected loss of $d_\eta(X, \mathbf{Y})$ given $\mathbf{Y}$. The risk function of d_η is then

$$R(\theta, d_\eta) = \int Q(d\mathbf{y}) R(\theta, d_\eta; \mathbf{y}). \tag{8.6.14}$$

We have

$$\begin{aligned} R(\theta, d_\eta; \mathbf{Y}) & \\ &= \int_{-\infty}^{\infty} W(d_\eta(x, \mathbf{Y}) - \theta)p(x - \theta \mid \mathbf{Y})\, dx \\ &= \int_{-\infty}^{\eta} W(x + \psi_2(\mathbf{Y}) - \theta)p(x - \theta \mid \mathbf{Y})\, dx \\ &\quad + \int_{\eta}^{\infty} W(x + \psi_1(\mathbf{Y}) - \theta)p(x - \theta \mid \mathbf{Y})\, dx \\ &= \int_{-\infty}^{\eta-\theta} W(x + \psi_2(\mathbf{Y}))p(x \mid \mathbf{Y})\, dx + \int_{\eta-\theta}^{\infty} W(x + \psi_1(\mathbf{Y}))p(x \mid \mathbf{Y})\, dx. \end{aligned} \tag{8.6.15}$$

Let

$$\rho(\mathbf{Y}) = \inf_{\psi} \int_{-\infty}^{\infty} W(x + \psi(\mathbf{Y}))p(x \mid \mathbf{Y})\, dx. \tag{8.6.16}$$

Then from (8.6.15),

$$R(\theta, d_\eta; \mathbf{Y}) = \rho(\mathbf{Y}) + \int_{-\infty}^{\eta-\theta} [W(x + \psi_2(\mathbf{Y})) - W(x + \psi_1(\mathbf{Y}))]p(x \mid \mathbf{Y})\, dx. \tag{8.6.17}$$

Let

$$\Delta(x, \mathbf{Y}) = W(x + \psi_2(\mathbf{Y})) - W(x + \psi_1(\mathbf{Y})). \tag{8.6.18}$$

Since $\psi_1(\mathbf{Y}) \leqslant \psi_2(\mathbf{Y})$ a.s., we obtain for all $\mathbf{Y}$ outside a null set that $\Delta(x, \mathbf{Y})$ as a function of x changes sign at most once, as x ranges over $(-\infty, \infty)$. This is due to the strict monotonicity (8.6.6). Furthermore,

$$\lim_{x\to-\infty} \Delta(x, \mathbf{Y}) \leqslant 0, \qquad \lim_{x\to\infty} \Delta(x, \mathbf{Y}) \geqslant 0. \tag{8.6.19}$$

Moreover,

$$\int_{-\infty}^{\infty} \Delta(x, \mathbf{Y})p(x \mid \mathbf{Y})\, dx = 0 \quad \text{a.s.} \tag{8.6.20}$$

Let $\xi(\mathbf{Y})$ be the point at which $\Delta(x, \mathbf{Y})$ changes its sign. Thus $\Delta(X, \mathbf{Y}) \leqslant 0$ for all $X < \xi(\mathbf{Y})$ and $\Delta(X, \mathbf{Y}) \geqslant 0$ for all $X \geqslant \xi(\mathbf{Y})$. Hence if $\eta - \theta < \xi(\mathbf{Y})$, then

$$\int_{-\infty}^{\eta-\theta} \Delta(x, \mathbf{Y})p(x \mid \mathbf{Y})\, dx \leqslant 0.$$

If $\eta - \theta \geqslant \xi(\mathbf{Y})$ we have

$$\begin{aligned}\int_{-\infty}^{\eta-\theta} \Delta(x, \mathbf{Y})p(x \mid \mathbf{Y})\, dx &= \int_{-\infty}^{\xi(Y)-0} \Delta(x, \mathbf{Y})p(x \mid \mathbf{Y})\, dx + \int_{\xi(Y)}^{\eta-\theta} \Delta(x, \mathbf{Y})p(X \mid \mathbf{Y})\, dx \\ &\leqslant \int_{-\infty}^{\infty} \Delta(x, \mathbf{Y})p(x \mid \mathbf{Y})\, dx = 0.\end{aligned} \tag{8.6.21}$$

From (8.6.21) and (8.6.22) we obtain, since

$$\int_{\xi(Y)}^{\eta-\theta} \Delta(x, \mathbf{Y})p(x \mid \mathbf{Y})\, dx \geqslant 0,$$

$$\int_{-\infty}^{\eta-\theta} \Delta(x, \mathbf{Y})p(X \mid \mathbf{Y})\, dx \leqslant 0, \qquad \text{all } -\infty < \theta < \infty. \tag{8.6.22}$$

Hence from (8.6.17) and (8.6.22), $R(\theta, d_\eta) \leqslant \rho$ for all θ, $-\infty < \theta < \infty$, where $\rho = \int Q(d\mathbf{y})\rho(\mathbf{y})$. We show now that strict inequality holds for all θ, $-\infty < \theta < \infty$. On the set A $\psi_1(\mathbf{Y}) < \psi_2(\mathbf{Y})$. The strict monotonicity property, (8.6.6), implies that $\delta(x, \mathbf{Y}) = 0$ for at most one point x. Hence, on A

$$\int_{-\infty}^{\eta-\theta} \Delta(x, \mathbf{Y})p(x \mid \mathbf{Y}) < 0, \qquad \text{all } -\infty < \theta < \infty.$$

Or,

$$R(\theta, d_\eta; \mathbf{Y}) < \rho(\mathbf{Y}), \quad \text{all} \quad -\infty < \theta < \infty, \quad \mathbf{Y} \in A. \tag{8.6.23}$$

Finally, since $Q(A) > 0$,

$$R(\theta, d_\eta) < \rho, \quad \text{for all } \theta, \quad -\infty < \theta < \infty. \tag{8.6.24}$$ (Q.E.D.)

The result of Theorem 8.6.2 can be extended to randomized equivariant estimators as well. For details, see Farrel [1].

The theorems of the previous sections, concerning the admissibility of various translation equivariant estimators of a single location parameter, have been generalized by Farrel [1] under the following conditions:

(i) for all real numbers c,

$$\int Q(\mathrm{d}\mathbf{y}) \int_{-\infty}^{\infty} W(x + c)p(x \mid \mathbf{y})\, \mathrm{d}x < \infty; \tag{8.6.25}$$

(ii) Except on a set N of a Q measure zero, $\int_{-\infty}^{\infty} W(x + c)p(x \mid \mathbf{Y})\, \mathrm{d}x$ assumes its minimum at the *single* point $\psi_0(\mathbf{Y})$.

(iii) The function $W(t)$ is bounded, or uniformly continuous, or convex, or satisfies $\sup_{x \leqslant 0} W(x) < \infty$; $W(x)$ is convex in $x > 0$.

(iv) There exists a real number p, $1 < p \leqslant 2$, such that

$$\int Q(\mathrm{d}\mathbf{y}) \int_{-\infty}^{\infty} |x|^p\, W(x)p(x \mid \mathbf{Y})\, \mathrm{d}x < \infty. \tag{8.6.26}$$

If $W(x)$ is convex, then

$$\int Q(\mathrm{d}\mathbf{y}) \int_{-\infty}^{\infty} |W'(d + x)|\, |x|^p\, p(x \mid \mathbf{y})\, \mathrm{d}x < \infty, \tag{8.6.27}$$

for all real numbers d. If $W(x)$ is partly bounded and partly convex then

$$\int Q(\mathrm{d}\mathbf{y}) \int_{0}^{\infty} |W'(d + x)|\, |x|^p\, p(x \mid \mathbf{y})\, \mathrm{d}x < \infty, \tag{8.6.28}$$

for all $d \geqslant 0$.

(v) There exists a function $\lambda(\theta)$, $0 < \lambda(\theta) \leqslant K < \infty$, such that

$$\alpha = \lim_{\theta \to -\infty} \lambda(\theta) > 0, \qquad \beta = \lim_{\theta \to \infty} \lambda(\theta) > 0. \tag{8.6.29}$$

Moreover, except on a set of Q measure zero,

$$\int_{-\infty}^{\infty} W(c - x + \theta)p(\theta \mid \mathbf{Y})\lambda(x - \theta)\,d\theta$$

assumes its minimum value at a *single* finite point $\hat{\theta}(X, \mathbf{Y})$.

Theorem 8.6.3. (Farrel [1].) *Under (i)–(v), if $W(t)$ is bounded, then $\hat{\theta}(X, \mathbf{Y})$ is almost admissible. If $Q(\mathbf{Y})$ is a degenerate distribution (puts its total mass on a single point) then $\hat{\theta}(X, \mathbf{Y})$ is almost admissible. If $W(t)$ is not bounded and the support of $Q(\mathbf{Y})$ contains more than one point then, if $p(x \mid \mathbf{Y})$ vanishes off a compact set, $\hat{\theta}(X, \mathbf{Y})$ is almost admissible.*

The proof of his theorem is omitted. The condition that $\lambda(\theta)$ is bounded (see (v)) can be replaced by a weaker restriction when $W(x) = x^2$. It is impossible, however, to prove the almost admissibility of $\hat{\theta}(X, \mathbf{Y})$ when $\lambda(\theta)$ is an arbitrary function. This is illustrated by the following example of Farrel [1].

Example 8.10. Let $W(x) = W(-x)$ be strictly convex and satisfy conditions (i)–(v) for all x, $-\infty < x < \infty$. Let $\lambda(\theta) = e^{\theta}$. If $\hat{\theta}_{\lambda}(X, \mathbf{Y})$ is a generalized Bayes estimator with respect to λ, the corresponding Bayes risk is

$$\text{(8.6.30)} \quad \int_{-\infty}^{\infty} W(\hat{\theta}_{\lambda}(X, \mathbf{Y}) - \theta)p(X - \theta \mid \mathbf{Y}) \cdot e^{\theta}\,d\theta$$

$$+\int_{-\infty}^{\infty} W(\hat{\theta}_{\lambda}(X, \mathbf{Y}) - X + \tau)p(\tau \mid \mathbf{Y})e^{X-\tau}d\tau$$

$$= e^{X}\int_{-\infty}^{\infty} W(\hat{\theta}_{\lambda}(X, \mathbf{Y}) - X - \tau)p(-\tau \mid \mathbf{Y})e^{\tau}\,d\tau \quad \text{a.s.}$$

As in (v), assume that $\int_{-\infty}^{\infty} W(c - X + \theta)p(\theta \mid \mathbf{Y})e^{X-\theta}\,d\theta$ assumes its minimum at a single point. Then $\hat{\theta}_{\lambda}(X, \mathbf{Y})$ is unique for each $(X, \mathbf{Y})$. In particular, if we consider in (8.6.30) the point $(0, \mathbf{Y})$, we obtain that

$$\text{(8.6.31)} \qquad \hat{\theta}_{\lambda}(X, \mathbf{Y}) = X + \hat{\theta}_{\lambda}(0, \mathbf{Y}) = X + \psi_{\lambda}(\mathbf{Y}).$$

Hence $\hat{\theta}_{\lambda}(X, \mathbf{Y})$ is a translation equivariant estimator. According to (ii), let $\psi_0(\mathbf{Y})$ be the single point at which $\int_{-\infty}^{\infty} W(x + c)p(x \mid \mathbf{Y})\,dx$ is minimized. We immediately obtain that if $\hat{\theta}_0(X, \mathbf{Y})$ is a generalized Bayes estimator against $\lambda_0(\theta) \equiv 1$, then $\hat{\theta}_0(X, \mathbf{Y}) = X + \psi_0(\mathbf{Y})$. Both $\hat{\theta}_{\lambda}(X, \mathbf{Y})$ and $\hat{\theta}_0(X, \mathbf{Y})$ are minimax estimators within $\mathfrak{D}_I$, since the risk function of every $d \in \mathfrak{D}_I$ is independent of θ (see Theorem 6.5.1). Hence from Theorem 8.6.2 we imply that if $\hat{\theta}_{\lambda}(X, \mathbf{Y})$ is not equivalent to $\hat{\theta}_0(X, \mathbf{Y})$, $\hat{\theta}_{\lambda}(X, \mathbf{Y})$ is not almost admissible. Since $W(x)$ satisfies (i)–(v) and is strictly convex, we have

$$\text{(8.6.32)} \qquad \int_{-\infty}^{\infty} W'(x + \psi_0(\mathbf{Y}) - \theta)p(x - \theta \mid \mathbf{Y})\,d\theta = 0 \quad \text{a.s.,}$$

where $W'(X + \psi_0(\mathbf{Y}) - \theta)$ is the derivative of $W(t)$ at the point $t = X + \psi_0(\mathbf{Y}) - \theta$. It is not difficult to verify that

$$\text{(8.6.33)} \qquad \int_{-\infty}^{\infty} W'(x + \psi_0(\mathbf{Y}) - \theta)\phi(x - \theta \mid \mathbf{Y})e^{\theta}\, d\theta > 0 \quad \text{a.s. } [Q].$$

On the other hand,

$$\text{(8.6.34)} \qquad \int_{-\infty}^{\infty} W'(x + \psi_\lambda(\mathbf{Y}) - \theta)p(x - \theta \mid \mathbf{Y})e^{\theta}\, d\theta = 0 \quad \text{a.s.}$$

It follows from (8.6.33) and (8.6.44) that

$$\text{(8.6.35)} \qquad W'(x + \psi_0(\mathbf{Y}) - \theta) > W'(x + \psi_\lambda(\mathbf{Y}) - \theta) \quad \text{a.s.} \quad [Q].$$

Finally, since $W(t)$ is strictly convex,

$$\text{(8.6.36)} \qquad \psi_0(\mathbf{Y}) > \psi_\lambda(\mathbf{Y}) \quad \text{a.s.} \quad [Q].$$

Hence, $\hat{\theta}_\lambda(X, \mathbf{Y})$ is not almost admissible. ■

We remark here that in the cases discussed in this section, if $W(t)$ is either bounded or strictly convex, the absolute continuity of a conditional distribution of $X - \theta$, given $\mathbf{Y}$, implies that every almost admissible estimator is admissible. This property has been discussed in Section 8.4.

Farrel extended Stein's results concerning the admissibility of the best equivariant estimator of a single location parameter, under squared-error loss functions, and proved the following theorem.

Theorem 8.6.4. (Farrel [1].) *Suppose that $\lambda(\theta)$ is a non-negative Borel measurable function satisfying, for some constants $c_1 > 0$, $c_2 \geqslant 0$, and α, $0 < \alpha < 1$, and for all $-\infty < \theta < \infty$, $\lambda(\theta) \leqslant c_1 + c_2 |\theta|^{\alpha}$. Suppose that there exists a value β, $1 + \alpha < \beta \leqslant 2$, for which*

$$\text{(8.6.37)} \qquad \int Q(d\mathbf{y}) \int_{-\infty}^{\infty} |x|^{2+\alpha+\beta}\, p(x \mid \mathbf{y})\, dx < \infty.$$

Let $\Omega = \{\theta;$ for all $\epsilon > 0$, $\int_{\theta-\epsilon}^{\theta+\epsilon} \lambda(\theta)\, d\theta > 0\}$. Then if $\hat{\theta}_\lambda(X, \mathbf{Y})$ is a generalized Bayes estimator with respect to $\lambda(\theta)$, and $W(t) = t^2$, then $\hat{\theta}_\lambda(X, \mathbf{Y})$ is admissible, for $\theta \in \Omega$.

The results of Katz [1] concerning the admissibility of the minimax estimator of θ for $\Theta = [0, \infty)$ which were presented in Section 8.3, were also generalized by Farrel. We do not discuss this generalization here. The reader can find it in Farrel's paper [1]. We present instead some interesting results of Cohen [1] on the admissibility of linear estimators of a vector valued location parameter, in the multivariate normal case, under a quadratic loss function.

Let $\mathbf{X}$ be a random vector having a p-variate normal distribution $\mathcal{N}(\boldsymbol{\theta}, \mathbf{I})$, where the location parameter $\boldsymbol{\theta} = (\theta_1, \ldots, \theta_p)'$ is unknown, $-\infty < \theta_i < \infty$, $i = 1, \ldots, p$. Let $\mathbf{A}$ be a $p \times p$ matrix independent of $\mathbf{X}$. A linear estimator of $\boldsymbol{\theta}$ is $\hat{\boldsymbol{\theta}}_A = \mathbf{AX}$. We consider the problem of estimating $\boldsymbol{\theta}$ linearly with the squared-error loss function $L(\boldsymbol{\theta}, \mathbf{A}) = \|\mathbf{AX} - \boldsymbol{\theta}\|^2$. Cohen's main result is formulated in the following theorem.

Theorem 8.6.5. (Cohen [1].) *The linear estimator* $\boldsymbol{\theta} = \mathbf{AX}$ *is an admissible estimator of* θ *if and only if* A *is symmetric and the characteristic roots of* A, *say* α_i $(i = 1, \ldots, p)$, *satisfy the inequality* $0 \leqslant \alpha_i \leqslant 1$, *for all* $i = 1, \ldots, p$, *with equality at one for at most two of the roots.*

Before proving this theorem we remark that the unbiased estimator of $\boldsymbol{\theta}$, $\tilde{\boldsymbol{\theta}} = \mathbf{X}$, is inadmissible when $p \geqslant 3$, as shown in Example 8.8, since the symmetric matrix of the linear transformation $\mathbf{X} \to \mathbf{X}$ is $\mathbf{A} = \mathbf{I}$, with all p characteristic roots equal to one. According to Cohen's theorem, at most two of the characteristic roots could be equal to one. We start by proving the following lemma.

Lemma 8.6.6. (Cohen [1].) *If* $\mathbf{A}$ *is any matrix* (independent of x) *and* $\mathbf{P}$ *is any orthogonal matrix, then* $\hat{\boldsymbol{\theta}}_2 = (\mathbf{P'AP})\mathbf{X}$ *is an admissible linear estimator of* $\boldsymbol{\theta}$ *if and only if* $\hat{\boldsymbol{\theta}}_1 = \mathbf{AX}$ *is an admissible linear estimator of* $\boldsymbol{\theta}$.

Proof. Since $\mathbf{X} \sim N(\boldsymbol{\theta}, \mathbf{I})$, $\mathbf{Y} = \mathbf{PX} \sim N(\mathbf{P}\boldsymbol{\theta}, \mathbf{I})$ for any orthogonal matrix P. Thus $\mathbf{AX}$ is an admissible estimator of $\boldsymbol{\theta}$ if and only if $\mathbf{APX}$ is an admissible estimator of $\mathbf{P}\boldsymbol{\theta}$. Moreover, since $L(\boldsymbol{\theta}, \mathbf{AX}) = \|\mathbf{AX} - \boldsymbol{\theta}\|^2$, $L(\mathbf{P}\boldsymbol{\theta}, \mathbf{PBX}) = L(\boldsymbol{\theta}, \mathbf{BX})$ for every B. In particular if $\mathbf{B} = \mathbf{P'AP}$ we obtain that $\mathbf{P'APX}$ is an admissible estimator of $\boldsymbol{\theta}$ if and only if $\mathbf{PBX} = \mathbf{APX}$ is an admissible estimator of $\mathbf{P}\boldsymbol{\theta}$. Combining the two results, we obtain that $\mathbf{AX}$ is an admissible estimator of $\boldsymbol{\theta}$ if and only if $\mathbf{P'APX}$ is an admissible estimator of $\boldsymbol{\theta}$. (Q.E.D.)

Proof of Theorem 8.6.5. (i) Suppose that $\mathbf{A}$ is a (real) symmetric matrix; then there exists an orthogonal matrix $\mathbf{P}$ such that $\mathbf{A} = \mathbf{P'DP}$, where $\mathbf{D}$ is a diagonal matrix. Hence, according to the previous lemma, $\mathbf{AX}$ is admissible if and only if $(\mathbf{P'DP})\mathbf{X}$ is admissible. Therefore, without loss of generality, we can assume that $\mathbf{A}$ is a diagonal matrix, with diagonal elements α_i $(i = 1, \ldots, p)$. The estimator $\hat{\boldsymbol{\theta}} = \mathbf{AX}$ is an admissible estimator of $\boldsymbol{\theta}$, when $0 \leqslant \alpha_i < 1$ for all $i = 1, \ldots, p$, since $\mathbf{AX}$ is in this case the unique Bayes estimator of $\boldsymbol{\theta}$ against a prior distribution $H(\boldsymbol{\theta})$ with

$$H(d\boldsymbol{\theta}) = (2\pi)^{-p/2} \prod_{i=1}^{p} \beta_i^{1/2} \exp\left\{\frac{-\beta_i\theta_i^2}{2}\right\} \prod_{i=1}^{n} d\theta_i, \tag{8.6.38}$$

where

$$\beta_i = \frac{(1 - \alpha_i)}{\alpha_i}, \qquad i = 1, \ldots, p.$$

If any of the characteristic roots α_i is zero, $H(\boldsymbol{\theta})$ assigns $\{\theta_i = 0\}$ probability one. The risk function of $\mathbf{AX}$, where $\mathbf{A}$ is a diagonal matrix, is given by

$$R(\boldsymbol{\theta}, p) = \sum_{i=1}^{p} [\alpha_i^2 + \theta_i^2(1 - \alpha_i)^2]. \tag{8.6.39}$$

Hence if any α_i is negative then $\alpha_i^2 + \theta_i(1 - \alpha_i)^2 > \theta_i^2$; and by setting $\alpha_i = 0$ we reduce the risk. Similarly, if $\alpha_i > 1$ we reduce the risk by setting $\alpha_i = 1$. This implies that if any of the characteristic roots is either negative or greater than one $\mathbf{AX}$ is inadmissible. Suppose that the first k diagonal elements of $\mathbf{A}$ are equal to one and $k \geqslant 3$. Write $R(\boldsymbol{\theta}, \mathbf{A}) = R_1(\boldsymbol{\theta}, \mathbf{A}) + R_2(\boldsymbol{\theta}, \mathbf{A})$. Then $R_1(\boldsymbol{\theta}, \mathbf{A}) = k$ and $R_2(\boldsymbol{\theta}, \mathbf{A}) = \sum_{i=k+1}^{p} [\alpha_i^2 + \theta_i^2(1 - \alpha_i)^2]$. The estimator $\mathbf{AX}$ is equal in this case to $\hat{\boldsymbol{\theta}} = (\mathbf{X}^{(1)}, \mathbf{A}^*\mathbf{X}^{(2)})$, where $\mathbf{X}^{(1)} = (X_1, \ldots, X_k)'$, $\mathbf{X}^{(2)} = (X_{k+1}, \ldots, X_p)'$, and $\mathbf{A}^*$ is a diagonal matrix of order $(p - k) \times (p - k)$, whose diagonal elements are in $(0, 1)$. As shown in Example 8.8, $\hat{\boldsymbol{\theta}}$ is inadmissible since the estimator $\hat{\boldsymbol{\theta}}^* = (\mathbf{X}^{(1)}(1 - (k - 2)/\|\mathbf{X}^{(1)}\|^2), \mathbf{A}^*\mathbf{X}^{(2)})'$ has a smaller risk for all $\boldsymbol{\theta}$. Indeed, $R(\boldsymbol{\theta}, \hat{\boldsymbol{\theta}}^*) < k + \sum_{i=k+1}^{p} (\alpha_i^2 + \theta_i^2(1 - \alpha_i)^2)$. This proves the necessity of the condition that at most two of the characteristic roots of $\mathbf{A}$ are equal to one. To prove the sufficiency of the condition we assume that $\mathbf{A}$ is a diagonal matrix, $\alpha_1 = \alpha_2 = 1$ and $0 \leqslant \alpha_i < 1$ for $i = 3, \ldots, p$. We prove the sufficiency by negation. If $\mathbf{AX}$ is inadmissible, there exists another estimator $(B_1(\mathbf{X}), \ldots, B_p(\mathbf{X}))'$ with a risk function $R(\boldsymbol{\theta}, \mathbf{B})$ such that, for all $\boldsymbol{\theta}$,

$$\begin{aligned} R(\boldsymbol{\theta}, \mathbf{B}) &= \int [(B_1(\mathbf{X}) - \theta_1)^2 + (B_2(\mathbf{X}) - \theta_2)^2]\, dN(\mathbf{X} \mid \boldsymbol{\theta}, \mathbf{I}) \\ &\quad + \sum_{i=3}^{p} \int (B_i(\mathbf{X}) - \theta_i)^2\, dN(\mathbf{X} \mid \boldsymbol{\theta}, \mathbf{I}) \leqslant 2 + \sum_{i=3}^{p} [\alpha_i^2 + \theta_i^2(1 - \alpha_i)^2], \end{aligned} \tag{8.6.40}$$

with a strict inequality for some θ_0. Consider a prior distribution of θ such that $(\theta_3, \ldots, \theta_p)$ has a prior density

$$h(\theta_3, \ldots, \theta_p) = \prod_{i=3}^{p} \left(\frac{\beta_i}{2\pi}\right)^{1/2} \exp\{-\tfrac{1}{2}\beta_i\theta_i^2\}. \tag{8.6.41}$$

Integrating $(\theta_3, \ldots, \theta_p)$ from both sides of (8.6.40) according to the density (8.6.41), we obtain

(8.6.42)

$$\begin{aligned} &\int h(\theta_3, \ldots, \theta_p) \int ((B_1(\mathbf{X}) - \theta_1)^2 + (B_2(\mathbf{X}) - \theta_2)^2)\, dN(\mathbf{X} \mid \boldsymbol{\theta}, \mathbf{I}) \prod_{i=3}^{p} d\theta_i \\ &\quad + \int h(\theta_3, \ldots, \theta_p) \sum_{j=3}^{p} (B_j(\mathbf{X}) - \theta_j)^2\, dN(\mathbf{X} \mid \boldsymbol{\theta}, \mathbf{I}) \prod_{i=3}^{p} d\theta_i \leqslant 2 + \sum_{i=3}^{p} \alpha_i. \end{aligned}$$

Indeed, according to the prior density (8.6.41),

$$E\{\alpha_i^2 + \theta_i^2(1 - \alpha_i)^2\} = \alpha_i^2 + \frac{1}{\beta_i}(1 - \alpha_i)^2 = \alpha_i, \qquad i = 3, \ldots, p. \tag{8.6.43}$$

Now since $\sum_{i=3}^{p} \alpha_i$ is the Bayes risk associated with (8.6.41), the second term on the left-hand side of (8.6.42) is greater or equal to $\sum_{i=3}^{p} \alpha_i$. It follows that

(8.6.44)

$$\int h(\theta_3, \ldots, \theta_p) \prod_{i=3}^{p} d\theta_i \int [(B_1(\mathbf{X}) - \theta_1)^2 + (B_2(\mathbf{X}) - \theta_2)^2]\, dN(\mathbf{X} \mid \boldsymbol{\theta}, \mathbf{I})$$
$$= \int [(B_1(\mathbf{X}) - \theta_1)^2 + (B_2(\mathbf{X}) - \theta_2)^2]\, dN(\mathbf{X} \mid \boldsymbol{\theta}^*, \boldsymbol{\Sigma}) \leqslant 2,$$

where $\boldsymbol{\theta}^* = (\theta_1, \theta_2, 0, \ldots, 0)'$ and $\boldsymbol{\Sigma}$ is a diagonal matrix with $\Sigma_{11} = \Sigma_{22} = 1$ and $\Sigma_{ii} = (1 + \beta_i)/\beta_i$, $i = 3, \ldots, p$. Since $\boldsymbol{\Sigma}$ is known, $(\mathbf{X}_1, \mathbf{X}_2)$ is a sufficient statistic for $\mathcal{N}(\boldsymbol{\theta}^*, \boldsymbol{\Sigma})$. Hence we can take $B_1(\mathbf{X}) = B_1(\mathbf{X}_1) \neq \mathbf{X}_1$ and $B_2(\mathbf{X}) = B_2(\mathbf{X}_2) \neq \mathbf{X}_2$. Thus from (8.6.42) we deduce that (8.6.44) holds for all (θ_1, θ_2) with a strict inequality for some (θ_1^0, θ_2^0). This implies that $(\mathbf{X}_1, \mathbf{X}_2)$ is an inadmissible estimator of (θ_1, θ_2), which contradicts the result of James and Stein [1] reported previously. This completes the proof of (i).

(ii) If $\mathbf{A}$ is asymmetric, then the estimator $\mathbf{A}^*\mathbf{X}$, with

$$\mathbf{A}^* = \mathbf{I} - [(\mathbf{A} - \mathbf{I})'(\mathbf{A} - \mathbf{I})]^{1/2}, \tag{8.6.45}$$

strictly dominates the estimator $\mathbf{AX}$.

The risk function of $\mathbf{AX}$ is

$$\begin{aligned} R(\boldsymbol{\theta}, \mathbf{A}) &= E_{\theta}\{(\mathbf{AX} - \boldsymbol{\theta})'(\mathbf{AX} - \boldsymbol{\theta})\} \\ &= \operatorname{tr}\{\mathbf{A}'\mathbf{A}\} + \boldsymbol{\theta}'(\mathbf{A} - \mathbf{I})'(\mathbf{A} - \mathbf{I})\boldsymbol{\theta}. \end{aligned} \tag{8.6.46}$$

Since $(\mathbf{A}^* - \mathbf{I})'(\mathbf{A}^* - \mathbf{I}) = (\mathbf{A} - \mathbf{I})'(\mathbf{A} - \mathbf{I})$, it remains to prove that $\operatorname{tr}\{\mathbf{A}^{*\prime}\mathbf{A}^*\} < \operatorname{tr}\{\mathbf{A}'\mathbf{A}\}$. Furthermore, since

$$(\mathbf{A}^* - \mathbf{I})'(\mathbf{A}^* - \mathbf{I}) = (\mathbf{A} - \mathbf{I})'(\mathbf{A} - \mathbf{I}),$$

it is sufficient to prove that

$$\operatorname{tr}\{\mathbf{I} - \mathbf{A}\} < \operatorname{tr}[(\mathbf{I} - \mathbf{A})'(\mathbf{I} - \mathbf{A})]^{1/2}. \tag{8.6.47}$$

But (8.6.47) holds for any matrix $\mathbf{I} - \mathbf{A}$ which is asymmetric. This completes the proof of (ii). (Q.E.D.)

8.7. ADMISSIBILITY IN THE GENERAL SEQUENTIAL CASE

Brown generalized in his paper [2] many of the results discussed in the previous sections to the case of sequential sampling. Due to space restrictions and to the extraordinary length of most of the proofs, we confine the discussion in this section to one theorem only. This theorem can be considered as a generalization of Stein's theorem concerning the admissibility of the best translation equivariant estimator of a single location parameter. It can be considered also as a generalization of Farrel's theorem (Theorem 8.6.2). The treatment of the admissibility problem in sequential models requires the introduction of several notions. We start, therefore, with setting the model and defining the required notions.

Let $X, Y_1, Y_2, \ldots$ be a sequence of random variables. The variables $Y_1, Y_2, \ldots$ are identically distributed and independent. Let $\mathcal{A}$ be the smallest Borel sigma-field generated by X and $\mathcal{B}^{(n)} = \sigma\{Y_1, \ldots, Y_n\}$ designate the smallest Borel sigma-field generated by $Y_1, \ldots, Y_n$. For each $n \geqslant 1$ we let $p_n(x - \theta \mid Y_1, \ldots, Y_{n-1}) \in \mathcal{B}^{(n)}$ be the conditional density of X, given $Y_1, \ldots, Y_{n-1}$. We assume that the conditional distribution of X, given $Y_1, \ldots, Y_{n-1}$, is absolutely continuous; that is,

$$\int_{-\infty}^{\infty} p_n(x - \theta \mid Y_1, \ldots, Y_{n-1})\, dx = 1 \quad \text{a.s.,} \tag{8.7.1}$$

all θ, $-\infty < \theta < \infty$. The parameter θ is a 1-dimensional location parameter.

Let $\mathcal{C}^{(n)} = \sigma\{\mathcal{A}(\bigcup_{j=0}^{n-1} \mathcal{B}^{(j)})\}$ be the smallest Borel sigma-field on $\mathcal{A}(\bigcup \mathcal{B}^{(j)})$, where $\mathcal{B}^{(0)} = \{\phi, \Omega\}$; $\mathcal{C}^{(\infty)} = \sigma\{\mathcal{A}(\bigcup_{j=0}^{\infty} \mathcal{B}^{(j)})\}$. A (well-defined) *stopping variable* is a function $S(X, \mathbf{Y})$, $\mathbf{Y} = (Y_1, Y_2, \ldots)$, measurable $\mathcal{C}^{(\infty)}$, such that $\{S(X, \mathbf{Y}) = n\} \in \mathcal{C}^{(n)}$, and $P_\theta[S(X, \mathbf{Y}) = n$ for some n finite$] = 1$, for all θ, $-\infty < \theta < \infty$. A sequential estimation procedure is a pair $(S(X, \mathbf{Y}), \hat{\theta}(X, \mathbf{Y}))$ such that $\{S(X, \mathbf{Y}) = n, \hat{\theta}(X, \mathbf{Y})\} \in \mathcal{C}^{(n)}$. That is, if the observed random variable is $(X, Y_1, \ldots, Y_{n-1})$, the estimator $\hat{\theta}(X, \mathbf{Y})$ depends only on X and on $(Y_1, \ldots, Y_{n-1})$, for each $n \geqslant 2$. If $S(X, \mathbf{Y}) = 0$, $\hat{\theta}(X, \mathbf{Y})$ is a constant independent of the random variables $(X, \mathbf{Y})$ If $S(X, \mathbf{Y}) = 1$ then $\hat{\theta}(X, \mathbf{Y}) = \tilde{\theta}(X)$.

Let $Q(\mathrm{d}y)$ be a probability measure on $(\mathcal{Y}, \mathcal{B})$ where $\mathcal{Y}$ is the sample space of $\mathbf{Y}$ and $\mathcal{B} = \sigma\{\mathbf{Y}\}$. Let $P(\cdot \mid \mathbf{Y})$ designate the conditional probability measure induced by $p(x \mid \mathbf{Y})$ when $\theta = 0$. We say that a function $g\colon (\mathcal{X}, \mathcal{Y}) \to \mathcal{R}^{(1)}$ (the real line) is *finitely measurable* if, for any given $\epsilon > 0$, there exists an $n(\epsilon) < \infty$ and a $\mathcal{C}^{(n)}$ measurable set A_n such that

$$\int P[A_n \mid \mathbf{Y} = y] Q(\mathrm{d}y) \geqslant 1 - \epsilon \tag{8.7.2}$$

and $g(X, \mathbf{Y})$ *restricted* to A_n, $\mathcal{C}^{(n)} \mid A_n$, is measurable. [The symbol $\mathcal{C}_r^{(n)} \mid A_n$ designates the smallest Borel sigma-field generated by $(X, \mathbf{Y})$ which is restricted to A_n]. If g is non-negative and finitely measurable, then

$$\int g(x, y) P(\mathrm{d}x \mid \mathbf{Y} = y) Q(\mathrm{d}y) = \lim_{n\to\infty} \int Q(\mathrm{d}y) \int_A g(x, y) p_n(x \mid \mathbf{Y} = y)\, \mathrm{d}x \leqslant \infty, \tag{8.7.3}$$

where $\{A_n\}$ is a sequence satisfying (8.7.2).

For the problem of sequentially estimating the single location parameter θ we adopt a loss function $W(\hat{\theta}(X, \mathbf{Y}), S(X, \mathbf{Y}), \mathbf{Y})$ such that, for each $n \geqslant 1$, $W(t, n, \mathbf{Y})$ is $\mathfrak{C}^{(n)}$-measurable. We assume that $W(\cdot, \cdot, \cdot) \geqslant 0$ and $W(\cdot, \infty, \cdot) = \infty$. The risk function $R(\theta, \delta)$ of any sequential procedure $\delta = (S(x, \mathbf{Y}), \hat{\theta}(x, \mathbf{Y}))$ such that $S(X, \mathbf{Y}) < \infty$ a.s. $[P, Q]$ is

(8.7.4)

$$R(\theta, \delta) = \int Q(\mathrm{d}y) \int W(\hat{\theta}(x, y) - \theta, S(x, y), y)\, \mathrm{d}P[X - \theta \leqslant x \mid \mathbf{Y} = y].$$

A sequential procedure δ is translation equivariant if

$$\hat{\theta}(X, \mathbf{Y}) = X + \psi(\mathbf{Y})$$

and

$$S(X, \mathbf{Y}) = S^*(\mathbf{Y}).$$

We assume that the class of all equivariant sequential procedures $\mathfrak{D}_I$ contains at least one procedure δ having a finite risk. It is easy to show that, for each $\delta \in \mathfrak{D}_I$, $R(\theta, \delta)$ is independent of θ. We let

(8.7.5)
$$R_0 = R(\theta_0) = \inf_{d \in \mathfrak{D}_I} R(\theta, \delta).$$

Brown introduced then the following two general conditions:

(i) If $\{\delta_i\}$ is a sequence of procedures in $\mathfrak{D}_I$ such that $R(\delta_i) \to R(\delta_0)$, then the components of δ_i satisfy

(8.7.6)
$$\hat{\theta}_i(X, \mathbf{Y}) \to X + \psi_0(\mathbf{Y})$$

in probability $[Q]$, where

$$\hat{\theta}_0(X, \mathbf{Y}) = X + \psi_0(\mathbf{Y});$$

and

(8.7.7)
$$S_i(X, \mathbf{Y}) \to S_0^*(\mathbf{Y}).$$

(ii)

$$\int_0^\infty \mathrm{d}\lambda \left\{ \sup_{\delta = (\psi, S^*) \in \mathfrak{D}_I} \int Q(\mathrm{d}y) \int_{-\lambda}^{\lambda} P(\mathrm{d}x \mid \mathbf{Y} = y) \right.$$
$$\left. \times\, [W(x + \psi_0(y), s^*(y), y) - W(x + \psi(y), s^*(y), y)] \right\} < \infty.$$

Theorem 8.7.1. (Brown [2].) *Suppose that* $R_0 < \infty$. *If (i) and (ii) are satisfied and if, furthermore,*

$$\int Q(\mathrm{d}y) \int |xW(x + \psi_0(y), S_0^*(y), y)| \, P(\mathrm{d}x \mid \mathbf{Y} = y) < \infty, \tag{8.7.8}$$

then the best equivariant procedure δ_0 *is admissible within the class of procedures* $\mathcal{S}_1$ *which are based on at least one observation. If, in addition,*

$$\sup_t \inf_y W(t, 0, y) > R_0, \tag{8.7.9}$$

then δ_0 *is admissible.*

The proof is very long and we therefore provide only it's main points. We remark first that (i) implies that the best translation equivariant estimator is unique. Thus (i) is *stronger* than the requirement (in Farrel's theorem) that the best equivariant estimator be unique. We remark also that (8.7.8) is, roughly speaking, a requirement that $P(x - \theta \mid \mathbf{Y})$ will have a moment of an order larger by one than $\int W(t, n, y) P(\mathrm{d}x \mid y)$ for each n. This condition is in the same spirit of Stein's condition, (8.4.4), and that of Fox and Rubin (Theorem 8.4.3). Brown [2] explored the cases under which both (i) and (ii) are fulfilled.

Proof of Theorem 8.7.1. (Only schematized.) Suppose that δ is a sequential estimation procedure with a risk function $R(\theta, \delta)$ such that $R(\theta, \delta) \leqslant R_0$ for all θ, $-\infty < \theta < \infty$. In order to establish the admissibility of δ_0 we have to show that $R(\theta, \delta) \equiv R_0$ for *all* θ, $-\infty < \theta < \infty$. For this purpose we have to show that

$$\int Q(\mathrm{d}y) \int_{-\infty}^{\infty} \Big\{ W(z + \psi_0(y), S_0^*(y), y) - W(z + \psi(x, y), S(x, y), y) \Big\} P(\mathrm{d}z \mid \mathbf{Y} = y) = 0 \quad \text{a.s.} \tag{8.7.10}$$

If $\delta(X, \mathbf{Y})$ dominates $\delta_0(X, \mathbf{Y})$, there exists an L, $0 < L < \infty$, such that

$$\int_{-L}^{L} \mathrm{d}\theta \int Q(\mathrm{d}y) \int P(\mathrm{d}z \mid \mathbf{Y} = y) \times \{W(z + \psi_0(y), S_0^*(y), y) - W(z + \psi(x, y), S(x, y), y)\} \geqslant 0. \tag{8.7.11}$$

Since $W \geqslant 0$ and $R(\theta, \delta) \geqslant 0$, we can interchange the order of integration in (8.7.11) to obtain

$$\int Q(\mathrm{d}y) \int_{-\infty}^{\infty} \mathrm{d}x \int_{x-L}^{x+L} \{W(z + \psi_0(y), S_0^*(y), y) - W(z + \psi(x, y), S(x, y), y)\} P(\mathrm{d}z \mid \mathbf{Y} = y) \geqslant 0. \tag{8.7.12}$$

The integral on the left-hand side of (8.7.12) is decomposed into the sum of five terms in the following manner.

$$\int Q(\mathrm{d}y)\int_{-\infty}^{\infty}\mathrm{d}x\int_{x-L}^{x+L}P(\mathrm{d}z\mid \mathbf{Y}=y)\{\ \} \tag{8.7.13}$$

$$=\int Q(\mathrm{d}y)\Bigg[\int_{-L/2}^{L/2}\mathrm{d}x\int_{-L/2}^{L/2}P(\mathrm{d}z\mid \mathbf{Y}=y)+\int_{-3L/2}^{-L/2}\mathrm{d}x\int_{-L/2}^{L+x}P(\mathrm{d}z\mid \mathbf{Y}=y)$$

$$+\int_{L/2}^{3L/2}\mathrm{d}x\int_{x-L}^{L/2}P(\mathrm{d}z\mid \mathbf{Y}=y)+\int_{L/2}^{\infty}P(\mathrm{d}z\mid \mathbf{Y}=y)\int_{z-L}^{z+L}\mathrm{d}x$$

$$+\int_{-\infty}^{-L/2}P(\mathrm{d}z\mid \mathbf{Y}=y)\int_{z-L}^{z+L}\mathrm{d}x\Bigg]\{\ \}.$$

We then prove that

$$\left(\int_{L/2}^{\infty}P(\mathrm{d}z\mid \mathbf{Y}=y)\int_{z-L}^{z+L}\mathrm{d}x+\int_{-\infty}^{-L/2}P(\mathrm{d}z\mid \mathbf{Y}=y)\int_{Z-L}^{Z+L}\mathrm{d}x\right)\{\ \} \tag{8.7.14}$$

$$\leqslant 4\int Q(\mathrm{d}y)\int_{|z|>L/2}|z|\,W(z+\varphi_0(y),S_0^*(y),y)P(\mathrm{d}z\mid \mathbf{Y}=y)\to 0,$$

as $L\to\infty$. To obtain this convergence to zero we need (8.7.8). Secondly, we can prove that

$$\limsup_{L\to\infty}\int Q(\mathrm{d}y)\int_{-3L/2}^{-L/2}\{\ \}P(\mathrm{d}z\mid \mathbf{Y}=y)\leqslant 0, \tag{8.7.15}$$

and

$$\limsup_{L\to\infty}\int Q(\mathrm{d}y)\int_{L/2}^{3L/2}\{\ \}P(\mathrm{d}z\mid \mathbf{Y}=y)\leqslant 0. \tag{8.7.16}$$

From (8.7.12), (8.7.14), (8.7.15), and (8.7.16) we obtain that

$$\varliminf_{L\to\infty}\int Q(\mathrm{d}y)\int_{-L/2}^{L/2}\mathrm{d}x\int_{-L/2}^{L/2}\{\ \}P(\mathrm{d}z\mid \mathbf{Y}=y). \tag{8.7.17}$$

On the other hand, we show that

$$(8.7.18)\quad \lim_{\lambda\to\infty} \int_{-\lambda}^{\lambda} dx \int Q(dy) \int_{-\lambda}^{\lambda} \{\quad\} P(dz \mid \mathbf{Y} = y) \leqslant \int_{-\infty}^{\infty} dx \int Q(dy) \{\quad\} P(dz \mid \mathbf{Y} = y).$$

For a fixed value of X we can consider the function $\psi(X, \mathbf{Y})$ as dependent only on $\mathbf{Y}$. Therefore, $\hat{\theta}(X, \mathbf{Y}) = X + \psi(X, \mathbf{Y})$ is translation equivariant. Hence since (i) implies that δ_0 is the unique best equivariant estimator, the term in $\{\quad\}$ on the right-hand side of (8.7.18) is not positive. This and (8.7.17) imply (8.7.10) and the required result. (Q.E.D.)

8.8. ON THE ADMISSIBILITY OF THE SAMPLE MEAN AMONG POLYNOMIAL UNBIASED EQUIVARIANT ESTIMATORS

In this section we investigate the conditions under which the sample mean is an admissible estimator of a single location parameter θ, $-\infty < \theta < \infty$, for the squared-error loss. We have proven that if $X_1, \ldots, X_n$ are i.i.d., and $E_\theta\{X\} = \theta$ for all θ, then the Pitman estimator $\hat{\theta} = \bar{X} - E\{\bar{X} \mid X_2 - X_1, \ldots, X_n - X_1\}$ is an admissible best translation equivariant estimator of a single location parameter θ; $\bar{X}$ is the sample mean. Kagan, Linnik, and Rao [1] proved that if the sample size is $n \geqslant 3$, the Pitman estimator θ coincides a.s. with the sample mean $\bar{X}$ if and only if the distribution function of X_i $(i = 1, \ldots, n)$ is normal, $\mathcal{N}(\theta, \sigma^2)$, where σ^2 is known. Thus if the distribution function of X is not normal, the sample mean is inadmissible. There are, however, subclasses of estimators within which the sample mean is admissible. For example, if we consider the subclass of linear unbiased estimators, the sample mean is the minimum variance linear unbiased estimator; $\bar{X}$ is obviously admissible within this subclass, irrespective of the distribution of X. (As long as it is a single location parameter distribution.) Let us designate by $\mathfrak{D}^{(1)}$ the subclass of all linear unbiased estimators. Linnik raised, in connection with the above result, the problem of whether it is possible to construct a monotone sequence of subclasses $\mathfrak{D}^{(1)} \subset \mathfrak{D}^{(2)} \subset \cdots$ such that the sample mean $\bar{X}$ is admissible within $\mathfrak{D}^{(k)}$ if and only if the first $(k + 1)$ moments of X coincide with those of some family of normal distributions. In the present section we present the solution of Kagan to this problem. Kagan [1] has shown that if $n \geqslant 3$, such a monotone sequence of subclasses is obtained by defining $\mathfrak{D}^{(k)}$ to be the subclass of all equivariant estimators $\bar{X} + \psi^{(k)}(X_2 - X_1, \ldots, X_n - X_1)$ where $E_\theta\{\psi^{(k)}(X_2 - X_1, \ldots, X_n - X_1)\} = 0$ for all θ, and $\psi^{(k)}(X_2 - X_1, \ldots, X_n - X_1)$ is a polynomial in $(X_2 - X_1, \ldots, X_n - X_1)$ of degree not exceeding k. Thus let

$$(8.8.1)\quad \psi^{(k)}(X_2 < X_1, \ldots, X_n < X_1) = c + \sum_{j=2}^{n} a_j(X_j - X_1)$$
$$+ \sum_{j_1,j_2=2}^{n} a_{j_1 j_2}(X_{j_1} - X_1)(X_{j_2} - X_1) + \cdots + \sum_{j_1,\ldots,j_k=2}^{n} a_{j_1\ldots j_k} \prod_{v=1}^{k} (X_{j_v} - X_1)$$

with sets of coefficients c, $\{a_j\}$, $\{a_{j_1 j_2}\}, \ldots, \{a_{j_1\ldots j_k}\}$ so that,

$$(8.8.2)\quad E_\theta\{\psi^{(k)}(X_2 - X_1, \ldots, X_n - X_1)\} = 0, \quad \text{all } \theta, \quad -\infty < \theta < \infty.$$

We assume that the common distribution function of the i.i.d. random variables $X_1, \ldots, X_n$ has a moment of order $2k$.

The set of all polynomials $P(X_1, \ldots, X_n)$ of degree not exceeding k, with the inner product $(P_1, P_2) = E_0\{P_1 P_2\}$, forms a Hilbert space $L_k^{(2)}$. Let $\Lambda_k = \{\psi^{(k)}(X_2 - X_1, \ldots, X_n - X_1);$ (8.8.1) and (8.8.2) are satisfied$\}$. The set Λ_k is a subspace of the above Hilbert space. Let $\mathcal{F}_k$ be the smallest Borel sigma-field generated by Λ_k. The conditional expectation operator $E_0\{\cdot \mid \mathcal{F}_k\}$ can be considered as a (right) projection of polynomials P in the Hilbert space on Λ_k. That is, if $P \in L_k^{(2)}$ then $E_0\{P \mid \mathcal{F}_k\} \in \Lambda_k$, and

$$(8.8.3)\quad (E_0\{P \mid \mathcal{F}_k\}, P - E_0\{P \mid \mathcal{F}_k\}) = 0,$$

for all $P \in L_k^{(2)}$. Thus $\|P - E_0\{P \mid \mathcal{F}_k\}\|^2$ is minimal.

Lemma 8.8.1. (Kagan [1].) *Let*

$$(8.8.4)\quad t^{(k)}(X_1, \ldots, X_n) = \bar{X} - E_0\{\bar{X} \mid \mathcal{F}_k\}.$$

Then, for each θ, $-\infty < \theta < \infty$,

$$(8.8.5)\quad E_\theta\{t^{(k)}(X_1, \ldots, X_n)\} = \theta,$$

and

$$(8.8.6)\quad E_\theta\{[t^{(k)}(X_1, \ldots, X_n) - \theta]^2\} \leqslant E_\theta\{(\bar{X} - \theta)^2\}.$$

Equality holds in (8.8.6) if and only if

$$E_0\{\bar{X} \mid \mathcal{F}_k\} = 0 \quad \text{a.s.}$$

Proof.

$$(8.8.7)\quad E_\theta\{E_0\{\bar{X} \mid \mathcal{F}_k\}\} = 0, \quad \text{for all } \theta,$$

due to (8.8.2) and the definition of $\mathcal{F}_k$. Hence $t^{(k)}(X_1, \ldots, X_n)$ is an unbiased estimator of θ, which belongs to $\mathcal{D}^{(k)}$. Since $t^{(k)}(X_1, \ldots, X_n) = \bar{X} - E_0\{\bar{X} \mid \mathcal{F}_k\}$ is orthogonal to any function of Λ_k, we obtain that

$$(8.8.8)\quad E_\theta\{t^{(k)}(X_1, \ldots, X_n) E_0\{\bar{X} \mid \mathcal{F}_k\}\} = 0, \quad \text{all } \theta.$$

Hence

$$(8.8.9)\quad E_\theta\{(\bar{X}-\theta)^2\} = E_\theta\{[t^{(k)}(X_1, \ldots, X_n) - \theta]^2\} + E_0\{(E_0\{\bar{X} \mid \mathcal{F}_k\})^2\}, \quad \text{all } \theta, \quad -\infty < \theta < \infty.$$

Equation 8.8.9 implies (8.8.6). Moreover, $E_0\{(E_0\{\bar{X} \mid \mathcal{F}_k\})^2\} = 0$ if and only if $E_0\{\bar{X} \mid \mathcal{F}_k\} = 0$ a.s. (Q.E.D.)

The important lemma is the following.

Lemma 8.8.2. (Kagan [1].) *Let $n \geqslant 3$. Then $E_0\{\bar{X} \mid \mathcal{F}_k\} = 0$ a.s. if and only if the first $(k+1)$ moments of the distribution function $F(x)$ coincides with the corresponding moments of some normal distribution.*

Proof. We provide only a sketch of the proof. Let $\mathcal{M}_k$ be the subspace of $L_k^{(2)}$ generated by all polynomials of degree $\leqslant k$ of the form (8.8.1). Obviously, $\Lambda_k \subset \mathcal{M}_k$. If $\psi^{(k)} \in \mathcal{M}_k$ and $E_\theta\{\psi^{(k)}\} = \varphi(\theta)$, then $\psi^{(k)} - \varphi(\theta) \in \Lambda_k$. Moreover, $E_0\{c\bar{X}\} = 0$ for all c. Hence $E_0\{\bar{X} \mid \mathcal{F}_k\} = 0$ a.s. if and only if $E_0\{\bar{X} \mid \mathcal{G}_k\} = 0$ a.s., where $\mathcal{G}_k$ is the Borel sigma-field generated by $\mathcal{M}_k$; $\mathcal{F}_k \subset \mathcal{G}_k$. It is more convenient to establish the result working with $E_0\{\bar{X} \mid \mathcal{G}_k\}$. Thus we have to prove that $E_0\{\bar{X} \mid \mathcal{G}_k\} = 0$ a.s. if and only if

$$(8.8.10)\quad E_0\{X^j\} = \begin{cases} 0, & \text{all odd } j, \quad 1 \leqslant j \leqslant k+1, \\ (j-1)!\,\sigma^j, & \text{all even } j, \quad 2 \leqslant j \leqslant k+1, \end{cases}$$

for some σ, $0 < \sigma < \infty$. The proof proceeds by induction on k. For $k = 1$ it is trivially true. Assume that it is true for all $\nu = 1, \ldots, k-1$, $k \geqslant 2$. Since $\mathcal{G}_k \supset \mathcal{G}_{k-1}$, $E_0\{\bar{X} \mid \mathcal{G}_k\} = 0$ a.s. implies $E_0\{\bar{X} \mid \mathcal{G}_{k-1}\} = 0$ a.s. Hence we have to show only that the $(k+1)$-th moment of X, under $\theta = 0$, is equal to the $(k+1)$-th moment of some normal distribution. Since $\mathcal{M}_{k-1}$ and the set of functions $\{\prod_{v=1}^{k} (X_{j_v} - X_1);\ j_1, \ldots, j_k = 2, \ldots, n\}$ generate the subspace $\mathcal{M}_k$, the condition

$$(8.8.11)\quad E_0\{\bar{X} \mid \mathcal{G}_k\} = 0 \quad \text{a.s.}$$

is equivalent to the set of conditions

$$(8.8.12)\quad \begin{cases} E_0\{\bar{X} \mid \mathcal{G}_{k-1}\} = 0 \quad \text{a.s.} \\ E_0\left\{\bar{X} \prod_{v=1}^{k} (X_{j_v} - X_1)\right\} = 0, \quad \text{all } j_1, \ldots, j_k = 2, \ldots, n. \end{cases}$$

We can write

$$(8.8.13)\quad \prod_{v=1}^{k} (X_{j_v} - X_1) = \prod_{v=1}^{s} (X_{i_v} - X_1)^{\alpha_v}$$

where $i_1, \ldots, i_s$ are pairwise distinct; $\alpha_1, \ldots, \alpha_s$ non-negative integers such that $\alpha_1 + \cdots + \alpha_s = k$. Thus for each $j_1, \ldots, j_k = 2, \ldots, n$ we have

(8.8.14)

$$E_0\left\{\bar{X}\prod_{v=1}^{k}(X_{j_v} - X_1)\right\} = \frac{1}{n}\sum_{j_1=0}\cdots\sum_{j_s=0}(-1)^j\binom{j_1}{\alpha_1}\cdots\binom{j_s}{\alpha_s}\mu_{j_1}\cdots\mu_{j_s}\mu_{k+1-j},$$

where $j_1 + \cdots + j_s = j$, $\mu_j = E_0\{X^j\}$. We have to consider separately then the case of $k - 1$ odd and the case of $k - 1$ even. Applying the induction hypothesis, $\mu_j = 0$ for all odd $j = 1, \ldots, k - 1$, and we obtain after some algebraic manipulation that $\mu_{k+1} = 0$. If $k - 1$ is even, we obtain from the induction hypothesis that $\mu_{k+1} = k!\,\sigma^{k+1}$. The details can be found in Kagan's paper [1]. We remark that the condition of $n \geqslant 3$ is necessary to assure that subspace $\mathcal{M}_k$ contains the function $\prod_{v=1}^{s}(X_{j_v} - X_1)$, with $\sum_{v=1}^{s}\alpha_v = k$.
(Q.E.D.)

The two lemmas proven imply the main theorem.

Theorem 8.8.3. (Kagan [1].) *Let the family of distribution functions $F(x - \theta)$, $-\infty < \theta < \infty$, have moments of order 2k. Then for samples of size $n \geqslant 3$, the sample mean $\bar{X}$ is an admissible of θ within the subclass $\mathfrak{D}^{(k)}$ if and only if the first $k + 1$ moments of $F(x)$ coincide with the corresponding moments of some normal distribution.*

We comment that Theorem 8.8.3 implies the result of Kagan, Linnik, and Rao [1] because the class of all unbiased estimators $\mathfrak{D}_u$ contains $\bigcup_{k=1}^{\infty}\mathfrak{D}^{(k)}$. Therefore, since every distribution whose moments coincide with those of a normal distribution is normal, $\bar{X}$ is an admissible unbiased estimator of θ if and only if $F(x)$ is a normal distribution. Kagan extended in his paper [1] the results reported here to cases of vector valued location parameters also.

8.9. ADMISSIBILITY OF ESTIMATORS AND PROCEDURES IN SAMPLING FINITE POPULATIONS

In recent years, the problem of the admissibility of various sampling procedures and estimators of finite population parameters has been the subject of a significant number of papers. A general review of some of these papers can be found in Solomon and Zacks [1]. We present here some of the results of Godambe and Joshi [1], Joshi [1; 4], Godambe [4], and Hanurav [2].

A finite population $\mathcal{U}$ is a set $\{u_1, \ldots, u_N\}$ of N units, which are identifiable. Every unit has a certain characteristic, which is represented by a real or vector valued function $x(u)$. The values $x_i = x(u_i)$, $i = 1, \ldots, N$, are unknown. We therefore consider the vector of unknown population values

$x = (x_1, \ldots, x_N)'$ as a parametric point in a Euclidean space $\mathcal{X}$. A parametric function $\theta(x)$ is a function from $\mathcal{X}$ to Θ, where Θ will be considered an interval in an appropriate Euclidean space. In the papers mentioned above the parametric function under consideration is the population total $T(x) = x'1_N$.

A sample s from $\mathcal{U}$ is any *ordered* finite sequence of units of $\mathcal{U}$; that is,

$$s = \langle u_i, \ldots, u_{i_{n(s)}} \rangle,$$

where $n(s)$ designates the sample size; $n(s)$ can be fixed or random. Since sampling could be with replacement, $n(s)$ can be as large as we wish. The effective sample size, $\nu(s)$, is the number of different units of $\mathcal{U}$ in s. Obviously, $\nu(s) \leqslant N$.

Let ζ designate the class of *all* ordered sequences of units of $\mathcal{U}$; ζ is called the sample space. The rule of choosing a sample s from ζ is represented by a probability function $p(s)$ on ζ, according to which a sample s is chosen from ζ with probability $p(s)$. The triplet $D = (\mathcal{U}, \zeta, p(\cdot))$ is called a *sampling design.*

Let (s, X) designate a sample s on ζ and a sequence X of the values $x(u)$ associated with the units u in s. Given a sample (s, X) and a design $D = (\mathcal{U}, \zeta, p(\cdot))$, the *likelihood function* of the parametric points x of $\mathcal{X}$ is

$$\text{(8.9.1)} \qquad L(x \mid (s, X)) = \begin{cases} cp(s), & \text{if } x_i = X_i, i \in s, \\ 0, & \text{otherwise,} \end{cases}$$

where $0 < c < \infty$ is an arbitrary constant. The likelihood function $L(x \mid (s, X))$ is the basis for an estimation procedure. The relationship of $L(x \mid (s, X))$ to Bayesian and non-Bayesian estimation procedures has been studied in various studies (see Godambe [4], Basu [3], Ericson [1], Zacks [9], and others). We introduce now the notions of estimators and their admissibility in relation to a sampling design D.

An *estimator* of a parametric function $\theta(x)$ is a function $e(s, X)$ from $(\zeta \times \mathcal{X})$ to Θ, which depends on x only through the values of $x(u)$ in X; that is, an estimator is a function of the units chosen and of their observed x-values. Given a specified loss function $W(\theta, e(s, X), n(s))$, the risk function associated with the sampling designs D and the estimator e is

$$\text{(8.9.2)} \qquad R(\theta; D, e) = \sum_{s \in \zeta} p(s) W(\theta, e(s, X), \eta(s)).$$

Let $\mathcal{E}$ be a class of estimators. An estimator e in $\mathcal{E}$ is said to be admissible within $\mathcal{E}$, with respect to a given design D and a loss function $W(\cdot, \cdot, \cdot)$, if there exists no other estimator e^* in $\mathcal{E}$ which strictly dominates e; that is,

$$\text{(8.9.3)} \qquad R(\theta; D, e^*) \leqslant R(\theta; D, e), \quad \text{for all } \theta \in \Theta,$$

with a strict inequality at some $\theta_0 \in \Theta$. An estimator e in $\mathcal{E}$ can be admissible with respect to a design D, but inadmissible with respect to another design D'. Thus Joshi [4] introduced the concept of *uniform admissibility* of an estimator e. This property states that the estimator is admissible with respect to all sampling designs. Since the sample size may play an important role here, Joshi's definition of uniform admissibility is as follows. A *sampling strategy* is a pair (D, e); (D_1, e_1) is called uniformly admissible if there exists no other sampling strategy (D_2, e_2) such that

(i) $E\{n(s) \mid D_2\} \leqslant E\{n(s) \mid D_1\}$,

and

(ii) $R(\theta; D_2, e_2) \leqslant R(\theta; D_1, e_1)$, for all $\sigma \in \Theta$, with strict inequality for some θ_0. The concept of uniform admissibility was also studied by Godambe [4].

Let $\pi(u)$ designate the inclusion probability of a unit from $\mathcal{U}$; $\pi(u) = \sum_{\{s:u\in s\}}, p(s)$. Given a design D, the Horvitz-Thompson unbiased estimator of the population total $T(x)$ is

$$\bar{e}(s, X) = \sum_{u\in s} \frac{x(u)}{\pi(u)}. \tag{8.9.4}$$

To prove that this estimator is unbiased we notice that

$$\sum_{s\in\zeta} p(s)\bar{e}(s, X) = \sum_{s\in\zeta} p(s) \frac{\sum_{u\in s} x(u)}{\sum_{\{s:u\in s\}} p(s)}$$
$$= \sum_{u\in\mathcal{U}} x(u) \frac{\sum_{\{s:u\in s\}} p(s)}{\sum_{\{s:u\in s\}} p(s)} = T(x).$$

The Horvitz-Thompson estimator occupies a central role in the theory and practice of sampling survey. We notice that it is *not* a Bayes estimator, since no Bayes estimator is a function of the sampling design. This is immediately verified from Bayes's theorem and the form of the likelihood function (8.9.1). The Horvitz-Thompson estimator in a fixed sample design is a special case of a linear estimator $e_1(s, X) = \sum_{u\in s} b(u)x(u)$ with

$$b(u) \geqslant 1, \qquad u \in \mathcal{U}, \quad \text{and} \quad \sum_{u\in\mathcal{U}} \frac{1}{b(u)} = n, \tag{8.9.5}$$

where the sample size n is fixed (independent of s in ζ). Indeed, if $A = \{s; n(s) \neq n\}$ and $\sum_{s\in A} p(s) = 0$, then $\sum_{s\in\mathcal{U}} p(s) = n$. Define then $b(u) = 1/\pi(u)$ and $e_1(s, X)$ is reduced to $\bar{e}(s, X)$. We now prove the following theorem.

Theorem 8.9.1. (Godambe and Joshi [1].) *For any fixed sample size design D, every estimator $e_1(s, X)$ satisfying (8.9.5), with n equal to the sample size, is* *admissible within the subclass of all linear estimators of $T(x)$ for the squared-error loss.*

Proof. Suppose there exists a linear estimator $e(s, X) = \sum_{u \in s} \beta(u, s)x(u)$ such that

$$\sum_{s \in \zeta} p(s)[e(s, X) - T(x)]^2 \leqslant \sum_{s \in \zeta} p(s)[e_1(s, X), -T(x)]^2, \tag{8.9.6}$$

for all x, with a strict inequality for some x_0. Let

$$\begin{aligned} h(s, X) &= \sum_{u \in s} [\beta(u, s) - b(u)]x(u) \\ &= \sum_{u \in s} \alpha(u, s)x(u). \end{aligned} \tag{8.9.7}$$

Substituting (8.9.7) in (8.9.6) we obtain

$$\sum_{s \in \zeta} p(s)h^2(s, X) + \sum_{s \in \zeta} p(s)[e_1(s, X) - T(x)]h(s, X) \leqslant 0. \tag{8.9.8}$$

Since (8.9.8) should hold for any x in $\mathfrak{X}$, the left-hand side of (8.9.8) is a seminegative quadratic form in x. Therefore the coefficients of $x^2(u)$ in the left-hand side of (8.9.8) must be nonpositive. Substituting (8.9.7) in (8.9.8) we obtain

$$\sum_{\{s: u \in s\}} p(s)\alpha^2(u, s) + \sum_{\{s: u \in s\}} p(s)\alpha(u, s)[b(u) - 1] \leqslant 0. \tag{8.9.9}$$

If we let $\delta(u) = \sum_{\{s: u \in s\}} p(s)\alpha(u, s)$, we can write (8.9.9) in the form

$$\sum_{\{s: u \in s\}} p(s)[\alpha(u, s) - \delta(u)]^2 + \pi(u)\delta^2(u) + \delta(u)[b(u) - 1] \leqslant 0. \tag{8.9.10}$$

Since $b(u) \geqslant 1$ we must have from (8.9.10) that

$$\delta(u) \leqslant 0, \quad \text{all} \quad u \in \mathcal{U}. \tag{8.9.11}$$

Consider now a special point in $\mathfrak{X}$, namely, $x^{(1)}$ such that $x^{(1)}(u) = 1/b(u)$. For this point, the estimator attains the value

$$e_1(s, X^{(1)}) = \sum_{u \in s} d(u) \frac{1}{b(u)} = n. \tag{8.9.12}$$

Moreover, according to (8.9.5), $T(x^{(1)}) = n$. Hence $e_1(s, X^{(1)}) - T(x^{(1)}) = 0$ and we obtain from (8.9.6) that, for all samples s such that $p(s) > 0$, $e(s, X^{(1)}) = 0$. Therefore $h(s, X^{(1)}) = 0$ with probability one. Furthermore, since

$$\begin{aligned} \sum_{s \in \zeta} p(s)h(s, X^{(1)}) &= \sum_{u \in \mathcal{U}} \delta(u)x^{(1)}(u) \\ &= \sum_{u \in \mathcal{U}} \delta(u) \frac{1}{b(u)}, \end{aligned} \tag{8.9.13}$$

we obtain

$$\sum_{u \in \mathcal{U}} \frac{(u)}{b(u)} = 0. \tag{8.9.14}$$

Since $\sum_{u \in \mathcal{U}} 1/b(u) = n$, $b(u) > 0$ for all $u \in \mathcal{U}$. Hence $\delta(u) = 0$ for each $u \in \mathcal{U}$. This implies that $\alpha(u, s) = 0$ for all s such that $p(s) > 0$. Hence $e(s, X) = e_1(s, X)$ a.s., for each x in $\mathcal{X}$. Therefore strict inequality cannot hold in (8.9.6) and $e_1(s, X)$ is admissible. (Q.E.D.)

As a corollary from the present theorem we deduce that the Horvitz-Thompson estimator is admissible within the subclass of linear estimators, whenever D is a fixed sample size design. Godambe and Joshi [1] show, by way of a counter-example, that the condition of a fixed sample size is necessary for the admissibility of $\bar{e}(s, X)$ within the subclass of linear estimators. Joshi [4] established that if the design is not necessarily a fixed sample size one, then the estimator

$$e^*(s, X) = \frac{N}{n(s)} \sum_{u \in s} x(u) \tag{8.9.15}$$

is admissible within the subclass of linear estimators, under the squared-error loss. The above results were extended by Joshi [4]. He proved that the Horvitz-Thompson estimator of $T(x)$ is admissible whenever the design is a fixed sample size design in the class of all measurable estimators, under the squared-error loss. He has established similar results for ratio estimators.

PROBLEMS

Section 8.1

1. Compare the risk function of the estimator D_n^2, defined in Example 8.1, to that of the minimum risk equivariant estimator.

2. Consider the case of a finite family of probability measures $\mathcal{P} = \{P_1, P_2\}$. Consider the function $g(P_i) = i$, $i = 1, 2$. What is the minimax estimator of $g(P_i)$ based on an observation X, with the loss function

$$L(g, \hat{g}) = \begin{cases} 1, & \text{if } g = 1, \hat{g} = 2, \\ 2, & \text{if } g = 2, \hat{g} = 1, \\ 0, & \text{otherwise?} \end{cases}$$

Is this minimax estimator admissible?

3. Apply Theorem 8.1.2 to derive admissible estimators of the parameter θ of the family of gamma distributions $\mathcal{G}(1/\theta, 1)$, $0 < \theta < \infty$, for a quadratic loss function.

4. Let X have a binomial distribution $\mathcal{B}(n; \theta)$, $0 < \theta < 1$. Consider the problem of estimating θ, with a quadratic loss function $L(\theta, \hat{\theta}) = \min\{(\theta - \hat{\theta})^2/\theta^2, 2\}$. Show that the unique minimax estimator is $\hat{\theta}(X) \equiv 0$. Is this estimator admissible? [This example, offered by Professor H. Rubin (see Ferguson [1], p. 97), illustrates a minimax estimator which is of no value.]

5. Let X have a negative binomial distribution N.B. (ψ, ν), $0 < \psi < 1$, ν known. Consider the loss function $L(\psi, \hat{\psi}) = (\psi - \hat{\psi})^2/\psi(1 - \psi)$.

(i) Derive the Bayes estimators against the prior beta (a, b) distributions. Are these estimators admissible?
(ii) Is the M.L.E. a generalized Bayes estimator? Is it a minimax estimator?

Section 8.2

6. Complete Example 8.4 by proving that $\tilde{\sigma}_n^2 = \max\{\sigma_0^2, S_n/(n + 2)\}$ has a uniformly smaller mean square error than $S_n/(n + 2)$ if $\Theta = \{\theta; \sigma \geqslant \sigma_0, 0 < \sigma_0 < \infty\}$.

7. Let $X_1, \ldots, X_n$ be i.i.d. random variables having a common extreme value distribution, with density $f(x - \alpha)$, where $f(x) = \exp\{-x - e^{-x}\}$, $-\infty < \alpha < \infty$. Prove that $(1/n)\sum_{i=1}^{n} e^{-X_i}$ is an admissible estimator of $e^{-\alpha}$ for the quadratic loss $L(e^{-\alpha}, d) = (d - e^{-\alpha})^2 e^{2\alpha}$. Is this estimator minimax?

8. Use the method of Karlin [4] to prove the almost admissibility of the Bayes equivariant estimators of the common mean of two normal distributions, which has been provided in Example 7.8. (See Zacks [7].)

Section 8.3

9. Let X have a gamma distribution $\mathcal{G}(1/\theta, 1)$, where $\Theta = \{\theta; \theta \geqslant \theta_0, 0 < \theta_0 < \infty\}$. Determine an almost admissible estimator of θ for a squared-error loss.

10. Consider the problem of estimating the variance components of A.O.V. Model II. The basic model has been formulated in Problem 18, Chapter 2. Assume that we know the values of μ and of σ^2. Provide an admissible estimator of τ^2 for the squared-error loss. (The commonly used unbiased estimator of τ^2 can assume negative values and is obviously inadmissible.)

11. Consider the N.B. (ψ, ν) distribution with density

$$g(x \mid \psi, \nu) = \frac{\Gamma(x + \nu)}{\Gamma(\nu)\Gamma(x + 1)} \psi^x (1 - \psi)^\nu, \qquad x = 0, 1, \ldots$$

where ν is specified, and $0 < \psi_0 \leqslant \psi < 1$. Determine an admissible estimator of $E_{\psi,\nu}\{X\} = \nu\psi/(1 - \psi)$.

Section 8.4

12. Let $X_1, \ldots, X_n$ be i.i.d. random variables having a location-parameter exponential distribution with a density

$$f(x; \alpha) = \exp\{-(x - \alpha)\}, \quad \alpha \leqslant x < \infty, \quad -\infty < \alpha < \infty.$$

(i) Show that the Pitman estimator of α, $\hat{\alpha}_n = X_{(1)} - 1/n$, is an admissible estimator for a squared-error loss.

(ii) Derive an admissible estimator of α, which is the analog of the Pitman estimator, for the bilinear convex loss function $L(\theta, d) = a(\theta - d)^+ + b(\theta - d)^-$ where $0 < a, b < \infty$, $x^+ = \max(0, x)$, $x^- = -\min(0, x)$.

13. Consider the family of logistic distributions, with densities

$$f(x; \alpha) = \frac{\exp\{-(x - \alpha)\}}{[1 + \exp\{-(x - \alpha)\}]^2}, \quad -\infty < x < \infty, \quad -\infty < \alpha < \infty.$$

(i) Determine the Pitman estimator of α.

(ii) Is the Pitman estimator admissible for a squared-error loss?

(iii) Determine an admissible estimator for the bilinear loss function $L(\theta, d) = a(\theta - d)^+ + b(\theta - d)^-$.

Section 8.5

14. Using a method similar to that of Example 8.9, prove that all the equivariant estimators of the variance components of the A.O.V. Model II (Problem 18, Chapter 2) are inadmissible for a squared-error loss. The group is of the affine transformations. (See Zacks [6].)

15. Consider a discrete distribution according to which

$$P_\theta[X = x] = \begin{cases} \frac{1}{2}, & \text{if} \quad x = \theta + 1, \\ \frac{1}{2}, & \text{if} \quad x = \theta - 1, \end{cases}$$

where $-\infty < \theta < \infty$. Consider the problem of estimating θ with the loss function

$$L(\theta, d) = \begin{cases} |\theta - d|, & \text{if} \quad |\theta - d| \leqslant 1, \\ 1, & \text{if} \quad |\theta - d| > 1. \end{cases}$$

(i) Determine a uniformly minimum risk translation equivariant estimator of θ.

(ii) What is the risk of a best translation equivariant estimator?

(iii) Show that the estimator $d(x) = (x + 1)I_{(-\infty,0)}(x) + (x - 1)I_{(0,\infty)}(x)$ has a uniformly smaller risk than the minimum risk translation equivariant estimator (see Blackwell [2]). (This is another example of a best equivariant estimator which is inadmissible.)

Section 8.6

16. Prove Lemma 8.6.1.

17. Give an example of an admissible estimator of a location parameter θ, which is constructed according to Theorem 8 6.4.

18. Given an example of an admissible linear estimator of a vector location parameter in the $\mathcal{N}(\boldsymbol{\theta}, \mathbf{I})$ case, for a squared-error loss, $L(\boldsymbol{\theta}, \mathbf{A}) = \|\mathbf{AX} - \boldsymbol{\theta}\|^2$.

Section 8.8

19. Consider n i.i.d. random variables having a location parameter exponential distribution, that is, $X \sim \alpha + G(1, 1)$. For $k \geqslant 2$, what is the form of the equivariant estimator $t^{(k)}(\mathbf{X}_n)$?

20. Construct the estimator $t^{(2)}(\mathbf{X}_n)$ for n, $n \geqslant 3$, i.i.d. random variables having a Laplacian distribution, with density

$$f(x; \alpha) = \tfrac{1}{2} \exp\{-|x - \alpha|\}, \quad -\infty < x < \infty, \quad -\infty < \alpha < \infty.$$

REFERENCES

Basu [3], Blackwell [2]; Blyth [1]; Brown [2]; Cohen [1]; Ericson [1]; Farrel [1]; Ferguson [1]; Fox and Rubin [1]; Girshick and Savage [1]; Godambe [4], [5]; Godambe and Joshi [1]; Hanurav [1], [2]; Hodges and Lehmann [2]; James and Stein [1]; Joshi [1], [4]; Kagan [1]; Kagan, Linnik, and Rao [1]; Karlin, [4]; Katz [1]; Perng [1]; Sacks [1]; Solomon and Zacks [1]; Stein [4], [5], [7], [8]; von Neumann and Morgenstern [1]; Wald [6]; Zacks [9].

CHAPTER 9

Testing Statistical Hypotheses

9.1. MULTIPLE HYPOTHESES TESTING—BAYES AND MINIMAX PROCEDURES FOR THE FIXED SAMPLE CASE

Let $X_1, \ldots, X_n$ be i.i.d. random variables, having a common distribution function $F_\theta(x)$, $\theta \in \Theta$; Θ is a specified interval in a Euclidean k-space, $E^{(k)}$. The sample size n is fixed. We let T designate the minimal sufficient statistic for the family $\mathfrak{F} = \{F_\theta;\ \theta \in \Theta\}$; T is some r-dimensional vector, $1 \leqslant r \leqslant n$. In case $\mathfrak{F}$ admits only a trivial sufficient statistic, the dimension of T is $r = n$. Since we do not consider in this section procedures with varying sample size, we consider only decision functions which depend on $(X_1, \ldots, X_n)$ only through T.

The problem of multiple hypotheses testing can be described in the following terms. A finite partition of the parameter space Θ is specified, that is, $\{\Theta_1, \ldots, \Theta_m\}$. The statistician observes T and has to decide to which one of the m subsets θ belongs. The decision that $\theta \in \Theta_j$ is interpreted as the acceptance of the hypothesis A_j: $\theta \in \Theta_j$ $(j = 1, \ldots, m)$ and the rejection of the other $(m - 1)$ alternative hypotheses. The decision is performed by a randomized test function $\phi(T)$, which is a probability vector $\phi(T) = (\phi_1(T), \ldots, \phi_m(T))'$, $\phi_j(T) \geqslant 0$, $j = 1, \ldots, m$; $\sum_{j=1}^m \phi_j(T) = 1$. The class $\mathfrak{D}$ of all randomized test functions is the $(M - 1)$-dimensional simplex of all probability vectors $\phi = (\phi_1, \ldots, \phi_m)'$. Non-negative functions $L_j(\theta)$, $j = 1, \ldots, m$, are defined on Θ. The function $L_j(\theta)$ designates the loss associated with the acceptance of H_j, when θ is the value of the parameter. If $\mathfrak{J}$ denotes the Borel sigma-field on Θ, then $L_j(\theta) \in \mathfrak{J}$ for all $j = 1, \ldots, m$.

We remark that in the classical Neyman-Pearson theory of testing two (composite) hypotheses, $m = 2$, and

$$L_j(\theta) = \begin{cases} 0, & \text{if } \theta \in \Theta_j, \quad j = 1, 2, \\ 1, & \text{otherwise} \end{cases} \tag{9.1.1}$$

The classical theory is concerned with the derivation of "optimal" procedures of testing. The optimality criterion is to maximize the *power* of a decision function ϕ under a constraint on the *size* of the test. The reader is most certainly familiar with these notions (see Lehmann [2], Ch. 3). Here we treat the testing problem in a more general fashion. We characterize, for certain families of distribution functions, the class of all Bayes procedures, the class of all admissible procedures, minimax procedures, etc.

Let $f(t; \theta)$ designate the density function of T with respect to a measure $\mu(\mathrm{d}t)$ under θ. Let $H(\theta)$ designate a prior distribution on Θ. The risk function associated with a test function ϕ is

$$R(\theta, \phi) = \sum_{j=1}^{m} L_j(\theta) \int \phi_j(t) f(t; \theta) \mu(\mathrm{d}t). \tag{9.1.2}$$

Since $0 \leqslant \phi_j(t) \leqslant 1$ for all $j = 1, \ldots, m$, obviously $0 \leqslant R(\theta, \phi) < \infty$ for all $\theta \in \Theta$. Therefore, the prior risk associated with $H(\theta)$ and ϕ is

$$R(H, \phi) = \sum_{j=1}^{m} \int \phi_j(t)\mu(\mathrm{d}t) \int H(\mathrm{d}\theta) L_j(\theta) f(t; \theta). \tag{9.1.3}$$

We assume that, for each $j = 1, \ldots, m$,

$$R_j(t) = \int H(\mathrm{d}\theta) L_j(\theta) f(t; \theta) < \infty \quad \text{a.s. } [\mu], \tag{9.1.4}$$

and that

$$\int \mu(\mathrm{d}t) R_j(t) < \infty. \tag{9.1.5}$$

This implies that $R(H, \phi) < \infty$ for all $\phi \in \mathfrak{D}$. A test function ϕ^H is called *Bayes* against H if it minimizes (9.1.3). It is easy to verify that

$$\phi_j^H(T) = \begin{cases} 1, & \text{if } R_j(T) = \min\limits_{i=1,\ldots,m} \{R_i(T)\}, \\ 0, & \text{otherwise.} \end{cases} \tag{9.1.6}$$

We notice that a Bayes procedure, against any prior distribution, is (a) not necessarily unique and (b) does not require randomization.

Example 9.1. A random variable X has an $\mathcal{N}(\theta, 1)$ distribution; $\Theta = (-\infty, \infty)$. The partition of Θ is to $\Theta_{-1} = (-\infty, -1)$, $\Theta_0 = [-1, 1]$, $\Theta_1 = (1, \infty)$. We assume that θ has a prior $\mathcal{N}(0, \tau^2)$ distribution. The loss functions are zero-one; that is,

$$L_{-1}(\theta) = \begin{cases} 0, & \text{if } \theta < -1, \\ 1, & \text{otherwise.} \end{cases}$$

$$L_0(\theta) = \begin{cases} 0, & \text{if } |\theta| \leqslant 1, \\ 1, & \text{otherwise.} \end{cases}$$

$$L_1(\theta) = \begin{cases} 0, & \text{if } \theta > 1, \\ 1, & \text{otherwise.} \end{cases}$$

Since the marginal distribution of X is $\mathcal{N}(0, 1 + \tau^2)$, these loss functions will induce a symmetric test procedure. We notice that

$$f(x;\theta)H(d\theta) = n(x \mid 0, 1 + \tau^2)\, n\left(\theta \,\Big|\, x\frac{\tau^2}{1+\tau^2}, \frac{\tau^4}{1+\tau^2}\right) d\theta,$$

where $n(\cdot \mid \xi, \sigma^2)$ is the density function of $\mathcal{N}(\xi, \sigma^2)$. Thus, according to (9.1.4),

$$\text{(9.1.7)} \quad R_{-1}(X) = n(X \mid 0, 1 + \tau^2)\int_{-1}^{\infty} n\left(\theta \,\Big|\, X\frac{\tau^2}{1+\tau^2}, \frac{\tau^4}{1+\tau^2}\right) d\theta$$

$$= n(X \mid 0, 1 + \tau^2)\Phi\left(\frac{1 + X^*}{\sigma}\right),$$

where $X^* = X(\tau^2/(1 + \tau^2))$ and $\sigma^2 = \tau^4/(1 + \tau^2)$. Similarly,

$$\text{(9.1.8)} \quad R_0(X) = n(X \mid 0, 1 + \tau^2)\left(\int_{-\infty}^{-1} + \int_{1}^{\infty}\right) n(\theta \mid X^*, \sigma^2)\, d\theta$$

$$= n(X \mid 0, 1 + \tau^2)\left[2 - \Phi\left(\frac{1 + X^*}{\sigma}\right) - \Phi\left(\frac{1 - X^*}{\sigma}\right)\right].$$

Furthermore,

$$R_1(X) = n(X \mid 0, 1 + \tau^2)\Phi\left(\frac{1 - X^*}{\sigma}\right).$$

For the comparison of $R_j(X)$, $j = -1, 0, 1$, we can disregard the common factor $n(X \mid 0, 1 + \tau^2)$. Thus let $R_j^*(X) = R_j(X)/n(X \mid 0, 1 + \tau^2)$. $R_0^*(X)$ is symmetric in X^* with a minimum at $X = 0$. Furthermore,

$$\text{(9.1.9)} \qquad \frac{d}{dX^*} R_0^*(X) = -\frac{1}{\sigma}\left[\varphi\left(\frac{1 + X^*}{\sigma}\right) - \varphi\left(\frac{1 - X^*}{\sigma}\right)\right],$$

where $\varphi(\cdot)$ is the standard normal density.

For all $X^* < 0$, $|1 + X^*| < 1 - X^*$. Hence $\varphi((1 + X^*)/\sigma) > \varphi((1 - X^*)/\sigma)$ for all $X^* < 0$. It follows from (9.1.9) that $R_0^*(X)$ is monotone decreasing on $(-\infty, 0)$, with $\lim_{X \to -\infty} R_0^*(X) = 1$. The function $R_{-1}^*(X)$ is monotone increasing with $\lim_{X \to -\infty} R_{-1}^*(X) = 0$ and $\lim_{X \to \infty} R_{-1}^*(X) = 1$. Hence there exists a unique point ξ_{-1} in $(-\infty, 0)$ at which $R_{-1}^*(X) = R_0^*(X)$. Symmetrically, $\xi_1 = -\xi_{-1}$ is the unique point in $(0, \infty)$ at which $R_0^*(X) = R_1^*(X)$. Thus the sample space of X, $\mathfrak{X} = (-\infty, \infty)$, is partitioned into three subsets, $(-\infty, \xi_{-1})$, $[\xi_{-1}, \xi_1]$, (ξ_1, ∞). If $X \in (-\infty, \xi_{-1})$, we accept A_{-1}; if $X \in [\xi_{-1}, \xi_1]$, we accept A_0; and if $X \in (\xi_1, \infty)$, we accept A_1. ■

The above example illustrates a special case in the general class of monotone procedures which will be discussed later on. The discussion of monotone procedures is preceded by some general theorems on the existence of minimax procedures.

Theorem 9.1.1. (Wald [6].) *Let Θ be a parameter space and $\mathfrak{D}$ a class of decision functions. Let $R(\theta, d)$ be a risk function on $\Theta \times \mathfrak{D}$. Assume the following:*

(i) *$R(\theta, d)$ is bounded below in d for each θ;*
(ii) *Θ is weakly semiseparable in the sense of Wald* (see Section 8.1);
(iii) *$\mathfrak{D}$ is weakly compact in the sense of Wald* (see Section 8.1);
(iv) *for each $d_1, d_2 \in \mathfrak{D}$ and α, $0 < \alpha < 1$, there exists $d_\alpha \in \mathfrak{D}$ such that*

$$R(\theta, d_\alpha) \leqslant \alpha R(\theta, d_1) + (1 - \alpha) R(\theta, d_2). \tag{9.1.10}$$

Let $\mathcal{H}$ be the class of all prior distributions on Θ, which concentrate on a finite number of points; then

$$\inf_{\mathfrak{D}} \sup_{\mathcal{H}} R(H, d) = \sup_{\mathcal{H}} \inf_{\mathfrak{D}} R(H, d), \tag{9.1.11}$$

and there exists a decision function d^* in $\mathfrak{D}$ such that

$$\sup_{\mathcal{H}} R(H, d^*) = \inf_{\mathfrak{D}} \sup_{\mathcal{H}} R(H, d). \tag{9.1.12}$$

Proof. Since Θ is weakly separable, there exists a sequence $\{\theta^i\} \subset \Theta$ such that

$$\overline{\lim_{i \to \infty}} R(\theta^i, d) \geqslant R(\theta, d), \quad \text{all } (\theta, d).$$

Let $\mathcal{H}_n$ be the family of all prior distributions on Θ, which are concentrated on $\{\theta_1, \ldots, \theta_n\}$. The convexity assumption, (9.1.10), implies that there exists a nonrandomized minimax decision function in $\mathfrak{D}$ with respect to $\mathcal{H}_n$. We choose a sequence $\{d^n\} \in \mathfrak{D}$ so that, for each n,

$$\sup_{\mathcal{H}_n} R(H, d^{(n)}) \leqslant \inf_{\mathfrak{D}} \sup_{\mathcal{H}} R(H, d) + \frac{1}{n}. \tag{9.1.13}$$

Since $\mathfrak{D}$ is weakly compact, there exists a subsequence $\{d^{(n_j)}\} \subset \{d^{(n)}\}$ and $d^0 \in \mathfrak{D}$ so that

$$\underline{\lim}_{j \to \infty} R(H, d^{(n_j)}) \geqslant R(H, d^0), \qquad \text{all } H \in \mathcal{H}. \tag{9.1.14}$$

Hence for each m,

$$\begin{aligned} \sup_{\mathcal{H}_m} R(H, d^0) &\leqslant \sup_{\mathcal{H}_m} \underline{\lim}_{j \to \infty} R(H, d^{(n_j)}) \\ &\leqslant \underline{\lim}_{j \to \infty} \sup_{\mathcal{H}_m} R(H, d^{(n_j)}) \\ &\leqslant \underline{\lim}_{j \to \infty} \sup_{\mathcal{H}_{n_j}} R(H, d^{(n_j)}), \end{aligned} \tag{9.1.15}$$

since for large enough j, $\mathcal{H}_m \subset \mathcal{H}_{n_j}$. Furthermore, according to the definition of $d^{(n)}$ and (9.1.15), we obtain

$$\text{(9.1.16)} \qquad \sup_{\mathcal{H}_m} R(H, d^0) \leqslant \varliminf_{j\to\infty} \left[\inf_{\mathfrak{D}} \sup_{\mathcal{H}_{n_j}} R(H, d) + \frac{1}{n_j} \right]$$

$$\leqslant \varliminf_{j\to\infty} \sup_{\mathcal{H}_{n_j}} \inf_{\mathfrak{D}} R(H, d) \leqslant \sup_{\mathcal{H}} \inf_{\mathfrak{D}} R(H, d).$$

On the other hand,

$$\text{(9.1.17)} \qquad \lim_{m\to\infty} \sup_{\mathcal{H}_m} R(H, d^0) = \lim_{m\to\infty} \sup_{1\leqslant i\leqslant m} R(\theta^i, d^0)$$

$$= \sup_{i\geqslant 1} R(\theta^i, d^0) \geqslant \varlimsup_{i\to\infty} R(\theta^i, d^0)$$

$$\geqslant R(\theta, d^0) \text{ for all } \theta,$$

due to the weak-compactness. Hence from (9.1.16) and (9.1.17) we obtain

$$\text{(9.1.18)} \qquad \lim_{m\to\infty} \sup_{\mathcal{H}_m} R(H, d^0) = \sup_{\Theta} R(\theta, d^0).$$

Finally,

$$\text{(9.1.19)} \qquad \sup_{\mathcal{H}} \inf_{\mathfrak{D}} R(H, d) \geqslant \lim_{m\to\infty} \sup_{\mathcal{H}_m} R(H, d^0)$$

$$= \sup_{\Theta} R(\theta, d^0) = \sup_{\mathcal{H}} R(\theta, d^0)$$

$$\geqslant \inf_{\mathfrak{D}} \sup_{\mathcal{H}} R(H, d).$$

This implies (9.1.11). (Q.E.D.)

To apply Theorem 9.1.1 for multiple testing we have to show that the class $\mathfrak{D}$ of all probability vectors $\phi = (\phi_1, \ldots, \phi_m)'$ is weakly compact in the sense of Wald, and that Θ is weakly semiseparable. This is sufficient, since the risk function

$$R(\theta, \phi) = \sum_{i=1}^{m} L(\theta) \int \phi_i(X) f(X; \theta) \mu(\mathrm{d}X)$$

is bounded below; $R(\theta, \phi)$ is linear in ϕ on $\mathfrak{D}$. Indeed,

$$R(\theta, \alpha\phi^{(1)} + (1 - \alpha)\phi^{(2)}) = \alpha R(\theta, \phi^{(1)}) + (1 - \alpha) R(\theta, \phi^{(2)}).$$

Hence (9.1.10) holds.

Define a metric on Θ as follows.

$$\text{(9.1.20)} \quad \rho(\theta_1, \theta_2) = \sum_{i=1}^{m} |L_i(\theta_1) - L_i(\theta_2)| + \int |f(x; \theta_1) - f(x; \theta_2)| \, \mu(\mathrm{d}x).$$

Let $\mathfrak{L}$ be the linear space spanned by $\{(L_1(\theta), \ldots, L_m(\theta)); \Theta\}$; $\mathfrak{L} \subset E^{(m)}$. Define the metric $\rho_1(\cdot, \cdot)$ on $\mathfrak{L}$ as

$$\text{(9.1.21)} \qquad \rho_1(\mathbf{L}^{(1)}, \mathbf{L}^{(2)}) = \sum_{i=1}^{m} |L_i^{(1)} - \mathbf{L}_i^{(2)}|, \quad \mathbf{L}^{(1)}, \mathbf{L}^{(2)} \in \mathfrak{L}.$$

The space $\mathfrak{L}$ is separable in the ρ_1 metric. Furthermore, let

$$\mathfrak{L}_1(\mu, \mathcal{B}) = \left\{f; f \in \mathcal{B} \text{ and } \int |f(x)|\, \mu(\mathrm{d}x) < \infty\right\},$$

where $\mathcal{B}$ is the Borel sigma-field on $\mathfrak{X}$. We define the norm $\|f\|_{\mathfrak{L}_1} = \int |f(x)|\, \mu(\mathrm{d}x)$; and $\rho_2(f^{(1)}, f^{(2)}) = \|f^{(1)} - f^{(2)}\|_{\mathfrak{L}_1}$. The space $\mathfrak{L}_1$ is separable in the ρ_2 metric. It follows from (9.1.20) that

$$\rho(\theta_1, \theta_2) = \rho_1(\mathbf{L}^{(1)}, \mathbf{L}^{(2)}) + \rho_2(f_{\theta_1}, f_{\theta_2}), \quad \text{where} \quad \mathbf{L}^{(i)} = \mathbf{L}(\theta_i).$$

Hence Θ is separable in the ρ metric. Thus for any $\theta_0 \in \Theta$ there exists $\{\theta_{n_i}\} \subset \Theta$ so that $\rho(\theta_0, \theta_n) \to 0$. This implies that $\rho_1(\mathbf{L}(\theta_1), \mathbf{L}(\theta_2)) \to 0$ and $\rho_2(f_{\theta_1}, f_{\theta_2}) \to 0$. Thus

(9.1.22)
$$\begin{aligned}
&\lim_{j\to\infty} |R(\theta_0, \phi) - R(\theta_{n_j}, \phi)| \\
&= \lim_{j\to\infty} \left| \sum_{i=1}^{m} \left[L_i(\theta_{n_j}) \int f(x; \theta_{n_j}) \phi_i(x) \mu(\mathrm{d}x) \right.\right. \\
&\qquad \left.\left. - L_i(\theta_0) \int f(x; \theta_0)\phi_i(x)\mu(\mathrm{d}x) \right] \right| \\
&= \lim_{j\to\infty} \left| \sum_{i=1}^{m} [L_i(\theta_{n_j}) - L_i(\theta_0)] \int \phi_i(x) f(x; \theta_{n_j})\mu(\mathrm{d}x) \right. \\
&\qquad \left. + \sum_{i=1}^{m} L_i(\theta_0) \int \phi_i(x)[f(x; \theta_{n_j}) - f(x; \theta_0)]\mu(\mathrm{d}x) \right| \\
&\leqslant \lim_{j\to\infty} \left\{ \sum_{i=1}^{m} |L_i(\theta_{n_j}) - L_i(\theta_0)| \right. \\
&\qquad \left. + \sum_{i=1}^{m} L_i(\theta_0) \int \mu(\mathrm{d}x)\, |f(x; \theta_{n_j}) - f(x; \theta_0)| \right\} \\
&= \lim_{j\to\infty} \rho_1(L(\theta_{n_j}), L(\theta_0)) + \lim_{j\to\infty} \rho_2(f_{\theta_{n_j}}, f_{\theta_j}) = 0, \quad \text{for all} \quad \phi \in \mathfrak{D}.
\end{aligned}$$

This proves that Θ is weakly semiseparable in the sense of Wald. It remains to show that $\mathfrak{D}$ is weakly compact.

Let $\{f^{(j)}\}$ be a countable dense subset of $\mathfrak{L}_1(\mu, \mathcal{B})$, and let $\{\phi_n\} \subset \mathfrak{D}$ be a given sequence. Since $0 \leqslant \phi_n \leqslant 1$ for each n, $\int \phi_n(x) f^{(j)}(x)\mu(\mathrm{d}x)$ is bounded for every $n \geqslant 1$ and each j. Let $\{\phi_{n_1}\} \subset \{\phi_n\}$ such that $\int \phi_{n_1}(x) f^{(1)}(x)\mu(\mathrm{d}x)$ converges to $F(f^{(1)})$. Consider $\int \phi_{n_1}(x) f^{(2)}(x)\mu(\mathrm{d}x)$. Since this sequence of reals is bounded, we can find a subsequence $\{\phi_{n_2}\} \subset \{\phi_{n_1}\}$ such that

$$\lim_{n_2\to\infty} \int \phi_{n_2}(x) f^{(2)}(x)\mu(\mathrm{d}x) = F(f^{(2)}).$$

Proceeding in this way we can determine a subsequence $\{m^*\} \subset \{\bigcap_{j=1}^{\infty} \{n_j\}\}$ such that

$$\text{(9.1.23)} \quad \lim_{m^* \to \infty} \int \phi_{m^*}(x) f^{(j)}(x)\mu(\mathrm{d}x) = F(f^{(j)}), \quad \text{all} \quad j = 1, 2, \ldots .$$

Since $\{f^{(j)}\}$ is dense in $\mathfrak{L}_1(\mu, \mathfrak{B})$, the sequence $\int \phi_{m^*}(x) f(x)\mu(\mathrm{d}x)$ converges for all $f \in \mathfrak{L}_1(\mu, \mathfrak{B})$. Let $F(f) = \lim_{m^* \to \infty} \int \phi_{m^*}(x) f(x)\mu(\mathrm{d}x)$; $F(f)$ is a linear functional on $\mathfrak{L}_1(\mu, \mathfrak{B})$, and

$$\text{(9.1.24)} \qquad \sup_{\|f\| \leqslant 1} |F(f)| \leqslant 1.$$

Indeed, for any $f \in \mathfrak{L}_1(\mu, \mathfrak{B})$,

$$|F(f)| \leqslant \overline{\lim_{m^* \to \infty}} \int |\phi_{m^*}(x)|\,|f(x)|\,\mu(\mathrm{d}x)$$
$$\leqslant \int |f(x)|\,\mu(\mathrm{d}x) = \|f\|_{\mathfrak{L}_1}.$$

This implies (9.1.24). Hence $F(f)$ is a linear bounded functional on $\mathfrak{L}_1(\mu, \mathfrak{B})$. According to the Riesz-Fisher theorem (see Naimark [1], p. 89) there exists a measurable function $\psi(X)$ such that

$$\text{(9.1.25)} \qquad F(f) = \int \psi(x) f(x)\mu(\mathrm{d}x).$$

Finally, the class $\mathfrak{D}$ of all probability vectors $\phi = (\phi_1, \ldots, \phi_m)'$, $0 \leqslant \phi_i(X) \leqslant 1$, is compact. Hence substituting above $f(X; 0)$ for $f(X)$, given a sequence $\{\phi_i^{(n)}; n \geqslant 1\}$, there exists a subsequence $\{\phi_i^{(n_j)}; j \geqslant 1\} \subset \{\phi_i^{(n)}; n \geqslant 1\}$ and ϕ_i in $\mathfrak{D}$ such that

$$\text{(9.1.26)} \qquad \lim_{j \to \infty} \int \phi_i^{(n_j)}(x) f(x; \theta)\mu(\mathrm{d}x) = \int \phi_i(x) f(x; \theta)\mu(\mathrm{d}x),$$

$i = 1, \ldots, m$. This implies that

$$\text{(9.1.27)} \qquad \lim_{j \to \infty} R(\theta, \phi^{(n_j)}) = R(\theta, \phi), \qquad \text{al} \ \ \theta \in \Theta.$$

This proves the weak-compactness of $\mathfrak{D}$. Thus we have proven the following theorem.

Theorem 9.1.2. *For the multiple hypotheses testing with a risk function (9.1.2) there exists a minimax procedure in the class $\mathfrak{D}$ of all randomized decision functions.*

We show in the next section that in our search for minimax procedures we can restrict attention to monotone procedures.

9.2. MONOTONE PROCEDURES OF TESTING—THE FIXED SAMPLE CASE

The theory of multiple testing for families of distribution functions having the monotone likelihood ratio (M.L.R.) property was developed by Rubin and Karlin [1], Karlin [1], and then extended by Karlin [2; 3] to general classes of Pólya type distributions. We present now the main results of these papers for the two-hypotheses case. The reader can find a nice exposition of the basic results for the n-hypotheses case, $m \geqslant 2$, in Ferguson [1], Section 6.1, and in Rubin and Karlin [1].

9.2.1. PRELIMINARIES

Let $\mathfrak{F} = \{F_\theta; \theta \in \Theta\}$ be a family of distribution functions of a (real valued) random variable, and assume that Θ is an interval on the real line. Let $f(x; \theta)$ designate the density of F_θ with respect to μ. The family $\mathfrak{F}$ is said to possess the *monotone likelihood ratio* (M.L.R.) property if, for any θ_1, θ_2 such that $\theta_1 < \theta_2$, the likelihood ratio $f(X; \theta_2)/f(X; \theta_1)$ is monotone increasing in a statistic $T(X)$. In this section we assume that $T(X) = X$. Thus the M.L.R. property is

$$f(X_1; \theta_1)f(X_2; \theta_2) - f(X_1; \theta_2)f(X_2; \theta_1) \geqslant 0 \tag{9.2.1}$$

for all $X_1 < X_2$ and $\theta_1 < \theta_2$.

The 1-parameter exponential family is an M.L.R. one. There are other examples of such families.

We introduce now some terminology. Let $g(\theta)$ be a function on Θ. A point θ_0 is called a *change point* of the *first kind* if, for all $\theta' < \theta_0 < \theta''$ sufficiently close to θ_0, $g(\theta')g(\theta'') \leqslant 0$. A change point θ_0 is of the *second kind* if, for all $\theta' < \theta_0 < \theta''$ sufficiently close to θ_0, $g(\theta')g(\theta_0)g(\theta'') \leqslant 0$. Points of change of the second kind enable us to use functions $g(\theta)$ with discrete points of discontinuity, or step functions.

Let $\pi = \{\theta_1, \ldots, \theta_M\}$ be partition points for Θ, $-\infty \leqslant \theta_1 < \theta_2 < \cdots < \theta_M \leqslant \infty$. We designate by $N_\pi(g)$ the number of sign changes in the ordered sequence $\langle g(\theta_1), \ldots, g(\theta_M)\rangle$. The number of sign changes in g is defined as

$$V(g) = \sup_\pi N_\pi(g). \tag{9.2.2}$$

If $V(g) = n$ there are n points of change to the function g.

A point of change of the first kind, θ_0, is called a *nodal* point of g, if $g(\theta_0) = 0$ and $g(\theta')g(\theta'') < 0$ for all $\theta' < \theta_0 < \theta''$ sufficiently close to θ_0.

The *spectrum* of a measure μ on $(\mathfrak{X}, \mathfrak{B})$ is the set of all X's such that if $X \in W$ and W is an open set, then $\mu(W) > 0$. For example, if P is a probability measure on $(\mathfrak{X}, \mathfrak{B})$, with a corresponding Poisson distribution, then the spectrum of P consists of the set of non-negative integers. We designate by $\mathcal{S}(\mu)$ the spectrum of μ. Let $\mathcal{C}(g)$ designate the set of all change points of g, and $\mathfrak{Z}(g)$ the set of zero points of g.

In the following lemma we assume that $f(X; \theta) = e^{\theta X}\beta(\theta)$.

Lemma 9.2.1. (Karlin [1].) *Let H be a distribution function on Θ and g a measurable function such that $f(X; \theta)g(\theta)$ is H-integrable a.s. $[\mu]$. If $V(g) = n$ and $f(X; \theta)$ is a 1-parameter exponential density, then*

$$f(X) = \int f(X; \theta)g(\theta)H(d\theta) \tag{9.2.3}$$

has at most n zeros, counting nonmodal zeros twice, or is identically zero; $f(X)$ is identically zero only if $\mathcal{S}(H) \subset \mathcal{C}(g) \subset \mathfrak{Z}(g)$.

Proof. Let $\theta_1 \leqslant \ldots \leqslant \theta_n$ designate the change points of g. The quantity $\theta_i = \theta_{i+1}$ only when these are change points of the second kind. For uniqueness, assume that $(-1)^n g(\theta) \geqslant 0$ for all $\theta < \theta_1$. The product $\prod_{i=1}^n (\theta - \theta_i)$ and $g(\theta)$ change signs at the same change points. Therefore

$$\prod_{i=1}^{n} (\theta - \theta_i)g(\theta) \geqslant 0, \qquad \text{all } \theta \in \Theta. \tag{9.2.4}$$

Hence

$$\int_\Theta f(X; \theta) \prod_{i=1}^{n} (\theta - \theta_i)g(\theta)H(d\theta) \geqslant 0. \tag{9.2.5}$$

Consider now the 1-parameter exponential case, that is, $f(X; \theta) = e^{\theta X}\beta(\theta)$. We obtain

$$K(X) = \int e^{\theta X}\beta(\theta) \prod_{i=1}^{n} (\theta - \theta_i)g(\theta)H(d\theta) > 0, \tag{9.2.6}$$

unless $\mathcal{S}(H) \subset (\mathcal{C}(g) \cup \mathfrak{Z}(g))$. Assuming that

$$\int |\theta|^n \, e^{\theta X}\beta(\theta) \, |g(\theta)| \, H(d\theta) < \infty, \tag{9.2.7}$$

we can differentiate $f(X)$ under the integral sign. We can prove then, by induction on n, that

(9.2.8)

$$K(X) = e^{\theta_1 X} \frac{d}{dX}\left\{e^{-\theta_n X + \theta_{n-1} X} \frac{d}{dX}\left\{\ldots \frac{d}{dX} e^{-\theta_2 X + \theta_1 X}\left\{\frac{d}{dX} e^{-\theta_1 X} f(X)\right\}\right\}\right\},$$

where the three dots show n-fold iterated differentiation. Let

$$D_j(X) = e^{-\theta_j X + \theta_{j-1} X} \frac{d}{dX} \{D_{j-1}(X)\}, \qquad j = 2, \ldots, \tag{9.2.9}$$

where $D_1(X) = e^{-\theta_1 X} f(X)$.

If $K(X) > 0$, we obtain from (9.2.8) that $(d/dX)D_n(X) > 0$. Hence $D_n(X)$ is a strictly increasing function of X. According to Rolle's theorem, $D_n(X)$ has *at most* one modal zero. Let ξ_1 be a zero of $D_n(X)$. If $X < \xi_1$ then $D_n(X) < 0$, and $D_{n-1}(X)$ is strictly decreasing. At the same time $D_{n-1}(X)$ is strictly increasing on (ξ_1, ∞). Hence, again by Rolle's theorem, $D_{n-1}(X)$ has *at most* two zeros, one at $(-\infty, \xi_1)$ and one at (ξ_1, ∞). Proceeding in this manner, we obtain that $D_{n-j}(X)$ has *at most* $j + 1$ zeros, for all $j = 0, \ldots, n - 1$. Finally, since $D_1 = e^{-\theta_1 X} f(X), f(X)$ has *at most* n zeros, counting twice the nonmodal zeros. (Q.E.D.)

We can also show that sign $\{f(-\infty)\}$ = sign $\{g(-\infty)\}$, and that if $V(f) = n$, then $f(X)$ and $g(\theta)$ change signs in the same order.

9.2.2. TESTING TWO HYPOTHESES

In the present subsection we concentrate on the problem of testing two hypotheses. We restrict attention to distribution functions belonging to the 1-parameter exponential family. Generalizations will be mentioned at the end.

Let $L_1(\theta)$ and $L_2(\theta)$ designate the loss functions associated with the two possible actions. Let $\Delta(\theta) = L_1(\theta) - L_2(\theta)$. We assume the following:

(i) $\Delta(\theta)$ changes sign *at most* n times over Θ;
(ii) $\Delta(\theta)$ has at most a countable set of discontinuities;
(iii) all the zeros of $\Delta(\theta)$ are modal;
(iv) $\mathfrak{Z}(\Delta) \subset \mathfrak{C}(\Delta)$.

Define

$$\Delta_H(X) = \int \Delta(\theta)\beta(\theta)e^{\theta X} H(d\theta). \tag{9.2.10}$$

According to (9.1.6), the Bayes decision against H is to accept the first hypothesis if $\Delta_H(X) < 0$; the second hypothesis if $\Delta_H(X) > 0$; and decide arbitrarily if $\Delta_H(X) = 0$. Assuming that $\mathcal{S}(H)$ is not contained in $\mathfrak{C}(\Delta)$, we obtain from Lemma 9.2.1 that $\Delta_H(X)$ changes sign *at most* n times. From (i)–(iv) we deduce that all the change points of $\Delta_H(X)$ are zero points. Let us designate these points by ξ_i $(i = 1, \ldots, n)$. We have $\xi_1 \leqslant \xi_2 \leqslant \cdots \leqslant \xi_n$. Furthermore, let $\xi_0 \equiv -\infty$, $\xi_{n+1} \equiv +\infty$. We assume also that sign $\{\Delta_H(X)\}$ $= -1$ for all $X < \xi_1$.

In the case of two hypotheses, the test function $\phi(X)$ can be taken as the first component of the probability vector $\boldsymbol{\phi}(X) = (\phi_1(X), \phi_2(X))$. Thus according to (9.1.6) and Lemma 9.2.1, the Bayes test function against H is of the general form

$$(9.2.11) \quad \phi^H(X) = \begin{cases} 1, & \text{if } \xi_{2j} \leqslant X \leqslant \xi_{2j+1}, \quad j = 0, 1, \ldots, \left[\frac{n}{2}\right]. \\ 0, & \text{if } \xi_{2j+1} < X < \xi_{2j+2}. \end{cases}$$

When $X = \xi_k$ $(k = 1, \ldots, n)$ the Bayes test function is arbitrary. There is no loss of optimality if we agree to let $\phi^H(\xi_k) = 1$ for all $k = 1, \ldots, n$. We summarize the above results in the following theorem.

Theorem 9.2.2. (Karlin [1].) *Under the above assumption, the Bayes testing procedure against any prior distribution H is uniquely determined, except for at most n points, and consists of at most* $(n + 1)$ *disjoint intervals on which the hypothesis is either accepted or rejected.*

The points at which the Bayes procedure is arbitrary are the (at most) n points of change of $\Delta_H(X)$. However, if we proceed according to the above convention, our Bayes procedures are unique. We assumed above that $\mathcal{S}(H)$ is not contained in $\mathcal{C}(\Delta)$. If $\mathcal{S}(H) \subset \mathcal{C}(\Delta)$, all procedures are Bayes against that H. Thus nontrivial Bayes procedures are characterized by at most *n critical points* $\xi_1, \ldots, \xi_n$, which are the change points (or zeros) of $\Delta_H(X)$. These n critical points define a partition of the real line to $(n + 1)$ disjoint intervals. On each interval only one of the two actions is to be taken. All procedures of testing which define some finite partition of the real line, so that on each of the disjoint intervals only one of the actions is decided upon, are called *monotone* procedures. All Bayes procedures are thus either arbitrary or monotone procedures. We designate by $\mathcal{M}_n$ the subclass of all monotone procedures which are determined by *at most n* critical points. We prove now that if $\Delta(\theta)$ has n sign changes, and $f(X; \theta) = e^{\theta X}\beta(\theta)$, then $\mathcal{M}_n$ is an essentially complete class. That is, we can consider only monotone procedures in $\mathcal{M}_n$.

Theorem 9.2.3. (Karlin [1].) *Under (i)–(iv) on* $\Delta(\theta)$, *and assuming that* $f(X; \theta) = e^{\theta X}\beta(\theta)$, $\mathcal{M}_n$ *is essentially complete.*

Proof. Let $\theta_1, \theta_2, \ldots$ be a dense subset of Θ containing *all* points of discontinuity of $\Delta(\theta)$. Let $\{\theta_1, \ldots, \theta_M\}$ be the first $M \geqslant 2$ points of this sequence. For each θ_i $(i = 1, \ldots, M)$ we have a risk of

$$(9.2.12) \qquad R_i(\phi) = L_2(\theta_i) + \Delta(\theta_i) \int \phi(x) e^{x\theta_i} \beta(\theta_i) \mu(\mathrm{d}x).$$

Let S_M designate the set of all M-dimensional vectors $(R_1(\phi), \ldots, R_M(\phi))$ where ϕ transverses over [0, 1]. The set S_M is a compact convex set in the Euclidean M-space. Indeed, the class of all functions ϕ, $0 \leqslant \phi(X) \leqslant 1$, is compact in the topology of weak* convergence† relative to $\mathfrak{L}_1(\mu, \mathcal{B})$. The mapping φ: $[0, 1] \to S_M$ is continuous in the weak* topology and the Euclidean topology. The continuous range of a compact set is itself compact. Hence S_M is compact. The convexity of S_M is implied from the convexity of $\{\phi; 0 \leqslant \phi \leqslant 1\}$. As mentioned in Section 8.1, the subclass of all Bayes procedures ϕ^H, against priors which concentrate on $\{\theta_1, \ldots, \theta_M\}$, is a complete class. Let $\mathfrak{D}_M$ designate this subclass. Given any procedure $\phi^* \notin \mathfrak{D}_M$, choose an *admissible* procedure $\phi_M{}^0$ in $\mathfrak{D}_M$ such that $R_i(\phi_M{}^0) \leqslant R_i(\phi^*)$ for all $\lambda = 1, \ldots, M$; $\phi_M{}^0$ is Bayes against some prior which concentrates on $\{\theta_1, \ldots, \theta_M\}$ and is therefore, according to Theorem 9.2, a monotone procedure. Since $\mathfrak{D}_M$ is complete, (9.2.12) implies that

$$\text{(9.2.13)} \quad \Delta(\theta_i)\int \phi_M{}^0(x)e^{\theta_i x}\beta(\theta_i)\mu(\mathrm{d}x) \leqslant \Delta(\theta_i)\int \phi^*(x)e^{\theta_i x}\beta(\theta_i)\mu(\mathrm{d}x),$$

for all $i = 1, \ldots, M$. The monotone class $\mathcal{M}_n$ is closed in the weak* topology; that is, for every sequence $\{\phi_M\} \subset \mathcal{M}_n$ we can find a $\phi_0 \in \mathcal{M}_n$ so that, for all $\theta_i \in \{\theta_1, \theta_2, \ldots\}$,

$$\text{(9.2.14)} \quad \lim_{M\to\infty} \int \phi_M(x)e^{\theta_i x}\beta(\theta_i)\mu(\mathrm{d}x) = \int \phi_0(x)e^{\theta_i x}\beta(\theta_i)\mu(\mathrm{d}x).$$

If $\phi_M{}^0 \xrightarrow{W^*} \phi^0$ in the sense of (9.2.14), then

$$\text{(9.2.15)} \quad \int \phi^0(x)e^{\theta_i x}\beta(\theta_i)\mu(\mathrm{d}x) \leqslant \int \phi^*(x)e^{\theta_i x}\beta(\theta_i)\mu(\mathrm{d}x).$$

At all points of θ not in $\{\theta_1, \theta_2, \ldots\}$ the function $\Delta(\theta)$ is continuous. Therefore we can find a subsequence in $\{\theta_{i_k}\} \subset \{\theta_i\}$ such that $\theta_{i_k} \to \theta$ as $k \to \infty$ and, according to (9.2.12), $R_{i_k}(\phi) \to R_\theta(\phi)$ for all ϕ, where $R_\theta(\phi) = L_2(\theta) + \Delta(\theta)\int \phi(x)e^{\theta x}\beta(\theta)\mu(\mathrm{d}x)$. In particular, this holds for ϕ^0 and ϕ^*. Hence, since $R_{i_k}(\phi^0) \leqslant R_{i_k}(\phi^*)$ for all $\{\theta_{i_k}\}$,

$$\text{(9.2.16)} \quad R_\theta(\phi^0) \leqslant R_\theta(\phi^*), \quad \text{all} \quad \theta \in \Theta.$$

This proves that $\mathcal{M}_n$ is essentially complete. (Q.E.D.)

The proof of the essential completeness of $\mathcal{M}_1$, when $\Delta(\theta)$ has only one point of change, is significantly simpler. For this proof see Lehmann [2], p. 72, and Ferguson [1], p. 286.

We prove that *all* procedures in $\mathcal{M}_n$ are admissible, which means that $\mathcal{M}_n$ is a minimal complete class. For this purpose we show first the following lemma.

† For a discussion of weak* topology see Simmons [1], pp. 232–235.

Lemma 9.2.4. (Karlin [1].) *If* $\phi^{(1)}$ *and* $^{(2)}$ *are both in* $\mathcal{M}_n$ *then*

$$D(\phi^{(1)}, \phi^{(2)}) = \int e^{\theta X}[\phi^{(1)}(x) - \phi^{(2)}(x)]\mu(\mathrm{d}x) \tag{9.2.17}$$

has at most $(n - 1)$ *zeros, counting multiplicities.*

Proof. The proof consists of two steps. In the first one we show that $\phi^{(1)}(X) - \phi^{(2)}(X)$ has at most $(n \dot{-} 1)$ change points. Then from Lemma 9.2.1 we obtain that (9.2.17) has at most $(n - 1)$ zeros, counting multiplicities. Thus let $\xi_1 \leqslant \cdots \leqslant \xi_n$ be the change points of $\phi^{(1)}(X)$, and $\eta_1 \leqslant \cdots \leqslant \eta_n$ the change points of $\phi^{(2)}(X)$. If $\xi_1 \leqslant \eta_1$ then $\phi^{(1)}(X) - \phi^{(2)}(X) \leqslant 0$ for all $X < \xi_2$. This follows directly from (9.2.11). Hence the first change point of $\phi^{(1)} - \phi^{(2)}$ is not smaller than ξ_2. Let ζ_1 designate the least value of X such that $X \geqslant \xi_2$ *and* $\phi^{(1)}(X) - \phi^{(2)}(X) > 0$; ζ_1 is the first change point of $\phi^{(1)} - \phi^{(2)}$. Let ξ_i $(i \geqslant 3)$ be the least change point of $\phi^{(1)}$ which is greater or equal to ξ_3. Since $\phi^{(2)}(X)$ is in $[0, 1]$, $\phi^{(1)}(X) - \phi^{(2)}(X) \geqslant 0$ for all $\zeta_1 \leqslant X < \xi_i$. Hence the second change point of $\phi^{(1)} - \phi^{(2)}$ is not smaller than ξ_i. In the same manner we show that the third change point of $\phi^{(1)} - \phi^{(2)}$ is not smaller than ξ_{i+j} $(j \geqslant 1)$. Proceeding in this manner we obtain that the total number of changes in $\phi^{(1)} - \phi^{(2)}$ is at most $(n - 1)$. If $\xi_1 > \eta_1$, then $\phi^{(1)}(X) - \phi^{(2)}(X) \geqslant 0$ for all $X < \xi_2$. Thus $\zeta_1 \geqslant \xi_2$. The same argument as above establishes the required result. (Q.E.D.)

We remark that if $\phi \in \mathcal{M}_n$ and $\phi^* \notin \mathcal{M}_n$, then

$$\int (\phi(x) - \phi^*(x)) f(x;\, \theta)\mu(\mathrm{d}x)$$

changes signs at most n times. We now prove the following theorem.

Theorem 9.2.5. (Karlin [1].) *If* $\Delta(\theta)$ *satisfies (i)–(iv) and if* $\phi^* \notin \mathcal{M}_n$, *there exists a procedure* ϕ^0, $\phi^0 \in \mathcal{M}_n$, *which strictly dominates* ϕ^*, *that is,*

$$R(\theta, \phi^0) \leqslant R(\theta, \phi^*), \quad \textit{for all} \quad \theta \in \Theta,$$

with equality only at the n points of change of $\Delta(\theta)$. *Furthermore, if* $\tilde{\phi} \in \mathcal{M}_n$ *satisfies* $R(\theta, \tilde{\phi}) \leqslant R(\theta, \phi^0)$, *then* $\tilde{\phi} = \phi^0$ *a.s.*

Proof. Let $\theta_1 \leqslant \cdots \leqslant \theta_n$ be the points of change of $\Delta(\theta)$. Consider the case $\theta_1 < \theta_2 < \cdots < \theta_n$. Let $I_i = (\theta_{i-1}, \theta_i)$, $i = 1, \ldots, n - 1$, where $\theta_0 \equiv -\infty$ and $\theta_{n+1} \equiv \infty$. The risk function of a procedure ϕ is

$$R(\theta, \phi) = L_2(\theta) + \Delta(\theta) \int \phi(x) f(x; \theta)\mu(\mathrm{d}x). \tag{9.2.18}$$

If $\phi^* \notin \mathcal{M}_n$ we have, since $\mathcal{M}_n$ is essentially complete, a procedure $\phi^0 \in \mathcal{M}_n$ so that $R(\theta, \phi^0) \leqslant R(\theta, \phi^*)$ for *all* $\theta \in \Theta$. Thus from (9.2.18),

$$\Delta(\theta)\beta(\theta) \int [\phi^0(x) - \phi^*(x)] e^{\theta x} \mu(\mathrm{d}x) \leqslant 0, \qquad \text{all } \theta \in \Theta. \tag{9.2.19}$$

If $\theta \in I_1$ then $\Delta(\theta) < 0$ and $\int (\phi^0 - \phi^*)e^{\theta x}\mu(dx) \geqslant 0$. If $\theta \in I_2$ then $\Delta(\theta) > 0$ and $\int (\phi^0 - \phi^*)e^{\theta x}\mu(dx) \leqslant 0$, etc. Since $\int (\phi^0(x) - \phi^*(x))e^{\theta x}\mu(dx)$ is a continuous function of θ,

$$\int (\phi^0(x) - \phi^*(x))e^{\theta_i x}\mu(dx) = 0, \qquad i = 1, \ldots, n, \tag{9.2.20}$$

at all the change points of Δ, that is, on $\mathcal{C}(\Delta) = \{\theta_1, \ldots, \theta_n\}$. Since $\int (\phi^0 - \phi^*)f\mu$ vanishes *at most n* times, (9.2.20) implies that

$$\int (\phi^0(x) - \phi^*(x))f(x; \theta)\mu(dx) \neq 0 \text{ if } \theta \notin \mathcal{C}(\Delta).$$

Furthermore, $\Delta(\theta)\beta(\theta) \neq 0$ if $\theta \notin \mathcal{C}(\Delta)$. Hence

$$R(\theta, \phi^0) - R(\theta, \phi) = \Delta(\theta)\beta(\theta)\int [\phi^0(x) - \phi^*(x)]e^{\theta x}\mu(dx) < 0 \tag{9.2.21}$$

for all $\theta \notin \mathcal{C}(\Delta)$.

Finally, if $\tilde{\phi} \in \mathcal{M}_n$ and $R(\theta, \tilde{\phi}) \leqslant R(\theta, \phi^0)$ for all $\theta \in \Theta$, then

$$\Delta(\theta)\beta(\theta)\int (\tilde{\phi}(x) - \phi^0(x))e^{\theta x}\mu(dx) \leqslant 0, \qquad \text{all } \theta \in \Theta. \tag{9.2.22}$$

As in (9.2.20),

$$\int (\tilde{\phi}(x) - \phi^0(x))e^{\theta x}\mu(dx) = 0, \qquad \theta \in \mathcal{C}(\Delta). \tag{9.2.23}$$

But according to Lemma 9.2.4, $\int (\tilde{\phi} - \phi^0)f\mu$ has *at most* $(n - 1)$ zeros. Hence $\tilde{\phi}(X) = \phi^0(X)$ a.s. $[\mu]$. (Q.E.D.)

Corollary 9.2.6. *Under assumptions (i)–(iv), $\mathcal{M}_n$ is a minimal complete class, and every monotone procedure in $\mathcal{M}_n$ is admissible.*

Proof. According to Theorem 9.2.5, if $\phi^* \notin \mathcal{M}_n$ there exists $\phi^0 \in \mathcal{M}_n$ which *strictly* dominates ϕ^*. Hence $\mathcal{M}_n$ is a complete class. Moreover, since there are no procedures in $\mathcal{M}_n$ that are strictly dominated by other procedures in $\mathcal{M}_n$, this subclass is minimal complete. Finally, any procedure in $\mathcal{M}_n$ cannot be dominated by a procedure not in $\mathcal{M}_n$ since the relationship of dominance is transitive. In other words, suppose that $\phi^0 \in \mathcal{M}_n$ and $\phi^* \notin \mathcal{M}_n$, and assume that $R(\theta, \phi^*) \leqslant R(\theta, \phi^0)$ for all $\theta \in \Theta$. According to Theorem 9.2.5, there exists a procedure $\tilde{\phi}$ in $\mathcal{M}_n$ such that $R(\theta, \tilde{\phi}) \leqslant R(\theta, \phi^*)$ for all θ, with equality on at most n points. Hence $R(\theta, \tilde{\phi}) \leqslant R(\theta, \phi^0)$ with equality on at most n points. This contradicts Theorem 9.2.5. (Q.E.D.)

Karlin [1] proved that, given any monotone procedure in $\mathcal{M}_n$, say with n critical points, it is Bayes against some prior distribution $H(\theta)$ concentrated on the n zero points of $\Delta(\theta)$. He provided the formula for determining such a prior distribution H. An extension of the results of this section to families of distribution functions of the Pólya type has been given by Karlin [2; 3]. Rubin and Karlin [1] studied the form of minimax procedures for testing two hypotheses under slightly more general assumptions than those of this section. We notice that if $\Delta(\theta)$ has n points of change, all minimax procedures are monotone, since $\mathcal{M}_n$ is minimal complete. Rubin and Karlin proved that there exists a minimax strategy for nature which consists of a prior distribution H^* concentrated on two points, provided $L_1(\theta)$ and $L_2(\theta)$ do not grow too fast. In the special case that $\Delta(\theta)$ has only one zero, every Bayes procedure in $\mathcal{M}_1$ is characterized by one ciritical point, say X_0. Thus $L_1(\theta) = 0$ if $\theta \leqslant \theta_0$ and $L_2(\theta) = 0$ if $\theta \geqslant \theta_0$. The risk under X_0 is

$$R(\theta, X_0) = \begin{cases} L_1(\theta)\displaystyle\int_{-\infty}^{X_0} f(x;\theta)\mu(\mathrm{d}x), & \text{if } \theta \geqslant \theta_0, \\ L_2(\theta)\displaystyle\int_{X_0}^{\infty} f(x;\theta)\mu(\mathrm{d}x), & \text{if } \theta \leqslant \theta_0. \end{cases} \tag{9.2.24}$$

Assume that $L_1(\theta)$ is continuous; then the risk function is continuous and a minimax value of X_0 is such that the two humps of $R(\theta, X_0)$ are equal. This is illustrated in the following example.

Example 9.2. Let $\mathcal{F}$ be the family of normal distributions $\mathcal{N}(\theta, 1)$, $-\infty < \theta < \infty$. Consider the problem of testing $A_0\colon \theta \leqslant 0$, versus $A_1\colon \theta > 0$ (these are two composite hypotheses). The distribution $\mathcal{N}(\theta, 1)$ belongs to a 1-parameter exponential family.

Consider the loss functions

$$L_0(\theta) = \begin{cases} 0, & \text{if } \theta \leqslant 0 \\ \theta, & \text{if } \theta > 0 \end{cases}$$

$$L_1(\theta) = \begin{cases} |\theta|, & \text{if } \theta \leqslant 0 \\ 0, & \text{if } \theta > 0. \end{cases}$$

The risk function associated with a monotone procedure, specified by X_0, is

$$R(\theta, X_0) = \begin{cases} \theta\Phi(X_0 - \theta), & \text{if } \theta > 0 \\ |\theta|\,[1 - \Phi(X_0 - \theta)], & \text{if } \theta < 0. \end{cases}$$

In particular, for $X_0 = 0$ we get

$$R(\theta, 0) = |\theta|\,\Phi(-|\theta|), \quad \text{all} \quad -\infty < \theta < \infty.$$

This risk function is *independent* of the sign of θ. Hence $\sup\limits_{0<\theta<\infty} \theta\Phi(-\theta) = \sup\limits_{-\infty<\theta<0} |\theta|\,\Phi(-|\theta|)$. It follows that X_0 is minimax. ■

9.3. EMPIRICAL BAYES TESTING OF MULTIPLE HYPOTHESES

As in Section 6.9 consider a case of repetitive experiment represented by a sequence of independent random vectors $(X_1, \theta_1), \ldots, (X_n, \theta_n), \ldots$. The observed random variable is X_n $(n = 1, 2, \ldots)$ while θ_n is unobserved. Furthermore, we assume that for each $n \geqslant 1$ the conditional density function of X_n, given θ_n, with respect to μ, is $f(x; \theta_n)$. All $\theta_n \in \Theta$ and there exists an unknown prior distribution on Θ, $H(\theta)$, which is common to all $\theta_1, \theta_2, \ldots$. A number m of hypotheses $(m \geqslant 2)$ are specified, that is, $A_i; \theta \in \Theta_i$, $i = 1, \ldots, m$. After each observation we have to take one of m actions. The i-th action, $a_i^{(n)}$, is to accept the hypothesis, $A_i^{(n)}: \theta_n \in \Theta_i$ $(i = 1, \ldots, m)$. The method of empirical Bayes testing provides a decision function whose prior risk converges to the Bayes risk. In this section we provide some of the results of Robbins [4] and Samuel [1] concerning empirical Bayes testing procedures.

Let $L_i(\theta)$, $\theta \in \Theta$, $i = 1, \ldots, m$, be the loss function associated with action $a_i^{(n)}$ $(i = 1, \ldots, m)$, $n \geqslant 1$. We define

$$(9.3.1) \qquad \Delta_i^{(H)}(X) = \int [L_i(\theta) - L_1(\theta)] f(X; \theta) H(\mathrm{d}\theta).$$

We assume that $L_i(\theta)$ satisfies,

$$(9.3.2) \qquad \int |L_i(\theta) - L_1(\theta)|\, f(X; \theta) H(\mathrm{d}\theta) < \infty, \qquad i = 2, \ldots, m.$$

The Bayes procedure is to accept A_i if

$$(9.3.3) \quad \Delta_i^{(H)}(X) = \min\,\{0, \Delta_2^{(H)}(X), \ldots, \Delta_m^{(H)}(X)\}, \qquad i = 1, \ldots, m.$$

If the minimum (9.3.3) is attained for several values of i, we can accept any one of the corresponding hypotheses arbitrarily. When the prior distribution H is unknown, the empirical Bayes approach is to provide a sequence

$$\{\Delta_i^{(n)}(x; \mathbf{X}_{n-1});\ n \geqslant i, 1 = 2, \ldots, m\}$$

such that

$$(9.3.4) \quad \operatorname*{p.lim}_{n\to\infty} \Delta_i^{(n)}(x; \mathbf{X}_{n-1}) = \Delta_i^{(H)}(x), \quad \text{a.s. } [\mu], \quad i = 1, \ldots, m.$$

The decision rule is to accept, at the n-th test, the hypothesis $A_i^{(n)}$ which corresponds to the minimal $\Delta_i^{(n)}(x; \mathbf{X}_{n-1})$. Later on we show how to estimate the prior distribution $H(\theta)$ consistently by an estimator $H_n(\theta \mid \mathbf{X}_{n-1})$. We can define then, for $i = 2, \ldots, m$,

$$(9.3.5) \qquad \Delta_i^{(n)}(x; \mathbf{X}_{n-1}) = \int [L_i(\theta) - L_1(\theta)] f(x; \theta) H_n(\mathrm{d}\theta \mid \mathbf{X}_{n-1}).$$

The proof that such an estimation rule will attain the Bayes risk is based on Theorem 6.9.1. Indeed, if $L_i(\theta) - L_1(\theta)$ is a continuous bounded function of θ, then by the Helly-Bray lemma,

$$\operatorname*{p.lim}_{n\to\infty} \Delta_i^{(n)}(x; \mathbf{X}_{n-1}) = \Delta_i^{(H)}(x), \quad \text{for each } i = 2, \ldots, m.$$

Example 9.3. This example was given by Robbins [4]. Let $X \sim P(\theta)$ be a Poisson random variable with mean θ; $\Theta = (0, \infty)$. There is a prior distribution $H(\theta)$ on Θ that is unknown. We wish to test two hypotheses: $A_1: \theta \leq \theta^*$ versus $A_2: \theta > \theta^*$. The loss functions considered are

$$L_1(\theta) = \begin{cases} 0, & \text{if } \theta \leqslant \theta^*, \\ \theta - \theta^*, & \text{otherwise,} \end{cases}$$

and

$$L_2(\theta) = \begin{cases} \theta^* - \theta, & \text{if } \theta \leqslant \theta^*, \\ 0, & \text{otherwise.} \end{cases}$$

Thus

$$\begin{aligned} \Delta^{(H)}(x) &= \int_0^\infty [L_2(\theta) - L_1(\theta)] e^{-\theta} \frac{\theta^x}{x!} H(d\theta) \\ &= \theta^* \int_0^\infty \frac{\theta^x}{x!} e^{-\theta} H(d\theta) - (x + 1) \int_0^\infty \frac{\theta^{x+1}}{(x + 1)!} e^{-\theta} H(d\theta). \end{aligned}$$

The marginal density of X with respect to the prior distribution H is

$$f_H(x) = \int_0^\infty \frac{\theta^x}{x!} e^{-\theta} H(d\theta), \qquad x = 0, 1, \ldots .$$

Accordingly,

$$\Delta^{(H)}(x) = \theta^* f_H(x) - (x + 1) f_H(x + 1).$$

The approach is to provide a consistent estimator of $f_H(x)$. Such an estimator could be, for example,

$$\hat{f}^{(n)}(x; \mathbf{X}_n) = \frac{1}{n} \sum_{j=1}^{n} U(x; X_j),$$

where

$$U(x; X_j) = \begin{cases} 1, & \text{if } x = X_j, \\ 0, & \text{otherwise.} \end{cases}$$

The sequence $\{\hat{f}^{(n)}; n \geqslant 1\}$ obeys the strong law of large numbers, and $\hat{f}^{(n)} \to f_H$ a.s. Finally, if $\phi^{(n)}(\mathbf{X}_n)$ is the probability of accepting $A_1^{(n)}$, at the n-th test, the empirical Bayes sequence is

$$\phi^{(n)}(\mathbf{X}_n) = \begin{cases} 1, & \theta^* \geqslant \dfrac{(X_n + 1)\hat{f}^{(n)}(X_n + 1; \mathbf{X}_n)}{\hat{f}^{(n)}(X_n; \mathbf{X}_n)} \\ 0, & \text{otherwise.} \end{cases}$$

We remark that $\hat{f}^{(n)}(X_n; \mathbf{X}_n) \geqslant 1/n$ for all $\mathbf{X}_n = (X_1, \ldots, X_n)'$. Therefore $\phi^{(n)}$ is well defined for all n. ■

Samuel [1] gave corresponding formulae for the empirical Bayes testing of two composite hypotheses in the geometric and the negative binomial cases.

Robbins [4] discusses the following method of estimating the unknown prior distribution $H(\theta)$. Let $X_1, X_2, \ldots$ be a sequence of i.i.d. random variables with a common distribution $F_H(x)$. For each real x, define

$$P_n(x) = \frac{\{\text{number of terms } x_1, \ldots, x_n \text{ which are} \leqslant x\}}{n}. \tag{9.3.6}$$

The function $P_n(x)$ is the empirical distribution of X. Let $\mathcal{H}$ be any class of prior distribution functions on Θ that contains H. Define the distance function between any two distributions of x as

$$\rho(F_1, F_2) = \sup_x |F_1(x) - F_2(x)|. \tag{9.3.7}$$

Let

$$D_n = \inf_{H \in \mathcal{H}} \sup_x |P_n(x) - F_H(x)|. \tag{9.3.8}$$

A sequence of estimators $H_n(\theta \mid \mathbf{X}_n)$ of $H(\theta)$ will be any sequence in $\mathcal{H}$ satisfying

$$\rho(P_n, F_{H_n}) \leqslant D_n + \epsilon_n, \tag{9.3.9}$$

where $\{\epsilon_n\}$ is any sequence of positive constants $\epsilon_n \downarrow 0$.

A sequence of estimators $\{H_n(\theta \mid \mathbf{X}_n), n \geqslant 1\}$ is called *effective* if

$$P\left[\lim_{n\to\infty} H_n(\theta \mid \mathbf{X}_n) = H(\theta), \text{ at every continuity point of } H(\theta)\right] = 1. \tag{9.3.10}$$

Theorem 9.3.1. (Robbins [4].) *Assume the following:*

(i) *for every fixed x, $F_\theta(x)$ is a continuous function on Θ;*
(ii) *the limits $\lim_{\theta\to-\infty} F_\theta(x) = F_{-\infty}(x)$ and $\lim_{\theta\to\infty} F_\theta(x) = F_\infty(x)$ exist for every x;*
(iii) *neither $F_{-\infty}$ nor F_∞ is a distribution function;*
(iv) *if H_1 and H_2 are any two prior distributions on Θ such that $F_{H_1} = F_{H_2}$, then $H_1 = H_2$.*

Then the sequence $\{H_n(\theta \mid \mathbf{X}_n); n \geqslant 1\}$ is effective for the class $\mathcal{H}$ of all prior distributions on Θ.

Proof. According to the Glivenko-Cantelli theorem (see Fisz [1], p. 391), the empirical distribution $P_n(x)$ converges strongly to $F_H(x)$, uniformly in x; that is,

$$P\left[\lim_{n\to\infty} \rho(P_n, F_H) = 0\right] = 1. \tag{9.3.11}$$

Hence due to the triangular inequality and (9.3.9),

$$\begin{aligned}\rho(F_{H_n}, F_H) &\leqslant \rho(F_{H_n}, P_n) + \rho(P_n, F_H)\\ &\leqslant D_n + \epsilon_n + \rho(P_n, F_H)\\ &\leqslant 2\rho(P_n, F_H) + \epsilon_n.\end{aligned}$$

Thus $\lim_{n\to\infty} \rho(F_{H_n}, F_H) = 0$ a.s. This implies that

$$\lim_{n\to\infty} \int F_\theta(x)H_n(\mathrm{d}\theta \mid \mathbf{X}_n) = \int F_\theta(x)H(\mathrm{d}\theta) \quad \text{a.s.,} \tag{9.3.12}$$

uniformly in x, $-\infty < x < \infty$. Let $\{H_{n_j}(\theta \mid \mathbf{X}_{n_j}); j \geqslant 1\}$ be a subsequence, converging weakly to $H^*(\theta)$; $H^*(\theta)$ is possibly a defective d.f., that is, $H^*(\infty) - H^*(-\infty) \leqslant 1$. According to the Helly-Bray theorem, since $F_\theta(x)$ is a continuous function of θ, for each fixed x,

$$\begin{aligned}\lim_{j\to\infty} &\int F_\theta(x)H_{n_j}(\mathrm{d}\theta \mid \mathbf{X}_{n_j})\\ &= \int F_\theta(x)H^*(\mathrm{d}\theta) + H^*(-\infty)F_{-\infty}(x) + (1 - H^*(\infty))F_\infty(x).\end{aligned} \tag{9.3.13}$$

Since $F_\theta(x)$ is a distribution function for each θ, $-\infty < \theta < \infty$, $F_{-\infty}(x)$ and $F_\infty(x)$ are between zero and one, monotone nondecreasing in x, and right continuous in x. The Lebesgue dominated convergence theorem implies that

$$\lim_{x\to-\infty} \lim_{j\to\infty} \int F_\theta(x)H_{n_j}(d\theta \mid \mathbf{X}_{n_j}) = \lim_{x\to-\infty} \int F_\theta(x)H^*(d\theta) = 0. \tag{9.3.14}$$

Hence

$$H^*(-\infty)F_{-\infty}(-\infty) + (1 - H^*(\infty))F_\infty(-\infty) = 0. \tag{9.3.15}$$

Similarly, by letting $x \to \infty$ in (9.3.13), we obtain

$$H^*(-\infty)F_{-\infty}(\infty) + (1 - H^*(\infty))F_\infty(\infty) = 0. \tag{9.3.16}$$

Hence if $H^*(-\infty) > 0$ and $H^*(\infty) < 1$, then (9.3.15) implies that $F_{-\infty}(-\infty) = 0$, $F_\infty(-\infty) = 0$; and (9.3.16) implies that $F_{-\infty}(+\infty) = 1$ and $F_\infty(\infty) = 1$. This contradicts (iii). Similarly, if we assume that $H^*(-\infty) = 0$ and $H^*(\infty) < 1$, or $H^*(-\infty) > 0$ and $H^*(\infty) = 1$, we contradict (iii). Hence $H^*(-\infty) = 0$ and $H^*(\infty) = 1$. This proves that H^* is a proper distribution. Finally, from (iv) and (9.3.13) we obtain that $H_{n_j}(\theta \mid X_{n_j}) \xrightarrow{W} H^*(\theta)$ (at all the continuity points θ of H^*) a.s. This proves (9.3.10). (Q.E.D.)

The following example is given by Robbins [4].

Example 9.4. Let $\mathcal{F} = \{F_\theta;\, \theta \in \Theta\}$ be a location parameter family of distributions, that is, $F_\theta(x) = F(x - \theta)$; Θ is the real line. The marginal distribution of X is

$$F_H(x) = \int F(x - \theta)H(d\theta), \qquad -\infty < x < \infty.$$

The convolution of F_θ and H is thus $F_H(x)$. As is well known, the characteristic function of the convolution of two d.f.'s is the product of the corresponding characteristic function. Let $\phi_F(t)$ designate the characteristic function of the d.f. F. Hence for any two prior distributions $H_1, H_2 \in \mathcal{H}$, $F_{H_1} = F_{H_2}$ implies that $\phi_F(t)\phi_{H_1}(t) = \phi_F(t)\phi_{H_2}(t)$. Thus if $\phi_F(t) \neq 0$ for all t, $\phi_{H_1}(t) = \phi_{H_2}(t)$ for all t. This in turn implies that $H_1(\theta) = H_2(\theta)$. If, in addition, (i)–(iii) are satisfied, Theorem 9.3.1 holds and the sequence $\{H_n(\theta \mid \mathbf{X}_n);\, n \geqslant 1\}$ is effective.

If $F_\theta(x)$ is the $\mathcal{N}(\theta, 1)$ distribution $F_\theta(x) = \Phi(x - \theta)$. This is a continuous function of θ for each fixed x. Moreover,

$$\lim_{\theta \to \infty} \Phi(x - \theta) = 0, \text{ for each } x, \qquad -\infty < x < \infty;$$

$$\lim_{\theta \to \infty} \Phi(x - \theta) = 1, \text{ for each } x, \qquad -\infty < x < \infty.$$

Hence (ii) and (iii) are satisfied. ■

9.4. BAYES SEQUENTIAL TESTING—GENERAL THEORY

Consider as before the problem of testing m hypotheses, $A_i\colon \theta \in \Theta_i$, $i = 1, \ldots, m$. The m loss functions $L_i(\theta)$ are defined on Θ; $X_1, X_2, \ldots$ is a sequence if i.i.d. random variables having a common distribution F_θ, $\theta \in \Theta$. We assume here that $\{\Theta_1, \ldots, \Theta_m\}$ is a partition of Θ. The value θ, which determines F_θ, is generated from Θ according to a specified prior distribution H. The problem is to decide in which of the m subsets θ lies. The sequential procedure of testing these multiple hypotheses is different from the fixed sample procedure in the following important feature. The statistician is allowed to decide whether to observe the above sequence of random variables or not, and after each observation he has the option of terminating the sampling and making his final decision (which one of the specified m hypotheses to accept). In order to make a meaningful decision to stop sampling, the statistician has to consider on the one hand the cost of sampling, and on the other hand the amount of information on θ he expects to draw from additional observations. Thus let $K_n(X_1, \ldots, X_n)$ be the cost of observing $(X_1, \ldots, X_n)$. This cost might be a function of the observed values, as is the case when X represents the time duration of the observation (lifetime of a piece of equipment); or $K_n(X_1, \ldots, X_n)$ could depend only on n, that is, $K_n(X_1, \ldots, X_n) = K(n)$. We further assume the following:

(i) $K_{n+1}(\mathbf{X}_{n+1}) \geqslant K_n(\mathbf{X}_n)$ a.s., all $n \geqslant 0$;
(ii) $\lim_{n \to \infty} K_n(\mathbf{X}_n) = \infty$ a.s.

As explained in Section 6.8, every sequential procedure is determined by two rules, a stopping rule and a decision rule. We have designated the stopping rule by $s(H, \mathbf{X})$ and the decision rule by $\delta(H, \mathbf{X})$, where $\mathbf{X}$ is the sequence of random variables $X_1, X_2, \ldots$. We specify here the following class of sequential procedure.

1. The function $\delta(H, \mathbf{X})$ is always the Bayes decision function against the prior distribution H, which is applied after sampling is terminated. That is, if the sample size is $N(s) = n$, then, given $(H, \mathbf{X}_n)$,

$$\delta(H, \mathbf{X}) = (\phi_1^{(H)}(\mathbf{X}_n), \ldots, \phi_m^{(H)}(\mathbf{X}_n))',$$

where $\phi_i^{(H)}(\mathbf{X}_n)$ is obtained by substituting $\mathbf{X}_n = (X_1, \ldots, X_n)'$ for T in (9.1.6).

2. The class of stopping rules $s(H, \mathbf{X})$ is the class of all randomized rules, which can be represented in the following manner. For each $n \geqslant 0$, let $\mathcal{F}_n$ designate the Borel sigma-field generated by $(\theta, X_1, \ldots, X_n)$, $\mathcal{F}_0 = \mathcal{C}$. Obviously, $\mathcal{F}_0 \subset \mathcal{F}_1 \subset \mathcal{F}_2 \subset \cdots$. Let $\mathcal{S}$ be the class of all sequences $\mathbf{s} = (s_1, s_2, \ldots)$ where $s_{n+1} = s_{n+1}(H, X_1, \ldots, X_n) \in \mathcal{F}_n$, $n \geqslant 1$, and $s_1 = s_1(H)$, $0 \leqslant s_{n+1}(H, X_1, \ldots, X_n) \leqslant 1$; s_{n+1} is the probability of observing X_{n+1}. Define

$$J_{n+1}(H, \mathbf{X}_n) = \begin{cases} 1, & \text{w.p. } s_{n+1}(H, \mathbf{X}_n) \\ 0, & \text{w.p. } 1 - s_{n+1}(H, \mathbf{X}_n). \end{cases} \tag{9.4.1}$$

The stopping rule, represented by a sequence $\mathbf{s}$ of $\mathcal{S}$, specifies that sampling terminates at the least integer $n \geqslant 0$ for which $J_{n+1} = 0$. The corresponding sample size, which is a stopping (random) variable, is defined as

$$N(\mathbf{s}) = \text{least } n, \quad n \geqslant 0, \quad \text{for which } \; J_{n+1}(H, \mathbf{X}_n) = 0. \tag{9.4.2}$$

Given $(H, \mathbf{X}_n)$, let $H(\theta \mid \mathbf{X}_n)$ designate the posterior distribution of θ. Let $F_H(x_1, \ldots, x_n)$ designate the joint marginal (mixed) distribution of $(X_1, \ldots, X_N)$ given H. Let $\rho_0(H, \mathbf{X}_n)$ designate the Bayes risk, corresponding to the Bayes decision, given $(H, \mathbf{X}_n)$, $N(s) = n$, that is,

$$\rho_0(H, \mathbf{X}_n) = \min_{i=1,\ldots,M} \left\{ \int L_i(\theta) H(\mathrm{d}\theta \mid \mathbf{X}_n) \right\}. \tag{9.4.3}$$

The prior risk associated with a stopping rule $\mathbf{s}$ and a Bayes decision rule is

$$R(H, \mathbf{s}) = \sum_{n=1}^{\infty} \int [K_n(\mathbf{x}_n) + \rho_0(H, \mathbf{x}_n)] \prod_{v=0}^{n-1} s_{v+1}(H, \mathbf{x}_v) \times [1 - s_{n+1}(H, \mathbf{x}_n)] F_H(d\mathbf{x}_n) + (1 - s_1(H))\rho_0(H), \tag{9.4.4}$$

where $F_H(d\mathbf{x}_n) = F_H(dx_1, \ldots, dx_n)$; $X_0 \equiv 0$. A stopping rule $\mathbf{s}^0$ is called Bayes (or optimal) against H if

$$R(H, \mathbf{s}^0) = \inf_{\mathbf{s} \in \mathcal{S}} R(H, \mathbf{s}). \tag{9.4.5}$$

We assume that $R(H, \mathbf{s}^0) < \infty$.

It is generally not a simple task to determine the Bayes stopping rule $\mathbf{s}^0$ for a specified prior distribution. In certain special cases the optimal stopping rule gives a procedure with a fixed sample size. We study now some general properties of $\mathbf{s}^0$.

Let $\mathcal{S}^{(M)}$ be the subclass of all stopping rules which are *truncated* at $N = M$; that is, $s_{j+1}(H, \mathbf{X}_j) = 0$, for all $j \geqslant M$ and all $\mathbf{s} \in \mathcal{S}^{(M)}$. We show first how to determine a Bayes truncated stopping rule $\mathbf{s}^{(M)} \in \mathcal{S}^{(M)}$, for which

$$R(H, \mathbf{s}^{(M)}) = \inf_{\mathbf{s} \in \mathcal{S}^{(M)}} R(H, \mathbf{s}). \tag{9.4.6}$$

According to the method of dynamic programming we determine $\mathbf{s}^{(M)}$ sequentially in the following manner. First we construct recursively $M + 1$ functions

$$\begin{aligned} \rho_0^{(M)}(H, \mathbf{X}_M) &= \rho_0(H, \mathbf{X}_M), \\ \rho_j^{(M)}(H, \mathbf{X}_{M-j}) &= \min \{\rho_0(H, \mathbf{X}_{M-j}), \\ E\{\Delta(Y; \mathbf{X}_{M-j}) &+ \rho_{j-1}^{(M)}(H, (\mathbf{X}_{M-j}, Y)) \mid \mathcal{F}_{M-j}\}\}, \qquad j = 1, \ldots, M, \end{aligned} \tag{9.4.7}$$

where $E\{\cdot \mid \mathcal{F}_{M-j}\}$ is the conditional expectation of the function in brackets, with respect to the distribution

$$F(y \mid \mathcal{F}_{M-j}) = \int F_\theta(y) H(d\theta \mid \mathbf{X}_{M-j}), \qquad j = 0, \ldots, M, \tag{9.4.8}$$

and

$$\Delta(Y; \mathbf{X}_{M-j}) = K_{M-j+1}(\mathbf{X}_{M-j}, Y) - K_{M-j}(\mathbf{X}_{M-j}). \tag{9.4.9}$$

We remark that $\rho_j^{(M)}(H, \mathbf{X}_{M-j})$, $j = 0, \ldots, M$, is the minimal expected Bayes risk, after $M - j$ observations, if we follow the optimal policy throughout the remaining decisions. The functions $\rho_j^{(M)}(H, \mathbf{X}_{M-j})$ for $j = 1, \ldots, M$ are also called in the literature the *preposterior Bayes risk* (see Raiffa and Schlaifer [1]).

Let

(9.4.10)
$$N^{(M)} = \text{least non negative integer } n, \quad n \leqslant M, \quad \text{for which} \quad \rho_{M-n}^{(M)}(H, \mathbf{X}_n) = \rho_0(H, \mathbf{X}_n);$$

$N^{(M)}$ is the sample size associated with the Bayes truncated sampling rule $\mathbf{s}^{(M)}$. It is a stopping variable, that is, $\{N^{(M)} = n\} \in \mathcal{F}_n$ for all $n \geqslant 0$. The Bayes stopping rule is $\mathbf{s}^{(M)} = (s_1^{(M)}, s_2^{(M)}, \ldots)$ where

$$s_j^{(M)} \equiv s_j^{(M)}(H, \mathbf{X}_{j-1}) = \begin{cases} 1, & \text{if } j \leqslant N^{(M)}, \\ 0, & \text{if } j > N^{(M)}. \end{cases} \tag{9.4.11}$$

For each $M \geqslant 1$, the function $\rho_M^{(M)}(H)$ is the prior Bayes risk associated with H and $\mathcal{S}^{(M)}$. Let $\mathcal{S}_1^{(M)}$, $M \geqslant 1$, designate the subclass of $\mathcal{S}^{(M)}$ in which *at least* one observation is taken. We then have

$$\rho_M^{(M)}(H) = \min\left\{\rho_0(H), \inf_{\mathbf{s}\in\mathcal{S}_1^{(M)}} R(H, \mathbf{s})\right\}. \tag{9.4.12}$$

Since $\mathcal{S}_1^{(M)} \subset \mathcal{S}_1^{(M+1)}$ for all $M \geqslant 1$,

$$\inf_{\mathbf{s}\in\mathcal{S}_1^{(M+1)}} R(H, \mathbf{s}) \leqslant \inf_{\mathbf{s}\in\mathcal{S}_1^{(M)}} R(H, \mathbf{s}). \tag{9.4.13}$$

We obtain from (9.4.12) and (9.4.13) that $\rho_M^{(M)}(H) \geqslant \rho_{M+1}^{(M+1)}(H)$ for all $M \geqslant 1$. Furthermore, since $\rho_M^{(M)}(H) \geqslant 0$ for all $M \geqslant 1$, we obtain that

$$\rho^{(\infty)}(H) = \lim_{M\to\infty} \rho_M^{(M)}(H) \tag{9.4.14}$$

exists.

Let $\rho(H)$ designate the prior Bayes risk with respect to the class $\mathcal{S}$ of all stopping rules, that is, $\rho(H) = R(H, \mathbf{s}^0)$. The function $\rho(H)$ satisfies the functional equation

$$\rho(H) = \min\{\rho_0(H), E_H\{K_1(Y) + \rho(H_Y)\}\}, \tag{9.4.15}$$

where $H_Y(\theta) \equiv H(\theta \mid Y)$ is the posterior distribution of θ given (H, Y). Indeed, if $\mathcal{S}_1$ designates the class of all stopping rules whose sample size is at least one, then

$$\rho(H) = \min\left\{\rho_0(H), \inf_{\mathbf{s}\in\mathcal{S}_1} R(H, \mathbf{s})\right\}. \tag{9.4.16}$$

Define for each $\mathbf{s} \in \mathcal{S}_1$ the derived stopping rule $\mathbf{s}^+$ in the following manner: If $\mathbf{s} = (s_1(H), s_2(H, X_1), \ldots)$, then $\mathbf{s}^+ = (s_2(H, X_1), s_3(H, X_1, X_2), \ldots)$. We then have, for each $\mathbf{s} \in \mathcal{S}_1$,

$$R(H, \mathbf{s}) = E_H\{K_1(Y) + R(H_Y, \mathbf{s}^+)\}, \quad \mathbf{s} \in \mathcal{S}_1. \tag{9.4.17}$$

It follows from (9.4.17) and Fatou's lemma that

$$\inf_{\mathbf{s}\in\mathcal{S}_1} R(H, \mathbf{s}) \geqslant E_H\left\{K_1(Y) + \inf_{\mathbf{s}\in\mathcal{S}_1} R(H_Y, \mathbf{s}^+)\right\} \tag{9.4.18}$$
$$= E_H\left\{K_1(Y) + \inf_{\mathbf{s}\in\mathcal{S}_1} R(H_Y, \mathbf{s})\right\}$$
$$= E_H\{K_1(Y) + \rho(H_Y)\}.$$

It follows from (9.4.16) and (9.4.18) that

$$\rho(H) \geqslant \min\{\rho_0(H), E_H\{K_1(Y) + \rho(H_Y)\}\}$$
$$\geqslant \inf_{\mathbf{s}\in\mathcal{S}} R(H, \mathbf{s}) = \rho(H).$$

This establishes the functional equation (9.4.15). In the same manner we show that, for each $M \geqslant 1$,

$$\rho_{M+1}^{(M+1)}(H) = \min\{\rho_0(H), E_H\{K_1(Y) + \rho_M^{(M)}(H_Y)\}\}. \tag{9.4.19}$$

If

$$E_H\{\rho_1^{(1)}(H_Y)\} < \infty \tag{9.4.20}$$

then, according to the Lebesgue monotone convergence theorem and the monotonicity of $\rho_M^{(M)}(H)$, $M \geqslant 1$,

$$\rho^{(\infty)}(H) = \lim_{M\to\infty} \rho_{M+1}^{(M+1)}(H) \tag{9.4.21}$$
$$= \min\left\{\rho_0(H), \lim_{M\to\infty} E_H\left\{K_1(Y) + \rho_M^{(M)}(H_Y)\right\}\right\}$$
$$= \min\{\rho_0(H), E_H\{K_1(Y) + \rho^{(\infty)}(H_Y)\}\}.$$

It follows that the functional $\rho^{(\infty)}(H)$ satisfies the functional equation (9.4.15). Thus we have proven the following theorem.

Theorem 9.4.1. *Given a prior distribution H, if*

$$E_H\{K_1(Y)\} < \infty,$$

and

$$E_H\{\rho_1^{(1)}(H_Y)\} < \infty,$$

then

$$\rho(H) = \lim_{M\to\infty} \rho_M^{(M)}(H).$$

Theorem 9.4.1 implies that, for any given $\epsilon > 0$, there exists an $M(\epsilon)$ such that the Bayes truncated stopping rules $\mathbf{s}^M$, for every $M \geqslant M(\epsilon)$, are ϵ-Bayes stopping rules for the untruncated problem.

Theorem 9.4.2. *If* $\rho(H) = \lim_{M\to\infty} \rho_M^{(M)}(H)$, *and if for all* $n \geqslant M_0$ *the following inequality holds,*

(9.4.22) $$\rho_0(H, \mathbf{X}_n) - E\{\rho_0(H, (\mathbf{X}_n, Y)) \mid \mathcal{F}_n\} \leqslant E\{\Delta(Y; \mathbf{X}_n) \mid \mathcal{F}_n\} \quad \text{a.s.,}$$

then $\rho_{M_0}^{(M_0)}(H) = \rho(H)$ *and the truncated Bayes sequential rule* $\mathbf{s}^{(M_0)}$ *is optimal.*

Proof. Let $M \geqslant M_0$, and consider the Bayes truncated stopping rule $\mathbf{s}^{(M)}$. Due to (9.4.7) and (9.4.22) we have

(9.4.23) $$\rho_{M-j}^{(M)}(H, X_j) = \rho_0(H, X_j), \qquad j = M_0, \ldots, M.$$

This implies that

(9.4.24) $$\rho_M^{(M)}(H) = \rho_{M_0}^{(M_0)}(H), \qquad \text{all } M \geqslant M_0.$$

Finally,

$$\lim_{M\to\infty} \rho_M^{(M)}(H) = \rho_{M_0}^{(M_0)}(H) = \rho(H).$$ (Q.E.D.)

Then we prove the following theorem.

Theorem 9.4.3. *If for each* $n \geqslant 1$ *the posterior risk* $\rho_0(H, \mathbf{X}_n)$ *is independent of* $\mathbf{X}_n$, *that is,* $\rho_0(H, \mathbf{X}_n) = \rho_0(H, n)$ a.s. *for each* $n \geqslant 1$, *then the Bayes sequential rule is a fixed sample rule with a sample size* n^0, *where*

(9.4.25) $n^0 =$ *smallest integer* n, $n \geqslant 0$, *which minimizes* $\tilde{R}(n) =$
$$\rho_0(H, n) + E_H\{K_n(\mathbf{X}_n)\}.$$

Proof. According to the definition of n^0, and since $\rho_0(H, n) = \rho_0(H, \mathbf{X}_n)$ a.s. for each $n \geqslant 1$, (9.4.22) is satisfied for all $n \geqslant n^0$. Hence $\mathbf{s}^{(n^0)}$ is a Bayes sequential rule. If $n^0 \leqslant 1$, the theorem is proven. Assume that $n^0 > 1$. It remains to show that $\mathbf{s}^{(n)^0}$ is a fixed sample procedure; that is,

$$S_{j+1}^{(n^0)}(H, \mathbf{X}_j) = 1 \qquad \text{a.s., for all } j = 0, \ldots, n^0 - 1.$$

Since $n^0 > 1$,

(9.4.26)

$$\rho_0(H, j) + E_H\{K_j(\mathbf{X}_j)\} > \rho_0(H, n_0) + E_H\{K_{n_0}(\mathbf{X}_{n_0})\}, \quad \text{for all } j < n^0$$

$$\rho_0^{(n^0)}(H, X_{n^0}) = \rho_0(H, n^0) \quad \text{a.s.}$$

$$\rho_1^{(n^0)}(H, X_{n^0-1}) = \min\{\rho_0(H, n^0 - 1), E\{\rho_0(H, n^0) + \Delta(Y; \mathbf{X}_{n^0-1}) \mid \mathcal{F}_{n^0-1}\}\}.$$

But from (9.4.26), $\rho_0(H, n^0 - 1) > \rho_1^{(n^0)}(H, X_{n^0-1})$ a.s. By induction on j, $j = 1, \ldots, n^0$, we prove that

(9.4.27) $$\rho_0(H, n^0 - j) > \rho_j^{(n^0)}(H, \mathbf{X}_{n^0-j}) \quad \text{a.s.,} \quad j = 1, \ldots, n_0.$$

This proves that the Bayes sequential truncated rule $\mathbf{s}^{(n^0)}$ will terminate sampling, with probability one, only at $N = n^0$. (Q.E.D.)

For more details about the general theory of Bayes sequential decision rules, see Blackwell and Girshick [1], Ch. 9; Ferguson [1], Ch. 7. As previously mentioned, the actual determination of the Bayes stopping rule in nontrivial cases is very difficult, especially when the hypotheses are composite (the corresponding subsets of Θ contain each more than one point). For a numerical example of a determination of Bayes truncated stopping rules in the normal case, see Goode [1]. More powerful analytic methods for approximating the Bayes stopping rule will be discussed later on.

9.5. THE WALD SEQUENTIAL PROBABILITY RATIO TEST FOR TWO SIMPLE HYPOTHESES

The Wald sequential probability ratio test (S.P.R.T.) of two simple hypotheses (both Θ_1 and Θ_2 are single-point sets) is a sequential test whose stopping rule belongs to $\mathcal{S}_1$ and is specified by two constants (a, b), $0 < a, b < \infty$. Thus, given n observations, we form the sum

$$S_n = \sum_{i=1}^{n} \log \frac{f(X_i; \theta_2)}{f(X_i; \theta_1)}, \qquad n \geqslant 1, \tag{9.5.1}$$

where $\theta_1 \in \Theta_1$ and $\theta_2 \in \Theta_2$, and $f(x; \theta)$ is the density function of F_θ with respect to μ. The S.P.R.T. requires *at least one* observation. The stopping rule says that sampling is continued as long as

$$-b < S_n < a.$$

If sampling terminates with N observations, hypothesis A_1 is accepted if $S_N \leqslant -b$; and A_2 is accepted if $S_N \geqslant a$. We prove now that the stopping variable N is finite with probability one; and that the tails of its distribution decrease to zero in a geometric rate.

Lemma 9.5.1. *If for a given θ, $P_\theta[Z = 0] < 1$ where $Z = \log f(X; \theta_2)/f(X; \theta_1)$, there exists a ρ, $0 < \rho < 1$, and a c, $0 < c < \infty$, such that*

$$P_\theta[N > n] \leqslant c\rho^n, \quad \text{for all } n \geqslant n_0. \tag{9.5.2}$$

Proof. If $P_\theta[Z = 0] < 1$, there exists $\epsilon > 0$, $p > 0$, such that $P_\theta[|Z| > \epsilon] > p$. If $\text{Var}_\theta\,\{Z\} = 0$, the distribution of Z is concentrated on a single point Z_0, $Z_0 \neq 0$, and the result is trivial, since $N \leqslant$ least integer n for which $n\,|Z_0| \geqslant \max\,(a, b)$. Assume that $\sigma^2 = \text{Var}_\theta\,\{Z\} > 0$. By the central limit theorem, the asymptotic distribution of S_n is normal. Hence for any $\epsilon > 0$ there exists $R_1(\epsilon)$ so that, if $r \geqslant R_1(\epsilon)$,

$$P_\theta[|S_r| > c] \geqslant P_\theta\{|r\mu + \sqrt{r}\,\sigma N(0, 1)| > c\} - \epsilon, \tag{9.5.3}$$

where $\mu = E_\theta\{Z\}$. Furthermore,

$$(9.5.4)\quad P_\theta[|r\mu + \sqrt{r}\,\sigma N(0, 1)| > c] = P_\theta\left[|N(\sqrt{r}\,\mu, \sigma^2)| > \frac{c}{\sqrt{r}}\right] \to 1$$

as $r \to \infty$, for every c, $0 < c < \infty$.

Finally, for every r sufficiently large, the definition of N implies that

$$(9.5.5)\quad P_\theta[N > mr] \leqslant P_\theta\{-b < S_r < a, |S_{2r} - S_r| < a + b, \ldots, |S_{mr} - S_{(m-1)r}| < ab.\}.$$

Let $p_2 = P_\theta\{|S_r| > a + b\}$. For r sufficiently large, $p_2 > 0$. Moreover, $S_{jr} - S_{(j-1)r}$ are i.i.d. for every $j = 1, 2, \ldots$. Hence from (9.5.5),

$$(9.5.6)\qquad P_\theta[N > mr] < (1 - p_2)^{m-1},$$

for r sufficiently large. Set $\rho = (1 - p_2)^{1/r}$ and $n = mr$. Then for n sufficiently large,

$$(9.5.7)\qquad P_\theta[N > n] \leqslant \rho^{(rm-1)} = c\rho^n,$$

where $c = \rho^{-r} = (1 - p_2)^{-1/r}$.

(Q.E.D.)

Corollary 9.5.2. *Assume that* $P_\theta[Z = 0] < 1$*: then,*

$$(9.5.8)\qquad \begin{aligned} &P_\theta[N < \infty] = 1, \\ &E_\theta\{\exp\{tN\}\} < \infty, \quad \textit{for some} \quad t > 0, \end{aligned}$$

and

$$(9.5.9)\qquad E_\theta\{N^k\} < \infty, \quad \textit{for all} \quad k = 0, 1, \ldots .$$

The proof is left to the reader.

Lemma 9.5.3. (Wald [4].) *If* $E_\theta\{[Z]\} < \infty$, *then*

$$(9.5.10)\qquad E_\theta\{S_N\} = E_\theta\{N\}E_\theta\{Z\}.$$

This is a special case of Lemma 4.4.1 and is implied from (9.5.8).

As in Chapter 4, we denote by $M_\theta(t)$ the moment generating function of Z.

Lemma 9.5.4. (Chernoff [2].) *Under the following conditions,*

(i) $E_\theta\{Z\} \neq 0$,
(ii) $P_\theta\{Z > 0\} > 0$, $P_\theta\{Z < 0\} > 0$,
(iii) $M_\theta(t) < \infty$, *for all* t,

there exists a $t_0 \neq 0$ such that $M_\theta(t_0) = 1$, and sign $(t_0) = -$sign (μ).

Proof. According to (ii) there exist p_1, p_2, and $\epsilon > 0$ such that

$$p_1 = P_\theta[Z > \epsilon] > 0. \tag{9.5.11}$$
$$p_2 = P_\theta[Z < -\epsilon] > 0.$$

Hence

$$M_\theta(t) = E_\theta\{e^{tZ}\} > e^{t\epsilon}P_1[Z > \epsilon] = p_1 e^{t\epsilon} \to \infty \qquad \text{as } t \to \infty. \tag{9.5.12}$$

Similarly,

$$M_\theta(t) > e^{-t\epsilon}P_\theta[Z < -\epsilon] \to \infty \quad \text{as} \quad t \to -\infty. \tag{9.5.13}$$

Moreover, $M_\theta(0) = 1$ and

$$\frac{d}{dt} M_\theta(t) = E_\theta\{Ze^{tZ}\}, \tag{9.5.14}$$

and, according to (ii),

$$\frac{d^2}{dt^2} M_\theta(t) = E_\theta\{Z^2 e^{tZ}\} > 0. \tag{9.5.15}$$

Thus $M_\theta(t)$ is a strictly convex function of t, which increases to ∞ as $|t| \to \infty$. Hence by the Rolle theorem there exists a value t_0 such that $M_\theta(t_0) = 1$, $t_0 \neq 0$. Moreover, $M_\theta(t)$ has a unique minimum at t^*, $|t^*| < |t_0|$, and $M_\theta(t^*) < 1$. If $\mu = E_\theta\{Z\} > 0$ we obtain from (9.5.14) that $M_\theta(t)$ is strictly increasing at $t = 0$, which implies that $t_0 < 0$. Conversely, if $\mu < 0$ then $\mu_0 > 0$. (Q.E.D.)

We introduce now the defective distribution

$$K_n^{(\theta)}(z) = \begin{cases} P_\theta[S_1 \leqslant z], & n = 1, \\ P_\theta[-b < S_1 < a, \ldots, -b < S_{n-1} < a, S_n \leqslant z], & n \geqslant 2. \end{cases} \tag{9.5.16}$$

Define the generating function

$$G(t, u) = \sum_{n=0}^{\infty} u^n \int_{-b}^{a} e^{tz} K_n^{(\theta)}(dz), \qquad 0 < |u| < 1. \tag{9.5.17}$$

It is easy to prove that

$$\int_{-b}^{a} e^{tz} K_n^{(\theta)}(dz) \leqslant \int_{-b}^{a} e^{tz}\, dP_\theta[S_n \leqslant z]. \tag{9.5.18}$$

Hence if $E_\theta\{Z\} \neq 0$, $P_\theta[Z < 0] > 0$, $P_\theta[Z > 0] > 0$, there exists $t^* \neq 0$ at which $M_\theta(t)$ is minimum, and

$$\int_{-b}^{a} e^{tz}\, dP_\theta[S_n \leqslant z] = \int_{-b}^{a} e^{(t-t^*)z + t^* z}\, dP_\theta[S_n \leqslant z] \tag{9.5.19}$$
$$\leqslant \begin{cases} e^{(t-t^*)a}(M_\theta(t^*))^n & \text{if } t \geqslant t^*, \\ e^{-b(t-t^*)}(M_\theta(t^*))^n, & \text{if } t < t^*. \end{cases}$$

Thus under the assumptions of the previous lemma, $|G(t, u)| < \infty$ for all $|u| < 1$ and $-\infty < t < \infty$. We now prove Wald's celebrated *fundamental identity*.

Theorem 9.5.5. (Wald [3].) *Let N denote the sample size of an S.P.R.T., defined by the stopping boundaries $(-b, a)$, $0 < a, b < \infty$. Assume that* (i)–(iii) *of Lemma 9.5.4 hold. Then*

$$E_\theta\{e^{tS_N} \cdot (M_\theta(t))^{-N}\} = 1, \qquad \text{all } t \text{ s.t. } M_\theta(t) > 1. \tag{9.5.20}$$

Proof. For all u, $0 < |u| < 1$, and all t, $-\infty < t < \infty$, we have

$$\begin{aligned} E_\theta\{e^{tS_N}u^N\} &= \sum_{n=1}^{\infty} u^n \left\{\int_{-\infty}^{-b} + \int_a^{\infty}\right\} e^{tz} K_n^{(\theta)}(dz) \\ &= \sum_{n=1}^{\infty} u^n \int_{-\infty}^{\infty} e^{tz} K_n^{(\theta)}(dz) - \sum_{n=1}^{\infty} u^n \int_{-b}^{a} e^{tz} K_n^{(\theta)}(dz) \\ &= \sum_{n=1}^{\infty} u^n \int_{-\infty}^{\infty} e^{tz} K_n^{(\theta)}(dz) - [G(t, u) - 1]. \end{aligned} \tag{9.5.21}$$

Furthermore, $S_n = S_{n-1} + Z$ where Z is independent of S_{n-1}. Thus

$$\begin{aligned} \int_{-\infty}^{\infty} e^{tz} K_n^{(\theta)}(dz) &= \int_{-b}^{a} \left[\int_{-\infty}^{\infty} e^{t(z-x)}\, dP_\theta[Z_n \leqslant z - x]\right] e^{tx} K_{n-1}^{(\theta)}(dx) \\ &= M_\theta(t) \int_{-b}^{a} e^{tx} K_{n-1}^{(\theta)}(dx). \end{aligned} \tag{9.5.22}$$

From (9.5.22) we obtain

$$\begin{aligned} \sum_{n=1}^{\infty} u^n \int_{-\infty}^{\infty} e^{tz} K_n^{(\theta)}(dz) &= uM_\theta(t) \sum_{n=0}^{\infty} u^n \int_{-b}^{a} e^{tx} K_n^{(\theta)}(dx) \\ &= uM_\theta(t) G(t, u). \end{aligned} \tag{9.5.23}$$

Hence for all u, $0 < |u| < 1$, and all t, $-\infty < t < \infty$,

$$E_\theta\{e^{tS_N}u^N\} = 1 + [uM_\theta(t) - 1]G(t, \mu). \tag{9.5.24}$$

Consider a value of t for which $M_\theta(t) > 1$ and substitute in (9.5.24) $u = (M_\theta(t))^{-1}$. This yields the fundamental identity (9.5.20). (Q.E.D.)

The Wald fundamental identity can be used to derive certain relationships between the moments of S_N and those of N. It is useful also for the following derivation.

Let $\pi(\theta)$ designate the probability of accepting A_1 under θ. This is the probability under θ of $\{S_n \leqslant -b\}$. The fundamental identity can be written as

$$\begin{aligned} &\pi(\theta) E_\theta\{e^{S_N t}(M_\theta(t))^{-N} \mid S_N \leqslant -b\} \\ &\qquad + (1 - \pi(\theta)) E_\theta\{e^{S_N t}(M_\theta(t))^{-N} \mid S_N \geqslant a\} = 1, \end{aligned} \tag{9.5.25}$$

for all t s.t. $M_\theta(t) > 1$. Let $t_0 \neq 0$ be the point at which $M_\theta(t_0) = 1$. Suppose, for the sake of the argument, that $\mu < 0$, $t_0 > 0$. Then in a closed neighborhood of t values greater than t_0, that is, on the interval $(t_0, t_0 + \delta)$, $\delta > 0$, (9.5.25) holds. Since the left-hand side of (9.5.25) is a continuous function of t, and since it is finite on the compact set $[t_0, t_0 + \delta]$, we obtain by approaching t_0 from the right that

$$\pi(\theta)E_\theta\{e^{t_0 S_N} \mid S_N \leqslant -b\} + (1 - \pi(\theta))E_\theta\{e^{t_0 S_N} \mid S_N \geqslant a\} = 1. \tag{9.5.26}$$

Neglecting the overshoot of S_N with respect to the boundaries we obtain the approximations

$$E_\theta\{e^{t_0 S_N} \mid S_N \leqslant -b\} \approx e^{-t_0(\theta)b}, \tag{9.5.27}$$

and

$$E_\theta\{e^{t_0 S_N} \mid S_N \geqslant a\} \approx e^{t_0(\theta)a}. \tag{9.5.28}$$

From (9.5.26)–(8.5.31) we obtain the approximation

$$\pi(\theta) \approx \frac{e^{t_0(\theta)a} - 1}{e^{t_0(\theta)a} - e^{-t_0(\theta)b}}. \tag{9.5.29}$$

Similarly, from (9.5.10) we obtain the approximation when

$$E_\theta\{N\} \approx \frac{1}{\mu}\left[a\,\frac{1 - e^{-t_0(\theta)b}}{e^{t_0(\theta)a} - e^{-t_0(\theta)b}} + b\,\frac{e^{t_0(\theta)a} - 1}{e^{t_0(\theta)a} - e^{-t_0(\theta)b}}\right]. \tag{9.5.30}$$

Example 9.5. Let $X_1, X_2, \ldots$ be i.i.d. random variables having an $\mathcal{N}(\theta, 1)$ distribution. The two simple hypotheses are $A_1\colon \theta = -1$, $A_2\colon \theta = 1$.

$$Z = \log\frac{f(X; 1)}{f(X; -1)} = 2X.$$

A transitive sufficient statistic (see Section 2.8) is $T_n = \sum_{i=1}^n X_i$. We also see that $S_n = \sum_{i=1}^n Z_i = 2T_n$. We therefore base the S.P.R.T. of T_n. Sampling continues as long as $-b < T_n < a$ for some $0 < a, b < \infty$. The m.g.f. of X is

$$G_\theta(t) = \exp\{\tfrac{1}{2}t^2 + \theta t\}.$$

Therefore

$$M_\theta(t) = \exp\{2t^2 + 2\theta t\}.$$

It follows that $t_0(\theta) = -\theta$. Thus

$$\pi(\theta) \approx \frac{e^{-\theta a} - 1}{e^{-\theta a} - e^{+\theta b}}$$

(9.5.31)

$$E_\theta\{N\} \approx \frac{1}{2\theta}\left[a\,\frac{1 - e^{\theta b}}{e^{-\theta a} - e^{\theta b}} + b\,\frac{e^{-\theta a} - 1}{e^{-\theta a} - e^{\theta b}}\right]. \quad \blacksquare$$

Let α and β designate the error probabilities of Type I and Type II, respectively, associated with an S.P.R.T. $(-b, a)$. That is,

$$\alpha = P_{\theta_1}[S_N \geqslant a], \tag{9.5.32}$$
$$\beta = P_{\theta_2}[S_N \leqslant -b].$$

Let $A(\alpha, \beta)$ and $B(\alpha, \beta)$ be functions of α, β such that

$$0 < B(\alpha, \beta) < 1 < A(\alpha, \beta) < \infty,$$

and if we substitute $a(\alpha, \beta) = \log A(\alpha, \beta)$ and $-b(\alpha, \beta) = \log B(\alpha, \beta)$, then (9.5.32) is satisfied. In other words, $(-b(\alpha, \beta), a(\alpha, \beta))$ determines an S.P.R.T. with the prescribed error probabilities α and β. It is generally quite laborious to determine the exact boundaries $(-b(\alpha, \beta), a(\alpha, \beta))$, which may be even randomized ones (see Wijsman [2]). Wald [4] suggested the following approximation.

Theorem 9.5.6. (Wald [4].) *If an S.P.R.T. is defined by* $(\log B, \log A)$, *where* $0 < B < 1 < A < \infty$, *then the error probabilities* α, β *satisfy*

$$A \leqslant \frac{1-\beta}{\alpha}, \qquad B \geqslant \frac{\beta}{1-\alpha}. \tag{9.5.33}$$

If we set $A' = (1-\beta)/\alpha$ *and* $B' = \beta/(1-\alpha)$, *the corresponding error probabilities* α', β' *satisfy*

$$\alpha' \leqslant \frac{\alpha}{1-\beta}, \qquad \beta' \leqslant \frac{\beta}{1-\alpha}, \tag{9.5.34}$$

and if $\alpha + \beta < 1$ *then* $\alpha' + \beta' \leqslant \alpha + \beta$.

Proof. Let ξ, $0 < \xi < 1$, be any prior probability that A_1 is true. Let $Q_1(\xi)$ be the prior probability that A_2 is true *and* A_2 is accepted.

$$\begin{aligned} Q_1(\xi) &= P[A_2 \text{ is accepted} \mid A_2 \text{ is true}]P_\xi[A_2] \\ &= (1-\xi)(1-\beta). \end{aligned} \tag{9.5.35}$$

For each $N = n$, the posterior probability of A_2, given X_n, is

$$1 - \xi(X_n) = \frac{(1-\xi)\prod_{i=1}^n f(X_i; \theta_2)}{\xi \prod_{i=1}^n f(X_i; \theta_1) + (1-\xi)\prod_{i=1}^n f(X_i; \theta_2)}. \tag{9.5.36}$$

Let $\lambda(\mathbf{X}_n) = \prod_{i=1}^n f(X_i; \theta_2)/f(X_i; \theta_1)$. Let

$$f_\xi(\mathbf{X}_n) = \xi \prod_{i=1}^n f(X_i; \theta_1) + (1-\xi)\prod_{i=1}^n f(X_i; \theta_2). \tag{9.5.37}$$

For each $n \geqslant 1$, $\lambda(X_n) \geqslant A$ if and only if $(1-\xi)A/[\xi + (1-\xi)A] \leqslant 1 - \xi(X_n)$. This is proven from (9.5.36) by showing that $ax/(1+ax)$ is monotone increasing in x for each a, $0 < a < \infty$. Thus since

$$Q_1(\xi) = \sum_{n=1}^{\infty} \int_{\{\lambda(\mathbf{X}_n)\geqslant A\}} (1 - \xi(\mathbf{X}_n)) \prod_{i=1}^{n} f_\xi(x_i)\mu(\mathrm{d}x_i), \tag{9.5.38}$$

we have

$$\begin{aligned} Q_1(\xi) &\geqslant \frac{(1-\xi)A}{\xi + (1-\xi)A} \sum_{n=1}^{\infty} \int_{\{\lambda(\mathbf{X}_n)\geqslant A\}} \prod_{i=1}^{n} f_\xi(x_i)\mu(\mathrm{d}x_i) \\ &= \frac{(1-\xi)A}{\xi + (1-\xi)A} [\xi\alpha + (1-\xi)(1-\beta)]. \end{aligned} \tag{9.5.39}$$

From (9.5.35) and (9.5.42) we obtain

$$(1-\beta)[\xi + (1-\xi)A] \geqslant A[\xi\alpha + (1-\xi)(1-\beta)], \text{ all } 0 < \xi < 1. \tag{9.5.40}$$

This yields that $A \leqslant (1-\beta)/\alpha$. In a similar fashion we prove that $B \geqslant \beta/(1-\alpha)$.

Let $A' = (1-\beta)/\alpha$ and $B' = \beta/(1-\alpha)$. Partition $\mathfrak{X}^{(n)}$, which is the sample space of the first n observations, to the three sets

$$\begin{aligned} \mathcal{A}_n &= \left\{\mathbf{X}_n;\ \frac{\beta}{1-\alpha} < \lambda(\mathbf{X}_1) < \frac{1-\beta}{\alpha}, \ldots, \frac{\beta}{1-\alpha} < \lambda(\mathbf{X}_{n-1}) < \frac{1-\beta}{\alpha}, \lambda(\mathbf{X}_n) \leqslant \frac{\beta}{1-\alpha}\right\}, \\ \mathcal{R}_n &= \left\{\mathbf{X}_n;\ \frac{\beta}{1-\alpha} < \lambda(\mathbf{X}_1) < \frac{1-\beta}{\alpha}, \ldots, \frac{\beta}{1-\alpha} < \lambda(\mathbf{X}_{n-1}) < \frac{1-\beta}{\alpha}, \lambda(\mathbf{X}_n) \geqslant \frac{1-\beta}{\alpha}\right\}, \\ \mathcal{C}_n &= \mathfrak{X}^{(n)} - \mathcal{A}_n - \mathcal{R}_n. \end{aligned} \tag{9.5.41}$$

The set $\mathcal{A}_n$ is the set of all sample points at which A_1 is accepted at the n-th trial, using the S.P.R.T. $(\log B', \log A')$. The set $\mathcal{R}_n$ is the set of all sample points at which A_2 is accepted at the n-th trial with that S.P.R.T. The error probabilities corresponding to $(\log B', \log A')$ are

$$\begin{aligned} \alpha' &= \sum_{n=1}^{\infty} \int_{\mathcal{R}_n} \prod_{i=1}^{n} f(x_i; \theta_1)\mu(\mathrm{d}x_i) \\ &\leqslant \frac{\alpha}{1-\beta} \sum_{n=1}^{\infty} \int_{\mathcal{R}_n} \prod_{i=1}^{n} f(x_i; \theta_2)\mu(\mathrm{d}x_i) \\ &= \frac{\alpha}{1-\beta}(1-\beta') \leqslant \frac{\alpha}{1-\beta}. \end{aligned} \tag{9.5.42}$$

Similarly,

$$\beta' = \sum_{n=1}^{\infty} \int_{\mathcal{A}_n} \prod_{i=1}^{n} f(x_i; \theta_2)\mu(\mathrm{d}x_i) \tag{9.5.43}$$

$$\leqslant \frac{\beta}{1-\alpha} \int_{\mathcal{A}_n} \prod_{i=1}^{n} f(x_i; \theta_1)\mu(\mathrm{d}x_i) = \frac{\beta}{1-\alpha}(1-\alpha') \leqslant \frac{\beta}{1-\alpha}.$$

This proves (9.5.34). Finally, consider the inequalities

(i) $\dfrac{\alpha'}{1-\beta'} \leqslant \dfrac{\alpha}{1-\beta}$,

(ii) $\dfrac{\beta}{1-\alpha'} \leqslant \dfrac{\beta}{1-\alpha}$.

Multiplying (i) by $(1-\beta')(1-\beta)$ and (ii) by $(1-\alpha')(1-\alpha)$ and adding both sides we derive

$$\alpha'(1-\beta) + \beta'(1-\alpha) \leqslant \alpha(1-\beta') + \beta(1-\alpha'). \tag{9.5.44}$$

From (9.5.44) we obtain, by simple algebraic simplification, $\alpha' + \beta' \leqslant \alpha + \beta$. (Q.E.D.)

Wald suggested the use of the limits $(-b', a')$ for a specified pair of (α, β). The extent of the deviation of α' from α and β' from β has been investigated by Wald [4].

9.6. BAYESIAN OPTIMALITY OF THE WALD SEQUENTIAL PROBABILITY RATIO TEST

Consider the case of two simple hypotheses $A_1\colon \theta = \theta_1$, $A_2\colon \theta = \theta_2$ (θ can be considered a discrete random variable with a two-point prior distribution). Let ξ be the prior probability of $\{\theta = \theta_1\}$. Suppose that the loss functions due to erroneous decisions are

$$L_1(\theta) = \begin{cases} 0, & \text{if } \theta = \theta_1, \\ r_1, & \text{if } \theta = \theta_2, \end{cases} \tag{9.6.1}$$

and

$$L_2(\theta) = \begin{cases} r_2, & \text{if } \theta = \theta_1, \\ 0, & \text{if } \theta = \theta_2. \end{cases} \tag{9.6.2}$$

Suppose that the cost of sampling is 1 (loss unit) per observations; that is,

$$K_n(\mathbf{X}_n) = n, \quad \text{all} \quad n = 1, 2, \ldots \quad \text{and } all \quad \mathbf{X}_n. \tag{9.6.3}$$

On the basis of the prior probability ξ without observing X, the Bayes decision function is

$$d^0(\xi) = \begin{cases} \theta_1, & \text{if} \quad \xi \geqslant \dfrac{r_1}{r_1 + r_2}, \\ \theta_2, & \text{if} \quad \xi < \dfrac{r_1}{r_1 + r_2}. \end{cases} \tag{9.6.4}$$

If $d^0(\xi) = \theta_1$ we accept A_1; otherwise we accept A_2. The associated prior risk is

$$\rho_0(\xi) = \xi r_2 I_{(0,r^*)}(\xi) + (1 - \xi) r_1 I_{[r^*,1]}(\xi), \tag{9.6.5}$$

$0 < \xi < 1$, where $r^* = r_1/(r_1 + r_2)$ and $I_A(\xi)$ designates the indicator function of the interval A. The function $\rho_0(\xi)$ is concave on [0, 1]. If we terminate sampling at $N = n$, the prior probability ξ is converted into the posterior probability

$$\xi(\mathbf{X}_n) = \frac{\xi f(\mathbf{X}_n; \theta_1)}{\xi f(\mathbf{X}_n; \theta_2) + (1 - \xi) f(\mathbf{X}_n; \theta_2)}, \tag{9.6.6}$$

and the Bayes decision function $d^0(\xi)$ is evaluated at $\xi = \xi(\mathbf{X}_n)$. The associated Bayes risk is $\rho_0(\xi(\mathbf{X}_n))$.

Let $\rho(\xi)$ designate the prior risk associated with the optimal (Bayes) stopping rule. According to (9.4.15), $\rho(\xi)$ satisfies the functional equation

$$\rho(\xi) = \min \{\rho_0(\xi), 1 + E\{\rho(\xi(X)) \mid \mathcal{F}_1\}\}. \tag{9.6.7}$$

The function $\rho_0(\xi)$ is bounded by max (r_1, r_2). Hence $E\{\rho_1^{(1)}(\xi(X)) \mid \mathcal{F}_1\} < \infty$ and, according to Theorem 9.4.1,

$$\rho(\xi) = \lim_{M \to \infty} \rho_M^{(M)}(\xi).$$

We have also seen that $\rho(\xi) = \min \{\rho_0(\xi), \rho_1(\xi)\}$ where

$$\rho_1(\xi) = \inf_{\mathbf{s} \in \mathcal{S}_1} R(\xi, \mathbf{s}), \qquad 0 < \xi < 1. \tag{9.6.8}$$

For any given $\mathbf{s}$, $R(\xi, \mathbf{s})$ is linear in ξ. Hence for any $0 < \xi_1 < \xi_2 < 1$ and $0 < \omega < 1$,

$$\begin{aligned} \rho_1(\xi_1 \omega + \xi_2(1 - \omega)) &= \inf_{\mathbf{s} \in \mathcal{S}_1} \{\omega R(\xi_1, \mathbf{s}) + (1 - \omega) R(\xi_2, \mathbf{s})\} \\ &\geqslant \omega \inf_{\mathbf{s} \in \mathcal{S}_1} R(\xi_1, \mathbf{s}) + (1 - \omega) \inf_{\mathbf{s} \in \mathcal{S}_1} R(\xi_2, \mathbf{s}) \\ &= \omega \rho_1(\xi_1) + (1 - \omega) \rho_1(\xi_2). \end{aligned} \tag{9.6.9}$$

This proves that $\rho_1(\xi)$ is a concave function of ξ on [0, 1]. Hence there exist two critical values $\xi_L(r_1, r_2)$ and $\xi_U(r_1, r_2)$, defined as

$$\xi_L(r_1, r_2) = \min \{r^*, \sup \{\xi;\ \rho_1(\xi) > \xi r_2\}\},$$

(9.6.10)

$$\xi_U(r_1, r_2) = \max \{r^*, \inf \{\xi;\ \rho_1(\xi) > (1 - \xi) r_1\}\}.$$

Obviously, $0 < \xi_L(r_1, r_2) \leqslant \xi_U(r_1, r_2) < 1$. The Bayes sequential stopping rule $\mathbf{s}^0$ yields

(9.6.11) $N(\mathbf{s}^0)$ = least non-negative integer n, such that

$$\xi(\mathbf{X}_n) \leqslant \xi_L(r_1, r_2) \text{ or } \xi(\mathbf{X}_n) \geqslant \xi_U(r_1, r_2),$$

where $\xi(X_0) \equiv \xi$. We now verify that for any a, b, $0 < a, b < \infty$, the S.P.R.T. specified by the boundaries $(-b, a)$ corresponds to a Bayes sequential procedure against some ξ, $0 < \xi < 1$; r_1, r_2, $0 < r_1, r_2 < \infty$.

The stopping rule of an S.P.R.T. with boundaries $(-b, a)$ leads to

(9.6.12) $$N = \text{least } n \geqslant 1 \text{ such that } \lambda(\mathbf{X}_n) \leqslant B \text{ or } \lambda(\mathbf{X}_n) \geqslant A,$$

where $\lambda(\mathbf{X}_n) = f(\mathbf{X}_n; \theta_2)/f(\mathbf{X}_n; \theta_1)$ and $B = e^{-b}$, $A = e^{a}$. Thus for any ξ, $0 < \xi < 1$, define

(9.6.13) $$\xi_L = \left(1 + \frac{1 - \xi}{\xi} A\right)^{-1}, \qquad \xi_U = \left(1 + \frac{1 - \xi}{\xi} B\right)^{-1}.$$

Obviously, since $0 < B < 1 < A < \infty$,

(9.6.14) $$0 < \xi_L < \xi < \xi_U < 1.$$

We now show that there exist $0 < r_1, r_2 < \infty$ such that $\xi_L = \xi_L(r_1, r_2)$ and $\xi_U = \xi_U(r_1, r_2)$ for the given value of ξ. Since (9.6.14) means that the Bayes sequential stopping rule, for these (ξ, r_1, r_2), belongs to $\mathcal{S}_1$, $N = N(\mathbf{s}^0)$ a.s. Also, according to (9.6.4), the S.P.R.T. accepts A_1 if and only if $d^0(\xi(\mathbf{X}_n)) = \theta_1$. Thus the existence of r_1 and r_2, having the above-mentioned property, proves that the S.P.R.T. is a Bayes sequential procedure.

Lemma 9.6.1.

(i) *For a fixed* r_1, $\xi_U(r_1, r_2)$ *is a decreasing function of* r_2.
(ii) *For a fixed* r_2, $\xi_L(r_1, r_2)$ *is an increasing function of* r_1.

Proof. (i) Let $\alpha(\mathbf{s})$ and $\beta(\mathbf{s})$ be the error probabilities of Type I and II, respectively, under stopping rule $\mathbf{s}$, and let $N(\mathbf{s})$ be the corresponding stopping variable. Then

(9.6.15) $$R(\xi, \mathbf{s}) = \xi[r_2\alpha(\mathbf{s}) + E_{\theta_1}\{N(\mathbf{s})\}] + (1 - \xi)[r_1\beta(\mathbf{s}) + E_{\theta_2}\{N(\mathbf{s})\}].$$

For a fixed value of r_1, $R(\xi, \mathbf{s})$ is an increasing function of r_2 for each $(\xi, \mathbf{s})$. Hence $\rho_1(\xi)$ is an increasing function of r_2 for a fixed r_1 for each ξ. Finally, assuming that $\xi_U(r_1, r_2) > r^*$, $\xi_U(r_1, r_2)$ is the intersection point of $\rho_1(\xi)$ and of $(1 - \xi)r_1$. Hence if r_1 is fixed, $r_2' > r_2$ implies that $\xi_U(r_1, r_2') \geqslant \xi_U(r_1, r_2)$. In a similar manner we prove (ii). (Q.E.D.)

Lemma 9.6.2.

(i) $\xi_L(r_1, r_2) \to 0$ *as* $r_2 \to \infty$ *uniformly in* $r_1 \leqslant r_2$ *on a bounded set;*
(ii) $\xi_U(r_1, r_2) \to 1$ *as* $r_1 \to \infty$ *uniformly in* $r_1 \geqslant r_2$.

Proof. (i) If r_1 is bounded, then from the definition of $\xi_L(r_1, r_2)$,

$$\xi_L(r_1, r_2) \leqslant \frac{r_1}{r_1 + r_2} \to 0 \qquad \text{as } r_2 \to \infty.$$

If r_1 can approach infinity also we prove (i) in the following manner. Let $\mathbf{s}_n$ be a fixed sample stopping rule with $N = n$. Let α_n and β_n designate the respective error probabilities. The Bayes decision function $d^0(\xi(\mathbf{X}_n))$ is equivalent to the most powerful Neyman-Pearson test of A_1 against A_2, with some α_n and β_n. In Lemma 4.5.2 we have proven that, in the fixed sample case,

$$\lim_{n\to\infty}\left\{\frac{1}{n}\ln\left(\frac{1}{\alpha_n}\right)\right\} = I(\theta_2, \theta_1). \tag{9.6.16}$$

We obtain a similar order of magnitude for β_n, by relabeling θ_1, θ_2, α_n, and β_n. Thus both α_n and β_n decrease to zero at a geometric rate as $n \to \infty$. For a given $\eta > 0$ choose n sufficiently large so that $\alpha_n, \beta_n < \eta$. The prior risk associated with fixed sample procedure $\mathbf{s}_n$ is

$$\begin{aligned} R(\xi, \mathbf{s}_n) &= n + \xi r_2 \alpha_n + (1 - \xi) r_1 \beta_n \\ &\leqslant n + 2\eta \max(r_1, r_2). \end{aligned} \tag{9.6.17}$$

Therefore, assuming that $r_2 \geqslant r_1$,

$$\rho(\xi) = \inf_{\mathbf{s}\in\mathfrak{S}} R(\xi, \mathbf{s}) \leqslant \rho_1(\xi) = \inf_{\mathbf{s}\in\mathfrak{S}_1} R(\xi, \mathbf{s}) \leqslant n + 2\eta r_2, \tag{9.6.18}$$

for all ξ, $0 < \xi < 1$. Let $\xi_L^{(n)}$ be the intersection point of $n + 2\eta r_2$ and ξr_2. That is,

$$\xi_L^{(n)} = \frac{n}{r_2} + 2\eta \to 2\eta \qquad \text{as } r_2 \to \infty. \tag{9.6.19}$$

According to (9.6.18),

$$\xi_L(r_1, r_2) \leqslant \xi_L^{(n)} \to 2\eta \qquad \text{as } r_2 \to \infty, \qquad \text{all } 0 < \xi < 1. \tag{9.6.20}$$

Finally, since η is arbitrary, $\lim_{r_2\to\infty} \xi_L(r_1, r_2) = 0$. Since $\xi_L^{(n)}$ is defined for $r_2 \geqslant r_1$ the convergence is uniform on every finite interval of r_1 values.

For the proof of (ii), reverse the roles of r_1 and r_2. (Q.E.D.)

Lemma 9.6.3. *The function* $\rho(\xi) \equiv \rho(\xi, r_1, r_2)$ *is a continuous function of* (ξ, r_1, r_2) *over the domain* $0 < \xi < 1$; $0 < r_1, r_2 < \infty$.

Proof. Let $R(\xi, r_1, r_2; \mathbf{s}) \equiv R(\xi, \mathbf{s})$ be the prior risk function of any stopping rule $\mathbf{s}$. This function is displayed in (9.6.15). Hence if (ξ, r_1', r_2') is a point in a close neighborhood of (ξ, r_1, r_2), that is, $|r_1' - r_1| < \delta_1$ and $|r_2' - r_2| < \delta_2$, then

$$(9.6.21)\quad \begin{aligned} &|R(\xi, r_1', r_2'; \mathbf{s}) - R(\xi, r_1, r_2; \mathbf{s})| \\ &\qquad = |\xi\alpha(\mathbf{s})(r_2' - r_2) + (1 - \xi)\beta(\mathbf{s})(r_1' - r_1)| \\ &\qquad \leqslant |r_2' - r_2| + |r_1' - r_1| < \delta_1 + \delta_2. \end{aligned}$$

Hence $R(\xi, r_1, r_2; \mathbf{s})$ is a continuous function of (r_1, r_2) for each ξ and $\mathbf{s}$. Furthermore, if $\max\{E_{\theta_1}\{N(\mathbf{s})\}, E_{\theta_2}\{N(\mathbf{s})\}\} \leqslant N^*$, then

$$(9.6.22)\quad |R(\xi', r_1', r_2'; \mathbf{s}) - R(\xi, r_1, r_2; \mathbf{s})| \leqslant |r_1' - r_1| + |r_2' - r_2| + N^*|\xi' - \xi|.$$

Thus if $\mathcal{S}(N^*)$ is the class of stopping rules,

$$\mathbf{s};\ \max\{E_{\theta_1}\{N(\mathbf{s})\}, E_{\theta_2}\{N(\mathbf{s})\}\} \leqslant N^*,$$

then $R(\xi, r_1, r_2; \mathbf{s})$ is a continuous function of (ξ, r_1, r_2) for each $\mathbf{s} \in \mathcal{S}(N^*)$. We have shown that the Bayes truncated stopping rule $\mathbf{s}^{(N^*)}$ is ϵ-Bayes for N^* sufficiently large; $\mathbf{s}^{(N^*)} \in \mathcal{S}(N^*)$ since $N(\mathbf{s}^{(N^*)}) \leqslant N^*$ a.s. for θ_1 and θ_2. Thus

$$(9.6.23)\quad \begin{aligned} &|\rho(\xi', r_1', r_2') - \rho(\xi, r_1, r_2)| \\ &\quad = |\rho(\xi', r_1', r_2') - R(\xi', r_1', r_2'; \mathbf{s}^{(N^*)}) + R(\xi', r_1', r_2'; \mathbf{s}^{(N^*)}) \\ &\qquad - R(\xi, r_1, r_2; \mathbf{s}^{(N^*)}) + R(\xi, r_1, r_2; \mathbf{s}^{(N^*)}) - \rho(\xi, r_1, r_2)| \\ &\quad \leqslant |\rho(\xi', r_1', r_2') - R(\xi', r_1', r_2'; \mathbf{s}^{(N^*)})| + |R(\xi', r_1', r_2'; \mathbf{s}^{(N^*)}) \\ &\qquad - R(\xi, r_1, r_2; \mathbf{s}^{(N^*)})| + |R(\xi, r_1, r_2; \mathbf{s}^{(N^*)}) - \rho(\xi, r_1, r_2)|. \end{aligned}$$

Choose N^* sufficiently large so that $\mathbf{s}^{(N^*)}$ is ϵ-Bayes for both (ξ, r_1, r_2) and (ξ', r_1', r_2'). Since $\mathbf{s}^{(N^*)} \in \mathcal{S}(N^*)$ we can choose (ξ', r_1', r_2') close enough to (ξ, r_1, r_2) so that

$$(9.6.24)\quad |R(\xi', r_1', r_2'; \mathbf{s}^{(N^*)}) - R(\xi, r_1, r_2; \mathbf{s}^{(N^*)})| < \epsilon.$$

Finally, from (9.6.23) and (9.6.24),

$$(9.6.25)\quad |\rho(\xi', r_1', r_2') - \rho(\xi, r_1, r_2)| < 3\epsilon.$$

(Q.E.D.)

Lemma 9.6.4. *Both* $\xi_L(r_1, r_2)$ *and* $\xi_U(r_1, r_2)$ *are continuous in* (r_1, r_2) *for* $0 < r_1, r_2 < \infty$.

Proof. Let

(9.6.26) $$D(\xi, r_1, r_2) = \rho_0(\xi, r_1, r_2) - \rho_1(\xi, r_1, r_2),$$

and

(9.6.27)
$$\psi(\xi, r_1, r_2) = D(\xi, r_1, r_2)\left\{\left|\frac{r_1}{r_1 + r_2} - \xi\right| + \max\left[0, D\left(\frac{r_1}{r_1 + r_2}, r_1, r_2\right)\right]\right\}.$$

The Bayes risk is designated by $\rho_0(\xi, r_1, r_2)$ when $N = 0$, and

$$\rho_1(\xi, r_1, r_2) = \inf_{s \in \mathcal{S}_1} R(\xi, r_1, r_2; \mathbf{s}).$$

The function $\psi(\xi, r_1, r_2)$ can assume the value 0 only at $\xi_L(r_1, r_2)$ and $\xi_U(r_1, r_2)$. Furthermore, $\psi(\xi, r_1, r_2)$ is continuous in (ξ, r_1, r_2). Fix ξ, and let $\{(r_1^{(n)}, r_2^{(n)}); n \geqslant 1\}$ be a sequence converging to (r_1^0, r_2^0). We wish to show that $\xi_L(r_1^{(n)}, r_2^{(n)}) \to \xi_L(r_1^0, r_2^0)$ and $\xi_U(r_1^{(n)}, r_2^{(n)}) \to \xi_U(r_1^0, r_2^0)$. Let $\xi_L^{(n)}(r_1^{(n)}, r_2^{(n)})$ and $\xi_U^{(n)}(r_1^{(n)}, r_2^{(n)})$ be the values given by (9.6.10) when $r_1 = r_1^{(n)}$ and $r_2 = r_2^{(n)}$. For each n, $n \geqslant 1$, we have

$$\psi(\xi_L^{(n)}(r_1^{(n)}, r_2^{(n)}), r_1^{(n)}, r_2^{(n)}) = 0,$$

and

$$\psi(\xi_U^{(n)}(r_1^{(n)}, r_2^{(n)}), r_1^{(n)}, r_2^{(n)}) = 0.$$

If $\{\xi_L^{(n)}(r_1^{(n)}, r_2^{(n)}); n \geqslant 1\}$ has a limit point, we have to show that $\lim_{n\to\infty} \xi_L^{(n)}(r_1^{(n)}, r_2^{(n)}) = \xi_L(r_1^0, r_2^0)$. Due to the continuity of $\psi(\xi, r_1, r_2)$,

(9.6.28) $$\psi(\lim_{n\to\infty} \xi_L^{(n)}(r_1^{(n)}, r_2^{(n)}), r_1^0, r_2^0) = 0.$$

Hence

$$\lim_{n\to\infty} \xi_L^{(n)}(r_1^{(n)}, r_2^{(n)}) = \xi_L(r_1^0, r_2^0).$$

It remains to show that $\xi_L(r_1^0, r_2^0) > 0$. Assume that $\xi_L(r_1^0, r_2^0) < r_1/(r_1 + r_2)$; $\xi_L^{(n)}(r_1^{(n)}, r_2^{(n)})$ is the point of intersection of $r_2^{(n)}\xi$ and of $\rho_1(\xi)$. Since $\rho_1(\xi) \geqslant 1$,

(9.6.29) $$\xi_L^{(n)}(r_1^{(n)}, r_2^{(n)}) \geqslant \frac{1}{r_2^{(n)}}, \qquad \text{all } n \geqslant 1.$$

Hence

(9.6.30) $$\xi_L(r_1^0, r_2^0) \geqslant \frac{1}{r_2^0} > 0.$$

In a similar manner we show that $\xi_U(r_1, r_2)$ is a continuous function of (r_1, r_2). (Q.E.D.)

Theorem 9.6.5. *Given any* $0 < \xi_L \leqslant \xi_U < 1$, *there exist* r_1 *and* r_2, $0 < r_1$, $r_2 < \infty$, *such that*

$$\xi_L(r_1, r_2) = \xi_L \quad \text{and} \quad \xi_U(r_1, r_2) = \xi_U.$$

Proof. Consider the complex plane $z = r_1 + ir_2$ and the complex valued function

$$F(r_1, r_2) = \xi_L(r_1, r_2) + i\xi_U(r_1, r_2). \tag{9.6.31}$$

By Lemma 9.6.4, $F(r_1, r_2)$ is continuous in (r_1, r_2). Consider the square in the (r_1, r_2)-plane with vertices $(1, 1)$, $(1, K)$, $(K, 1)$, and (K, K). The transformation $F(r_1, r_2)$ transforms it to a triangle T in the (ξ_L, ξ_U)-plane, whose sides are the line segments connecting $(1/(1 + K), 1/(1 + K)) \to (K/(1 + K), K/(1 + K))$, $(K/(1 + K), K/(1 + K)) \to (1/(1 + K), K/(1 + K))$, $(1/(1 + K), K/(1 + K)) \to (1/(1 + K), 1/(1 + K))$. By choosing K large enough the Jordan region inscribed by the triangle will contain any given point (ξ_*, ξ^*) with $0 < \xi_* \leqslant \xi^* < 1$. The closed region $\mathfrak{R}$ inscribed by the above square is transformed onto the closed region T inscribed by that triangle. Therefore $F^{-1}(\xi_*, \xi^*) \subset \mathfrak{R}$ and there exists a point $(r_1, r_2) \in \mathfrak{R}$ such that $F(r_1, r_2) = (\xi_L, \xi_U)$. (Q.E.D.)

Conclusion. Given any S.P.R.T. $(-b, a)$, $0 < b$, $a < \infty$, there exists a (ξ, r_1, r_2) for which the S.P.R.T. $(-b, a)$ is a Bayes sequential procedure.

Theorem 9.6.6. (Wald [4].) *Let* N *be the stopping variable of an S.P.R.T. with error probabilities* α *and* β. *Let* $\mathbf{s}$ *be an arbitrary stopping rule in* $\mathcal{S}_1$ *with error probabilities* $\alpha(\mathbf{s})$, $\beta(\mathbf{s})$ *and stopping variable* $N(\mathbf{s})$. *If* $\alpha(\mathbf{s}) \leqslant \alpha$ *and* $\beta(\mathbf{s}) \leqslant \beta$, *then*

$$E_{\theta_i}\{N\} \leqslant E_{\theta_i}\{N(\mathbf{s})\}, \qquad i = 1, 2. \tag{9.6.32}$$

Proof. Given an S.P.R.T. $(-b, a)$ with error probabilities α and β, let (ξ, r_1, r_2) be the prior probability and the loss values against which it is Bayes. Let $R(\xi, \mathbf{s})$ be the corresponding prior risk of $\mathbf{s}$ in $\mathcal{S}_1$. Then according to (9.6.15),

$$\begin{aligned}\xi[r_2\alpha + E_{\theta_1}\{N\}] &+ (1 - \xi)[r_1\beta + E_{\theta_2}\{N\}] \\ &\leqslant \xi[r_2\alpha(\mathbf{s}) + E_{\theta_1}\{N(\mathbf{s})\}] + (1 - \xi)[r_1\beta(\mathbf{s}) + E_{\theta_2}\{N(\mathbf{s})\}].\end{aligned} \tag{9.6.33}$$

Hence

$$\begin{aligned}\xi r_2(\alpha - \alpha(\mathbf{s})) + (1 - \xi)r_1(\beta - \beta(\mathbf{s})) &\leqslant \xi E_{\theta_1}\{N(\mathbf{s}) - N\} \\ &\quad + (1 - \xi)E_{\theta_2}\{N(\mathbf{s}) - N\}.\end{aligned} \tag{9.6.34}$$

Since $\alpha(\mathbf{s}) \leqslant \alpha$ and $\beta(\mathbf{s}) \leqslant \beta$, the left-hand side of (9.6.34) is non-negative for each ξ. Letting $\xi \to 0$, we obtain

$$0 \leqslant r_1(\beta - \beta(\mathbf{s})) \leqslant E_{\theta_2}\{N(\mathbf{s}) - N\}. \tag{9.6.35}$$

Similarly, letting $\xi \to 1$, we obtain

$$0 \leqslant r_2(\alpha - \alpha(\mathbf{s})) \leqslant E_{\theta_1}\{N(\mathbf{s}) - N\}. \tag{9.6.36}$$

Finally, since $0 < r_1, r_2 < \infty$, (9.6.35) and (9.6.36) yield (9.6.32). (Q.E.D.)

For further reading on the optimality of the S.P.R.T., see Matthes [1], Wald and Wolfowitz [1], Wolfowitz [5], Ferguson [1], and Burkholder and Wijsman [1].

9.7. SOME CONTINUOUS TIME PROCESS ANALOGS

Continuous time process analogs of the S.P.R.T. are useful decision procedures either when a process is observed continuously or as an approximating model for discrete time processes. One of the earliest papers in which these ideas are explored is that of Dvoretzky, Kiefer, and Wolfowitz [1]. The results and methodology of this paper have been applied in many subsequent studies. We therefore present here their main ideas. A novel application of these ideas is found in the paper of Epstein and Sobel [1], which we present also. DeGroot [2] approached the problem of testing whether the mean of a normal distribution, μ, is nonpositive against the hypothesis that it is positive by deriving the minimax test for a corresponding Wiener process. We also provide these results in this section. Finally, we give several extensions of previous theorems of Wald to the Wiener process case, following the recent results of Hall [2].

9.7.1. S.P.R.T. FOR CONTINUOUS TIME PROCESSES

In certain cases one can observe X continuously in time and obtain a sample function of a stochastic process $\{X(t); t \geqslant 0\}$. In such cases a generalization of Wald's S.P.R.T. to a continuous time decision process can have several advantages, as will be indicated later.

For example, suppose an r.v. X has a density belonging to a 1-parameter exponential family, that is,

$$f(x; \theta) = h(x) \exp\{U(x)\psi_1(\theta) + \psi_2(\theta)\},$$

where $\psi_1(\theta)$ is strictly increasing in θ. Suppose we test two simple hypotheses A_i: $\theta = \theta_i$, $i = 1, 2$, by means of Wald's S.P.R.T. We have

(9.7.1)

$$S_n = \sum_{i=1}^{n} \log \frac{f(X_i; \theta_2)}{f(X_i; \theta_1)} = (\psi_1(\theta_2) - \psi_1(\theta_1)) \sum_{i=1}^{n} U(X_i) + n[\psi_2(\theta_2) - \psi_1(\theta_1)].$$

Thus the S.P.R.T. $(-b, a)$ is equivalent to the sequential test in which one continues as long as

(9.7.2)
$$-\frac{b}{\psi_1(\theta_2)-\psi_1(\theta_1)} - n\frac{\psi_2(\theta_2)-\psi_2(\theta_1)}{\psi_1(\theta_2)-\psi_1(\theta_1)} < \sum_{i=1}^{n} U(X_i) < \frac{a}{\psi_1(\theta_2)-\psi_1(\theta_1)} - n\frac{\psi_2(\theta_2)-\psi_2(\theta_1)}{\psi_1(\theta_2)-\psi_1(\theta_1)}.$$

Define $Y(t)$ to be a continuous analog (stochastic process) of $Y_n = \sum_{i=1}^{n} U(X_i)$. The acceptance and rejection limits can be considered also as functions of t, $0 < t < \infty$, and thus we obtain a continuous generalization of Wald's S.P.R.T. The acceptance and rejection lines of this sequential test are, respectively,

$$A(t) = -\frac{b}{\Delta} - t\frac{\Gamma}{\Delta},$$

and

$$R(t) = \frac{a}{\Delta} - t\frac{\Gamma}{\Delta},$$

where $\Delta = \psi_1(\theta_2) - \psi_1(\theta_1)$, and $\Gamma = \psi_2(\theta_2) - \psi_2(\theta_1)$. Define the stopping time

(9.7.3)
$$T = \inf\{t;\, t \geqslant 0,\, t \notin (A(t), R(t))\}$$

and the decision rule if $Y(T) = A(T)$ accept A_1; if $Y(T) = R(T)$ accept A_2. We thus consider the following decision problem.

Let $\{X_1(t); t \geqslant 0\}$ and $\{X_2(t); t \geqslant 0\}$ be two different stochastic processes. A statistician observes continuously, beginning at $t = 0$, a process $\{X(t); t \geqslant 0\}$ which is either $\{X_1(t); t \geqslant 0\}$ or $\{X_2(t); t \geqslant 0\}$. The objective is to decide "as soon as possible" which process is observed. Let α, β be assigned error probabilities of the first and second kind. Let T be the stopping time. The objective is to minimize $E_i\{T\}$, $i = 1, 2$, under the constraint that the error probabilities do not exceed α and β, respectively.

We assume that $X_1(t)$ and $X_2(t)$ have specified density functions $f_i(X; t)$, $i = 1, 2$. Let

(9.7.4)
$$Z(t) = \log\frac{f_2(X(t); t)}{f_1(X(t); t)}, \qquad Z(0) \in 0.$$

We assume that $\{Z(t); t \geqslant 0\}$ is a process of *stationary independent increments*. The theory outlined below has two important examples: testing hypotheses about the parameter of a continuous Poisson process; and testing hypotheses about the mean of a Wiener process.

The stopping time T of the present procedure is well defined, and $E_i\{T^k\} < \infty$ $(i = 1, 2)$ for all $k = 0, 1, \ldots$. This is implied from the results we have established on the stopping variable N of the discrete S.P.R.T.

Let $E\{Z(1)\} = \mu$ and suppose $\mu \neq 0$. Wald's theorem is extended to

$$E_\theta\{Z(T)\} = \mu E_\theta\{T\}. \tag{9.7.5}$$

The extension of the fundamental identity is

$$E_\theta\{e^{uZ(T)}(M_\theta(u))^{-T}\} = 1 \tag{9.7.6}$$

where $M_\theta(u) = E_\theta\{e^{uZ(1)}\} \geqslant 1$. These extensions are proven easily by applying Doob's results on continuous martingales ([2], Ch. 7, Theorem 11.8) and will be discussed also in Section 9.7.4.

Example 9.6. Let $X(t)$ be a stationary Poisson process with

$$P_\lambda[X(t) = j] = e^{-\lambda t}\frac{(\lambda t)^j}{j!}, \qquad j = 0, 1, \ldots. \tag{9.7.7}$$

The two hypotheses under consideration are

$$A_1: \lambda = \lambda_1, \qquad A_2: \lambda = \lambda_2, \qquad 0 < \lambda_1 < \lambda_2 < \infty.$$

From the log-likelihood ratio of the discrete case we base the S.P.R.T. on the analogous stochastic process

$$Z(t) = -t(\lambda_2 - \lambda_1) + \left(\log\left(\frac{\lambda_2}{\lambda_1}\right)\right)X(t). \tag{9.7.8}$$

We choose appropriate boundaries $(-b, a)$, $0 < a$, $b < \infty$, and define the stopping time

$$T = \inf\{t; t \geqslant 0, Z(t) \notin (-b, a)\}. \tag{9.7.9}$$

The process $X(t)$ is a jump process with $X(t + 0) - X(t - 0) \leqslant 1$, with probability 1, for all t. Hence $Z(t + 0) - Z(t) \leqslant \log(\lambda_2/\lambda_1)$, with probability 1. We observe that whenever the process $Z(t)$ hits the lower boundary $-b$, $Z(T) = -b$. On the other hand, if $Z(t)$ crosses the boundary a, $Z(T) = a$. But $a > Z(T - 0) \geqslant a - \log(\lambda_2/\lambda_1)$. Let

$$M_\lambda(u) = E_\lambda\{e^{uZ(1)}\} = \exp\left\{-(\lambda_2 - \lambda_1)u - \lambda\left(1 - \left(\frac{\lambda_2}{\lambda_1}\right)^u\right)\right\}.$$

Thus if $u(\lambda) \neq 0$ designates the value of u for which $M_\lambda(u) = 1$, we have

$$u(\lambda) = -\frac{\lambda}{\lambda_2 - \lambda_1} + \frac{1}{\lambda_2 - \lambda_1}\left(\frac{\lambda_2}{\lambda_1}\right)^{u(\lambda)}. \tag{9.7.10}$$

From the extended fundamental identity, (9.7.6), we obtain that the probability of accepting A_1 under λ is

$$\pi(\lambda) = \frac{E_\lambda\{e^{u(\lambda)Z(T)} \mid Z(T) \geqslant a\} - 1}{E_\lambda\{e^{u(\lambda)Z(T)} \mid Z(T) \geqslant a\} - e^{-u(\lambda)b}}. \tag{9.7.11}$$

If $u(\lambda) > 0$,

$$e^{u(\lambda)a} \leqslant E_\lambda\{e^{u(\lambda)Z(T)} \mid Z(T) \geqslant a\} \leqslant e^{u(\lambda)[a+\log(\lambda_2/\lambda_1)]} \tag{9.7.12}$$

We obtain from (9.7.10)–(9.7.12) that when $u(\lambda) > 0$,

$$\frac{e^{au(\lambda)} - 1}{[(\lambda_2 - \lambda_1)u(\lambda) + \lambda]e^{au(\lambda)} - e^{-bu(\lambda)}} \leqslant \pi(\lambda) \leqslant \frac{[(\lambda_2 - \lambda_1)u(\lambda) + \lambda]e^{u(\lambda)a} - 1}{e^{u(\lambda)a} - e^{-bu(\lambda)}}. \tag{9.7.13}$$

If $u(\lambda) < 0$, the above inequalities reverse. The exact formula of $\pi(\lambda)$ was derived by Dvoretzky, Kiefer, and Wolfowitz [1].

9.7.2. SEQUENTIAL LIFE TESTS

We now present the application of the above results to sequential life testing in the exponential case, as given in the article of Epstein and Sobel [1]. We restrict attention to an experimental set-up with replacement. In such an experiment, n identical units are put into operation at time $t_0 = 0$. As soon as a unit fails it is replaced by a new unit. The process is observed continuously and the failure time points are recorded. The experiment is terminated as soon as the r-th failure occurs, if it has not been stopping previously; $1 \leqslant r \leqslant n$. Assuming that the interfailure time, in each unit, is a random variable $X \sim G(1/\theta, 1)$, θ is the mean time between failures (M.T.B.F.). The objective is to test the two hypotheses, $A_1\colon \theta \geqslant \theta_1$, $A_2\colon \theta \leqslant \theta_2$, $0 < \theta_2 < \theta_1 < \infty$. We consider a continuous version of an S.P.R.T. with error probabilities α and β.

Let $0 \leqslant t_1 \leqslant \cdots \leqslant t_r$ denote the epochs of the first r failures. The total life of the n units up to t_r is $T_{n,r} = nt_r$. Now define the corresponding Poisson process on $(0, t_r]$, namely,

$$X_{n,r}(t) = \text{total number of failures in } (0, t], \quad 0 < t \leqslant t_r. \tag{9.7.14}$$

We define the stopping time, for specified values of $0 < a, b < \infty$, as

$$\tau_{n,r} = \min\left\{t_r, \sup\left\{t; t > 0, \frac{nt - h_0}{s} < X_{n,r}(t) < \frac{nt + h_1}{s}\right\}\right\} \tag{9.7.15}$$

where

(9.7.16)

$$h_0 = \frac{b}{(1/\theta_2) - (1/\theta_1)}, \qquad h_1 = \frac{a}{(1/\theta_2) - (1/\theta_1)}, \qquad s = \frac{\log(\theta_1/\theta_2)}{(1/\theta_2) - (1/\theta_1)}.$$

Hypothesis A_1 is accepted if $X_{n,r}(\tau_{n,r}) = (1/s)(n\tau_{n,r} - h_0)$; A_2 is accepted if $X_{n,r}(\tau_{n,r}) \geqslant (1/s)(n\tau_{n,r} + h_1)$. It is easy to check that $\tau_{n,r}$ corresponds to a continuous time version of an S.P.R.T.

The total life $T_{n,r}(t)$ corresponding to $X_{n,r}(t)$ is given by $T_{n,r}(t) = nt$ if $t_j \leqslant t < t_{j+1}$ and $X_{n,r}(t) = j, j = 0, 1, \dots, r-1$.

If we designate by $\pi(\theta)$ the probability of accepting A_1 under θ, then due to the governing Poisson process with $\lambda = 1/\theta$ we obtain from (9.7.10) and (9.7.11) the following (approximative) parametric representation:

$$\pi(\theta) = \frac{e^{ah} - 1}{e^{ah} - e^{-bh}},$$

(9.7.17)

$$\theta = \frac{(\theta_1/\theta_2) - 1}{h(1/\theta_2 - 1/\theta_1)}.$$

According to (9.7.8), we can write

(9.7.18)
$$E_\theta\{X_{n,r}(\tau_{n,r})\} \log \frac{\theta_1}{\theta_2} - \left(\frac{1}{\theta_2} - \frac{1}{\theta_1}\right) E_\theta\{T_{n,r}(\tau_{n,r})\}$$
$$= \pi(\theta) E_\theta\{Z_{n,r}(\tau_{n,r}) \mid Z_{n,r}(\tau_{n,r}) \leqslant -b\} + (1 - \pi(\theta)) \times E_\theta\{Z_{n,r}(\tau_{n,r}) \mid Z_{n,r}(\tau_{n,r}) \geqslant a\},$$

where

$$Z_{n,r}(t) = X_{n,r}(t) \log \left(\frac{\theta_1}{\theta_2}\right) - \left(\frac{1}{\theta_2} - \frac{1}{\theta_1}\right) T_{n,r}(t).$$

The total life at the time of stopping is designated by $T_{n,r}(\tau_{n,r})$; $X_{n,r}(\tau_{n,r})$ is the total number of failures at the time of stopping. We notice that if we set $b = -\log(\beta/(1-\alpha))$ and $a = \log((1-\beta)/\alpha)$, neglecting the possible overshoot of $X_{n,r}(t)$ over the upper boundary, we obtain the approximation

(9.7.19)
$$E_\theta\{X_{n,r}(\tau_{n,r})\} \log \left(\frac{\theta_1}{\theta_2}\right) - \left(\frac{1}{\theta_2} - \frac{1}{\theta_1}\right) E_\theta\{T_{n,r}(\tau_{n,r})\}$$
$$\simeq \pi(\theta) \log \frac{\beta}{1-\alpha} + (1 - \pi(\theta)) \log \frac{1-\beta}{\alpha}.$$

We establish the following analog of the Wald theorem, namely,

(9.7.20)
$$E_\theta\{T_{n,r}(\tau_{n,r})\} = \theta E_\theta\{X_{n,r}(\tau_{n,r})\}.$$

That is, the expected total life at stopping is equal to the M.T.B.F. multiplied by the expected total number of failures at stopping. From (9.7.19) and (9.7.20), we obtain

(9.7.21)
$$E_\theta\{X_{n,r}(\tau_{n,r})\} \simeq \begin{cases} \dfrac{h_1 - \pi(\theta)(h_0 + h_1)}{s - \theta}, & s \neq \theta, \\[2ex] \dfrac{h_0 h_1}{s^2}, & s = \theta. \end{cases}$$

We now prove (9.7.20).

Suppose that we continue with the experiment of life testing (with replacement of failing units) indefinitely. The stopping time $\tau_{n,r}$ is considered now as the first passage time from the continuation region outside. Let $R = X_{n,r}(\tau_{n,r})$, and fix an integer N larger than R. Then

$$t_N = \tau_{n,r} + (t_{R+1} - \tau_{n,r}) + (t_{R+2} - t_{R+1}) + \cdots + (t_N - t_{N-1}).$$

Due to the well-known characteristics of the Poisson process, the interfailure times $(t_{R+2} - t_{R+1}), \ldots, (t_N - t_{N-1})$ are i.i.d. random variables having the negative exponential distribution with mean θ/n; that is, $t_{j+1} - t_j \sim G(n/\theta, 1)$, $j \geqslant R + 1$. Furthermore, given $\tau_{n,r}$, $(t_{R+1} - \tau_{n,r})$ is independent of $\{(t_{j+1} - t_j); j \geqslant R + 1\}$, having the same $\mathcal{G}(n/\theta, 1)$ distribution. Hence we obtain the following equalities:

$$E_\theta\{t_N\} = N\frac{\theta}{n}, \tag{9.7.22}$$

and

$$E_\theta\{t_N\} = E_\theta^{(N)}\{\tau_{n,r}\} + (N - E_\theta^{(N)}\{R\})\frac{\theta}{n}. \tag{9.7.23}$$

Hence for each $N > R$ we have

$$E_\theta^{(N)}\{\tau_{n,r}\} = \frac{\theta}{n} E_\theta^{(N)}\{R\}. \tag{9.7.24}$$

Finally, since $\tau_{n,r} \leqslant t_r$ and $R = X_{n,r}(\tau_{n,r}) \leqslant r$, and since $E_\theta\{t_r\} = r(\theta/n)$, we obtain from the Lebesgue dominated convergence theorem that

$$\begin{aligned} E_\theta\{\tau_{n,r}\} &= \lim_{N\to\infty} E_\theta^{(N)}\{\tau_{n,r}\} = \frac{\theta}{n}\lim_{N\to\infty} E_\theta^{(N)}\{R\} \\ &= \frac{\theta}{n} E_\theta\{X_{n,r}(\tau_{n,r})\}. \end{aligned} \tag{9.7.25}$$

Moreover, since $T_{n,r}(\tau_{n,r}) = n\tau_{n,r}$, (9.7.20) is implied from (9.7.25).

9.7.3. MINIMAX SEQUENTIAL TESTS FOR WIENER PROCESSES

Consider first the following problem: $X_1, X_2, \ldots$ is a sequence of i.i.d. random variables having a common $\mathcal{N}(\mu, 1)$ distribution. We have to test the hypotheses $A_0\colon \mu \leqslant 0$, $A_1\colon \mu > 0$. The loss functions due to erroneous decisions are

$$L_0(\mu) = \begin{cases} 0, & \text{if } \mu \leqslant 0, \\ c\mu^r, & \text{if } \mu > 0, \end{cases}$$

and

$$L_1(\mu) = \begin{cases} c\,|\mu|^r, & \text{if } \mu \leqslant 0, \\ 0, & \text{if } \mu > 0. \end{cases}$$

The problem of determining a Bayes sequential procedure for a given prior distribution $H(\mu)$ is not a simple one. This problem will be discussed in Section 9.8. We have proven that every S.P.R.T. is a Bayes sequential test procedure for testing two simple hypotheses, say $A_1^*: \mu = -\delta$ versus $A_2^*: \mu = \delta$, for some $\delta > 0$, against some (ξ, r_0, r_1), where ξ is the prior probability of A_1^*. We do not know, however, what happens when $\delta \to 0$, and we test the composite hypotheses $A_0: \mu \leqslant 0$, $A_1: \mu > 0$. We prove here, following DeGroot [2], that a minimax sequential procedure is a proper symmetric S.P.R.T. We approach the problem by considering a continuous time analog to the given sequence.

Let $\mathcal{W}_\mu = \{X(t); t \geqslant 0\}$ be a stationary Wiener process with $X(0) \equiv 0$. This is a process of independent increments, with

$$X(t) - X(s) \sim N(\mu(t-s), t-s), \quad \text{all} \quad t > s > 0. \tag{9.7.26}$$

The class of symmetric S.P.R.T. for $\mathcal{W}_\mu$ is the class of all procedures, according to which $X(t)$ is observed as long as it is in the interval $(-d, d)$ for some $d > 0$. Thus we define the stopping time

$$T_d = \inf\{t; t > 0, |X(t)| \geqslant d\}. \tag{9.7.27}$$

If $X(T_d) = d$ we accept A_1 and if $X(T_d) = -d$ we accept A_0. This test procedure is an extension of a symmetric S.P.R.T. to Wiener processes. We remark that an S.P.R.T. of $A_{-1}^*: \mu = -1$ versus $A_1^*: \mu = 1$ is based on the sequence of partial sums $S_n = \sum_{i=1}^n X_i$ distributed like $\mathcal{N}(n\mu, n)$. This is the marginal distribution of $X(n)$ of the Wiener process $\mathcal{W}_\mu$. Writing $Z = \log (f(X(1), \mu)/f(X(1), -\mu)) = 2\mu X(1)$, $-\infty < \mu < \infty$, we obtain that the m.g.f. $M_\mu(t)$ of Z attains the value 1 at $t_0(\mu) = -\mu \neq 0$. Hence by the extended fundamental identity, (9.7.6), the probability of accepting A_0 under μ, in the symmetric S.P.R.T. for $\mathcal{W}_\mu$, is

$$\pi(\mu) = \frac{e^{-2d\mu} - 1}{e^{-2d\mu} - e^{2d\mu}} = (1 + e^{2d\mu})^{-1}, \qquad -\infty < \mu < \infty. \tag{9.7.28}$$

Here we obtain an exact determination of $\pi(\mu)$, since the sample functions $X(t)$ of $\mathcal{W}_\mu$ are continuous in t, with probability 1. Furthermore, for the extended Wald theorem, Theorem 9.7.5, we obtain

$$\begin{aligned} E_\mu\{T_d\} &= \frac{1}{\mu}\{-\pi(\mu)d + (1 - \pi(\mu))d\} \\ &= \frac{d(e^{2\mu d} - 1)}{\mu(e^{2\mu d} + 1)}. \end{aligned} \tag{9.7.29}$$

This is the expected time to termination. It is simple to show that $\lim_{\mu \to 0} E_\mu\{T_d\} = E_0\{T_d\} = d^2/2$. The risk function associated with a symmetric S.P.R.T. is

$$R(\mu, d) = kE_\mu\{T_d\} + \pi(\mu)L_1(\mu) + (1 - \pi(\mu))L_0(\mu). \tag{9.7.30}$$

(Here k is the sampling cost per time unit.) Substituting (9.7.28) and (9.7.29) in (9.7.30) we obtain

$$R(\mu, d) = k\frac{d}{\mu}\cdot\frac{e^{2d\mu}-1}{e^{2d\mu}+1} + c\,|\mu|^r\frac{e^{2d|\mu|}}{1+e^{2d|\mu|}}. \tag{9.7.31}$$

In order to determine the minimax value of d, say d^*, it is convenient to make the following transformations:

$$\begin{aligned} d &= \left(\frac{c}{k}\right)^{(r+2)^{-1}}\eta, \\ \mu &= \left(\frac{k}{c}\right)^{(r+2)^{-1}} m. \end{aligned} \tag{9.7.32}$$

The risk function can then be written as

$$\begin{aligned} R(m, \eta) &= \frac{m^r}{e^{m\eta}+1} + \frac{\eta(e^{m\eta}-1)}{2m(e^{m\eta}+1)}, \qquad m \geqslant 0 \\ &= R(-m, \eta), \qquad m < 0. \end{aligned} \tag{9.7.33}$$

We wish to find values m^*, η^* such that

$$R(m^*, \eta^*) = \inf_\eta R(m^*, \eta) = \sup_m R(m, \eta^*). \tag{9.7.34}$$

Because of the symmetry in m of $R(m, \eta)$, we can restrict attention to $m \geqslant 0$.

The solution provided by DeGroot assumes that $r \leqslant 2$. In this case, m^* and η^* should satisfy the simultaneous equations

$$\begin{aligned} r(1+e^{-m\eta}) - m\eta &= \frac{\eta^2}{m^r}\left[\frac{\sinh(m\eta)}{m\eta - 1}\right], \\ \frac{m^{r+1}}{\eta} &= \frac{\sinh(m\eta)}{m\eta+1}. \end{aligned} \tag{9.7.35}$$

The values (m^*, η^*), which are the roots of (9.7.35), exist and are unique. Transforming m^* and η^* back to μ^* and d^* we obtain the minimax symmetric S.P.R.T. We show that this procedure is minimax at large; d^* is obviously Bayes against the prior distribution which assigns $\mu = \mu^*$ probability $\frac{1}{2}$ and $\mu = -\mu^*$ probability $\frac{1}{2}$. Furthermore, due to the symmetry of $R(\mu, \delta)$ with respect to μ, the Bayes risk against this prior distribution is $\rho^*(d^*) = R(|\mu^*|, d^*)$. Finally, $R(|\mu^*|, d^*) = \sup_\mu R(\mu, d^*)$. Hence, by Theorem 6.5.2, d^* is a minimax procedure.

9.7.4. ON WALD'S EQUATION FOR WIENER PROCESSES

In Section 9.7.1 we cited the result of Dvoretsky, Kiefer, and Wolfowitz [1] concerning the extension of Wald's theorem (Theorem 9.5.3) to continuous parameter processes. Recent papers by Robbins and Samuel [1] and by Brown [1] provide general results on Wald's equation in discrete time processes. These results have been generalized by Hall [2] to Wiener processes.

Let $\mathcal{W}_\mu$ be a Wiener process with drift μ, $-\infty < \mu < \infty$, and variance $\sigma^2 = 1$ per time unit. A function T on $(\mathfrak{X}, \mathfrak{B}, P)$ is called a *stopping time* if T is a random variable assuming non-negative values and the event $\{T \leqslant t\}$ is $\mathcal{F}_t$-measurable where, for $0 < t < \infty$, $\mathcal{F}_t$ is a subfield of $\mathfrak{B}$, generated by $\{X(s); 0 < s \leqslant t\}$.

A process $\{Y_t, \mathcal{F}_t; t \geqslant 0\}$ is called a supermartingale if $E\,|Y_t| < \infty$, and

$$E\{Y_t \mid \mathcal{F}_s\} \leqslant Y_s \quad \text{a.s.,} \quad \text{all} \quad 0 < s \leqslant t. \tag{9.7.36}$$

The fundamental identity for the S.P.R.T., (9.5.23), gives in the case of normally distributed random variables

$$E_\theta\{e^{-tS_N}\} \leqslant 1, \quad \text{all} \quad 0 \leqslant t \leqslant 2\theta. \tag{9.7.37}$$

The proof of (9.5.23) restricts (9.3.37) to stopping variables of the S.P.R.T. We now extend (9.7.37) to a general stopping time for a Wiener process $\mathcal{W}_\mu$, using martingale theory.

Theorem 9.7.1. (Hall [2].) *Let T be a stopping time on a Wiener process $\mathcal{W}_\mu$; then if $\mu > 0$,*

$$E_\mu\{\exp(-uX(T))\} \leqslant 1, \quad \textit{all} \quad 0 \leqslant u \leqslant 2\mu, \tag{9.7.38}$$

and $E_\mu\{X^r(T)\}$ exists $(\leqslant \infty)$ for every positive integer r.

Proof. We first notice that $\{e^{-uX(t)}, \mathcal{F}_t; t \geqslant 0\}$ is a supermartingale whenever $0 \leqslant u \leqslant 2\mu$. Indeed, $X(t) - X(s)$ is independent of $X(s)$, for all $0 < s < t$. Hence for all $0 < s \leqslant t$,

$$\begin{aligned} (9.7.39)\quad & E_\mu\{\exp(-uX(t)) \mid \mathcal{F}_s\} \\ &= e^{-uX(s)}E_\mu\{\exp\{-u(X(t) - X(s))\}\} \\ &= \exp\{-uX(s) + \tfrac{1}{2}u^2(t-s)^2 - \mu u(t-s)\} \quad \text{a.s.} \end{aligned}$$

But if $0 \leqslant u \leqslant 2\mu$, then $\exp\{\frac{1}{2}u^2(t-s)^2 - \mu u(t-s)\} \leqslant 1$. Hence

$$E_\mu\{\exp\{-uX(t)\} \mid \mathcal{F}_s\} \leqslant \exp\{-uX(s)\} \text{ a.s.}$$

Consider now the stopping time $T_n = \min\{n, T\}$. Another supermartingale is $\{\exp\{-uX(T_n)\}, \mathcal{F}_n; n \geqslant 0\}$. (See Doob [1], Theorem 11.6.) Hence, for all $0 \leqslant u \leqslant 2\mu$,

$$E_\mu\{e^{-uX(T_n)}\} \leqslant E_\mu\{e^{-uX(0)}\} = 1. \tag{9.7.40}$$

Finally, by the Lebesgue monotone convergence theorem,

$$(9.7.41)\qquad E_{\mu}\{e^{-uX(T)}\} = E_{\mu}\left\{\lim_{n\to\infty} e^{-uX(T_n)}\right\} = \lim_{n\to\infty} E_{\mu}\{e^{-uX(T^n)}\} \leqslant 1.$$

The existence of $E_{\mu}\{X^r(T)\}$ is immediately implied. (Q.E.D)

Returning again to the fundamental identity of the S.P.R.T., one obtains by differentiating (9.5.23) with respect to t (see Wald [3]),

$$(9.7.42)\qquad \frac{d}{dt} E\{e^{tS_N}(M(t))^{-N}\} = E\{S_N e^{tS_N}(M(t))^{-N}\} - M'(t)E\{Ne^{tS_N}(M(t))^{-N-1}\}.$$

For $t = 0$ we obtain from (9.7.42)

$$E\{S_N\} - \mu E\{N\} = 0.$$

Second-order differentiation yields

$$(9.7.43)\qquad E\{S_N{}^2\} - 2\mu E\{NS_N\} + \mu E\{N(N+1)\} - E\{N\}E\{Z^2\} = 0,$$

and so on. These results have been generalized to the Wiener process $\mathfrak{W}_{\mu}$ and any stopping time T, as in the following theorem.

Theorem 9.7.2. (Hall [2].) *Suppose that T is a stopping time on $\mathfrak{W}_{\mu}$.*

(i) *If $E_{\mu}\{T^k\} < \infty$ for some positive integer k then all terms on the left-hand side of (9.7.44) exist and are finite, and*

$$(9.7.44)\qquad E_{\mu}\{X^m(T)\} + a_{m1}E_{\mu}\{TX^{m-2}(T)\} + \cdots + a_{mk}E_{\mu}\{T^kX^{m-2k}(T)\} = 0,$$

for $m = 1, 2, \ldots, 2k$.

(ii) *If $E_{\mu}\{T^{(k+1)/2}\} < \infty$ for some positive integer k, and if $E_{\mu}\{|X(T)|^{2k+1}\} < \infty$, then all terms on the left-hand side of (9.7.44) exist and are finite and (9.7.44) holds for $m = 1, \ldots, 2k+1$.*

For a proof, see Hall [2].

Theorem 9.7.2 implies that if $E_{\mu}\{T^m\} < \infty$ for some positive integer m, then $E_{\mu}\,|X(T)|^m < \infty$ and

$$(9.7.45)\qquad \sum_{j=0}^{m/2} a_{mj} \sum_{i=0}^{m-2j} \binom{m-2j}{i} (-\mu)^{m-2j-i} E_{\mu}\{X^i(T)T^{m-i-j}\} = 0.$$

Formula 9.7.45 provides us with an upper bound to the variance of any stopping rule T, defined on a Wiener process $\mathfrak{W}_{\mu}$, such that $E_{\mu}\{T^2\} < \infty$ and $E_{\mu}\{X^2(T)\} < \infty$. Application of these results will be provided later on. We remark here that Theorem 9.7.2 does not imply the result of Robbins and

Samuel [1], since they provided a proof of the Wald equation $E_\mu\{X(T)\} = \mu E_\mu\{T\}$, which does not require the condition that $E_\mu |T| < \infty$ and $E|X(T)| < \infty$. However, their result pertains to the discrete time case. Hall [2] proved this result for Wiener processes also.

9.8. MORE ON THE DETERMINATION OF THE BAYES SEQUENTIAL STOPPING RULE FOR MULTIPLE TESTING

In Section 9.4 we presented some general theory of Bayes sequential procedures for testing multiple (possibly composite) hypotheses. We have shown that under some mild conditions (Theorem 9.4.1) we can approximate the Bayes procedure by a Bayes truncated stopping rule, for a sufficiently large M. In practice it may often be very difficult to determine even the Bayes truncated stopping rule. In this section we provide some approximations for the case of multiple *simple* hypotheses, and in particular we provide the entire solution for two simple hypotheses and symmetric loss matrices. This approximation is based on an important observation of Whittle [1]. The present exposition follows Shiryaev [1].

Consider again the recursive functional equation, (9.4.7). We restrict attention in the present discussion to simple hypotheses, and $\Delta(Y; \mathbf{X}_n) = c$ for all $(\mathbf{X}_n, Y)$, $n = 0, 1, \ldots$. Let $\xi = (\xi_1, \ldots, \xi_m)$ designate the prior probability vector. The posterior Bayes risk, when stopping at $N = n$, is

$$\rho_0(\xi, \mathbf{X}_n) = \inf_{1 \leqslant i \leqslant m} \sum_{j=1}^{m} L_i(\theta_j)\xi_j(\mathbf{X}_n), \tag{9.8.1}$$

where $0 \leqslant L_i(\theta_j) < \infty$ is the loss due to accepting A_i when θ_j is true $(i, j = 1, \ldots, m)$. Since $c = c\sum_{j=1}^m \xi_j(\mathbf{X}_n)$, both $\rho_0(\xi, \mathbf{X}_n)$ and c are homogeneous functions of the first degree over the simplex Σ of probability vectors $\xi = (\xi_1, \ldots, \xi_m)$; $\xi_j \geqslant 0$ $(j = 1, \ldots, m)$, $\sum_{j=1}^m \xi_j = 1$.

For each $M \geqslant 1$, $1 \leqslant n \leqslant M$,

$$\rho_n^{(M)}(\xi, \mathbf{X}_{M-n}) = \min\{\rho_0(\xi, \mathbf{X}_n), c + E\{\rho_{n-1}^{(M)}(\xi, (\mathbf{X}_{M-n}, Y)) \mid \mathcal{F}_{M-n}\}\}. \tag{9.8.2}$$

We can prove by induction on n, $n = 0, \ldots, M$, that $\rho_n^{(M)}(\xi, \mathbf{X}_{M-n})$ is a homogeneous function of the first degree. This observation was first announced by Whittle [1]. It plays a major role in the search for the optimal stopping rule. All the loss components $L_i(\theta_j)$ are bounded. Hence $\lim_{M\to\infty} \rho_n^{(M)}(\xi) = \rho(\xi)$, where $\rho(\xi)$ is the prior risk of the Bayes nontruncated stopping rule. Therefore the minimal prior risk $\rho(\xi)$ is also a homogeneous functional on Σ, of the first degree; $\rho(\xi)$ is a solution of the functional equation (9.4.5). We have shown that the simplex $\not\Sigma$ is partitioned to two regions $\mathcal{C}$ and $\mathcal{T}$ where

(9.8.3)
$$\mathcal{C} = \{\xi;\ \xi \in \not\Sigma \text{ and } \rho(\xi) < \rho_0(\xi)\}$$
$$\mathcal{T} = \{\xi;\ \xi \in \not\Sigma \text{ and } \rho(\xi) \geqslant \rho_0(\xi)\}.$$

Region $\mathcal{C}$ is the continuation region of the Bayes sequential stopping rule. As long as $\xi(\mathbf{X}_n)$ belongs to $\mathcal{C}$, sampling is continued. Region $\mathcal{T}$ is the termination or stopping region. Now if $\xi \in \mathcal{C}$, then the following integral equation is satisfied:

(9.8.4)
$$\rho(\xi) = c + \int \rho(\xi(y)) \sum_{i=1}^{m} \xi_i f_i(y)\mu(\mathrm{d}y).$$

However, since $\rho(\xi)$ is homogeneous of the first degree in ξ, we can write

(9.8.5)
$$\rho(\xi) = c + \int \rho(\xi_1 f_1(y), \ldots, \xi_m f_m(y))\mu(\mathrm{d}y)$$

for all $\xi \in \mathcal{C}$.

We have to solve (9.8.5) under the supplementary condition that $\rho(\xi) = \rho_0(\xi)$ for all $\xi \in \mathcal{T}$.

Let $\mathcal{C} = \|c_{ij};\ i, j = 1, \ldots, m\|$ be a matrix of coefficients having the property

(9.8.6)
$$\sum_{j=1}^{m} c_{ij} = 0, \qquad \text{for each } i = 1, \ldots, m,$$

and

(9.8.7)
$$\sum_{j=1}^{m} c_{ij} I(i, j) = c, \qquad \text{for each } i = 1, \ldots, m,$$

where $I(i, i) = 0$ for all $i = 1, \ldots, m$.

(9.8.8)
$$I(i, j) = \int \left(\log \frac{f_i(x)}{f_j(x)}\right) f_i(x)\mu(\mathrm{d}x), \qquad i, j = 1, \ldots, m,$$

is the Kullback-Leibler information number. Then the function

(9.8.9)
$$\rho^*(\xi) = \sum_{i=1}^{m} \sum_{j=1}^{m} c_{ij}\xi_i \log \xi_j$$

is a particular solution of the inhomogeneous equations of (9.8.5). This can be easily verified. Define

(9.8.10)
$$G(\beta) = \int \prod_{i=1}^{m} f_i^{\beta_i}(y)\mu(\mathrm{d}y).$$

The Kullback-Leibler information numbers can be derived from $G(\beta)$ according to the formula

(9.8.11)
$$I(i, j) = \frac{\partial}{\partial \beta_i} G(\beta) - \frac{\partial}{\partial \beta_j} G(\beta) \bigg|_{\beta = e^{(i)}},$$

where $e^{(i)} = (e_1^{(i)}, \ldots, e_m^{(i)})'$, $e_i^{(i)} = 1$, and $e_j^{(i)} = 0$ for all $j \neq i$. The general solution of the homogeneous equation

$$\rho(\xi) = \int \rho(\xi_1 f_1(y), \ldots, \xi_m f_m(y))\mu(\mathrm{d}y) \tag{9.8.12}$$

is given by

$$\tilde{\rho}(\xi) = \int \prod_{i=1}^{m} \xi_i^{\nu_i} \phi(\mathrm{d}\nu), \tag{9.8.13}$$

where $\nu = (\nu_1, \ldots, \nu_m)$ (possibly with complex components) is a solution to

$$G(\beta) = 1$$

under the constraint $\sum_{i=1}^{m} \nu_i = 1$; $\phi(\mathrm{d}\nu)$ is an appropriate measure. Indeed,

(9.8.14)

$$\begin{aligned}\int \mu(\mathrm{d}y)\tilde{\rho}(\xi_1 f_1(y), \ldots, \xi_m f_m(y)) &= \int \phi(\mathrm{d}\nu) \prod_{i=1}^{m} \xi_i^{\nu_i} \int \mu(\mathrm{d}y) \prod_{j=1}^{m} f_j^{\nu_j}(y) \\ &= \int \phi(\mathrm{d}\nu) \prod_{i=1}^{m} \xi_i^{\nu_i} G(\nu) = \tilde{\rho}(\xi).\end{aligned}$$

The solution of the integral equation, (9.8.4), is expressed thus by the formula

$$\rho(\xi) = \sum_{i=1}^{m} \sum_{j=1}^{m} c_{ij}\xi_i \log \xi_j + \int \prod_{i=1}^{m} \xi_i^{\nu_i} \phi(\mathrm{d}\nu), \tag{9.8.15}$$

where $\nu'1 = 1$, $G(\nu) = 1$, and (c_{ij}) satisfy (9.8.6) and (9.8.7). The quantity $\rho(\xi)$ is not yet determined from (9.8.15) since $\phi(\mathrm{d}\nu)$ is not specified. The coefficients c_{ij}, $i, j = 1, \ldots, m$, have still to be determined. This requires supplementary conditions. One such condition is the boundary condition $\rho(\xi) = \rho_0(\xi)$ on $\mathcal{C}$. Let $\partial\mathcal{C}$ be the boundary of $\mathcal{C}$. The Bayes stopping rule says that N is the least integer n, $n \geqslant 0$, for which $\xi(\mathbf{X}_n)$ crosses the boundary $\partial\mathcal{C}$. Whittle [1] conjectured that if $\xi(\mathbf{X}_n) \in \partial\mathcal{C}$ then one may confine attention only to the real roots of $G(\nu) = 1$. In many cases the (9.8.15) can be determined with the aid of the following side conditions, called the *merging risks*; that is,

$$\frac{\partial}{\partial \xi_i} \rho(\xi) = \frac{\partial}{\partial \xi_i} \rho_0(\xi), \qquad i = 1, \ldots, m. \tag{9.8.16}$$

In the next section we show that these conditions should hold on the boundary in the case of certain continuous time processes.

We now show how we can completely specify the solution in the case of two simple hypotheses ($m = 2$) and symmetric loss $L_1(\theta_2) = L_2(\theta_1) = r$.

Neglecting the overshoot across the boundaries, we consider only the real roots ν_1, ν_2 of $1 = G(\nu) = \int f_1^{\nu_1}(x) f_2^{\nu_2}(x)\mu(\mathrm{d}x)$ such that $\nu_1 + \nu_2 = 1$.

Obviously we have the two pairs of roots

$$\nu^{(1)} = (1, 0), \qquad \nu^{(2)} = (0, 1).$$

For $\nu^{(1)}$ we obtain $\xi^{\nu_1^{(1)}}(1-\xi)^{\nu_2^{(1)}} = \xi$. For $\nu^{(2)}$ we have $\xi^{\nu_1^{(2)}}(1-\xi)^{\nu_2^{(2)}} = 1-\xi$. Let $\phi(\nu)$ be a measure concentrated on the two points $\nu^{(1)}$ and $\nu^{(2)}$. Then

$$\tilde{\rho}(\xi) = \phi_1\xi + \phi_2(1-\xi). \tag{9.8.17}$$

According to (9.8.6) and (9.8.7) we can choose the coefficients c_{ij}, $i, j = 1, 2$, to be

$$c_{11} = -c_{12}, \qquad c_{12} = \frac{c}{I(1, 2)},$$

and

$$c_{22} = -c_{21}, \qquad c_{21} = \frac{c}{I(2, 1)}.$$

Hence

$$\begin{aligned}\rho^*(\xi) &= c_{11}\xi \log \xi + c_{12}\xi \log(1-\xi) \\ &\quad + c_{21}(1-\xi)\log\xi + c_{22}(1-\xi)\log(1-\xi) \\ &= \frac{c}{I(1, 2)}\xi \log\frac{1-\xi}{\xi} + \frac{c}{I(2, 1)}(1-\xi)\log\frac{\xi}{1-\xi}.\end{aligned} \tag{9.8.18}$$

The risk function is $\rho(\xi) = \rho^*(\xi) + \tilde{\rho}(\xi)$; ϕ_1 and ϕ_2 are still undetermined. To simplify the presentation, suppose that $c/(I(1, 2)) = c/(I(2, 1)) = k$. The Bayes risk $\rho_0(\xi)$ is

$$\rho_0(\xi) = \begin{cases} r\xi, & \text{if } 0 < \xi \leqslant \frac{1}{2}, \\ r(1-\xi), & \text{if } \frac{1}{2} \leqslant \xi < 1. \end{cases}$$

Hence

$$\frac{\partial}{\partial\xi}\rho_0(\xi) = \begin{cases} r, & \text{if } 0 < \xi < \frac{1}{2}, \\ -r, & \text{if } \frac{1}{2} < \xi < 1. \end{cases} \tag{9.8.19}$$

On the other hand,

$$\begin{aligned}\frac{\partial}{\partial\xi}\rho(\xi) &= \frac{\partial}{\partial\xi}\{\rho^*(\xi) + \tilde{\rho}(\xi)\} \\ &= 2k\left[\log\frac{1-\xi}{\xi} - \frac{1}{1-\xi}\right] + \frac{k}{\xi(1-\xi)} + \phi_1 - \phi_2.\end{aligned} \tag{9.8.20}$$

In Section 9.6 we saw that if the sample size of the Bayes sequential stopping rule is $N \geqslant 1$ a.s., there exist two constants $0 < \xi_L < \xi_U < 1$ such that sampling continues as long as $\xi(\mathbf{X}_n) \in (\xi_L, \xi_U)$. Thus in the present case the boundary of $\mathcal{C}$ is $\{\xi_L, \xi_U\}$. Since $0 < \xi_L < \frac{1}{2} < \xi_U < 1$, we obtain from

(9.8.19) and (9.8.20) the system of equations

$$
\begin{aligned}
&2\left[\log\frac{1-\xi_L}{\xi_L}-\frac{1}{1-\xi_L}\right]+\frac{1}{\xi_L(1-\xi_L)}+\frac{1}{k}(\phi_1-\phi_2)=R,\\
&2\left[\log\frac{1-\xi_U}{\xi_U}-\frac{1}{1-\xi_U}\right]+\frac{1}{\xi_U(1-\xi_U)}+\frac{1}{k}(\phi_1-\phi_2)=-R,
\end{aligned}
\tag{9.8.21}
$$

where $R = r/k$. Subtracting one from the other we obtain

$$
\begin{aligned}
2\left[\log\frac{(1-\xi_L)\xi_U}{\xi_L(1-\xi_U)}-\left(\frac{1}{1-\xi_L}-\frac{1}{1-\xi_U}\right)\right]&\\
+\left(\frac{1}{\xi_L(1-\xi_L)}-\frac{1}{\xi_U(1-\xi_U)}\right)&=2R.
\end{aligned}
\tag{9.8.22}
$$

Due to the symmetry of all the cost components, $\xi_U = 1 - \xi_L$. Substituting this relationship in (9.8.22), we obtain

$$
2\log\psi_L+\left(\psi_L-\frac{1}{\psi_L}\right)=R,
\tag{9.8.23}
$$

where $\psi_L = (1 - \xi_L)/\xi_L$. There exists a unique solution to (9.8.23), which determines the optimal boundary $(\xi_L, 1 - \xi_L)$.

9.9. BAYES SEQUENTIAL STOPPING RULES AND THE STEPHAN FREE BOUNDARY PROBLEM

The Bayes sequential stopping rule for testing m hypotheses was specified in Section 9.4 in terms of the Bayes risk $\rho_0(H, \mathbf{X}_n)$ and the minimal pre-posterior risk $\rho(H, \mathbf{X}_n)$. The stopping variable N of the Bayes sequential procedure is the least non-negative integer n for which $\rho_0(H, \mathbf{X}_n) \leqslant \rho(H, \mathbf{X}_n)$. Thus we can specify the Bayes sequential stopping rule in terms of the random walk of $\rho_0(H, \mathbf{X}_n)$ and its first passage time out of the continuation region.

The sequence $\{\rho_0(H, \mathbf{X}_n), \mathcal{F}_n; n \geqslant 0\}$ is a positive supermartingale, where $\mathcal{F}_n$ is the Borel sigma-field generated by $(\theta, \mathbf{X}_n)$; $\mathcal{F}_0 \subset \mathcal{F}_1 \subset \cdots$. Indeed, for each $n \geqslant 0$ we obtain from Fatou's lemma

$$
\begin{aligned}
E\{\rho_0(H,(\mathbf{X}_n,Y))\mid\mathcal{F}_n\}&=E\left\{\min_{1\leqslant j\leqslant m}\int L_j(\theta)H(d\theta\mid\mathbf{X}_n,Y)\mid\mathcal{F}_n\right\}\\
&\leqslant\min_{1\leqslant j\leqslant m}E\left\{\int L_j(\theta)H(d\theta\mid\mathbf{X}_n,Y)\mid\mathcal{F}_n\right\}\\
&=\min_{1\leqslant j\leqslant m}\int L_j(\theta)E\{H(d\theta\mid\mathbf{X}_n,Y)\mid\mathcal{F}_n\},
\end{aligned}
\tag{9.9.1}
$$

where $H(\theta \mid \mathbf{X}_n, Y)$ is the posterior d.f. of θ, given $(\mathbf{X}_n, Y)$. The conditional expectation in (9.9.1) is with respect to the conditional marginal distribution of Y, given $\mathscr{F}_n$. It is immediate to prove that

$$E\{H(\mathrm{d}\theta \mid \mathbf{X}_n, Y) \mid \mathscr{F}_n\} = H(\mathrm{d}\theta \mid \mathbf{X}_n) \quad \text{a.s.} \tag{9.9.2}$$

Hence from (9.9.1) and (9.9.2) we obtain that

$$E\{\rho_0(H, (\mathbf{X}_n, Y)) \mid \mathscr{F}_n\} = \rho_0(H, \mathbf{X}_n) \quad \text{a.s.}, \quad n \geqslant 0. \tag{9.9.3}$$

From this we immediately obtain that the minimal preposterior risk $\rho(H, \mathbf{X}_n)$, $n \geqslant 0$, generates a supermartingale $\{\rho(H, \mathbf{X}_n), \mathscr{F}_n; n \geqslant 0\}$, too. We show now that these supermartingale sequences can be embedded under certain conditions in proper Wiener processes. Thus the Bayes sequential stopping rule for the Wiener process in which the supermartingale sequence $\{\rho_0(H, \mathbf{X}_n), \mathscr{F}_n; n \geqslant 0\}$ is embedded provides continuous time approximation to the Bayes sequential stopping rule for the original process. The embedding method shown here was demonstrated by Skorokhod [1] for martingale sequences and developed recently for submartingales or supermartingales by Hall [1]. We provide here the essential definitions and the main embedding theorem.

A sequence $\{Y_n, \mathscr{F}_n; n = 1, 2, \ldots\}$ is called an *embedable supermartingale* if there exists a constant $\theta < 0$ such that $\{e^{-\theta Y_n}, \mathscr{F}_n; n \geqslant 0; e^{-\theta Y_0} \equiv 1\}$ is a supermartingale; θ is called the exponent of the sequence.

In the proof of Theorem 9.7.1. we have shown that if $\mathcal{W}_\mu = \{X(t); t \geqslant 0\}$ is a Wiener process, then $\{e^{-uX(t)}, \mathscr{F}_t; t \geqslant 0\}$ is a supermartingale for all $0 \leqslant u \leqslant 2\mu$ or $-2\mu \leqslant u \leqslant 0$. We have also mentioned that if T is any stopping time for the Wiener process $\mathcal{W}_\mu$, and $T_n = \min\{n, T_n\}$, then $\{e^{-uX(T_n)}, \mathscr{F}_n; n \geqslant 0\}$ is a supermartingale. These results are used in the proof of the following theorem.

Theorem 9.9.1. (Hall [1].) *Let* $\mathcal{W}_\mu = \{X(t); t \geqslant 0, X(0) \equiv 0\}$ *be a Wiener process. There exists a sequence* $\{T_n\}$ *of successive stopping times for* $\mathcal{W}_\mu$ *for which*

$$\mathfrak{L}(X(T_n)) = \mathfrak{L}(Y_n), \qquad n = 1, 2, \ldots, \tag{9.9.4}$$

if and only if $\{Y_n, \mathscr{F}_n; n \geqslant 1\}$ *is an embedable supermartingale sequence for some* $\{\mathscr{F}_n; n \geqslant 1\}$, *with exponent* 2μ. *Moreover,*

$$E_\mu\{T_n\} = \frac{1}{\mu} E\{Y_n\} \qquad (\leqslant \infty). \tag{9.9.5}$$

For a proof, see Hall [1], Theorem 4.1. The notation $\mathfrak{L}(X(T_n))$ means the distribution law of $X(t)$ at the random time T_n. This theorem can be used in the following manner. Given a positive supermartingale $\{\rho_0(H, \mathbf{X}_n), \mathcal{F}_n; n \geqslant 0\}$, let $\mu < 0$, $Y_n = -(1/2\mu) \ln \rho_n$, where $\rho_n \equiv \rho_0(H, \mathbf{X}_n)$. According to Theorem 9.9.1 there exists a sequence $\{T_n\}$ of successive stopping times for a Wiener process $\mathcal{W}_\mu$ so that

$$\mathcal{L}(X(T_n)) = \mathcal{L}(Y_n), \qquad n = 0, 1, \ldots .$$

We have thus embedded the random sequence $\{Y_n, \mathcal{F}_n; n \geqslant 0\}$ in a Wiener process $\mathcal{W}_\mu$. The optimal stopping time for the Wiener process $\mathcal{W}_\mu$, in which $\{Y_n, \mathcal{F}_n; n \geqslant 0\}$ is embedded, yields an approximating Bayes stopping variable for the original sequence $\{\rho_0(H, \mathbf{X}_n)\}$.

Example 9.7. Consider the problem of testing two simple hypotheses A_0: $\theta = 0$ and A_1: $\theta = 1$. Let $X_1, X_2, \ldots$ be a sequence of i.i.d. random variables with density functions $f_i(x)$, $i = 0, 1$, with respect to a dominating measure μ. We have proven in Section 9.6 that the Bayes sequential stopping rule is to continue sampling as long as the posterior probability of A_0, say $\xi(\mathbf{X}_n)$, is in some interval (ξ_L, ξ_U). Thus we consider the random walk of the posterior probability function $\xi(\mathbf{X}_n)$. Instead we can consider, as in the Wald S.P.R.T., the random walk of the likelihood-ratio $\lambda(\mathbf{X}_n) = f_1(\mathbf{X}_n)/f_0(\mathbf{X}_n)$. It is not difficult to verify that $\{\lambda(\mathbf{X}_n), \mathcal{F}_n; n \geqslant 1\}$ is a supermartingale, under A_0. Furthermore, if $L_n = \ln \lambda(\mathbf{X}_n)$ then $\lambda(\mathbf{X}_n) = \exp\{-\theta L_n\}$, with $\theta = -1$. Thus $\{L_n, \mathcal{F}_n; n \geqslant 1\}$ is an embedable sequence with an exponent -1. This sequence can be embedded in a Wiener process $\mathcal{W}_{-1/2}$.

We determine the optimal boundaries of the continuation region for testing $\mathcal{W}_{-1/2}$ against $\mathcal{W}_{1/2}$, calculate after each observation the value of $L_n = \ln \lambda(\mathbf{X}_n)$, $n \geqslant 1$, and stop at the least n for which L_n crosses outside the continuation region. ■

We now develop the theory of Bayes sequential stopping rules for Wiener processes $\mathcal{W}_\mu$ when the drift μ of $\mathcal{W}_\mu$ is assigned a prior normal distribution $\mathcal{N}(\mu_0, \sigma_0^2)$. As shown by Chernoff [7], the posterior distribution of μ, given the value $X(t)$ of $\mathcal{W}_\mu$ at time point t, is $\mathcal{N}(Y(s), s)$, where

$$Y(s) = \frac{\mu_0\sigma_0^{-2} + X(t)}{\sigma_0^{-2} + t}, \tag{9.9.6}$$

and

$$s = (t + \sigma_0^{-2})^{-1}. \tag{9.9.7}$$

Furthermore $Y(s)$ is a realization of a Wiener process with drift zero in the $-s$ scale (a backward process) originating at $(\mu_0, \sigma_0^2) = (y_0, s_0)$. The Wiener process $Y(s)$ is observed. The statistician may stop at any time s, $s > 0$. The corresponding Bayes risk when stopping at s is $\rho_0(Y(s), s)$. The problem is to find a stopping rule which will minimize the expected cost. Thus if $\tilde{s}$

designates a stopping time for $\{Y(s), \mathcal{F}_s; 0 < s \leqslant s_0\}$, the problem is to find $\tilde{s}^0$ such that

$$E\{\rho_0(Y(\tilde{s}^0), \tilde{s}^0) \mid Y(s_0) = y_0\} = \inf_{\tilde{s}} E\{\rho_0(Y(\tilde{s}), \tilde{s}) \mid Y(s_0) = y_0\}. \tag{9.9.8}$$

Let

$$\tilde{\rho}(y, s) = E\{\rho_0(Y(\tilde{s}^0), \tilde{s}^0) \mid Y(s) = y\}, \qquad 0 < s \leqslant s_0.$$

Then in analogy to the functional equation, (9.4.15), we have (see Grigelionis and Shiryaev [1])

$$\tilde{\rho}(y, s) = \min\{\rho_0(y, s), E\{\tilde{\rho}(Y(\tilde{s}_\epsilon), \tilde{s}_\epsilon) \mid Y(s) = y\}\} \tag{9.9.9}$$

where for the small value of ϵ, $\epsilon > 0$,

$$\tilde{s}_\epsilon = \sup\{s'; s' < s, |Y(s') - y| \geqslant \epsilon\}; \tag{9.9.10}$$

that is, $\tilde{s}_\epsilon$ is the first exit time of $Y(s)$ from the interval $(y - \epsilon, y + \epsilon)$. The functional equation (9.9.9) means that the optimal stopping rule $\tilde{s}^0$ is to continue observing $Y(s)$ as long as $\rho_0(Y(s), s) > \tilde{\rho}(Y(s), s)$. In order to find the optimal stopping time $\tilde{s}^0$, we have to determine $\tilde{\rho}(y, s)$, and the continuation set $\mathcal{C}$ on which $\tilde{\rho}(y, s) < \rho_0(y, s)$.

Following Chernoff [7], we show now that under some mild conditions the minimal Bayes risk function $\tilde{\rho}(y, s)$ satisfies within the continuation region $\mathcal{C}$ the heat-equation

$$\frac{\partial}{\partial s} \tilde{\rho}(y, s) = \frac{1}{2} \frac{\partial^2}{\partial y^2} \tilde{\rho}(y, s). \tag{9.9.11}$$

Let $h > 0$ be sufficiently small so that the points $(Y(s + h), s + h)$ and $(Y(s), s)$ are both within the continuation region. This assumption is warranted, since the Wiener process $\{Y(s): s > 0\}$ is almost surely continuous. Since both points are within the continuation region, (9.9.9) implies that

$$\tilde{\rho}(y, s + h) = E\{\tilde{\rho}(Y(s), s) \mid Y(s + h) = y\}. \tag{9.9.12}$$

Since $\{Y(s): s > 0\}$ is a standard Wiener process, we can expand the right-hand side of (9.9.12) in the form

$$\begin{aligned} &E\{\tilde{\rho}(Y(s), s) \mid Y(s + h) = y\} \\ &= E\{\tilde{\rho}(y + \sqrt{h}\, U, s)\} \\ &= E\{\tilde{\rho}(y, s) + \sqrt{h}\, U \tilde{\rho}_y(y, s) + \tfrac{1}{2} h U^2 \tilde{\rho}_{yy}(y, s) + o_p(h)\}, \text{ as } h \to 0; \end{aligned} \tag{9.9.13}$$

where $U \sim N(0, 1)$. Equation 9.9.13 yields under mild conditions on $\tilde{\rho}(y, s)$,

$$E\{\tilde{\rho}(Y(s), s) \mid Y(s + h) = y\} = \tilde{\rho}(y, s) + \frac{h}{2} \tilde{\rho}_{yy}(y, s) + o(h), \tag{9.9.14}$$

as $h \to 0$. The heat-equation (9.9.11) is obtained from (9.9.12) and (9.9.14). Indeed,

$$\frac{\partial}{\partial s}\tilde{\rho}(y, s) = \lim_{h \searrow 0}\left\{\frac{1}{h}\left[\tilde{\rho}(y, s+h) - \tilde{\rho}(y, s)\right]\right\} \tag{9.9.15}$$

$$= \lim_{h \searrow 0}\left\{\frac{1}{h}\left[\tfrac{1}{2}h\tilde{\rho}_{yy}(y, s) + o(h)\right]\right\}.$$

In addition, following Chernoff [7], we show that on the boundary Γ of $\mathcal{C}$ the function $\tilde{\rho}(y, s)$ satisfies

$$\frac{\partial}{\partial y}\tilde{\rho}(y, s) = \frac{\partial}{\partial y}\rho_0(y, s), \qquad (y, s) \in \Gamma. \tag{9.9.16}$$

Equation 9.9.16 was used in Section 9.8 in order to show an explicit determination of the Bayes sequential stopping rule. In order to demonstrate the boundary condition (9.9.16) suppose that (y_1, s_1) is a point on Γ such that

$$\tilde{\rho}(y, s_1) = \rho_0(y, s_1) \quad \text{for all} \quad y \geqslant y_1, \tag{9.9.17}$$

and

$$\tilde{\rho}(y, s_1) \leqslant \rho_0(y, s_1) \quad \text{for all} \quad y \leqslant y_1. \tag{9.9.18}$$

Equation (9.9.16) obviously holds at (y_1, s_1) for right derivatives. When $y < y_1$, (9.9.18) implies that

$$\frac{\partial^-}{\partial y}\tilde{\rho}(y_1, s_1) \geqslant \frac{\partial^-}{\partial y}\rho_0(y_1, s_1), \tag{9.9.19}$$

where $\partial^-/\partial y$ denotes the left derivative.

Now, for s_1 such that $s_1 + \delta \leqslant s_0$, $\tilde{\rho}(y_1, s_1 + \delta)$ is the minimal expected risk at $s_1 + \delta$, $Y(s_1 + \delta) = y_1$. Suppose that we consider at time point s_1 to proceed according to a stopping rule which observes $Y(s)$ for at least δ units, and then proceed in an optimal manner. Then

$$\tilde{\rho}(y_1, s_1 + \delta) \leqslant E\{\tilde{\rho}(y_1 + \sqrt{\delta}\, U, s_1)\} \tag{9.9.20}$$

where $U \sim N(0, 1)$. Taylor expansion yields

(9.9.21)

$$\tilde{\rho}(y_1 + \sqrt{\delta}\, U, s_1) = \tilde{\rho}(y_1, s_1) + \sqrt{\delta}\, U \frac{\partial^+}{\partial y}\tilde{\rho}(y_1, s_1) + o(\sqrt{\delta}), \qquad U > 0$$

$$= \tilde{\rho}(y_1, s_1) + \sqrt{\delta}\, U \frac{\partial^-}{\partial y}\tilde{\rho}(y_1, s_1) + o(\sqrt{\delta}), \qquad U < 0,$$

as $\delta \to 0$. Hence from (9.9.17) and (9.9.20),

(9.9.22)

$$E\{\tilde{\rho}(y_1 + \sqrt{\delta}\, U, s_1)\} = \tilde{\rho}(y_1, s_1) + \sqrt{\delta}\left\{\int_0^\infty u\varphi(u)\, du \cdot \frac{\partial^+}{\partial y}\rho_0(y_1, s_1)\right.$$

$$\left. + \int_{-\infty}^0 u\varphi(u)\, du \frac{\partial^-}{\partial y}\tilde{\rho}(y_1, s_1)\right\} + o(\sqrt{\delta})$$

$$= \tilde{\rho}(y_1, s_1) + \frac{\sqrt{\delta}}{\sqrt{(2\pi)}}\left\{\frac{\partial^+}{\partial y}\rho_0(y_1, s_1) - \frac{\partial^-}{\partial y}\tilde{\rho}(y_1, s_1)\right\} + o(\sqrt{\delta}), \qquad \text{as } \delta \to 0.$$

The density function of $N(0, 1)$ is designated by $\varphi(u)$. Hence from (9.9.20) and (9.9.22) we deduce that if $[\tilde{\rho}(y_1, s_1 + \delta) - \tilde{\rho}(y_1, s_1)]/\delta$ is bounded, then

$$\frac{\partial^+}{\partial y}\rho_0(y_1, s_1) - \frac{\partial^-}{\partial y}\tilde{\rho}(y_1, s_1) \geqslant 0. \tag{9.9.23}$$

Finally, from (9.9.29) and (9.9.23) we obtain that

$$\frac{\partial}{\partial y}\rho_0(y_1, s_1) = \frac{\partial^+}{\partial y}\rho_0(y_1, s_1) \geqslant \frac{\partial^-}{\partial y}\tilde{\rho}(y_1, s_1) \tag{9.9.24}$$

$$\geqslant \frac{\partial^-}{\partial y}\rho_0(y_1, s_1) = \frac{\partial}{\partial y}\rho_0(y_1, s_1).$$

This proves (9.9.16) at (y_1, s_1). A more rigorous proof is given by Grigelionis and Shiryaev [1].

The problem of determining a solution $\tilde{\rho}(y, s)$ to the heat equation, (9.9.11), and a boundary Γ such that $\tilde{\rho}(y, s) = \rho_0(y, s)$ for $(y, s) \in \mathcal{C}$ and $\tilde{\rho}(y, s)$ satisfies the smooth-pasting condition, (9.9.16), on Γ is known as the *Stephan free boundary problem*. We have shown that a solution of the optimization problem is a solution of the corresponding Stephan free boundary problem. An important question is whether a solution of the Stephan problem is necessarily a solution of the optimization problem. The answer is negative, unless additional conditions are imposed. Chernoff [7] has proven that if both $\tilde{\rho}(y, s)$ and $\rho_0(y, s)$ have bounded derivatives up to the third order, a solution of the Stephan problem is also a solution of the optimization problem.

Generally when the posterior risk function $\rho_0(y, s)$ is not a very simple one, the solution of the Stephan free boundary problem is very difficult. Bounds and asymptotic approximations to the solution of the optimization problem for certain sequential analysis problems were developed by Breakwell and Chernoff [1], Chernoff [6], Chernoff and Ray [2], and others. See also Dynkin and Yushkevich [1], Ch. 3, 4.

9.10. ASYMPTOTICALLY OPTIMAL TEST PROCEDURES

Several significant papers have been published on the asymptotic optimality of certain sequential testing procedures of composite hypotheses. Among these we mention the papers of Chernoff [3]; A. E. Albert [1]; Bessler [1]; Schwarz [1]; Kiefer and Sacks [1]; and Bickel and Yahav [1]. The papers of Bickel and Yahav [1] and of Schwarz [1] deal entirely with sequential testing, while the other papers also consider relevant experimental design problems. The paper of Kiefer and Sacks [1] generalizes the results of Chernoff [3], Albert [1], Bessler [1], and Schwarz [1]. The study of Bickel and Yahav [1] provides some stronger results concerning pointwise optimal procedures. Their pointwise asymptotic theory was discussed in Section 6.10 in the context of asymptotically pointwise optimal estimation procedures. We do not discuss their results in this section and the reader is referred to [1]. All the studies mentioned here consider asymptotic procedures in the large sample sense, as the cost per observation $c \to 0$. We start the discussion with the theory for testing two composite hypotheses which are not separated by a zone of indifference. In the second part we discuss the problem of testing composite hypotheses with an indifference zone.

Consider two composite hypotheses A_1 and A_2, corresponding to two disjoint sets Θ_1 and Θ_2 in some Euclidean parameter space Θ, where $\Theta_1 \cup \Theta_2 = \Theta$. Let $\{f(x; \theta); \theta \in \Theta\}$ be a family of density functions with respect to a sigma-finite measure μ. The function $L_1(\theta)$ is the loss function corresponding to the decision to accept A_1, and $L_2(\theta)$ the loss function for accepting A_2. Let $H(\theta)$ designate a prior distribution on Θ.

Let $X_1, X_2, \ldots$ be a sequence of i.i.d. random variables having a common density function $f(x; \theta)$. Suppose that c is the cost per observation (independent of $X_1, X_2, \ldots$) and $0 < c < 1$. Whenever sampling terminates, a Bayes decision is made for the acceptance of A_1 or A_2. For a stopping rule we consider the simplified rule: *stop at the smallest integer* n, $n \geq 0$, *for which* $\rho_0(H, \mathbf{X}_n) \leq c$, where $\rho_0(H, \mathbf{X}_n)$ is the posterior Bayes risk, $X_0 \equiv 0$ and $\rho_0(\mathbf{X}_0) \equiv \rho_0$ is the prior Bayes risk.

Let $\mathbf{s}^*(c)$ designate this stopping rule which, combined with the Bayes acceptance decision function, can be rephrased in the following terms.

Accept A_1 if

$$\int L_1(\theta) f(x_n; \theta) H(d\theta) \leq c \int L_2(\theta) f(x_n; \theta) H(d\theta); \tag{9.10.1}$$

accept A_2 if

$$\int L_1(\theta) f(x_n; \theta) H(d\theta) > \frac{1}{c} \int L_2(\theta) f(x_n; \theta) H(d\theta); \tag{9.10.2}$$

take an additional observation otherwise. The function $f(\mathbf{X}_n; \theta)$ designates the joint density of $(X_1, \ldots, X_n)$ given θ. In cases of m composite hypotheses ($m > 2$) the stopping rule $\mathbf{s}^*(c)$ remains the same. We determine $\rho_0(H, \mathbf{X}_n)$ sequentially and terminate sampling at the first $n, n \geqslant 0$, for which $\rho_0(H, \mathbf{X}_n) \leqslant c$. We now establish that $\mathbf{s}^*(c)$ is asymptotically Bayes, as $c \to 0$. That is, if $R_c(H, \mathbf{s}^*(c))$ is the prior risk associated with $\mathbf{s}^*(c)$ and H, and if $\rho_c(H)$ is the Bayes risk, $\rho_c(H) = \inf_{\mathbf{s} \in S} R_c(K, \mathbf{s})$, then

$$\lim_{c \to 0} \frac{R_c(H, \mathbf{s}^*(c))}{\rho_c(H)} = 1.$$

To attain this objective we first present several lemmas. All these lemmas and the ensuing theorem are taken from Kiefer and Sacks [1].

$$\underline{L}_i = \inf_\theta L_i(\theta), \qquad \bar{L}_i(\theta) = \sup_\theta L_i(\theta).$$

As in Chapter 4 we let $Z(\theta, \varphi) = \log f(X; \theta)/f(X; \varphi)$ and designate by $I(\theta, \varphi)$ the Kullback-Leibler information number. We have proven that $I(\theta, \varphi) > 0$ for all $\theta \neq \varphi$. The question, however, is how $I(\theta, \varphi)$ behaves when φ is very close to θ. This is an important question when we test two composite hypotheses, when Θ_1 and Θ_2 have a part of their boundaries in common. There is a danger that if $\theta \in \Theta_1$ and $\varphi \in \Theta_2$, but both are very close to the common boundary, the information available in the sample to discriminate between Θ_1 and Θ_2 will not be sufficient to obtain an efficient test. We therefore assume the following

A.1.

(9.10.3) $$0 \leqslant \underline{L}_i \leqslant \bar{L}_i < \infty, \qquad i = 1, 2.$$

A.2.

(9.10.4) $$I^{(1)} = \inf_{\theta \in \Theta_1} \inf_{\varphi \in \Theta_2} I(\theta, \varphi) > 0,$$

(9.10.5) $$I^{(2)} = \inf_{\theta \in \Theta_2} \inf_{\varphi \in \Theta_1} I(\theta, \varphi) > 0.$$

$I(\theta, \varphi)$ and $I_1(\theta) = \inf_{\varphi \in \Theta_2} I(\theta, \varphi)$ are both continuous on Θ_1. Similarly, $I(\theta, \varphi)$ and $I_2(\theta) = \inf_{\varphi \in \Theta_1} I(\theta, \varphi)$ are both continuous on Θ_2.

A.3. For each $\theta \in \Theta_1$ and $\varphi \in \Theta_2$ the following hold:

(i) $E_\theta\{Z^2(\theta, \varphi)\} < \infty$;
(ii) $E_\varphi\{Z^2(\varphi, \theta)\} < \infty$;
(iii) $\lim_{\rho \downarrow 0} E_\theta\{[\log \sup_{|\varphi' - \varphi| \leqslant \rho} f(X; \varphi') - \log f(X; \varphi)]^2\} = 0$;

(iv) $\lim_{\rho\downarrow 0} E_\varphi\{[\log \sup_{|\theta'-\theta|\leqslant\rho} f(X;\theta') - \log f(X;\theta)]^2\} = 0;$

(v) $\lim_{\theta'\to\theta} E_\theta\{Z^2(\theta, \theta')\} = 0;$

(vi) $\lim_{\varphi'\to\varphi} E_\varphi\{Z^2(\varphi, \varphi')\} = 0.$

Let $N^*(c)$ denote the stopping variable associated with the above stopping rule. The following lemma is an extension of Chernoff's results [3].

Lemma 9.10.1. (Kiefer and Sacks [1].) *Suppose that A.1–A.3 hold.*

(i) *If* Θ_1 *is compact, then for each* $\theta_0 \in \Theta_1$

$$E_{\theta_0}\{N^*(c)\} \leqslant \frac{(1+o(1))\,|\log c|}{I_1(\theta_0)}, \qquad \text{as } c \to 0. \tag{9.10.6}$$

(ii) *If* Θ_2 *is compact, then for each* $\varphi_0 \in \Theta_2$

$$E_{\varphi_0}\{N^*(c)\} \leqslant \frac{(1+o(1))\,|\log c|}{I_2(\varphi_0)}, \qquad \text{as } c \to 0. \tag{9.10.7}$$

The proof of this lemma is very laborious and is not provided. The reader can find a proof in Kiefer and Sacks [1], Lemma 2. We mention that the convergence of the $o(1)$ term on the right-hand side of (9.10.6) depends on θ_0 and is not uniform in θ on Θ_1. Similarly, the convergence of the $o(1)$ term on the right-hand side of (9.10.7) depends on φ_0. In order to obtain uniform convergence of $o(1)$ to zero, we have to impose in addition to A.1–A.3 the continuity condition stated below under A.4, and the condition that both Θ_1 and Θ_2 are compact (see [1], Lemma 3). We remark further that the expected sample size required to test A_1 versus A_2 according to the stopping rule $\mathbf{s}^*(c)$ is of order $|\log c|$, as $c \to 0$, and is inversely proportional to the minimal information number. The above assumption of continuity is formally stated as follows:

A.4. *The functions* $E_\theta\{Z^2(\theta, \varphi)\}$ *and* $E_\theta\{[\log \sup_{|\varphi'-\varphi|\leqslant\rho} f(x;\varphi') - \log f(x;\varphi)]^2\}$ *are continuous on* Θ_1 *for each* φ *and* ρ, $0 < \rho$ *sufficiently small. Similarly,* $E_\varphi\{Z^2(\varphi, \theta)\}$ *and* $E_\varphi\{[\log \sup_{|\theta'-\theta|\leqslant\rho} f(x;\theta') - \log f(x;\theta)]^2\}$ *are continuous on* Θ_2 *for each* θ *and* ρ. *Moreover,* $E_\theta\{Z^2(\theta, \varphi)\}$ *is jointly continuous in* θ *and* φ.

Theorem 9.10.2. (Kiefer and Sacks [1].) *If* Θ_1 *and* Θ_2 *are compact and A.1–A.4 are satisfied, then* $\mathbf{s}^*(c)$ *is asymptotically optimal.*

The proof of this theorem is also deleted. From Theorem 9.10.2 the following corollary can be drawn.

Corollary 9.10.3. (Kiefer and Sacks [1].) *If* $\mathbf{s}'(c)$ *is any stopping rule with an associated stopping variable* $N'(c)$, *and if*

$$\sup_{\theta\in\Theta_1} R_c(\theta, \mathbf{s}'(c)) + \sup_{\varphi\in\Theta_2} R_c(\varphi, \mathbf{s}'(c)) \leqslant Kc\,|\log c| \tag{9.10.8}$$

for some constant K, $0 < K < \infty$, *then*

$$E_\theta\{N'(c)\} \geqslant \frac{(1 + o(1)\,|\log c|}{I_1(\theta)} \qquad \textit{for all } \theta \in \Theta_1, \tag{9.10.9}$$

$$E_\varphi\{N'(c)\} \geqslant \frac{(1 + o(1)\,|\log c|}{I_2(\varphi)} \qquad \textit{for all } \varphi \in \Theta_2. \tag{9.10.10}$$

The maximal error probabilities are shown to be approximately of the order of magnitude $O(c\,|\log c|)$, as $c \to 0$, which is also the order of magnitude of the (prior) Bayes risk, $\rho_c(H)$.

Schwarz [1] provided a very elegant characterization of the asymptotic (as $c \to 0$) shape of the Bayes continuation region, when the density functions $f(x; \theta)$, $\theta \in \Theta$ belong to a 1-parameter exponential family, and the parametric sets Θ_1 and Θ_2 are separated by an "indifference region" Θ_0, so that $L_i(\theta) = 0$ for all $\theta \in \Theta_0$, $i = 1, 2$. Thus let

$$f(x; \theta) = \exp\{\theta x - b(\theta)\}, \qquad -\infty < x < \infty,$$

be the density function of X with respect to a sigma-finite measure μ. The statistic $T_n = \sum_{i=1}^n X_i$ is a minimal sufficient statistic for each $n \geqslant 1$, with a density function

$$f_n(t; \theta) = \exp\{\theta t - nb(\theta)\}.$$

The natural parameter space Θ, as defined in previous sections, is an interval on the real line over which $\int_{-\infty}^{\infty} e^{\theta x}\mu(dx) < \infty$, $\theta \in \Theta$. The function $b(\theta)$ is defined over Θ and satisfies

$$E_\theta\{X\} = b'(\theta),$$

$$\operatorname{Var}_\theta\{X\} = b''(\theta).$$

Hence if the distribution of X is nondegenerate, $b(\theta)$ is a convex function on Θ. Since $b'(\theta)$ is strictly increasing, there is a unique inverse $\mu^{-1}(\theta)$ for $E_\theta\{X\} = \mu = b'(\theta)$.

Consider the composite hypothesis $A_1: \theta \leqslant \theta_1$ versus the composite hypothesis $A_2: \theta \geqslant \theta_2$, where $-\infty < \theta_1 < \theta_2 < \infty$. The interval $\Theta_0 = (\theta_1, \theta_2)$ is an indifference region. Let $L_i(\theta)$, $i = 1, 2$, be the loss functions associated with accepting A_i. We assume that $L_i(\theta)$ is *bounded* and that $L_i(\theta) = 0$ if $\theta \in (\Theta_0 \cup \Theta_i)$. The (posterior) Bayes risk $\rho_0(H, \mathbf{X}_n)$ depends on $\mathbf{X}_n$ through T_n; we therefore designate it by $\hat{R}(n, T_n)$. The continuation region for a stopping rule $\mathbf{s}^*(c)$ defines for each n an interval $C_n^*(c)$ such that if $T_n \in C_n^*(c)$, an additional observation is taken. The continuation region is designated by $\mathcal{C}^*(c)$. This is the union of the corresponding intervals $C_n^*(c)$.

Let

$$\mathcal{C}^*(\gamma) = \{(n, T_n);\ \hat{R}(n, T_n) \geqslant \gamma\}, \qquad 0 < \gamma < \infty, \tag{9.10.11}$$

be a family of continuation for which sampling is continued as long as $\hat{R}(n, T_n) \geqslant \gamma$. In the following theorem we provide an upper and lower bound for the Bayes continuation region $\mathcal{B}(c)$.

Theorem 9.10.4. (Schwarz [1].) *For all c sufficiently small,* $0 < c < c_0$,

$$\mathcal{C}^*(c) \supset \mathcal{B}(c) \supset \mathcal{C}^*\left(\frac{3}{\Delta(\theta_1, \theta_2)}\, c\, |\log c|\right), \tag{9.10.12}$$

where

$$\Delta(\theta_1, \theta_2) = b(\theta_1) + b(\theta_2) - 2b\left(\frac{\theta_1 + \theta_2}{2}\right).$$

Remark. The quantity $\Delta(\theta_1, \theta_2)$ is non-negative for all θ_1, θ_2, $\theta_1 < \theta_2$, since $b(\theta)$ is convex.

Proof. As proven in Section 9.4, the Bayes continuation region is

$$\mathcal{B}(c) = \{(n, T_n);\ \hat{R}(n, T_n) > c + E\{\rho(n + 1, T_n + X) \mid \mathcal{F}_n\}\},$$

where $\rho(n, T_n)$ is the minimal expected Bayes risk, and $\mathcal{F}_n$ is the Borel sigma subfield generated by (θ, T_n). Hence every point (n, T_n) which belongs to $\mathcal{B}(c)$ belongs also to $\mathcal{C}^*(c)$. This proves the left-hand side of (9.10.12). To verify the right-hand side of (9.10.12) we show that if $(n, T_n) \in \mathcal{C}^*((3/\Delta(\theta_1, \theta_2))c\,|\log c|)$ there exists a procedure (possibly a fixed sample one) that leads to taking at least one more observation, and whose expected posterior risk is less than $3c\,|\log c|/\Delta(\theta_1, \theta_2)$. If this is the case, then every point (n, T_n) which belongs to $\mathcal{C}^*(3c\,|\log c|/\Delta(\theta_1, \theta_2))$ belongs also to $\mathcal{B}(c)$.

Consider a fixed sample procedure with a sample size N, where N is the least integer greater or equal to $2\,|\log c|/\Delta(\theta_1, \theta_2)$. Let $\bar{L}_1 = \sup_\theta L_1(\theta)$ and $\bar{L}_2 = \sup_\theta L_2(\theta)$. Suppose that one uses the following decision function: Accept A_2 if

$$\frac{f_N(T_N;\theta_2)}{f_N(T_N;\theta_1)} \geqslant \frac{\bar{L}_2}{\bar{L}_1}. \tag{9.10.13}$$

Let

$$\phi(T_N) = \begin{cases} 1, & \text{if } A_2 \text{ is accepted,} \\ 0, & \text{otherwise.} \end{cases}$$

We prove now that the risk function of this fixed sample procedure is smaller than $3c\ |\log c|/\Delta(\theta_1, \theta_2)$ for all θ in Θ_1 or in Θ_2, provided c is sufficiently small.

The risk function of the above fixed sample procedure is

(9.10.14)

$$R_N^*(\theta) = Nc + L_1(\theta)\int(1 - \phi(t))f_N(t;\theta)\mu_N(dt) + L_2(\theta)\int\phi(t)f_N(t;\theta)\mu_N(dt)$$
$$= Nc + L_1(\theta)\pi_N^*(\theta) + L_2(\theta)[1 - \pi_N^*(\theta)],$$

where

$$\pi_N^*(\theta) = \int(1 - \phi(t))f_N(t;\theta)\mu_N(dt)$$

is the probability of accepting A_1 under θ.

We assume that $L_1(\theta) = 0$ if θ belongs to Θ_1 or Θ_0 $(i = 1, 2)$. Hence we can write

$$(9.10.15)\quad R_N^*(\theta) = Nc + L_1(\theta)I_{\Theta_2}(\theta)\pi_N^*(\theta) + L_2(\theta)I_{\Theta_1}(\theta)[1 - \pi_N^*(\theta)],$$

where $I_{\Theta_i}(\theta)$ is the indicator function of Θ_i $(i = 1, 2)$. We notice that

$$(9.10.16)\qquad 1 - \pi_N^*(\theta) \leqslant 1 - \pi_N^*(\theta_1) \quad \text{if} \quad \theta \in \Theta_1,$$

and

$$(9.10.17)\qquad \pi_N^*(\theta) \leqslant \pi_N^*(\theta_2) \quad \text{if} \quad \theta \in \Theta_2.$$

This is implied from the M.L.R. property of the 1-parameter exponential family (see Lehmann [2], p. 72). Hence from (9.10.15)–(9.10.17) we obtain the inequality

$$(9.10.18)\qquad R_N^*(\theta) \leqslant Nc + \bar{L}_1\pi_N^*(\theta_2) + \bar{L}_2(1 - \pi_N^*(\theta_1)),$$

all $\theta \in (\Theta_1 \cup \Theta_2)$. Proceeding with the right-hand side of (9.10.18) we write, according to (9.10.13),

$$(9.10.19)\quad \bar{L}_1[1 - \phi(t)]f_N(t;\theta_2) + \bar{L}_2\phi(t)f_N(t;\theta_1)$$
$$= \min\{\bar{L}_1 f_N(t;\theta_2), \bar{L}_2 f_N(t;\theta_1)\} \leqslant (\bar{L}_1\bar{L}_2)^{1/2}[f_N(t;\theta_1)f_N(t;\theta_2)]^{1/2}.$$

Furthermore, as is easily verified,

$$(9.10.20)\qquad \int[f_N(t;\theta_1)f_N(t;\theta_2)]^{1/2}\mu_N(dt) = \exp\left\{-\frac{N}{2}\Delta(\theta_1, \theta_2)\right\}.$$

Hence from (9.10.18)–(9.10.20) we obtain

$$(9.10.21)\qquad R_N^*(\theta) \leqslant Nc + (\bar{L}_1\bar{L}_2)^{1/2}\exp\left\{-\frac{N}{2}\Delta(\theta_1, \theta_2)\right\},$$

for all θ. We remark that if θ belongs to the indifference region Θ_0, $L_i(\theta) = 0$ and $R_N^*(\theta) = Nc$. Hence (9.10.21) holds for all θ.

According to the definition of N,

$$(N-1)c < \frac{2c\,|\log c|}{\Delta(\theta_1, \theta_2)} \tag{9.10.22}$$

and

$$\exp\left\{-\frac{N}{2}\Delta(\theta_1, \theta_2)\right\} \leqslant c. \tag{9.10.23}$$

Hence from (9.10.21)–(9.10.23) we obtain

$$R_N^*(\theta) < \frac{2c\,|\log c|}{\Delta(\theta_1, \theta_2)} + c(1 + (\bar{L}_1\bar{L}_2)^{1/2}). \tag{9.10.24}$$

Let

$$c_0 = \exp\{-\Delta(\theta_1, \theta_2)(1 + (\bar{L}_1\bar{L}_2)^{1/2})\}.$$

Then for all c, $0 < c \leqslant c_0$,

$$c(1 + (\bar{L}_1\bar{L}_2)^{1/2}) < \frac{c\,|\log c|}{\Delta(\theta_1, \theta_2)}. \tag{9.10.25}$$

Hence $R_N^*(\theta) < 3c\,|\log c|/\Delta(\theta_1, \theta_2)$ for all θ. Finally, if $(j, T_j) \in \mathcal{C}^*(3c\,|\log c|/\Delta(\theta_1, \theta_2))$ for all $j = 1, \ldots, n$, we know that the sample size of the above fixed sample procedure, N, is at least $n + 1$. The posterior risk associated with that fixed sample procedure is $E\{R_N^*(\theta) \mid \mathcal{F}_n\}$. This posterior risk is obviously smaller than $3c\,|\log c|/\Delta(\theta_1, \theta_2)$. (Q.E.D.)

In Theorem 9.10.2 we specified the general conditions under which the stopping rule represented by the continuation region $\mathcal{C}^*(c)$ is asymptotically Bayes. Schwarz's theorem yields a stronger result in the 1-parameter exponential case and a region of indifference separating Θ_1 and Θ_2. We proved that $\mathcal{B}(c) \subset \mathcal{C}^*(c)$ and provided an inner bound for $\mathcal{B}(c)$, for c sufficiently small. The region $\mathcal{C}^*(3c\,|\log c|/\Delta(\theta_1, \theta_2))$ grows to infinity in each direction as $c \to 0$. Thus the Bayes continuation region $\mathcal{B}(c)$ grows to infinity in each direction and is represented asymptotically by $\mathcal{C}^*(c)$. It is generally very difficult to draw the boundary of $\mathcal{C}^*(c)$. Schwarz [1] therefore suggested considering the asymptotic shape of a continuation region based on a generalized probability likelihood ratio test, which (as will be shown) is also asymptotically optimal. Thus define the likelihood ratio statistics

$$\lambda_i(n, T_n) = \frac{(\xi)\operatorname{ess\,sup}_{\theta\in\Theta_i}(\exp\{\theta T_n - nb(\theta)\})}{(\xi)\operatorname{ess\,sup}_{\theta\in\Theta}(\exp\{\theta T_n - nb(\theta)\})}, \qquad i = 1, 2. \tag{9.10.26}$$

Here ξ designates the prior probability measure on $(\Theta, \mathcal{T})$ corresponding to the prior distribution H on Θ. The quantity $(\xi)\operatorname{ess\,sup}_{\theta\in\Theta_i}(\cdot)$ designates the almost sure supremum of $(\cdot)$, over the set Θ_i, with respect to the prior probability measure ξ. Due to the monotonicity of the logarithmic function, we have

(9.10.27)
$$\log \lambda_i(n, T_n) = (\xi) \operatorname{ess\,sup}_{\theta \in \Theta_i} \{\theta T_n - nb(\theta)\} - (\xi) \operatorname{ess\,sup}_{\theta \in \Theta} \{\theta T_n - nb(\theta)\}.$$

Therefore the function $\log \lambda_i(n, T_n)$ is homogeneous of the first order. That is,

$$\log \lambda_i(\alpha n, \alpha T_n) = \alpha \log \lambda_i(n, T_n), \qquad i = 1, 2, \tag{9.10.28}$$

for arbitrary α, $\alpha > 0$.

Define now the family of continuation sets for all γ, $0 < \gamma < 1$,

$$\Lambda(\gamma) = \{(n, T_n); \log \gamma_i(n, T_n) \geqslant \log \gamma, i = 1, 2\}. \tag{9.10.29}$$

The set $\Lambda(1/e)$ is denoted by $\mathcal{B}_0$. For every $\gamma > 0$ we can obtain $\Lambda(\gamma)$ by a homothetic transformation of $\mathcal{B}_0$. In this transformation each coordinate of the points of $\mathcal{B}_0$ is multiplied by the same positive constant, say α. We denote this transformation by $\mathcal{B}_0\alpha$. Thus

$$\Lambda(\gamma) = \mathcal{B}_0 |\log \gamma|, \qquad 0 < \gamma < 1. \tag{9.10.30}$$

Indeed,

$$\mathcal{B}_0 = \{(n, T_n); \log \lambda_i(n, T_n) \geqslant -1, i = 1, 2\}. \tag{9.10.31}$$

Therefore

$$\begin{aligned} \mathcal{B}_0 |\log \gamma| &= \left\{(n, T_n); \log \lambda_i\left(\frac{n}{|\log \gamma|}, \frac{T_n}{|\log \gamma|}\right) \geqslant -1, i = 1, 2\right\} \\ &= \{(n, T_n); |\log \gamma|^{-1} \log \lambda_i(n, T_n) \geqslant -1, i = 1, 2\} \\ &= \{(n, T_n); \log \lambda_i(n, T_n) \geqslant -|\log \gamma|, i = 1, 2\} \\ &= \Lambda(\gamma), \qquad 0 < \gamma < 1. \end{aligned} \tag{9.10.32}$$

Let

$$\Lambda_i(\gamma) = \{(n, T_n); \log \lambda_i(n, T_n) \geqslant \log \gamma\}; \tag{9.10.33}$$

$\Lambda(\gamma)$ is the intersection of $\Lambda_1(\gamma)$ and $\Lambda_2(\gamma)$. We denote by Λ_i the region $\Lambda_i(e^{-1})$, $i = 1, 2$. Thus $\mathcal{B}_0 = \Lambda_1 \cap \Lambda_2$.

According to (9.10.27) and (9.10.31),

$$\begin{aligned} \Lambda_1 = \Big\{(n, T_n); (\xi) \operatorname{ess\,sup}_{\theta \in \Theta} (\theta \bar{X} - b(\theta)) \\ \leqslant \frac{1}{n} + (\xi) \operatorname{ess\,sup}_{\theta \in \Theta_1} (\theta \bar{X} - b(\theta))\Big\}. \end{aligned} \tag{9.10.34}$$

For a given sample mean $\bar{X}$, $\theta\bar{X} - b(\theta)$ is a concave function, having a unique maximum at $\theta(\bar{X}) = \mu^{-1}(\bar{X})$, which is the root of the equation $b'(\theta) = \bar{X}$. The function $b'(\theta)$ is strictly increasing, and therefore $\theta(\bar{X})$ is a strictly increasing function of $\bar{X}$. For values of $\bar{X}$ such that $\theta(\bar{X}) \in \Theta_1$ we obtain that, if Θ_1 is contained in the support of ξ,

(9.10.35)
$$(\xi) \operatorname*{ess\,sup}_{\theta\in\Theta_1} (\theta\bar{X} - b(\theta)) = (\xi) \operatorname*{ess\,sup}_{\theta\in\Theta} (\theta\bar{X} - b(\theta)) = \bar{X}\theta(\bar{X}) - b(\theta(\bar{X})).$$

For such an $\bar{X}$ the inequality in (9.10.34) holds for every $n \geqslant 1$, and the entire ray $\theta\bar{X}$, $\theta \geqslant 0$, is contained in Λ_1. Since $\theta(\bar{X})$ is strictly increasing with $\bar{X}$, and $\Theta_1 = (-\infty, \theta_1]$, if $\theta(\bar{X}) \geqslant \theta_1$ and θ_1 belongs to the support of ξ,

$$(\xi) \operatorname*{ess\,sup}_{\theta\in\Theta_1} (\theta\bar{X} - b(\theta)) = \theta_1\bar{X} - b(\theta_1). \tag{9.10.36}$$

Let k_1 be a real value such that $\theta(k_1) = \theta_1$, that is, $k_1 = b'(\theta_1)$. Thus for all $\bar{X} > k_1$ we can present the boundary of Λ_1 as

$$\partial\Lambda_1 = \left\{(n, T_n);\ (\xi) \operatorname*{ess\,sup}_{\theta\in\Theta} (\theta T_n - nb(\theta)) = 1 + \theta_1 T_n - nb(\theta_1)\right\}. \tag{9.10.37}$$

Here we assume that $\Theta_1 \subset \operatorname{supp} \xi$, where $\operatorname{supp} \xi$ denotes the support of ξ. In a similar manner, we obtain a value k_2, $k_2 > k_1$, such that for all $k_2 = b'(\theta_2)$ the boundary of Λ_2 is, when $\Theta_2 \subset \operatorname{supp} \xi$,

$$\partial\Lambda_2 = \left\{(n, T_n);\ (\xi) \operatorname*{ess\,sup}_{\theta\in\Theta} (\theta T_n - nb(\theta)) = 1 + \theta_2 T_n - nb(\theta_2)\right\}. \tag{9.10.38}$$

Thus from (9.10.37) and (9.10.38) we can characterize $\mathcal{B}_0 = \Lambda_1 \cap \Lambda_2$ as

(9.10.39)
$$\mathcal{B}_0 = \left\{(n, T_n);\ 1 + \min_{i=1,2} \{\theta_i T_n - nb(\theta_i)\} \geqslant (\xi) \operatorname*{ess\,sup}_{\theta\in\Theta} (\theta T_n - nb(\theta))\right\},$$

assuming that $(\Theta_0 \cup \Theta_1) \subset \operatorname{supp} \xi$.

Theorem 9.10.5. (Schwarz [1].) *Let* **T** *denote the set of all* (n, T_n) *for which* n *is a positive integer. Let* $\mathcal{C}^*(\gamma)$ *be the region defined in* (9.10.11). *Then*

$$\mathcal{C}^*(\gamma) = (\mathcal{B}_0 \,|\log \gamma| + 0(|\log \gamma|) \cap \mathbf{T} \quad \text{as} \quad \gamma \to 0. \tag{9.10.40}$$

Proof. Let us characterize the points in the (n, T_n)-plane at which the rays of expected values $(n, nb'(\theta))$ intersect the boundary of $\mathcal{C}^*(\gamma)$.

If (n, T_n) is a point on the boundary at which A_2 is accepted, we have the associated Bayes risk

$$\hat{R}(n, T_n) = \frac{\int I_{\Theta_1}(\theta) \exp\{\theta T_n - nb(\theta)\} L_2(\theta) H(d\theta)}{\int \exp\{\theta T_n - nb(\theta)\} H(d\theta)}, \tag{9.10.41}$$

where $H(\theta)$ is the prior distribution of θ. If (n, T_n) is a point of intersection of the ray $(n, nb'(\theta))$ with the boundary of $\mathcal{C}^*(\gamma)$, at which A_2 is accepted, we have

$$\gamma = \frac{\int g^n(\theta)M(d\theta)}{\int e^n(\theta)H(d\theta)}, \tag{9.10.42}$$

where

$$\begin{aligned} g(\theta) &= I_{\Theta_1}(\theta) \exp\{\theta b'(\theta) - b(\theta)\}, \\ e(\theta) &= \exp\{\theta b'(\theta) - b(\theta)\}, \\ M(d\theta) &= L_2(\theta)H(d\theta). \end{aligned} \tag{9.10.43}$$

Following Loève [1], p. 160, if X is a random variable whose r-th absolute moment, with respect to a d.f. F, is finite, we define the norm of X in the corresponding L_r space as

$$\|X\|_r^{(F)} = \{E^{(F)}|X^r|\}^{1/r}.$$

Thus (9.10.42) can be written in the form

$$\gamma^{1/n} = \frac{\|g\|_n^{(M)}}{\|e\|_n^{(H)}}. \tag{9.10.44}$$

For each n, γ is bounded away from zero. Hence $\lim_{\gamma\to 0} n = \infty$. Otherwise $\lim_{\gamma\to 0} \gamma^{1/n} = 0$. As proven in Loève [1], p. 160,

$$\lim_{n\to\infty} \|X\|_n^{(F)} = (F) \text{ ess sup } |X|. \tag{9.10.45}$$

Let $n = t(\gamma)\,|\log \gamma|$. Then $\gamma^{1/n} = e^{-1/t(\gamma)}$ and we obtain for $\tau = \lim_{\gamma\to 0} t(\gamma)$,

$$e^{-1/\tau} = \frac{(M) \text{ ess sup } \{I_{\Theta_1}(\theta) \exp\{\theta b'(\theta) - b(\theta)\}\}}{(H) \text{ ess sup } \{\exp\{\theta b'(\theta) - b(\theta)\}\}}. \tag{9.10.46}$$

Since the null sets of M and H coincide,

(9.10.47)

$$\tau^{-1} = \left[(H) \operatorname*{ess\,sup}_{\Theta} (\theta b'(\theta) - b(\theta)) - (H) \operatorname*{ess\,sup}_{\Theta_1} (\theta b'(\theta) - b(\theta))\right].$$

Substituting $\bar{X}$ for $b'(\theta)$ and n for τ we obtain from (9.10.47) the boundary of Λ_1, $\partial\Lambda_1$, as given in (9.10.37). In a similar manner we show that the part of the boundary of $\mathcal{C}^*(\gamma)$ on which A_1 is accepted is (asymptotically as $c \to 0$) like the boundary $\partial\Lambda_2$. (Q.E.D.)

We now prove the main theorem.

Theorem 9.10.6. (Schwarz [1].) *The continuation region of the Bayes stopping rule*, $\mathcal{B}(c)$, *satisfies*

$$\mathcal{B}(c) = [\mathcal{B}_0\,|\log c| + o(|\log c|)] \cap \mathbf{T} \quad \text{as} \quad c \to 0. \tag{9.10.48}$$

Proof. According to Theorem 9.10.5,

$$\mathcal{C}^*(c) = [\mathcal{B}_0 \,|\log c| + o(|\log c|)] \cap \mathbf{T} \quad \text{as} \quad c \to 0.$$

Similarly,

$$(9.10.49) \quad \mathcal{C}^*(3c\,|\log c|/\Delta(\theta_1, \vartheta_2)) = \left[\mathcal{B}_0 \left| \log \frac{3c}{\Delta(\theta_1, \theta_2)} \,|\log c| \right| + o\left(\left| \log \frac{3c}{\Delta(\theta_1, \theta_2)} \,|\log c| \right|\right)\right] \cap \mathbf{T} \quad \text{as} \quad c \to 0.$$

But

$$(9.10.50) \quad \mathcal{B}_0 \left(\log \frac{3c}{\Delta(\theta_1, \theta_2)} \,|\log c|\right)^{-1} = \mathcal{B}_0 \,|\log c| + \mathcal{B}_0 \,|\log |\log c|| + \mathcal{B}_0 \log \frac{\Delta(\theta_1, \theta_2)}{3}.$$

Substituting (9.10.50) in (9.10.49) and noticing that

$$\log |\log c| = o(|\log c|) \quad \text{as} \quad c \to 0,$$

and

$$\log \left(\frac{\Delta(\theta_1, \theta_2)}{3}\right) = O(1) \quad \text{as} \quad c \to 0,$$

we obtain

$$(9.10.51) \quad \mathcal{C}^*\left(\frac{3c}{\Delta(\theta_1, \theta_2)} \,|\log c|\right) = [\mathcal{B}_0 \,|\log c| + o(|\log c|)] \cap \mathbf{T},$$

as $c \to 0$. Finally, according to (9.10.12) the Bayes continuation region $\mathcal{B}(c)$ is bounded by $\mathcal{C}^*(c)$ and $\mathcal{C}^*((3c/\Delta(\theta_1, \theta_2))\,|\log c|)$. (Q.E.D.)

The asymptotic shape of $\mathcal{B}(c)$ is thus given by that of $\mathcal{B}_0 \,|\log c|$ and is represented in the $(n/|\log c|, T_n/|\log c|)$-plane by $\mathcal{B}_0$. Schwarz [1] provided explicit formulae for the boundary of $\mathcal{B}_0$, when the support of ξ is the whole real line for several cases of interest. Kiefer and Sacks [1] generalized and extended the above results of Schwarz [1] for testing two composite hypotheses which are separated by an indifference zone. In the model treated by Kiefer and Sacks we cannot obtain a geometric representation similar to that of Schwarz, since they did not confine attention to the 1-parameter exponential family.

PROBLEMS

Section 9.1

1. Consider a finite population $\mathcal{U}$ consisting of N units. A simple random sample of size n, $1 \leqslant n < N$, is drawn from $\mathcal{U}$ without replacement. Let M be the

number of units in $\mathcal{U}$ having a certain attribute, and let T be the observed number of units in the sample having the specified attribute. Consider the problem of testing the hypotheses:

$$A_1: M \leqslant \theta_1 N, \qquad A_2: \theta_1 N < M \leqslant \theta_2 N, \qquad A_3: \theta_2 N < M \leqslant N,$$

where $0 < \theta_1 < \theta_2 < 1$ are specified rationals so that $\theta_1 N$ and $\theta_2 N$ are integers. The loss functions for erroneous decisions are

$$\begin{aligned} L_1(M) &= c_1(M - \theta_1 N) I_{\{M > \theta_1 N\}}, \\ L_2(M) &= c_2(\theta_1 N - M) I_{\{M \leqslant \theta_1 N\}} \\ &\qquad + c_2(M - \theta_2 N) I_{\{M > \theta_2 N\}}, \\ L_3(M) &= c_3(\theta_2 N - M) I_{\{M < \theta_2 N\}}. \end{aligned}$$

(i) Determine the Bayes test procedure for the prior binomial distribution $\mathcal{B}(N; \varphi)$ of M.

(ii) Determine the Bayes test procedure for the prior uniform distribution of M.

2. Consider the problem of testing two simple hypotheses, that is, $\Theta = \{\theta_1, \theta_2\}$, $A_1: \theta = \theta_1$, $A_2: \theta = \theta_2$. Show that the most powerful Neyman-Pearson test (see Lehmann [2], p. 65) is equivalent to some Bayes test.

3. Let X have a location parameter logistic distribution with a density $f(x - \theta)$, $-\infty < \theta < \infty$, where

$$f(x) = \frac{e^{-x}}{(1 + e^{-x})^2}, \qquad -\infty < x < \infty.$$

Consider the three hypotheses

$$A_{-1}: \theta < -\theta_0, \qquad A_0: -\theta_0 \leqslant \theta \leqslant \theta_0, \qquad A_1: \theta \geqslant \theta_0; \qquad 0 < \theta_0 < \infty.$$

Furthermore, suppose that the loss functions are the step functions

$$\begin{aligned} L_{-1}(\theta) &= I_{[-\theta_0,\theta_0]}(\theta) + 2I_{(\theta_0,\infty)}(\theta), \\ L_0(\theta) &= I_{(-\infty,-\theta_0)}(\theta) + I_{(\theta_0,\infty)}(\theta), \\ L_1(\theta) &= 2I_{(-\infty,-\theta_0)}(\theta) + I_{[-\theta_0,\theta_0]}(\theta). \end{aligned}$$

(i) Determine the form of the Bayes tests.

(ii) What is a minimax test? (See Ferguson [1], p. 290.)

(iii) Are the assumptions of Theorem 9.1.1 fulfilled?

Section 9.2

4. Consider the problem of testing the following hypotheses about the mean θ of a normal distribution $\mathcal{N}(\theta, 1)$: $A_{-1}: \theta < -1$; $A_0: -1 \leqslant \theta \leqslant 1$; $A_1: \theta > 1$. Suppose that the loss functions are

$$\begin{aligned} L_{-1}(\theta) &= \tfrac{1}{2}(\theta + 1)^2 I_{(-1,\infty)}(\theta), \\ L_0(\theta) &= (1 + \theta^2)^{-1}, \\ L_1(\theta) &= \tfrac{1}{2}(\theta - 1)^2 I_{(-\infty,1)}(\theta). \end{aligned}$$

(i) Derive the Bayes test against the prior $\mathcal{N}(0, \tau^2)$ distribution of θ. Is it a monotone procedure?

(ii) What is an essentially complete class of test procedures?

5. Let $X \sim G(1/\theta, 1)$, $0 < \theta < \infty$. Let $A_i: \theta = \theta_i$ $(i = 1, \ldots, k)$, where $0 < \theta_1 < \cdots < \theta_k < \infty$ are specified. The prior $H(\theta)$ is a discrete distribution concentrated on $\{\theta_1, \ldots, \theta_k\}$. The loss function is a zero-one loss. Derive the Bayes test procedure. (See Ferguson [1], p. 293.)

6. Consider the *k-sample slippage problem.* In this problem we consider k independent random variables $X_1, \ldots, X_n$ having (generalized) density functions $f_1(x), \ldots, f_k(x)$. We formulate $k + 1$ hypotheses:

$$A_0: f_1(x) = \cdots = f_k(x) = \varphi(x);$$

$$A_1: f_2(x) = \cdots = f_k(x) = \varphi(x) \neq f_1(x) = \psi(x);$$

$$\vdots$$

$$A_i: f_1(x) = \cdots = f_{i-1}(x) = f_{i+1}(x) = \cdots = f_k(x) = \varphi(x) \neq \psi(x) = f_i(x), \; i = 2, \ldots, k - 1;$$

$$\vdots$$

$$A_k: f_1(x) = \cdots = f_{k-1}(x) = \varphi(x) \neq \psi(x) = f_k(x).$$

Consider the (permutation) invariant prior distribution, which assigns the hypothesis A_0 a prior probability $1 - k\xi$ and A_i $(i = 1, \ldots, k)$ the equal prior probability ξ, $0 < \xi < 1/k$, and a zero-one loss function. Derive the Bayes invariant test. Is this test a minimax test? (See Ferguson [1], pp. 299–301.)

Section 9.3

7. Derive an empirical Bayes test for the negative binomial case N.B. (ψ, ν), ν known, and $0 < \psi < 1$, under the assumption that $\theta = \nu\psi/(1 - \psi)$ has a prior gamma distribution $\mathcal{G}(\tau, 1)$, $0 < \tau < \infty$, τ unknown. The hypotheses to test are

$$A_0: 0 < \theta \leqslant \theta_0, \qquad A_1: \theta_0 < \theta < \infty.$$

The loss function is as in Example 9.3. (See Samuel [1].)

8. Consider the problem of an empirical Bayes testing for the mean of an exponential distribution $\mathcal{G}(1/\theta, 1)$, $0 < \theta < \infty$. The two composite hypotheses are

$$A_0: 0 < \theta \leqslant \theta_0, \qquad A_1: \theta_0 < \theta < \infty.$$

The loss functions are zero-one. Derive an empirical Bayes test when the family $\mathcal{H}$ of prior distributions of θ are such that $1/\theta \sim G(\tau, 1)$, $0 < \tau < \infty$.

Section 9.4

9. Let $X_1, X_2, \ldots$ be a sequence of i.i.d. random variables having a discrete distribution,

$$X_1 = \begin{cases} -1, & \text{w.p. } \dfrac{\theta}{2} \\ 0, & \text{w.p. } 1 - \theta \\ 1, & \text{w.p. } \dfrac{\theta}{2}. \end{cases}$$

Consider the problem of testing the two composite hypotheses $A_0\colon \theta \leqslant \theta_0$, $A_1\colon \theta > \theta_0$, $0 < \theta_0 < 1$. The loss functions are

$$L_0(\theta) = \frac{(\theta - \theta_0)^2}{\theta(1 - \theta)} I_{(\theta_0, 1)}(\theta),$$

and

$$L_1(\theta) = \frac{(\theta_0 - \theta)^2}{\theta(1 - \theta)} I_{(0, \theta_0)}(\theta).$$

For a prior rectangular distribution of θ, that is, $\theta \sim \mathcal{R}(0, 1)$, what is $\rho_M^{(M)}(H)$ for $M \geqslant 1$? Is $R(H, \mathbf{s}^0) = \inf_{\mathbf{s} \in S} R(H, s) < \infty$?

10. Let $X_1, X_2, \ldots$ be a sequence of i.i.d. random variables. $X_1 \sim G(1/\theta, 1)$. Consider the problem of testing $A_0\colon 0 < \theta \leqslant \theta_0$ and $A_1\colon \theta_0 < \theta < \infty$ with a zero-one loss function. Derive the truncated Bayes sequential stopping rule for $M = 3$ against the $\mathcal{G}(1, 1)$ prior distribution of $1/\theta$.

11. Suppose that we have a multiple hypotheses testing problem, which is invariant under a group of transformations $\mathcal{G}$ (see the definition of invariant tests in Lehmann [2], Ch. 6). How would you determine the optimal invariant sequential procedure? (See Ferguson [1], p. 340.)

Section 9.5

12. Let $X_1, X_2, \ldots$ be a sequence of i.i.d random variables, $X_1 \sim N(\theta, 1)$. Suppose that the two simple hypotheses are $A_{-1}\colon \theta = -1$, $A_1\colon \theta = 1$.

(i) Define the stopping variable N for a Wald S.P.R.T. $(-b, a)$ with boundary constants $0 < a, b < \infty$.
(ii) Determine recursive formulae for the exact computation of $P_\theta[N = n]$.

13. Let $X_1, X_2, \ldots$ be a sequence of i.i.d. random variables, $X_1 \sim G(1/\theta, \nu)$, ν known, $0 < \theta < \infty$. Define the random variable $Z = \log (f(X; \theta_2)/f(X; \theta_1))$, $0 < \theta_1 < \theta_2 < \infty$. Determine the m.g.f. of Z, $M_\theta(t)$.

14. Let $X_1, X_2, \ldots$ be a sequence of i.i.d. random variables, $X_1 \sim \mathcal{R}(0, \theta)$, $0 < \theta < \infty$. Consider an S.P.R.T. of the two simple hypotheses $A_1\colon \theta = 1$ and $A_2\colon \theta = 2$.

(i) Show that the S.P.R.T. has the following stopping variable:

$$N = \text{least integer } n,\ n \geqslant 1 \text{ such that } X_n \geqslant 1 \text{ or } N = M,$$

where M is a (fixed) finite positive constant. Furthermore, show that the acceptance rule is: Accept A_1 if and only if $N = M$.

(ii) Prove that the error probabilities are $\alpha = P[\text{rejecting } A_1 \mid \theta = 1] = 0$, and $\beta = P[\text{accepting } A_1 \mid \theta = 2] = 2^{-M}$.

(iii) Prove that $E\{N \mid \theta = 1\} = M$ and $E\{N \mid \theta = 2\} = 2(1 - 2^{-M})$.

15. Consider the Wald S.P.R.T. for testing two simple hypotheses concerning the probability of success, θ, in a sequence of Bernoulli trials. Let $A_1: \theta = \theta_1$, $A_2: \theta = 1 - \theta_1$.

(i) Determine the S.P.R.T. $(-b, a)$.

(ii) Show that for certain values of a and b, the acceptance probability $\pi(\theta)$ of A_1 can be determined exactly for every θ by reducing the S.P.R.T. to a gambler ruin problem (see Feller [1], Ch. 14).

Section 9.6

16. Compare the proof of Wald and Wolfowitz [1] and Wolfowitz [5] on the optimality of the S.P.R.T. to that of Section 9.6.

Section 9.7

17. Let $X_1, X_2, \ldots$ be a sequence of i.i.d. random variables, $X_1 \sim N(0, \sigma^2)$, $0 < \sigma < \infty$.

(i) Construct an S.P.R.T. for testing $A_0: \sigma = 1 - \delta$ against $A_1: \sigma = 1 + \delta$, $0 < \delta < 1$, so that the error probabilities will be approximately α and β, $0 < \alpha, \beta < 1$, respectively. Determine $E_{1-\delta}\{N\}$ and $E_{1+\delta}\{N\}$, approximately.

(ii) Since the sufficient sequence $S_n = \sum_{i=1}^n X_i^2$ has an asymptotically normal distribution, and since $E_\sigma\{N\}$ is of order $0(1/\delta)$ as $\delta \to 0$, we can approximate N by a stopping time $T(\delta)$ of a Wiener process. Find the appropriate Wiener process and the corresponding $E_\sigma\{T(\delta)\}$ and $\pi(\sigma)$.

18. Let $X_1, X_2, \ldots$ be a sequence of i.i.d. random variables, $X_1 \sim N(\mu, \sigma^2)$, $\mu > 0$. Let $S_n = \sum_{i=}^n X_i$ and consider the stopping variable

$$N = \text{least integer } n, \quad n \geqslant 1, \quad \text{for which} \quad S_n \geqslant a + b_n + cn^2, \quad 0<a,\ b, \quad c<\infty.$$

Approximate the sequence S_n by a Wiener process and use (9.7.45) to approximate $E_{\mu,\sigma}\{N^3\}$.

Section 9.8

19. Prove by induction on n that, for each M, $\rho_n^{(M)}(\xi, \mathbf{X}_{N-n})$ is a homogeneous function of the first degree.

20. Determine the Bayes sequential procedure for testing two simple hypotheses, with a symmetric loss, $c = 1$, $r = 5$, for the following cases:

(i) $X \sim N(\theta, \sigma^2)$; A_0: $\theta = 0$, $\sigma = 1$, A_1: $\theta = 1$, $\sigma = 2$.
(ii) $X \sim G(1/\theta, 1)$; A_0: $\theta = 1$, A_1: $\theta = 2$.

REFERENCES

Albert [1]; Albert [1], [2]; Bessler [1]; Bickel and Yahav [1]; Blackwell and Girshick [1]; Breakwell and Chernoff [1]; Brown [1]; Chernoff [2], [3], [4], [6], [7]; Chernoff and Ray [2]; DeGroot [2]; Doob [1]; Dvoretzky, Kiefer, and Wolfowitz [1]; Dynkin [2]; Dynkin and Yushkevich [1]; Epstein and Sobel [1]; Ferguson [1]; Goode [1]; Grigelionis and Shiryaev [1]; Hall [1], [2]; Karlin [1], [2], [3]; Kemperman [2]; Kiefer and Sacks [1]; Lehmann [2]; Loève [1]; Matthes [1]; Naimark [1]; Page [1]; Raiffa and Schlaifer [1]; Robbins [4]; Robbins and Samuel [1]; Rubin and Karlin [1]; Samuel [1]; Schwarz [1]; Shiryaev [1]; Simmons [1] Skorokhod [1]; Wald [3], [4]; Wald and Wolfowitz [1]; Wijsman [2]; Whittle [1]; Wolfowitz [5].

CHAPTER 10

Confidence and Tolerance Intervals

10.1 MOST ACCURATE CONFIDENCE INTERVALS—THE 1-PARAMETER CASE

Let $\mathcal{F} = \{F_\theta;\ \theta \in \Theta\}$ be a 1-parameter family of distribution functions of a random variable X; Θ is an interval on the real line. A subinterval of Θ, $[\underline{\theta}(X), \bar{\theta}(X)]$ whose limits depend on the observed random variable X (properly measurable statistics) is called a γ confidence interval, $0 < \gamma < 1$, if

$$(10.1.1)\qquad P_\theta\{\theta \in [\underline{\theta}(X), \bar{\theta}(X)]\} \geqslant \gamma, \quad \text{all} \quad \theta \in \Theta.$$

The specified value γ on the right-hand side of (10.1.1) is called the *confidence level*; $\underline{\theta}(X)$ and $\bar{\theta}(X)$ are called lower and upper *confidence limits*. We notice that $[\underline{\theta}(X), \bar{\theta}(X)]$ is a random interval in Θ. The distribution law of $\{\underline{\theta}(X), \bar{\theta}(X)\}$ is determined by that of the random variable X.

In certain cases we may be interested in specifying only a lower or an upper confidence limit for θ. In these cases we let $\bar{\theta}(X) = \infty$ a.s. or $\underline{\theta}(X) = -\infty$ a.s. The corresponding intervals are called, respectively, lower and upper γ confidence intervals. In this section we discuss a method of constructing confidence intervals based on families of test functions. These confidence intervals carry, as will be shown, certain optimum properties which characterize the corresponding families of test functions. The theory presented here was originated by Neyman [2] and can be found in various sources. The reader is especially referred to Lehmann [2]. We start with a discussion of lower confidence intervals. The same theory applies for upper confidence intervals.

Confidence limits are generally not unique. For example, suppose that $X_1, X_2, \ldots, X_n$ are i.i.d. having a common gamma distribution $\mathcal{G}(1/\theta, 1)$, $0 < \theta < \infty$. A minimal sufficient statistic is $T_n = \sum_{i=1}^n X_i$, $T_n \sim 2\theta\chi^2[2n]$.

It is easy to verify that $P_\theta\{\theta \geqslant T_n/2\chi_\beta^2[2n]\} = 1 - \beta$ for all θ. Thus $\underline{\theta}(\mathbf{X}_n) = T_n/2\chi^2_{1-\gamma}[2n]$ is a γ lower confidence limit for θ. Let $X_{(1)} = \min_{1\leqslant i\leqslant n} \{X_i\}$, $X_{(1)} \sim (2\theta/n)\chi^2[2]$. Hence $\tilde{\underline{\theta}}(\mathbf{X}_n) = (n/2)X_{(1)}/\chi^2_{1-\gamma}[2]$ is also a γ lower confidence limit for θ. The question is which of these two lower confidence limits we should use. Intuitively, we could argue that $\tilde{\underline{\theta}}(\mathbf{X}_n)$ is not a function of a minimal sufficient statistic; $\underline{\theta}(\mathbf{X}_n)$ is expected, therefore, to be (in some sense) a more informative lower confidence limit than $\tilde{\underline{\theta}}(\mathbf{X}_n)$. We present now a theory of uniformly most accurate (U.M.A.) lower confidence limits, according to which one can choose an optimal lower confidence limit (if it exists).

A γ lower confidence limit for θ, $\underline{\theta}(X)$, is called *uniformly most accurate* if

$$P_\theta[\theta' \geqslant \underline{\theta}(X)] \leqslant P_\theta[\theta' \geqslant \tilde{\underline{\theta}}(X)], \quad \text{all} \quad \theta' < \theta, \tag{10.1.2}$$

and all θ, $\theta' \in \Theta$, where $\underline{\theta}(\tilde{X})$ is any other γ lower confidence limit. That is, the probability that a U.M.A. lower confidence interval covers points smaller than θ is minimal uniformly in θ. We show a strong connection between the theory of U.M.A. lower confidence limits and minimum risk confidence limits.

Lemma 10.1.1. (Lehmann [2].) *Let $L(\theta, d(X))$ be a loss function associated with a γ lower confidence limit $d(X)$. Assume that $L(\theta, d(X))$ is a nonincreasing function of $d(X)$ for each θ; and $L(\theta, d(X)) = 0$ for $d(X) \geqslant \theta$. Let $\underline{\theta}(X)$ be a U.M.A. γ lower confidence limit. Then*

$$E_\theta\{L(\theta, \underline{\theta}(X))\} \leqslant E_\theta\{L(\theta, d(X))\}, \quad \text{all} \quad \theta \in \Theta. \tag{10.1.3}$$

Proof. Consider the two distribution functions

$$G(y; \theta) = \begin{cases} P_\theta[y \geqslant \underline{\theta}(X)]/P_\theta[\theta \geqslant d(X)], & y < \theta, \\ 1, & y \geqslant \theta, \end{cases}$$

and

$$\tilde{G}(y; \theta) = \begin{cases} P_\theta[y \geqslant d(X)]/P_\theta[\theta \geqslant d(X)], & y < \theta, \\ 1, & y \geqslant \theta. \end{cases}$$

Since $\underline{\theta}(X)$ is a U.M.A. lower confidence limit, (10.1.2) implies that $G(y; \theta) \leqslant \tilde{G}(y; \theta)$ for all $-\infty < y < \infty$ and each θ.

It is easy to prove (see Lehmann [2], p. 112) that if $\phi(y)$ is any decreasing function of y, which is G and $\tilde{G}$-integrable, then

$$\int \phi(y)G(dy; \theta) \leqslant \int \phi(y)\tilde{G}(dy; \theta). \tag{10.1.4}$$

In particular, for $\phi(y) = L(\theta, y)$ we obtain, since $L(\theta, y) = 0$ for all $y \geqslant \theta$,

$$E_\theta\{L(\theta, \underline{\theta}(X))\} = \int L(\theta, y)\, dP_\theta[\underline{\theta}(X) \leqslant y] \tag{10.1.5}$$

$$= P_\theta[d(X) \leqslant \theta] \int L(\theta, y) G(dy; \theta)$$

$$\leqslant P_\theta[d(X) \leqslant \theta] \int L(\theta, y) \tilde{G}(dy; 0)$$

$$= \int L(\theta, y)\, dP_\theta[d(X) \leqslant y]$$

$$= E_\theta\{L(\theta, d(X))\}$$

(Q.E.D.)

Thus a U.M.A. γ lower confidence limit minimizes the risk of underestimating θ uniformly in θ. We now provide a method of constructing U.M.A. γ lower confidence limits which is based on *uniformly most powerful* (U.M.P.) tests of composite hypotheses (if exist).

Consider a simple hypothesis $A_1: \theta = \theta_1$ versus a (possibly composite) hypothesis $A_2: \theta \in \Theta_2$. The quantity θ_1 is not an element of Θ_2. For each value of θ_1, $-\infty < \theta_1 < \infty$, partition the sample space $\mathfrak{X}$ into $\mathcal{A}(\theta_1)$ and $\mathcal{R}(\theta_1)$, where $\mathcal{A}(\theta_1)$ is the set of all x values in $\mathfrak{X}$ at which A_1 is accepted. The set $\{\mathcal{A}(\theta); \theta \in \Theta\}$ is a family of acceptance regions for A_1. For each $x \in \mathfrak{X}$ define the subset of Θ, $S(x) = \{\theta; \theta \in \Theta, x \in \mathcal{A}(\theta)\}$. Let $\alpha = 1 - \gamma$ be the level of significance of the test based on $\mathcal{A}(\theta)$. That is,

$$P_\theta[X \in \mathcal{R}(\theta)] = \alpha, \quad \text{all} \quad \theta \in \Theta. \tag{10.1.6}$$

Then according to the definition of $S(x)$,

$$P_\theta[\theta \in S(X)] = P_\theta[X \in \mathcal{A}(\theta)] = \gamma, \quad \text{all} \quad \theta \in \Theta. \tag{10.1.7}$$

The family $S_\gamma = \{S(X); X \in \mathfrak{X}\}$ is a *family of γ confidence sets* in Θ. In other words, every family of acceptance regions $\mathcal{A}_\alpha = \{\mathcal{A}(\theta); \theta \in \Theta\}$ of the simple hypothesis A_1, at level of significance α, corresponds to a family S_γ, $\gamma = 1 - \alpha$, of confidence sets in Θ.

A family of acceptance regions $\mathcal{A}_\alpha$ is called *uniformly most powerful* (U.M.P.) at level α if

$$P_\varphi[X \in \mathcal{R}(\theta)] \geqslant P_\varphi[X \in \tilde{\mathcal{R}}(\theta)], \quad \text{all} \quad \varphi \in A_2(\theta), \tag{10.1.8}$$

and all $\theta \in \Theta$, where $\{(\tilde{\mathcal{A}}(\theta), \tilde{\mathcal{R}}(\theta)); \theta \in \Theta\}$ is any other family of acceptance and rejection regions for the hypothesis A_1; and $A_2(\theta)$ is the parametric set corresponding to the alternative hypothesis, A_2.

The concept of uniformly most accurate (U.M.A.) lower confidence limits is generalized to suit the present context of confidence sets. We say that S_γ is a *uniformly most accurate family of* γ *confidence sets* if

$$P_\varphi[\theta \in S(X)] \leqslant P_\varphi[\theta \in \tilde{S}(X)], \quad \text{all} \quad \varphi \in A_2(\theta), \tag{10.1.9}$$

and all $\theta \in \Theta$, where $\tilde{S}_\gamma = \{\tilde{S}(X); X \in \mathfrak{X}\}$, is any other family of γ confidence sets. We immediately obtain the following lemma.

Lemma 10.1.2. (Lehmann [2].) *If* S_γ *is a family of* γ *confidence sets corresponding to a U.M.P. family* $\mathcal{A}_\alpha$, $\alpha = 1 - \gamma$, *of acceptance regions, then* S_γ *is U.M.A.*

Proof. Let $\mathcal{A}_\alpha$ be any family of acceptance regions for A_1 and $\tilde{S}_\gamma$ the corresponding family of γ confidence sets, $\gamma = 1 - \alpha$. Since $\mathcal{A}_\alpha$ is U.M.P. we obtain from (10.1.7) and (10.1.8) that

$$\begin{aligned} P_\varphi[\theta \in S(X)] &= 1 - P_\varphi[X \in \mathcal{R}(\theta)] \\ &\leqslant 1 - P_\varphi[X \in \tilde{\mathcal{R}}(\theta)] = P_\varphi[\theta \in \tilde{S}(X)], \quad \text{all} \quad \varphi \in A_2(\theta), \end{aligned} \tag{10.1.10}$$

and all $\theta \in \Theta$. Hence S_γ is U.M.A. (Q.E.D.)

Corollary 10.1.3. *If* $\mathcal{F} = \{f(x; \theta); \theta \in \theta\ \Theta\}$ *is a family of density functions which is M.L.R. in* $T(X)$, *and if the distribution function* $H(t; \theta)$ *of* $T(X)$ *is continuous in* t *for each* θ, *then*

(i) *there exist a U.M.A.* γ *lower confidence limit* $\underline{\theta}(T)$ *for all* γ, $0 < \gamma < 1$;
(ii) *if the root* $\hat{\theta}_\gamma(T)$ *of the equation* $H(T; \theta) = \gamma$ *belongs to* Θ, *then* $\underline{\theta}(T) = \hat{\theta}_\gamma(T)$.

Proof. Consider the hypotheses $A_1: \theta = \theta_1$ versus $A_2: \theta \in A_2(\theta_1) = (\theta_1, \infty)$. As proven in Lehmann [2], since $\mathcal{F} = \{f(x; \theta); \theta \in \Theta\}$ is an M.L.R. in T, and $H(t; \theta)$ is continuous in t, a family of uniformly most powerful acceptance regions, at level α, is given by the intervals

$$\mathcal{A}(\theta) = \{T; T \leqslant c_{1-\alpha}(\theta)\}, \qquad \theta \in \Theta, \tag{10.1.11}$$

where $c_{1-\alpha}(\theta)$ is the $(1 - \alpha)$-fractile of $H(t; \theta)$. Furthermore,

$$P_\varphi[T \leqslant c_{1-\alpha}(\theta)] < P_\theta[T \leqslant c_{1-\alpha}(\theta)], \quad \text{all} \quad \theta < \varphi. \tag{10.1.12}$$

It follows that $c_{1-\alpha}(\theta)$ is a strictly increasing function of θ for each $\alpha, 0 < \alpha < 1$.

Define the statistic

$$\hat{\theta}_\gamma(T) = \text{root of the equation}; \quad c_{1-\alpha}(\theta) = T. \tag{10.1.13}$$

The function $\hat{\theta}_\gamma(T)$ is an increasing function of T; and $c_{1-\alpha}(\hat{\theta}_\gamma(T)) = T$. Hence $\hat{\theta}_\gamma(T)$ is a γ lower confidence limit for θ, $\gamma = 1 - \alpha$. Indeed,

$$P_\theta[\theta \geqslant \hat{\theta}_\gamma(T)] = P_\theta[c_{1-\alpha}(\theta) \geqslant T] = 1 - \alpha = \gamma, \tag{10.1.14}$$

for all $\theta \in \Theta$. Finally, from the previous lemma, since $\mathcal{A}_\alpha = \{\mathcal{A}(\theta); \theta \in \Theta\}$ is U.M.P. at level α, $S_\gamma = \{[\theta; \theta \geqslant \hat{\theta}_\gamma(T)]; T \in T(\mathfrak{X})\}$ is a U.M.A. family of lower confidence intervals, with $\gamma = 1 - \alpha$. (Q.E.D.)

Example 10.1. We have mentioned previously the case of n i.i.d. random variables, distributed like $G(1/\theta, 1)$, $0 < \theta < \infty$. We derived two families of γ lower confidence limits,

$$\underline{\theta}(X_n) = \frac{T_n}{2\chi^2_{1-\gamma}[2n]} \quad \text{and} \quad \tilde{\underline{\theta}}(\mathbf{X}_n) = \frac{nX_{(1)}}{2\chi^2_{1-\gamma}[2]}.$$

The family of gamma distributions $\{\mathcal{G}(1/\theta, 1); 0 < \theta < \infty\}$ is an M.L.R. one. The U.M.P. test of $A_1: \theta = \theta_1$ versus $A_2: \theta > \theta_1$, for all $0 < \theta_1 < \infty$ at level α $(0 < \alpha < 1)$ is represented by the family $\mathcal{A}_\alpha$ of acceptance regions $A(\theta_1) = \{T_n; T_n \leqslant 2\theta_1\chi^2_{1-\alpha}[2n]\}$. Thus the family of U.M.A. γ lower confidence intervals is $S_\gamma = \{S_\gamma(T_n); 0 < T_n < \infty\}$ where $S_\gamma(T_n) = \{\theta; \theta \geqslant \hat{\theta}_\gamma(T_n)\}$, $\hat{\theta}_\gamma(T_n) = T_n/2\chi_\gamma{}^2[2n]$. The function $\hat{\theta}_\gamma(T_n)$ coincides with $\underline{\theta}(\mathbf{X}_n)$. Hence

$$P_{\theta'}[\theta \geqslant \hat{\theta}_\gamma(T_n)] < P_{\theta'}[\theta \geqslant \tilde{\underline{\theta}}(\mathbf{X}_n)], \quad \text{all} \quad 0 < \theta < \theta' < \infty,$$

where $\underline{\theta}(\mathbf{X}_n) = nX_{(1)}/2\chi_\gamma{}^2[2]$. ■

If X is a discrete random variable, we cannot apply Corollary 10.1.3 directly even if its distribution belongs to an M.L.R. family, since the distribution of the statistic T_n is a step function. In order to overcome this difficulty we consider randomized test functions, which have the same power as any specified test function. Thus let $\phi(X)$ be a specified test function, which is the probability of rejecting A_1 given X. Let R be a random variable independent of X, having a rectangular distribution on $[0, 1]$, and consider the test function $\phi^*(X, R) = J(R; \phi(X))$ where

$$J(X; Y) = \begin{cases} 1, & \text{if} \quad X \leqslant Y, \\ 0, & \text{if} \quad X > Y. \end{cases}$$

Then, for all θ,

$$E_\theta\{\phi^*(X, R)\} = P_\theta[R \leqslant \phi(X)] = E_\theta\{\phi(X)\}. \tag{10.1.15}$$

This shows that the power functions of $\phi^*(X, R)$ and of $\phi(X)$ are the same. When X is an integer-valued random variable we can represent (X, R) by $Y = X + R$. There is a one-to-one correspondence between Y and (X, R), since $X = [Y]$ and $R = Y - [Y]$ where $[x]$ is the largest integer that does not exceed x. The distribution function of $Y = X + R$ is absolutely continuous, and we can apply Corollary 10.1.3 to obtain a U.M.A. γ lower confidence limit for the parameter θ of a discrete random variable X having an M.L.R. distribution. This lower confidence limit is randomized, in the sense that it depends not only on the observed random variable but also on an independent rectangular random variable R (which can be obtained from a table of random numbers).

Theorem 10.1.4. (Zacks [10].) *Let $\mathfrak{F} = \{f(x; \theta); \theta \in \Theta\}$ be an M.L.R. family of density functions of an integer valued discrete random variable, X. Let R be an independent random variable having a rectangular distribution on* (0, 1). *Then the U.M.A. γ lower confidence limit for θ is the root $\underline{\theta}$ of the equation*

$$RF(X; \underline{\theta}) + (1 - R)F(X - 1; \theta) = \gamma, \tag{10.1.16}$$

where $F(X; \theta)$ is the distribution function of X.

Proof. Without loss of generality, assume that $\mathfrak{F}$ is an M.L.R. in X and that the values assumed by X are the non-negative integers. Thus $f(x; \theta) > 0$ for all $x = 0, 1, \ldots$ and all $\theta \in \Theta$. Let $F^{-1}(\gamma; \theta)$ designate the γ-fractile of $F(x; \theta)$, that is,

$$F^{-1}(\gamma; \theta) = \text{least non-negative integer } j, \text{ such that } F(j; \theta) \geqslant \gamma. \tag{10.1.17}$$

Let

$$\theta_j = \inf \{\theta; F^{-1}(\gamma; \theta) = j\}, \qquad j = 0, 1, \ldots . \tag{10.1.18}$$

Let $c_\gamma(\theta)$ be the γ-fractile of the distribution of $Y = X + R$. We now prove that

$$c_\gamma(\theta) = j + \frac{\gamma - F(j - 1; \theta)}{f(j; \theta)}, \qquad \theta_j \leqslant \theta < \theta_{j+1}, \tag{10.1.19}$$

where $F(-1; \theta) \equiv 0$ for all θ. Indeed, for every $\theta \in \Theta$ and each $j = 0, 1, \ldots,$

$$\begin{aligned} P_\theta[X + R \leqslant j + \phi] &= P_\theta[X \leqslant j, R \leqslant \phi] \\ &+ P_\theta[X \leqslant j - 1, R > \phi] = \phi F(j; \theta) + (1 - \phi)F(j - 1; \theta]. \end{aligned} \tag{10.1.20}$$

Let $\phi_\gamma(\theta)$ be the root ϕ of the equation

$$\phi F(j; \theta) + (1 - \phi)F(j - 1; \theta) = \gamma. \tag{10.1.21}$$

It is easily verified that

$$\phi_\gamma(\theta) = \frac{\gamma - F(j - 1; \theta)}{f(j; \theta)}, \qquad \theta_j \leqslant \theta < \theta_{j+1}. \tag{10.1.22}$$

We notice also that if $\theta \in [\theta_j, \theta_{j+1})$, then $F(j - 1; \theta) < \gamma$. Finally, the γ lower confidence limit for θ is the root of the equation

$$X + R = c_\gamma(\theta), \qquad \theta_X \leqslant \theta < \theta_{X+1}. \tag{10.1.23}$$

It is easy to check that (10.1.23) is equivalent to (10.1.16). Since the distribution function of $X + R$ is absolutely continuous, Corollary 10.1.3 implies that the root of (10.1.16) is a U.M.A. γ lower confidence limit.

(Q.E.D.)

Example 10.2. Let $X_1, X_2, \ldots, X_n$ be i.i.d. random variables having a common binomial distribution $\mathcal{B}(N, \theta)$; N is a known positive integer, θ an unknown parameter, $0 < \theta < 1$. A minimal sufficient statistic is $T_n = \sum_{i=1}^{n} X_i$; $T_n \sim \mathcal{B}(nN, \theta)$. The family $\mathcal{F} = \{\mathcal{B}(nN, \theta); 0 < \theta < 1\}$ is an M.L.R. in T_n.

Let $B(k; Nn, \theta)$ designate the distribution function of T_n. We utilize the well-known relationship

$$(10.1.24) \quad B(k; Nn, \theta) = I_{1-\theta}(Nn - k, k + 1), \qquad k = 0, \ldots, Nn - 1,$$

where $Ix(p, q)$, $0 < p, q < \infty$, is the incomplete beta-function ratio. In addition, $Ix(p, q)$ is the distribution function of beta (p, q) at the point x, $0 < x < 1$.

Let $I(\beta; p, q)$ designate the β-fractile of beta (p, q). Then, according to (10.1.24), the γ-fractile of $\mathcal{B}(Nn, \theta)$ is

$$(10.1.25) \quad B^{-1}(\gamma; Nn, \theta) = \text{least non-negative integer } k \text{, for which } I(\gamma; Nn - k, k + 1) \leqslant 1 - \theta.$$

Finally, from (10.1.16), the U.M.A. γ lower confidence limit for θ is $\hat{\theta}_\gamma(T_n + R)$, where

$$(10.1.26) \quad \hat{\theta}_\gamma(T_n + R) = \text{root of the equation:}$$

$$RI_{1-\theta}(Nn - T_n, T_n + 1) + (1 - R)I_{1-\theta}(Nn - T_n + 1, T_n) = \gamma,$$

if $T_n = 1, \ldots, Nn - 1$. If $T_n = 0$, the lower confidence limit is $\hat{\theta}_\gamma(0) \equiv 0$. If $T_n = Nn$ the lower confidence limit is $\hat{\theta}_\gamma(Nn) = (1 - \gamma)^{1/Nn}$. ∎

If lower and upper confidence limits are required the method of construction is similar to the previous one. We let $\mathcal{A}_\alpha$ be a family of acceptance regions of the simple hypothesis A_1: $\theta = \theta_1$ versus the composite hypothesis A_2: $\theta \neq \theta_1$, at level α. We define as before

$$S(x) = \{\theta; x \in A(\theta), A(\theta) \in \mathcal{A}_\alpha\}.$$

The family $S_\gamma = \{S(X); X \in \mathfrak{X}\}$ is a family of two-sided γ confidence sets. In the case of exponential families of distribution functions, $\mathcal{A}_\alpha$ is a convex collection of intervals; and correspondingly $S(X)$ are intervals $[\underline{\theta}(X), \bar{\theta}(X)]$.

Example 10.3. Suppose that $X_1, X_2, \ldots, X_n$ are i.i.d. random variables having a common normal distribution $\mathcal{N}(0, \sigma^2)$, $0 < \sigma^2 < \infty$. A minimal sufficient statistic is $T_n = \sum_{i=1}^{n} X_i^2$; $T_n \sim \sigma^2\chi^2[n]$. It is easy to verify that

$\mathcal{A}_\alpha = \{A(\sigma); 0 < \sigma < \infty\}$, where

$$A(\sigma) = \{T_n; \sigma^2\chi^2_{\alpha/2}[n] \le T_n \le \sigma^2\chi^2_{1-(\alpha/2)}[n]\},$$

is a family of acceptance intervals at level α. Therefore $\underline{\sigma}^2(T_n) = T_n/\chi^2_{1-(\alpha/2)}[n]$ and $\bar{\sigma}^2(T_n) = T_n/\chi^2_{\alpha/2}[n]$ are lower and upper confidence limits for σ at level $\gamma = 1 - \alpha$. Are these confidence limits optimal? ∎

In Lemma 10.1.2 we have proven that if $\mathcal{A}_\alpha$ is a family of U.M.P. acceptance regions, then the corresponding family of confidence sets S_γ is U.M.A. In many cases U.M.P. tests do not exist. However, if U.M.P. tests exist among the unbiased or invariant tests, the corresponding confidence sets are called U.M.A. unbiased or invariant, respectively. We describe these types of confidence intervals in the following section.

10.2. CONFIDENCE INTERVALS FOR CASES WITH NUISANCE PARAMETERS

In cases where the family of distribution functions depends on more than one parameter and the objective is to determine confidence sets to single parameters (one or each of them) or to subvectors of parameters, we consider the problem as one with nuisance parameters. Problems of estimation with nuisance parameters have been considered in previous chapters. In this section we discuss some procedures for determining U.M.A. unbiased and univariant confidence limits for models with nuisance parameters.

Example 10.4. Let $X_1, X_2, \ldots, X_n$ be i.i.d. having a common normal distribution $\mathcal{N}(\mu, \sigma^2)$ where both μ and σ are unknown, $-\infty < \mu < \infty$, $0 < \sigma^2 < \infty$. The problem of determining confidence limits for μ is a problem with a nuisance parameter. The common practice is to consider the two-sided $(1 - \alpha)$ confidence limits for μ,

$$(\bar{X} - t_{1-(\alpha/2)}[n-1]s/n^{1/2}, \bar{X} + t_{1-(\alpha/2)}[n-1]s/\sqrt{n})$$

where $t_\gamma[\nu]$ is the γ-fractile of the t-distribution with ν degrees of freedom; $\bar{X} = \sum X_i/n$, $s^2 = \sum_{i=1}^n (X_i - \bar{X})^2/(n-1)$. The commonly used two-sided $(1-\alpha)$ confidence limits for σ^2 are $\underline{\sigma}^2 = Q/\chi^2_{1-(\alpha/2)}[n-1]$, $\bar{\sigma}^2 = Q/\chi^2_{\alpha/2}[n-1]$, where $Q = \sum (X_i - \bar{X})^2$. These limits are obtained from acceptance intervals which yield equal tail probabilities. ∎

Let X be a random variable having a distribution function $F(x; \theta, \nu)$, $\theta \in \Omega$, $\nu \in \Theta$. We consider confidence sets for θ, where ν is a nuisance parameter. A set $S(X)$ in Ω is called a γ confidence set for θ if

$$P_{\theta,\nu}[\theta \in S(X)] \geqslant \gamma \quad \text{for all} \quad \theta \in \Omega, \nu \in \Theta. \tag{10.2.1}$$

Consider a family $\{S(X); X \in \mathfrak{X}\}$ of confidence sets which corresponds to a family $\mathcal{A}_\alpha$ of acceptance regions of a *composite* hypothesis A_1 versus a composite hypothesis A_2. Suppose that $A_1(\theta_1)$ is a parametric representation of the set of parametric points corresponding to A_1 and $A_2(\theta_1)$ a parametric representation of the set corresponding to A_2; θ_1 is a point in Ω. Then (10.2.1) is extended to

$$(10.2.2)\quad P_{\theta,\nu}[\theta_1 \in S(X)] \geqslant \gamma, \quad \text{all} \quad \theta \in A_1(\theta_1), \quad \text{all} \quad \theta_1 \in \Omega, \quad \text{all} \quad \nu \in \Theta.$$

A family of γ confidence sets S_γ is called *unbiased* if, in addition to (10.2.2),

$$(10.2.3)\quad P_{\theta,\nu}[\theta_1 \in S(X)] \leqslant \gamma, \quad \text{all} \quad \theta \in A_2(\theta_1), \quad \text{all} \quad \theta_1 \in \Omega, \quad \text{all} \quad \nu \in \Theta.$$

That is, whenever the true parameter point (θ, ν) does not belong to $A(\theta_1) \times \Theta$, then the coverage probability of θ_1 is at most γ. The family S_γ is called *strictly unbiased* if strict inequality holds in (10.2.3).

In cases where Ω is an interval on the real line; Θ is an interval in a k-dimensional Euclidean space; and if A_1: $\theta \leqslant \theta_1$, ν arbitrary, A_2: $\theta > \theta_1$, ν arbitrary, then $A_1(\theta_1) = \{\theta; \theta \leqslant \theta_1\}$, $A_2(\theta_1) = \{\theta > \theta_1\}$, and (10.2.2) and (10.2.3) in cases of two-sided confidence intervals reduce to

$$(10.2.4)\qquad P_{\theta,\nu}[\underline{\theta}(X) \leqslant \theta_1 \leqslant \bar{\theta}(X)] \geqslant \gamma, \quad \text{all} \quad \theta \in A_1(\theta_1),$$

all $\theta_1 \in \Omega$, all $\nu \in \Theta$; and

$$(10.2.5)\qquad P_{\theta,\nu}[\underline{\theta}(X) \leqslant \theta_1 \leqslant \bar{\theta}(X)] \leqslant \gamma, \quad \text{all} \quad \theta \in A_2(\theta_1),$$

all $\theta_1 \in \Omega$, and all $\nu \in \Theta$. A family S_γ of unbiased γ confidence sets is called *uniformly most accurate unbiased* (U.M.A.U.) if, subject to (10.2.2), its confidence sets $S(X)$ minimize the coverage probabilities

$$P_{\theta,\nu}[\theta_1 \in S(X)], \quad \text{all} \quad \theta \in A_2(\theta_1)$$

uniformly in $\theta_1 \in \Omega$ and $\nu \in \Theta$.

To obtain a family of U.M.A.U. γ confidence sets we construct a family $\mathcal{A}_\alpha$, $\alpha = 1 - \gamma$, of uniformly most powerful unbiased (U.M.P.U.) tests of $A_1(\theta_1)$ versus $A_2(\theta_1)$, $\theta_1 \in \Omega$, ν arbitrary. If $A(\theta)$, $\theta \in \Omega$, are the acceptance regions of the U.M.P.U. family $\mathcal{A}_\alpha$, then $S(X) = \{\theta; X \in A(\theta)\}$ are the corresponding confidence sets of the U.M.A.U. family of S_γ. The reader can find in Lehmann [2], Ch. 4 and Ch. 5, a discussion of the theory of U.M.P.U. tests and the related U.M.A.U. confidence sets. We reproduce here the main results for the exponential family of distributions. Starting from the 1-parameter exponential case, let

$$f(x; \theta) = \beta(\theta)e^{\theta x}, \qquad -\infty < \theta < \infty.$$

As proven in Lehmann [2], p. 126, there exists a U.M.P.U. test of the hypothesis $A_1: \theta = \theta_1$ versus $A_2: \theta \neq \theta_1$. This test is given by the test function

$$\phi(X) = \begin{cases} 1, & \text{if } X < c_\alpha^{(1)}(\theta_1) \quad \text{or } X > c_\alpha^{(2)}(\theta_1), \\ \gamma_i, & \text{if } X = c_\alpha^{(i)}(\theta_1), \quad i = 1, 2, \\ 0, & \text{otherwise.} \end{cases} \tag{10.2.6}$$

The critical values $c_\alpha^{(1)}(\theta_1)$ and $c_\alpha^{(2)}(\theta_1)$ are determined from the two conditions

$$E_{\theta_1}\{\phi(X)\} = \alpha, \tag{10.2.7}$$

$$E_{\theta_1}\{X\phi(X)\} = \alpha E_{\theta_1}\{X\}. \tag{10.2.8}$$

Condition 10.2.7 is obvious. Condition 10.2.8 is obtained from the condition of unbiasedness, according to which the power function of an unbiased test of A_1 versus A_2 should attain a minimum value at $\theta = \theta_1$. The power function of a test function $\phi(X)$ is

$$\psi(\theta) = \beta(\theta) \int \phi(x) e^{\theta x} \mu(\mathrm{d}x).$$

Hence

$$\begin{aligned} \psi'(\theta) &= \beta'(\theta) \int \phi(x) e^{\theta x} \mu(\mathrm{d}x) + \beta(\theta) \int x \phi(x) e^{\theta x} \mu(\mathrm{d}x) \\ &= \frac{\beta'(\theta)}{\beta(\theta)} E_\theta\{\phi(X)\} + E_\theta\{X\phi(X)\}. \end{aligned} \tag{10.2.9}$$

In particular, if $\tilde{\phi}(X) = \alpha$ a.s., $\psi(\theta) = \alpha$ for all θ. Hence

$$\begin{aligned} 0 &= \frac{\beta'(\theta)}{\beta(\theta)} E_\theta\{\tilde{\phi}(X)\} + E_\theta\{X\tilde{\phi}(X)\} \\ &= \frac{\beta'(\theta)}{\beta(\theta)} + E_\theta\{X\}. \end{aligned} \tag{10.2.10}$$

Substituting (10.2.10) in (10.2.9), we obtain

$$\psi'(\theta) = E_\theta\{X\phi(X)\} - E_\theta\{X\}E_\theta\{\phi(X)\}. \tag{10.2.11}$$

A necessary condition for a test function ϕ to minimize $\psi(\theta)$ at $\theta = \theta_1$ is that $\psi'(\theta_1) = 0$. Hence we imply (10.2.8). We infer from this result that the two-sided confidence intervals for σ^2 given in Example 10.3 are not U.M.V.U.

We turn now to the multiparameter exponential case. We present the density function of the observed random variable

$$f(x; \theta, \nu) = \beta(\theta, \nu) \exp\{\theta x + \nu' T(x)\} \tag{10.2.12}$$

where ν is a k-dimensional vector ($k \geqslant 1$) and $T(X)$ a corresponding k-dimensional random vector; $(X, T(X))$ is a minimal sufficient statistic;

$\theta \in \Omega$; Ω, is an interval on the real line; and $\nu \in \Theta$, which is an interval in a k-dimensional Euclidean space. The conditional density of X, given T, is of the 1-parameter exponential type

$$f_{X|T}(x; \theta) = k(\theta, T) \exp\{\theta x\}. \tag{10.2.13}$$

Let $A_1: \theta = \theta_1$, ν arbitrary, $A_2: \theta \neq \theta_1$, ν arbitrary, be two composite hypotheses. If the family $\mathfrak{F}_1^T$ of the distribution functions of the statistic T is boundedly complete when $\theta = \theta_1$, then every unbiased test $\phi(X; T)$ of A_1 versus A_2 should satisfy the conditions

$$E_{\theta_1}\{\phi(X; T) \mid T\} = \alpha \quad \text{a.s.}, \tag{10.2.14}$$

$$E_{\theta_1}\{X\phi(X; T) \mid T\} = \alpha E_{\theta_1}\{X \mid T\} \quad \text{a.s.} \tag{10.2.15}$$

Conditions 10.2.14 and 10.2.15 are analogous to (10.2.7) and (10.2.8). We therefore consider a family of acceptance regions $\mathcal{A}_\alpha(T)$ which is based on the conditional distributions of X given T. The U.M.P.U. test of A_1 versus A_2 is of the same structure as (10.2.6). The difference is that $c_\alpha^{(i)}(\theta_1, T)$ and $\gamma_i(\theta_1, T)$, $i = 1, 2$, are functions of the observed value of T and determined so that both (10.2.14) and (10.2.15) are satisfied. The reader can find many interesting examples in Lehmann's book [2], Ch. 4 and Ch. 5.

Example 10.5. In this example we provide a family of U.M.A.U. γ two-sided confidence intervals for the reliability function of a two-component system in series. The results presented in this example were derived by Lentner and Buehler [1].

Let $X_1, X_2, \ldots, X_{n_1}$ be i.i.d. random variables distributed like $G(1/\theta_1, 1)$. Let $Y_1, \ldots, Y_{n_2}$ be i.i.d. random variables distributed like $G(1/\theta_2, 1)$, $0 < \theta_1, \theta_2 < \infty$; $\{X_1, \ldots, X_{n_1}\}$ and $\{Y_1, \ldots, Y_{n_2}\}$ are mutually independent. Suppose that X represents the life-time of a component of Type I, Y represents the lifetime of a component of Type II. These two components are connected in series. The reliability of this two-component system is the probability that the minimal lifetime will exceed a specified value τ; that is,

$$R(\tau) = P_{\theta_1,\theta_2}\{\min(X, Y) \geqslant \tau\} = \exp\left\{-\tau\left(\frac{1}{\theta_1} + \frac{1}{\theta_2}\right)\right\}. \tag{10.2.16}$$

Let $\omega = (1/\theta_1 + 1/\theta_2)$. We derive U.M.A.U. confidence limits for ω, based on the given observations. These limits will induce U.M.A.U. confidence limits for $R(\tau)$, since $R(\tau)$ is a monotone function of ω.

A minimal sufficient statistic is (T_1, T_2), where $T_1 = \sum_{i=1}^{n_1} X_i$, $T_2 = \sum_{j=1}^{n_2} Y_j$; $T_i \sim G(1/\theta_i, n_i)$, $i = 1, 2$. Both T_1 and T_2 are independent. The joint density of T_1, T_2 is

$$f(t_1, t_2; \theta_1, \theta_2) = \frac{t_1^{n_1-1} t_2^{n_1-1}}{\Gamma(n_1)\Gamma(n_1)\theta_1^{n_1}\theta_2^{n_2}} \exp\left\{-\frac{t_1}{\theta_1} - \frac{t_2}{\theta_2}\right\}.$$

Make the transformation

$$\begin{cases} T_1 = T_1 \\ U = T_2 - T_1 \end{cases} \tag{10.2.17}$$

The joint density of T_1 and U is

$$f_{T_1,U}(t, u; \theta_2, \omega) = \lambda e^{u/\theta_2} t^{\nu_1}(t-u)^{\nu_2} e^{-\omega t}, \tag{10.2.18}$$

where $\lambda^{-1} = \Gamma(n_1)\Gamma(n_2)\theta_1^{n_1}\theta_2^{n_2}$. The marginal density of U is

$$f_U(u; \theta_2, \omega) = \lambda A(u; \omega) e^{u/\theta_2}, \tag{10.2.19}$$

where

$$A(u;\omega) = \begin{cases} e^{-\omega u}\omega^{-\nu_1-\nu_2-1}\sum_{i=0}^{\nu_1}\binom{\nu_1}{i}(\nu_1+\nu_2-i)!\,(\omega u)^i, & u > 0 \\ \omega^{-\nu_1-\nu_2-1}\sum_{i=0}^{\nu_2}\binom{\nu_2}{i}(\nu_1+\nu_2-i)!\,(-\omega u)^i, & u < 0. \end{cases} \tag{10.2.20}$$

Finally, the conditional density of T_1 given U is

$$f_{T_1|U}(t; \omega) = \frac{t^{\nu_1}(t-U)^{\nu_2}}{A(U;\omega)} e^{-\omega t}, \quad t \geqslant \max(U, 0) \tag{10.2.21}$$

Thus the conditional distribution of T_1 given U belongs to a 1-parameter exponential family. We can obtain the U.M.A.U. γ confidence limits for ω by the method described earlier. ■

10.3. EXPECTED LENGTH OF CONFIDENCE INTERVALS

In Sections 10.1 and 10.2 we discussed optimal confidence intervals from the point of view of minimizing the coverage probability of points in the domain of the alternative hypotheses. Thus if we considered lower confidence intervals, the objective was to minimize the probability of underestimating θ. In two-sided cases we discussed unbiased confidence intervals which minimize the probability of covering parameter points different from the true parameter point. Other criteria of optimality of confidence intervals have been adopted in various studies. An important class of criteria is associated with the expected length of two-sided confidence intervals. Pratt [2] has shown that the criterion of minimizing the expected length of a confidence interval is related to the criterion of minimizing the probability of covering false values in the parameter space. However, the minimization of the expected length of a confidence interval generally depends on the unknown true value of θ. Pratt explored the possibility of minimizing some weighted expectation of the expected length. As is shown in Section 10.7, Pratt's criterion is not the same as the Bayesian method, although there are apparent similarities. In Example 10.6 we illustrate Pratt's method on a confidence interval for the mean of a normal distribution. This example is taken from Pratt's paper [2]. Madansky [1] provided an example in which minimizing the expected length of a confidence interval is not a good criterion. This is illustrated in Example 10.7. To overcome possible difficulties, Harter [1] investigated the criteria of minimizing, for two-sided confidence intervals,

$$E_\theta\{|\underline{\theta}(X) - \theta| + |\bar{\theta}(X) - \theta|\},$$

or

$$E_\theta\{(\underline{\theta}(X) - \theta)^2 + (\bar{\theta}(X) - \theta)^2\}.$$

As in Section 10.1, assume that the parameter space Θ is an interval in the real line. Let S_γ be a family of γ confidence sets for θ. We introduce a measure $m(\cdot)$ of the confidence sets $S(X)$ of S_γ. Thus if $\{S(X); X \in \mathfrak{X}\}$ is a family of finite intervals, that is, $S(X) = (\underline{\theta}(X), \bar{\theta}(X))$, $m(\cdot)$ could be the Lebesgue linear measure $m(S(X)) = \bar{\theta}(X) - \underline{\theta}(X)$. In this case $m(\mathrm{d}\theta) = \mathrm{d}\theta$, and $m(S(X))$ is the (random) length of $S(X)$. In cases where $S(X)$ is a one-sided confidence interval of an infinite length, for example, $S(X) = (\underline{\theta}(X), \infty)$, $m(\cdot)$ could be finite (like a probability measure).

Lemma 10.3.1. (Pratt [2].) *If $m(\cdot)$ is a nonatomic measure on $(\Theta, \mathcal{C})$ and S is a family of $\mathcal{C}$-measurable γ confidence sets in Θ, then*

$$\int m(S(x)) f(x; \theta)\mu(\mathrm{d}x) = \int_{\{\theta' \neq \theta\}} P_\theta[\theta' \in S(X)] m(\mathrm{d}\theta'), \tag{10.3.1}$$

for all $\theta \in \Theta$.

Remark.

(i) The left-hand side of (10.3.1) is the expected measure of $S(X)$, under F_θ. The right-hand side of (10.3.1) is the expected coverage probability of false parameter points under the measure $m(\cdot)$.

(ii) If $m(\mathrm{d}\theta)$ is a prior probability measure on $(\Theta, \mathcal{C})$ then the right-hand side of (10.3.1) is the prior expectation of coverage probabilities of false parameter points.

(iii) The field $\mathcal{C}$ is the Borel sigma-field generated by Θ.

Proof. Since $m(S(x)) \geqslant 0$ we have

(10.3.2)

$$\begin{aligned}\int m(S(x)) f(x; \theta)\mu(\mathrm{d}x) &= \int F_\theta(\mathrm{d}x) \int I_{S(x)}(\theta') m(\mathrm{d}\theta') \\ &= \int m(\mathrm{d}\theta') \int I_{S(x)}(\theta') F_\theta(\mathrm{d}x) = \int m(\mathrm{d}\theta') P_\theta[\theta' \in S(x)].\end{aligned}$$

Finally, since $m(\mathrm{d}\theta')$ is nonatomic,

$$(10.3.3)\qquad \int m(\mathrm{d}\theta')P_\theta[\theta' \in S(x)] = \int_{\{\theta \neq \theta'\}} m(\mathrm{d}\theta')P_\theta[\theta' \in S(x)],$$

all $\theta \in \Theta$. (Q.E.D.)

From Lemma 10.3.1 we deduce that if S_γ is a family of U.M.A.U. γ confidence intervals and $m(\mathrm{d}\theta) = \mathrm{d}\theta$, the expected length of $S(X) = (\underline{\theta}(X), \bar{\theta}(X))$ is minimal. Indeed, $m(S(x)) = \bar{\theta}(x) - \underline{\theta}(x)$, and the left-hand side of (10.3.1) is the expected length, under F_θ, of $S(X)$. Since S_γ is U.M.A.U., $P_\theta[\theta' \in S(X)]$ is minimal, with respect to all families of unbiased γ confidence intervals uniformly in $\theta \neq \theta'$. Thus, for example, if $X \sim N(\theta, 1)$, $\{(X - \Phi^{-1}(1 - (\alpha/2)), X + \Phi^{-1}(1 - (\alpha/2)); X \in \mathfrak{R}\}$ is a U.M.A.U. family of $(1 - \alpha)$ confidence intervals for θ. The length of these confidence intervals is fixed at $2\Phi^{-1}(1 - (\alpha/2))$. This is the minimal expected length of two-sided confidence intervals among the families of unbiased ones. As will be illustrated later, the expected length of the confidence intervals can be locally reduced by applying the following method.

Suppose that $H(\theta)$ is a (prior) distribution function on Θ, and let $f_H(x)$ be the marginal (mixed) density of X under H; that is,

$$f_H(x) = \int f(x; \theta)H(\mathrm{d}\theta).$$

Consider the family $\mathcal{A}_\alpha^{(H)}$ of acceptance regions of the hypothesis $A_1(\theta_1)$: $\theta = \theta_1$ versus the alternative hypothesis $A_2(H)$. This alternative hypothesis specifies that the observed random variable has the density $f_H(x)$. For each value of θ_1 there exists (by the Neyman-Pearson fundamental lemma) a most powerful test of $A_1(\theta_1)$ versus $A_2(H)$, with a test function.

$$(10.3.4)\qquad \phi^{(H,\theta_1)}(X) = \begin{cases} 1, & \text{if } \dfrac{f_H(X)}{f(X; \theta_1)} > c_\alpha(\theta_1), \\ \gamma_\alpha(\theta_1), & \text{if } \dfrac{f_H(X)}{f(X; \theta_1)} = c_\alpha(\theta_1), \\ 0, & \text{otherwise.} \end{cases}$$

The quantity $c_\alpha(\theta_1)$ is determined so that the test function satisfies $E_\theta\{\phi^{(H,\theta_1)}(X)\} = \alpha$; $0 < \alpha < 1$. The corresponding acceptance regions of $\mathcal{A}_\alpha^{(H)}$ are

$$(10.3.5)\qquad A(\theta) = \{X; f_H(X) < c_\alpha f(X; \theta)\} \cup \{(X, R); f_H(X) \leqslant c_\alpha f(X; \theta) \quad \textit{and} \quad R \leqslant 1 - \gamma_\alpha(\theta)\},$$

where R is independent of X, having a rectangular distribution on $[0, 1]$. The corresponding family of confidence sets $S_\gamma^{(H)}$ has the property that its confidence sets have minimal expected $m(\cdot)$ measure, under H; that is,

$$E_H\{m(S(X))\} = \int H(d\theta) \int F_\theta(dx) m(S(x))$$

is minimal. This result can be used to obtain locally minimal expected length γ confidence intervals, by considering a d.f. $H(\theta)$ which is concentrated at one point. This is exhibited in the following example, taken from Pratt [2].

Example 10.6. Let $X \sim N(\theta, 1)$, where $-\infty < \theta < \infty$. We have remarked that $X \pm \Phi^{-1}(1 - (\alpha/2))$, where $\gamma = 1 - \alpha$, are upper and lower γ confidence limits for θ, which constitute a U.M.A.U. family. Consider the family of tests, at level α, of the hypotheses $A_1(\theta_1)$: $\theta = \theta_1$ versus A_2: $\theta = 0$. Here $H(\theta)$ is a distribution function concentrated on $\{\theta = 0\}$. The family of acceptance intervals for $A_1(\theta)$, $-\infty < \theta < \infty$, is

$$(10.3.6)\qquad A_1(\theta) = \begin{cases} & X \geqslant \theta - \Phi^{-1}(1-\alpha), \quad \text{if} \quad \theta > 0, \\ X; & X \leqslant \theta + \Phi^{-1}(1-\alpha), \quad \text{if} \quad \theta < 0, \\ & \text{arbitrary}, \qquad\qquad\quad \text{if} \quad \theta = 0, \end{cases}$$

where $\Phi^{-1}(1 - \alpha)$ is the $(1 - \alpha)$-fractile of the standard normal d.f., $\Phi(\cdot)$. It is easy to check that the corresponding γ confidence intervals are

$$(10.3.7)\quad S(X) = [\min\{0, X - \Phi^{-1}(1-\alpha)\}, \max\{0, X + \Phi^{-1}(1-\alpha)\}].$$

This family of confidence intervals has the minimal expected length at $\theta = 0$. We readily find that the expected length of $S(X)$ is

$$(10.3.8)\quad \Lambda(\theta) = (\Phi^{-1}(1-\alpha) + \theta)\Phi(\Phi^{-1}(1-\alpha) + \theta) + (\Phi^{-1}(1-\alpha) - \theta) \\ \times \Phi(\Phi^{-1}(1-\alpha) - \theta) + \varphi(\Phi^{-1}(1-\alpha) + \theta) + \varphi(\Phi^{-1}(1-\alpha) - \theta).$$

The function $\Lambda(\theta)$ is, as intuitively expected, an even function of θ. At $\theta = 0$ we have

$$(10.3.9)\qquad \Lambda(0) = 2[(1-\alpha)\Phi^{-1}(1-\alpha) + \varphi(\Phi^{-1}(1-\alpha))];$$

$\Lambda(0)$ is smaller than $2\Phi^{-1}(1 - (\alpha/2))$, which is the length of the U.M.A.U. confidence interval. On the other hand, as $|\theta| \to \infty$, $\Lambda(\theta) \to \infty$, whereas the U.M.A.U. confidence interval has a fixed length. Pratt [2] provided examples of this kind for the parameter θ of binomial distributions also. In [3] we find a derivation of the shortest expected-length confidence intervals, when $H(\theta)$ is a normal distribution. ■

Expression 10.3.1 holds for families of two-sided confidence intervals based on corresponding families of tests of $A_1(\theta_1)$: $\theta = \theta_1$ versus $A_2(\theta_1)$: $\theta \neq \theta_1$ and a nonatomic measure $m(\cdot)$. The implication that a family of most powerful tests of $A_1(\theta_1)$ versus $A_2(\theta_1)$ induces a family of confidence intervals with minimal expected length holds only if $A_1(\theta_1)$ and $A_2(\theta_1)$ are two-sided tests, as above, and the corresponding confidence intervals are two-sided. In the one-sided case we construct most powerful tests of $A_1(\theta_1)$: $\theta \leqslant \theta_1$[resp $\theta \geqslant \theta_1$] versus $A_2(\theta_1)$: $\theta > \theta_1$[resp $\theta < \theta_1$] to obtain the most accurate lower (resp upper) confidence limits. The expected value of $m(\cdot)$ can be expressed in the form

$$\text{(10.3.10)} \quad \int m(S(x))f(x;\theta)\mu(dx) = \int_{\{\theta'>\theta\}} P_\theta[\theta' \in S(X)]m(d\theta') + \int_{\{\theta'<\theta\}} P_\theta[\theta' \in S(X)]m(d\theta').$$

We see that in the U.M.A. lower (resp upper) interval case, the first (resp second) term on the right-hand side of (10.3.10) is minimized while the second (resp first) term may not be. Accordingly, in the one-sided case, uniformly most accurate lower or upper confidence intervals may not yield confidence intervals with the shortest expected length. We now provide an example of such a case. This example was published by Madansky [1].

Example 10.7. Let $X_1, X_2, \ldots, X_n$ be i.i.d. random variables distributed like $G(1/\theta, 1)$, $0 < \theta < \infty$. As shown in Example 10.1, if $T_n = \sum_{i=1}^n X_i$, then $[0, 2T_n/\chi^2_{1-\gamma}[2n]]$ is a U.M.A. γ upper confidence interval for θ. The expected length of this U.M.A. interval is $\theta/(\chi^2_{1-\gamma}[2n]/2n)$. We exhibit another γ confidence interval with a smaller expected length, for certain values of n and $\alpha = (1 - \gamma)$.

Since the distribution of X is negative exponential with mean θ, the distribution of $\exp\{-(1/\theta)X\}$ is uniform on [0, 1]. Let $X_{(1)} \leqslant \cdots \leqslant X_{(n)}$ be the order statistic. The distribution of $\exp\{-(1/\theta)X_{(i)}\}$ is like that of the $(n - i + 1)$-th order statistic from a uniform distribution on [0, 1]. It is easy to verify that

$$\exp\left\{-\frac{1}{\theta}X_{(i)}\right\} \sim \text{beta}\,(n - i + 1, i).$$

Hence for every ξ, $0 < \xi < 1$,

$$\text{(10.3.11)} \qquad P\left\{\exp\left\{-\frac{1}{\theta}X_{(i)}\right\} \leqslant \xi\right\} = I_\xi(n - i + 1, i),$$

where $I_\xi(p, q)$ designates the incomplete beta function ratio. Equation 10.3.11 is also equivalent to

$$\text{(10.3.12)} \qquad P\left\{-\frac{X_{(i)}}{\ln \xi} \geqslant \theta\right\} = I_\xi(n - i + 1, i).$$

Setting the right-hand side of (10.3.12) equal to γ, we obtain that $[0, -(X_{(i)}/\ln \xi)]$ is a γ upper confidence interval for θ, where i and ξ are chosen so that

$$I_\xi(n - i + 1, i) = \gamma. \tag{10.3.13}$$

The expected value of $X_{(i)}$ is

$$E_\theta\{X_{(i)}\} = \theta \sum_{j=1}^{i} \frac{1}{n - j + 1}. \tag{10.3.14}$$

Hence the expected length of the γ confidence interval $\tilde{S}(\mathbf{X}) = [0, (X_{(i)}/-\ln \xi)]$ is $(\theta/-\ln \xi) \sum_{j=1}^{i} (n - j + 1)^{-1}$. Madansky [1] has shown that for $\gamma = 0.3$, $\xi = 0.99$, and $n = 120$, the expected length of $\tilde{S}(\mathbf{X})$ is 0.829θ. On the other hand, the expected length of the U.M.A. γ confidence interval is 0.956θ. ■

We have discussed the reason for the phenomenon illustrated in the above example. Lehmann [2], p. 182, argues that for such reasons the criterion of minimizing the expected length of one-sided confidence intervals is not adequate. In the above example, $P_\theta[\theta' \in S(X)] \leqslant P_\theta[\theta' \in \tilde{S}(X)]$ for all $\theta' > \theta$ and all θ, where $S(X)$ is the U.M.A. upper confidence interval. Hence according to (10.3.10),

$$\int_0^\theta P_\theta[\theta' \in S(X)]\, d\theta' > \int_0^\theta P_\theta[\theta' \in \tilde{S}(X)]\, d\theta', \qquad \text{all } \theta. \tag{10.3.15}$$

Let $\bar{\theta}(X)$ and $\tilde{\theta}(X)$ denote the upper confidence limits of $S(X)$ and $\tilde{S}(X)$, respectively. The probability of covering a false value by a γ confidence interval $[0, d(X)]$ is

$$\begin{aligned} P_\theta[\theta' \leqslant d(X)] &= \gamma P_\theta[\theta' \leqslant d(X) \mid d(X) \geqslant \theta] \\ &\quad + (1 - \gamma) P_\theta[\theta' \leqslant d(X) \mid d(X) < \theta]. \end{aligned} \tag{10.3.16}$$

Hence from (10.3.15) and (10.3.16) we obtain

(10.3.17)

$$\int_0^\theta P_\theta[\theta' \leqslant \bar{\theta}(X) \mid \bar{\theta}(X) < \theta]\, d\theta' > \int_0^\theta P_\theta[\theta' \leqslant \tilde{\theta}(X) \mid \tilde{\theta}(X) < \theta]\, d\theta',$$

for all θ. The last inequality means that if a value θ' is chosen at random from the interval $(0, \theta)$, the conditional probability that it will be covered by the U.M.A. confidence interval $S(X)$, given $\{\bar{\theta}(X) < \theta\}$, is greater than the conditional probability that $\tilde{S}(X)$ will cover θ', given $\{\tilde{\theta}(X) < \theta\}$. From this point of view the family of U.M.A. upper confidence intervals is superior to the family $\tilde{S}_\gamma(X)$, even if its expected-intervals length could sometimes be greater than that of $\tilde{S}_\gamma(X)$. Madansky [1] discusses this point with the notions of "expected excess," $E_\theta\{d(X) - \theta \mid d(X) \geqslant \theta\}$, and "expected shortage," $E_\theta\{\theta - d(X) \mid d(X) \leqslant \theta\}$, of one-sided confidence intervals. The criteria of Harter, mentioned at the beginning of the section, are designed to overcome these difficulties. Criterion (i) in the one-sided upper interval case, $S(X) = (-\infty, \bar{\theta}(X)]$, is to minimize

$$(10.3.18)\quad \begin{aligned} E_\theta\{|\bar{\theta}(X) - \theta|\} &= P_\theta\{\bar{\theta}(X) \geqslant \theta\}E\{\bar{\theta}(X) - \theta \mid \bar{\theta}(X) \geqslant \theta\} \\ &\quad + P_\theta\{\bar{\theta}(X) < \theta\}E\{\theta - \bar{\theta}(X) \mid \bar{\theta}(X) < \theta\} \\ &= \gamma E_\theta\{\bar{\theta}(X) - \theta \mid \bar{\theta}(X) \geqslant \theta\} \\ &\quad + (1-\gamma)E_\theta\{\theta - \bar{\theta}(X) \mid \bar{\theta}(X) < \theta\}. \end{aligned}$$

This is a convex combination of the expected excess and the expected shortage of an upper confidence interval.

10.4. TOLERANCE INTERVALS

Let X be an observed random variable having a distribution function F_θ, which belongs to a family $\mathfrak{F}$. In this section we consider parametric families $\mathfrak{F} = \{F_\theta;\, \theta \in \Theta\}$ where Θ is an interval in a Euclidean k-space. The objective is to determine, on the basis of the observed sample value of X, a β-content interval of the unknown distribution F_θ with a confidence level γ; $0 < \beta, \gamma < 1$. In other words, let $F^{-1}(\xi;\theta)$ be a ξ-fractile of the distribution function F_θ. We have, for every $0 \leqslant \xi_1 < \xi_2 \leqslant 1$,

$$(10.4.1)\quad P_\theta\{F^{-1}(\xi_1;\theta) \leqslant X \leqslant F^{-1}(\xi_2;\theta)\} \geqslant \xi_2 - \xi_1, \quad \text{all} \quad \theta \in \Theta.$$

If $\beta = \xi_2 - \xi_1$ we say that $[F^{-1}(\xi_1;\theta), F^{-1}(\xi_2;\theta)]$ is a β-content interval of F_θ. Let Y be a random variable independent of X and with the same d.f. F_θ. We wish to determine an interval $[\underline{L}_{\beta,\gamma}(X), \bar{L}_{\beta,\gamma}(X)]$ such that

$$(10.4.2)\qquad P_\theta\{P_\theta\{\underline{L}_{\beta,\gamma}(X) \leqslant Y \leqslant \bar{L}_{\beta,\gamma}(X) \mid X\} \geqslant \beta\} \geqslant \gamma,$$

for all θ.

If (10.4.2) is satisfied for all θ,

$$(10.4.3)\qquad P_\theta\left\{\int_{\underline{L}_{\beta,\gamma}(X)}^{\bar{L}_{\beta,\gamma}(X)} dF_\theta(y) \geqslant \beta\right\} \geqslant \gamma, \qquad \text{all } \beta.$$

To assure (10.4.2) we can determine $\xi_1 = \frac{1}{2}(1-\beta)$ and $\xi_2 = \frac{1}{2}(1+\beta)$ so that

$$(10.4.4)\quad P_\theta\{\underline{L}_{\beta,\gamma}(X) \leqslant F^{-1}(\xi_1;\theta);\, \bar{L}_{\beta,\gamma}(X) \geqslant F^{-1}(\xi_2;\theta)\} \geqslant \gamma, \quad \text{all} \quad \theta \in \Theta.$$

We can interpret (10.4.4) to mean that with confidence probability not smaller than γ, we can predict that at least a β-proportion of future observations from F_θ will fall in the interval $[\underline{L}_{\beta,\gamma}(X), \bar{L}_{\beta,\gamma}(X)]$. This interval is called a (β, γ) *two-sided tolerance interval.* If we set $\xi_1 = 0$ and $\xi_2 = \beta$, then $F^{-1}(0, \theta) \equiv -\infty$ and $\underline{L}_{\beta,\gamma}(X) \equiv -\infty$. The interval $(-\infty, \bar{L}_{\beta,\gamma}(X))$ is called a (β, γ) *upper tolerance interval.* In the case of upper or lower (β, γ) tolerance intervals the problem is to determine a γ lower or γ upper confidence limit for the β-fractile, $F^{-1}(\beta; \theta)$, of the d.f. F_θ. In this section we show how the theory of confidence intervals can be used to determine tolerance intervals. We start with a couple of examples and then discuss the criterion of *uniformly most accurate* (β, γ) tolerance intervals, for M.L.R. families of density functions. This will be followed by some remarks on β *expectation tolerance intervals.* These are intervals $(\underline{E}_\beta(X), \bar{E}_\beta(X))$ such that if X, Y are i.i.d., then

$$\text{(10.4.5)} \quad E_\theta\{P_\theta\{\underline{E}_\beta(X) \leqslant Y \leqslant \bar{E}_\beta(X) \mid X\}\} = P_\theta\{\underline{E}_\beta(X) \leqslant Y \leqslant \bar{E}_\beta(X)\} \geqslant \beta.$$

Example 10.9. Let $X_1, X_2, \ldots, X_n$ be i.i.d. having a normal distribution $\mathcal{N}(\mu, \sigma^2)$. Both μ and σ are unknown; $-\infty < \mu < \infty$, $0 < \sigma < \infty$. A minimal sufficient statistic is $(\bar{X}, Q)$, where $\bar{X} = (1/n) \sum_{i=1}^n X_i$, $Q = \sum_{i=1}^n (X_i - \bar{X})^2$. We now establish an upper (β, γ) tolerance interval. We are seeking a statistic $\bar{L}_{\beta,\gamma}(\bar{X}, Q)$ such that

$$\text{(10.4.6)} \qquad P_{\mu,\sigma}\{\bar{L}_{\beta,\gamma}(\bar{X}, Q) \geqslant \mu + u_\beta\sigma\} \geqslant \gamma, \quad \text{all} \quad (\mu, \sigma),$$

where $u_\beta = \Phi^{-1}(\beta)$ is the β-fractile of $\mathcal{N}(0, 1)$. Let $\bar{L}_{\beta,\gamma}(\bar{X}, Q) = \bar{X} + \lambda_{\beta,\gamma} Q^{1/2}$; then

$$\bar{L}_{\beta,\gamma}(\bar{X}, Q) \geqslant \mu + u_\beta\sigma \Leftrightarrow \bar{X} - \mu - u_\beta\sigma \geqslant -\lambda_{\beta,\gamma} Q^{1/2}$$

Furthermore, $\bar{X} - \mu - u_\beta\sigma \sim N(-u_\beta\sigma, \sigma^2/n)$; $Q^{1/2} \sim \sigma\sqrt{(\chi^2[n-1])}$; $\bar{X}$ and Q are independent. Hence

$$\text{(10.4.7)} \quad P_{\mu,\sigma}\{\bar{X} + \lambda_{\beta,\gamma} Q^{1/2} \geqslant \mu + u_\beta\sigma\} = P_{\mu,\sigma}\left\{\frac{N(0, 1) + \sqrt{n}\, u_\beta}{(\chi^2[n-1]/(n-1))^{1/2}} \leqslant \lambda_{\beta,\gamma}\sqrt{(n(n-1))}\right\}.$$

Furthermore,

$$\frac{N(0, 1) + \delta}{(\chi^2[\nu]/\nu)^{1/2}} \sim t[\nu; \delta]$$

where $t[\nu; \delta]$ designates a noncentral t with ν degrees of freedom and parameter of noncentrality δ. Hence from (10.4.7), setting $\lambda_{\beta,\gamma}\sqrt{(n(n-1))} = t_\gamma[n-1; \sqrt{n}\, u_\beta]$, we have

$$\text{(10.4.8)} \qquad \bar{L}_{\beta,\gamma}(\bar{X}, Q) = \bar{X} + t_\gamma[n-1; \sqrt{n}\, u_\beta] \frac{\hat{\sigma}}{\sqrt{n}},$$

where $\hat{\sigma}^2 = Q/(n-1)$ is the sample variance.

Values of $t_\gamma[n-1; \sqrt{n}\, u_\beta]$ for certain values of (β, γ) can be compiled from the tables of Resnikoff and Lieberman [1]. We now derive the β-expectation upper tolerance limit. According to (10.4.5), the β-expectation upper tolerance limit, $\bar{E}_\beta(\bar{X}, Q)$, should satisfy

$$P_{\mu,\sigma}\{Y \leqslant \bar{E}_\beta(\bar{X}, Q)\} \geqslant \beta, \quad \text{all} \quad (\mu, \sigma). \tag{10.4.9}$$

Set $\bar{E}_\beta(\bar{X}, Q) = \bar{X} + a_\beta Q^{1/2}$; then since Y is independent of $(\bar{X}, Q)$,

$$\frac{Y - \bar{X}}{Q} \sim \left(1 + \frac{1}{n}\right)^{1/2} \frac{N(0,1)}{\sqrt{(\chi^2[n-1])}} \sim \frac{1}{\sqrt{(n-1)}}\left(1 + \frac{1}{n}\right)^{1/2} t[n-1], \tag{10.4.10}$$

where $t[\nu]$ is the central t-statistic with ν degrees of freedom. Hence setting

$$a_\beta = \frac{1}{\sqrt{(n-1)}}\left(1 + \frac{1}{n}\right)^{1/2} t_\beta[n-1], \tag{10.4.11}$$

we obtain

$$P_{\mu,\sigma}\left\{Y \leqslant \bar{X} + t_\beta[n-1]\hat{\sigma}\left(1 + \frac{1}{n}\right)^{1/2}\right\} \geqslant \beta, \qquad \text{all } (\mu, \sigma). \tag{10.4.12}$$

Hence

$$\bar{E}_\beta(\bar{X}, Q) = \bar{X} + t_\beta[n-1]\hat{\sigma}\left(1 + \frac{1}{n}\right)^{1/2}. \tag{10.4.13}$$

It is easy to verify that

$$\lim_{n\to\infty} \bar{E}_\beta(\bar{X}, Q) = \mu + u_\beta \sigma \quad \text{a.s.} \tag{10.4.14}$$

■

Example 10.10 In Example 10.2 we derived U.M.A. γ lower confidence limits for θ in the binomial case $\mathcal{B}(N, \theta)$, $0 < \theta < 1$, N known. We derive here a (β, γ) upper tolerance limit for this distribution. We prove later that this upper tolerance limit is U.M.A., in a sense to be defined. This example is taken from the paper of Zacks [10].

As in Example 10.2, let $X_1, X_2, \ldots, X_n$ be i.i.d. random variables having a common distribution $\mathcal{B}(N, \theta)$. Let $T_n = \sum_{i=1}^n X_i$, $T_n \sim \mathcal{B}(nN, \theta)$. In analogy to Example 10.2 we can show that a γ upper confidence limit for θ is

(10.4.15) $\bar{\theta}_\gamma(T_n + R) =$ root of the equation:

$$(1 - R)I_{1-\theta}(Nn - T_n + 1, T_n) + RI_{1-\theta}(Nn - T_n, T_n + 1) = 1 - \gamma,$$

if $T_n = 1, \ldots, Nn - 1$, and $\bar{\theta}_\gamma(R) = \theta_0$, $\bar{\theta}_\gamma(Nn - 1 + R) = 1$, where R is an independent random variable having a rectangular distribution on (0, 1); $\theta_0 = 1 - (1 - \gamma)^{1/Nn}$. Now, since the family of binomial distributions

$\mathcal{B}(N, \theta)$ is an M.L.R., the β-fractile of $\mathcal{B}(N, \theta)$, $B^{-1}(\beta; N, \theta)$, is a non-decreasing function of θ for each β, $0 < \beta < 1$ (see Lehmann [2], p. 74). That is,

$$B^{-1}(\beta; N, \theta) \leqslant B^{-1}(\beta; N, \theta'), \quad \text{all} \quad \theta < \theta'. \tag{10.4.16}$$

Hence $\theta \leqslant \bar{\theta}_\gamma(T_n + R)$ implies that $B^{-1}(\beta; N, \theta) \leqslant B^{-1}(\beta; N, \bar{\theta}(T_n + R))$.

Therefore,

$$\begin{aligned} P_\theta\{B^{-1}(\beta; N, \theta) &\leqslant B^{-1}(\beta; N, \bar{\theta}_\gamma(T_n + R))\} \\ &\geqslant P_\theta\{\theta \leqslant \bar{\theta}_\gamma(T_n + R)\} = \gamma, \quad \text{all} \quad \theta \in (0, 1). \end{aligned} \tag{10.4.17}$$

This means that

$$\bar{L}_{\beta,\gamma}(T_n + R) = B^{-1}(\beta; N, \bar{\theta}_\gamma(T_n + R)) \tag{10.4.18}$$

is a (β, γ) upper tolerance limit for $\mathcal{B}(N, \theta)$. We remark that $B^{-1}(\beta; N, \theta)$ is obtained from (10.1.25) by replacing Nn by N. The upper tolerance limit $\bar{L}_{\beta,\gamma}(T_n + R)$ is a randomized one. Numerical examples indicate that in many cases it is not sensitive to R (see Zacks [10]). ■

The concept of uniformly most accurate confidence intervals can be extended to tolerance intervals in the following manner.

Let $\bar{L}_{\beta,\gamma}(X)$ be a (β, γ) upper tolerance limit for a 1-parameter family of distribution functions, $\mathcal{F} = \{F_\theta; \theta \in \Theta\}$; $\bar{L}_{\beta,\gamma}(X)$ is called *uniformly most accurate* (U.M.A.) if for every other (β, γ) upper tolerance limit $L^*_{\beta,\gamma}(X)$ and each $\theta' > \theta$ such that $F^{-1}(\beta; \theta') > F^{-1}(\beta; \theta)$,

$$P_\theta\{\bar{L}_{\beta,\gamma}(X) \geqslant F^{-1}(\beta; \theta')\} \leqslant P_\theta\{L^*_{\beta,\gamma}(X) \geqslant F^{-1}(\beta; \theta')\}, \tag{10.4.19}$$

for all $\theta \in \Theta$ and each β, $0 < \beta < 1$. In a similar manner we define U.M.A. lower tolerance limits.

Theorem 10.4.1. (Zacks [10].) *If $\mathcal{F}$ is an M.L.R. family of 1-parameter distribution functions and $K_\gamma(X)$ is a U.M.A. γ upper [resp lower] confidence limit for θ, then $\bar{L}_{\beta,\gamma}(X) = F^{-1}(\beta; K_\gamma(X))$ is a U.M.A. (β, γ) upper [resp lower] tolerance limit.*

Proof. We provide here the proof of the theorem for upper tolerance limits and the discrete case. The proof for lower tolerance limits is similar. In the absolutely continuous case the proof is somewhat simpler and is left to the reader. We further assume, without loss of generality, that the discrete random variable X attains only non-negative integers. An extension of the present proof to the general case is immediate.

Let X be the observed random and R an independent random variable having a rectangular distribution over $[0, 1]$; $X + R$ has an absolutely continuous d.f. Let $K_\gamma(X + R)$ be the U.M.A. γ upper confidence limit for θ. We prove that $L_{\beta,\gamma}(X + R) = F^{-1}(\beta; K_\gamma(X + R))$ is a U.M.A. (β, γ) upper tolerance limit. Suppose there exists a (β, γ) upper tolerance limit $L^*_{\beta,\gamma}(X + R)$ such that, for some $\theta' < \theta''$, $F^{-1}(\beta; \theta') < F^{-1}(\beta; \theta'')$ and

$$P_{\theta'}\{L^*_{\beta,\gamma}(X + R) \geqslant F^{-1}(\beta; \theta'')\} < P_{\theta'}\{L_{\beta,\gamma}(X + R) \geqslant F^{-1}(\beta; \theta'')\} \tag{10.4.20}$$

For each $j = 0, 1, \ldots$ define

$$\theta_j = \sup \{\theta; F^{-1}(\beta; \theta) = j\}. \tag{10.4.21}$$

Since $\mathfrak{F}$ is an M.L.R. family, $\theta_j < \theta_{j+1}$ for all $j = 0, 1, \ldots$. Suppose that $\theta'' \in (\theta_v, \theta_{v+1}]$ for some $v \geqslant 1$. Since $F^{-1}(\beta; \theta') < F^{-1}(\beta; \theta'')$, $\theta' \in (\theta_r, \theta_{r+1}]$ for some $r < v$. According to the definition of $L_{\beta,\gamma}(X + R)$,

$$L_{\beta,\gamma}(X + R) \geqslant F^{-1}(\beta; \theta'') \Leftrightarrow K_\gamma(X + R) > \theta_v. \tag{10.4.22}$$

Thus

$$P_{\theta'}\{L_{\beta,\gamma}(X + R) \geqslant F^{-1}(\beta; \theta'')\} = P_{\theta'}\{K_\gamma(X + R) \geqslant \theta_v\}. \tag{10.4.23}$$

We have written $K_\gamma(X + R) \geqslant \theta_v$ on the right-hand side of (10.4.23) since the distribution of $X + R$ is absolutely continuous.

Define the statistic,

$$\theta^*_{\beta,\gamma}(X + R) = \sup \{\theta; L^*_{\beta,\gamma}(X + R) = F^{-1}(\beta; \theta)\}. \tag{10.4.24}$$

Then for all $\theta \in (\theta_v, \theta_{v+1}]$,

$$\theta^*_{\beta,\gamma}(X + R) \geqslant \theta_{v+1} \Leftrightarrow L^*_{\beta,\gamma}(X + R) \geqslant F^{-1}(\beta; \theta). \tag{10.4.25}$$

Furthermore,

$$\begin{aligned} \gamma \leqslant P_\theta\{L^*_{\beta,\gamma}(X + R) \geqslant F^{-1}(\beta; \theta)\} &= P_\theta\{\theta^*_{\beta,\gamma}(X + R) \geqslant \theta_{v+1}\} \\ &= P_\theta\{\theta^*_{\beta,\gamma}(X + R) \geqslant \theta\}, \quad \text{all } \theta, \quad \theta \in (\theta_v, \theta_{v+1}]. \end{aligned} \tag{10.4.26}$$

Thus $\theta^*_{\beta,\gamma}(X + R)$ is a γ upper confidence limit for θ. Finally, according to (10.4.25),

$$P_{\theta'}\{L^*_{\beta,\gamma}(X + R) \geqslant F^{-1}(\beta; \theta'')\} = P_{\theta'}\{\theta^*_{\beta,\gamma}(X + R) \geqslant \theta_v + \epsilon\}, \tag{10.4.27}$$

for all $0 < \epsilon \leqslant \theta_{v+1} - \theta_v$. Hence from (10.4.20), (10.4.23), and (10.4.27),

$$P_{\theta'}\{\theta^*_{\beta,\gamma}(X + R) \geqslant \theta_v + \epsilon\} < P_{\theta'}\{K_\gamma(X + R) \geqslant \theta_v + \epsilon\}, \tag{10.4.28}$$

for some ϵ, $\epsilon > 0$ sufficiently small; $\theta_v + \epsilon > \theta'$ since $\theta' \leqslant \theta_v$. This contradicts the assumption that $K_\gamma(X + R)$ is a U.M.A. γ upper confidence limit.

(Q.E.D.)

The (β, γ) upper tolerance limit of Example 10.10 is U.M.A. In Example 10.7 we have shown that $-X_{(i)}/\ln \beta$ is also a γ upper confidence limit for the mean, θ, of $\mathcal{G}(1/\theta, 1)$, where

$$i = \text{least } j, \quad j = 1, \ldots, n, \quad \text{such that} \quad I_\beta(n - j + 1, j) \geqslant \gamma.$$

From this we immediately obtain that $X_{(i)}$ is a (β, γ) lower tolerance limit for $\mathcal{G}(1/\theta, 1)$. Indeed,

$$P_\theta\left[\frac{1}{\theta}\int_{X_{(i)}}^{\infty} e^{-y/\theta}\,\mathrm{d}y = \beta\right] \geqslant \gamma.$$

This tolerance limit is not, however, U.M.A. The U.M.A. γ lower confidence limit for θ is $2T_n/\chi_\gamma^2[2n]$, where $T_n = \sum_{i=1}^n X_i$. We notice that if $\underline{L}_{\beta,\gamma}(T_n) = (-\ln\beta)2T_n/\chi_\gamma^2[2n]$ then

$$P_\theta\left[\frac{(-\ln\beta)2T_n}{\chi_\gamma^2[2n](-\ln\beta)} \leqslant \theta\right] = P_\theta\left[\exp\left\{-\frac{1}{\theta}\underline{L}_{\beta,\gamma}(T_n)\right\} = \beta\right]$$

$$= P_\theta\left\{\frac{1}{\theta}\int_{\underline{L}_{\beta,\gamma}(T_n)}^{\infty} e^{-y/\theta}\,\mathrm{d}y = \beta\right\} = \gamma.$$

Thus $\underline{L}_{\beta,\gamma}(T_n)$ is a (β, γ) lower tolerance limit for $\mathcal{G}(1/\theta, 1)$. Since $2T_n/\chi_\gamma^2[2n]$ is a U.M.A. lower confidence limit, $\underline{L}_{\beta,\gamma}(T_n)$ is a U.M.A. lower tolerance limit.

Another criterion mentioned in the literature (see Goodman and Madansky [1]) is the one of minimizing the expected length of a (β, γ) tolerance interval. As we have demonstrated and discussed in the previous section, if Θ is an interval on the real line with a finite lower limit, say a, and if $\bar{\theta}_\gamma(X)$ is a U.M.A. γ upper confidence limit for θ, the expected length of the corresponding interval, that is, $E_\theta\{\bar{\theta}_\gamma(X) - a\}$, is not necessarily minimal. If we consider a two-sided U.M.A.U. γ confidence interval for $F^{-1}(\beta; \theta)$ we will obtain, in the case of M.L.R. distributions, a confidence interval of minimal expected length. However, such an interval has no clear interpretation from the point of view of tolerance limits. We are actually looking for an interval $\underline{L}[_{\beta,\gamma}(X), \bar{L}_{\beta,\gamma}(X)]$ of a minimal expected length. We should provide *simultaneously*, therefore, a lower confidence limit for $F^{-1}(\alpha_1; \theta)$ and an upper confidence limit for $F^{-1}(1 - \alpha_2; \theta)$, $0 < \alpha_1 < \alpha_2 < 1$ and $\beta = 1 - \alpha_1 - \alpha_2$, such that the expected distance between these two confidence limits is minimal. This is a different problem.

Goodman and Madansky [1] considered tolerance intervals which are called *most stable*. Most stable (β, γ) tolerance intervals are ones which maximize the probabilities

$$P_\theta\left\{1 - \alpha_1 < \int_{\underline{L}_{\gamma,\beta}(X)}^{\bar{L}_{\beta,\gamma}(X)} \mathrm{d}F_\theta(y) < 1 - \alpha\right\},$$

and

$$P_\theta\left\{1 - \alpha < \int_{\underline{L}_{\beta,\gamma}(X)}^{\bar{L}_{\beta,\gamma}(X)} dF_\theta(y) < 1 - \alpha_2\right\},$$

for all $\theta \in \Theta$ and all $0 \leqslant \alpha_2 < \alpha < \alpha_1 \leqslant 1$. In the one-sided case, say upper tolerance intervals, they obtain most stable limits by considering the U.M.A. confidence intervals for θ. In two-sided cases the situation is significantly more complicated. They introduce further *ad hoc* criteria to obtain some meaningful tolerance limits.

Fraser and Guttman [1] approached the problem of determining an optimal β-expectation tolerance interval from the following point of view. Suppose that $X_1, X_2, \ldots, X_n$ are the observed i.i.d. random variables, and suppose that Y is independent of $X_1, \ldots, X_n$ but identically distributed. Let $T(\mathbf{X}_n) = T_n$ be a (minimal) sufficient statistic for $\mathfrak{F}$. Let $S_\beta(T_n)$ be a β-expectation tolerance interval, that is,

$$P_\theta\{Y \in S_\beta(T_n)\} \geqslant \beta, \quad \text{all} \quad \theta \in \Theta. \tag{10.4.29}$$

Let $I_{S_\beta(T_n)}(y)$ be the indicator function of $S_\beta(T_n)$. Then

$$P_\theta\{Y \in S_\beta(T_n)\} = E_\theta\{E\{I_{S_\beta(T_n)}(Y) \mid T_n\}\}, \quad \text{all} \quad \theta. \tag{10.4.30}$$

We can consider $\phi(Y \mid T_n) = 1 - I_{S_\beta(T_n)}(Y)$ as a test function of the two composite hypotheses,

$$A_1\colon F^\theta_{X_1,X_2,\ldots,X_n,Y}(x_1, \ldots, x_n, y) = \prod_{i=1}^n F_{X_i}{}^\theta(x_i) \cdot F_Y{}^\theta(y) \tag{10.4.31}$$

versus

$$A_2\colon F^\theta_{X_1,\ldots,X_n,Y}(x_1, \ldots, x_n, y) = \prod_{i=1}^n F_{X_i}{}^\theta(x_i) \cdot G_Y{}^\theta(y), \tag{10.4.32}$$

for a properly chosen d.f. $G_Y{}^\theta(y)$, $\theta \in \Theta$.

Thus the problem is to determine, for a properly chosen $G_Y{}^\theta(\cdot)$, in optimum conditional test function, given T_n, at level of significance $\alpha = 1 - \beta$. If $S_\beta(T_n)$ is the acceptance region of a most powerful test of A_1 versus A_2, then $S_\beta(T_n)$ is called a *most powerful* β-expectation tolerance interval. Uniformly most powerful β-expectation tolerance intervals generally do not exist. However, if we restrict attention to uniformly most powerful unbiased or invariant test functions then we call the corresponding β-expectation tolerance intervals uniformly most powerful unbiased or invariant, respectively. Fraser and Guttman [1] prove that if $X_1, X_2, \ldots, X_n$, Y are i.i.d. such as $N(\mu, \sigma^2)$, then the interval $[\bar{X} - t_{1-(\alpha/2)}[n-1]\hat{\sigma}(1 + 1/n)^{1/2}, \ \bar{X} + t_{1-(\alpha/2)}[n-1]\hat{\sigma}(1 + 1/n)^{1/2}]$ is a uniformly most powerful invariant and unbiased $(1 - \alpha)$-expectation tolerance interval, where $\bar{X}$ and

$\hat{\sigma}$ are the sample mean and sample standard deviation based on $\{X_1, \ldots, X_n\}$. In Example 10.9 we have derived a one-sided β-expectation tolerance interval in the normal case. This interval is U.M.P.U. (or invariant). Guttman [1] derived U.M.P.U. (or invariant) β-expectation tolerance intervals for negative exponential distributions and double-exponential distributions.

10.5. DISTRIBUTION FREE CONFIDENCE AND TOLERANCE INTERVALS

In Example 10.7 we have derived γ-lower confidence and tolerance limits for the negative exponential distribution, which are functions of the order statistics. The method of deriving this kind of confidence and tolerance limits is common to all absolutely continuous distribution functions. We therefore call this type of confidence and tolerance limits *distribution free* limits. The theory of distribution free confidence intervals was developed in several studies. Among these we mention Wilks [2; 3]; Fraser [1; 3]; Fraser and Guttman [1]; Kemperman [1]; Epstein [1]; Barlow and Proschan [2]; Tukey [1; 2]; Scheffé and Tukey [1]; Goodman and Madansky [1]; Hanson and Koopmans [1]; and Robbins [1]. An expository introduction to this theory is given by Walsh [1]. We treat the case of absolutely continuous distributions. Thus let $\mathfrak{F}$ be the class of all absolutely continuous distributions. Suppose that $X_1, \ldots, X_n$ are i.i.d. random variables having a common d.f. F, $F \in \mathfrak{F}$. Let $X_{(1)} \leqslant X_{(2)} \leqslant \cdots \leqslant X_{(n)}$ be the order statistic. This statistic is minimal sufficient. The function $F(X)$ has a rectangular distribution on [0, 1]. Let $X_p(F)$ designate the p-the fractile of F; that is,

$$X_p(F) = \inf \{x; F(x) \geqslant p\}, \qquad 0 < p < 1.$$

A (β, γ) upper tolerance limit for $\mathfrak{F}$ is a statistic $\overline{L}_{\beta,\gamma}(X)$ satisfying

$$P_F\{\overline{L}_{\beta,\gamma}(X) \geqslant X_\beta(F)\} \geqslant \gamma, \tag{10.5.1}$$

for all $F \in \mathfrak{F}$. Since $F(X_{(i)}) \sim \beta(i, n - i + 1)$, $i = 1, \ldots, n$, $X_{(i^0)}$ is a (β, γ) upper tolerance limit, where

$$i^0 = \text{least integer } j, \quad j = 1, \ldots, n, \quad \text{such that} \quad I_{1-\beta}(n - j + 1, j) \geqslant \gamma. \tag{10.5.2}$$

Such a distribution free tolerance limit was derived in Example 10.7 for the negative exponential case. In the case of (β, γ) lower tolerance limits we have

$$\underline{L}_{\beta,\gamma}(X) = X_{(i^0)}, \quad \text{where} \quad i^0 = \text{maximal } j, \quad j = 1, \ldots, n, \quad \text{such that} \quad I_{1-\beta}(j, u - j + 1) \geqslant \gamma. \tag{10.5.3}$$

Or

$$(10.5.4)\quad i^0 = \text{maximal } j, \quad j = 1, \ldots, n, \quad \text{such that} \quad I_\beta(n - j + 1, j) \leqslant 1 - \gamma.$$

In the case of two-sided tolerance intervals we construct lower and upper tolerance limits in the following manner. Let $1 \leqslant i < j \leqslant n$ and $X_{(i)}$, $X_{(j)}$ be the corresponding order statistics. We have to determine i, j such that

$$(10.5.5)\qquad P\{F(X_{(j)}) - F(X_{(i)}) \geqslant \beta\} \geqslant \gamma, \quad \text{all} \quad F \in \mathfrak{T}.$$

Let $Y_{(i)} = F(X_{(i)})$; $Y_{(1)} \leqslant \cdots \leqslant Y_{(n)}$ is the order statistic of i.i.d. rectangular (0, 1) random variables. Furthermore, since $X_1, \ldots, X_n$ are i.i.d., the d.f. of $Y_{(j)} - Y_{(i)}$ is like that of $Y_{(j-i)}$, for all i, j; $1 \leqslant i < j \leqslant n$. (See Fraser [5], p. 150–151.) Hence

$$\begin{aligned}(10.5.6)\quad P\{F(X_{(j)}) - F(X_{(i)}) \geqslant \beta\} &= P\{Y_{(j-i)} \geqslant \beta\} \\ &= \frac{1}{B(j - i, n - j + i + 1)} \int_\beta^1 u^{j-i-1}(1 - u)^{n-j+i}\, du \\ &= 1 - I_\beta(j - i, n - j + i + 1) \\ &= \sum_{v=0}^{j-i+1} \binom{n}{v} \beta^v (1 - \beta)^{n-v}.\end{aligned}$$

Thus for specified values of β and γ we can determine from tables of the binomial distribution the smallest sample size n and the least value of $j - i$ so that

$$\sum_{v=0}^{j-i+1} \binom{n}{v} \beta^v (1 - \beta)^{n-v} \geqslant \gamma.$$

Ordinarily $j = n$ and $i = 1$. Walsh [1] discusses the determination of two-sided confidence limits for the median of a distribution when it is assumed, in addition to the assumption of absolute continuity, that the d.f. is symmetric. Tables are also provided.

The above tolerance limits have the disadvantage that (10.5.6) is satisfied only for $n \geqslant n(\beta, \gamma)$. Hanson and Koopmans [1] provided distribution free (β, γ) upper (lower) tolerance intervals for all $n, n \geqslant 2$, when the class $\mathcal{F}$ of distribution functions considered is a subclass of $\mathfrak{T}$, with the property of increasing failure rate. Thus we assume that every distribution function F in $\mathcal{F}$ is absolutely continuous, and the function

$$(10.5.7)\qquad h_F(x) = \frac{f(x)}{1 - F(x)}, \qquad -\infty < x < \infty,$$

where $f(x)$, the corresponding density function, is an increasing function of x. A distribution function F in $\mathfrak{F}$ is called an I.F.R. distribution. The hazard or failure rate of F at x is designated by $h_F(x)$. We notice that $h_F(x) = -(\mathrm{d}/\mathrm{d}x)\ln(1 - F(x))$ almost everywhere. Thus if $h_F(x)$ is an increasing function, $-\ln(1 - F(x))$ is a strictly convex function of x. Hanson and Koopmans have named upper tolerance limits for F in $\mathfrak{F}$ *log-convex* (L.C.) tolerance limits.

Let H be a continuous strictly monotonic function on $[0, 1]$. Let $\mathfrak{F}_H$ be the subclass of all distribution functions in $\mathfrak{F}$ such that the composite function $HF(\cdot)$ is *convex*. Let $X_{(1)} \leqslant \cdots \leqslant X_{(n)}$ be the order statistic of a sample of i.i.d. random variables. The type of upper tolerance limit considered by Hanson and Koopmans is $X_{(n-k-j)} + b(X_{(n-k)} - X_{(n-k-j)})$ such that $1 \leqslant k + j \leqslant n - 1, 0 \leqslant k \leqslant n - 2$.

Lemma 10.5.1. (Hanson and Koopmans [1]). *Let H be a strictly increasing continuous function on* $[0, 1]$, *and let η be a point in the range of H. Let $F \in \mathfrak{F}_H$. Then, for every η and every $b \geqslant 1$,*

$$(10.5.8)\quad P_F\{F(X_{(n-k-j)} + b(X_{(n-k)} - X_{(n-k-j)})) \geqslant H^{-1}(\eta)\} \geqslant P\{bH(U_{(n-k)}) - (b - 1)H(U_{(n-k-j)}) \geqslant \eta\}, \quad \text{all} \quad F \in \mathfrak{F}_H,$$

where $U_{(1)} \leqslant \cdots \leqslant U_{(n)}$ is the order statistic of a sample of n i.i.d. random variables having a rectangular distribution on $[0, 1], 0 \leqslant k < k + j \leqslant n - 1$.

Proof. Since H is strictly increasing,

$$(10.5.9)\quad P_F\{F(X_{(n-k-j)} + b(X_{(n-k)} - X_{(n-k-j)})) \geqslant H^{-1}(\eta)\} = P_F\{HF(X_{(n-k-j)} + b(X_{(n-k)} - X_{(n-k-j)}) \geqslant \eta\}, \quad \text{all} \quad F \in \mathfrak{F}_H.$$

The function $HF(\cdot)$ is strictly convex since $F \in \mathfrak{F}_H$; F is continuous on $(-\infty, \infty)$ and therefore $HF(\cdot)$ is continuous. Since $HF(\cdot)$ is strictly convex,

$$(10.5.10)\quad HF(X_{(n-k-j)} + b(X_{(n-k)} - X_{(n-k-j)})) > HF(X_{(n-k-j)}) + b[HF(X_{(n-k)}) - HF(X_{(n-k-j)})].$$

Furthermore, since $F(X_{(i)}) \sim U_{(i)}$, $i = 1, \ldots, n$, we obtain from (10.5.10) that

$$(10.5.11)\quad HF(X_{(n-k-j)} + b(X_{(n-k)} - X_{(n-k-j)})) > bH(U_{(n-k)}) - (b - 1)H(U_{(n-k-j)}).$$

Finally, from (10.5.11) we obtain, for every η in the range of H,

$$(10.5.12)\quad P\{HF(X_{(n-k-j)} + b(X_{(n-k)} - X_{(n-k-j)})) \geqslant \eta\} \geqslant P\{bH(U_{(n-k)}) - (b - 1)H(U_{(n-k-j)}) \geqslant \eta\}.$$

Equations 10.5.12 and 10.5.9 imply (10.5.8) for all $F \in \mathfrak{F}_H$. (Q.E.D.)

Define, for $0 < \beta < 1$, $b \geqslant 1$,

$$\text{(10.5.13)} \qquad \gamma(b) \equiv \gamma(b; n, k, j, \beta) = P\{U_{(k+1)} \leqslant (1-\beta)^{1/b} U_{(k+j+1)}^{(b-1)/b}\}.$$

Lemma 10.5.2. (Hanson and Koopmans [1].)

$$\text{(10.5.14)} \qquad P_F\{F(X_{(n-k-j)} + b(X_{(n-k)} - X_{(n-k-j)}) \geqslant \beta\} \geqslant \gamma(b),$$

all $F \in \mathcal{F}_H$; $0 \leqslant k < j + k \leqslant n - 1$, $0 < \beta < 1$.

Proof. Let $H(x) = -\ln(1-x)$, $0 < x < 1$. The function $H(x)$ is strictly increasing continuous on $(0, 1)$; $H(U_{(i)}) \sim Y_{(i)}$, $i = 1, \ldots, n$, where $Y_{(1)} \leqslant \cdots \leqslant Y_{(n)}$ is the order statistic of n i.i.d. random variables having a common exponential distribution $G(x) = 1 - e^{-x}$, $x \geqslant 0$. Let $\eta = -\ln(1-\beta)$; $\beta = H^{-1}(\eta)$; $0 < \eta < \infty$. According to (10.5.8),

$$\begin{aligned}\text{(10.5.15)} \quad & P_F\{F(X_{(n-k-j)} + b(X_{(n-k)} - X_{(n-k-j)})) \geqslant \beta\} \\ & \geqslant P\{bY_{(n-k)} - (b-1)Y_{(n-k-j)} \geqslant -\ln(1-\beta)\} \\ & = P\{-b\ln(1-U_{(n-k)}) + (b-1)\ln(1-U_{(n-k-j)}) \geqslant -\ln(1-\beta)\}, \\ & F \in \mathcal{F}_H.\end{aligned}$$

Moreover, $1 - U_{(n-k)} \sim U_{(k+1)}$, $k = 0, \ldots, n-1-j$. Hence we obtain from (10.5.15) that

$$\begin{aligned}\text{(10.5.16)} \quad & P_F\{F(X_{(n-k-j)} + b(X_{(n-k)} - X_{(n-k-j)})) \geqslant \beta\} \\ & \geqslant P\{U_{(k+1)}^b \leqslant (1-\beta)U_{(k+j+1)}^{b-1}\} = \gamma(b), \qquad \text{all } F \in \mathcal{F}_H. \quad (Q.E.D.)\end{aligned}$$

If we set b_γ so that $\gamma(b_\gamma) = \gamma$, then $X_{(n-k-j)} + b(X_{(n-k)} - X_{(n-k-j)})$ is a (β, γ) upper tolerance limit for $\mathcal{F}_H$. Tables of b_γ for $\beta = 0.5, 0.75, 0.90$, and 0.95; $\gamma = 0.90, 0.95$, and 0.99; $n = 2(1)44$ were given by Hanson and Koopmans.

Barlow and Proschan [2] derived another system of distribution free tolerance limits for families of I.F.R. distributions. They restrict attention to the function $H(x) = -\ln(1-x)$. If $G(x)$ is the d.f. of $G(1, 1)$, then $G^{-1}(p) = -\ln(1-p)$, $0 < p < 1$. Hence they restrict attention to the family $\mathcal{F}_{G^{-1}}$ of I.F.R. distributions. These are the distribution functions F such that $G^{-1}F(x)$ is a convex function of x. Furthermore, Barlow and Proschan consider only positive random variables, that is, $F(0) = 0$ for all $F \in \mathcal{F}_{G^{-1}}$.

Let $X_{(1)} \leqslant \cdots \leqslant X_{(n)}$ be the order statistic and consider the "total life" statistic up to the r-th order statistic $X_{(r)}$, that is, $T_{n,r} = \sum_{i=1}^{r} (n-i+1) \times (X_{(i)} - X_{(i-1)})$, where $X_{(0)} \equiv 0$. We have discussed this statistic in Example 2.3 of Section 2.1. For each $i = 1, \ldots, n$, $X_{(i)} - X_{(i-1)}$ is distributed like the minimal value in a sample of $(n-i+1)$ i.i.d. random variables having the common distribution F. The summands of $T_{n,r}$ are independent. In the negative exponential case, $F_\theta(x) = 1 - e^{-x/\theta}$, $x \geqslant 0$, $0 < \theta < \infty$, $(n - i + 1)(X_{(i)} - X_{(i-1)}) \sim \theta G(1, 1)$, and $T_{n,r} \sim (\theta/2)\chi^2[2r]$. Hence

$$(10.5.17) \qquad P_\theta\left\{\frac{2T_{n,r}}{\chi_\gamma^2[2r]} \leqslant \theta\right\} = \gamma \qquad \text{for all } \theta, \qquad 0 < \theta < \infty.$$

Since

$$(10.5.18) \qquad \frac{1}{\theta}\int_{\underline{L}_{\beta,\gamma}(T_{n,r})}^{\infty} \exp\left\{-\frac{y}{\theta}\right\} dy = \exp\left\{-\frac{1}{\theta}\underline{L}_{\beta,\gamma}(T_{n,r})\right\},$$

a (β, γ) lower tolerance limit in the exponential case is therefore

$$(10.5.19) \qquad \underline{L}_{\beta,\gamma}(T_{n,r}) = \frac{2(-\ln \beta)T_{n,r}}{\chi_\gamma^2[2r]}.$$

We derive now a similar form of a (β, γ) lower tolerance limit for any I.F.R. distribution in $\mathcal{F}_{G^{-1}}$. Since $G^{-1}F(\cdot)$ is a convex function,

$$(10.5.20) \qquad G^{-1}F\left(\sum_{i=1}^{r} a_i X_{(i)}\right) \leqslant \sum_{i=1}^{r} a_i G^{-1}F(X_{(i)}),$$

for any $a_i \geqslant 0$, $i = 1, \ldots, r$; $\sum_{i=1}^r a_i \leqslant 1$. We have already shown that $G^{-1}F(X_{(i)}) \sim Y_{(i)}$, $i = 1, \ldots, r$, where $Y_{(1)} \leqslant \cdots \leqslant Y_{(n)}$ is the order statistic of n i.i.d. random variables with the exponential d.f. G. Furthermore,

$$(10.5.21) \qquad \sum_{i=1}^{r} a_i X_{(i)} = \sum_{i=1}^{r} A_i(X_{(i)} - X_{(i-1)}), \qquad X_0 \equiv 0,$$

where $A_i = \sum_{j=i}^r a_j$. Thus from (10.5.20) and (10.5.21) we obtain

$$(10.5.22) \qquad F\left(\sum_{j=1}^{r} A_j(X_{(j)} - X_{(j-1)})\right) \leqslant G\left(\sum_{j=1}^{r} A_j(Y_{(j)} - Y_{(j-1)})\right)$$

where $F \in \mathcal{F}_{G^{-1}}$, $0 \leqslant A_j \leqslant 1$, $j = 1, \ldots, r$. In particular, if we substitute

$$(10.5.23) \qquad A_j = \min\left\{1, \frac{2(-\ln \beta)}{\chi_\gamma^2[2r]}(n - j + 1)\right\}$$

we obtain from the stochastic inequality (10.5.22),

$$(10.5.24) \qquad P\left\{F\left(\sum_{i=1}^{r} A_i(X_{(i)} - X_{(i-1)})\right) \leqslant 1 - \beta\right\}$$
$$\geqslant P\left\{G\left(\sum_{i=1}^{r} A_i(Y_{(i)} - Y_{(i-1)})\right) \leqslant 1 - \beta\right\}$$
$$\geqslant P\left\{\frac{2(-\ln \beta)}{\chi_\gamma^2[2p]}\sum_{i=1}^{r}(n - i + 1)(Y_{(i)} - Y_{(i-1)}) \leqslant G^{-1}(1 - \beta)\right\} = \gamma.$$

Theorem 10.5.3. (Barlow and Proschan [2].) *If* $F \in \mathcal{F}_{G-1}$ *then a* (β, γ) *lower tolerance limit is*

$$L^*_{\beta,\gamma}(T_{n,r}) = \min\left\{\frac{1}{n}, \frac{2(-\ln \beta)}{\chi_\gamma^2[2r]}\right\} T_{n,r}, \tag{10.5.25}$$

where $T_{n,r} = \sum_{i=1}^r (n - i + 1)(X_{(i)} - X_{(i-1)})$.

The proof of this theorem is based on (10.5.25), with

$$A_i = \min\left\{\frac{1}{n}, \frac{2(-\ln \beta)}{\chi_\gamma^2[2r]}\right\}(n - i + 1), \qquad i = 1, \ldots, r.$$

Barlow and Proschan call these tolerance limits "conservative tolerance limits," since they are deflated by multiplying $T_{n,r}$ by $\min\{1/n, 2(-\ln \beta)/\chi_\gamma^2[2r]\}$. In the exponential case these limits will be very inefficient, because $\min\{1/n, 2(-\ln \beta)/\chi_\gamma^2[2r]\}$ could be considerably smaller than $2(-\ln \beta)/\chi_\gamma^2[2r]$ when n is even of a medium size. However, if one knows that F is an exponential distribution, that is, $F_\theta(x) = G(x/\theta)$, then even the lower limit $(2(-\ln \beta)/\chi_\gamma^2[2r])T_{n,r}$ is inefficient, unless $n = r$. (It is an efficient lower tolerance limit if a life-testing experiment is terminated right after the r-th failure.) Barlow and Proschan [2] have not provided any efficiency comparison between their conservative tolerance limits and the log-convex tolerance limits of Hanson and Koopmans [1]. Barlow and Proschan do give however, upper tolerance limits for I.F.R. distributions and also discuss tolerance limits for I.F.R.A. distributions. A distribution function F, with $F(0) = 0$, is called an I.F.R.A. (increasing failure rate average) distribution if

$$h(t) = \frac{1}{t}\int_0^t f(\tau)(1 - F(\tau))^{-1}\, d\tau$$

is an increasing function of t. They prove the following theorem.

Theorem 10.5.4. (Barlow and Proschan [2].) *If F is an I.F.R.A. distribution with* $F(0) = 0$, *then*

$$P_F\left\{\min\left\{\frac{1}{n}, \frac{2(-\ln \beta)}{\chi_\gamma^2[2r]}\right\} X_{(1)} \leqslant F^{-1}(1 - \beta)\right\} \geqslant \gamma. \tag{10.5.26}$$

That is, $\min\{1/n, 2(-\ln \beta)/\chi_\gamma^2[2r]\}X_{(1)}$ is a (β, γ) lower tolerance limit for an I.F.R.A. distribution F, $F(0) = 0$. For a proof of this theorem see [2]. Similar results are given there for decreasing failure rate (D.F.R.) and decreasing failure rate average (D.F.R.A.) distributions.

For more general distribution free tolerance regions, see Fraser [3], Tukey [1], and Kemperman [1].

10.6. BAYES AND FIDUCIAL CONFIDENCE AND TOLERANCE INTERVALS

Interval estimators can be derived also in a Bayesian framework, in the following manner. Let X be an observed random variable having a distribution function F_θ, $\theta \in \Theta$, and an associated probability space $(\mathfrak{X}, \mathfrak{B}, P_\theta)$. Let Θ be an interval in an Euclidean k-space, and let $\mathfrak{C}$ be the smallest Borel sigma-field on Θ. Suppose that $H(\theta)$ is a prior distribution of θ, defined on the Euclidean k-space. Let $H(\theta \mid X)$ be the posterior distribution corresponding to $F_\theta(x)$ and $H(\theta)$. If θ is real, the γ-fractile of $H(\theta \mid X)$ designated by $H^{-1}(\gamma \mid X)$ is a statistic having the property

$$(10.6.1) \quad H^{-1}(\gamma \mid X) = \text{least real number } \eta \text{ such that } P[\theta \leqslant \eta \mid H, X] \geqslant \gamma,$$

where $P[T \mid H, X]$ designates the posterior probability of T, $T \in \mathfrak{C}$, given (H, X). Having observed the random variable X, the Bayesian who has chosen the prior probability $H(\cdot)$ predicts that, with a confidence (posterior) probability γ, θ does not exceed $H^{-1}(\gamma \mid X)$. Thus $H^{-1}(\gamma \mid X)$ is called a *γ-Bayes upper confidence limit* for θ under H (see Lindley [2], p. 35). In the same manner we can define Bayesian lower confidence limits as well as two-sided limits. If θ is not real but vector valued we can determine a *γ-Bayes confidence region* for θ, which is a Borel set $S_\gamma(X)$ in $\mathfrak{C}$, such that

$$(10.6.2) \quad \int_{S_\gamma(X)} H(d\theta \mid X) \geqslant \gamma.$$

We see that the Bayesian confidence intervals (or regions) differ conceptually from the non-Bayesian confidence intervals (regions). The non-Bayesian confidence regions are sets which cover the true (but unknown) parameter points independently of any prior assumptions. Therefore the confidence probability attached to the non-Bayesian confidence region is a total probability on $(\mathfrak{X}, \mathfrak{B})$ and not a conditional one. The Bayesian confidence probability, on the other hand, is a posterior probability of θ, given (H, X), which is specific to the prior distribution assumed and to the observed value of X. It is a conditional probability on $(\Theta, \mathfrak{C})$, given (H, X). From the Bayesian point of view, after X has been observed there is no meaning to total probabilities on $(\mathfrak{X}, \mathfrak{B})$. One should consider only posterior probabilities on $(\Theta, \mathfrak{C})$.

In a similar manner we can define the notion of a Bayesian tolerance region. Suppose that X is a random variable (real) and let $F^{-1}(\beta; \theta)$ be the β-fractile of its distribution. A *(β, γ)-Bayes upper tolerance limit* for F_θ, given (H, X), say $\bar{L}_{\beta,\gamma}(H, X)$, is a statistic satisfying (see Aitchison [1])

$$(10.6.3) \quad P\{F^{-1}(\beta; \theta) \leqslant \bar{L}_{\beta,\gamma}(H, X) \mid H, X\} \geqslant \gamma.$$

If F_θ is an M.L.R. distribution, $F^{-1}(\beta; \theta)$ is a nondecreasing function of θ foı each β, $0 < \beta < 1$. Hence

$$(10.6.4)\quad P\{F^{-1}(\beta; \theta) \leqslant F^{-1}(\beta; \bar{\theta}_\gamma(H, X)) \mid H, X\} \geqslant P\{\theta \leqslant \bar{\theta}_\gamma(H, X) \mid H, X\} \geqslant \gamma,$$

where $\bar{\theta}_\gamma(H, X) = H^{-1}(\gamma \mid X)$ is the γ upper Bayes confidence limit for θ. Thus in the M.L.R. case the (β, γ)-Bayes upper tolerance limit for F_θ, given (H, X), is

$$(10.6.5)\qquad \bar{L}_{\beta,\gamma}(H, X) = F^{-1}(\beta; \bar{\theta}_\gamma(H, X)).$$

In analogy to previous definitions we can also define a *β-expectation Bayes upper tolerance limit* as the β-fractile of the marginal distribution of a future observation on Y, given (H, X). In other words, let X and Y be i.i.d. random variables with a common distribution function F_θ. Given (H, X), let

$$F(y \mid H, X) = \int F_\theta(y) H(d\theta \mid X)$$

be the marginal d.f. of Y, mixed with respect to the posterior distribution of θ, given (H, X). Then

$$(10.6.6)\qquad P\{Y \leqslant F^{-1}(\beta \mid H, X) \mid H, X\} \geqslant \beta,$$

where $F^{-1}(\beta \mid H, X)$ is the β-fractile of $F(y \mid H, X)$. We notice that

$$(10.6.7)\quad P\{Y \leqslant F^{-1}(\beta \mid H, X) \mid H, X\} = E\left\{\int_{-\infty}^{F^{-1}(\beta \mid H, X)} F_\theta(dy) \mid H, X\right\} = \int H(d\theta \mid X) \int_{-\infty}^{F^{-1}(\beta \mid H, X)} F_\theta(dy).$$

This is the justification for calling $F^{-1}(\beta \mid H, X)$ a β-expectation Bayes upper tolerance limit. We provide two examples.

Example 10.11. (i) Suppose $X_1, X_2, \ldots, X_n$ are i.i.d. random variables with a common normal distribution $\mathcal{N}(\theta, 1)$, $-\infty < \theta < \infty$. Suppose that the prior distribution of θ is $\mathcal{N}(\mu_0, \tau^2)$. A minimal sufficient statistic is $\bar{X} = \sum_{i=1}^n X_i/n$. The posterior distribution of θ, given $\bar{X}$ is

$$\mathcal{N}((\bar{X}n + \mu_0\tau^{-2})(n + \tau^{-2})^{-1}, (n + \tau^{-2})^{-1}).$$

Hence a γ-Bayes upper confidence limit for θ, given $\bar{X}$, is

$$(10.6.8)\qquad \bar{\theta}_\gamma(\bar{X}) = \frac{\bar{X}}{1 + (n\tau^2)^{-1}} + \frac{\mu_0}{1 + n\tau^2} + \Phi^{-1}(\gamma)(n + \tau^{-2})^{-1/2}.$$

Since $F^{-1}(\beta; \theta) = \theta + \Phi^{-1}(\beta)$, the (β, γ)-Bayes upper tolerance limit is

$$(10.6.9)\quad \bar{L}_{\beta,\gamma}(\bar{X}) = \frac{\bar{X}}{1 + 1/n\tau^2} + \frac{\mu_0}{1 + n\tau^2} + \Phi^{-1}(\beta) + \Phi^{-1}(\gamma)(n + \tau^{-2})^{-1}.$$

The marginal distribution of Y, given $\bar{X}$, is $\mathcal{N}((\bar{X}n + \mu_0\tau^{-2})(n + \tau^{-2})^{-1}, 1 + (n + \tau^{-2})^{-1})$. Hence the β-expectation Bayes upper tolerance limit is

$$(10.6.10)\quad L_\beta^*(\bar{X}) = \frac{\bar{X}}{1 + 1/n\tau^2} + \frac{\mu_0}{1 + n\tau^2} + \Phi^{-1}(\beta)\left[1 + \frac{1}{n + \tau^{-2}}\right]^{1/2}.$$

We notice that when $\tau^2 \to \infty$ both $\bar{\theta}_\gamma(\bar{X})$, $\bar{L}_{\beta,\gamma}(\bar{X})$ and $L_\beta^*(\bar{X})$, approach the corresponding non-Bayesian limits. The non-Bayesian limits are also improper Bayesian limits, against a "uniform" prior distribution on the whole real line. These are types of "fiducial" limits which will be discussed later in this section.

(ii) We now derive a (β, γ)-Bayes upper tolerance limit for the $\mathcal{N}(\theta, \sigma^2)$ case, when both θ and σ^2 are unknown. This example in a slightly different form was provided by Aitchison [1].

Given a sample $X_1, \ldots, X_n$ of i.i.d. random variables with a common $\mathcal{N}(\theta, \sigma^2)$ distribution, the (β, γ)-Bayes upper tolerance limit is a function of the minimal sufficient statistic $(\bar{X}, Q)$, where $\bar{X} = \sum_{i=1}^n X_i/n$, $Q = \sum_{i=1}^n (X_i - \bar{X})^2$. We adopt a prior distribution for (θ, σ), according to which the conditional prior distribution of θ, given σ, is $\mathcal{N}(\mu_0, \rho\sigma^2)$, and the prior distribution of $(2\sigma^2)^{-1}$ is the gamma distribution $\mathcal{G}(\zeta, \nu)$.

Let $\omega = (2\sigma^2)^{-1}$. Then the posterior distribution of θ, given $(\bar{X}, \omega)$, is the normal distribution $\mathcal{N}(\tilde{\mu}, \rho/2\omega(1 + n\rho))$, where

$$(10.6.11)\qquad \tilde{\mu} = \bar{X}\left(1 + \frac{1}{\rho n}\right)^{-1} + \mu_0(1 + n\rho)^{-1}.$$

We notice that the marginal distribution of $\bar{X}$, given ω (mixed with respect to the conditional prior distribution of θ given ω), is $\mathcal{N}(\mu_0, (1/2\omega)(1 + n\rho))$. Since $\bar{X}$ and Q are independent, given (θ, ω), the posterior distribution of ω, given $\bar{X}$, Q, is the gamma distribution $\mathcal{G}(V, (n/2) + \nu)$, where

$$(10.6.12)\qquad V = Q + \frac{n}{1 + \rho n}(\bar{X} - \mu_0)^2 + \zeta.$$

From the above results we obtain that

$$(10.6.13)\quad \mathcal{L}\left((\theta - \tilde{\mu})\left(\frac{2\omega(1 + n\rho)}{\rho}\right)^{1/2} \,\middle|\, \bar{X}, \omega\right) = \mathcal{N}(0, 1) \quad \text{a.s. } [\omega],$$

where $\mathcal{L}(\cdot \mid \bar{X}, \omega)$ designates the posterior distribution law. Hence $(\theta - \tilde{u}) \times (2\omega(1 + n\rho)/\rho)^{1/2}$ is posteriorly independent of ω, given $(\bar{X}, Q)$. The

β-fractile of $\mathcal{N}(\theta, \sigma^2)$ is $\theta + u_\beta\sigma$, where $u_\beta = \Phi^{-1}(\beta)$. We notice that the inequality

$$\theta + u_\beta/\sqrt{(2\omega)} \leqslant \bar{L}_{\beta,\gamma}(\bar{X}, Q)$$

is equivalent to

(10.6.14)
$$\sqrt{(2\omega)}\,(\theta - \tilde{\mu})\left(\frac{1 + n\rho}{\rho}\right)^{1/2} + u_\beta\left(\frac{1 + n\rho}{\rho}\right)^{1/2} \leqslant (L - \tilde{\mu})\left(\frac{(2\omega)(1 + n\rho)}{\rho}\right)^{1/2},$$

$L \equiv \bar{L}_{\beta,\gamma}(\bar{X}, Q)$. Furthermore, if 2ν is an integer then

$$\text{(10.6.15)} \qquad \mathcal{L}(\omega \mid \bar{X}, Q) = \mathcal{G}\left(V, \frac{n}{2} + \nu\right) = \mathcal{L}\left(\frac{1}{2V}\chi^2[n + 2\nu]\right).$$

It follows that in this case

$$\text{(10.6.16)} \quad \mathcal{L}\left(\frac{\sqrt{(2\omega)}\,(\theta - \tilde{\mu})\left(\dfrac{1 + n\rho}{\rho}\right)^{1/2} + u_\beta\left(\dfrac{1 + n\rho}{\rho}\right)^{1/2}}{\left(\dfrac{2\omega V}{n + 2\nu}\right)^{1/2}} \,\middle|\, \bar{X}, Q\right)$$

$$= \mathcal{L}\left(t\left[n + 2\nu;\, u_\beta\left(\frac{1 + n\rho}{\rho}\right)^{1/2}\right]\right)$$

Hence from (10.6.14) and (10.6.16),

$$\text{(10.6.17)} \quad P\left\{\theta + \frac{u_\beta}{\sqrt{(2\omega)}} \leqslant \bar{L}_{\beta,\gamma}(\bar{X}, Q) \mid \bar{X}, Q\right\}$$

$$= P\left\{t\left[n + 2\nu;\, u_\beta\left(\frac{1 + n\rho}{\rho}\right)^{1/2}\right] \leqslant \frac{(\bar{L}_{\beta,\gamma}(\bar{X}, Q) - \tilde{\mu})\left(\dfrac{1 + n\rho}{\rho}\right)^{1/2}}{\left(\dfrac{V}{n + 2\nu}\right)^{1/2}} \,\middle|\, \bar{X}, Q\right\}$$

Finally, setting

$$\text{(10.6.18)} \quad \bar{L}_{\beta,\gamma}(\bar{X}, Q) = \tilde{\mu} + t_\gamma\left[n + 2\nu;\, u_\beta\left(\frac{1 + n\rho}{\rho}\right)^{1/2}\right]\left(\frac{\rho V}{(1 + n\rho)(n + 2\nu)}\right)^{1/2},$$

we obtain that

$$\text{(10.6.19)} \qquad P[\theta + u_\beta\sigma \leqslant \bar{L}_{\beta,\gamma}(\bar{X}, Q) \mid \bar{X}, Q] \geqslant \gamma.$$

Therefore $\bar{L}_{\beta,\gamma}(\bar{X}, Q)$ given by (10.6.18) is a (β, γ)-Bayes upper tolerance limit against the prior distribution specified above. We notice that as $\rho \to \infty$, $\tilde{\mu} \to \bar{X}$ a.s. and $V \to Q + \zeta$ a.s. If in addition we let $\zeta \to 0$ and $\nu \to -\frac{1}{2}$, we obtain that

$$\lim_{\rho \to \infty} \lim_{\substack{\zeta \to 0 \\ \nu \to -1/2}} \bar{L}_{\beta,\gamma}(\bar{X}, Q) = \bar{X} + t_\gamma[n-1; u_\beta\sqrt{n}]\frac{\hat{\sigma}}{\sqrt{n}},$$

which is the non-Bayesian (β, γ) upper tolerance limit, (10.4.8). From (10.6.13) and (10.6.15) we also obtain that a γ-Bayes upper confidence limit for θ is

$$\text{(10.6.20)} \qquad \bar{\theta}_\gamma(\bar{X}, Q) = \tilde{\mu} + t_\gamma[n + 2\nu]\left(\frac{\rho V}{(1 + n\rho)(n + 2\nu)}\right)^{1/2}. \qquad ■$$

Example 10.12. Let $X_1, X_2, \ldots, X_n$ be i.i.d. random variables having a common Poisson distribution $\mathcal{P}(\lambda)$ with mean λ, $0 < \lambda < \infty$; $T_n = \sum_{i=1}^n X_i \sim \mathcal{P}(n\lambda)$. Using the relationship

$$\text{(10.6.21)} \qquad e^{-\lambda}\sum_{j=0}^{c}\frac{\lambda^j}{j!} = P\{\chi^2[2c+2] \geqslant 2\lambda\}, \qquad c = 0, 1, \ldots,$$

we define the β-fractile of $\mathcal{P}(\gamma)$ as

$$\text{(10.6.22)} \quad P^{-1}(\beta; \gamma) = \text{least non-negative integer } c \text{ for which } P\{\chi^2[2c+2] \geqslant 2\gamma\} \geqslant \beta.$$

Since $\mathcal{P}(\lambda)$ is an M.L.R. distribution, $P^{-1}(\beta; \lambda)$ is a nondecreasing function of λ for each β, $0 < \beta < 1$. We now derive the γ-Bayes upper confidence limit for γ, against the prior gamma distribution $\mathcal{G}(\Lambda, \nu)$ of λ. Given T_n, the posterior distribution of λ is $\mathcal{G}(\Lambda + n, T_n + \nu)$. Let $G^{-1}(\gamma; p, q)$ designate the γ-fractile of $\mathcal{G}(p, q)$. Then $G^{-1}(\gamma; \Lambda + n, T_n + \nu)$ is a γ-Bayes upper confidence limit for λ, and $P^{-1}(\beta; G^{-1}(\gamma; \Lambda + n, T_n + \nu))$ is a (β, γ)-Bayes upper tolerance limit. ■

When the distribution functions of $\mathcal{F} = \{F_\theta; \theta \in \Theta\}$ depends on a 1-(real) parameter, the γ-Bayes upper (lower) confidence limit for θ is uniquely determined for each prior distribution H of θ. Thus in these cases there is no question of the optimum Bayes upper (lower) confidence limits. However, if we seek two-sided confidence intervals there are generally different intervals with the required coverage (posterior) probability. In this case we may adopt the criterion of choosing a confidence interval with *the smallest length*. For example, if we wish to determine a two-sided Bayes confidence interval for θ the posterior distribution of θ given $\bar{X}$ is $\mathcal{N}(\bar{X}(1 + 1/n\tau^2)^{-1}, (1 + 1/n\tau^2)^{-1})$. It is easy to verify that the shortest Bayes confidence interval, at level γ, is $\bar{X}(1 + 1/n\tau^2)^{-1} \pm u_{(1+\gamma)/2}(1 + 1/n\tau^2)^{-1/2}$. If we consider Bayes tolerance regions for k-dimensional ($k \geqslant 2$) vectors of random variables there is no unique solution. In these cases there is a role for a theory of optimum Bayes tolerance regions. Aitchison [1] discusses this point and suggests ascribing a loss function to the various families of tolerance intervals and choosing an interval which minimizes the associated posterior risk. The reader is referred to the study of Thatcher [1] concerning Bayes tolerance intervals in the binomial case, which yields, for two particular prior distributions, prediction rules similar to the "Laplace rule of succession" (see Feller [1], p. 113).

The subject of "fiducial interval estimators" occupies a significant place in the development of the theory of statistical inference. Fiducial intervals were introduced by Fisher [4] and used by many other statisticians. Many papers have been devoted to the discussion of certain apparent differences between fiducial intervals and confidence intervals and to the interpretation of fiducial intervals. In particular, see Dempster [1] and Stein [6].

In Section 7.5 we discussed the topic of minimum risk fiducial estimators. Here we discuss fiducial intervals in the same context. Thus, given a family $\mathcal{F} = \{F_\theta;\, \theta \in \Theta\}$ of distribution functions defined on a sample space $\mathcal{X}$ of an r.v. X, we designate by $\mathcal{G}$ a group of 1:1 transformations on $\mathcal{X}$; $\mathcal{G}$ is locally compact and $\mathcal{G} \leftrightarrow \Theta$. For each $\theta \in \Theta$ we designate by $[\theta]$ the corresponding element of $\mathcal{G}$. We represent X by the pair (T, V) where V is a maximal invariant statistic with respect to $\mathcal{G}$. We further assume that for each θ of Θ

$$F_\theta(\mathrm{d}x) = f([\theta]^{-1}t \mid V)\lambda(\mathrm{d}v)m(\mathrm{d}t),$$

where $m(\mathrm{d}t)$ is left invariant. The probability element of the fiducial distribution on Θ was defined in (7.4.6) as

$$g(\theta \mid (T, V))\, \mathrm{d}\theta = f([\theta]^{-1}T \mid V)\, \Delta([T])\nu(\mathrm{d}[\theta]).$$

Whenever such a structural model exists we define the *γ-fiducial limits* for θ as the γ-Bayes confidence limits for θ, against the (improper Bayes) prior invariant measure $\nu(d[\theta])$. In a similar manner we define the *(β, γ)-fiducial tolerance regions* as regions which are (β, γ)-Bayes tolerance regions against $\nu(d[\theta])$. Thus if $g(\theta \mid (T, \nu))$ is a fiducial density with respect to a group $\mathcal{G}$ and if θ is real, then a γ-fiducial upper limit for θ is a statistic $\bar{\theta}_\gamma(T, V)$ satisfying

$$\int_{-\infty}^{\theta_\gamma(T,V)} g(\theta \mid (T, V))\, \mathrm{d}\theta \geqslant \gamma. \tag{10.6.23}$$

Similarly, if $\psi(B \mid (T, V))$ designates the fiducial probability measure of a Borel set $B \in \mathscr{C}$, given (T, V), a (β, γ)-fiducial upper tolerance limit in the real case, $\bar{L}_{\beta,\gamma}(T, V)$, satisfies

$$\psi\{\bar{L}_{\beta,\gamma}(T, V) \geqslant F^{-1}(\beta; \theta) \mid (T, V)\} \geqslant \gamma. \tag{10.6.24}$$

In the following example we illustrate some fiducial limits.

Example 10.13. Consider the structural model of Example 7.4. We have that $X_1, X_2, \ldots, X_n$ are i.i.d. having a common exponential distribution, with parameters (μ, σ),

$$f(X; \mu, \sigma) = \frac{1}{\sigma} \exp\left\{-\frac{x-\mu}{\sigma}\right\}, \qquad x \geqslant \mu,$$

$-\infty < \mu < \infty$; $0 < \sigma < \infty$. The corresponding order statistic is $X_{(1)} \leqslant \cdots \leqslant X_{(n)}$. A minimal sufficient statistic is $(X_{(1)}, s_X)$, where $s_X = \sum_{i=2}^{n} (X_{(i)} - X_{(1)})/(n-1)$. The group $\mathscr{G}$ is of affine transformations, $X_i \to a + cX_i$ $(i = 1, \ldots, n)$. We have shown in Example 7.5 that the fiducial density of (μ, σ), given $T_X = (X_{(1)}, s_X)$, is given by (7.4.16); namely,

$$g(\mu, \sigma \mid T_X) = \frac{n(n-1)^{n-1}}{\Gamma(n-1)} s_X^{n-1} \sigma^{-(n+1)} \exp\left\{-n\frac{X_{(1)}-\mu}{\sigma} - (n-1)\frac{s_X}{\sigma}\right\},$$

where $\mu \leqslant X_{(1)}$ and $0 < \sigma < \infty$. Integrating $g(\mu, \sigma \mid T_X)$ with respect to μ, we get that the fiducial density of σ, given T_X, is

(10.6.25)

$$g(\sigma \mid T_X) = \frac{(n-1)^{n-1}}{\Gamma(n-1)} s_X^{n-1} \sigma^{-n} \exp\left\{-(n-1)\frac{s_X}{\sigma}\right\}, \qquad 0 < \sigma < \infty.$$

The γ-fractile of this fiducial distribution is the γ upper fiducial limit for σ. If we designate this limit by $\bar{\sigma}_\gamma(T_X)$ we obtain from (10.6.25) that

$$\bar{\sigma}_\gamma(T_X) = \frac{1}{G^{-1}(1-\gamma; (n-1)s_x, n-1)}. \tag{10.6.26}$$

We notice that $\bar{\sigma}_\gamma(T_X)$ is an equivariant upper limit of σ, with respect to $\mathscr{G}$. Indeed, $\bar{\sigma}_\gamma(T_X)$ satisfies the equation

$$\frac{(n-1)^{n-1}}{\Gamma(n-1)} s_X^{n-1} \int_0^{\bar{\sigma}_\gamma(T_X)} \sigma^{-n} \exp\left\{-(n-1)\frac{s_X}{\sigma}\right\} d\sigma = \gamma. \tag{10.6.27}$$

If $X_i \to a + cX_i$ $(i = 1, \ldots, n)$ for some a, $-\infty < a < \infty$, and c, $0 < c < \infty$, then $s_X \to cs_X$, and since $\bar{\sigma}_\gamma(T_X) = D(s_X)$ depends only on s_X,

$$\text{(10.6.28)}\quad \frac{(n-1)^{n-1}}{\Gamma(n-1)} s_X^{n-1} \int_0^{D(s_X)} \sigma^{-n} \exp\left\{-(n-1)\frac{s_X}{\sigma}\right\} d\sigma$$

$$\to \frac{(n-1)^{n-1}}{\Gamma(n-1)} s_X^{n-1} \int_0^{D(cs_X)} (c^{n-1}\sigma^{-n}) \exp\left\{-(n-1)\frac{cs_X}{\sigma}\right\} d\sigma.$$

Letting $\omega = \sigma/c$, we obtain

$$\text{(10.6.29)}\quad \frac{(n-1)^{n-1}}{\Gamma(n-1)} s_X^{n-1} \int_0^{D(cs_X)} (c^{n-1}\sigma^{-n}) \exp\left\{-(n-1)\frac{cs_X}{\sigma}\right\} d\sigma$$

$$= \frac{(n-1)^{n-1}}{\Gamma(n-1)} s_X^{n-1} \int_0^{(1/c)D(cs_X)} \omega^{-n} \exp\left\{-(n-1)\frac{s_X}{\omega}\right\} d\omega.$$

If we equate the right-hand side of (10.6.29) with the left-hand side of (10.6.28) we obtain

$$\text{(10.6.30)}\quad D(cs_X) = cD(s_X), \quad \text{all} \quad 0 < c < \infty.$$

We also notice that the γ upper fiducial limit for σ is the uniformly most accurate *invariant* γ *upper confidence* limit for σ, with respect to the group $\mathcal{G}$ of affine transformations.

We now derive the (β, γ) upper fiducial tolerance limits.

The β-fractile of F_θ is

$$\text{(10.6.31)}\quad F^{-1}(\beta; \mu, \sigma) = \mu - \sigma \ln(1 - \beta).$$

We consider a statistic of the form $\bar{L}_{\beta,\gamma}(T_X) = X_{(1)} + \lambda_{\beta,\gamma} s_X$. To find $\lambda_{\beta,\gamma}$ we start from the general definition,

$$\text{(10.6.32)}\quad \psi\{\mu - \sigma \ln(1-\beta) \leqslant X_{(1)} + \lambda_{\beta,\gamma} s_X \mid T_X\} \geqslant \gamma.$$

Furthermore,

(10.6.33)

$$\psi(\mu - \sigma \ln(1-\beta) \leqslant X_{(1)} + \lambda_{\beta,\gamma} s_X \mid T_X) = \frac{n(n-1)^{n-1}}{\Gamma(n-1)} s_X^{n-1} \int_0^\infty d\sigma$$

$$\times\, \sigma^{-(n+1)} e^{-(n-1)\frac{s_X}{\sigma}} \int_{-}^{X_{(1)}+\lambda_{\beta,\gamma}s_X+\sigma\ln(1-\beta)} \exp\left\{-n\frac{X_{(1)}-\mu}{\sigma}\right\} d\mu.$$

Making the change of variables

$$u = n\frac{X_{(1)} - \mu}{\sigma}$$

and

$$z = \frac{1}{\sigma},$$

we obtain

$$\begin{aligned}(10.6.34)\quad &\psi\{\mu - \sigma \ln(1-\beta) \leqslant X_{(1)} + \lambda_{\beta,\gamma} s_X \mid (X_{(1)}, s_X)\} \\ &= (1-\beta)^{-n} \frac{(n-1)^{n-1} s_X^{n-1}}{\Gamma(n-1)} \int_0^\infty z^{n-2} \exp\{-[(n-1) + n\lambda_{\beta,\gamma}] s_X z\}\, dz \\ &= (1-\beta)^{-n}\left[1 + \frac{n}{n-1}\lambda_{\beta,\gamma}\right]^{-n}.\end{aligned}$$

Equating (10.6.34) to γ we derive

$$(10.6.35)\qquad \lambda_{\beta,\gamma} = \frac{n-1}{n}[(\gamma^{1/n-1}(1-\beta))^{-1} - 1].$$

Since $0 < \gamma, \beta < 1$, $\gamma_{\beta,\gamma} > 0$. Thus

$$\bar{L}_{\beta,\gamma}(T_X) = X_{(1)} + ((n-1)/n)[(\gamma^{1/(n-1)}(1-\beta))^{-1} - 1] s_X$$

is a (β, γ) upper fiducial tolerance limit for $\mathfrak{F}$ with respect to the group of affine transformations. ■

We conclude this section with a few comments concerning the admissibility of confidence intervals. The definition of admissibility of confidence intervals was formulated by Godambe [2]. In his formulation, a system $\mathcal{S} = \{S(X) = (a(X), b(X)), X \in \mathfrak{X}\}$ is called *admissible* if there exists no other system, say $\mathcal{S} = \{S^*(X) = (a^*(X), b^*(X)), X \in \mathfrak{X}\}$ such that

$$(10.6.36)\qquad b^*(X) - a^*(X) \leqslant b(X) - a(X) \quad \text{a.s.}$$

and

$$(10.6.37)\qquad P_\theta[\theta \in (a^*(X), b^*(X))] \geqslant P_\theta[\theta \in (a(X), b(X))],$$

for almost all θ, with a strict inequality for a non-null set of θ values. Joshi [3] reformulated this definition, with the more stringent requirement that (10.6.31) holds for *all* θ with a strict inequality for at least one θ. Joshi's definition is in agreement with our previous definitions of admissibility, while Godambe's definition is more like almost admissibility.

Theorem 10.6.1. (Joshi [3].) *Let $X_1, \ldots, X_n$ be i.i.d. random variables having an absolutely continuous distribution function F_θ, with a density function $f(x; \theta)$*

which depends on a real parameter. Let $\mathcal{F} = \{F_\theta;\ \theta \in \Theta\}$ *and assume the following:*

(i) $\mathcal{F}$ *admits a sufficient statistic* $T(X_1, \ldots, X_n)$, *with a density function* $p(t - \theta)$, $-\infty < \theta < \infty$;

(ii) $p(t)$ *is strictly decreasing for* $t \geqslant 0$ *and strictly increasing for* $t \leqslant 0$, *and* $p(t)$ *is continuous for all* t *and*

$$\int_0^\infty dt_1 \left[\int_{t_1}^\infty p(t)\, dt + \int_{-\infty}^{-t_1} p(t)\, dt\right] < \infty; \tag{10.6.38}$$

(iii) $f(x; \theta)$ *is continuous in* x, *for all* $-\infty < x < \infty$ *and all* θ, $-\infty < \theta < \infty$;

(iv) $V_1(\mathbf{X}_n)$ *and* $V_2(\mathbf{X}_n)$ *are non-negative statistics distributed independently of* $T(\mathbf{X}_n)$ *and* θ, *and* $p(-V_2(\mathbf{X}_n)) = p(V_1(\mathbf{X}_n))$. *Furthermore, if* $V(\mathbf{X}_n) = \max\{V_1(\mathbf{X}_n), V_2(\mathbf{X}_n)\}$ *then* $\operatorname{Var}\{V(\mathbf{X}_n)\} < \infty$. *Then the system of confidence intervals*

$$\{(T(\mathbf{X}_n) - V_1(\mathbf{X}_n),\ T(\mathbf{X}_n) + V_2(\mathbf{X}_n));\ \mathbf{X}_n \in \mathfrak{X}^{(n)}\}$$

is admissible in the sense of Joshi.

We do not provide the proof of this theorem, which is very long and highly technical. We mention a simple example instead. Suppose that $X_1, \ldots, X_n$ are i.i.d. distributed like $\mathcal{N}(\theta, 1)$; $-\infty < \theta < \infty$; $T_n(\mathbf{X}_n) = \bar{X}_n = 1/n \sum_{i=1}^n X_i$. Set $V_1(\mathbf{X}_n) = V_2(\mathbf{X}_n) = \Phi^{-1}(\beta)$, $\frac{1}{2} < \beta < 1$. It is easy to verify that all the conditions of the above theorem are satisfied and $(\bar{X} - \Phi^{-1}(\beta), \bar{X} + \Phi^{-1}(\beta))$ is an admissible confidence interval of level $\gamma = 2\beta - 1$. In Example 10.6 we have shown that this two-sided confidence interval does not have a uniformly shortest expected length. However, it is admissible and minimax. In a series of four consecutive papers [4; 5; 6; 7], which followed [3], Joshi extended and generalized the above theorem to cases of sampling from finite populations. Joshi [7] proved that the usual confidence sets for the mean of a univariate or a bivariate normal distribution are admissible. Joshi [5] established a result analogous to that of Stein (Example 8.8), according to which the usual confidence regions for the mean of a p-variate normal distribution ($p \geqslant 3$) are inadmissible.

10.7. THE PROBLEM OF FIXED-WIDTH CONFIDENCE INTERVALS—SEVERAL SPECIAL CASES

The precision of a family of interval estimators is generally measured in terms of the length of the intervals and their coverage probabilities. Confidence intervals $S_\gamma(X)$ are called *fixed-width confidence intervals* if $S_\gamma(X) = (\hat{\theta}(X) - \delta, \hat{\theta}(X) + \delta)$ and

$$P_\theta[\theta \in (\hat{\theta}(X) - \delta, \hat{\theta}(X) + \delta)] \geqslant \gamma, \quad \text{all} \quad \theta \in \Theta. \tag{10.7.1}$$

In the general case θ could be a real parameter or some real functional on the family $\mathcal{F}$ of distribution functions. The problem of fixed-width confidence intervals is that of determining whether such a family exists and providing the sampling scheme which will guarantee that a proper statistic $\hat{\theta}(X)$ will satisfy (10.7.1). It is well known that in certain interval estimation problems there exist no such fixed-width confidence intervals under certain sampling procedures. The first example of this nature was apparently given by Dantzig [1] who proved that, in the normal case, if the variance is unknown there does not exist a fixed-width confidence interval for the mean, under a fixed size sampling procedure. We provide a simple proof of this result later on. A few years later Stein [2] showed that a fixed-width confidence interval for the mean of a normal distribution can be constructed if sampling is performed in two stages and the size of the second stage sample is a random variable that depends on the observed values of the first stage. As proven in Section 10.8, the problem of fixed-width confidence intervals often has a solution in a two-stage sampling scheme. There are cases, however, in which no sampling scheme exists, with a predetermined number of stages, that can yield a fixed-width confidence interval. Purely sequential schemes are required. Results of this kind were published by Blum and Rosenblatt [2], Farrel [3], and others. Moreover, there are cases for which there exists no sampling scheme that can guarantee fixed-width confidence intervals. In this section we provide several examples of such cases. This section is partitioned into three subsections. In the first subsection we provide the classical two-stage sampling procedure of Stein. In the second subsection we prove that a fixed-width confidence interval for the mean of any I.F.R. distribution cannot be attained under a fixed sample size procedure. We show that all the moments of an I.F.R. distribution, with $F(0) = 0$, can be estimated by a fixed-width confidence interval if there are at least two stages of sampling. In the third subsection we show an interesting case in which no proper sampling procedure can guarantee a fixed-width confidence interval.

10.7.1. THE STEIN TWO-STAGE SAMPLING PROCEDURE

Let $X_1, X_2, \ldots$ be a sequence of i.i.d. random variables having a common normal distribution $\mathcal{N}(\mu, \sigma^2)$. Both μ and σ are unknown; $-\infty < \mu < \infty$, $0 < \sigma < \infty$. We show first that there exists no fixed-width confidence interval for μ if the sample size n is fixed. For a fixed sample size n, $(\bar{X}_n, Q_n)$ is a minimal sufficient statistic; $\bar{X}_n = (1/n) \sum_{i=1}^n X_i$, $Q_n = \sum_{i=1}^n (X_i - \bar{X}_n)^2$. We have to show that there exists no statistic $\hat{\mu}(\bar{X}_n, Q_n)$ such that

$$(10.7.2) \qquad P_{\mu,\sigma}\{|\mu - \hat{\mu}(\bar{X}_n, Q_n)| < \delta\} \geqslant \gamma, \quad \text{for all} \quad (\mu, \sigma),$$

where δ, $0 < \delta < \infty$, is specified and $\gamma > 0$. Indeed,

(10.7.3)

$$\sup_{\hat{\mu}} \inf_{\mu,\sigma} P_{\mu\sigma}\{|\mu - \hat{\mu}(\bar{X}_n, Q_n)| < \delta\} \leqslant \underline{\lim}_{\sigma\to\infty} \inf_{\mu} \sup_{\hat{\mu}} P_{\mu,\sigma}\{|\mu - \hat{\mu}(\bar{X}_n, Q_n)| < \delta\}.$$

We have shown in Example 6.7 that, for a given value of σ, the minimax midpoint statistic is $\hat{\mu}(\bar{X}_n, Q_n) = \bar{X}_n$; that is,

$$\text{(10.7.4)} \quad \inf_{\mu} \sup_{\hat{\mu}} P_{\mu,\sigma}\{|\mu - \hat{\mu}(\bar{X}_n, Q_n)| < \delta\} = 2\Phi\left(\frac{\delta}{\sigma}\sqrt{n}\right) - 1.$$

Hence

$$\text{(10.7.5)} \quad \sup_{\hat{\mu}} \inf_{\mu,\sigma} P_{\mu,\sigma}\{|\mu - \hat{\mu}(\bar{X}_n, Q_n)| < \delta\} \leqslant \lim_{\sigma\to\infty}\left[2\Phi\left(\frac{\delta}{\sigma}\sqrt{n}\right) - 1\right] = 0.$$

This proves the statement that there exists no fixed-width confidence interval for μ in the fixed sample size case.

Stein [2] has proven that the following two-stage sampling procedure yields a fixed-width confidence interval for μ.

Stage I. Choose a sample of a fixed size n_1, $n_1 \geqslant 2$, and determine its sufficient statistic $(\bar{X}_{n_1}, Q_{n_1})$.

Stage II. Take an additional sample of (a random size)

$$\text{(10.7.6)} \quad N_2 = \text{least integer } n \text{ greater or equal to } \left[F_\gamma[1, n_1 - 1]\frac{s_{n_1}^2}{\delta^2} - n_1\right]^+,$$

where $a^+ = \max(0, a)$, $s_{n_1}^2 = Q_{n_1}/(n_1 - 1)$; and $F_\gamma[1, n_1 - 1]$ is the γ-fractile of the $F[1, n_1 - 1]$ distribution. We designate by δ half of the prescribed length of the confidence interval. The confidence interval determined after the second stage is $I_\delta(\mathbf{X}_N) = [\bar{X}_N - \delta, \bar{X}_N + \delta]$ where $N = n_1 + N_2$. We remark that the sample size of the second stage, N_2, is a random variable that can assume the value zero with a positive probability. Whenever $N_2 = 0$ sampling terminates after the first stage. It is easy to prove that $P_{\mu,\sigma}\{N_2 < \infty\} = 1$ for all (μ, σ). Thus N_2 is a proper random variable. Furthermore, N_2 depends only on $s_{n_1}^2$ and is therefore independent of $\bar{X}_{n_1}$ and of $(X_{n_1+1}, \ldots, X_N)$ (when $N_2 \geqslant 1$). Hence the conditional distribution of $\bar{X}_N$, given N, is $\mathcal{N}(\mu, \sigma^2 N^{-1})$. Thus

$$\begin{aligned}
\text{(10.7.7)} \quad & P_{\mu,\sigma}\{|\bar{X}_N - \mu| < \delta\} \\
&= E\{P_{\mu,\sigma}\{|\bar{X}_N - \mu| < \delta \mid N\}\} \\
&= 2E\left\{\Phi\left(\frac{\delta}{\sigma}N^{1/2}\right)\right\} - 1 \geqslant 2E\left\{\Phi\left(\frac{\delta}{\sigma}\cdot\frac{s_{n_1}}{\delta}t_{\gamma'}[n_1 - 1]\right)\right\} - 1 \\
&= 2P\left\{\frac{N(0, 1)}{(\chi^2[n_1 - 1]/(n_1 - 1))^{1/2}} \leqslant t_{\gamma'}[n_1 - 1]\right\} - 1 = \gamma, \qquad \text{all } (\mu, \sigma),
\end{aligned}$$

where $\gamma' = (1 + \gamma)/2$ and $N(0, 1)$ is independent of $\chi^2[n_1 - 1]$. Thus the average probability of the fixed-width interval $I_\delta(\mathbf{X}_N)$ is at least γ *for all* (μ, σ); that is, $I({}_\delta\mathbf{X}_N)$ is a fixed-width confidence interval with confidence level γ. A different solution to the problem here has been provided by Graybill [1]. Graybill's solution is somewhat less efficient than Stein's solution. This problem of estimating the mean of a normal distribution was studied also by Weiss [1]. In Section 10.9 we discuss the Stein two-stage sampling procedure further.

10.7.2. TWO-STAGE ESTIMATION IN THE CLASS OF I.F.R. DISTRIBUTIONS

Consider the class $\mathfrak{F}$ of I.F.R. distribution functions discussed in Section 10.5. We restrict attention to distribution function F in $\mathfrak{F}$ such that $F(0) = 0$. Suppose we wish to estimate the expected value of $\mu(F) = \int_0^\infty x\, F(dx)$. In this subsection we derive the results of Blum and Rosenblatt [3]. We start by showing that there exist no fixed-width confidence intervals for $\mu(F)$, $F \in \mathfrak{F}$, if the sample size n is fixed. We prove first the following slightly generalized version of a theorem of Blum and Rosenblatt.

Lemma 10.7.1. (Blum and Rosenblatt [3].) *Suppose that $\mathfrak{F}$ is a family of absolutely continuous distributions depending on a scale parameter θ, that is, $F_\theta(x) = F(x/\theta)$, $0 < \theta < \infty$, and having a density function $(1/\theta)f(x/\theta)$, continuous in θ. Then there exists no fixed-width confidence interval for θ, say $[\hat{\theta}(X), \hat{\theta}(X) + l]$, $l > 0$, based on one observation.*

Proof. The proof is by negation. Suppose that $[\hat{\theta}(X), \hat{\theta}(X) + l]$ is a fixed-width γ confidence interval; that is,

$$P_\theta[\hat{\theta}(X) \leqslant \theta \leqslant \hat{\theta}(X) + l] \geqslant \gamma, \quad \text{all } \theta, \quad 0 < \theta < \infty. \tag{10.7.8}$$

Hence for any fixed λ, $0 < \lambda < \infty$,

$$P_{\lambda\theta}[\hat{\theta}(X) \leqslant \lambda\theta \leqslant \hat{\theta}(X) + l] \geqslant \gamma, \quad \text{all } \theta, \quad 0 < \theta < \infty.$$

We notice that if the distribution function of X is $F(x/\theta)$ then the distribution function of $Y = \lambda X$ is $F(y/\lambda\theta)$. Hence

(10.7.9)

$$P_\theta[\hat{\theta}(\lambda X) \leqslant \lambda\theta \leqslant \hat{\theta}(\lambda X) + l] = P_\theta\left[0 \leqslant \theta - \frac{\hat{\theta}(\lambda X)}{\lambda} \leqslant \frac{l}{\lambda}\right] \geqslant \gamma, \qquad \text{all } \theta.$$

Let $\mathfrak{X}$ be the sample space of X and $\mathfrak{B}$ its corresponding Borel sigma-field. Let $P_\theta(\cdot)$ be a probability measure on $(\mathfrak{X}, \mathfrak{B})$ induced by F_θ. The function

$P_\theta(\cdot)$ is a continuous function of θ for each $B \in \mathcal{B}$. Indeed, for $B \in \mathcal{B}$,

$$P_\theta(B) = \frac{1}{\theta}\int I_B(x) f\left(\frac{x}{\theta}\right) \mathrm{d}x = \int I_B(\theta x) f(x)\, \mathrm{d}x, \tag{10.7.10}$$

where $f(x)$ is the density function corresponding to $F(x)$. By the Lebesgue dominated convergence theorem,

$$\begin{aligned} &\overline{\lim_{d\to 0}}\, |(\theta + d)P_{\theta+d}(B) - \theta P_\theta(B)| \\ &\leqslant \overline{\lim_{d\to 0}} \int_\Lambda I_B(u\theta) f(u) \left| \frac{f[u(\theta/(\theta + \mathrm{d}))]}{f(u)} - 1 \right| \mathrm{d}u \\ &= \int_\Lambda I_B(u\theta)\, \overline{\lim_{d\to 0}} \left| \frac{f[u(\theta/(\theta + \mathrm{d}))]}{f(u)} - 1 \right| \cdot f(u)\, \mathrm{d}u = 0, \end{aligned} \tag{10.7.11}$$

where $\Lambda = \{u; f(u) > 0\}$ is the support of $F(x)$.

Let N be an integer such that $\gamma N \geqslant 2$. Since $P_\theta(B)$ is continuous in θ, we can find N points $\theta_1, \theta_2, \ldots, \theta_N$ such that, for any $B \in \mathcal{B}$,

$$|P_{\theta_i}(B) - P_{\theta_j}(B)| \leqslant \frac{1}{2N}, \quad \text{all } i \neq j = 1, \ldots, N. \tag{10.7.12}$$

Choose λ so large that $l/\lambda < \frac{1}{2} \min_{i\neq j} |\theta_i - \theta_j|$. The events $B_i = \{0 \leqslant \theta_i - ((\hat{\theta}(\lambda X))/\lambda) \leqslant l/\lambda\}$, $i = 1, \ldots, N$, are mutually exclusive. Furthermore, from (10.7.12)

$$P_{\theta_1}(B_i) \geqslant \gamma - \frac{1}{2N}, \quad \text{all } i = 1, \ldots\ N. \tag{10.7.13}$$

Finally,

$$P_{\theta_1}\left(\bigcup_{i=1}^N B_i\right) = \sum_{i=1}^N P_{\theta_1}(B_i) \geqslant N\gamma - \tfrac{1}{2} \geqslant \tfrac{3}{2}. \tag{10.7.14}$$

This is a contradiction! (Q.E.D.)

As a corollary to this lemma we infer that there exist no fixed-width confidence intervals for the class of I.F.R. distributions when the sample size is fixed. Indeed, the negative exponential densities satisfy the conditions of the lemma. We show now that there exist fixed-width confidence intervals if sampling is at two-stages. Let $\xi_p(F)$ be the p-fractile of an I.F.R. distribution F. We have shown in Section 10.5 (Lemma 10.5.2) that a γ upper confidence limit for $\xi_p(F)$ exists for each n. This is a (p, γ) upper tolerance limit for $\mathcal{F}$. Let us designate this limit by $\bar{L}_{p,\gamma}(\mathbf{X}_n)$. Thus, given a first-stage sample of size n_1, compute $\bar{L}_{p,\gamma}(\mathbf{X}_{n_1})$. We have

$$P_F\{\bar{L}_{p,\gamma}(\mathbf{X}_{n_1}) \geqslant \xi_p(F)\} \geqslant \gamma, \quad \text{all} \quad F \in \mathcal{F}. \tag{10.7.15}$$

Barlow and Proschan [5], p. 27, prove that if $F \in \mathcal{F}$ then

$$1 - F(t) \leqslant e^{-\beta_F(p)t}, \quad \text{all } t \geqslant \xi_p(F), \tag{10.7.16}$$

where $\beta_F(p) = -\ln(1-p)/\xi_p(F)$. Thus let

$$\hat{\beta}(p; \mathbf{X}_{n_1}) = -\frac{\ln(1-p)}{\bar{L}_{p,\gamma}(\mathbf{X}_{n_1})}, \tag{10.7.17}$$

and define

$$\hat{G}(t; \mathbf{X}_{n_1}) = \begin{cases} 1, & \text{if } t \leqslant \bar{L}_{p,\gamma}(\mathbf{X}_{n_1}) \\ \exp\{-t\hat{\beta}(p; \mathbf{X}_{n_1})\}, & \text{if } t > \bar{L}_{p,\gamma}(\mathbf{X}_{n_1}). \end{cases} \tag{10.7.18}$$

Then form (10.7.15) and (10.7.16)

$$P_F\{1 - F(t) \leqslant G(t; \mathbf{X}_{n_1})\} \geqslant \gamma, \quad \text{all} \quad F \in \mathcal{F}. \tag{10.7.19}$$

Let $\mu_r(F)$ designate the r-th moment of F. The variance of X^r is

$$\begin{aligned} \operatorname{Var}_F\{X^r\} &= \mu_{2r}(F) - \mu_r^2(F) \\ &\leqslant \int_0^\infty x^{2r} F(\mathrm{d}x) = 2r \int_0^\infty x^{2r-1}[1 - F(x)]\,\mathrm{d}x \\ &\leqslant 2r \int_0^\infty x^{2r-1} G(x)\,\mathrm{d}x. \end{aligned} \tag{10.7.20}$$

Let

$$\begin{aligned} H_r(\mathbf{X}_{n_1}) &= 2r \int_0^\infty x^{2r-1} \hat{G}(x; X_{n_1})\,\mathrm{d}x \\ &= (\bar{L}_{p,\gamma}(X_{n_1}))^{2r} \\ &\quad + \frac{2r\Gamma(2r)}{(\hat{\beta}(p; X_{n_1}))^{2r}}[1 - P\{\chi^2[4r] \leqslant 2\bar{L}_{p,\gamma}(\mathbf{X}_{n_1})\hat{\beta}(p; \mathbf{X}_{n_1})\}]. \end{aligned} \tag{10.7.21}$$

If $\operatorname{Var}_F(X^r)$ is known, $X_1, \ldots, X_n$ are i.i.d. random variables having a common I.F.R. distribution F, and if $\hat{\mu}_r(\mathbf{X}_n) = (1/n)\sum_{i=1}^n X_i^r$, then from the Tchebytchev inequality,

$$P_F\{|\hat{\mu}_r(\mathbf{X}_n) - \mu_r| < \delta\} \geqslant 1 - \frac{\operatorname{Var}_F\{X^r\}}{\delta^2}. \tag{10.7.22}$$

Hence if

$$n_2 = \text{least integer greater or equal to } \frac{\operatorname{Var}_F\{X^r\}}{\alpha\delta^2},$$

then

$$P_F\{|\hat{\mu}(\mathbf{X}_{n_2}) - \mu| < \delta\} \geqslant 1 - \alpha.$$

Accordingly, define

(10.7.24) $$N_2 = \text{least integer greater or equal to } \frac{H_r(\mathbf{X}_{n_1})}{\alpha\delta^2}.$$

From (10.7.20) and (10.7.23) we deduce that

(10.7.25) $$P_F\{N_2 \geqslant n_2\} \geqslant \gamma, \quad \text{all} \quad F \in \mathcal{F}.$$

Consider now the following two-stage procedure.

Stage I. Choose a value of n_1, and observe $\{X_1, \ldots, X_{n_1}\}$. Compute $\bar{L}_{p,\gamma}(\mathbf{X}_{n_1})$ for some $0 < p, \gamma < 1$; $\hat{\beta}(p; \mathbf{X}_{n_1}); H_r(\mathbf{X}_{n_1})$ and N_2.

Stage II. Observe an additional N_2 random variables $Y_1, \ldots, Y_{N_2}$, independently of $\{X_1, \ldots, X_{n_1}\}$, and compute $\hat{\mu}_r(\mathbf{Y}_{N_2}) = (1/N_2) \sum_{i=1}^{n} Y_i^r$.

Lemma 10.7.2. (Blum and Rosenblatt [3].) *If $\mathcal{F}$ is the class of I.F.R. absolutely continuous distributions, with $F(0) = 0$, then under the above two-stage sampling procedure*

(10.7.26) $$P_F\{|\hat{\mu}_r(\mathbf{Y}_{N_2}) - \mu_r| < \delta\} \geqslant \gamma(1 - \alpha), \quad \textit{all} \quad F \in \mathcal{F}; \quad r = 1, 2, \ldots .$$

Proof. The conditional distribution of $\{Y_1, \ldots, Y_{N_2}\}$ given the Borel sigma-field generated by $\{X_1, \ldots, X_{n_1}\}$, say $\mathcal{B}^{(n_1)}$, is like that of N_2 i.i.d. random variables having a common d.f. Therefore,

(10.7.27) $$P_F\{|\hat{\mu}_r(\mathbf{Y}_{N_2}) - \mu_r| < \delta \mid N_2 \geqslant n_2\} \geqslant 1 - \alpha, \quad \text{all} \quad F \in \mathcal{F}.$$

Finally, according to (10.7.25) and (10.7.27),

(10.7.28) $$\begin{aligned} &P_F\{|\hat{\mu}_r(\mathbf{Y}_{N_2}) - \mu_r| < \delta\} \\ &\quad \geqslant P_F\{|\hat{\mu}_r(\mathbf{Y}_{N_2}) - \mu_r| < \delta \mid N_2 \geqslant n_2\} P_F\{N_2 \geqslant n_2\} \geqslant \gamma(1 - \alpha), \end{aligned}$$

all $F \in \mathcal{F}$. [For more details, see (10.8.31)]. (Q.E.D.)

Thus we exhibited a two-stage sampling procedure and an associated interval estimator $I_\delta(\mathbf{Y}_{N_2}) = (\hat{\mu}_r(\mathbf{Y}_{N_2}) - \delta, \ \hat{\mu}_r(\mathbf{Y}_{N_2}) + \delta)$ which is a fixed-width confidence interval for the r-th moment of an I.F.R. distribution function.

10.7.3. A CASE OF AN ESTIMATION PROBLEM FOR WHICH A FIXED-WIDTH CONFIDENCE INTERVAL PROCEDURE DOES NOT EXIST

Bahadur and Savage [1] exhibited an estimation problem for which there exist no fixed-width confidence intervals for the mean under any sampling procedure which terminates with probability one. They considered the family $\mathcal{F}$ of *all* distribution functions F for which $E_F\{|X|\} < \infty$. This is a very rich family of distributions and we are not very astonished at this kind of result.

We exhibit here an interesting recent result of Blum and Rosenblatt [4], which shows that in the family of m-dependent normal random variables there exists no sampling procedure under which we can attain fixed-width confidence intervals for the mean. More specifically, let $\cdots X_{-2}, X_{-1}, X_0, X_1, X_2, \ldots$ be a doubly infinite sequence of independent random variables, identically distributed like $N(0, 1)$. Let

$$Y_n = \mu + X_n + Z_{n,m}, \qquad n = 1, 2, \ldots, \tag{10.7.29}$$

where

$$Z_{n,m} = \begin{cases} 0, & \text{if } m = 0, \\ \dfrac{1}{\sqrt{m}} \displaystyle\sum_{j=1}^{m} X_{n-j}, & \text{if } m \geqslant 1. \end{cases} \tag{10.7.30}$$

The parameters μ, $-\infty < \mu < \infty$, and m, $m = 0, 1, \ldots$ are both unknown. We show that there exists no sampling procedure that terminates with probability one under which we can determine a fixed-width confidence interval for μ.

The variables $Y_1, Y_2, \ldots$ form a stationary m-dependent normal sequence, that is, $Y_n \sim N(\mu, 2)$ for all $n = 1, 2, \ldots$ and

$$\operatorname{cov}(Y_n, Y_{n+j}) = \begin{cases} 1 - \dfrac{j}{m} + \dfrac{1}{\sqrt{m}}, & j = 0, 1, \ldots, m, \\ 0, & j = m + 1, \ldots, \end{cases} \tag{10.7.31}$$

for *all n*.

Let M be a fixed positive integer. Let $P_{\mu,m}^{(M)}(\cdot)$ designate the probability measure on $(\mathfrak{X}^{(M)}, \mathfrak{B}^{(M)})$ which is induced by the joint normal distribution of $\mathbf{Y}_M = (Y_1, \ldots, Y_M)$. The space $\mathfrak{X}^{(M)}$ is the sample space of $\mathbf{Y}_M$ and $\mathfrak{B}^{(M)}$ the corresponding Borel sigma-field. Let $P_{\mu,m|z}^{(M)}(\cdot)$ designate the probability measure on $(\mathfrak{X}^{(M)}, \mathfrak{B}^{(M)})$ induced by the *conditional* joint normal distribution of $\mathbf{Y}_M$, given $\{Z_{1,m} = z\}$.

Lemma 10.7.3. (Blum and Rosenblatt [4].) *For each fixed positive integer M,*

$$\lim_{m\to\infty} \sup_{B\in\mathfrak{B}^{(M)}} \sup_{z\in I} \sup_{\mu} |P_{\mu,m|z}^{(M)}(B) - P_{\mu+z,0}^{(M)}(B)| = 0, \tag{10.7.32}$$

where I is any fixed bounded interval of reals.

Proof. It is easy to show that if $m > M$, the conditional distribution of $(Y_1, \ldots, Y_M)$, given $\{Z_{1,m} = z\}$, is multivariate normal with mean $(\mathbf{1}_M - (1/m)\mathbf{W}_M)z$, where $\mathbf{1}_M' = (1, \ldots, 1)$ and $\mathbf{W}_M' = (0, 1, \ldots, M-1)$, and covariance matrix $\Sigma_M(z)$ whose (i, j)-th element is

$$\begin{aligned} \sigma_{ij}^{(M)}(z) = \delta_{ij} + 1 + |i - j|\, m^{-1} + (1 - \delta_{ij}) m^{-1/2} \\ - [1 - (j-1)m^{-1}][1 - (i-1)m^{-1}], \end{aligned} \tag{10.7.33}$$

where δ_{ij} is the Kronecker delta. Hence as $n \to \infty$ the mean vector approaches $z1_M$ and the covariance matrix $\Sigma_M(z)$ approaches the identity matrix I_M. Furthermore, it is simple to show that this approach is uniform on any finite interval I of z values.

We notice that, in order to prove (10.7.32), it is sufficient to prove that

$$\lim_{m\to\infty} \sup_{B\in\mathfrak{B}^{(M)}} \sup_{z\in I} |P^{(M)}_{0,m|z}(B) - P^{(M)}_{0,z}(B)| = 0. \tag{10.7.34}$$

Indeed, if $B \in \mathfrak{B}^{(M)}$ and $B_\mu = B + \mu 1_M$ then

$$P^{(M)}_{\mu,m|z}(B) = P^{(M)}_{0,m|z}(B_\mu), \quad \text{all } \mu, \ -\infty < \mu < \infty; \tag{10.7.35}$$

and

$$P^{(M)}_{z+\mu,m}(B) = P^{(M)}_{z,m}(B_\mu), \quad -\infty < \mu < \infty. \tag{10.7.36}$$

We now prove (10.7.34). For any given $\epsilon > 0$, let $I^{(M)}_\epsilon$ be an M-dimensional interval such that

$$\int_{I_\epsilon^{(M)}} P^{(M)}_{z,0}(dy_1, \ldots, dy_M) > 1 - \frac{\epsilon}{4}. \tag{10.7.37}$$

Since $P^{(M)}_{0,m|z}(\cdot) \to P^{(M)}_{z,0}(\cdot)$ uniformly on finite intervals, we can choose $m_0(\epsilon, M)$ sufficiently large so that for all $m \geqslant m_0(\epsilon, M)$,

$$\int_{I_\epsilon^{(M)}} P^{(M)}_{0,m|z}(dy_1, \ldots, dy_M) > 1 - \frac{\epsilon}{4}. \tag{10.7.38}$$

For each event $B \in \mathfrak{B}^{(m)}$,

(10.7.39)

$$\begin{aligned}
|P^{(M)}_{0,m|z}(B) - P^{(M)}_{z,0}(B)| &\leqslant \int_B |n(y \mid 0, m, z) - n(y \mid z, 0)|\, dy \\
&= \left(\int_{B\cap I_\epsilon^{(M)}} + \int_{B\cap \bar{I}_\epsilon^{(M)}}\right) |n(y \mid 0, m, z) - n(y \mid z, 0)|\, dy \\
&\leqslant \int_{B\cap I_\epsilon^{(M)}} |n(y \mid 0, m, z) - n(y \mid z, 0)|\, dy + \frac{\epsilon}{2},
\end{aligned}$$

where $n(y \mid \mu, m, z)$ designates the joint normal density function corresponding to $P^{(M)}_{\mu,m|z}(\cdot)$ and $n(y \mid \mu, m)$ the joint normal density function corresponding to $P^{(M)}_{\mu,m}(\cdot)$. Finally, since $n(y \mid 0, m, z) \to n(y \mid z, 0)$ uniformly on any finite z interval I, and since $n(y \mid 0, m, z)$ is a uniformly bounded function, we can choose m so large that

$$\text{(10.7.40)} \quad \int_{B \cap I_\epsilon^{(M)}} |n(y \mid 0, m, z) - n(y \mid z, 0)| \, dy$$

$$\leqslant \int_{I_\epsilon^{(M)}} |n(y \mid 0, m, z) - n(y \mid z, 0)| \, dy < \frac{\epsilon}{2}$$

uniformly in $z \in I$. Equation 10.7.33 is now implied from (10.7.39) and (10.7.40). (Q.E.D.)

Lemma 10.7.4. (Blum and Rosenblatt [4].) *If there is a sampling scheme which terminates with probability one, that is,* $P_{\mu,m}[N < \infty] = 1$ *for all* μ, $-\infty < \mu < \infty$, *and all* m, $m = 0, 1, \ldots$, *then there exists a bounded set* K *which is contained in the complement of* $[-l, l]$, *where* $2\Phi(l) - 1 < 1 - 3\alpha$ *for some* $0 < \alpha < \frac{1}{3}$ *such that*

$$\frac{1}{\sqrt{(2\pi)}} \int_K e^{-\frac{1}{2}x^2} \, dx \geqslant 2\alpha.$$

Furthermore, for each positive integer i *there exists a positive integer* b_i *satisfying*

$$\text{(10.7.41)} \qquad P_{\mu,0}\{N \leqslant b_i\} > 1 - \frac{1}{i}, \quad \text{all } \mu \in K.$$

Proof. For each b and i, $P_{\mu,0}\{N \leqslant b\}$ is a continuous function of μ. Hence the set of μ values $B_i = \{\mu; P_{\mu,0}\{N \leqslant b_i\} > 1 - 1/i\}$ is an open set. Therefore B_i and $\bar{B}_i$ are measurable sets, as well as $A_i = \{\mu; P_{\mu,0}\{N \leqslant b_i\} < 1 - 1/i\}$ and its complement $\bar{A}_i$. Choose u_α so that

$$\text{(10.7.42)} \qquad \int_{-u_\alpha}^{-l} \varphi(u) \, du + \int_l^{u_\alpha} \varphi(u) \, du = \frac{5}{2}\alpha,$$

where $\varphi(u) = (1/\sqrt{(2\pi)})e^{-\frac{1}{2}u^2}$. Fix i, and consider the measurable set

$$\text{(10.7.43)} \quad K_{bi} = \left\{\mu \in [-u_\alpha, -l] \cup [l, u_\alpha]; P_{\mu,0}[N \leqslant b] > 1 - \frac{1}{i}\right\};$$

K_{bi} is nondecreasing in b, and $\lim_{b\to\infty} K_{bi} = [-u_\alpha, -l] \cup [l, u_\alpha]$. Therefore, for sufficiently large value of b, say b_i, the $N(0, 1)$ measure of $K_{b_i,i}$ is

$$\text{(10.7.44)} \qquad \int_{K_{b_i,i}} \varphi(u) \, du > \frac{5}{2}\alpha - \frac{\alpha}{2^{i+1}}, \quad i = 1, 2, \ldots.$$

We notice that $b_1 < b_2 < \cdots$. Define $K = \bigcup_{i=1}^{\infty} K_{b_i,i} = K_{b_i,1}$. Then

$$\int_K \varphi(u)\,du > \frac{5}{2}\alpha - \frac{\alpha}{4} = \frac{9}{4}\alpha > 2\alpha. \tag{10.7.45}$$

Finally, if $\mu \in K$ then $\mu \in K_{b_i,i}$ for *all* $i = 1, 2, \ldots$. This implies (10.7.41). (Q.E.D.)

We now prove the main theorem.

Theorem 10.7.5. (Blum and Rosenblatt [4].) *Let $l > 0$ and $0 < \alpha < \frac{1}{3}$ be given. Under the condition that $2\Phi(l) - 1 < 1 - 3\alpha$ if $Y_n = \mu + X_n + Z_{n,m}$, there does not exist a sampling procedure that terminates with probability one for all μ real and $m = 0, 1, \ldots$, under which a fixed-width confidence interval for μ is attainable.*

Proof. The proof is by negation. Suppose there exists a sampling plan, which terminates with probability one and under which the interval estimator $I_\delta(\mathbf{Y}_N) = (\hat{\mu}(\mathbf{Y}_N) - \delta,\ \hat{\mu}(\mathbf{Y}_N) + \delta)$ is a confidence interval, that is,

$$P_{\mu,m}[\mu \in I_\delta(\mathbf{Y}_N)] \geqslant \gamma > 0 \quad \text{for all} \quad (\mu, m). \tag{10.7.46}$$

Then for $\alpha = 1 - \gamma$ we have

$$\begin{aligned} \alpha &\geqslant P_{0,m}[0 \notin I_\delta(\mathbf{Y}_N)] \\ &\geqslant \int_K P_{0,m}[0 \notin I_\delta(\mathbf{Y}_N) \mid Z_{1,m} = z] \cdot dP_{0,m}(Z_{1,m} \leqslant z). \end{aligned} \tag{10.7.47}$$

By the definition of the measurable set K of Lemma 10.7.4, if $z \in K$ then $\{z \in I_\delta(\mathbf{Y}_N)\}$ implies that $\{0 \notin I_\delta(\mathbf{Y}_N)\}$. Hence

$$\begin{aligned} &\int_K P_{0,m}[0 \notin I_\delta(\mathbf{Y}_N) \mid Z_{1,m} = z]\, dP_{0,m}(Z_{1,m} \leqslant z) \\ &\qquad \geqslant \int_K P_{0,m}[z \in I_\delta(\mathbf{Y}_N) \mid Z_{1,m} = z]\, dP_{0,m}(Z_{1,m} \leqslant z). \end{aligned} \tag{10.7.48}$$

Therefore from (10.7.47) and (10.7.48)

$$\int_K P_{0,m}[z \in I_\delta(\mathbf{Y}_N) \mid Z_{1,m} = z]\, dP_{0,m}(Z_{1,m} \leqslant z) \leqslant \alpha. \tag{10.7.49}$$

According to Lemma 10.7.3, for each $\epsilon > 0$ and any i there exists an integer $m(b_i, \epsilon, K)$ sufficiently large so that, for all $m \geqslant m(b_i, \epsilon, K)$,

(10.7.50)

$$|P_{0,m}[z \in I_\delta(\mathbf{Y}_N),\, N \leqslant b_i \mid Z_{1,m} \leqslant z] - P_{z,0}[z \in I_\delta(\mathbf{Y}_N),\, N \leqslant b_i]| < \epsilon.$$

Furthermore

$$(10.7.51)\quad P_{z,\delta}[z \in I_\delta(\mathbf{Y}_N)] - P_{z,\delta}[z \in I_\delta(\mathbf{Y}_N),\, N \leqslant b_i] = P_{z,0}[z \in I_\delta(\mathbf{Y}_N),\, N > b_i] < \frac{1}{i}, \quad \text{for all } z \in K.$$

Hence for all $z \in K$,

$$(10.7.52)\quad |P_{0,m}[z \in I_\delta(\mathbf{Y}_N),\, N \leqslant b_i \mid Z_{1,m} = z] - P_{z,0}[z \in I_\delta(\mathbf{Y}_N)]| < \epsilon + \frac{1}{i}.$$

From (10.7.49) and (10.7.52) we obtain

$$(10.7.53)\quad \int_K \left(P_{z,0}[z \in I_\delta(\mathbf{Y}_N)] - \left(\epsilon + \frac{1}{i}\right)\right) \mathrm{d}P_{0,m}(Z_{1,m} \leqslant z) \leqslant \alpha,$$

for all $m \geqslant m(b_i, \epsilon, K)$. That is, for all $m \geqslant m(b_i, \epsilon, K)$,

$$(10.7.54)\quad \int_K P_{z,0}[z \in I_\delta(\mathbf{Y}_N)]\, \mathrm{d}P_{0,m}(Z_{1,m} \leqslant z) \leqslant \alpha + \left(\epsilon + \frac{1}{i}\right) P_{0,m}(Z_{1,m} \in K).$$

On the other hand, from (10.7.46) and Lemma 10.7.4,

$$(10.7.55)\quad \int_K P_{z,0}[z \in I_\delta(\mathbf{Y}_N)]\, \mathrm{d}P_{0,m}(Z_{1,m} \leqslant z) \geqslant (1-\alpha)\int_K \mathrm{d}P_{0,m}(Z_{1,m} \leqslant z) = 2\alpha(1-\alpha).$$

Finally, since for all $0 < \alpha < \frac{1}{3}$, $\alpha < 2\alpha(1-\alpha)$, (10.7.54) and (10.7.55) are contradictory (ϵ and i^{-1} can be chosen sufficiently small). (Q.E.D.)

10.8. GENERAL EXISTENCE THEOREMS FOR MULTISTAGE AND SEQUENTIAL FIXED-WIDTH CONFIDENCE ESTIMATION

In the previous section we presented two examples of cases requiring at least two stages of sampling and discussed special two-stage sampling schemes which solve the problem of fixed-width confidence intervals. In this section we state and prove several theorems, which establish necessary and sufficient conditions for the existence of certain multistage or sequential sampling schemes for fixed-width interval estimation. The results stated here have been published by Singh [1] and by Blum and Rosenblatt [1]. We start with a general formulation of the problem and of the sampling procedures.

Let $X_1, X_2, \ldots$ be a sequence of i.i.d. random variables having a common distribution F, belonging to a family $\mathcal{F}$. The objective is to estimate a functional $\theta(F)$ on $\mathcal{F}$. We consider here $\theta(F)$ which are real or vector valued.

Suppose that $(X_1, X_2, \ldots, X_N)$, where N is fixed or random, is an observed sample. The problem is to provide an interval estimator for $\theta(F)$, namely, $I_\delta(\mathbf{X}_N) = [\hat{\theta}(\mathbf{X}_N) - \delta, \hat{\theta}(\mathbf{X}_N) + \delta]$ such that

$$P_F\{\theta(F) \in I_\delta(\mathbf{X}_N)\} \geqslant \gamma > 0, \quad \text{all} \quad F \in \mathcal{F}. \tag{10.8.1}$$

We have seen that an interval estimator $I_\delta(\mathbf{X}_N)$ that satisfies (10.8.1) can be constructed only if sampling is performed in a proper manner. Since X_1, $X_2, \ldots$ are i.i.d., the only crucial sampling design problem is the problem of the stopping rule. We distinguish between fixed sample, multistage, and purely sequential stopping rules. The fixed sample stopping rules are those which specify the sample size N before observations commence; N is in these cases a fixed parameter. In a multistage stopping rule the number of stages, say m, $m \geqslant 2$, is specified at the beginning, and so is the first sample size n_1. The stopping rule specifies also $(m - 1)$ measurable functions $J_2(\cdot), \ldots, J_m(\cdot)$ which give the sample sizes of consecutive stages and are defined inductively in the following manner. Let $\mathbf{X}_{n_1}^{(1)}$ be the observed random vector at the first stage, and let $\mathcal{B}^{(1)}$ be the Borel sigma-field generated by $\mathbf{X}_{n_1}^{(1)}$. The function $J_2(\cdot)$ is a non-negative integer valued function of $\mathbf{X}_{n_1}^{(1)}$, which is $\mathcal{B}^{(1)}$-measurable. Let $N_2 = n_1 + J_2$, where J_2 is the random value that $J_2(\cdot)$ realizes. Let $(\mathbf{X}_{n_1}^{(1)}, \mathbf{X}_{J_2}^{(2)})$ be the observed random vector of the two stages and $\mathcal{B}^{(2)}$ the corresponding Borel sigma-field. The function $J_3(\cdot)$ is a function of $(\mathbf{X}_{n_1}^{(1)}, \mathbf{X}_{J_2}^{(2)})$ and is $\mathcal{B}^{(2)}$-measurable. Generally, for every $k = 1, 2, \ldots, m - 1$, let $\mathbf{Y}_{N_k} = (\mathbf{X}_{n_1}^{(1)}, \ldots, \mathbf{X}_{J_k}^{(k)})'$ be the observed random vector of the first k-stage. The function $J_{k+1}(\cdot)$ is a function of $\mathbf{Y}_{N_k}$ and is $\mathcal{B}^{(k)}$-measurable. The stopping rule associated with the specified functions $\{J_2(\cdot), \ldots, J_m(\cdot)\}$ is the following: Terminate sampling after the k-th stage $(k = 1, \ldots, m - 1)$ where k is the least integer value in $\{1, \ldots, m - 1\}$ for which $J_{k+1}(\cdot) = 0$; or after the m-th stage. In a purely sequential scheme the number of stages m is not specified ahead, and each stage consists of one observation only. The stopping rule, described in previous chapters, prescribes a measurable function N such that $\{N = n\} \in \mathcal{B}^{(n)}$; N is a stopping variable. In the case of an m-stage procedure, $N = N^{(m)} = n_1 + J_2 + \cdots + J_m$. We say that the stopping variable N is *proper* or *closed* if sampling terminates with probability one, that is, $P_F\{N < \infty\} = 1$ for all $F \in \mathcal{F}$. Let $\mathcal{P}$ be the family of probability measures on $(\mathcal{X}, \mathcal{B})$ induced by d.f.'s F in $\mathcal{F}$.

Given n random variables $X_1, \ldots, X_n$ from the above sequence of i.i.d. random variables, the associated sample space is $\mathcal{X}^{(n)}$ and the Borel sigma-field is $\mathcal{B}^{(n)}$. If P and Q are two probability measures in $\mathcal{P}$, we define the distance function

$$d^{(n)}(P, Q) = \sup_{B \in \mathcal{B}^{(n)}} |P(B) - Q(B)|, \quad n \geqslant 1, \tag{10.8.2}$$

and

$$d(P, Q) = \sup_{B \in \mathcal{B}} |P(B) - Q(B)|. \tag{10.8.3}$$

If density functions exist with respect to a sigma-finite measure μ, then

(10.8.4)

$$d^{(n)}(P, Q) = \tfrac{1}{2}\int_{\mathfrak{X}^{(n)}} |p(x_1, \ldots, x_n) - q(x_1, \ldots, x_n)|\, \mu(dx_1, \ldots, dx_n),$$

where $p(\cdot)$ and $q(\cdot)$ are the density functions corresponding to P and Q.

Lemma 10.8.1. (Singh [1].) *If N is a proper stopping variable and $\hat{\theta}(\mathbf{X}_N)$ is a statistic which is $\mathcal{B}^{(N)}$-measurable and satisfies*

$$P\left[|\hat{\theta}(\mathbf{X}_N) - \theta(P)| \leqslant \frac{\delta}{2}\right] \geqslant 1 - \alpha, \quad \text{all } P \in \mathcal{T}, \tag{10.8.5}$$

then there exists a $\mathcal{B}^{(N)}$-measurable function $Z^{(N)}$ such that

$$P[|Z^{(N)} - \theta(P)| < \delta] \geqslant 1 - 2\alpha, \quad P \in \mathcal{T}. \tag{10.8.6}$$

Proof. Let $H_N(y) = P[\hat{\theta}(\mathbf{X}_N) \leqslant y \mid \mathcal{B}^{(N)}]$ designate the conditional distribution of $\hat{\theta}(\mathbf{X}_N)$, given $\mathcal{B}^{(N)}$. Let

$$K^{(N)} = \min\left\{k;\, H_N\left(\frac{k}{2}\,\delta\right) \geqslant \tfrac{1}{2}\right\}; \tag{10.8.7}$$

$K^{(N)}$ is an integer valued random variable which is $\mathcal{B}^{(N)}$-measurable. Moreover,

$$\{K^{(N)} = k\} = \left\{H_N\left(\frac{k-1}{2}\,\delta\right) < \frac{1}{2}\right\} \cap \left\{H_N\left(\frac{k}{2}\,\delta\right) \geqslant \frac{1}{2}\right\}. \tag{10.8.8}$$

Define the $\mathcal{B}^{(N)}$-measurable random variable $Z^{(N)} = (\delta/2)K^{(N)}$ where $Z^{(N)}$ is the median of $H_N(y)$. According to the hypothesis of the theorem,

$$\begin{aligned} &P\left\{|\hat{\theta}(\mathbf{X}_N) - \theta(P)| < \frac{\delta}{2}\right\} \\ &\quad = E_P\left\{P\left\{|\hat{\theta}(\mathbf{X}_N) - \theta(P)| < \frac{\delta}{2} \,\middle|\, \mathcal{B}^{(N)}\right\}\right\} \geqslant 1 - \alpha, \quad \text{for all } P \in \mathcal{T}. \end{aligned} \tag{10.8.9}$$

Furthermore, $P[|\hat{\theta}(\mathbf{X}_N) - \theta(P)| < \delta/2 \mid \mathcal{B}^{(N)}]$ is a random variable. We have generally that if V is a random variable such that $P[0 \leqslant V \leqslant 1] = 1$ then, for every ϵ, $0 < \epsilon < 1$,

$$P[V > \epsilon] \geqslant \frac{E\{V\} - \epsilon}{1 - \epsilon}. \tag{10.8.10}$$

Hence for $\epsilon = \frac{1}{2}$ we obtain from (10.8.9) and (10.8.10) that

$$P\left\{P\left[|\hat{\theta}(\mathbf{X}_N) - \theta(P)| < \frac{\delta}{2} \,\middle|\, \mathcal{B}^{(N)}\right] > \frac{1}{2}\right\} \geqslant \frac{1 - \alpha - \frac{1}{2}}{1 - \frac{1}{2}} = 1 - 2\alpha, \tag{10.8.11}$$

for all $P \in \mathcal{F}$. Finally, $P[|\hat{\theta}(\mathbf{X}_N) - \theta(P)| < \delta/2 \mid \mathcal{B}^{(N)}] > \frac{1}{2}$ implies that $|Z^{(N)} - \theta(P)| < \delta$. Hence

$$P[|Z^{(N)} - \theta(P)| < \delta] \geqslant 1 - 2\alpha. \tag{10.8.12}$$

(Q.E.D.)

Theorem 10.8.2. (Singh [1].) *Let N be a proper stopping variable. Then for every $\epsilon > 0$ and every $P \in \mathcal{F}$ there is a $\delta > 0$ such that $d^{(N)}(P, Q) < \epsilon$ if $d^{(1)}(P, Q) < \delta$; δ does not depend on P if $\lim_{k\to\infty} P[N > k] = 0$ uniformly over $\mathcal{F}$.*

Proof. For a given $\epsilon > 0$ and $P \in \mathcal{F}$, let k be a number sufficiently large so that

$$P[N > k] < \frac{\epsilon}{2}. \tag{10.8.13}$$

Such a k exists, since N is a proper stopping variable. We now show that for this k we can choose a $\delta > 0$, $\delta \leqslant d_0(k)$, so that

$$\mathrm{d}^{(N)}(P, Q) \leqslant P(N > k) + \frac{\epsilon}{2}, \tag{10.8.14}$$

for all $P, Q \in \mathcal{F}$, such that $d^{(1)}(P, Q) < \delta$. Since k is a fixed number, we can choose δ, $\delta > 0$, such that $d^{(1)}(P, Q) < \delta$ implies that $d^{(k)}(P, Q) < \epsilon/2(k + 1)$. Let $B \in \mathcal{B}^{(N)}$. Without loss of generality, assume that $N \geqslant 1$ a.s. Then the Q measure of B is

$$\begin{aligned} Q(B) &= \sum_{n=1}^{\infty} Q\{B \mid N = n\}Q(N = n) \\ &\geqslant \sum_{n=1}^{k} E_Q\{I_{B_n}(\mathbf{X}_n)I_{\{N=n\}}(\mathbf{X}_n)\}. \end{aligned} \tag{10.8.15}$$

Since $\{N = n\} \in \mathcal{B}^{(n)}$, we obtain from (10.8.15) and the fact that $d^{(k)}(P, Q) < \epsilon/2(k + 1)$ that

$$\begin{aligned} Q(B) &\geqslant \sum_{n=1}^{k}\left[E_P\{I_{B_n}(\mathbf{X}_n)I_{\{N=n\}}(\mathbf{X}_n)\} - \frac{\epsilon}{2(k + 1)}\right] \\ &\geqslant P(B) - P(N > k) - \frac{\epsilon}{2}, \quad \text{all } B \in \mathcal{B}^{(N)}. \end{aligned} \tag{10.8.16}$$

Hence

$$|P(B) - Q(B)| \leqslant P(N > k) + \frac{\epsilon}{2}, \quad B \in \mathcal{B}^{(N)}. \tag{10.8.17}$$

This implies (10.8.14) and that $d^{(N)}(P, Q) < \epsilon$. Finally, since $\lim_{k\to\infty} P(N > k) = 0$ uniformly over $\mathcal{T}$, we can choose k to satisfy (10.8.13) independently of P. Consequently, δ depends only on ϵ. (Q.E.D.)

Theorem 10.8.3. (Singh [1].) *If $\hat{\theta}(X)$ is a $\mathcal{B}$-measurable statistic such that*

$$P\left[|\hat{\theta}(X) - \theta(P)| < \frac{\delta}{2}\right] \geqslant 1 - \alpha, \quad \text{all } P \in \mathcal{T}, \tag{10.8.18}$$

then $|\theta(P) - \theta(Q)| < \delta$ whenever $d(P, Q) < 1 - 2\alpha$ for $P, Q \in \mathcal{T}$.

Proof. Let

$$Z = \begin{cases} 1 & \text{if } |\hat{\theta}(X) - \theta(P)| < \dfrac{\delta}{2}, \\ 0, & \text{otherwise.} \end{cases}$$

Then

$$\begin{aligned} |E_P\{Z\} - E_Q\{Z\}| &\leqslant \sup_{\psi} |E_P\{\psi(X)\} - E_Q\{\psi(X)\}| \\ &= d(P, Q), P, Q \in P, \end{aligned} \tag{10.8.19}$$

where $\psi; \mathfrak{X} \to [0, 1]$. Therefore

$$E_Q\{Z\} \geqslant E_P\{Z\} - d(P, Q). \tag{10.8.20}$$

Furthermore,

$$\begin{aligned} &Q\left\{|\hat{\theta}(X) - \theta(P)| \leqslant \frac{\delta}{2}, |\hat{\theta}(X) - \theta(Q)| \leqslant \frac{\delta}{2}\right\} \\ &\quad\geqslant Q\left\{|\hat{\theta}(X) - \theta(P)| \leqslant \frac{\delta}{2}\right\} + Q\left\{|\hat{\theta}(X) - \theta(Q)| \leqslant \frac{\delta}{2}\right\} - 1 \\ &\quad\geqslant ((1 - \alpha) - d(P, Q)) + (1 - \alpha) - 1 \\ &\quad= 1 - 2\alpha - d(P, Q). \end{aligned} \tag{10.8.21}$$

If $d(P, Q) < 1 - 2\alpha$ then

$$Q\left\{|\hat{\theta}(X) - \theta(P)| \leqslant \frac{\delta}{2}, |\hat{\theta}(X) - \theta(Q)| \leqslant \frac{\delta}{2}\right\} > 0. \tag{10.8.22}$$

But (10.8.22) is possible only if $|\theta(P) - \theta(Q)| \leqslant \delta$. (Q.E.D.)

In the following theorem we establish necessary conditions for the existence of interval estimators which are, under certain sampling procedures, fixed-width confidence intervals.

Theorem 10.8.4. (Singh [1].) *Let (Θ, D) be a metric space with one-to-one mapping $\theta \to P_\theta$, from Θ to $\mathcal{S}$.*

(i) *If $\theta \to P_\theta$ from $(\Theta, D) \to (\mathcal{S}, d^{(1)})$ is continuous, then the existence of a fixed-width confidence interval $I_\delta(\mathbf{X}_N)$ for $\hat{h}(\theta)$ under a sequential procedure implies that $\hat{h}(\theta)$ is a continuous function on (Θ, D).*

(ii) *If $\theta \to P_\theta$ is uniformly continuous then the existence of a fixed-width confidence interval for $\hat{h}(\theta)$, based on a fixed size sample, implies that $\hat{h}(\theta)$ is uniformly continuous on (Θ, D).*

Proof. (i) For every $\theta \in \Phi$ define $h(P)$ on $\mathcal{S}$ by $h(P_\theta) = \hat{h}(\theta)$. We have to show that the existence of a fixed-width confidence interval for $h(P_\theta)$, based on a sequential sampling procedure, implies that $h(P)$ is continuous on $(\mathcal{S}, d^{(1)})$. Let N be a proper stopping variable and $I_\delta(\mathbf{X}_N) = (\hat{\theta}(\mathbf{X}_N) - \delta, \hat{\theta}(\mathbf{X}_N) + \delta)$ an interval estimator. Suppose that $\hat{\theta}(\mathbf{X}_N)$ satisfies

$$(10.8.23) \qquad P\left\{|\hat{\theta}(\mathbf{X}_N) - h(P)| < \frac{\xi}{2}\right\} \geqslant \frac{2}{3}, \quad \text{all } P \in \mathcal{S}.$$

By Theorem 10.8.2 there exists $\eta > 0$ such that, for $Q \in \mathcal{S}$, $d^{(1)}(P, Q) < \eta$ implies that $d^{(N)}(P, Q) < \frac{1}{3}$. Consequently, from Theorem 10.8.3 we imply that $|h(P) - h(Q)| < \eta$. Thus $h(P_\theta) = \hat{h}(\theta)$ is continuous on (Θ, D).

(ii) If $\theta \to P_\theta$ is uniform on (Θ, D), a similar proof verifies that the existence of a fixed-width confidence interval, under a fixed sample size procedure, implies the uniform continuity of $h(P)$ on $(\mathcal{S}, d^{(1)})$. (Q.E.D.)

Example 10.14. (Singh [1].) Let $\mathcal{F}$ be the class of Poisson distributions; that is, $\mathcal{F} = \{P(\lambda); 0 < \lambda < \infty\}$. Let $h(P_\lambda) = \lambda$. Then

$$(10.8.24) \quad \mathrm{d}^{(1)}(P_{\lambda_1}, P_{\lambda_2}) = \tfrac{1}{2}\sum_{x=0}^{\infty}\left|\frac{e^{-\lambda_1}\lambda_1^{\,x}}{x!} - \frac{e^{-\lambda_2}\lambda_2^{\,x}}{x!}\right|$$

$$= \sum_{x=0}^{x_0}\left|\frac{e^{-\lambda_1}\lambda_1^{\,x}}{x!} - \frac{e^{-\lambda_2}\lambda_2^{\,x}}{x!}\right| = e^{|\lambda_1 - \lambda_2|} - 1,$$

where x_0 is the largest integer which does not exceed $(\lambda_2 - \lambda_1)/\ln(\lambda_2/\lambda_1)$. Hence $h(P_\lambda)$ is uniformly continuous on $(\mathcal{S}, d^{(1)})$. However, a fixed sample procedure of a fixed-width confidence interval estimation *does not exist.* ■

We investigate now the conditions under which there exists a two-stage sampling procedure which yields a fixed-width confidence interval. We start with a general formulation of the two-stage procedure discussed in Section 10.7.2. This formulation will give us some clues about the nature of some sufficient conditions.

Suppose that a two-stage procedure has a first sample of size n_1 (fixed). We consider a sequence $\{\hat{\theta}(\mathbf{X}_n); n \geqslant 1\}$ of consistent estimators of $\theta(P)$, that is, for any given δ, $\delta > 0$,

$$\lim_{n\to\infty} P\{|\hat{\theta}(\mathbf{X}_n) - \theta(P)| < \delta\} = 1, \quad \text{all } P \in \mathcal{T}. \tag{10.8.25}$$

Suppose that $\hat{\theta}(\mathbf{X}_n)$ has a finite variance and that

$$\text{Var}_P\,\{\hat{\theta}(\mathbf{X}_n)\} = \frac{D^2(P)}{n} < \infty, \quad \text{all } P \in \mathcal{T}. \tag{10.8.26}$$

If we can determine a β upper confidence limit for $D^2(P)$, say $\bar{D}_\beta{}^2(\mathbf{X}_{n_1})$, based on the first sample $(X_1, \ldots, X_{n_1})$, then we can determine a stopping variable N_2 so that

$$P\{|\hat{\theta}(X_{n_1+1}, \ldots, X_{n_1+N_2}) - \theta(P)| < \delta\} \geqslant \gamma, \quad \text{all} \quad P \in \mathcal{T}. \tag{10.8.27}$$

This can be done, as in Section 10.7.2, on the basis of the Tchebytchev inequality in the following manner. Let $\beta > \gamma$ and define

$$n_2(P) = \text{least integer } n \text{ such that } n \geqslant \frac{\beta D^2(P)}{\delta^2(\beta - \gamma)}. \tag{10.8.28}$$

In analogy to (10.8.28), define the stopping variable

$$N_2 = \text{least integer } n \text{ such that } n \geqslant \frac{\beta \bar{D}_\beta{}^2(X_{n_1})}{\delta^2(\beta - \gamma)}. \tag{10.8.29}$$

We notice that N_2 is a random variable which is $\mathcal{B}^{(n_1)}$-measurable. Moreover, since

$$P\{\bar{D}_\beta{}^2(\mathbf{X}_{n_1}) < \infty\} = 1, \quad \text{all} \quad P \in \mathcal{T}, \tag{10.8.30}$$

N_2 is a proper stopping variable. The conditional joint distribution of $(X_{n_1+1}, \ldots, X_{n_1+N_2})$, given $\{N_2 = k\}$, is like that of $(X_1, X_2, \ldots, X_k)$. Hence

(10.8.31)

$$\begin{aligned}
&P\{|\hat{\theta}(X_{n_1+1}, \ldots, X_{n_1+N_2}) - \theta(P)| < \delta\} \\
&\quad \geqslant P\{|\hat{\theta}(X_{n_1+1}, \ldots, X_{n_1+N_2}) - \theta(P)| < \delta \mid N_2 \geqslant n_2(P)\}P\{N_2 \geqslant n_2(P)\} \\
&\quad \geqslant \sum_{n=n_2(P)}^{\infty} P[N_2 = n]P\{|\hat{\theta}(X_{n_1+1}, \ldots, X_{n_1+N_2}) - \theta(P)| < \delta \mid N_2 = n\} \\
&\quad = \sum_{n=n_2(P)}^{\infty} P[N_2 = n]P\{|\hat{\theta}(X_1, \ldots, X_n) - \theta(P)| < \delta\} \\
&\quad \geqslant \left(1 - \frac{D^2(P)}{\delta^2 n_2(P)}\right)P[N_2 \geqslant n_2(P)] \geqslant \gamma, \quad \text{all } P \in \mathcal{T}.
\end{aligned}$$

This proves (10.8.27). Now we generalize these results. We first notice that the convergence in (10.8.25) is generally not a uniform one (which is the main reason for the nonexistence of a fixed sample procedure). Thus for specified values of γ, δ, $0 < \gamma < 1$, $0 < \delta < \infty$, and $P \in \mathfrak{T}$, we determine $n(\gamma, \delta, P)$ such that for all $n \geqslant n(\gamma, \delta, P)$

$$P[|\hat{\theta}(\mathbf{X}_n) - \theta(P)| < \delta] \geqslant \gamma.$$

We designate by $\mathfrak{T}_{m,\gamma,\delta}$ a subfamily of probability measures $\mathfrak{T}_{m,\gamma,\delta} \subset \mathfrak{T}$ such that $n(\gamma, \delta, P) \geqslant m$ for all $P \in \mathfrak{T}_{m,\gamma,\delta}$. Clearly if $m_1 < m_2$, then $\mathfrak{T}_{m_2,\gamma,\delta} \subset \mathfrak{T}_{m_1,\gamma,\delta}$.

The two-stage procedure which we described above has the following two important characteristics. Let

$$S_j = \{\mathbf{X}_{n_1}; N_2 \geqslant j\}, \qquad j = 0, 1, \ldots .$$

(i) Since N_2 is a proper stopping variable,

$$\lim_{j\to\infty} P[S_j] = 0, \quad \text{all } P \in \mathfrak{T}. \tag{10.8.32}$$

(ii) If for some β, $0 < \gamma < \beta < 1$, there exists a statistic $\bar{D}_\beta(\mathbf{X}_{n_1})$ such that $P[\bar{D}_\beta(\mathbf{X}_{n_1}) \geqslant D^2(P)] \geqslant \beta$ for all $P \in \mathfrak{T}$, then $P[N_2 \geqslant n_2(P)] \geqslant \beta$. Therefore, if $P \in \mathfrak{T}_{m,(\gamma/\beta),\delta}$ then $n_2(P) \geqslant m$ and

$$P[S_m] = P[N \geqslant m] \geqslant P[N_2 \geqslant n_2(P)] \geqslant \beta. \tag{10.8.33}$$

Properties (i) and (ii) define a general family of probability measures for which a fixed-width confidence interval estimator exists under a two-stage sampling procedure. This is stated in the following theorem.

Theorem 10.8.5. (Blum and Rosenblatt [1].) *Let $\mathfrak{T}$ be a family of probability measures on $(\mathfrak{X}, \mathfrak{B})$. Let $X_1, X_2, \ldots$ be a sequence of i.i.d. random variables having a common distribution in $\mathfrak{T}$. Let $\{\hat{\theta}(\mathbf{X}_n); n \geqslant 1\}$ be a sequence of consistent estimators of $\theta(P)$, $P \in \mathfrak{T}$. Assume the following:*

(i) *For a fixed positive integer, n_1, there exists a sequence $\{S_j; j \geqslant 1\}$ of $\mathfrak{B}^{(n_1)}$-measurable sets such that*

$$\lim_{j\to\infty} P[S_j] = 0, \quad \textit{all } P \in \mathfrak{T}; \tag{10.8.34}$$

where

$$P[S_j] = \int_{S_j} \prod_{v=1}^{n_1} F_P(dx_v), \quad j = 1, 2, \ldots \tag{10.8.35}$$

and $P \in \mathfrak{T}$.

(ii) *There exists a constant β, $0 < \gamma < \beta < 1$, such that*

$$\inf_{P\in\mathfrak{T}_{m,(\gamma/\beta),\delta}} P[S_m] \geqslant \beta, \quad \textit{all } m = 0, 1, \ldots, \tag{10.8.36}$$

where $S_0 \equiv \mathfrak{X}^{(n_1)}$ and $\mathfrak{T}_{0,(\gamma/\beta),\delta} \equiv \mathfrak{T}$.

Then there exists a two-stage sampling scheme with a γ fixed-width confidence interval for $\theta(P)$ whose first sample is of size n_1. The associated stopping variable for the size of the second-stage sample is

(10.8.37) $N_2 =$ *least non-negative integer n such that*

$$(X_1, \ldots, X_{n_1}) \in S_k, \qquad k = 0, \ldots, n \quad \textit{and} \quad (X_1, \ldots, X_{n_1}) \notin S_{n+1}.$$

Proof. Equation 10.8.34 guarantees that N_2 is a proper stopping variable. Furthermore, N_2 is $\mathcal{B}^{(n_1)}$-measurable, and therefore the event $\{N_2 \geqslant k\}$ and the events in the Borel sigma-field generated by $\{X_{n_1+1}, X_{n_1+2}, \ldots\}$ are independent, $k = 0, 1, \ldots$. The interval

$$\begin{aligned} I_\delta(\mathbf{X}_{n_1+N_2}) &= [\hat{\theta}(\mathbf{X}_{n_1+N_2}) - \delta, \hat{\theta}(\mathbf{X}_{n_1+N_2}) + \delta] \\ &= \bigcup_{j=0}^{\infty} \{[\hat{\theta}(\mathbf{X}_{n_1+j}) - \delta, \hat{\theta}(\mathbf{X}_{n_1+j}) + \delta], N_2 = j\} \end{aligned}$$

is measurable. We now show that $I_\delta(\mathbf{X}_{n_1+N_2})$ is a fixed-width confidence interval for $\theta(P)$, whenever $\hat{\theta}(\mathbf{X}_{n_1+N_2})$ depends only on $(X_{n_1+1}, \ldots, X_{n_1+N_2})$.

For each $P \in \mathcal{T}$ let $n_2(P) \equiv n(\gamma/\beta), \delta, P)$, which is the maximal positive integer j satisfying $P \in \mathcal{T}_{j,(\gamma/\beta),\delta}$. According to this definition of $n_2(P)$ and (10.8.36),

$$P[S_{n_2(P)}] \geqslant \beta, \quad \text{all} \quad P \in \mathcal{T}. \tag{10.8.38}$$

Furthermore, according to (10.8.37), if $(X_1, \ldots, X_{n_1}) \in S_{n_2(P)}$ then $N_2 \geqslant n_2(P)$. Hence

$$P[N_2 \geqslant n_2(P)] \geqslant P[S_{n_2(P)}] \geqslant \beta, \quad \text{all} \quad P \in \mathcal{T}. \tag{10.8.39}$$

Finally, the coaverage probability of $I_\delta(\mathbf{X}_{n_1+N_2})$ is

$$\begin{aligned} &P\{|\hat{\theta}(\mathbf{X}_{n_1+N_2}) - \theta(P)| < \delta\} \\ &\geqslant \sum_{n=n_2(P)}^{\infty} P[N_2 = n] P\{|\hat{\theta}(\mathbf{X}_{n_1+N_2}) - \theta(P)| < \delta \mid N_2 = n\} \\ &= \sum_{n=N_2(P)}^{\infty} P\{|\hat{\theta}(\mathbf{X}_n) - \theta(P)| < \delta] P[N_2 = n] \\ &\geqslant \frac{\gamma}{\beta} P[N_2 \geqslant n_2(P)] \geqslant \frac{\gamma}{\beta} \cdot \beta = \gamma, \quad \text{all } P \in \mathcal{T}. \end{aligned} \tag{10.8.40}$$

(Q.E.D.)

In the above mentioned study [1], Blum and Rosenblatt generalized these sufficient conditions of existence to cases of m-stage sampling ($m > 2$). We notice that there is no restriction on the sample size of the first stage n_1 only through conditions (i) and (ii) of Theorem 10.8.5. As shown by Abbott and Rosenblatt [1], there are cases (e.g., the family of normal distributions) in which we can take only one observation in the first stage. Such procedures are generally inefficient. To construct efficient fixed-width estimators we generally need purely sequential procedures. In the following sections we discuss such sequential procedures, which are asymptotically optimal.

10.9. ASYMPTOTICALLY EFFICIENT SEQUENTIAL FIXED-WIDTH CONFIDENCE INTERVALS FOR THE MEAN OF A NORMAL DISTRIBUTION

Let $X_1, X_2, \ldots$ be a sequence of i.i.d. random variables having a common $\mathcal{N}(\mu, \sigma^2)$ distribution, where both μ and σ are unknown; $-\infty < \mu < \infty$; $0 < \sigma < \infty$. We wish to determine a fixed-width confidence interval estimator for μ. In Section 10.7.1 we presented the Stein procedure, which solves the problem of fixed-width confidence intervals for the mean μ. However, the Stein procedure is not an efficient one in terms of the expected sample size. To show this we have to define first the notion of efficiency of a sampling procedure which yields fixed-width confidence intervals.

If σ^2 is known, there exists a fixed sample procedure with size $n_0 \equiv n_0(\delta, \sigma)$, where

$$n_0(\delta, \sigma) = \text{least integer } n \text{ such that } n \geqslant \frac{a^2\sigma^2}{\delta^2}, \quad a = \Phi^{-1}\left(\frac{1+\gamma}{2}\right), \quad 0 < \gamma < 1, \tag{10.9.1}$$

and an interval estimator $I_\delta(\mathbf{X}_{n_0}) = (\bar{X}_{n_0} - \delta, \bar{X}_{n_0} + \delta)$ such that $P_{\mu\sigma}\{\mu \in I_\delta(\mathbf{X}_{n_0})\} \geqslant \gamma$ for all (μ, σ). Wald and Stein [1] have proven that when σ^2 is known, all sequential (and in particular multistage) procedures which solve the fixed-width confidence intervals problem have an expected sample size greater than or equal to $n_0(\delta, \sigma)$. If σ^2 is unknown we can define the efficiency of a sampling procedure, with a proper stopping variable $N(\delta)$, to be the ratio of the expected sample sizes, that is, $n_0(\delta, \sigma)/E_{\mu,\sigma}\{N(\delta)\}$. A sampling procedure is called *asymptotically efficient* if

$$\lim_{\delta \to 0} \frac{n_0(\delta, \sigma)}{E_{\mu,\sigma}\{N(\delta)\}} \geqslant 1, \quad \text{all } (\mu, \sigma). \tag{10.9.2}$$

Returning to the Stein two-stage procedure, the stopping variable $N(\delta)$, for a first sample size of size n_1, is

$$N(\delta) = n_1 I_A\{s_{n_1}^2\} + \left(\left[\frac{t_{\gamma'}^2[n_1 - 1)s_{n_1}^2}{\delta^2}\right] + 1\right) I_{\bar{A}}\{s_{n_1}^2\}, \tag{10.9.3}$$

where

$$A = \left\{s_{n_1}^2; s_{n_1}^2 \leqslant \frac{n_1 \delta^2}{t_{\gamma'}^2[n_1 - 1]}\right\}, \quad \gamma' = \frac{1+\gamma}{2},$$

and $[x]$ is the integral part of x. Hence

(10.9.4)

$$E_\sigma\{N(\delta)\} \geqslant n_1 P\left\{\chi^2[n_1 - 1] \leqslant \frac{n_1(n_1 - 1)}{n_0(\delta, \sigma)\rho(\gamma, n_1)}\right\} + \frac{n_0(\delta, \sigma)\rho(\gamma, n_1)}{n_1 - 1} \int_{(n_1(n_1-1))/(n_0(\delta,\sigma)\rho(\gamma,n_1))}^{\infty} x \, dP\{\chi^2[n_1 - 1] \leqslant x\},$$

where

$$\rho(\gamma, n_1) = \frac{t_{\gamma'}^2[n_1 - 1]}{u_{\gamma'}^2}. \tag{10.9.5}$$

We notice that $t_{\gamma'}^2[n_1 - 1] > u_{\gamma'}^2$ for all $n_1 \geqslant 2$ and γ'. Hence $\rho(\gamma, n_1) > 1$ for each $0 < \gamma < 1$, $n_1 \geqslant 2$. Since

$$n_1 P\left\{\chi^2[n_1 - 1] \leqslant \frac{n_1(n_1 - 1)}{n_0(\delta, \sigma)\rho(\gamma, n_1)}\right\} \geqslant \frac{n_0(\delta, \sigma)\rho(\gamma, n_1)}{n_1 - 1} \int_0^{(n_1(n_1-1))/(n_0(\delta,\sigma)\rho(\gamma,n_1))} x \, dP\{\chi^2[n_1 - 1] \leqslant x\}, \tag{10.9.6}$$

we obtain from (10.9.4) that

$$\begin{aligned} E_\sigma\{N(\delta)\} &\geqslant \frac{n_0(\delta, \sigma)\rho(\gamma, n_1)}{n_1 - 1} E\{\chi^2[n_1 - 1]\} \\ &= n_0(\delta, \sigma)\rho(\gamma, n_1). \end{aligned} \tag{10.9.7}$$

Hence the relative efficiency function of the Stein procedure is

$$\frac{n_0(\delta, \sigma)}{E_\sigma\{N(\delta)\}} \leqslant \frac{1}{\rho(\gamma, n_1)}, \quad \text{all } (\mu, \sigma). \tag{10.9.8}$$

The procedure is not even asymptotically efficient since

$$\overline{\lim_{\delta \to 0}} \frac{n_0(\delta, \sigma)}{E_\sigma\{N(\delta)\}} \leqslant \frac{1}{\rho(\gamma, n_1)} < 1, \quad \text{all } (\mu, \sigma). \tag{10.9.9}$$

It is easy to derive from (10.9.3) that

$$\lim_{n_1 \to \infty} \lim_{\sigma \to \infty} \frac{n_0(\delta, \sigma)}{E_{\mu,\sigma}\{N(\delta)\}} \geqslant 1. \tag{10.9.10}$$

This means that for a sufficiently large initial sample size n_1 and very large σ^2 the Stein procedure is almost efficient. On the other hand, as $\sigma^2 \to 0$,

$$\lim_{\sigma \to 0} \frac{n_0(\delta, \sigma)}{E_\sigma\{N(\delta)\}} \leqslant \frac{1}{n_1 + \rho(\gamma, n_1)}. \tag{10.9.11}$$

This shows that the Stein procedure could be very inefficient for small values of σ. In several studies an attempt has been made to improve the efficiency of the Stein procedure by some modified two-stage procedures and by sequential procedures. In particular, see Seelbinder [1] and Moshman [1] for the proper selection of n_1 in a two-stage procedure. We devote the rest of this section to sequential procedures which are asymptotically efficient (as $\delta \to 0$) for all (μ, σ).

Consider the following sequential procedure. Take initially n_1, $n_1 \geqslant 2$, observations. Afterwards sample the random variables in a purely sequential manner. After each observation compute $s_n^2 = (1/(n-1)) \sum_{i=1}^{n} (X_i - \bar{X}_n)^2$, $n \geqslant n_1$. Terminate sampling at

(10.9.12) $N(\delta)$ = least non-negative integer n such that

$$n \geqslant \max\left\{n_1, \frac{a_n^2 s_n^2}{\delta^2}\right\},$$

where $\{a_n\}$ is a sequence of positive constants converging to $u_{\gamma'}$, $u_{\gamma'} = \Phi^{-1}(\gamma')$, $\gamma' = (1 + \gamma)/2$. After sampling is terminated, estimate μ by

$$I_\delta(\mathbf{X}_N) = (\bar{X}_{N(\delta)} - \delta, \bar{X}_{N(\delta)} + \delta).$$

Lemma 10.9.1. *The stopping variable $N(\delta)$ defined in* (10.9.12) *has the following properties:*

(i) *$N(\delta)$ is a proper stopping variable for each (μ, σ);*
(ii) *if $\delta_1 < \delta_2$, then $N(\delta_1) \geqslant N(\delta_2)$ a.s., and* $\lim_{\delta \to 0} N(\delta) = \infty$ *a.s.;*
(iii) *$E_\sigma\{N(\delta)\} < \infty$, all σ, $0 < \sigma < \infty$; $0 < \delta < \infty$;*
(iv) $\lim_{\delta \to 0} N(\delta)/n_0(\delta, \sigma) = 1$ *a.s., all σ, $0 < \sigma < \infty$;*
(v) $\lim_{\delta \to 0} E_\sigma\{N(\delta)\}/n_0(\delta, \sigma) = 1$, *all σ, $0 < \sigma < \infty$.*

Proof. (i) For each value of $\delta > 0$ and $0 < \sigma < \infty$, if $n \geqslant n_1$

$$\begin{aligned} P_\sigma\{N(\delta) > n\} &= P_\sigma\left\{\bigcap_{j=n_1}^{n} \left[s_j^2 > \frac{j\delta^2}{a_j^2}\right]\right\} \\ &\leqslant P_\sigma\left\{s_n^2 > \frac{n\delta^2}{a_n^2}\right\} = P\left\{\chi^2[n-1] > \frac{n(n-1)\delta^2}{a_n^2\sigma^2}\right\} \\ &\approx 1 - \Phi\left(\frac{n(n-1)\delta^2/a_n^2\sigma^2 - (n-1)}{\sqrt{(2(n-1))}}\right) \\ &= O(n^{-3/2}e^{-1/2 n^3}), \quad \text{as } n \to \infty. \end{aligned} \tag{10.9.13}$$

This order of magnitude is attained by noticing that $(1/\sqrt{n})(\chi^2[n] - n) \xrightarrow{d} N(0, 2)$ and $1 - \Phi(x) \approx (1/x)\varphi(x)$ as $x \to \infty$, where $\varphi(x)$ is the $N(0, 1)$ density function. Hence $\lim_{n\to\infty} P_\sigma\{N(\delta) > n\} = 0$ for each $\delta > 0$, $0 < \delta < \infty$.

(ii) If $\delta_1 < \delta_2$, then $\max\{n_1, (a_n^2 s_n^2)/\delta_1^2\} \geqslant \max\{n_1, (a_n^2 s_n^2)/\delta_2^2\}$ a.s. Hence $N(\delta_1) \geqslant N(\delta_2)$ a.s. Moreover, $\lim_{\delta\to 0} (a_n^2 s_n^2)/\delta^2 = \infty$ a.s. for each $n \geqslant 2$. This proves (ii).

(iii)
$$\begin{aligned} E_\sigma\{N(\delta)\} &= \sum_{n=n_1}^{\infty} n P_\sigma\{N(\delta) = n\} \\ &= n_1 P_\sigma\{N(\delta) \geqslant n_1\} + \sum_{j=n_1+1}^{\infty} P_\sigma\{N(\delta) \geqslant j\} \\ &\leqslant n_1 + \sum_{j=n_1+1}^{\infty} P_\sigma\{N(\delta) \geqslant j\}. \end{aligned}$$

From (10.9.13) we immediately obtain that $\sum_{j=n_1+1}^{\infty} P_\sigma\{N(\delta) \geqslant j\} < \infty$ for all $\delta > 0$, $0 < \sigma < \infty$.

(iv) $\lim_{n\to\infty} s_n^2 = \sigma^2$ a.s. Hence, since $\lim_{\delta\to 0} N(\delta) = \infty$ a.s., $\lim_{\delta\to 0} s^2_{N(\delta)} = \sigma^2$ a.s. Furthermore, $\lim_{\delta\to 0} a^2_{N(\delta)} = a^2$ a.s. Hence

$$\lim_{\delta\to 0} \frac{N(\delta)}{n_0(\delta, \sigma)} = 1 \quad \text{a.s.} \tag{10.9.14}$$

(v) From (10.9.14) and the Lebesgue dominated convergence theorem, in order to prove that

$$\lim_{\delta\to 0} \frac{E_\sigma\{N(\delta)\}}{n_0(\delta, \sigma)} = 1, \quad \text{all } \sigma, \quad 0 < \sigma < \infty, \tag{10.9.15}$$

it is sufficient to show that $N(\delta)/n_0(\delta, \sigma)$ is bounded by a δ-free integrable function. The definitions of $N(\delta)$ and of $n_0(\delta, \sigma)$ imply that if $s^2_{N(\delta)}$ is bounded above by a δ-free integrable function, (10.9.15) holds. We notice furthermore that $s^2_{N(\delta)}$ is bounded above by a δ-free integrable function if $W = \sup_{n\geqslant 2} s_n^2$ has a finite second moment. For each $n \geqslant 2$, $s_n^2 \leqslant 2\sigma^2 \sum_{i=1}^{n} U_i^2/n$, where $U_1, U_2, \ldots$ is a sequence of i.i.d. random variables distributed like $N(0, 1)$. Hence from the celebrated Wiener ergodic theorem (Wiener [1]), the existence of the second moment of U_1^2 implies that

$$E\left\{\left[\sup_{n\geqslant 1} \frac{1}{n} \sum_{i=1}^{n} U_i^2\right]^2\right\} < \infty. \tag{10.9.16}$$

(Q.E.D.)

We remark that Lemma 10.9.1 was proven by Starr [1] in a somewhat different, and in parts more complicated, manner. The above lemma establishes the asymptotic efficiency of the sequential procedure specified by the stopping variable, (10.9.12). In the following lemma we prove a stronger result.

Lemma 10.9.2. (Simons [1].) *If $N(\delta)$ is the stopping variable defined in* (10.9.12) *with $a_n = a$ for all $n \geqslant n_1$, then*

$$E_\sigma\{N(\delta)\} \leqslant n_0(\delta, \sigma) + n_1 + 1, \quad \text{all} \quad \delta > 0, \qquad 0 < \sigma < \infty. \tag{10.9.17}$$

Proof. According to (10.9.12), writing N for $N(\delta)$ we obtain

$$\begin{aligned} N - 1 &< \max\left\{n_1, 1 + \left[\frac{a^2}{\delta^2} s_{N-1}^2\right]\right\} \\ &\leqslant n_1 + 1 + \frac{a^2}{\delta^2}\cdot\frac{1}{N-2}\sum_{i=1}^{N-1}(X_i - \bar{X}_{N-1})^2 \\ &\leqslant n_1 + 1 + \frac{a^2}{\delta^2}\cdot\frac{1}{N-2}\sum_{i=1}^{N-1}(X_i - \mu)^2. \end{aligned} \tag{10.9.18}$$

From (10.9.18) we imply

$$\begin{aligned} E_\sigma\{(N-1)(N-2)\} &\leqslant (n_1 + 1)E_\sigma\{N - 2\} \\ &\quad + \frac{n_0(\delta, \sigma)}{\sigma^2} E_\sigma\left\{\sum_{i=1}^{N-1}(X_i - \mu)^2\right\}. \end{aligned} \tag{10.9.19}$$

Moreover,

$$E_\sigma\{(N-1)(N-2)\} \geqslant (E_\sigma\{N\})^2 - 3E_\sigma\{N\} + 2, \tag{10.9.20}$$

and from the Wald theorem (Lemma 4.4.1), since $E_\sigma\{N(\delta)\} < \infty$,

$$E_\sigma\left\{\sum_{i=1}^{N-1}(X_i - \mu)^2\right\} = \sigma^2 E_\sigma\{N - 1\}. \tag{10.9.21}$$

From (10.9.19)–(10.9.21) we derive the inequality

$$(E_\sigma\{N\})^2 - (3 + n_1 + n_0(\delta, \sigma))E_\sigma\{N\} + (2n_1 + n_0(\delta, \sigma) + 2) \leqslant 0. \tag{10.9.22}$$

Equation 10.9.17 is implied from (10.9.22). (Q.E.D.)

Actually we can obtain from (10.9.22) a somewhat stronger result than (10.9.17). Equation 10.9.17 means that the expected sample size is close to the minimal sample size required for attaining the specified coverage probability. There is, however, a danger in being too efficient in terms of the expected sample size. This is the danger that the corresponding interval estimator $I_\delta(\mathbf{X}_N)$ will not attain the specified coverage probability. This is indeed the case, as shown by a series of numerical computations published by Starr [1]. The numerical results of Starr [1] show, however, a good performance of the interval estimator $I_\delta(\mathbf{X}_N)$ over a wide range of parameter values. Simons [1] has proven that there exists, in the normal case, a finite integer k, which is independent of δ and of σ, such that an additional k observations will guarantee the prescribed coverage probability γ (k may depend on γ). In order to prove this important result, we start with a lemma of Robbins.

Lemma 10.9.3. (Robbins [3].) *Let $X_1, X_2, \ldots$ be a sequence of i.i.d. random variables distributed like $N(\mu, \sigma^2)$; $-\infty < \mu < \infty$, $0 < \sigma < \infty$. For any fixed n, $n \geqslant 2$, let $(s_2^2, \ldots, s_n^2)$ be the sample variances of the first 2, 3, ..., n observations. Let $\mathcal{B}^{(n)} = \mathcal{B}(s_2^2, \ldots, s_n^2)$ be the Borel sigma-field generated by $(s_2^2, \ldots, s_n^2)$. Let $\bar{X}_n$ be the sample mean of the first n observations. Then $\bar{X}_n$ is independent of $\mathcal{B}^{(n)}$ for each $n = 2, 3, \ldots$.*

Proof. Let $Y_i = (X_i - \mu)/\sigma$; $Y_i \sim N(0, 1)$ for all $i = 1, 2, \ldots$. Let $\mathbf{Y}_n = (Y_1, \ldots, Y_n)'$, and make the Helmert orthogonal transformation

$$\mathbf{U}_n = \mathbf{P}\mathbf{Y}_n,$$

where

$$\mathbf{P} = \begin{bmatrix} \dfrac{1}{\sqrt{n}} & \mathbf{1}'_n & & & \\ \dfrac{1}{\sqrt{2}} & -\dfrac{1}{\sqrt{2}} & & O'_{n-2} & \\ \dfrac{1}{\sqrt{6}} & \dfrac{1}{\sqrt{6}} & -\dfrac{2}{\sqrt{6}} & & O'_{n-3} \\ & & & \cdot & \\ & & & & \cdot \\ & & & & \quad\cdot \\ \dfrac{1}{\sqrt{(n(n-1))}}\,\mathbf{1}'_{n-1} & & & & -\dfrac{n-1}{\sqrt{(n(n-1))}} \end{bmatrix}.$$

The vector $\mathbf{1}'_k = (1, 1, \ldots, 1)$, $k = 2, \ldots, n$, is a k-dimensional vector of 1's, and $\mathbf{0}'_k = (0, 0, \ldots, 0)$ is a k-dimensional vector of 0's. Then

$$\mathbf{U}_n \sim N(\mu\mathbf{P}\mathbf{1}_n, I_n), \qquad \mathbf{P}\mathbf{1}_n = (\sqrt{n}, \mathbf{0}'_{n-1})'.$$

Thus $U_1, U_2, \ldots, U_n$ are independent, $U_1 \sim N(\mu\sqrt{n}, 1)$, and $U_i \sim N(0, 1)$, $i = 2, \ldots, n$. For each $i = 1, \ldots, n$, let $\mathbf{U}_{(i)} = (U_1, \ldots, U_i)'$, $\mathbf{Y}_{(i)} = (Y_1, \ldots, Y_i)'$, and $\mathbf{P}_{(i)}$ be the $i \times i$ upper left corner submatrix of $\mathbf{P}$. Then for every $i = 2, \ldots, n$,

$$(10.9.23) \qquad \begin{aligned} s_i^2 &= \frac{\sigma^2}{i-1}\,\mathbf{Y}'_{(i)}\left(I_{(i)} - \frac{1}{i}\,J_{(i)}\right)\mathbf{Y}_{(i)} \\ &= \frac{\sigma^2}{i-1}\,\mathbf{U}'_{(i)}P_{(i)}\left(I_{(i)} - \frac{1}{i}\,J_{(i)}\right)P'_{(i)}\mathbf{U}_{(i)}, \end{aligned}$$

where $I_{(i)}$ is the identity matrix of order i and $J_{(i)} = 1_i 1_i'$. Furthermore,

$$(10.9.24)\quad P_{(i)}\left(I_{(i)} - \frac{1}{i} J_{(i)}\right)P'_{(i)} = I_{(i)} - \frac{1}{i} P_{(i)} 1_i 1_i' \mathbf{P}'_{(i)}$$

$$= \left(\begin{array}{c|c} 0 & 0'_{i-1} \\ \hline 0_{i-1} & I_{i-1} \end{array}\right), \qquad i = 2, \ldots, n.$$

From (10.9.23) and (10.9.24) we obtain that

$$(10.9.25)\quad s_i^2 \sim \frac{\sigma^2}{i-1} \sum_{j=2}^{i} U_j^2 \sim \frac{\sigma^2}{i-1} \chi^2[i-1], \qquad i = 2, \ldots, n.$$

On the other hand,

$$\bar{X}_n \sim \mu + \frac{\sigma}{\sqrt{n}} U_1.$$

Finally, since U_1 is independent of $(U_2, \ldots, U_n)$, $\bar{X}_n$ is independent of $(s_2^2, \ldots, s_n^2)$ for each $n = 2, 3, \ldots$. (Q.E.D.)

We apply Lemma 10.9.3 for the evaluation of the coverage probability of the interval estimator $I_\delta(\mathbf{X}_{N(\delta)})$, where $N(\delta)$ is defined by (10.9.12). This coverage probability is given by

$$(10.9.26)\quad P_{\mu,\sigma}\{\mu \in I_\delta(X_{N(\delta)})\} = P_{\mu,\sigma}\{|\bar{X}_{N(\delta)} - \mu| < \delta\}$$

$$= \sum_{n=n_1}^{\infty} P_\sigma\{N(\delta) = n\}$$

$$\times P_{\mu,\sigma}\{|\bar{X}_{N(\delta)} - \mu| < \delta \mid N(\delta) = n\}.$$

But according to Lemma 10.9.3, since $\{N(\delta) = n\} \in \mathcal{B}^{(n)}$,

$$(10.9.27)\quad \mathcal{L}\left((\bar{X}_N - \mu)\frac{\sqrt{N}}{\sigma} \,\middle|\, N = n\right) = \mathcal{N}(0, 1) \quad \text{for all} \quad n.$$

Hence from (10.9.26) and (10.9.27),

$$(10.9.28)\quad P_{\mu,\sigma}\{\mu \in I_\delta(\mathbf{X}_{N(\delta)})\} = 2 \sum_{n=n_1}^{\infty} \Phi\left(\frac{\delta}{\sigma}\sqrt{n}\right) P_\sigma\{N(\delta) = n\} - 1.$$

Robbins [3] provided recursive formulae according to which $P_\sigma\{N(\delta) = n\}$ can be determined. These formulae were used by Starr [1] in the numerical evaluation of the coverage probabilities (10.9.28). We derive now the result of Simons [1] and show that there exists an integer k, independent of δ, σ, and μ, such that

$$(10.9.29)\qquad P_{\mu,\sigma}\{\mu \in I_\delta(\mathbf{X}_{N(\delta)+k})\} \geqslant \gamma, \quad \text{for all} \quad (\mu, \sigma).$$

We can say that the statistician has to "pay" the equivalent of an extra k observation for his ignorance of σ^2. Simons [1] derived (10.9.29) by considering, together with the above stopping variable $N(\delta)$, a reverse stopping variable $M(\delta)$, defined as

$$(10.9.30)\quad M(\delta) = \begin{cases} \text{maximal integer} \quad n \geqslant n_1 \quad \text{for which} \quad n < \dfrac{a_n^2 s_n^2}{\delta^2}, \\[2ex] n_1 - 1, \quad \text{if} \quad n \geqslant \dfrac{a_n^2 s_n^2}{\delta^2} \quad \text{for all} \quad n \geqslant n_1, \\[2ex] \infty, \qquad \text{if} \quad n < \dfrac{a_n^2 s_n^2}{\delta^2} \quad \text{infinitely often.} \end{cases}$$

We notice that if $n \geqslant (a_n^2 s_n^2)/\delta^2$ for all $n \geqslant N(\delta)$, then $M(\delta) = N(\delta) - 1$. Thus we expect that $M(\delta)$ will be close to $N(\delta)$ in many cases and give us supplementary information on $N(\delta)$. The reverse stopping variable $M(\delta)$ is related to *reverse martingale* sequences, which are defined in the following manner:

Let $(\mathfrak{X}, \mathfrak{B}, P)$ be a probability space. Let $\{\mathfrak{B}_i; \mathfrak{B}_i \subset \mathfrak{B}, i \in I\}$ be a nonincreasing sequence of sigma-subfields of $\mathfrak{B}$, that is, $\mathfrak{B}_i \supset \mathfrak{B}_{i+1}$ for all $i \in I$, where I is a continuous sequence of integers containing possibly $\pm\infty$. Then $\mathfrak{Z} = \{Z_i, \mathfrak{B}_i, i \in I\}$ is called a *reverse martingale* if, for all $i \in I$,

(i) Z_i is a $\mathfrak{B}_i$-measurable random variable;
(ii) $E\,|Z_i| < \infty$;
(iii) For all $A \in \mathfrak{B}_k$,

$$(10.9.31)\qquad \int_A Z_j \, dP = \int_A Z_k \, dP, \quad \text{all} \quad k \in I, \quad k \geqslant j.$$

Equation 10.9.31 can be restated in the form

$$(10.9.32)\qquad E\{Z_j \mid \mathfrak{B}_k\} = Z_k, \quad \text{for all} \quad k \geqslant j.$$

Thus if we reverse the order of the indices in I, we obtain from $\mathfrak{Z}$ a martingale.

Example 10.15. (i) Represent by $X_1, X_2, \ldots$ a sequence of i.i.d. random variables and by $U_{j_0}, U_{j_0+1}, \ldots$ a sequence of U-statistics for some $j_0 \geqslant 1$. (See Section 3.7.) If $E\{|U_{j_0}|\} < \infty$ and $\mathfrak{B}_j = \mathfrak{B}(U_j, U_{j+1}, \ldots)$, all $j \geqslant j_0$, then $\mathfrak{Z} = \{U_j, \mathfrak{B}_j, j \geqslant j_0\}$ is a reverse martingale.

(ii) The variables $X_1, X_2, \ldots$ are i.i.d. random variables distributed like $G(1/\theta, \nu)$; $\gamma, \theta > 0$. For any β, $0 < \beta < \infty$, let $T_n = \sum_{i=1}^n X_i$ and $Z_n = T_n^{\beta}/E\{T_n^{\beta}\}$. The nonincreasing sequence $\{\mathfrak{B}_n; n \geqslant 1\}$ is given by $\mathfrak{B}_n = \mathfrak{B}(Z_n, Z_{n+1}, \ldots)$. We show now that $\{Z_n, \mathfrak{B}_n, n \geqslant 1\}$ is a reverse martingale for each β, $0 < \beta < \infty$. For this purpose, it is sufficient to prove that

$$E\{Z_n \mid Z_{n+1}\} = Z_{n+1} \quad \text{a.s.,} \quad \text{all} \quad n \geqslant 1. \tag{10.9.33}$$

For each n we have, since $T_n \sim G(1/\theta, n\nu)$,

$$E_{\theta,\nu}\{T_n^{\beta}\} = \theta^{\beta} \frac{\Gamma(n\nu + \beta)}{\Gamma(n\nu)}. \tag{10.9.34}$$

It is easy to verify that

$$\mathfrak{L}\left(\frac{T_n}{T_{n+1}} \middle| T_{n+1}\right) = \beta(n\nu, \nu). \tag{10.9.35}$$

Hence

$$E\{T_n^{\beta} \mid T_{n+1}\} = T_{n+1}^{\beta} \frac{c_{n,\beta}}{c_{n+1,\beta}}, \quad n \geqslant 1, \tag{10.9.36}$$

where

$$c_{n,\beta} = \frac{\Gamma(n\nu + \beta)}{\Gamma(n\nu)}, \qquad n \geqslant 1.$$

Finally, from (10.9.34)–(10.9.36) we imply (10.9.33). ■

A simple generalization of a result of Doob [1], p. 300, yields that if $\mathfrak{Z}$ is a reverse martingale and M a reverse stopping variable, and if I has a first element i_0 (possible $-\infty$), then

$$E\{|Z_M|\} \leqslant E\{|Z_{i_0}|\} < \infty; \qquad E\{Z_M\} = E\{Z_{i_0}\}. \tag{10.9.37}$$

These results will be applied in the proof of the following lemma.

Lemma 10.9.4. (Simons [1].) *Let $N(\delta)$ and $M(\delta)$ be defined as in (10.9.12) and* (10.9.30), *respectively, with $a_n \equiv a$. Then, for all σ,*

$$\begin{aligned} E_{\sigma}\{M(\delta)\} &\leqslant n_0(\delta, \sigma) + (n_1 - 1)P_{\sigma}\{M(\delta) = n_1 - 1\} \\ &\leqslant n_0(\delta, \sigma) + n_1 - 1, \end{aligned} \tag{10.9.38}$$

and

$$E_{\delta}\{M(\delta)\} \geqslant n_0(\delta, \sigma) - 2 - \frac{2}{n_1}. \tag{10.9.39}$$

Proof. The event $\{M(\delta) = \infty\}$ has zero probability for all (μ, σ, δ). This is implied from the strong convergence of s_n^2 to σ^2. Thus from (10.9.30),

$$M(\delta) \leqslant \frac{a^2 s_{M(\delta)}^2}{\delta^2} + (n_1 - 1)I_{\{M(\delta)=n_1-1\}}, \tag{10.9.40}$$

where $I_{\{\cdot\}}$ is the indicator function of $\{\cdot\}$. It follows immediately that

$$E_{\sigma}\{M(\delta)\} \leqslant \frac{a^2}{\delta^2} E_{\sigma}\{s_{M(\delta)}^2\} + (n_1 - 1)P_{\sigma}\{M(\delta) = n_1 - 1\}. \tag{10.9.41}$$

Let $U_1, U_2, \ldots$ be a sequence of i.i.d. random variables distributed like $N(0, 1)$. We notice that $s_{n+1}^2 \sim (\sigma^2/n) \sum_{i=1}^{n} U_i^2$ for all $n = 1, 2, \ldots$. According to Example 10.15 (ii), with $\beta = 1$, we obtain that $\{s_n^2, \mathcal{B}_n^*; n \geqslant 1\}$ is a reverse martingale, where $s_1^2 \equiv s_2^2$, and $\mathcal{B}_n^* = \mathcal{B}(s_n^2, s_{n+1}^2, \ldots)$. Hence

$$E_\sigma\{s_{M(\delta)}^2\} = E_\sigma\{s_2^2\} = \sigma^2, \quad \text{all} \quad 0 < \sigma < \infty. \tag{10.9.42}$$

Substituting (10.9.24) in (10.9.41) we obtain (10.9.38). Define

$$M'(\delta) = \max\{n_1 + 1, M(\delta)\}; \tag{10.9.43}$$

$M'(\delta)$ is a reverse stopping variable. If $M(\delta) = n_1 - 1$ or $M(\delta) = n_1$, then $M'(\delta) = n_1 + 1$ and $M(\delta) + 2 \geqslant a^2 s_{M'(\sigma)}^2/\delta^2$. If $M(\delta) \geqslant n_1 + 1$ then $M'(\delta) = M(\delta)$ and

$$\begin{aligned} M(\delta) + 1 &\geqslant \frac{a^2}{\delta^2} s_{M(\delta)+1}^2 \\ &= \frac{a^2}{\delta^2} \frac{1}{M(\delta)} \sum_{i=1}^{M(\delta)+1} (X_i - \bar{X}_{M(\delta)+1})^2 \geqslant \frac{a^2}{\delta^2} \cdot \frac{1}{M(\delta)} \sum_{i=1}^{M(\delta)} (X_i - \bar{X}_{M(\delta)})^2 \\ &= \frac{a^2}{\delta^2} \cdot \frac{M(\delta) - 1}{M(\delta)} s_{M(\delta)}^2 \\ &= \frac{a^2}{\delta^2} s_{M(\delta)}^2 - \frac{a^2}{\delta^2} \cdot \frac{s_{M(\delta)}^2}{M(\delta)}. \end{aligned} \tag{10.9.44}$$

Furthermore, since $M(\delta) \geqslant n_1 + 1$,

$$\frac{a^2}{\delta^2} \frac{s_{M(\delta)}^2}{M(\delta)} \leqslant \frac{M(\delta) + 1}{M(\delta) - 1} = 1 + \frac{2}{M(\delta) - 1} \leqslant 1 + \frac{2}{n_1}. \tag{10.9.45}$$

Hence from (10.9.44) and (10.9.45) we obtain (10.9.39). (Q.E.D.)

Further results required for the proof of the main theorem follow.

Lemma 10.9.5. (Simons [1].) *For normally disrtibuted random variables*

$$E_\sigma\{M^\nu(\delta)\} \leqslant S^\nu + 0(S^{\nu-1}), \quad \text{as} \quad \delta \to 0, \nu = 1, 2, \ldots; \tag{10.9.46}$$

$$E_\sigma\{N^\nu(\delta)\} \leqslant S^\nu + 0(S^{\nu-1}), \quad \text{as} \quad \delta \to 0, \nu = 1, 2, \ldots; \tag{10.9.47}$$

$$E_\sigma\{M(\delta) - N(\delta)\} = 0(1), \quad \text{for any} \quad n_1 \geqslant 3; \tag{10.9.48}$$

$$E_\sigma\{M^\nu(\delta)\} \geqslant S^\nu + 0(S^{\nu-1}), \quad \text{as} \quad \delta \to 0, \nu = 1, 2, \ldots; \tag{10.9.49}$$

$$E_\sigma\{N^\nu(\delta)\} \geqslant S^\nu + 0(S^{\nu-1}), \quad \text{as} \quad \delta \to 0, \nu = 1, 2, \ldots; \tag{10.9.50}$$

where $S \equiv n_0(\delta, \sigma)$. Moreover, for θ, $0 < \theta < 1$;

$$P_\sigma\{N(\delta) \leqslant \theta S\} = 0(S^{-(n_1-1)/2}), \quad \text{as} \quad \delta \to 0; \tag{10.9.51}$$

$$E_\sigma\{M(\delta) - N(\delta) \mid N(\delta) = n\} \leqslant S + 1, \quad \text{for all} \quad n \geqslant n_1; \tag{10.9.52}$$

and for $n > \theta S$, $\theta > \frac{1}{2}$,

$$E_\sigma\{M(\delta) - N(\delta) \mid N(\delta) = n\} \leqslant K(\theta), \quad \textit{independently of } S. \tag{10.9.53}$$

We omit the proof of this lemma. The reader can find a detailed proof in Simons' paper. We now prove the main theorem.

Theorem 10.9.6. (Simons [1].) *For the stopping variable* $N(\delta)$ *defined by* (10.9.12), *with* $a_n \equiv a$ *and* $n_1 \geqslant 3$, *we have, for some finite integer* $k \geqslant 0$,

$$P_{\mu,\sigma}\{|\bar{X}_{N+k} - \mu| < \delta\} \geqslant \gamma, \quad \textit{all} \quad (\mu, \sigma) \quad \textit{and} \quad \delta > 0. \tag{10.9.54}$$

Proof. In (10.9.28) we have shown that

$$P_{\mu,\sigma}\{|\bar{X}_{N+k} - \mu| < \delta\} = 2E_\sigma\left\{\Phi\left(\frac{\delta}{\sigma}\sqrt{(N+k)}\right)\right\} - 1, \quad \text{all } 0 < \sigma < \infty. \tag{10.9.55}$$

Let $\lambda = \delta/\sigma$, and $g(x) = \Phi(\lambda\sqrt{x})$. Then

$$\begin{aligned} g'(x) &= \varphi(\lambda\sqrt{x})\frac{\lambda}{2\sqrt{x}}, \\ g''(x) &= -\lambda\frac{1 + \lambda^2 x}{4x^{3/2}}\varphi(\lambda\sqrt{x}), \end{aligned} \tag{10.9.56}$$

where $\varphi(\cdot)$ is the standard normal density. Hence expansion of $g(x)$ in a Taylor series about $x_0 = S = n_0(\delta, \sigma)$ yields, for arbitrary θ, $0 < \theta < 1$,

$$\begin{aligned} E_\sigma\{\Phi(\lambda\sqrt{(N+k)})\} &\geqslant E_\sigma\{\Phi(\lambda\sqrt{(N+k)})I_{\{N+k\geqslant\theta^2 S\}}\} \\ &\geqslant \Phi(a)P_\sigma\{N + k \geqslant \theta^2 S\} \\ &\quad + a\varphi(a)\frac{1}{2S}E_\sigma\{(N + k - S)I_{\{N+k>\theta^2 S\}}\} \\ &\quad - a(a^2\theta^2 + 1)\varphi(a\theta)(8\theta^3 S^2)^{-1} \\ &\quad \times E_\sigma\{(N + k - S)I_{\{N+k>\theta^2 S\}}\} \\ &\geqslant \Phi(a) + a\varphi(a)(2S)^{-1}E_\sigma\{N + k - S\} \\ &\quad - a(a^2\theta^2 + 1)\varphi(a\theta)(8\theta^3 S^2)^{-1} \\ &\quad \times E_\sigma\{(N + k - S)^2\} \\ &\quad + \left[-\Phi(a) + a\varphi(a)\frac{1 - \theta^2}{2}\right. \\ &\quad \left. + a(a^2\theta^2 + 1)\varphi(a\theta)\frac{(1 - \theta^2)^2}{8\theta^3}\right] \\ &\quad \times P_\sigma\{N + k \leqslant \theta^2 S\}. \end{aligned} \tag{10.9.57}$$

For small values of θ, $\theta > 0$, the coefficient of $P_\sigma\{N + k \leqslant \theta^2 S\}$ is positive. For these values of θ,

$$(10.9.58)\quad E_\sigma\{\Phi(\lambda\sqrt{(N + k)})\} \geqslant \Phi(a) + a\varphi(a)(2S)^{-1}(k + E_\sigma(N - S)) \\ -a(a^2\theta^2 + 1)\varphi(a\theta)(8\theta^3 S^2)^{-1}[k^2 + 2kE_\sigma\{N - S\} + E_\sigma\{(N - S)^2\}].$$

According to (10.9.47) and (10.9.50),

$$(10.9.59)\quad E_\sigma\{\Phi(\lambda\sqrt{(N + k)})\} - \Phi(a) \geqslant 0(S^{-2})k^2 \\ + [a\varphi(a)(2S)^{-1} + 0(S^{-2})]k + 0(S^{-1}) \\ = [a\varphi(a)(2S)^{-1} + 0(S^{-2})][0(S^{-1})k^2 + k + 0(1)].$$

Hence, for some *large* k,

$$E_\sigma\{\Phi(\lambda\sqrt{(N + k)})\} \geqslant \Phi(a) = \frac{1 + \gamma}{2}.$$

This and (10.9.55) imply (10.9.54). (Q.E.D.)

10.10. ASYMPTOTICALLY EFFICIENT SEQUENTIAL FIXED-WIDTH CONFIDENCE ESTIMATION OF THE MEAN IN A GENERAL CASE

Chow and Robbins [1] provided the basic asymptotic theory for sequential fixed-width confidence estimation in the general case. They considered the problem of estimating the mean of an unknown distribution, under the sole assumption that its second moment exists. Let $X_1, X_2, \ldots$ be a sequence of i.i.d. random variables having a common distribution function F belonging to the class $\mathcal{F}$ of *all* distribution functions having a finite second moment. Thus the results of Chow and Robbins are in a sense distribution free. The assumption of a finite second moment, in addition to the assumption that the random variables are i.i.d., assures the validity of the central limit theorem. That is, if $\mu(F) = E_F\{X\}$, $\sigma^2(F) = \mathrm{Var}_F\{X\}$,

$$(10.10.1)\qquad \lim_{n\to\infty} P\left\{\sqrt{n}\,\frac{\bar{X}_n - \mu(F)}{\sigma(F)} \leqslant x\right\} = \Phi(x), \quad \text{all} \quad F \in \mathcal{F},$$

where $\bar{X}_n$ is the sample mean. Thus we know that the interval estimator $I_\delta(\mathbf{X}_n) = [\bar{X}_n - \delta, \bar{X}_n + \delta]$ will cover $\mu(F)$, with probability not less than γ, $0 < \gamma < 1$, if $n \geqslant n(\delta, \gamma, F)$. A fixed sample procedure for which

$$(10.10.2)\qquad P\{\mu(F) \in I_\delta(\mathbf{X}_n)\} \geqslant \gamma, \quad \text{all} \quad F \in \mathcal{F},$$

does not exist. Indeed, in Section 10.8 we have shown some special cases in which there is no such fixed sample procedure. We do not yet have a general sampling procedure which guarantees (10.10.2) for all F and any positive δ. Chow and Robbins introduced the following *asymptotic* criteria.

A fixed-width confidence interval $I_\delta(\mathbf{X}_N) = (\bar{X}_N - \delta, \bar{X}_N + \delta)$, based on a stopping variable N, is said to be *asymptotically consistent* if

$$(10.10.3) \qquad \lim_{\delta \to 0} P\{\mu(F) \in I_\delta(X_{N(\delta)})\} \geqslant \gamma, \quad \text{all} \quad F \in \mathcal{F},$$

for some γ; $0 < \gamma < 1$.

A stopping variable $N(\delta)$ is called *asymptotically efficient* if $I_\delta(\underline{X}_{N(\delta)})$ is asymptotically consistent and

$$(10.10.4) \qquad \lim_{\delta \to 0} \frac{E_F\{N(\delta)\}\delta^2}{a^2\sigma^2(F)} = 1, \quad \text{all} \quad F \in \mathcal{F},$$

where $a = \Phi^{-1}((1 + \gamma)/2)$.

We see later that if $N(\delta) \nearrow \infty$ a.s., the sequence of sample means $\bar{X}_{N(\delta)}$ obeys the central limit theorem. In the normal case, if σ^2 is known then the minimal sample size required, for a coverage probability $\geqslant \gamma$, is $n_0(\delta, \sigma) = a^2\sigma^2/\delta^2$. Thus (10.10.4) coincides with our previous definition of asymptotic efficiency in the normal case, (10.9.2). Consider the stopping variable

$$(10.10.5) \qquad N(\delta) = \text{smallest integer} \quad n \quad \text{such that}$$

$$n \geqslant \frac{a_n^2}{\delta^2}\left\{\frac{1}{n}\left(1 + \sum_{i=1}^{n}(X_i - \bar{X}_n)^2\right)\right\},$$

where $\{a_n; n \geqslant 1\}$ is a sequence of positive constants converging to a. We notice that $N(\delta)$ is of the same form as (10.9.12). The only difference is that $1/n$ is added to the sample variance. This is done to avoid zero value of s_n^2 in discrete cases. The main result of Chow and Robbins [1] is that $I_\delta(\underline{X}_{N(\delta)})$ is asymptotically consistent and $N(\delta)$ an asymptotically efficient stopping variable. To prove it we start with the following lemma.

Lemma 10.10.1. (Chow and Robbins [1].) *Let* $\{V_n; n = 1, 2, \ldots\}$ *be a sequence of a.s. positive random variables and* $\lim_{n\to\infty} V_n = 1$ *a.s. Let* $f(n)$ *be any sequence of constants satisfying*

$$(10.10.6) \qquad f(n) > 0, \qquad \lim_{n\to\infty} f(n) = \infty, \qquad \lim_{n\to\infty} \frac{f(n)}{f(n-1)} = 1.$$

For each t, $t > 0$, *define*

$$(10.10.7) \qquad N(t) = \text{least integer } k \geqslant 1 \text{ such that } f(k) \geqslant tV_k.$$

Then $N(t)$ *is a proper, nondecreasing (in* t*) stopping variable. Moreover,*

$$\lim_{t\to\infty} N(t) = \infty \quad \text{a.s.}, \qquad \lim_{t\to\infty} E\{N(t)\} = \infty,$$

and

(10.10.8) $$\lim_{t\to\infty} \frac{f(N(t))}{t} = 1 \quad \text{a.s.}$$

If, in addition, we assume that $E_F\{\sup_{n\geq 1} V_n\} < \infty$ *then*

(10.10.9) $$\lim_{t\to 0} E_F\left\{\frac{f(N(t))}{t}\right\} = 1, \qquad F \in \mathcal{F}.$$

Proof. For each $t, t > 0$, $N(t)$ is a proper stopping variable, since $f(k) \to \infty$ and, on the other hand, $V_k \to 1$ a.s., as $k \to \infty$. From (10.10.7) we obtain immediately that if $t_1 < t_2$ then $N(t_1) \leqslant N(t_2)$ a.s. Moreover, $\lim_{t\to\infty} N(t) = \infty$ a.s., since, for each finite $k, f(k) < \infty$. Finally, since $V_{N(t)} \to 1$ a.s. as $t \to \infty$, and

$$V_{N(t)} \leqslant \frac{f(N(t))}{t} < \frac{f(N(t))}{f(N(t)-1)} V_{N(t)-1},$$

we obtain (10.10.8) by letting $t \to \infty$.

To prove (10.10.9), let $W = \sup_{n\geq 1} V_n$. Choose K so large that $f(n)/f(n-1) \leqslant 3/2$ for all $n \geqslant K$. Then if $N(t) \geqslant K$, we have

(10.10.10) $$\frac{f(N(t))}{t} = \frac{f(N(t))}{f(N(t)-1)} \cdot \frac{f(N(t)-1)}{t} < \tfrac{3}{2} V_{N(t)-1} \leqslant \tfrac{3}{2} W.$$

On the other hand, if $N(t) < K$ then

(10.10.11) $$\frac{f(N(t))}{t} \leqslant \max_{1\leqslant n<K} \frac{f(n)}{t} \leqslant f(1) + \cdots + f(K), \qquad t \geqslant 1.$$

Hence for all $t \geqslant 1$,

(10.10.12) $$\begin{aligned}\frac{f(N(t))}{t} &< \tfrac{3}{2} W I_{\{N(t)\geqslant K\}} + \sum_{i=1}^{K} f(i) \cdot I_{\{N(t)<K\}} \\ &\leqslant \tfrac{3}{2} W + \sum_{i=1}^{K} f(i).\end{aligned}$$

If $E_F\{W\} < \infty$, we obtain from the Lebesgue dominated convergence theorem and (10.10.8) that

(10.10.13) $$\lim_{t\to\infty} E_F\left\{\frac{f(N(t))}{t}\right\} = E_F\left\{\lim_{t\to\infty} \frac{f(N(t))}{t}\right\} = 1.$$

(Q.E.D.)

We notice that since $\sigma^2(F) < \infty$ for each $F \in \mathcal{F}$,

$$\lim_{n\to\infty} \frac{1}{\sigma^2(F)} \left\{ \frac{1}{n} + \frac{1}{n} \sum_{i=1}^{n} (X_i - \bar{X}_n)^2 \right\} = 1 \quad \text{a.s.} \tag{10.10.14}$$

Hence if we set $f(n) = n/a_n^2$, $t = \sigma^2/\delta^2$, and $V_n = (1/\sigma^2)(\sum_{i=1}^{n} (X_i - \bar{X})^2 + 1)/n$, then the stopping variable $N(\delta)$ defined in (10.10.5) coincides with (10.10.7). Thus we imply the following theorem.

Theorem 10.10.2. (Chow and Robbins [1].) *The stopping variable* $N(\delta)$ *defined in* (10.10.5) *is proper, and* $N(\delta) \nearrow \infty$ *a.s. as* $\delta \searrow 0$. *Furthermore,*

$$\lim_{\delta\to 0} \frac{\delta^2 N(\delta)}{a^2\sigma^2(F)} = 1 \quad a.s., \qquad all \quad F \in \mathcal{F}; \tag{10.10.15}$$

and if $E\left\{\sup_{n\geq 1} s_n^2\right\} < \infty$ *then*

$$\lim_{\delta\to 0} \frac{\delta^2 E_\sigma\{N(\delta)\}}{a^2\sigma^2(F)} = 1. \tag{10.10.16}$$

As we have seen in the previous section, if V_n can be represented as $(1/n)\sum_{i=1}^{n} U_i$, where U_i are i.i.d., and if $E\{U_1^2\} < \infty$, then by the Wiener ergodic theorem, $E\left\{\sup_{n\geq 1} V_n\right\} < \infty$ and (10.10.16) holds. This requires, in addition that $E_F\{X_i^4\} < \infty$. Chow and Robbins [1] give alternative conditions under which (10.10.16) holds. These conditions are stated in Lemma 10.10.3.

Lemma 10.10.3. (Chow and Robbins [1].) *If in addition to the conditions of Lemma* 10.10.1, $\lim_{n\to\infty} f(n)/n = 1$ *and*

$$E_F\{N(t)\} < \infty, \quad all \quad t > 0, \qquad F \in \mathcal{F}, \tag{10.10.17}$$

$$\limsup_{t\to\infty} \frac{E_F\{N(t)V_{N(t)}\}}{E_F\{N(t)\}} \leqslant 1, \qquad F \in \mathcal{F}, \tag{10.10.18}$$

and if there exists a sequence $\{g(n); n \geqslant 1\}$ *of constants such that*

$$g(n) > 0, \qquad \lim_{n\to\infty} g(n) = 1, \qquad V_n \geqslant g(n)V_{n-1}, \tag{10.10.19}$$

then

$$\lim_{t\to\infty} E_F\left\{\frac{N(t)}{t}\right\} = 1, \qquad F \in \mathcal{F}. \tag{10.10.20}$$

We omit the proof of this lemma.

Thus under the conditions of Lemmas 10.10.1 and 10.10.3, the stopping variable given by (10.10.5) is asymptotically efficient, provided it is asymptotically consistent. The asymptotic consistency of this stopping variable is implied from a theorem stated first by Anscombe [1] and proven more rigorously later by Rényi [1]. This theorem generalizes the central limit theorem to the case of a random number of terms in the standardized sums.

Theorem 10.10.4. (Anscombe [1]; Rényi [1]). *Suppose that $U_1, U_2, \ldots$ is a sequence of i.i.d. random variables; $E\{U_1\} = 0, E\{U_1^2\} = 1$. Let $S_n = \sum_{i=1}^n U_i$. Let $N(t)$ denote a positive integer valued random variable, $t > 0$, such that*

(10.10.21) $$p \cdot \lim_{t\to\infty} \frac{N(t)}{t} = C, \qquad 0 < C < \infty.$$

Then

(10.10.22) $$\lim_{t\to\infty} P\left\{\frac{1}{\sqrt{(N(t))}} S_{N(t)} \leqslant x\right\} = \Phi(x), \qquad -\infty < x < \infty.$$

Proof. Let ϵ, $0 < \epsilon < \frac{1}{5}$, be arbitrary. Choose $t_1 > 0$ such that if $t \geqslant t_1$, then

(10.10.23) $$P\{|N(t) - Ct| \geqslant Ct\epsilon\} \leqslant \epsilon.$$

Since

(10.10.24) $$P\{(N(t))^{-1/2} S_{N(t)} \leqslant x\} = \sum_{n=1}^{\infty} P\{n^{-1/2} S_n \leqslant x, N(t) = n\},$$

we obtain, for all $t \geqslant t_1$,

(10.10.25) $$\left| P[(N(t))^{-1/2} S_{N(t)} \leqslant x] - \sum_{\{|n-Ct|<\epsilon Ct\}} P\left\{\frac{S_n}{\sqrt{n}} \leqslant x, N(t) = n\right\}\right|$$
$$\leqslant \sum_{\{|n-Ct|>\epsilon Ct\}} P\left\{\frac{S_n}{\sqrt{n}} \leqslant x, N(t) = n\right\}$$
$$\leqslant P\{|N(t) - Ct| \geqslant \epsilon Ct\} < \epsilon.$$

Define $N_1 = [C(1-\epsilon)t]$, $N_2 = [C(1+\epsilon)t]$. Then for $|n - ct| \leqslant \epsilon Ct$,

(10.10.26) $$P\left\{\frac{S_n}{\sqrt{n}} \leqslant x, N(t) = n\right\} \leqslant P\{S_{N_1} < x\sqrt{N_2} + \rho, N(t) = n\},$$

where

(10.10.27) $$\rho = \max_{N_1 < n \leqslant N_2} \left| \sum_{N_1 < k \leqslant n} U_k \right|.$$

Similarly,

(10.10.28) $$P\left\{\frac{S_n}{\sqrt{n}} \leqslant x, N(t) = n\right\} \geqslant P\{S_{N_1} < x\sqrt{N_1} - \rho, N(t) = n\}.$$

According to the Kolmogorov inequality (see Fisz [1], p. 221),

$$(10.10.29) \qquad P\{\rho \geqslant \epsilon^{1/3}\sqrt{N_1}\} \leqslant \frac{N_2 - N_1}{N_1\epsilon^{2/3}} \leqslant 5\epsilon^{1/3}, \qquad t \geqslant \frac{1}{C\epsilon}.$$

Let $R = \{\rho < \epsilon^{1/3}\sqrt{N_1}\}$ and $E = \{|n - Ct| < Ct\epsilon\}$. Then,

$$(10.10.30) \quad P\left\{\frac{S_{N(t)}}{\sqrt{(N(t))}} \leqslant x\right\} \leqslant P\left\{\frac{S_{N_1}}{\sqrt{N_1}} \leqslant x\left(\frac{N_2}{N_1}\right)^{1/2} + \epsilon^{1/3}, R \cap E\right\} + 6\epsilon^{1/3},$$

and

$$(10.10.31) \quad P\left\{\frac{S_{N(t)}}{\sqrt{(N(t))}} \leqslant x\right\} \geqslant P\left\{\frac{S_{N_1}}{\sqrt{N_1}} \leqslant x - \epsilon^{1/3}, R \cap E\right\} - \epsilon.$$

Finally, it follows that

$$(10.10.32) \qquad P\left\{\frac{S_{N_1}}{\sqrt{N_1}} \leqslant x - \epsilon^{1/3}\right\} - 7\epsilon^{1/3}$$

$$\leqslant P\left\{\frac{S_{N(t)}}{\sqrt{(N(t))}} \leqslant x\right\}$$

$$\leqslant P\left\{\frac{S_{N_1}}{\sqrt{N_1}} \leqslant x\left(\frac{1 + 2\epsilon}{1 - 2\epsilon}\right)^{1/2} + \epsilon^{1/3}\right\} + 6\epsilon^{1/3}.$$

Letting $N_1 \to \infty$ and $\epsilon \to 0$, we obtain the required result from the central limit theorem. (Q.E.D.)

The stopping variable (10.10.5) and the interval estimator $[\bar{X}_N - \delta, \bar{X}_N + \delta]$ proposed by Chow and Robbins [1] are asymptotically optimal if no more than the existence of the fourth moment is assumed about the class $\mathcal{F}$ of distributions under consideration. Since the class $\mathcal{F}$ is very rich, it is not expected that any substantially better sampling variable can be proposed. However, if we restrict attention to a subclass of $\mathcal{F}$ on which more can be said, the above stopping variable and interval estimator may not be asymptotically optimal. This may be the case, for example, when the subclass of $\mathcal{F}$ under consideration is a parametric family of distribution functions that satisfies the regularity conditions of Section 4.3 and of Theorem 5.5.3. Under these conditions, the M.L.E. of θ is a B.A.N. estimator, with an asymptotic normal distribution, whose covariance matrix is the inverse of the Fisher information matrix. Thus a stopping variable based on the appropriate element of the Fisher information matrix, combined with an interval estimator $[\hat{\theta}(\mathbf{X}_N) - \delta, \hat{\theta}(\mathbf{X}_N) + \delta]$, where $\hat{\theta}(\mathbf{X}_n)$ is an M.L.E. of θ, may be asymptotically more efficient than the stopping variable (10.10.5) and the interval estimator based on the sample mean. Such a case was studied by Zacks [3] and is illustrated in the following example.

Example 10.16. Let $X_1, X_2, \ldots$ be a sequence of i.i.d. random variables $X_i \sim \exp\{N(\mu, \sigma^2)\}$, $-\infty < \mu < \infty$, $0 < \sigma < \infty$. That is, $\log X_i \sim N(\mu, \sigma^2)$. In Example 5.6 we verified that the M.L.E. of the mean $\xi = \exp\{\mu + \sigma^2/2\}$ is $\hat{\xi}_n = \exp\{\bar{Y}_n + (1/2n)Q_n\}$, where $Y_i = \log X_i$ $(i = 1, 2, \ldots)$, $\bar{Y}_n = (1/n)\sum_{i=1}^n Y_i$, $Q_n = \sum_{i=1}^n (Y_i - \bar{Y}_n)^2$. We also verified that $\hat{\xi}_n$ is a B.A.N. estimator of ξ, and $\sqrt{n}(\hat{\xi}_n - \xi) \xrightarrow{d} N[0, \xi^2\sigma^2(1 + (\sigma^2/2))]$. On the other hand, if $D^2 = \text{Var}\{X\}$ we know that $D^2 = \xi^2(e^{\sigma^2} - 1)$, and the sample mean satisfies $\sqrt{n}(\bar{X}_n - \xi) \xrightarrow{d} N(0, \xi^2(e^{\sigma^2} - 1))$. The sample mean, $\bar{X}_n$, is *not* B.A.N.

The stopping variable $N(\delta)$ given by (10.10.5) yields in the log-normal case

$$\lim_{\delta \to 0} \delta^2 N(\delta) = a^2\xi^2(e^{\sigma^2} - 1) \quad \text{a.s.,} \quad \text{all} \quad (\xi, \sigma), \tag{10.10.33}$$

and

$$\lim_{\delta \to 0} \delta^2 E_{\xi,\sigma}\{N(\delta)\} = a^2\xi^2(e^{\sigma^2} - 1), \qquad \text{all} \quad (\xi, \sigma). \tag{10.10.34}$$

On the other hand, since $Q_n/n \to \sigma^2$ a.s., as $n \to \infty$, we can define the stopping variable

$$\begin{aligned} \tilde{N}(d) = {} & \text{least non-negative integer } n \geqslant 2 \\ & \text{such that} \quad n \geqslant \frac{a_n^2}{\delta^2} \bar{X}_n^2 \hat{\sigma}_n^2(1 + \tfrac{1}{2}\hat{\sigma}_n^2), \end{aligned} \tag{10.10.35}$$

where $\hat{\sigma}_n^2 = Q_n/n$ and $\{a_n\}$ is a sequence of positive constants such that $a_n \to a$.

$$\bar{X}_n^2\hat{\sigma}_n^2\left(1 + \frac{\hat{\sigma}_n^2}{2}\right) \to \xi^2\sigma^2\left(1 + \frac{\sigma^2}{2}\right) \quad \text{a.s.} \quad \text{as} \quad n \to \infty.$$

Define $V_n = \bar{X}_n^2\hat{\sigma}_n^2(1 + \frac{1}{2}\hat{\sigma}_n^2)/\xi^2\sigma^2(1 + (\sigma^2/2))$. Then $V_n \to 1$ a.s. Let $f(n) = n/a_n^2$ and $t = \xi^2\sigma^2(1 + (\sigma^2/2))/\delta^2$. Then from (10.10.8) we obtain that

$$\lim_{\delta \to 0} \delta^2 \tilde{N}(\delta) = a^2\xi^2\sigma^2\left(1 + \frac{\sigma^2}{2}\right) \quad \text{a.s.} \tag{10.10.36}$$

Since moments of all orders of the log-normal distribution exist, it is easy to prove, applying the Wiener ergodic theorem, that $E\left\{\sup_{n \geqslant 2} \{\bar{X}_n^2\hat{\sigma}_n^2(1 + \frac{1}{2}\hat{\sigma}_n^2)\}\right\} < \infty$. Hence from (10.10.9) we obtain that

$$\lim_{\delta \to 0} \delta^2 E_{\xi,\sigma}\{\tilde{N}(\delta)\} = a^2\xi^2\sigma^2\left(1 + \frac{\sigma^2}{2}\right), \quad \text{all} \quad (\xi, \sigma). \tag{10.10.37}$$

The comparison of (10.10.34) and (10.10.37) shows that the stopping variable $\tilde{N}(\delta)$ is asymptotically more efficient than the stopping variable $N(\delta)$, provided $I_\delta(\mathbf{X}_{\tilde{N}})$ is asymptotically consistent, that is,

$$\lim_{\delta\to 0} P_{\xi,\sigma}\{\hat{\xi}_{\tilde{N}(\delta)} - \delta \leqslant \xi \leqslant \hat{\xi}_{\tilde{N}(\delta)} + \delta\} \geqslant \gamma,$$

for all (ξ, σ). This asymptotic consistency is implied from the asymptotic normality of the M.L.E. $\hat{\xi}_n$ and the following theorem of Anscombe [1]. ■

Theorem 10.10.5. (Anscombe [1].) *Let $\{Y_n; n \geqslant 1\}$ be a sequence of random variables satisfying the following conditions:*

(i) Convergence in law. *There exists a real number θ, a distribution function $F(x)$, and a sequence of reals $\{\omega_n\}$ such that*

$$\lim_{n\to\infty} P\{Y_n - \theta \leqslant \omega_n x\} = F(x), \tag{10.10.38}$$

at all the continuity points x of F.

(ii) Uniform continuity in probability. *Given any $\epsilon > 0$ and $\eta > 0$ there is a large v and a small positive c such that for any $n > v$,*

$$P\{|Y_{n'} - Y_n| < \epsilon\omega_n \quad \text{for all } n' \text{ such that } |n' - n| < cn\} > 1 - \eta \tag{10.10.39}$$

Let $\{n(t)\}$ be an increasing sequence of positive integers, $n(t) \nearrow \infty$; and let $N(t)$ be a proper stopping variable such that

$$\frac{N(t)}{n(t)} \to 1 \quad a.s. \quad as \quad t \to \infty. \tag{10.10.40}$$

Then

$$\lim_{t\to\infty} P\{Y_{N(t)} - \theta \leqslant \omega_{n(t)} x\} = F(x) \tag{10.10.41}$$

at all the continuity points x of F.

We omit the proof of this theorem. Anscombe [1] has shown that a sequence $\{\hat{\sigma}_n\}$ of M.L.E.'s, under the usual regularity conditions, satisfies (i) and (ii) of the above theorem. Hence

$$\sqrt{(\tilde{N}(\delta))}(\hat{\xi}_{\tilde{N}(\delta)} - \xi) \xrightarrow{d} N\left(0, \xi^2\sigma^2\left(1 + \frac{\sigma^2}{2}\right)\right).$$

This proves (10.10.37).

It is interesting to notice in Example 10.15 that if the nuisance parameter σ^2 is known, the M.L.E. of ξ is $\hat{\xi}_n(\sigma) = \exp\{\bar{Y}_n + \sigma^2/2\}$. The asymptotic distribution of $\sqrt{n}(\hat{\xi}_n(\sigma) - \xi)$ is $\mathcal{N}(0, \xi^2\sigma^2)$. Thus if σ^2 is unknown, the M.L.E. of ξ and the corresponding stopping variable $\tilde{N}(\delta)$ cannot attain the same asymptotic efficiency as in the case of a known σ^2. It has therefore been suggested (Nadas [1], Khan [1]) to define the notion of asymptotic efficiency, in cases of regular families of distribution functions, in terms of the asymptotic normal distribution of the corresponding M.L.E. This is done in the following manner.

Let $\mathcal{F} = \{F_\theta; \theta \in \Theta\}$ be a regular family of distribution functions, where Θ is an interval of a k-dimensional Euclidean space. Let $f(x; \theta)$, $\theta = (\theta_1, \ldots, \theta_k)'$ be the corresponding density functions. We denote by $I(\theta) = \|I_{ij}(\theta); i, j = 1, \ldots, k\|$ the Fisher information matrix and by $I^{ij}(\theta)$, $i, j = 1, \ldots, k$, the elements of the inverse matrix $I^{-1}(\theta)$. Let $\hat{\theta}_n$ denote the M.L.E. of θ. Under the conditions of Theorem 5.3.1, $\hat{\theta}_n \to \theta$ a.s. as $n \to \infty$. Suppose that we wish to construct a fixed-width interval estimator for $\theta^{(1)} \equiv \theta_1$. We define, for a given proper stopping variable N, the interval estimator

(10.10.42) $$I_\delta(\mathbf{X}_N) = (\hat{\theta}_N^{(1)} - \delta, \hat{\theta}_N^{(1)} + \delta),$$

where $\hat{\theta}_N^{(1)}$ is the M.L.E. of θ_1. $I_\delta(\mathbf{X}_N)$ is called asymptotically consistent if

(10.10.43) $$\lim_{\delta \to 0} P_\theta\{\theta^{(1)} \in I_\delta(\mathbf{X}_N)\} \geqslant \gamma, \quad \text{all} \quad \theta \in \Theta.$$

A stopping variable N for which (10.10.43) holds is called *asymptotically efficient* if

(10.10.44) $$\lim_{\delta \to 0} \frac{\delta^2 E_\theta\{N(\delta)\}}{a^2 I^{11}(\theta)} = 1, \quad \text{all} \quad \theta \in \Theta.$$

As in Example 10.15, Khan [1] suggested the stopping variable

(10.10.45) $$S(\delta) = \text{least integer } n,\ n \geqslant 1, \quad \text{such that} \quad n \geqslant \frac{a_n^{\,2}}{\delta^2} I^{11}(\hat{\theta}_n),$$

where $\{a_n\}$ is a sequence of positive constants converging to a.

Theorem 10.10.6. (Khan [1].) *If $\mathcal{F}$ is a k-parameter family of distribution satisfying the strong consistency conditions of the M.L.E. sequence $\{\hat{\theta}_n; n \geqslant 1\}$ and the B.A.N. conditions of $\{\hat{\theta}_n; n \geqslant 1\}$, then $S(\delta)$ defined by (10.10.45) is a proper stopping variable satisfying*

(10.10.46) $$S(\delta) \nearrow \infty \quad a.s. \quad as \quad \delta \searrow 0; \qquad \lim_{\delta \to 0} E_\theta\{S(\delta)\} = \infty.$$

Moreover,

(10.10.47) $$\lim_{\delta \to 0} \frac{\delta^2 S(\delta)}{a^2 I^{11}(\theta)} = 1 \quad a.s., \quad all \quad \theta \in \Theta.$$

The interval estimator $I_\delta(X_{S(\delta)})$ *is asymptotically consistent,*

$$\lim_{\delta\to 0} P_\theta[\theta^{(1)} \in I_\delta(X_{S(\delta)})] \geqslant \gamma, \quad all \quad \theta \in \Theta;$$

and $S(\delta)$ *is asymptotically efficient,*

$$\lim_{\delta\to 0} \frac{\delta^2 E_\theta\{S(\delta)\}}{a^2 I^{11}(\theta)} = 1, \quad all \quad \theta \in \Theta. \tag{10.10.48}$$

The proof is based on the previous lemmas and theorems.

PROBLEMS

Section 10.1

1. Let $X_1, \ldots, X_n$ be i.i.d. random variables having a negative binomial distribution N.B. (ψ, ν), ν known, $0 < \psi < 1$. Determine a γ upper U.M.A. confidence limit for ψ.

Section 10.2

2. Let $X_1, \ldots, X_n$ be i.i.d. random variables having an $\mathcal{N}(\mu_1, \sigma_1^2)$ distribution; and let $Y_1, \ldots, Y_n$ be i.i.d. random variables (independent of the X's), having an $\mathcal{N}(\mu_2, \sigma_2^2)$ distribution. Determine a U.M.A. unbiased γ upper confidence limit for $\rho = \sigma_2^2/\sigma_1^2$.

3. Let $X_1, \ldots, X_n$ be i.i.d. random variables which have an absolutely continuous d.f. Let $\mathcal{F}$ be the family of all absolutely continuous d.f.'s on the real line. Determine a U.M.A. γ lower confidence limit for the median of the distribution, with respect to $\mathcal{F}$. (See Lehmann [2], p. 116.)

4. Let $X_1, \ldots, X_n$ be i.i.d. random variables having a common rectangular distribution $\mathcal{R}(\theta_1, \theta_2)$, $-\infty < \theta_1 < \theta_2 < \infty$.

(i) If $\theta_1 = 0$, what is a γ upper confidence limit for θ_2? Is it a U.M.A. confidence limit?
(ii) If θ_1 and θ_2 are unknown, determine a set $S(X_n)$ in the space of (θ_1, θ_2) for which $P_{\theta_1,\theta_2}[(\theta_1, \theta_2) \in S(X_n)] \geqslant \gamma$ for *all* θ_1, θ_2. (Such a set is called a simultaneous confidence region.)

5. Let $X_1, \ldots, X_n$ be i.i.d. random p-dimensional vectors having a common p-variate normal distribution $\mathcal{N}(\xi, \Sigma)$.

(i) Determine a γ confidence region for ξ when Σ is known.
(ii) Determine a γ confidence region for ξ when Σ is unknown.
(iii) Are the confidence sets determined in (i) and (ii) optimal in some sense?

6. Determine two-sided confidence limits for each of the variance components of an A.O.V. Model II. (The statistical model is specified in Problem 18, Chapter 2.)

7. Let X be an r.v. having an $\mathcal{N}(\mu, \sigma^2)$ distribution. Show that for any γ, $0 < \gamma < 1$, there exists a K, $0 < K < \infty$ (which may depend on γ), so that

$$P_{\mu,\sigma}[X - K|X| \leqslant \mu \leqslant X + K|X|] \geqslant \gamma, \quad \text{all} \quad (\mu, \sigma).$$

(This is an interesting example of a confidence interval for the mean of a normal distribution based on one observation (see Machol and Rosenblatt [1]).

Section 10.3

8. (i) Following Pratt [2], derive a two-sided γ confidence interval for the parameter θ of a binomial distribution which, at $\theta = \frac{1}{2}$, has a smaller expected length than that of a U.M.A.U. one.
   (ii) Derive a similar system of confidence intervals for the parameter λ of a Poisson distribution.

9. Let $X_1, \ldots, X_n$ be a sequence of i.i.d. random variables having a Cauchy distribution with a location parameter θ, $-\infty < \theta < \infty$.

   (i) Determine a confidence interval estimator for θ based on the sample median M_n.
   (ii) Determine a method for a numerical computation of the asymptotically U.M.A.U. confidence intervals.

Section 10.4

10. Prove Theorem 10.4.1 for the absolutely continuous case.

11. Derive U.M.A. (β, γ) upper tolerance limits for the Poisson case (see Zacks [10]).

12. Let $(x_1, Y_1), \ldots, (x_n, Y_n)$ be pairs of fixed real points x_i and random variables Y_i $(i = 1, \ldots, n)$ such that for each x_i, $Y_i \sim N(\alpha + \beta x_i, \sigma^2)$, $i = 1, \ldots, n$. Furthermore, $\{Y_1, \ldots, Y_n\}$ are independent.

   (i) Determine an upper (β, γ) tolerance limit for a future observation on Y at level $x = x_0$.
   (ii) Is the tolerance limit determined in (i) a U.M.A. one?

13. Study the problem of determining (β, γ) tolerance regions for a p-variate normal distribution (see John [1]).

14. Derive U.M.A.U. β-expectation tolerance intervals for the negative exponential distribution (see Guttman [1]).

Section 10.5

15. Let F be an absolutely continuous d.f. For a sample of size $n = 10$, for what values of β and γ, $0 < \beta, \gamma < 1$, is there a distribution free (β, γ) two-sided tolerance interval?

16. Determine the formula required for the computation of $\gamma(b)$, defined by (10.5.13).

17. Prove Theorem 10.5.4.

18. Given n i.i.d. observations from a Weibull distribution, with a density function

$$f(x; \alpha, \theta, \nu) = \frac{1}{\theta^{\nu}\Gamma(\nu)} x^{\nu\alpha-1} \exp\left\{-\frac{1}{\theta} x^{\alpha}\right\}, \qquad x \geqslant 0,$$

where $0 < \nu, \theta, \alpha < \infty$,

(i) show that if $\alpha > 1$ then F is an I.F.R. distribution;
(ii) determine a U.M.A. (β, γ) upper tolerance limit for the case of known values of α and ν;
(iii) determine a (β, γ) upper tolerance limit for the case where only ν is known.

Section 10.6

19. Let X be a random variable having a binomial distribution $\mathscr{B}(n; \theta)$. Let Y be a random variable, independent of X, having a binomial distribution $\mathscr{B}(k; \theta)$. On the basis of the observed value of X and a prior beta (a, b) distribution of θ determine

(i) a γ upper Bayes confidence limit for θ;
(ii) a (β, γ)-Bayes upper tolerance limit for Y.

20. Let $X_1, \ldots, X_n$ be i.i.d. random variables, $X_1 \sim N(0, \sigma_1^2)$, and $Y_1, \ldots, Y_n$ be i.i.d. random variables, $Y_1 \sim N(0, \sigma_2^2)$. The two samples are independent. Derive a γ-fiducial upper confidence limit for the variance ratio $\rho = \sigma_2^2/\sigma_1^2$.

21. Let $X_1, \ldots, X_n$ be i.i.d. random variables, $X_1 \sim N(\xi, \sigma_1^2)$. Let $Y_1, \ldots, Y_n$ be i.i.d. random variables, $Y_1 \sim N(\eta, \sigma_2^2)$. The two samples are independent. Derive the γ-fiducial limits for $R = \xi/\eta$, with respect to the proper group of transformations.

Section 10.7

22. Let $X_1, X_2, \ldots$ be a sequence of i.i.d. random variables having a common log-normal distribution, that is, $X_1 \sim \exp\{N(\mu, \sigma^2)\}$. Consider the problem of estimating $\xi = \exp\{\mu + \sigma^2/2\}$. The proportional closeness of an estimator $\hat{\xi}_N$ is defined as $P[|\hat{\xi}_N - \xi| < \lambda\xi]$, where λ is a specified positive real. Show that with a fixed sample procedure there exists no estimator $\hat{\xi}_N$ such that the proportional closeness is at least γ, $0 < \gamma < 1$, for all (ξ, σ^2). (See Zacks [3].)

23. Prove that the total sample size N in Stein's two-stage procedure is a proper random variable, that is, $P_{\mu,\sigma}\{N < \infty\} = 1$ for all (μ, σ).

24. Provide a two-stage sampling procedure and an estimator $\hat{\xi}_N$ so that in estimating the mean ξ of a log-normal distribution the proportional closeness will be at least γ, $0 < \gamma < 1$, for all (ξ, σ).

25. Show that if $\mathscr{F}$ is any family of d.f.'s depending on a location parameter of the translation type, that is, $F_\theta(x) = F(x - \theta)$, $-\infty < \theta < \infty$, there exists a fixed-width confidence interval, based on a fixed size sample.

26. Prove inequality (10.7.16).

27. Develop a two-stage sampling procedure for a fixed-width confidence interval of the scale parameter θ in the rectangular distributions $\mathscr{R}(0, \theta)$, $0 < \theta < \infty$.

Section 10.8

28. Prove the following theorem of Graybill [1]. Let $l(\mathbf{X}_n)$ be the (random) width of a confidence interval $(\underline{\theta}(\mathbf{X}_n), \bar{\theta}(\mathbf{X}_n))$ for a parameter θ, based on a sample $\mathbf{X}_n = (X_1, \ldots, X_n)$ of size n. Suppose that $l(\mathbf{X}_n)$ depends on n and on an unknown parameter τ (τ may be the parameter θ). Furthermore, suppose that there exists a (measurable) function $g(l; \tau, n)$ such that its distribution does not depend on any unknown parameter except n. Let $f(n)$ be such that $P[g(l; \tau, n) < f(n)] = \beta$, $0 < \beta < 1$. Let $\xi = \xi(\tau, n)$ be the solution of $g(l; \tau, n) = f(n)$, and suppose the following:

(i) $g(\xi; \tau, n)$ is increasing in ξ for each (τ, n).
(ii) $\xi(\tau, n)$ is increasing in τ for each n.
(iii) $\xi(\tau, n)$ is decreasing in n for each τ.

Moreover, let Z be a random variable [depending on $(X_1, \ldots, X_n)$] such that $P[t(Z) > t] = \beta$, for some function $t(Z)$ independent of θ, τ or n. Then if n is such that $\xi(t(Z), n) \leqslant d$ is saisfied, then $P[l(X_n) \leqslant d] \geqslant \beta^2$.

29. Apply Graybill's theorem, stated in the previous problem, to determine a two-stage procedure for a confidence interval of the mean μ of a normal distribution whose width does not exceed a fixed value d, $0 < d < \infty$. (See Graybill [1].)

30. Let $X_1, X_2, \ldots$ be a sequence of i.i.d. log-normal random variables, $X_1 \sim \exp\{N(\mu, \sigma^2)\}$, $0 < \mu, \sigma < \infty$. Design a two-stage sampling scheme for a fixed-width confidence interval estimator of the coefficient of variation $\tau = \sigma/\mu$ of the corresponding normal distribution. (See Koopmans, Owen, and Rosenblatt [1].)

Section 10.9

31. Consider the following three-stage procedure for a fixed-width interval estimator of the mean of a normal distribution:

(i) Take a random sample of n_1 observations. Compute the sample variance $s_{n_1}{}^2$. If $n_1 > (a^2/\delta^2)s_{n_1}{}^2$, terminate sampling. Otherwise, add an independent sample of size

$$N_2 = \max\left\{1, \frac{1}{2}\left(\frac{a^2}{\delta^2} s_{n_1}{}^2 - n_1\right)\right\}.$$

(ii) Compute the (pooled) sample variance $s^2_{N_1+N_2}$. If $n_1 + N_2 \geqslant (a^2 s^2_{N_1+N_2})/\delta^2$ terminate sampling. Otherwise, add

$$N_3 = \left[\frac{a^2 s^2_{n_1+N_2}}{\delta^2}\right] + 1 - (n_1 + N_2)$$

independent observations and terminate sampling. Let N be the total sampling size and the corresponding interval estimator $I_\delta(\mathbf{X}_N) = (\bar{X}_N - \delta, \bar{X}_N + \delta)$. What are the characteristics of this procedure?

Section 10.10

32. Let $X_1, X_2, \ldots$ be a sequence of i.i.d. random variables having a common $\mathcal{N}(\mu, \sigma^2)$ distribution, $0 < \mu, \sigma < \infty$.

    (i) Design a sequential procedure for a fixed-width, $l = 2\delta$, confidence interval estimation of the coefficient of variation $\tau = \sigma/\mu$.

    (ii) Define the asymptotic efficiency of the procedure in terms of the Fisher information, as a function of $\tau = \sigma/\mu$. Is your procedure asymptotically efficient (as $\delta \to 0$)?

33. Consider a sequence $X_1, X_2, \ldots$ of i.i.d. random variables $X_1 \sim G(1/\theta, 1)$. Provide a sequential procedure and an estimator of the reliability function $R(\tau) = e^{-\tau/\theta}$ such that

$$\lim_{\delta \to 0} P_\theta\{|\hat{R}_N(\tau) - R(\tau)| < \delta R(\tau)\} \geqslant \gamma, \quad \text{all } \theta, \quad 0 < \theta < \infty.$$

Is this procedure asymptotically efficient? Give bounds on $E_\theta\{N\}$ for each δ, $0 < \delta < 1$. (See Simons and Zacks [1].)

REFERENCES

Abbott and Rosenblatt [1]; Aitchison [1]; Anscombe [1]; Bahadur and Savage [1]; Barlow and Proschan [1], [2]; Blum and Rosenblatt [1], [2], [3], [4]; Chow and Robbins [1]; Dantzig [1]; Dempster [1]; Doob [1]; Epstein [1]; Farrel [3]; Feller [1]; Fisher [4], Fisz [1]; Fraser [1], [3], [5]; Fraser and Guttman [1]; Gabriel [1]; Godambe [2]; Goodman and Madansky [1]; Graybill [1]; Guttman [1]; Hanson and Koopmans [1]; Harter [1]; Joshi [4], [5], [6], [7]; Kemperman [1]; Khan [1]; Lehmann [2]; Lentner and Buehler [1]; Resnikoff and Lieberman [1]; Lindley [2]; Madansky [1]; Moshman [1]; Nadas [1]; Neyman [2]; Pratt [2], [3]; Rényi [1]; Robbins [1], [3]; Scheffé and Tukey [1]; Seelbinder [1]; Simons [1]; Singh [1]; Starr [1]; Stein [2], [6]; Stein and Wald [1]; Thatcher [1]; Tukey [1], [2]; Walsh [1]; Weiss [1]; Wiener [1]; Wilks [2], [3]; Zacks [3], [10].

References

J. H. Abbott, J. Rosenblatt

[1] Two-stage estimation with one observation on the first stage, *Ann. Inst. Statist. Math.* **14**: 229–235 (1963).

Om P. Aggarwal

[1] Some minimax invariant procedures for estimating cumulative distribution functions, *Ann. Math. Statist.*, **26**: 450–463 (1955).

J. Aitchison

[1] Bayesian tolerance regions, *Jour. Roy. Statist. Soc. B*, **26**: 161–175 (1964).

A. E. Albert

[1] The sequential design of experiments for infinitely many states of nature, *Ann. Math. Statist.*, **32**: 774–799 (1961).

G E. Albert

[1] On the computation of the sampling characteristics of a general class of sequential decision problems, *Ann. Math. Statist.*, **25**: 340–356 (1954).

[2] Accurate sequential tests on the mean of an exponential distribution, *Ann. Math. Statist.*, **27**: 460–470 (1956).

T. W. Anderson

[1] *Introduction to Multivariate Statistical Analysis*, John Wiley and Sons, New York, 1958.

F. J. Anscombe

[1] Large sample theory of sequential estimation, *Proc. Cambridge Philos. Soc.*, **48**: 600–607 (1952).

R. R. Bahadur

[1] Sufficiency and statistical decision functions, *Ann. Math. Statist.*, **25**: 423–462 (1954).

[2] Statistics and subfields, *Ann. Math. Statist.*, **26**: 490–497 (1955).

[3] Examples of inconsistency of maximum likelihood estimates, *Sankhya*, **20**: 207–210 (1958).

[4] On the asymptotic efficiency of tests and estimators, *Sankhya*, **22**: 229–252 (1960).

R. R. Bahadur, L. J. Savage

[1] The nonexistence of certain statistical procedures in nonparametric problems, *Ann. Math. Statist.*, **27**: 1115–1122 (1956).

K. S. Banerjee

[1] Singularity in Hotelling's weighing designs and a generalized inverse, *Ann. Math. Statist.*, **37**: 1021–1032 (1966).

[2] On nonrandomized fractional weighing designs, *Ann. Math. Statist.*, **37:** 1836–1841 (1966).

E. W. Barankin

[1] Locally best unbiased estimates, *Ann. Math. Statist.*, **20:** 477–501 (1949).

[2] Sufficient parameters: solution of the minimal dimensionality problem, *Ann. Inst. Stat. Math.*, **12:** 91–118 (1960).

E. W. Barankin, J. Gurland

[1] On asymptotically normal, efficient estimators: I. *University of California Publications in Statist.*, **1:** 89–130 (1951)

E. W. Barankin, M. Katz

[1] Sufficient statistics of minimal dimension, *Sankhya*, **21:** 217–246 (1959).

R. E. Barlow, F. Proschan

[1] *Mathematical Theory of Reliability*, John Wiley, New York, 1965.

[2] Tolerance and confidence limits for classes of distributions based on failure rate, *Ann. Math. Statist.*, **37:** 1593–1601 (1966).

G. A. Barnard

[1] Some logical aspects of the fiducial argument. *Jour. Roy. Statist. Soc. B*, **25:** 111–114 (1963).

V. D. Barnett

[1] Evaluation of the maximum-likelihood estimator where the likelihood equation has multiple roots, *Biometrika*, **53:** 151–165 (1966).

A. P. Basu

[1] Estimates of reliability for some distribution useful in life testing, *Technometrics*, **6:** 215–219 (1964).

D. Basu

[1] Contributions to Statistical Inference, Ph.D. Dissertation, Calcutta University, 1953.

[2] An inconsistency of the method of maximum likelihood, *Ann. Math. Statist.*, **26:** 144–145 (1955).

[3] Sufficiency in sample survey theory, *Sankhya* (to be published).

Bateman Manuscript Project

[1] Tables of Integral Transforms, Vol. 1, A. Erdelyi, ed., McGraw-Hill Book Co., New York, 1954.

R. E. Bechhofer, J. Kiefer, M. Sobel

[1] *Sequential Identification and Ranking Procedures*, The University of Chicago Press, Chicago, 1968.

A. Bhattacharyya

[1] On some analogues to the amount of information and their uses in statistical estimation, *Sankhya*, **8:** 1–14 (1946).

S. A. Bessler

[1] Theory and applications of sequential design of experiments, k-actions and infinitely many experiments, Tech. Rep. No. 55, Dept. of Statist., Stanford University, (1960).

P. J. Bickel, J. A. Yahav

[1] Asymptotically pointwise optimal procedures in sequential analysis, *Proc. Fifth Berkeley Symp. Math. Statist. Prob.*, **1:** 401–413 (1965).

[2] Asymptotically optimal Bayes and minimax procedures in sequential estimation *Ann. Math. Statist.*, **39:** 442–456 (1968).

D. Blackwell

[1] Comparison of experiments, *Proc. Second Berkeley Symp. Math. Stat. Prob.*, 93–102 (1951).

[2] On the translation parameter problem for discrete variables, *Ann. Math. Statist.*, **22:** 393–399 (1951).

D. Blackwell, M. A. Girshick

[1] *Theory of Games and Statistical Decisions*, John Wiley, New York, 1954.

J. R. Blum, J. Rosenblatt

[1] On multistage estimation, *Ann. Math. Statist.* **34:** 1452–1458 (1963).

[2] On some statistical problems requiring purely sequential sampling schemes, *Ann. Inst. Statist. Math*, **18:** 351–355 (1966).

[3] Fixed precision estimation in the class of IFR distributions, *Ann. Inst. Stat. Math.*, **21:** 1 (1969).

[4] On partial a-priori information in statistical inference, *Ann. Math. Statist.*, **38:** 1671–1678 (1967).

[5] On fixed precision estimation in time series, *Ann. Math. Statist.*, **40:** 1021–1032 (1969).

C. R. Blyth

[1] On minimax statistical decision procedures and their admissibility, *Ann. Math. Statist.*, **22:** 22–42 (1951).

J. Breakwell, H. Chernoff

[1] Sequential tests for the mean of a normal distribution II (large t) *Ann. Math. Statist.*, **35:** 162–173 (1964).

B. M. Brown

[1] Moments of a stopping rule related to the central limit theorem, Mimeo Series No. 159, Dept. of Statist., Purdue University, 1968.

L. D. Brown

[1] Sufficient statistics in the case of independent random variables, *Ann. Math. Statist.*, **35:** 1456–1474 (1964).

[2] On the admissibility of invariant estimators of one or more location parameters, *Ann. Math. Statist.*, **37:** 1087–1136 (1966).

H. D. Brunk

[1] On the estimation of parameters restricted by inequalities, *Ann. Math. Statist.*, **29:** 437–454 (1958).

D. L. Burkholder, R. H. Wijsman

[1] Optimum properties and admissibility of sequential tests, *Ann. Math. Statist.*, **34:** 1–17 (1963).

D. G. Chapman, H. Robbins

[1] Minimum variance estimation without regularity assumptions, *Ann. Math. Statist.*, **22:** 581–586 (1951).

H. Chernoff

[1] A measure of asymptotic efficiency for tests of a hypothesis based on the sum of observations, *Ann. Math. Statist.*, **23:** 493–507 (1952).

[2] Large sample theory; parametric case, *Ann. Math. Statist.*, **27:** 1–22 (1956).

[3] Sequential design of experiments, *Ann. Math. Statist.*, **30:** 755–770 (1959).

[4] Sequential tests for the mean of a normal distribution, *Proc. Fourth Berkeley Symp. Math. Statist. Prob.*, **1:** 79–91 (1961).
[5] A series of lectures on large sample theory, Stanford University, Stanford, California. 1963.
[6] Sequential tests for the mean of a normal distribution III (small t), *Ann. Math. Statist.*, **36:** 28–54 (1965).
[7] Optimal stochastic control, *Sankhya*, Series A, **30:** 221–252 (1968).

H. Chernoff, L. E. Moses

[1] *Elementary Decision Theory*, John Wiley, New York, 1959.

H. Chernoff, S. N. Ray

[1] A Bayes sequential sampling inspection plan, *Ann. Math. Statist.*, **36:** 1387–1407 (1965).

H. Chernoff, S. Zacks

[1] Estimating the current mean of a normal distribution which is subjected to changes in time, *Ann. Math. Statist.*, **35:** 999–1018 (1964).

C. L. Chiang

[1] On regular best asymptotically normal estimators, *Ann. Math. Statist.*, **27:** 336–351 (1956).

Y. S. Chow, H. Robbins

[1] On the asymptotic theory of fixed-width sequential confidence intervals for the mean, *Ann. Math. Statist.*, **36:** 457–462 (1965).

A. Cohen

[1] On estimating the mean and standard deviation of truncated normal distributions' *Jour. Amer. Statist. Assoc.*, **44:** 518–525 (1949).
[2] Estimating parameters of Pearson Type III populations from truncated samples, *Jour. Amer. Statist. Assoc.*, **45:** 411–425 (1950).
[3] Estimation of the Poisson parameters from truncated samples and from censored samples, *Jour. Amer. Statist. Assoc.*, **49:** 158–168 (1954).
[4] All admissible linear estimates of the mean vector, *Ann. Math. Statist.*, **37:** 458–463 (1966).

H. Cramér

[1] A contribution to the theory of statistical estimation *Aktuariestidskrift* **29:** 458–463 (1966).
[2] *Mathematical Methods in Statistics*, Princeton, 1948.
[3] *Random Variables and Probability Distributions,* (2d ed.), Cambridge Tracts in Math. and Math. Physics, Cambridge University Press, 1962.

W. Commins, Jr.

[1] Asymptotic variance as an approximation to expected loss for maximum likelihood estimates, Tech. Rep. No. 46, Contract N60nr-2S140 (NR-342-022), Dept. of Statist., Stanford University, (1959).

H. E. Daniels

[1] The asymptotic efficiency of a maximum likelihood estimator, *Proc. Fourth Berkeley Symp. Math. Statist. Prob.*, **1:** 151–164 (1961).

G. B. Dantzig

[1] On the nonexistence of tests of "Student's" hypothesis having power functions independent of σ^2, *Ann. Math. Statist.*, **11:** 186–192 (1940).

M. H. DeGroot

[1] Unbiased binomial sequential estimation, *Ann. Math. Statist.*, **30:** 80–101 (1959).
[2] Minimax sequential tests of some composite hypotheses, *Ann. Math. Statist.*, **31,** 1193–1200 (1960).
[3] Optimal allocation of observations, *Ann. Inst. Statist. Math.*, **18:** 13–28 (1966).

M. H. DeGroot, M. M. Rao

[1] Bayes estimation with convex loss, *Ann. Math. Statist.*, **34:** 839–846 (1963).

A. P. Dempster

[1] On the difficulties inherent in Fisher's fiducial argument, *Jour. Amer. Statist. Assoc.* **59:** 56–66 (1964).

J. L. Denny

[1] Minimal dimension of sufficient statistics. Unpublished doctoral dissertation, University of California, Berkeley (1962).
[2] On continuous sufficient statistics, *Ann. Math. Statist.*, **35:** 1229–1233 (1964).
[3] A continuous real-valued function on E^n almost everywhere 1-1, *Fundamenta Mathematicae*, **LV:** 95–99 (1964).
[4] Sufficient conditions for a family of probabilities to be exponential, *Proc. Nat. Acad. Sci.*, **57:** 1184–1187 (1967).
[5] Note on a theorem of Dynkin on the dimension of sufficient statistics, *Ann. Math. Statist.*, **40:** 1474–1476 (1969).

J. L. Doob

[1] Probability and statistics, *Trans. Amer. Math. Soc.*, **36:** 759–772 (1934).
[2] *Stochastic Processes*, John Wiley, New York, 1953.

A. Dvoretzky, J. Kiefer, J. Wolfowitz

[1] Sequential decision problems for processes with continuous time parameter, *Ann. Math. Statist.*, **24:** 254–264 (1953).

H. B. Dwight

[1] *Tables of Integrals and Other Mathematical Data* (4th ed.), The Macmillan Co. New York, 1961.

E. B. Dynkin

[1] Necessary and sufficient statistics for a family of probability distributions, *Uspeh Matem. Nauk.*, 6: English translation appeared in *Selected Translations in Math. Stat. and Prob.*, **1:** 1–740 (1961).
[2] *Markov Processes*, Academic Press, New York, 1965.

E. B. Dynkin, A. A. Yushkevich

[1] *Markov Processes: Theorems and Problems*, Plenum Press, New York, 1969.

S. Ehrenfeld

[1] Some experimental design problems in attribute life testing, *Jour. Amer. Statist. Assoc.* **57:** 668–679 (1962).

S. Ehrenfeld, S. Zacks

[1] Randomization and factorial experiments, *Ann. Math. Statist.*, **32:** 270–297 (1961).

B. E. Ellison

[1] Two theorems for inference about the normal distribution with applications in acceptance sampling, *Jour. Amer. Statist. Assoc.*, **59:** 89–95 (1964).

B. Epstein

[1] Tolerance limits based on life test data taken from an exponential distribution, *Indust. Qual. Cont.*, **17:** 10–11 (1960).

B. Epstein, M. Sobel

[1] Some theorems relevant to life testing from an exponential distribution, *Ann. Math. Statist.*, **25:** 373–381 (1954).
[2] Sequential life testing in the exponential case, *Ann. Math. Statist.*, **26:** 82–93 (1955).

W. A. Ericson

[1] Subjective Bayesian models in sampling finite populations, II: Stratification, *Symposium on the Foundations of Survey Sampling*, John Wiley, New York, 1969.

J. Fabius

[1] Asymptotic behavior of Bayes estimates, *Ann. Math. Statist.*, **35:** 846–856 (1964).

R. H. Farrel

[1] Estimators of a location parameter in the absolutely continuous case, *Ann. Math. Statist.*, **35:** 949–998 (1964).
[2] Bounded length confidence intervals for the p-point of a distribution function, III, *Ann. Math. Statist.*, **37:** 589–592 (1966).
[3] Weak limits of sequences of Bayes procedures in estimation theory. *Proc. Fifth Berkeley Symp. Math. Statist. Prob.*, **1:** 83–111 (1966).

A. V. Fend

[1] On the attainment of Cramér-Rao and Bhattacharyya bounds for the variances of an estimate, *Ann. Math. Statist.*, **30:** 381–388 (1959).

W. Feller

[1] *An Introduction to Probability Theory and its Applications*, Vol. I (2d ed.), John Wiley, New York, 1957.

T. S. Ferguson

[1] *Mathematical Statistics: A Decision Theoretic Approach*, Academic Press, New York, (1967).

R. A. Fisher

[1] The goodness of fit of regression formulae and distribution of regression coefficients, *Jour. Roy. Statist. Soc.*, **65:** 597–612 (1922).
[2] On the mathematical foundations of theoretical statistics, *Philos. Trans. Roy. Soc. A.*, **222:** 309–368 (1922).
[3] Theory of statistical estimation, *Proc. Camb. Phil. Soc.*, **22:** 700–715 (1925).
[4] Inverse probability, *Proc. Camb. Phil. Soc.*, **26:** 528–535 (1930).
[5] Two new properties of mathematical likelihood, *Proc. Roy. Soc. A.*, Series **A, 14:** 285–307 (1934).
[6] The fiducial argument in statistical inference, *Annals of Eugenics*, **6:** 391–398 (1935).
[7] *Statistical Methods and Scientific Inference*, Oliver and Boyd, Edinburgh, 1956.

M. Fisz

[1] *Probability Theory and Mathematical Statistics* (3d ed.), John Wiley, New York, 1963.

M. Fox, H. Rubin

[1] Admissibility of quantile estimates of a single location parameter, *Ann. Math. Statist.*, **35:** 1019–1031 (1964).

D. A. S. Fraser

[1] Sequentially determined statistically equivalent blocks, *Ann. Math. Statist.*, **22:** 372–381, (1951).
[2] Sufficient statistics and selection depending on the parameter, *Ann. Math. Statist.*, **23:** 417–425 (1952).
[3] Completeness of the order statistics, *Can. J. Math.*, **6:** 42 (1953).
[4] Non-parametric tolerance regions, *Ann. Math. Statist.*, **24:** 44–55 (1953).
[5] *Nonparametric Methods in Statistics*, John Wiley, New York, 1957.
[6] The fiducial method and invariance, *Biometrika*, **48:** 261–280 (1961).
[7] On sufficiency and the exponential family, *Jour. Roy. Statist. Soc.*, **25:** 115–123 (1963).
[8] Fiducial inference for location and scale parameters, *Biometrika*, **51:** 17–24 (1964).
[9] *The Structure of Inference*, John Wiley, New York, 1968.

D. A. S. Fraser, I. Guttman

[1] Tolerance regions, *Ann. Math. Statist.*, **27:** 162–179 (1957).

D. A. Freedman

[1] On the asymptotic behavior of Bayes' estimates in the discrete case, *Ann. Math. Statist.*, **34:** 1386–1403 (1963).

K. B. Gabriel

[1] Simultaneous test procedures—some theory of multiple comparisons, *Ann. Math. Statist.*, **40:** 224–250 (1969).

D. B. Gaver, Jr., D. G. Hoel

[1] Comparison of certain small-sample Poisson probability estimates, *Technometrics* (to be published).

M. A. Girshick, L. G. Savage

[1] Bayes and minimax estimates for quadratic loss function, *Proc. Second Berkeley Symp. Math. Statist. Prob.*, **1:** 53–74 (1951).

G. J. Glasser

[1] Minimum variance unbiased estimators of Poisson probabilities, *Technometrics*, **4:** 409–418 (1962).

V. P. Godambe

[1] A unified theory of sampling from finite populations, *Jour. Roy. Statist. Soc. B*, **17:** 268–278 (1955).
[2] Bayesian shortest confidence intervals and admissibility, *Bull. Inst. Internat. Statist.*, **33** Session, 1961.
[3] Bayesian sufficiency in survey sampling, *Inst. Statist. Math.*, **20:** 363–373 (1968).
[4] Admissibility and Bayes estimation in sampling finite populations, V, *Ann. Math. Statist.*, **40:** 672–676 (1969).
[5] Some aspects of the theoretical developments in survey sampling, *Symposium on the Foundation of Survey Sampling*, John Wiley, New York, 1969.

V. P. Godambe, V. M. Joshi

[1] Admissibility and Bayes estimation in sampling finite populations, I, *Ann. Math. Statist.*, **36:** 1707–1722 (1965).

A. J. Goldman, M. Zelen

[1] Weak generalized inverse and minimum variance linear unbiased estimation, *J. Res. Nat. Bur. Stand.*, **66B:** 151–172 (1964).

H. H. Goode

[1] Deferred decision theory, *Recent Developments in Information and Decision Processes*, R. E. Machol and P. Gray (eds.), The Macmillan Co., New York, 1962.

L. A. Goodman, A. Madansky

[1] Parameter free and non-parametric tolerance limits: the exponential distribution, *Technometrics*, **4**: 75–95 (1962).

J. J. Gort

[1] An extension of the Cramér-Rao inequality, *Ann. Math. Statist.*, **30**: 367–380 (1959).

F. A. Graybill

[1] Determining sample size for a specified width confidence interval, *Ann. Math. Statist.*, **29**: 282–287 (1958).

[2] *An Introduction to Linear Statistical Models*, McGraw-Hill, New York, 1961.

B. G. Greenberg, A. E. Sarhan

[1] Matrix inversion, its interest and application in analysis of data, *Jour. Amer. Statist., Assoc.*, **54**: 755–766.

B. I. Grigelionis, A. N. Shiryaev

[1] On Stephan's problem and optimal stopping rules for Markov processes, *Theory of Prob. and Its Appl.*, **9**: 541–558 (1966).

I. Guttman

[1] Optimum tolerance regions and power when sampling from some non-normal universes, *Ann. Math. Statist.*, **30**: 926–938 (1959).

W. J. Hall

[1] Embedding submartingales in Wiener processes with drift, with applications to sequential analysis, Tech. Rep. No. 34 (NSF GP-5705), Dept. of Statist., Stanford University, 1968.

[2] On Wald's equation for Wiener processes, Tech. Rep. No. 32 (NSF GP-5705 and U.S. Public Health Service Grant GM-10397, Dept. of Statist., Stanford University, 1968.

W. J. Hall, R. A. Wijsman, J. K. Ghosh

[1] The relationship between sufficiency and invariance with applications in sequential analysis, *Ann. Math. Statist.*, **36**: 575–614 (1965).

P. R. Halmos

[1] The theory of unbiased estimation, *Ann. Math. Statist.*, **17**: 34–43 (1946).

[2] *Measure Theory*, Van Nostrand, New York, 1950.

P. R. Halmos, L. J. Savage

[1] Application of the Radon-Nikodym theorem to the theory of sufficient statistics, *Ann. Math. Statist.*, **20**: 225–241 (1949).

J. Hannan

[1] Consistency of maximum likelihood estimation of discrete distributions, *Contributions to Probability and Statistics, Essays in Honor of Harold Hotelling*, Stanford University Press, Ch. 21, pp. 249–257, 1960.

D. L. Hanson, L. H. Koopmans

[1] Tolerance limits for the class of distributions with increasing hazard rates, *Ann. Math. Statist.*, **35**: 1571–1570 (1964).

T. V. Hanurav

[1] Some aspects of unified sampling theory, *Sankhya*, **A28**: 175–204 (1966).

[2] Hyper-admissibility and optimum estimators for sampling finite populations, *Ann. Math. Statist.*, **39:** 621–642 (1968).

H. L. Harter

[1] Criteria for best substitute interval estimators, with an application to the normal distribution, *J. Amer. Statist. Assoc.*, **59:** 1133–1140 (1964).

J. Hartigan

[1] Invariant prior distributions, *Ann. Math. Statist.*, **35:** 836–845 (1964).

J. L. Hodges, E. L. Lehmann

[1] Some problems in minimax estimation, *Ann. Math. Statist.*, **21:** 182–197 (1950).
[2] Some applications of the Cramér-Rao inequality, *Second Berkeley Symp. Math. Statist. Prob.*, **1:** 13–22 (1951).
[3] The use of previous experience in reaching statistical decisions, *Ann. Math. Statist.*, **23:** 396–407 (1952).

R. V. Hogg, A. T. Craig

[1] Some results on unbiased estimation, *Sankhya*, *Series A*, **24:** 333–338 (1962).

R. B. Hora, R. J. Buehler

[1] Fiducial theory and invariant estimation, *Ann. Math. Statist.*, **37:** 643–656 (1965).
[2] Fiducial theory and invariant prediction, *Ann. Math. Statist.*, **38:** 795–801 (1967).

H. Hotelling

[1] Some improvements in weighing and making other types of measurements, *Ann. Math. Statist.*, **15:** 297–303 (1944).

G. A. Hunt, C. Stein

[1] Most stringent tests of statistical hypotheses, unpublished, 1945.

V. S. Huzurbazar

[1] On a property of distributions admitting sufficient statistics, *Biometrika*, **36:** 71–74 (1949).

E. Jahnke, F. Emde

[1] *Tables of Functions with Formulas and Curves* (4th ed.), Dover Publications, New York, 1945.

W. James, C. Stein

[1] Estimation with quadratic loss, *Proc. Fourth Berkeley Symp. Math. Statist. I. Prob.*, **2:** 361–379 (1960).

H. Jeffreys

[1] *Theory of Probability* (3d ed.), Oxford University Press, 1960.

S. John

[1] A tolerance region for multivariate normal distributions, *Sankhya*, Series A, **24:** 363–368 (1962).

M. V. Johns, Jr.

[1] Non-parametric empirical Bayes procedures, *Ann. Math. Statist.*, **28:** 649–669 (1957).

V. M. Joshi

[1] Note on minimax design for cluster sampling, *Ann. Math. Statist.*, **39:** 278–281 (1968).
[2] Admissibility and Bayes estimation in sampling finite populations, II, III, *Ann. Math. Statist.*, **36:** 1723–1742 (1965).
[3] Admissibility of confidence intervals, *Ann. Math. Statist.*, **37:** 629–638 (1966).

[4] Admissibility and Bayes estimation in sampling finite populations, IV, *Ann. Math. Statist.*, **37:** 1658–1670 (1966).

[5] Inadmissibility of the usual confidence sets for the mean of a multivariate normal population, *Ann. Math. Statist.*, **38:** 629–638 (1967).

[6] Confidence intervals for the mean of a finite population, *Ann. Math. Statist.*, **38:** 1280–1307 (1967).

[7] Admissibility of the usual confidence sets for the mean of a univariate or bivariate normal populations, *Ann. Math. Statist.*, **40:** 1042–1067 (1969).

A. M. Kagan

[1] On the estimation theory of location parameters, *Sankyha*, Series A, **28:** 335–352 (1966).

A. Kagan, Yu. Linnik, C. R. Rao

[1] A characterization of normal law by an admissibility of a sample mean, *Sankhya*, Series A, **27:** 405–406 (1965).

B. K. Kale

[1] On the solution of the likelihood equation by iteration processes, *Biometrika*, **48:** 452–456 (1961).

[2] On the solution of the likelihood equation by iteration processes—the multiparametric case, *Biometrika*, **49:** 479–486 (1962).

[3] A note on the loss of information due to grouping of observations, *Biometrika*, **51:** 495–497 (1964).

[4] Approximations to the maximum-likelihood estimator using grouped data, *Biometrika*, **53:** 282–285 (1966).

G. Kallianpur

[1] von Mises functionals and maximum likelihood estimation, *Sankhya*, Series A, **25:** 149–158 (1963).

S. Karlin

[1] Decision theory for Pólya type distributions. Case of two actions, I, *Proc. Third Berkeley Symp. Math. Statist. Prod.*, **1:** 115–128 (1956).

[2] Pólya type distributions, II, *Ann. Math. Statist.*, **28:** 281–308 (1957).

[3] Pólya type distributions, III. Admissibility for multi-action problems, *Ann. Math. Statist.*, **29:** 406–436 (1958).

[4] Admissibility for estimation with quadratic loss, *Ann. Math. Statist.*, **29:** 406–436 (1958).

M. W. Katz

[1] Admissible and minimax estimates of parameters in truncated spaces, *Ann. Math. Statist.*, **32:** 136–142 (1961).

J. H. B. Kemperman

[1] Generalized tolerance limits, *Ann. Math. Statist.*, **27:** 180–186 (1956).

[2] A Wiener-Hopf type method for a general random Wald S.P.R.T. with two-sided boundaries, *Ann. Math. Statist.*, **35:** 1168–1193 (1964).

M. G. Kendall, W. R. Buckland

[1] *A Dictionary of Statistical Terms*, The International Statistical Institute, Oliver and Boyd, London, 1957.

R. A. Khan

[1] A general method of determining fixed-width confidence intervals, *Ann. Math. Statist.* **40:** 704–709 (1969).

J. Kiefer

[1] Sequential minimax estimation for the rectangular distribution with known range, *Ann. Math. Statist.*, **23:** 586–593 (1954).

[2] Invariance, minimax sequential estimation and continuous time processes, *Ann. Math, Statist.*, **28:** 573–601 (1957).

J. Kiefer, J. Wolfowitz

[1] Consistency of the maximum likelihood estimator in the presence of infinitely many incidental parameters, *Ann. Math. Statist.*, **27:** 887–906 (1956).

J. Kiefer, J. Sacks

[1] Asymptotically optimum sequential inference and design, *Ann. Math. Statist.*, **34:** 705–750 (1963).

T. Kitagawa

[1] The operational calculus and the estimation of functions of parameters admitting sufficient statistics, *Bull. Math. Statist.*, **6:** 95–108 (1956).

A. N. Kolmogorov

[1] Unbiased estimates, *Izvestia Akad, Nauk SSSR, Seriya Matematiceskava*, **14:** 303–326 (Amer. Math. Soc. Translation No. 98), 1950.

L. H. Koopmans, D. B. Owen, J. I. Rosenblatt

[1] Confidence intervals for the coefficient of variation for the normal and log-normal distributions, *Biometrika*, **51:** 25–32 (1964).

H. Kudō

[1] On minimax invariant estimates of the transformation parameter, *Natural Science Report of the Ochanomizea Univ.*, **6:** 31–73 (1955).

[2] On partial prior information and the property of parametric sufficiency, *Proc. Fifth Berkeley Symp. Math. Statist. Prob.*, **1:** 251–265 (1967).

S. Kullback

[1] *Information Theory and Statistics*, John Wiley, New York, 1959.

S. Kullback, R. A. Leibler

[1] On information and sufficiency, *Ann. Math. Statist.*, **22:** 79–86 (1951).

G. Kulldorf

[1] *Estimation from Grouped Data and Partially Grouped Samples*, John Wiley, New York, 1961.

L. LeCam

[1] On some asymptotic properties of maximum likelihood estimates and related Bayes estimates, *University of California Publ. in Statist.*, **1:** 277–330 (1953).

[2] On the asymptotic theory of estimation and testing hypotheses, *Proc. Third Berkeley Symp. Math. Statist. Prob.* V. **1:** 129–156 (1956).

[3] Sufficiency and approximate sufficiency, *Ann. Math. Statist.*, **35:** 1419–1455 (1964).

E. L. Lehmann

[1] *Notes on the Theory of Estimation*, University of California, 1950.

[2] *Testing Statistical Hypotheses*, John Wiley, New York, 1959.

E. L. Lehmann, H. Scheffé

[1] Completeness, similar regions and unbiased estimation, Part I, *Sankhya*, **10:** 305–340 (1950).

M. N. Lentner

[1] Generalized least-squares estimation of a subvector of parameters in randomized fractional factorial experiments, *Ann. Math. Statist.*, **40**: 1344–1352 (1969).

M. N. Lentner, R. J. Buehler

[1] Some inference about Gamma parameters with an application to a reliability problem, *J. Amer. Statist. Assoc.*, **58**: 670–677 (1963).

G. J. Lieberman, G. J. Resnikoff

[1] Sampling plans for inspection by variables, *Jour. Amer. Statist. Assoc.*, **50**: 457–516 (1955).

D. V. Lindley

[1] The use of prior probability distributions in statistical inference and decisions, *Fourth Berkeley Symp. Math. Statist. Prob.*, **1**: 453–468 (1961).

[2] *Introduction to Probability and Statistics: Part* 2, *Inference*, Cambridge University Press, 1965.

E. H. Lloyd

[1] Least squares estimation of location and scale parameters using order statistics, *Biometrika*, **39**: 88–95 (1952).

M. Loève

[1] Probability Theory (3d ed.), D. Van Nostrand, New York, 1963.

R. E. Machol, J. Rosenblatt

[1] Confidence interval based on single observation, *Proc. IEEF*, **54**: 1087–1088 (1966).

A. Madansky

[1] More on length of confidence intervals, *J. Amer. Statist. Assoc.*, **57**: 586–589 (1962).

T. A. Magness, J. B. McGuire

[1] Comparison of least-squares and minimum variance estimates of regression parameters, *Ann. Math. Statist.*, **33**: 462–470 (1962).

T. K. Matthes

[1] On the optimality of the sequential probability ratio test, *Ann. Math. Stat.*, **34**: 18–21 (1963).

V. S. Mikhalevich

[1] A Bayes test of two hypotheses concerning the mean of a normal process, *Visnik Kiius'kogo Universitétu*, **1**: 101–104 (1958). (In Ukranian.)

R. G. Miller

[1] *Simultaneous Statistical Inference*, McGraw-Hill, New York, 1966.

K. Miyasawa

[1] An empirical Bayes estimator of the mean of a normal population, *Bull. Inst. Internat. Statist.*, **38**: 181–188 (1961).

H. Morimoto, M. Sibuya

[1] Sufficient statistics and unbiased estimation of restricted selection parameters, *Sankhya*, Series A, **29**: 15–40 (1967).

J. Moshman

[1] A method for selecting the size of the initial sample in Stein's two sample procedure, *Ann. Math. Statist.*, **29**: 1271–1275 (1958).

F. Mosteller

[1] On some useful "inefficient" statistics, *Ann. Math. Statist.*, **17**: 377–408 (1946).

L. Nachbin

[1] *The Haar Integral* D. van Nostrand, Princeton, 1965.

A. J. Nades

[1] On the Asymptotic Theory of Estimating the Mean by Sequential Confidence Intervals of Prescribed Accuracy Ph.D. thesis Dept. of Statist., Colombia University, New York. 1967.

M. A. Naimark

[1] *Normed Rings* (English translation), Noordhoff-Groningen, The Netherlands, 1959.

J. Neyman

[1] Sur un teorems concerente le cosidette statistiche sufficienti. *Inst. Ital. Atti. Giorn.*, **6**: 320–334 (1935).

[2] Outline of the theory of statistical estimation based on the classical theory of probability, *Phil. Trans. Roy. Soc. London*, **236**: 333–380 (1937).

[3] Contributions to the theory of χ^2-test, *Proc. Berkeley Symp. Math. Statist. Prob.*, 239–273 (1949).

[4] Two breakthroughs in the theory of statistical decision making, *Rev. Inst. Internat. Statist.*, **30**: 11–27 (1962).

J. Neyman, E. L. Scott

[1] Consistent estimates based on partially consistent observations, *Econometrika*, **16**: 1–32 (1948).

H. W. Northan

[1] One likelihood adjustment may be inadequate, *Biometrics*, **12**: 79–81 (1956).

I. Olkin, J. W. Pratt

[1] Unbiased estimation of certain correlation coefficients, *Ann. Math. Statist.*, **29**: 201–211 (1958).

E. S. Page

[1] An improvement to Wald's approximation for some properties of sequential tests, *J. Roy. Statist. Soc. B* **16**: 136–139 (1954).

G. P. Patil, J. K. Wani

[1] Minimum variance unbiased estimation of the distribution function admitting a sufficient statistic, *Ann Inst. Stat. Math.*, **18**: 39–47 (1966).

E. S. Pearson, H. O. Hartley

[1] *Biometrika Tables for Statisticians, Vol.* 1, Cambridge University Press, 1958.

Shian-Koong Perng

[1] Inadmissibility of various good statistical procedures which are translation invariant, Tech. Rep. No. 16, Dept. of Statist., Michigan State University, 1967.

E. J. Pitman

[1] The estimation of the location and scale parameters of a continuous population of any given form, *Biometrika*, **30**: 391–421 (1939).

J. W. Pratt

[1] On a general concept of "In probability," *Ann. Math. Statist.*, **30**: 549–558 (1959).

[2] Length of confidence intervals, *J. Amer. Statist. Assoc.*, **56**: 260–272 (1961).

[3] Shorter confidence intervals for the mean of a normal distribution with known variance, *Ann. Math. Statist.*, **34**: 574–586 (1963).

H. Raiffa, R. Schlaifer

[1] *Applied Statistical Decision Theory*. Harvard University Press, Boston, 1961.

C. R. Rao

[1] Information and the accuracy attainable in the estimation of statistical parameters, *Bull. Calcutta Math. Soc.*, **37**: 81–91 (1945).
[2] Maximum likelihood estimation for multinomial distributions, *Sankhya*, **18**: 139–148 (1957).
[3] Asymptotic efficiency and limiting information, *Proc. Fourth Berkeley Symp. Math. Statist. Prob.*, **1**: 531–546 (1961).
[4] A note on a generalized inverse of a matrix with applications to problems of mathematical statistics, *J. Roy. Stat. Soc. B*, **24**: 152–158 (1962).
[5] Efficiency estimates and optimum inference procedures in large samples, *J. Roy. Stat. Soc.*, **24**: 46–72 (1962).
[6] Criteria of estimation in large samples, *Sankhya*, Series A, **25**: 189–206 (1963).
[7] *Linear Statistical Inference and Its Applications*, John Wiley, New York, 1965.

G. J. Resnikoff, G. J. Lieberman

[1] *Tables of the Noncentral t-Distribution*, Stanford University Press, Stanford, 1957.

A. Rényi

[1] On the asymptotic distribution of the sum of a random number of independent random variables, *Acta Math. Acad. Scien. Hung.*, **8**: 193–199 (1957).

H. Robbins

[1] On distribution free tolerance limits in random sampling, *Ann. Math. Statist.*, **15**: 214–216 (1944)
[2] The empirical Bayes approach to statistics, *Proc. Third Berkeley Symp. Math. Statist. Prob.*, **1**: 157–164 (1955).
[3] Sequential estimation of the mean of a normal population, *Probability and Statistics—The Herald Cramér Volume*, 235–245, Almquist and Wilsell, Uppsala, Sweden, 1959.
[4] The empirical Bayes approach to statistical decision problems, *Ann. Math. Statist.*, **35**: 1–20 (1964)

H. Robbins, E. Samuel

[1] An extension of a lemma of Wald, *J. Appl. Prob.*, **3**: 272–273 (1966).

H. L. Royden

[1] *Real Analysis*, Macmillan, New York, 1963.

H. Rubin, S. Karlin

[1] The theory of decision procedures for distributions with monotone likelihood ratio, *Ann. Math. Statist.*, **27**: 272–300 (1956).

J. R. Rutherford

[1] The empirical Bayes approach: estimating posterior quantiles, *Biometrika*, **54**: 672–675 (1967).

J. R. Rutherford, R. G. Krutchkoff

[1] The empirical Bayes approach: estimating the prior distribution, *Biometrika*, **54**: 326–328 (1967).

J. Sacks

[1] Generalized Bayes solutions in estimation problems, *Ann. Math. Statist.*, **34:** 751–768 (1963).

E. Samuel

[1] An empirical Bayes approach to the testing of certain parametric hypotheses. *Ann. Math Statist.*, **34:** 1370–1385 (1963).

A. E. Sarhan, B. G. Greenberg

[1] *Contributions to Order Statistics*, John Wiley, New York, 1962.

H. Scheffé

[1] *The Analysis of Variance*, John Wiley, New York, 1959.

H. Scheffé, J. W. Tukey

[1] Non-parametric estimation:. I Validation of order statistics, *Ann. Math. Statist.*, **16:** 187–192 (1945).

L. Schmetterer

[1] On the asymptotic efficiency of estimates, *Research Papers in Statistics*, Festschrift for J Neyman, F. N. David (ed.), John Wiley, New York, 1966.

G. Schwarz

[1] Asymptotic shapes of Bayes sequential testing regions, *Ann Math. Statist.*, **33:** 224–236 (1962).

B. M. Seelbinder

[1] On Stein's two-stage sampling scheme, *Ann. Math. Statist.*, **24:** 640–649 (1953).

J. Sethuraman

[1] Conflicting criteria of "goodness" of statistics, *Sankhya*, Series A, **23:** 187–190 (1961).

R. Shimizu.

[1] Remarks on sufficient statistics, *Ann. Inst. Statist. Math.*, **18:** 49–66 (1966).

A. N. Shiryaev

[1] Sequential analysis and controlled random processes, *Kibernetika*, **3:** 1–24 (1965) (in Russian).

Yu, A. Shreider, ed.

[1] *The Monte Carlo Method, The Method of Statistical Trials*, Pergamon Press, London, 1966.

G. F. Simmons

[1] *Introduction to Topology and Modern Analysis*, McGraw-Hill, New York, 1963.

G. Simons

[1] On the cost of not knowing the variance when making a fixed-width confidence interval for the mean, *Ann. Math. Statist.*, **39:** 1946–1952 (1968).

G. Simons, S. Zacks

[1] A sequential estimation of tail probabilities in exponential distributions with a prescribed proportional closeness, Tech. Rep., Dept. of Statist., Stanford University, 1967.

R. Singh

[1] Existence of bounded length confidence intervals, *Ann. Math. Statist.*, **34:** 1474–1485 (1963).

A. V. Skorokhod

[1] *Studies in the Theory of Random Processes* (translated from Russian in 1965), Addison-Wesley, Reading, 1961.

H. Solomon, S. Zacks

[1] Optimal design of sampling from finite populations: A critical review and development of new research areas, *Jour. Amer. Statist. Assoc.*, **65:** 653–677 (1970).

N. Starr

[1] The performance of a sequential procedure for the fixed width interval estimation of the mean, *Ann. Math. Statist.*, **37:** 36–50 (1966).

C. Stein

[1] Unpublished communication.
[2] A two-sample test for a linear hypothesis whose power is independent of the variance *Ann. Math. Statist.*, **16:** 243–258 (1945).
[3] Unbiased estimates with minimum variance, *Ann. Math. Statist.*, **21:** 406–415 (1950).
[4] A necessary and sufficient condition for admissibility, *Ann. Math. Statist.*, **26:** 518–522 (1955).
[5] Inadmissibility of the usual estimator for the mean of a multi-variate normal distribution, *Proc. Third Berkeley Symp. Math. Statist. Prob. V.* **1:** 197–206 (1956).
[6] An example of wide discrepancy between fiducial and confidence intervals, *Ann. Math. Statist.*, **30:** 877–880 (1959).
[7] The admissibility of the Pitman's estimator for a single location parameter, *Ann. Math. Statist.*, **30:** 970–979 (1959).
[8] Inadmissibility of the usual estimate for the variance of a normal distribution with unknown mean, *Annals Inst. Statist. Math.*, **16:** 155–160 (1964).

C. Stein, A. Wald

[1] Sequential confidence intervals for the mean of a normal distribution with known variance, *Ann. Math. Statist.*, **18:** 427–433 (1947).

M. Stone

[1] Right Haar measure for convergence in probability to quasi posterior distributions, *Ann. Math. Statist.*, **36:** 440–453 (1965).

E. Sverdrup

[1] The present state of the decision theory and Neyman-Pearson theory, *Review Int. Statist. Instit.*, **34:** 309–333 (1966).

P. S. Swamy

[1] On the amount of information supplied by censored samples of grouped observations in the estimation of statistical parameters, *Biometrika*, **49:** 245–249 (1962).
[2] On the amount of information supplied by truncated samples of grouped observations in the estimation of the parameters of normal populations, *Biometrika*, **50:** 207–213 (1963).

K. Takeuchi

[1] On the fallacy of a theory of Gunner Blom, *Rep. Statist. Appl. Res.*, JUSE, **9:** 34–35 (1961).

R. F. Tate

[1] Unbiased estimation: Functions of location and scale parameters, *Ann. Math. Statist.* **30:** 341–366 (1959).

A. R. Thatcher

[1] Relationships between Bayesian and confidence limits for predictions, *J. Roy. Statist. Soc. B.* **26:** 176–191 (1964).

J. W. Tukey

[1] Non-parametric estimation: II. Statistically equivalent blocks and tolerance regions—the continuous case, *Ann. Math. Statist.*, **18:** 529–539 (1947).
[2] Non-parametric estimation, III. Statistically equivalent blocks and multivariate tolerance regions—the discontinuous case, *Ann. Math. Statist.*, **19:** 30–39 (1948).

U.S Department of Commerce, N.B.S.

[1] Tables of the Bivariate Normal Distribution Function and Related Functions, *Applied Math. Series* 50, 1959.

J. von Neumann, O. Morgenstern

[1] *Theory of Games and Economic Behavior* (2d ed.), Princeton University Press, 1947.

A. Wald

[1] Asymptotically shortest confidence intervals, *Ann. Math. Statist.*, **13:** 127–137 (1942).
[2] On cumulative sums of random variables, *Ann. Math. Statist.*, **15:** 283–296 (1944).
[3] Differentiation under the expectation sign in the fundamental identity of sequential analysis, *Ann. Math. Statist.*, **17:** 493–497 (1946).
[4] *Sequential Analysis*, John Wiley, New York, 1947.
[5] Note on the consistency of the maximum likelihood estimate, *Ann. Math. Statist.*, **20:** 595–601 (1949).
[6] *Statistical Decision Functions*, John Wiley, New York, 1950.

A. Wald, J. Wolfowitz

[1] Optimum character of the sequential probability ratio test, *Ann. Math. Statist.*, **19:** 326–339 (1948).

A. M. Walker

[1] A note on asymptotic efficiency of an asymptotically normal estimator sequence, *J. Roy. Stat. Soc. B*, **25:** 193–200 (1963).

J. E. Walsh

[1] Nonparametric confidence intervals and tolerance regions, Chapter 8. *Contributions to Order Statistics*, A. E. Sarhan and B. G. Greenberg (eds.), John Wiley, New York, 1962.

M. T. Wasan

[1] Sequential optimum procedures for unbiased estimation of a binomial parameter, *Technometrics*, **6:** 259–272 (1964).

Y. Washio, H. Morimoto, N. Ikeda

[1] Unbiased estimation based on sufficient statistics, *Bull. of Math. Statist.*, **6:** 69–94 (1956).

L. Weiss

[1] On confidence intervals of given length for the mean of a normal distribution with unknown variance, *Ann. Math. Statist.*, **26:** 348–352 (1955).

L. Weiss, J. Wolfowitz

[1] Generalized maximum likelihood estimators, *Theory of Prob. and Its Appl.*, **11:** 58–81 (1966).

O. Wesler

[1] Invariance theory and a modified minimax principle, *Ann. Math. Statist.*, **30:** 1–20 (1959).

P. Whittle

[1] Some general results in sequential analysis, *Biometrika*, **51:** 123–141 (1964).

D. V. Widder

[1] *The Laplace Transform*, Princeton University Press, Princeton, 1946.

N. Wiener

[1] The ergodic theorem, *Duke Math. Jour.*, **5:** 1–18 (1939).

R. A. Wijsman

[1] On the theory of B.A.N. estimates, *Ann. Math. Stat.*, **30:** 185–191 (1959).

[2] Existence uniqueness and monotonicity of sequential probability ratio tests, *Ann. Math. Statist.*, **35:** 1541–1548 (1964).

[3] Cross-sections of orbits and their applications to densities of maximal invariants, *Proc. Fifth Berkeley Symp. Math. Statist. Prob.*, **1:** 389–400 (1967).

S. S. Wilks

[1] Shortest average confidence intervals from large samples, *Ann. Math. Statist.*, **9:** 166–175 (1938).

[2] On the determination of sample sizes for setting tolerance limits, *Ann. Math. Statist.*, **12:** 91–96 (1941).

[3] Statistical prediction with special reference to the problem of tolerance limits, *Ann. Math. Statist.*, **13:** 400–409 (1942).

J. Wolfowitz

[1] The efficiency of sequential estimates and Wald's equation for sequential processes, *Ann. Math. Statist.*, **18:** 215–230.

[2] On Wald's proof of the consistency of the maximum likelihood estimate, *Ann. Math. Statist.*, **20:** 601–602 (1949).

[3] Minimax estimates of the mean of a normal distribution with known variance, *Ann. Math. Statist.*, **21:** 218–230 (1950).

[4] Asymptotic efficiency of maximum likelihood estimators, *Theory of Prob. and Its Appl.*, **10:** 247–260 (1965).

[5] Remark on the optimum character of the sequential probability ratio test, *Ann. Math. Statist.*, **37:** 726–727 (1966).

S. Zacks

[1] On a complete class of linear unbiased estimators for randomized factorial experiments *Ann. Math. Statist.*, **34:** 769–779 (1963).

[2] Randomized fractional weighing designs, *Ann. Math. Statist.*, **37:** 1382–1395 (1966).

[3] Sequential estimation of the mean of a log-normal distribution having a prescribed proportional closeness, *Ann. Math. Statist.*, **37:** 1688–1696 (1966).

[4] Unbiased estimation of the common mean of two normal distributions based on small samples of equal size, *Jour. Amer. Statist. Assoc.*, **61:** 467–476 (1966).

[5] Bayesian design of single, double and sequential stratified sampling for estimating proportion in finite populations, *Technometrics*, **12:** 119–130 (1970).

[6] Bayes equivariant estimators of variance components, *Ann. Inst. Statist. Math.*, **22:** 27–40 (1970).

[7] Bayes and fiducial equivariant estimators of the common mean of two normal distributions, *Ann. Math. Statist.*, **41:** 59–69 (1970).

[8] Bayes sequential designs of stock levels, *Nav. Res. Log. Quart.*, **16:** 143–155 (1969).

[9] Bayes sequential designs of fixed size samples from finite populations, *Jour. Amer. Statist. Assoc.*, **64:** 1342–1349 (1969).

[10] Uniformly most accurate upper tolerance limits for monotone likelihood ratio families of discrete distributions, *Amer. Statist., Assoc.*, **64:** 307–316 (1970).

[11] Generalized least squares estimators for randomized fractional replication designs *Ann. Math. Statist.*, **35:** 696–704 (1964).

S. Zacks, Roy C. Milton

[1] Mean square errors of the best unbiased and maximum likelihood estimators of tail probabilities in normal distributions, Tech. Rep. No. 185, Dept. of Math. and Statist., University of New Mexico, 1969.

P. W. Zehna

[1] Invariance of maximum likelihood estimation, *Ann. Math. Statist.*, **37:** 755 (1966).

O. Zyskind

[1] On canonical forms, non-negative covariance matrices and best and simple least-squares linear estimators in linear models, *Ann. Math. Statist.*, **38:** 1092–1109 (1967).

Index

Applied Probability and Statistics (Continued)

GROSS and CLARK · Survival Distributions: Reliability Applications in the Biomedical Sciences
GROSS and HARRIS · Fundamentals of Queueing Theory
GUTTMAN, WILKS, and HUNTER · Introductory Engineering Statistics, *Second Edition*
HAHN and SHAPIRO · Statistical Models in Engineering
HALD · Statistical Tables and Formulas
HALD · Statistical Theory with Engineering Applications
HARTIGAN · Clustering Algorithms
HILDEBRAND, LAING, and ROSENTHAL · Prediction Analysis of Cross Classifications
HOEL · Elementary Statistics, *Fourth Edition*
HOLLANDER and WOLFE · Nonparametric Statistical Methods
HUANG · Regression and Econometric Methods
JAGERS · Branching Processes with Biological Applications
JESSEN · Statistical Survey Techniques
JOHNSON and KOTZ · Distributions in Statistics
Discrete Distributions
Continuous Univariate Distributions-1
Continuous Univariate Distributions-2
Continuous Multivariate Distributions
JOHNSON and KOTZ · Urn Models and Their Application: An Approach to Modern Discrete Probability Theory
JOHNSON and LEONE · Statistics and Experimental Design in Engineering and the Physical Sciences, Volumes I and II, *Second Edition*
KEENEY and RAIFFA · Decisions with Multiple Objectives
LANCASTER · An Introduction to Medical Statistics
LEAMER · Specification Searches: Ad Hoc Inference with Nonexperimental Data
McNEIL · Interactive Data Analysis
MANN, SCHAFER, and SINGPURWALLA · Methods for Statistical Analysis of Reliability and Life Data
MEYER · Data Analysis for Scientists and Engineers
OTNES and ENOCHSON · Digital Time Series Analysis
PRENTER · Splines and Variational Methods
RAO and MITRA · Generalized Inverse of Matrices and Its Applications
SARD and WEINTRAUB · A Book of Splines
SEARLE · Linear Models
THOMAS · An Introduction to Applied Probability and Random Processes
WHITTLE · Optimization under Constraints
WILLIAMS · A Sampler on Sampling
WONNACOTT and WONNACOTT · Econometrics
WONNACOTT and WONNACOTT · Introductory Statistics, *Third Edition*
WONNACOTT and WONNACOTT · Introductory Statistics for Business and Economics, *Second Edition*
YOUDEN · Statistical Methods for Chemists
ZELLNER · An Introduction to Bayesian Inference in Econometrics